DATE DUE

Demco, Inc. 38-293

FREEZING AND MELTING HEAT TRANSFER IN ENGINEERING

FREEZING AND MELTING HEAT TRANSFER IN ENGINEERING

Selected Topics on Ice-Water Systems and Welding and Casting Processes

Edited by

K. C. Cheng
Department of Mechanical Engineering
University of Alberta
Edmonton, Alberta, Canada

N. Seki
Professor Emeritus
Department of Mechanical Engineering
Hokkaido University, Sapporo, Japan

⦿**HEMISPHERE PUBLISHING CORPORATION**
A member of the Taylor & Francis Group

New York   Washington   Philadelphia   London

FREEZING AND MELTING HEAT TRANSFER IN ENGINEERING:
Selected Topics on Ice-Water Systems and Welding and Casting Processes

1 2 3 4 5 6 7 8 9 0   E B E B   9 8 7 6 5 4 3 2 1

Cover design by Debra Eubanks Riffe.
A CIP catalog record for this book is available from the British Library.

Library of Congress Cataloging-in-Publication Data

Freezing and melting heat transfer in engineering : selected topics on
  ice-water systems and welding and casting processes / editors, K. C.
  Cheng, N. Seki.
       p. cm.
    Includes bibliographical references and index.

    1. Welding.   2. Heat—Transmission.   3. Water-pipes—Freezing.
  4. Icing (Meteorology)   I. Cheng, K. C.   II. Seki, N. (Nobuhiro),
  date.
  TS227.2.F74   1991
  671.5'2—dc20   91-6932
  ISBN 0-89116-985-7   CIP

Contents

SPECIAL TOPICS (Casting and Welding Processes)

In Memory

Professor Robert Ridgeway Gilpin (1942–1982)

This book is dedicated to the memory of Dr. Robert R. Gilpin, Professor of Mechanical Engineering (1970–1982) at the University of Alberta, Edmonton, Alberta, Canada.

Dr. Gilpin was well known for his work on freezing of water (growth of dendritic ice, ice-band phenomena and modes of ice formation and flow blockage in water pipes), frost heave in freezing soils, regelation in ice, particle rejection and engulfment during freezing, wave formation at an ice-water interface in turbulent water flow and radiative heating in ice.

He was a fine human being as well as a brilliant scientist.

Contributors

Kazuo Aoki, Department of Mechanical Engineering, Nagaoka University of Technology, Nagaoka 940-21, Japan

R. S. Chen, Department of Power Mechanical Engineering, National Tsing Hua University, Hsinchu, Taiwan 30043, R.O.C.

K. C. Cheng, Department of Mechanical Engineering, University of Alberta, Edmonton, Alberta, Canada T6G 2G8

Steven F. Daly, U.S. Army Cold Regions Research and Engineering Laboratory, Hanover, New Hampshire 03755-1290, U.S.A.

S. Fukusako, Department of Mechanical Engineering, Hokkaido University, Sapporo 060, Japan

E. M. Gates, Formerly, Department of Mechanical Engineering, University of Alberta, Edmonton, Alberta, Canada T6G 2G8

Masaru Hattori, Department of Mechanical Engineering, Nagaoka University of Technology, Nagaoka 940-21, Japan

Yujiro Hayashi, Department of Mechanical Engineering, Kanazawa University, Kanazawa 920, Japan

Tetsuo Hirata, Department of Mechanical Engineering, Shinshu University, Nagano 380, Japan

G. J. Hwang, Department of Power Mechanical Engineering, National Tsing Hua University, Hsinchu, Taiwan 30043 R.O.C.

Hideo Inaba, Department of Mechanical Engineering, Okayama University, Okayama 700, Japan

K. Kimoto, Akita City Office, Akita O10, Japan

J.-M. Konrad, Department of Civil Engineering, Université Laval, Ste Foy, Quebec, Canada G1K 7P4

G. S. H. Lock, Department of Mechanical Engineering, University of Alberta, Edmonton, Alberta, Canada T6G 2G8

E. P. Lozowski, Division of Meteorology, Department of Geography, University of Alberta, Edmonton, Alberta, Canada T6G 2H4

Virgil J. Lunardini, U.S. Army Cold Regions Research and Engineering Laboratory, Hanover, New Hampshire 03755-1290, U.S.A.

Takahiro Ohrai, Seiken Co., Ltd. 2-11-16, Kawarayamachi, Chuo-ku, Osaka, Japan

Itsuo Ohnaka, Department of Materials Science and Processing, Faculty of Engineering, Osaka University, Suitashi, Osaka 565, Japan

L. Robillard, Department of Mechanical Engineering, Case Postale 6079, Succursale "A", Campus de l'Université de Montréal, Montéal, Québec, Canada H3C 3A7

Hakaru Saito, Department of Mechanical Engineering, Muroran Institute of Technology, Muroran 050, Japan

Takeo Saitoh, Department of Mechanical Engineering II, Tohoku University, Sendai 980, Japan

N. Seki, Department of Mechanical Engineerng, Hokkaido University, Sapporo 060, Japan

M. Sugawara, Department of Mechanical Engineering, Akita University, Akita 010, Japan

Ikuo Tokura, Department of Mechanical Engineering, Muroran Institute of Technology, Muroran 050, Japan

Chon L. Tsai, Department of Welding Engineering, The Ohio State University, Columbus, Ohio 43210, U.S.A.

P. Vasseur, Department of Mechanical Engineering, Case Postale 6079, Succursale "A", Campus de l'Université de Montréal, Québec, Canada H3C 3A7

Hideo Yamamoto, Seiken Co., Ltd., 2-11-16, Kawarayamachi, Chuo-ku, Osaka, Japan

Yin-Chao Yen, U.S. Army Cold Regions Research and Engineering Laboratory, Hanover, New Hampshire 03755, U.S.A.

W. Paul Zakrzewski, Division of Meteorology, Department of Geography, University of Alberta, Edmonton, Alberta, Canada T6G 2H4

Preface

Since the early 1960's, considerable progress has been made in freezing and melting heat transfer in engineering in general, and cold regions heat transfer in particular, due mainly to industrial activities relating to the natural resources development in the arctic regions. The idea of preparing a book dealing with selected topics on freezing and melting heat transfer was conceived in 1982, after the tragic death of Dr. R. R. Gilpin on February 25, 1982 while on a cross-country skiing trip at Lake Placid, New York, where he was spending a winter holiday. At the time, he was on a study leave from the University of Alberta in the Department of Chemical Engineering at Clarkson University in Potsdam, New York, where he was pursuing his research into regelation phenomena involving the organic crystal and the engineering physics of ice. He was only 39 years old.

This book is dedicated to the memory of Professor R. R. Gilpin [B.Sc. (1964), University of Alberta; M.Sc. (1965) and Ph.D (1968), California Institute of Technology] who made a pioneering contribution to freezing phenomena in ice-water systems during a rather short research and teaching career of 12 years at the University of Alberta. Dr. Gilpin was both a physicist and an engineer. His complete mastery of both disciplines was evident in his broad range of research interests, from high temperature plasma physics, magnetohydrodynamics, and atmospheric physical chemistry early in his research career, to his later focus on ice formation problems and frost heave phenomena in soil systems. His research was characterized by incisive physical understanding and insight of the phenomenon and simplified theoretical analysis describing the complex physical phenomenon. His experimental investigations were very ingenious in clarifying the physical phenomena.

This book is designed to provide a focus on the following topics: (1) heat transfer with freezing or melting, (2) ice formation problems, (3) snow melting and frost formation, (4) solidification and melting of solutions, (5) frazil ice, (6) ground freezing and frost heave, (7) atmospheric and marine icings of structures, and (8) special topics dealing with the analogous problems of casting and welding.

The authors of this book know Dr. R. R. Gilpin's works quite well. Thus, no attempt was made to find authors who are experts on many other important topics in freezing and melting heat transfer. The editors are fully aware of the limited scope of this work since the literature on this subject is currently very extensive. The review articles deal with recent advances and the basic concepts and physical understanding are emphasized. This work also serves as a source book.

K. C. Cheng
N. Seki

Acknowledgments

The editors wish to thank all authors for their ready acceptance of the invitation to write articles on their specialities and their time and effort in preparing the camera-ready manuscript. Their patience in a long wait for the publication is greatly appreciated. The preparation of this book was supported by a grant from the Central Research Fund of the University of Alberta. The editors wish to thank Hemisphere Publishing Corporation for publishing a book of this nature. It is a great pleasure to acknowledge Mrs. B. Bloedorn's skillful and able editorial assistance in preparing this book. Dr. Yin-Chao Yen and Prof. S. Fukusako provided several photographs used in this front matter.

Photographs

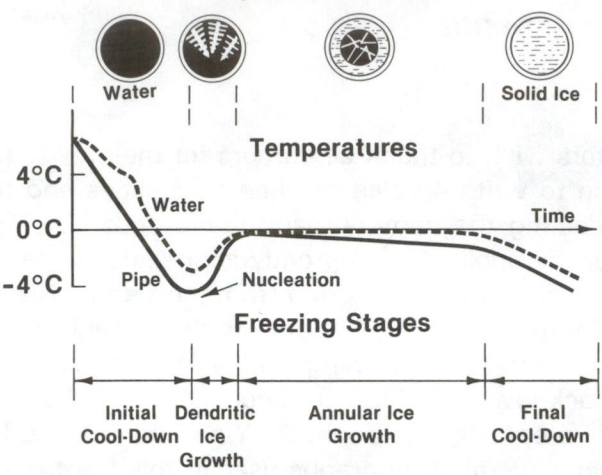

Pipe Cross Section

Water | | | Solid Ice |

Temperatures

4°C

Water

0°C

Time

−4°C

Pipe — Nucleation

Freezing Stages

Initial Cool-Down | Dendritic Ice Growth | Annular Ice Growth | Final Cool-Down

Freezing history of still water in a pipe (after R.R. Gilpin, 1977)

Flow and temperature characteristics in a cooled tube (courtesy of S. Fukusako)

Tube diameter = 64 mm, initial water temperature = 5°C, cooling rate = 0.27°C/min.
Left half: flow pattern; Right half: temperature distribution
Elapsed time: 1. 0 min. 2. 12 min.
   3. 16 min. 4. 48 min.

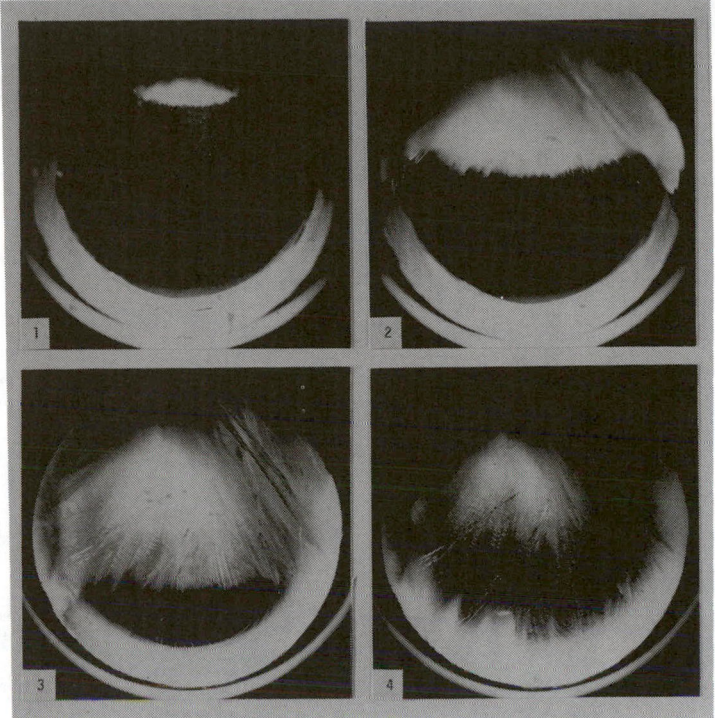

Occurrence of dendritic ice in a cooled tube (courtesy of S. Fukusako)
Tube diameter = 64 mm, initial water temperature = 5°C, cooling rate = 0.27°C/min.
Elapsed time: 1. 112 min. 00.12 sec.,
                 2. 112 min. 01.55 sec.,
                 3. 112 min. 02.25 sec.,
                 4. 112 min. 03.00 sec.

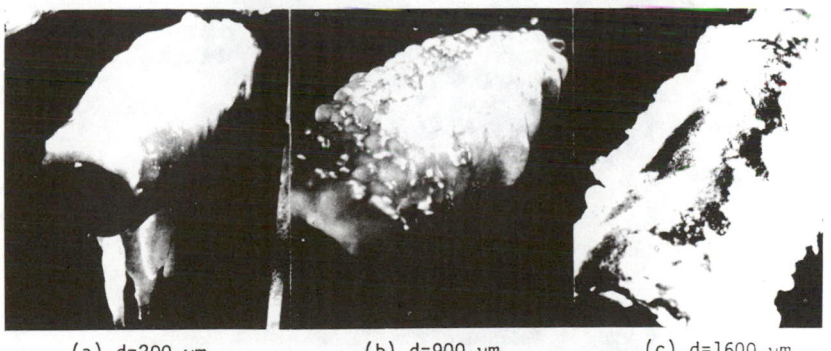

(a) d=200 μm   (b) d=900 μm   (c) d=1600 μm

Effect of liquid–droplet diameter on icing (courtesy of S. Fukusako)

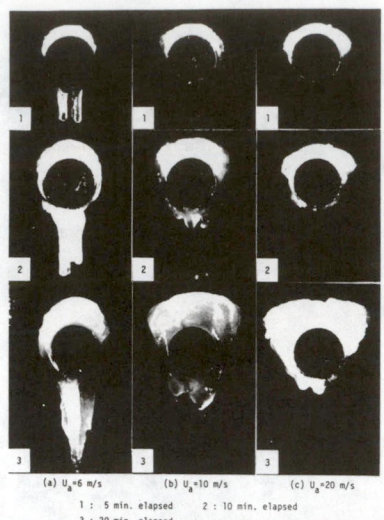

(a) U_a=6 m/s   (b) U_a=10 m/s   (c) U_a=20 m/s

1 : 5 min. elapsed   2 : 10 min. elapsed
3 : 20 min. elapsed

Effect of air velocity on icing (courtesy of S. Fukusako)

Icing on a microwave tower (after Charles C. Ryerson, U.S. Army CRREL)

Icing on antenna tower (after Charles C. Ryerson, U.S. Army CRREL)

Icing on transmission cable (after John W. Govoni, U.S. Army CRREL)

Icing on structures (after John W. Govoni, U.S. Army CRREL)

Air Hoar

Rime

Clear Ice

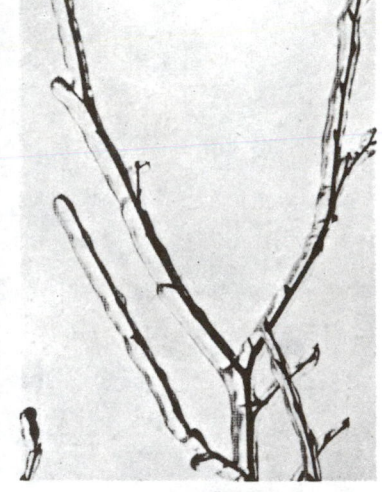

Glaze

Classification of icing (after Daisuke Koroiwa, 1958)

Soft rime (see page 618)

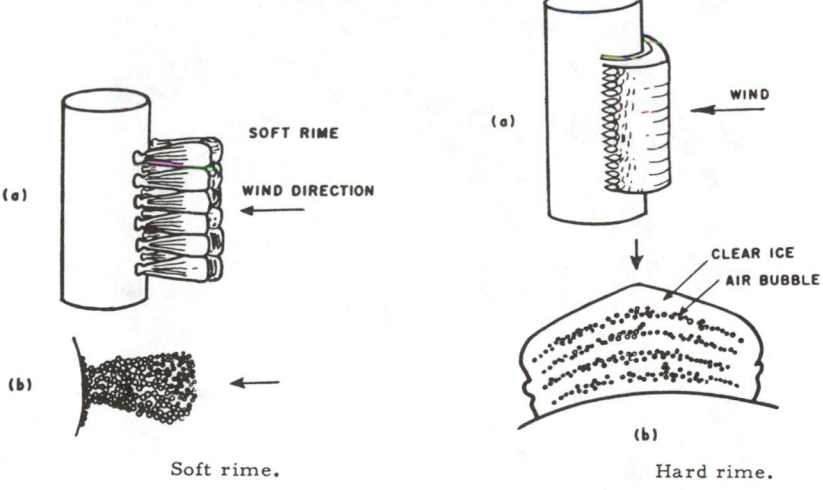

Soft rime.

Hard rime.

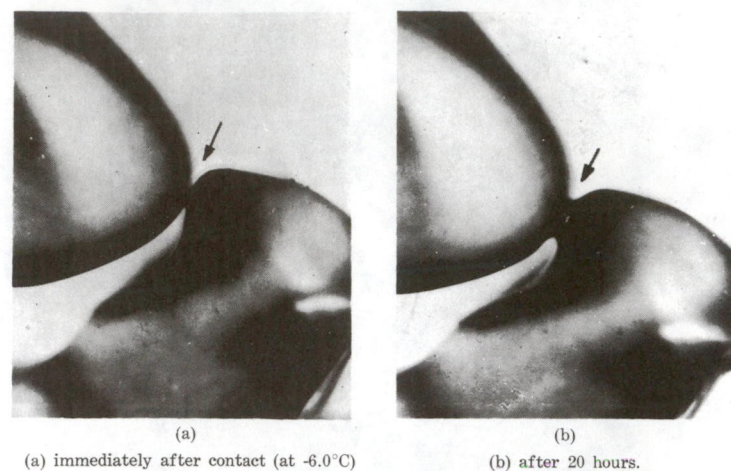

(a) immediately after contact (at -6.0°C)

(b) after 20 hours.

Formation of ice bridge between two particles of snow (after Daisuke Kuroiwa, 1958)

Magnified frazil ice crystals (courtesy of S.F. Daly, U.S. Army CRREL)

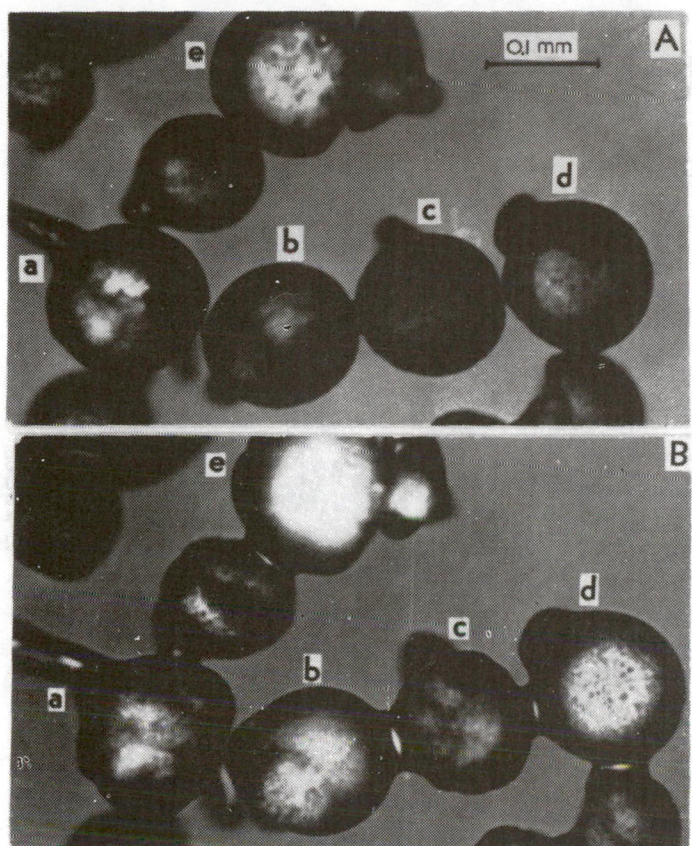

Sintering process of pure ice spheres taken by microscope with polarized light at −1.5°C.
(A) immediately after the sintering, (B) 165 min. after the beginning of sintering. Note: Change of crystallographic orientation of ice spheres (after Daisuke Kuroiwa, 1975).

Using river−ice to cool nuclear reactor (after U.S. Army CRREL)

Typical vertical support members with heat pipes for Trans–Alaska oil pipeline (courtesy of J.P. Zarling)

Heat pipes used to stabilize a railroad bed crossing ice rich ground (photo by D.W. Hayley)

a. Frazil   b. Frazil slush

c. Plate ice   d. Shore ice

e. Slush balls   f. Pancake ice

Forms of river and lake ice (after U.S. Army CRREL monograph on winter regime on rivers and lakes by B. Michel, 1971)

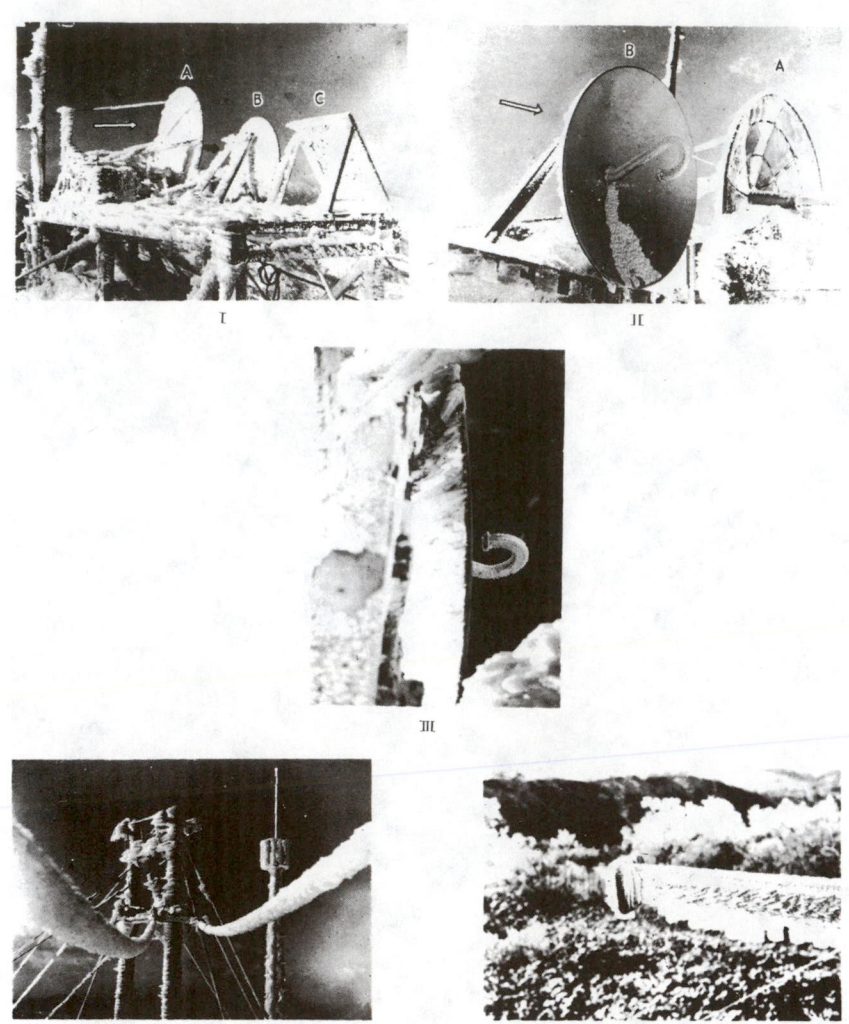

Ice accretion deposited on microwave antennas at Mt. Monbetsu, Hokkaido, Japan (after Daisuke Koroiwa, 1958)

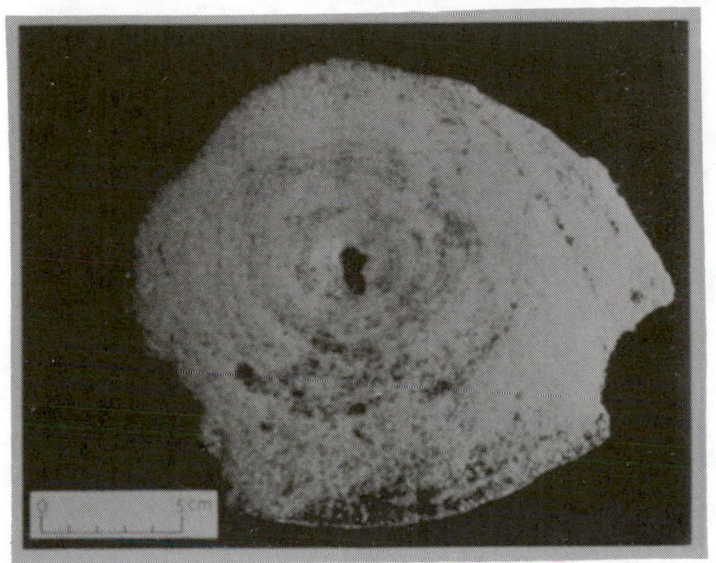

Annual ring pattern of snow accretion on electric transmission line (diameter = 5 mm, snow diameter = 18 cm) (after Kazuo Goto and Daisuke Kuroiwa, 1975)

Icing on ship (after U.S. Army CRREL)

A pingo (ground ice) on the coastal plain east of the MacKenzie River delta (after U.S. Army CRREL, Hanover, New Hampshire)

Courtesy Gulf Canada Resources

Molilpaq in the Beaufort Sea

Artificial island Molilpaq in the Beaufort Sea (after APOA Review, courtesy of Gulf Canada Resources)

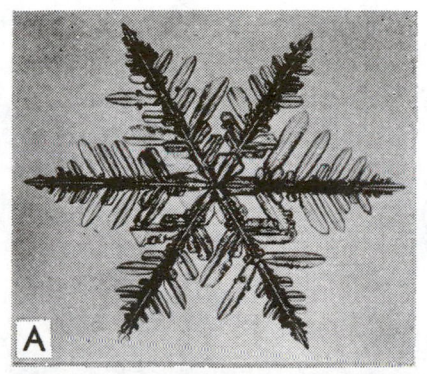

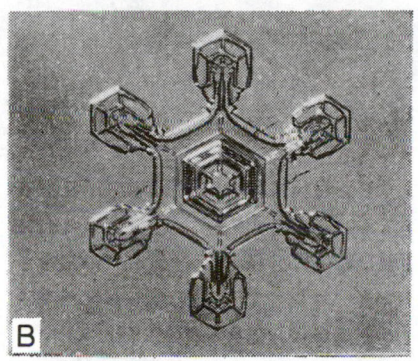

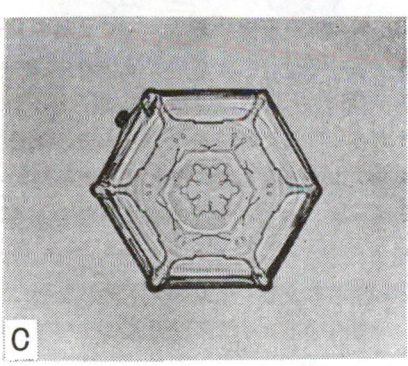

Typical snow crystals.
(A) A dendritic stellar snow crystal, (B) Hexagonal plate with plate extensions, (C) Hexagonal plate (after Daisuke Kuroiwa, 1976).

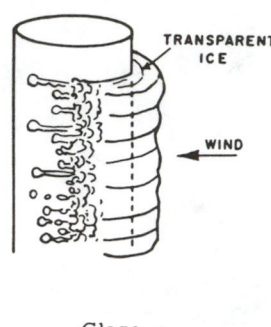

Glaze

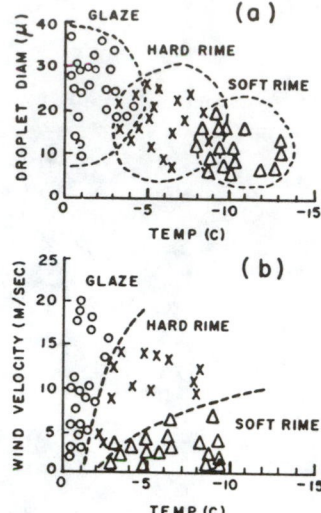

Relation between types of icing and meteorological conditions.

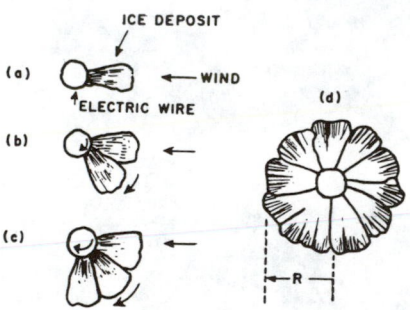

Growth process of icing on wires

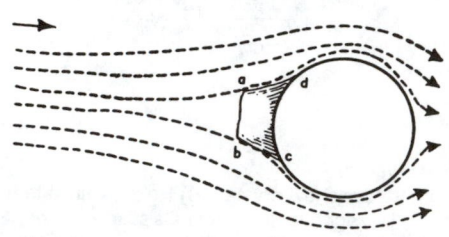

Ice deposit on a cylinder

Chapter 1

Introduction

K. C. CHENG

Department of Mechanical Engineering, University of Alberta,
Edmonton, Alberta, Canada T6G 2G8

N. SEKI
Department of Mechanical Engineering, Hokkaido University,
Sapporo, Japan

CONTENTS

1.   Brief Review and Classification of Freezing and Melting Heat Transfer Problems

Solidification and melting processes are of considerable importance and interest in a wide range of problems in technologies and geophysics. The subject of freezing and melting heat transfer dealing with historical background, scientific principles and technologies (fundamentals and applications) is at present very extensive. The freezing and melting phenomena occur in natural phenomena and in many applications in modern technologies ranging from ice formation to manufacture of crystals in space for semiconductors and computer chips.

The mathematical theory of heat conduction originated with Joseph Fourier's classical work, "The Analytical Theory of Heat (Théorie Analytique de la Chaleur) in 1822 and the first analytical work on phase-change problem was published by Lamé and Clapeyron in 1831. The mathematical theory of ice formation generally known as the classical Stefan problem originated with a famous series of lectures (unpublished) given by Neumann in Königsberg in the early 1860's and Stefan's classical paper in 1889. Since these classical works, very few exact analytical solutions exist and approximate and/or numerical techniques are required to obtain solutions of freezing and melting heat transfer problems.

The historical development relating to soil freezing, ice formation in rivers and oceans, and various solidification and melting problems in technologies require extensive treatment and the roots generally can be traced to the 19th century. It is thus seen that the history of solidification and melting engineering is about 150 years. It is also of interest to note that the subject was studied by applied mathematicians, crystal growers, metallurgists, engineers, meteorologists, geologists, geophysicists, physicists, chemists and others in agriculture, biology and medical science.

It is difficult to systematize or classify the existing materials (information) or prepare a complete list for freezing and melting phenomena. Specialized study areas include: (1) All solidification processes in materials science, (2) permafrost and freezing grounds, (3) atmospheric icings of structures, (4) aircraft and helicopter icings, (5) icings of ships and fishing vessels, (6) ice formation in rivers and lakes, (7) refrigeration and food processing, (8), ice formation in pipes, (9) frost formation and defrosting, (10) glaciology (snow and ice), (11) welding processes, (12) crystal growth, and (13) other industrial processes. Apparently, heat transfer, fluid mechanics, thermodynamics, and nucleation and interface kinetics are common to these freezing and melting phenomena. Freezing and melting phenomena can also be classified into two classes, namely natural phenomena in cold regions and industrial processes in modern technology. Depending on the nature of research and development, one may have mathematical, numerical, scientific or technical problems. The nature of works may be theoretical or experimental, but ultimately physical understanding and practical design information are required. The freezing and melting heat transfer problems can be classified in terms of temperature ranges such as (1) melting temperatures of metals or other materials (high temperature), (2) atmospheric temperature range in cold climates, (3) refrigeration temperature range and (4) cryogenic temperature range. Classification of the problems can also be based on solution techniques, physical problems (conduction–or convection–dominated problems) or geometrical shapes.

In order to show the scope and limitations of the present monograph, it is convenient to list application examples for freezing and melting phenomena or processes.

(1)   Applications in engineering include:

(1) melting and solidification of metals, alloys, glass, plastic and other materials, (2) crystal growth, (3) freeze coatings, (4) freezing and thawing of soils and foodstuffs (food preservation), (5) welding or soldering, (6) ablation of space vehicles and missiles under aerodynamic heating, (7) ice making and storage, (8) heat storage using latent heat materials, (9) artificial soil freezing for construction, (10) LNG and LPG tanks, (11) sludge treatment by freezing, (12) freezing of water pipes, freezing valve, (13) frost formation and defrosting in condensers, (14) desalination of sea water, (15) artificial permafrost storage using heat pipes, (16) application of heat pipes in cold regions, (17) snow melting, (18) atmospheric icing of structures, (19) aircraft and helicopter icing and deicing, (20) ship and fishing vessel icings, (21) construction of artificial islands in arctic seas, (22) ice formation in rivers, lakes and seas, (23) nuclear reactor safety, (24) coal liquefaction, (25) quick freezing techniques, (26) polymer processing and processing of thermoplastic materials, (27) icing and snow accretion on electric wires, (28) frost heaving, (29) frost formation in low–temperature insulation, (30) thermophysical properties at low temperatures, (31) desalination process.

(2)   Applications in agriculture, fisheries and food processing include:

(1) food preservation (freezing and thawing), (2) freeze–drying of foodstuffs, (3) freeze–dehydration, (4) freeze–concentration, (5) freeze–denaturation of protein, (6) cryomilling, (7) frost injury in plants, (8) wintering of insects, (9) freeze resistance of eggs.

(3)   Applications in medical science and biology include:

(1) freezing of blood, (2) hypothermia, (3) cryosurgery, (4) freeze–preservation, (5) extracellular freezing, (6) freeze tolerance in animals, (7) cryobiology.

The scope and overviews of recent literature on freezing and melting heat transfer can be found from books and review articles and these are listed below for reference.

(1)   Books

1.   Carlsaw, H.S. and Jaeger, J.C., 1965, Conduction of Heat in Solids, 2nd ed., Clarendon Press, Oxford.
2.   Crank, J., 1964, The Mathematics of Diffusion, Clarendon Press, Oxford.
3.   Rubinstein, L.I., 1971, The Stefan Problem, Translations of Mathematical Monographs, Vol. 27, American Mathematical Society, Providence, Rhode Island.
4.   Flemings, M.C., 1974, Solidification Processing, McGraw-Hill, New York.
5.   Ockendon, J.R. and Hodgkins, W.R., 1975, Moving-Boundary Problems in Heat Flow and Diffusion, Clarendon Press, Oxford.
6.   Wilson, D.G., Solomon, A.D. and Boggs, P.T., 1978, Moving-Boundary Problems, Academic Press, New York.
7.   Crank, J., 1984, Free- and Moving-Boundary Problems, Clarendon Press, Oxford.
8.   Kurz, W. and Fisher, D.J., 1986, Fundamentals of Solidification, Trans Tech Publications, Aedermannsdorf, Switzerland.
9.   Hill, J.M., 1987, One-Dimensional Stefan Problems: An Introduction, Longman Scientific and Technical, New York.

(2)   Review Articles

1.   Bankoff, S.G., 1964, Heat Conduction or Diffusion with Change of Phase, Adv. Chem. Engng., Vol. 5, pp. 75–150.
2.   Tiller, W.A., 1963, Principles of Solidification, J.J. Gilman, ed., Art and Science of Growing Crystals, Wiley.
3.   Boley, B.A., 1963, The Analysis of Problems of Heat Conduction and Melting, High Temperature Structure and Materials, Proc. 3rd Symp. on Naval Structural Mech. Pergamon Press.
4.   Goodman, T.R., 1964, Application of Integral Methods to Transient Non-Linear Heat Transfer, Adv. Heat Transfer, Vol. 1, pp. 51–122.
5.   Muehlbauer, J.C. and Sunderland, J.E., 1965, Heat Conduction with Freezing or Melting, Appl. Mech. Rev., Vol. 18, pp. 951–959.
6.   Boley, B.A., 1972, Survey of Recent Developments in the Fields of Heat Conduction in Solids and Thermo-Elasticity, Nuclear Engng. Design, Vol. 18, pp. 377–399.
7.   Mori, A. and Araki, K., 1976, Methods of Analysis of the Moving Boundary-Surface Problem, Int. Chem. Engng., Vol. 16, pp. 734–744.
8.   Rubinstein, L., 1979, The Stefan Problem: Comments on its Present State, J. Inst. Maths. Applics., Vol. 24, pp. 259–277.
9.   Viskanta, R., 1983, Phase-Change Heat Transfer, G.A. Lane, ed., Solar Heat Storage: Latent Heat Materials, Vol. 1, CRC Press, Inc., Boca Raton, Florida, pp. 153–222.
10.   Epstein, M. and Cheung, F.B., 1983, Complex Freezing-Melting Interfaces in Fluid Flow, Ann. Rev. Fluid Mech., Vol. 15, pp. 293–319.
11.   Cheung, F.B. and Epstein, M., 1984, Solidification and Melting in Fluid Flow, A.S. Mujumdar and R.A. Mashelkar, eds., Advances in Transport Processes, Vol. 3, John Wiley & Sons, New York.
12.   Viskanta, R., 1985, Natural Convection in Melting and Solidification, S. Kakac, W. Aung and R. Viskanta, eds., Natural Convection: Fundamentals and Applications, Hemisphere, Washington, pp. 845–877.
13.   Fukusako, S. and Seki, N., 1987, Fundamental Aspects of Analytical and Numerical Methods on Freezing and Melting Heat Transfer Problems, T.C. Chawla, ed., Ann. Rev. Numerical Fluid Mechanics and Heat Transfer, Vol. 1, Hemisphere, Washington, pp. 351–402.
14.   Viskanta, R., 1988, Heat Transfer During Melting and Solidification of Metals, ASME Journal of Heat Transfer, Vol. 110, pp. 1205–1219.

15.   Yao, L.S. and Prusa, J., 1989, Melting and Freezing, Adv. Heat Transfer, Vol. 19, Academic Press, pp. 1–95.
16.   Huppert, H.E., 1990, The Fluid Mechanics of Solidification, J. Fluid Mech., Vol. 212, pp. 209–240.

2.   COLD REGIONS HEAT TRANSFER

Since the early 1960's, one observes considerable progresses and advances in cold regions engineering and technology in connection with the exploration and development of natural resources such as crude oils and natural gases in the Arctic regions. The engineering activities in cold regions face many unique engineering and environmental problems. Arctic or cold environments provide many challenging freezing and melting phenomena (natural or artificial) in water, air, earth and biological systems. The subject of heat and mass transfer is of basic importance to cold regions engineering design, construction and operation and is characterized by multidisciplinary approach. Heat transfer occurs whenever a temperature difference exist and is a universal phenomena (like gravity) regardless of time, space or material. The natural phenomena relating to heat transfer are too numerous to be listed here.

In recent years, international conferences or symposia have been held regularly on such subjects as permafrost, ground freezing, atmospheric icing of structures, snow and ice (glaciology), ice problems, port and ocean engineering under Arctic conditions, offshore mechanics and Arctic engineering and other related topics. As well, such topics as Arctic heat transfer, frost and ice formation and related topics on freezing and melting heat transfer receive much attention in recent years. At present the literature on cold regions heat transfer is scattered in various journals, proceedings, and reports. Although limited in scope, the present monograph is designed to provide a focus and reviews on some aspects of the following topics from the common viewpoints of cold regions heat transfer:

(1)   Permafrost and ground freezing (freezing and melting in soil systems)
(2)   Atmospheric and marine icings of structures
(3)   River and lake ice (ice engineering)
(4)   Hot oil and chilled-gas pipelines in permafrost
(5)   Application and utilization of cold regions technology (foundations, construction, water supplies, sewage treatment, transportation, heat pipes and thermosyphons, and others)
(6)   Conventional freezing and melting heat transfer problems including metal castings and welding.

2.1   Subjects in Cold Regions Science and Engineering

In order to see the scope of the subjects in cold regions science and technology, it is useful to list the following classification used in a monograph series of the U.S. Army Cold Regions Research and Engineering Laboratory, Hanover, New Hampshire (CRREL monographs 1-A1, 1965 and 1-A2, 1966).

I.   Environment

        A.   General
                1.   Geology and Physiography
                2.   Permafrost
                3.   Climatology
                        a.   General and Northern Hemisphere 1
                        b.   Northern Hemisphere 2
                        c.   Southern Hemisphere
                        d.   Radioactive fallout in the Northern Hemisphere

4. Vegetation
   a. Patterns of vegetation
   b. Regional descriptions
   c. Utilization of vegetation

B. Regional
   1. The Antarctic ice sheet
   2. The Greenland ice sheet

II. Physical Science

A. Geophysics
   1. Heat exchange at the earth's surface
   2. Exploration geophysics

B. Physics and Mechanics of Snow as a Material

C. Physics and Mechanics of Ice
   1. Snow and ice on the earth's surface
   2. Ice as a material

D. Physics and Mechanics of Frozen Ground

III. Engineering

A. Snow Engineering
   1. Engineering properties of snow
   2. Construction
   3. Technology
      a. Explosions and snow
      b. Snow removal and ice control
      c. Blowing snow
      d. Avalanches
   4. Oversnow transportation

B. Ice Engineering

C. Frozen Ground Engineering
   1. Site exploration and excavation
   2. Buildings, utilities and dams
   3. Roads, railroads and airfields
   4. Foundations
   5. Sanitary engineering
      a. General and water supply
      b. Sewerage and waste disposal
   6. Cold weather construction

D. General

Here the cold regions are defined as those areas of the earth where freezing and thawing of soils is an essential consideration in engineering and operational difficulties due to freezing temperatures may occur. Cold regions engineering is characterized by cold climates and the related water–freezing and ice–melting phenomena (phase–change thermodynamic processes). Thus it is desirable to have classification of cold regions heat transfer problems based on physical phenomena in cold climates and the results are shown below in Table 1.

2.2 Freezing and Melting Phenomena in Cold Regions

In cold climates with air temperatures in the range $30°$ C $\sim -80°$ C, for example, water circulation in lithosphere, hydrosphere and atmosphere takes place in three phases, vapour, liquid and solid. Evaporation and condensation occur between water (liquid) and vapour (gas), and freezing and melting phenomena occur between water (liquid) and ice (solid). Sublimation evaporation and solidification can occur between ice (solid) and water vapour (gas). It is thus seen that freezing and melting phenomena involving conduction, convection, radiation can be very complicated with such effects as supersaturation, supercooling, superheating, materials, geometry and molecular effects. The natural freezing and melting phenomena have considerable effects on animals and plants, and engineering problems.

The development of natural resources and the increasing industrial activities in the arctic regions have stimulated considerable recent advances in both fundamental and applied research in cold regions heat transfer. Cold regions engineering has emerged as an important interdisciplinary study encompassing many fields of engineering and science and the subjects range from the environment to ice, snow and permafrost (perennially frozen ground). It is noted that about 14% of the land area on earth is underlain by permafrost and almost 10% of the whole ocean on the earth is perennially or seasonally covered with ice layers. Thermal effects on permafrost, frost heave phenomena and the heat exchange between permafrost and the atmosphere have been studied extensively in recent years by engineers in different fields of study. The subject of cold regions heat transfer is clearly emerged into a distinct, specialized field somewhat similar to other areas such as nuclear heat transfer.

One may appreciate the scope of the problems by classifying the ice formation problems as follows:

(1)   Icings in the atmosphere: snow, hail, fog, air hoar, rime, clear ice, glaze, super–cooling and freezing of water droplets, transmission wires, aircraft, helicopters, ships, engines, tall land–based structures.

(2)   Icings on the earth's surface: freezing of rivers, lakes, estuaries, harbours and arctic seas, glaciers, frazil ice, ice in plant and animal cells, snow removal and ice control.

(3)   Icings in the soil: permafrost, various ice forms in ground (pingo, ice wedge, patterned ground, massive ground ice), frost heave, freezing of soil around LNG tanks.

(4)   Stefan problem and its extensions and variations in industrial applications: freezing of water pipes, supercooling and formation of dendrites, frost formation, laminar or turbulent flow, free and forced convection effects, maximum water density effect.

In recent years, the classical Stefan problem (heat conduction with a moving boundary) has been extended to cases with real effects such as convection and maximum water density involving either external or internal flow.

The design of water supply and waste–water systems in cold regions must consider the possibility of freezing with and without main flow. A water pipe can be blocked by dendritic ice formation instead of the commonly understood annular ice growing inward even in the case of no main flow. For normal tap water supercooling up to $-7°$ C may occur before nucleation starts.

TABLE 1   COLD REGIONS HEAT TRANSFER – FREEZING PHENOMENA AND RELATED ENGINEERING PROBLEMS
(References: A. Higashi, S. Fukusako, see Chapter 2)

	Phenomena	Engineering Problems	Objects and Places	Related Fields
1.	Snow Fall, Deposited Snow	Electric wave and visibility obstruction, snow removal, avalanche, skiing, transportation	Roads, railroads, airfields, buildings, dams	Civil Eng. Mechanical Eng. Architecture Agriculture Transportation Eng.
2.	Icing, Snow Accretion, Atmospheric & Marine Icings	Navigational hazards, power & communication transmission lines, communication equipments, aircraft, helicopter, tall structure	Sea, ship, tower, traffic signals	Oceanography Navigation Communication Aeronautics
3.	Frost Formation	Loss of heat transfer effectiveness, energy loss	Heat exchangers	Mechanical Eng. Refrigeration Eng.
4.	Ice Formation	River & lake ice eng., water delivery system, ice jam, hydraulic power generation, frazil ice	Water mains, sewage & drain pipes, river, lake, sea	Mechanical Eng. Civil Eng.
5.	Land Ice	Skating, frost damages on water mains, drain & sewage pipes	Rivers, lakes, canals, hydraulic structures	Hydraulics River Eng.
6.	Sea Ice	Navigational hazards, artificial islands, pipelines	Port & ocean structures, embankment	Oceanography, Port & Ocean Eng.
7.	Artificial Ice	Ice making, freeze drying, food storage	Dehydrated foods	Refrigeration Eng. Mechanical Eng. Food Eng.
8.	Biological Ice	Plant frost damage, food preservation	Frozen foods, freezing of blood	Biology, Food Eng. Biomedical Eng.
9.	Frozen Ground, Permafrost	Frost heaving, freezing construction, thawing of permafrost	Artificial ground freezing, LNG ($-160°$ C) & LPG ($-42°$ C) tanks, oil & gas pipelines, roads, railways, airfields, buildings	Civil Eng. Environmental Eng.
10.	Glaciers, Icebergs	Water resources, climate change	Water utilization plan, hydraulic power	Hydrology Glaciology

Recent knowledge on river ice is summarized in several monographs (see Chapter 2) and an analytical model and some experimental data for the formation of ripples on the underside of river ice covers are available.

For steady state ice formation involving forced external or internal flow, the steady ice layer profile depends on the heat balance between the rate of heat conduction through ice layer and the rate of convective heat transfer between the flow and the interface. Some experimental studies on transition or turbulent flow over an ice surface in external or internal flows have been reported.

For pipelines transporting crude oil in arctic regions, solidification may occur during pump failure. Thus, the pipeline must be designed so that the flow can be restarted within a period of about 3 weeks. For the buried portion of the trans-Alaska pipeline, one must predict the melting of the surrounding permafrost for 30 years since the hot pipeline represents a heat source in the fragile permafrost regions. It is interesting to observe that over 120,000 heat pipes (two-phase closed thermosyphons) were used to protect the permafrost. For gas pipelines, the gas will be chilled to below the surrounding ground temperature in order to protect the continuous permafrost. However, in he discontinuous permafrost region, frost heaving may occur and this problem represents the greatest difficulty in the design of a gas pipeline.

Since research results on cold regions heat transfer in general and solidification and melting heat transfer in particular are widely scattered in the literature, it is clear that there is a critical need for survey on review papers bringing together all the currently available information on a special topic such as ice formation.

The fundamentals of cold regions engineering consist of snow and ice, freezing ground and frost heaving, ice mechanics, snow and ice accretion (atmospheric icings), river and lake ice, sea ice, mechanical properties of ice, thermophysical properties, and problems relating to life and industrial activities in the north. Sound engineering designs and practices must be based on complete understanding of physical phenomena involving freezing and melting heat transfer. It is noted that thermodynamics, heat transfer, fluid mechanics, applied mechanics, physics and others are basic subjects in cold regions engineering.

2.3   Sources of Technical Information in Cold Regions Heat Transfer

Technical information in cold regions heat transfer can be found in books, proceedings, review articles and papers in journals. The following sources are listed for reference.

International Conference or Symposium

1.   International Conference on Permafrost
2.   International Symposium on Ground Freezing
3.   International Workshops on Atmospheric Icing off Structures
4.   International Conference on Offshore Mechanics and Arctic Engineering
5.   IAHR (International Association for Hydraulic Research) Symposium on Ice Problems
6.   IAHR Ice Symposium
7.   International Conference on Port and Ocean Engineering Under Arctic Conditions
8.   International Symposium on Cold Regions Heat Transfer

Specialty Journals

1.   Cold Regions Science and Technology
2.   Journal of Glaciology (International Glaciological Society)
3.   Snow and Ice (Seppyo) (Japanese Society of Snow and Ice) (in Japanese)
4.   ASME - Journal of Offshore Mechanics and Arctic Engineering
                - Journal of Energy Resources Technology
5.   ASCE - Journal of Cold Regions Engineering

Monographs

1. Higashi, A., 1981, Fundamentals of Cold Regions Engineering Science, Kokon–Shoin Co. Ltd., Tokyo (in Japanese).
2. Lunardini, V.J., 1981, Heat Transfer in Cold Climates, Van Nostrand Reinhold Co., New York.
3. Johnston, G.H., ed., 1981, Permafrost – Engineering Design and Construction, John Wiley & Sons, New York.
4. Andersland, O.B. and Anderson, D.M., 1978, Geotechnical Engineering for Cold Regions, McGraw–Hill, New York.
5. Kinosita, S., ed., 1982, Ground Freezing – Its Control and Applications, Japan Soc. of Soil Mechanics and Foundation Engineering, Tokyo (in Japanese).
6. Kinosita, S., ed., 1982, Physics of Frozen Ground, Morikita Pub. Co., Tokyo (in Japanese).
7. ASCE Technical Council on Cold Regions Engineering Monograph Series, New York (see Chapter 2).
8. Michel, B., 1978, Ice Mechanics, Les Presses de L'Université Laval, Québec.
9. Colbeck, S.C., ed., 1980, Dynamics of Snow and Ice Masses, Academic Press.
10. Frost, W., ed., 1975, Heat Transfer at Low Temperatures, Plenum Press.
11. Ashton, G.D., ed., 1986, River and Lake Ice Engineering, Water Resources Publications, Littleton, CO.
12. Lunardini, V.J., 1988, Heat Conduction with Freezing or Thawing, CRREL Monograph 88–1, U.S. Army Cold Regions Research & Engineering Laboratory, 329 p.
13. Gray, D.M. and Male, D.H., eds., 1981, Handbook of Snow, Pergamon Press.
14. Lock, G.S.H., 1990, The Growth and Decay of Ice, Cambridge University Press.

Further details and the related review articles are listed in Chapter 2.

3. DR. R.R. GILPIN'S PUBLICATIONS RELATING TO FREEZING AND MELTING PHENOMENA AND HEAT TRANSFER

Dr. R.R. Gilpin had made significant and original contributions to freezing phenomena involving ice–water systems before his tragic death in 1982 at the age of only 39. His many technical papers cover such topics as freezing and melting heat transfer, jet cutting, gas dynamics and radiation, energy conservation, and the use of solar energy in industry and in building heating systems. Many of his publications dealing with freezing and melting heat transfer have since become standard references in later works by others and are widely quoted in the recent literature. His considerable influence on cold regions heat transfer is also evident in this monograph, but his publications relating to freezing phenomena in water pipes and frost heave model in freezing soil systems are considered as pioneering works.

Dr. Gilpin's unpublished final work at Clarkson University, Potsdam, New York is noteworthy and the following statement provided by W.R. Wilcox is quoted here for reference:

While at Clarkson University he made an important discovery on regelation. The classical picture of regelation in ice is that the force on the wire causes melting underneath it due to the volume change on melting. However, Dr. Gilpin theorized that a film of liquid water is always present between the wire and the ice. The force on the wire merely causes the water to flow around the wire. Dr. Gilpin discovered shortly before his death that regelation also occurs for organic compounds, for which the volume change on melting is the opposite of that of ice. Just as in regelation in ice, the velocity increased dramatically as the melting point was approached. Unfortunately, we have been unable to locate Dr. Gilpin's data. We plan to continue his pioneering experiments.

It is of interest to observe that Dr. Gilpin's work on "Modes of Ice Formation and Flow Blockage that occur While Filling a Cold Pipe", Cold Regions Science and Technology, Vol. 5, No. 2, Nov. 1981, pp. 163–171, is considered as a landmark breakthrough in the understanding of the physics of the freezing of flowing water in pipelines. This work renders obsolete the analyses and recommendations which are stated in pages 304–306, Chapter 17, "ASHRAE Handbook of Fundamentals", 1976. One clearly recognizes Dr. Gilpin's outstanding work.

Since his publications dealing with freezing and melting phenomena and heat transfer are scattered in various journals, his technical papers in this area are listed below for reference.

Freezing Phenomena and Convective Heat Transfer

1. Gilpin, R.R., 1975, Cooling of a Horizontal Cylinder of Water Through Its Maximum Density Point at 4° C, Int. J. Heat and Mass Transfer, Vol. 18, pp. 1307–1315.
2. Gilpin, R.R. and Wong, B.K., 1975, The Ground Temperature Regime and Its Relationship to Soil Properties and Ground Surface Cover, ASME Paper No. 75–WA/HT–98, presented at Winter Annual Meeting, Houston.
3. Gilpin, R.R. and Faulkner, M.G., 1976, Expansion Joints for Low Temperature Above Ground Water Piping Systems, Trans. ASHRAE, Vol. 82, Part 1, pp. 164–171.
4. Gilpin, R.R., 1976, The Influence of Natural Convection on Dendritic Ice Growth, J. of Crystal Growth, Vol. 36, pp. 101–108.
5. Gilpin, R.R. and Wong, B.K., 1976, 'Heat–Valve' Effects in the Ground Thermal Regime, ASME J. Heat Transfer, Vol. 98, pp. 537–542.
6. Gilpin, R.R., 1976, The Ground Temperature Boundary Condition Provided by a Moss Covered Surface, Paper No. 76–WA/Ht–61, ASME Winter Annual Meeting.
7. Gilpin, R.R., 1977, A Study of Pipe Freezing Mechanisms. In: Utilities Delivery in Arctic Regions, Environmental Protection Service, Environment Canada, Ottawa, ON, Report No. EPS 3–WP–77–1, pp. 207–220.
8. Gilpin, R.R. and Faulkner, M.G., 1976, Expansion Joints for Low Temperature Above Ground Water Piping Systems, Proc. Symp. on Utilities Delivery in Arctic Regions, Environmental Protection Service, Environment Canada.
9. Gilpin, r.R., 1977, Ice formation in Pipes, Proc. of 2nd Int. Symp. on Cold Regions Engineering, 12–14 August 1976, Cold Regions Engineers Professional Association, Fairbanks, AK, pp. 4–11.
10. Gilpin, R.R., 1977, The Effects of Dendritic Ice Formation in Water Pipes, Int. J. Heat and Mass Transfer, Vol. 20, pp. 693–699.
11. Gilpin, R.R., Robertson, R.B. and Singh, B., 1977, Radiative Heating in Ice, ASME J. Heat Transfer, Vol. 99, pp. 227–232.
12. Gilpin, R.R., 1977, The Effects of Cooling Rate on the Formation of Dendritic Ice in a Pipe with No Main Flow, ASME J. Heat Transfer, Vol. 99, pp. 419–424.
13. Gilpin, R.R., Imura, H. and Cheng, K.C., 1978, Experiments on the Onset of Longitudinal Vortices in Horizontal Blasius Flow Heated from Below, ASME J. Heat Transfer, Vol. 100, pp. 71–77.
14. Gilpin, R.R., 1978, A Study of Factors Affecting the Ice Nucleation Temperature in a Domestic Water Supply, The Canadian Journal of Chemical Engineering, Vol. 56, pp. 466–471.
15. Lipsett, A.W. and Gilpin, R.R., 1978, Laminar Jet Impingement Heat Transfer Including the Effects of Melting, Int. J. Heat Mass Transfer, Vol. 21, pp. 25–33.
16. Cheng, K.C., Takeuchi, M. and Gilpin, R.R., 1978, Transient Natural Convection in Horizontal Water Pipes with Maximum Density Effects and Supercooling, Numerical Heat Transfer, Vol. 1, pp. 101–115.
17. Imura, H., Gilpin, R.R. and Cheng, K.C., 1978, An Experimental Investigation of Heat Transfer and Buoyancy Induced Transition from Laminar Forced Convection to Turbulent Free Convection Over a Horizontal Isothermally Heated Plate, ASME J. Heat Transfer, Vol. 100, pp. 429–434.

18. Gilpin, R.R., Hirata, T. and Cheng, K.C., 1978, Longitudinal Vortices in a Horizontal Boundary Layer in Water Including the Effects of the Density Maximum at 4° C, Paper No. 78–HT–25, AIAA–ASME Thermophysics and Heat Transfer Conf., Palo Alto, Ca.

19. Gilpin, R.R. and Lipsett, A.W., 1978, Impingement Melting: Experiment and Numerical Simulation, Paper No. EN–8, Proc. Sixth Int. Heat Transfer Conf., Toronto, Vol. 3, pp. 43–47.

20. Gilpin, R.R., 1979, The Morphology of Ice Structure in a Pipe at or Near Transition Reynolds Numbers, AIChE Symp. Series, 189, Vol. 75, pp. 89–94.

21. Hirata, T., Gilpin, R.R., Cheng, K.C. and Gates, E.M., 1979, The Steady State Ice Layer Profile on a Constant Temperature Plate in Forced Convection Flow – I Laminar Regime, Int. J. Heat and Mass Transfer, Vol. 22, pp. 1425–1433.

22. Hirata, T., Gilpin, R.R. and Cheng, K.C., 1979, The Steady State Ice Layer Profile on a Constant Temperature Plate in Forced Convection Flow – II Turbulent Regime, Int. J. of Heat and Mass Transfer, Vol. 22, pp. 1435–1443.

23. Gilpin, R.R., 1979, A Study of Factors Effecting the Ice Nucleation Temperature in a Domestic Water Supply, Can. J. of Chem. Eng., Vol. 56, pp. 466–471.

24. Gilpin, R.R., 1979, A Model of the "Liquid–Like" Layer Between Ice and a Substrate with Application to Wire Regelation and Particle Migration, J. Colloid and Interface Sc., Vol. 68, pp. 235–251.

25. Gilpin, R.R., 1980, Theoretical Studies of Particle Engulfment, J., Colloid and Interface Sc., Vol. 74, pp. 44–63.

26. Gilpin, R.R., 1980, Wire Regelations at Low Temperatures, J. Colloid and Interface Sc., Vol. 17, pp. 435–448.

27. Gilpin, R.R., 1980, A Model for the Prediction of Ice Lensing and Frost Heave in Soils, Water Resources Research, Vol. 16, pp. 918–930.

28. Cheng, K.C. and Gilpin, R.R., 1980, Freezing Mechanism in Some Solidification Problems, Proc. Int'l. Symp. on Flow Visualization, Bochum, W. Germany, Flow Visualization II, pp. 497–502.

29. Gilpin, R.R., Hirata, T. and Cheng, K.C., 1980, Wave Formation and Heat Transferr at an Ice–Water Interface in the Presence of a Turbulent Flow, J. Fluid Mech., Vol. 99, pp. 616–640.

30. Gilpin, R.R., 1981, Modes of Ice Formation and Pipe Blockage that Occur While Filling a Cold Pipe, Cold Regions Sc. & Tech., Vol. 5, pp. 163–171.

31. Cheng, K.C., Inaba, H. and Gilpin, R.R., 1981, An Experimental Investigation of Ice Formation Around an Isothermally Cooled Cylinder in Cross–Flow, ASME J. Heat Transfer, Vol. 103, pp. 733–738.

32. Gilpin, R.R., 1981, Ice Formation in a Pipe Containing Flows in the Transition and Turbulent Regimes, ASME J. Heat Transfer, Vol. 103, pp. 363–368.

33. Gilpin, R.R., 1981, Surface Spreading as a Bio–Transport Mechanism, J. Biol. Phys., Vol. 9, pp. 109–132.

34. Gilpin, R.R., 1982, A Frost Heave Interface Condition for Use in Numerical Modelling, Proc. of the Fourth Canadian Permafrost Conf., The Roger J.E. Brown Memorial Volume, Associate Committee on Geotechnical Research, National Research Council of Canada, H.M. French, ed., pp. 459–465.

35. Gilpin, R.R., 1973–74, The Ablation of Ice by a Water Jet, Trans. CSME, Vol. 2, No. 2, pp. 91–96.

36. Cheng, K.C., Obata, T. and Gilpin, R.R., 1988, Bouyancy Effects on Forced Convection Heat Transfer in the Transition Regime of a Horizontal Boundary Layer Heated from Below, ASME J. of Heat Transfer, Vol. 110, pp. 596–603.

37. Cheng, K.C., Inaba, H. and Gilpin, R.R., 1988, Effect of Natural Convection on Ice Formation Around an Isothermally Cooled Horizontal Cylinder, ASME J. of Heat Transfer, Vol. 110, pp. 931–937.

38. Cheng, K.C., Obata, T. and Gilpin, R.R., 1988, An Experimental Investigation on Mixed Convection in Turbulent Boundary Layer Flow over a Horizontal Flat Plate at Constant Wall Temperature, ASME Procs. of 1988 National Heat Transfer Conference, HTD–96, Vol. 2, pp. 9–18.

1. Zukowski, E.E. and Gilpin, R.R., 1967, Large Amplitude Electrothermal Waves in a Nonequilibrium Plasma, Phys. of Fluids, Vol. 10, pp. 1974–1977.
2. Zukowski, E.E. and Gilpin, R.R., 1967, Electrothermal Instability, Proc. 8th Symp. on Engineering Aspects of Magnetohydrodynamics, Palo Alto, 1967.
3. Gilpin, R.R. and Zukowski, E.E., 1968, Experimental and Theoretical Studies of Electro-Thermal Waves, Proc. of the Ninth Symposium on Engineering Aspects of Magnetohydrodynamics, Tallahoma, pp. 116–125.
4. Gilpin, R.R. and Zukowski, E.E., 1969, Experimental and Theoretical Studies of Electrothermal Waves, AIAA, Vol. I, pp. 1438–1445.
5. Gilpin, R.R., 1970, Simplified Derivation of the Condition for Electrothermal Instability, Phys. of Fluids, Vol. 6, pp. 1642-1645.
6. Gilpin, R.R., 1970, Approximation to the Radiative Transport of Energy in an Optically Thick Line, UTIAS Report No. 150.
7. London, G., Gilpin, R.R., Schiff, H.I. and Weldge, K.H., 1971, Collisional Deactivation of 0('s) by 0_3 at Room Temperature, J. Chem. Phys., Vol. 54, pp. 4512–4513.
8. Weldge, K.H. and Gilpin, R.R., 1971, Time–of–Flight Spectroscopy of CO_2 Photodissociation in the Vacuum Ultraviolet Electron Emission from Cesium Surface by Metastable Singlet Oxygen Atom, J. Chem. Phys., Vol. 54, pp. 4224–4227.
9. Gilpin, R.R. and Weldge, K.H., 1971, Time–of–Flight Spectroscopy of Metastable Photodissociation Fragments – N_2O Dissociation in Vacuum UV, J. Chem. Phys., Vol. 55, pp. 975–978.
10. Gilpin, R.R., Schiff, H.I. and Weldge, K.H., Photodissociation of 0_3 in the Hartley Band, J. Chem. Phys., Vol. 55, pp. 1087–1094.
11. Gilpin, R.R., 1972, An Approximation to Radiative Transport in a Non–Grey Gas, J. AIAA, Vol. 10, pp. 1340–1342.
12. Gilpin, R.R., 1972, Some Analytical Solutions to a Non–linear Heat Transfer Problem, Phys. of Fluids, Vol. 15, pp. 1529–1531.
13. Gilpin, R.R., Heat Transfer in a Horizontal Layer with Radiative Heating, Trans. CSME, Vol. 1, No. 1, pp. 213–218.

Some aspects of Dr. R.R. Gilpin's work on freezing phenomena in water pipes are well summarized in the following article: Carey, K.L., 1982, The Freezing and Blocking of Water Pipes, Cold Regions Technical Digest, No. 82–1, U.S. Army Cold Regions Research and Engineering Laboratory.

4. CONTENTS OF A MONOGRAPH

The original goals were to present reasonably comprehensive overviews on freezing and melting heat transfer in engineering with main emphasis on ice–water systems in cold regions heat transfer. Freezing and melting phenomena are of considerable importance in a wide range of technological problems in refrigeration, cryogenics, materials processing, food processing, ice formation in water, air, earth and life relating to various industrial processes and activities, and various natural phenomena in cold regions. The topics are also closely related to low–temperature engineering and physics, and space technologies.

Since the subject is so extensive and the topics are so numerous, it is desirable to have a special monograph dealing with selected topics on freezing and melting phenomena relating to ice–water and related systems. It is of particular interest to observe that in recent years books and monographs dealing with two–phase flow and heat transfer (boiling and condensation phenomena) are numerous and certainly reflect the recent advances and progresses in power generation, refrigeration, cryogenics, and utilization of nuclear energy. In contrast, the number of monographs and books dealing with freezing and melting phenomena in cold regions is rather limited in spite of equal importance in technologies and engineering. It is hoped that the present monograph represents a small step in this direction. It is also of interest to observe that most heat transfer texts cover boiling and condensation, but solidification and melting are usually neglected.

The articles in this monograph are arranged under the following sections:

1. Historical and Recent Developments in Cold Regions Heat Transfer
2. Freezing and Melting Heat Transfer
3. Ice Formation
4. Convective Heat Transfer in Ice–Water and Convective Instability
5. Ice and Snow
6. Frost Formation
7. Solidification and Melting of Solutions
8. Frazil Ice
9. Ground Freezing and Frost Heave
10. Atmospheric and Marine Icings
11. Special Topics (Casting and Welding Processes)

Progress in science usually means generalization. It is useful to observe the relationships between the ice formation problems and the analogous problems in welding and casting processes, for example, from the viewpoints of physical phenomena and theoretical (numerical) analysis.

A few chapters in this monograph deal with special technical contributions, but most chapters are review articles on specific topics. This monograph can be regarded as a source book. The editors are fully aware of the shortcoming of this monograph and many important topics were not covered.

Frontispieces are included to illustrate the various ice formation phenomena and the related engineering problems in cold regions. This monograph provides a focus on a class of freezing and melting heat transfer problems in recent technological applications.

Since the title of the monograph is quite general in order to provide a focus on freezing and melting heat transfer problems in engineering, it is useful to list other relevant topics which are not covered in detail in this volume for future reference. These are:

(1) Regelation phenomena (organic crystal); (2) melting of snow cover and ice; (3) refreezing of partially melted snow; (4) freezing of underground pipes; (5) phase change heat transfer in permafrost; (6) natural convection during melting of ice in pure and saline water (iceberg utilization); 7) freezing of saline water; (8) frost formation and its effect on heating, ventilating and air conditioning equipment performance; (9) freezing of water unsaturated soil; (10) formation of saline permafrost; (10) artificial soil freezing in construction; (11) freezing and thawing of moist soil in the immediate vicinity of buildings; (12) applications of heat pipes and thermosyphons in cold regions; (13) thermophysical properties of ice, snow, sea ice, frost, pure and saline water, freezing soils, and other materials at low temperatures; (14) thermal insulation at low temperatures; (15) freezing of soils (porous media) saturated with water–salt solutions; (16) melting of ice and freezing of water in geophysical applications; (17) ice formation in rivers, lake and ocean; (18) surface energy between ice and liquid water.

HISTORICAL AND RECENT DEVELOPMENTS

Chapter 2

Historical and Recent Developments in the Research of Cold Regions Heat Transfer—Ice in Air, Water, and Earth

K. C. CHENG
Department of Mechanical Engineering, University of Alberta,
Edmonton, Alberta, Canada T6G 2G8

YIN-CHAO YEN
U.S. Army Cold Regions Research and Engineering Laboratory,
Hanover, New Hampshire 03755, USA

ABSTRACT

A brief review of historical and recent developments of ice
formation problems in air, water and earth was made covering such
subjects as atmospheric and marine icings of structures, permafrost and
ground freezing (frost heave), river and lake ice (frazil ice and
supercooling), arctic oil and gas pipelines, and heat transfer with
freezing or melting from the unified view-point of cold regions heat
transfer. An attempt was made to review the varied technical fields
involving ice formation phenomena from the common viewpoint of heat
transfer to show the scope and subjects of cold regions heat transfer
engineering.

CONTENTS

1. INTRODUCTION

Heat and gravity are universal phenomena. Since Newton's formulation of the universal law of gravity, the theory of gravity has needed only a minor modification by Einstein's general theory of relativity. Thus, the laws of mechanics are equally valid for planetary motion and all the motions in the gravitational field, for example. In contrast, heat phenomena are extremely complicated and many problems are still under investigations by scientists and engineers. The heat phenomena are characterized by temperature and temperatures ranging from the surface temperature (about 6000°K) of the sun and 0°K are of interest. On the other hand, human activities occur in the atmosphere and may also reach 1000 m or more below the surface of the earth. Such subjects as mathematics, mechanics, heat, light, sound, electricity and magnetism, fluid mechanics, thermodynamics and many others were developed to explain various physical phenomena in all fields of science and technology which may occur on the earth. The historical developments of various subjects in science and technology are well covered in various monographs.

It is of interest to observe that monographs on the historical developments of such subjects as strength of materials and elasticity, thermodynamics, heat, fluid mechanics, electricity and magnetism, for examples, are available; but at present, no monograph is available on the historical development of heat transfer science or engineering. Recently some aspects of heat transfer science and engineering were presented in [1]. Review and some observations on the historical development of heat transfer from Newton to Eckert for the years 1700 to 1960 are presented in [2,3]. In view of recent progress and advances in cold regions engineering relating to engineering activities in the arctic regions, it is of considerable interest to review the phase-change heat transfer phenomena in permafrost engineering or ground freezing, river and lake ice engineering, atmospheric icings of structures, and freezing and melting heat transfer in engineering from the viewpoints of historical perspectives, recent advances and ice formation. It is, indeed, in the earth's atmosphere, on the arctic ocean, in rivers and lakes and in the earth's ground that ice produces its most spectacular phenomena and effects and has the most direct influence on man. Sea ice, for example, is regarded as a product of interaction between the atmosphere and the ocean and similar remarks apply also to river and lake ices and various forms of ground ice.

The great advances in our understanding of freezing and melting phenomena are due not only to intensive and comprehensive works on many detailed problems, but principally to the pioneering achievements of a few exceptionally gifted researchers. These pioneers have opened up completely new fields for us and thus made it possible for our knowledge to advance in leaps and bounds in recent years. It has been pointed in the past that achievements of this kind by highly gifted researchers are of the utmost importance to progress in any scientific or engineering field.

The purpose of this paper is to present a historical review on freezing and melting heat transfer in general and some recent advances in the following topics:

1. Heat transfer with freezing or melting
2. Atmospheric and marine icings of structures
3. River and lake ice (frazil ice and supercooling)
4. Permafrost and ground freezing (frost action)
5. Hot-oil and chilled-gas pipelines in permafrost
6. Freezing phenomena in pipes
7. Applications of heat pipes and thermosyphons
8. Applications and utilization of cold regions technology

The scope of this review will be rather limited and serves only as the starting point for future studies. With the availability of modern textbooks, monographs on special topics, and journals and reports, present-day students, research workers and practicing engineers seldom feel the practical need to study or read the original pioneering papers relating to freezing or melting phenomena up to the end of the 19th century. However, it would be most interesting to trace the genesis of ideas and the development or evolution of concepts and theories in cold regions heat transfer engineering leading to the present state of the art of ice formation theories.

2. A BRIEF HISTORICAL REVIEW OF ICE FORMATION PHENOMENA IN AIR, WATER, AND EARTH

2.1 Freezing and Melting Heat Transfer

Newton, in his classical paper on "A Scale of the Degrees of Heat" published in the Philosophical Transactions in 1701, made a statement which involves the hypothesis that the rate of cooling of a body is proportional to its excess of temperature over the surrounding medium. This surmise has since been tested experimentally by Dulong and Petit and many others, and has been shown to be true within a small range of temperatures. Newton's only paper on heat is the source for Newton's Law of Cooling and the related problem is now known as Newton's cooling problem in heat transfer.

The fundamental laws of heat flow were first propounded by J. Biot (1804, 1816) and were given their definitive forms by Fourier (1807 and 1811). The mathematical theory of heat conduction was established by Joseph Fourier whose work on "The Analytical Theory of Heat" was published in Paris in 1822 [4]. Fourier's analysis introduced the Fourier series and integral which play so important a role in the theoretical treatment of all wave phenomena. Fourier's work on heat conduction was extended by Laplace (1820), Poisson (1821), Lamé (1839), C. Neumann (1864), C. Baer (1888) using Bessel functions, Wangerin (1875-76), Niven (1880), and many others to various heat conduction problems and with further development in applied mathematics, the mathematical theory of heat conduction was well established by the end of the 19th century. Fick's law for diffusion (1855), Darcy's law for conduction of fluids in porous media and Ohm's law for the conduction of electricity are mathematically analogous to Fourier's law of heat conduction and the related physical phenomena are somewhat analogous. Heat conduction theory has since found many applications in the physical, geological, biological and medical sciences. Theoretical results in heat conduction were well summarized by Carslaw in 1906 [5], 1921 [6], and Carslaw and Jaeger in 1959 [7]. Carslaw's work is considered to be classic and provides an excellent bibliography particularly for the references in the 19th century [6].

Josef Stefan (1835-1893), who is also known for his empirical discovery of Stefan-Boltzmann's law in thermal radiation in 1879, published four papers relating to heat conduction with phase change (melting and solidification) [8-11] which become the source of Stefan Problem. Stefan's work and the more general result known as Neumann's solution given by Franz Neumann in his lectures in the 1860's are presented in Chapter 11 of [7]. Currently three monographs on the Stefan Problem are available [12-14]. Luikov [15] notes that more than 50 original works on ground freezing problems were reported in the 19th century. He also notes that the Stefan problem was treated in 1831 by Lamé and Clapeyron, members of the Russian Academy of Sciences. The phase-change heat conduction problems are also reviewed by Ozisik [16].

The concept of specific heat is due to Joseph Black (1760) and his best known discovery was that of "latent heat" (1763). Black is credited with the clear distinction between heat and temperature. Black also explained the phenomenon of supercooling of water. In 1724, Fahrenheit described the supercooling of water. In 1772, Johan Carl Wilcke published a paper on the cold produced when snow melts giving the value of latent heat. He also presented a table of specific heats based on experiments in 1781. The discovery of specific and latent heat is discussed by McKie and Heathcote [17]. The most notable feature of the scientific study of heat in the 18th century is the experimental work in calorimetry by Black, Wilcke, Lavoisier, Laplace, Count Rumford and the eventual crowning experimental success of Joule in the 19th century.

Heat in the 17th and 18th centuries and the experimental study of heat and radiant energy in the 19th century are well reviewed in monographs such as R. Taton, History of Science (1964, 1965). The maximum density of water at 4°C, was demonstrated by Hope in 1805. He conclusively confirmed the 17th century observation that water expands as it nears its freezing point. The maximum density phenomenon of water was also studied by many physicists in the 19th century.

The history of the general equations of hydrodynamics was treated by Truesdell [18]. The Navier-Stokes equation for the motion of a viscous fluid was derived by Navier in 1827 based on molecular hypothesis and by Stokes in 1845. The modern exposition of the Navier-Stokes equations is based on Stokes' derivation. The energy equation for flowing fluids was derived by Fourier in 1820 and by Poisson in 1835, but neither of them gave any solution of it. In the early literature, the energy equation is known as Fourier-Poisson's equation. The intention here is to present a brief review on the historical landmarks in heat transfer relating to freezing and melting heat transfer; further details can be found in [2,3].

2.2 Terrestrial Heat Flow

Historical review of terrestrial heat flow is well presented by Bullard [19]. For brevity, his abstract will be quoted here and further details are referred to in his paper:

The first reference to high temperatures within the Earth known to the writer is by J.B. Morin in 1619; there is no known reference to the matter in the works of Agricola. Boyl, in 1671, raised many of the questions discussed in recent work. Little interest was taken in underground temperatures during the 18th century. The subject was energetically pursued by Committees of the British Association in 1868-1883 and 1935-1939. Measurements were first made at sea in 1950. In spite of the existence of about 2000 measurements of heat flow, the temperatures deep within the Earth are still very uncertain. Observations of magnetic variations combined with the study of samples from Moholes may give better estimates.

Terrestrial heat flow is the study of the outflow of heat from the earth's interior and deals with all geothermal problems. Geothermal processes play an important role in all theories about the origin, development, and surface features of the earth. A comprehensive study and bibliography on the subject of ground temperature can be found in [19,20].

2.3 Ice Physics

The history of the study of snow crystals in Europe is given by Hellmann (1893). Nakaya [21] also presented a brief historical account. The first sketch of snow crystals observed with the eye was made by Olaus Magnus about 1550. Kepler first pointed out that snow crystals have hexa-gonal symmetry. In 1635, Descartes published his sketches of snow crystals in Amsterdam [21]. Robert Hooke presented a number of sketches of snow and frost crystals observed by microscope in Micrographia published in 1665. Hook's work was followed by Friedrick Martens, Donat Rossetti (1681), William Scoresby (1820), Doi (1832) in Japan, and James Glaisher (1855). With the development of photo-micrography, research on snow crystal has made remarkable progress since the end of the 19th century and Bentley's book (1931) containing a large collection of photographs is most well-known. Snow crystals were studied later by Dobrowolski (1922) and Nakaya (1932). The first success in the artificial production of snow crystals was made by Nakaya in 1936. The unique properties of ice have interested scientists for hundreds of years and one may mention the experiments carried out by Bacon, Hooke, Faraday, and Kelvin. In recent years significant advances have been made in understanding many of the properties of ice in terms of its molecular structure providing insights into more complex molecular systems [22-26]. Dorsey's book on Properties of Ordinary Water Substance (1940, 1968) gives a nearly complete list of some 2000 references prior to 1938, including 600 specifically on ice. Ice is now attracting the interest of not only physicists and chemists, but also of biologists and engineers. It is of interest to note that much of the detailed understanding of the structure and properties of ice have been attained during the past 40 years [24]. The applied aspects of ice physics also continue to increase in importance to geophysics, atmospheric sciences, hydrology, ice engineering, and cryobiology.

2.4 Permafrost and Ground Freezing

The history of permafrost was reviewed by Williams [27,28] and Harris [29]. Although the existence of permafrost was known to the inhabitants of Siberia for centuries, not until 1836 did scientists of the western world take seriously reports of thick frozen ground existing under northern forest and grasslands [30]. During the 19th century, permafrost was studied actively in Siberia and the results of the research were applied in the development of the region. In contrast, systematic studies of perennially frozen ground were not undertaken in the United States and Canada until World War II. In 1895, I.V. Mushketov, with others, wrote the first "Instructions for Investigation of Frozen Soils in Siberia". Sumgin published in 1927 the first textbook entitled "Permafrost within the U.S.S.R.", which established the study of permafrost as an independent branch of science. The development of research institutes in the USSR, Japan, USA, Canada, France and China since 1930 and publications on permafrost research are described by Harris [29]. In 1943, Muller [31] summarized the available knowledge on permafrost for the U.S. Army, including Soviet literature. References on all aspects of permafrost research can be found in the Proceedings of the five International Conferences on Permafrost since 1963 (1st) and of the four Canadian Permafrost Conferences (1st in 1958, 4th in 1981).

The International Symposia on Ground Freezing were held regularly since 1978 and the proceedings are available. A brief history of geotechnical activities, and recent scientific research in the north relating to the Trans-Alaska Pipeline and the gas pipeline is given by Williams [27,28]. The notes given at the end of each chapter provide considerable insight into the history of the scientific study of the effect of ground freezing and thawing. The historical review on soil freezing and frost heaving with special applications to roads and railroads is given by Beskow [32]. The earth heaving phenomenon was first mentioned by Urban Hiarne in 1694. Frost heaving seems to have been well known by the middle of 1700, when a full explanation was given by E.O. Runeberg in 1765 [32]. Since 1960, artificial soil freezing methods for various underground construction works including the underground storage of a cold liquid (LNG) are quite common and one observes considerable research activities in both basic and applied areas. Simon Johansson (1914) was the first investigator to find that during freezing of a fine-grain soil, water flows by capillary action to the freezing layers and explained how frost heaving and frost boils occur [32]. The phenomenon became generally recognized in the 1930's [27]. Taber (1929, 1930) carried out careful investigations on frost heaving. Beskow also presented progress of Scandinavian soil frost research from 1935 to 1946 [32]. The Russian work is described by Tsytovich [33] and the origins and developments of the mechanics of frozen soils in the USSR from the beginning of this century are explained. Prior to the construction of the Alaska-Canada highay during World War II, little attention was paid in the United States to permafrost problems.

2.5 Atmospheric and Marine Icings of Structures
Atmospheric Icing

Aircraft and helicopter icings can occur on the ground as well as in the air and are caused by particular meteorological conditions with temperature in the range $0 \sim -40°C$. The ground icing problems are similar to those of ground transportation vehicles. The aircraft icing problem begins with the appearance of the modern aircraft and the physics of aircraft icing and the statistical data of atmospheric icing

conditions of various measured degrees of severity have been under
investigation since late 1920's. Work on aircraft icing was
particularly active in the period 1940 to 1960 and the physics of the
ice-accretion process, its causes (natural environments) and preventive
measures are well understood. The literature on aircraft icing is very
extensive and only a brief review will be made here.

The early literature on aircraft icing can be found in the
transactions of the American Society of Mechanical Engineers [34]
(1941-1954) and the Journal of the Aeronautical Science (1934-1958)
among many other sources. In 1930, Scott [35] reported the work on
"Ice Formation on Aircraft and Its Prevention" considering frost
formation from water vapor, and glaze and fog ice formation from
freezing water droplets. Simpson's paper [36] on "Ice Accretion on
Aircraft" in 1942 was considered one of the classics in the field.
Hardy [37] presents a comprehensive report on aircraft icing. Early
aircraft icing literature can be found in [38-54]. Geer [51] lists the
most important physical conditions of atmospheric ice in relation to
airplanes giving 71 references and discusses preventive measures.
Tribus [49] presents a report discussing the design problems in heated
surface anti-icing equipment, consolidating and compiling all of the
heat transfer data known to be available from past works. The analysis
of an unheated icing surface was made by Tribus [46] and Messinger
[54]. Experimental data [39] and analysis [44] of heat transfer over a
small cylinder in icing conditions were presented. A brief review of
early works on aircraft icing is given by Hardy [55] and Spencer [56].

In 1957, Brun [57] made a general survey on icing phenomenon (icing
conditions, mechanics of icing, thermodynamics of icing), and presented
brief remarks on airplane icing and defensive measures against it,
quoting 97 references. A more complete introductory documentation can
be found in [37, 58-60]. Brun's survey paper [57] also provides an
excellent source for bibliography up to 1957. In view of the
comprehensive nature of the report [57] on aircraft icing problems the
contents of the report [57] are listed in Table 1 to show the scope of
the review.

Some aspects of helicopter icing [61] and Canadian research in the
field of helicopter icing [62, 63] are discussed by Stallabrass. The
early studies in the laboratory and under natural icing conditions on
the icing of simple objects such as spheres and cylinders show that the
factors affecting rate of ice accretion are [44, 62]: air temperature,
air density, atmospheric water content, number and size distribution of
water droplets, wind velocity, shape and size of object on which ice is
accreting, velocity of droplet relative to the object, and time taken
for freezing to occur. Accretions of ice occur in two forms [64]:
glaze forms in wet-growth conditions with a surface temperature at 0°C
and rime forms in dry-growth conditions with a surface temperature
below 0°C. As the combination of temperature and water flux increases,
the density of rime increases from 0.01 Mg/m^3 for soft rime to a value
greater than 0.6 Mg/m^3 for hard rime. For the more severe conditions,
the properties of rime begin to merge with those of glaze.

Table 1   <u>Contents of AGARDograph 16 [57] on Icing Problems</u>

Chapter 1   E.A. Brun, General Survey, 97 references

Chapter 2   D. Fraser and K.G. Pettit, Icing Conditions to be Considered
in the Design of Protection Systems (The Large Scale
Collection of Icing Data, Results of Statistical
Measurements, Icing Requirements), 14 references

Chapter 3   D.G.A. Rendel and F.J. Bigg, The Measurements of Icing
Conditions (Measurement of Liquid Water Content, Measurement
of Droplet Diameter, Mesurement of Air Temperature,
Calibration of Instruments, Positioning of Instruments on
the Aircraft), 40 references

Chapter 4   R.B. Morrizon, J.A. Nicholls, R.E. Cullen, H.E. Stubbs,
Icing Wind Tunnel Tests (Conditions to be Simulated in an
Icing Wind tunnel, Design Considerations for an Icing Wind
Tunnel, Evaluation of the Results of Icing Wind Tunnels,
Brief Descriptions of Icing Wind Tunnel Facilities (NACA
Icing Tunnel, Lewis Flight Propulsion Laboratory, Cleveland,
Ohio, 1944; NRC Icing Wind Tunnel, Low Temperature
Laboratory, Ottawa, Canada; Univ. of Michigan Icing Wind
tunnel, Ann Arbor, Michigan; Mont Lachart Icing Tunnel,
France; Artington Icing Tunnel, Great Britain; Other Tunnels
(Lockhead, Boeing, Goodrich, The Installation on Mt.
Washington), 19 references

Chapter 5   G.C. Abel and J.K. Thompson, Flight Tests in Simulated Icing
Conditions (Water Spray Rigs, Water Tanker Aircraft).

Chapter 6   J.K. Thompson, Icing Flight Test Concepts (Test Directives,
Test Objectives, Instrumentation, Scope of Icing Tests,
Operational Problems)

In the case of small water droplets, around 10 and 20 microns in
diameter, the droplets are deflected by the air as it flows around the
aircraft so that only a fraction of these in its path strikes the
surface (collection efficiency). In the case of cylinders, the
collection efficiency can be calculated from the data of Langmuir and
Blodgett [65] who extended earlier works by G.I. Taylor and Glavert
[55]. Langmuir's theoretical work [65] on the trajectories of water
droplets in air passing a cylinder at high velocities is quoted very
often in the literature and is considered to be classic. Ice in the
atmosphere (ice nuclei, the origins of ice particles in clouds, the
growth and nature of ice particles in clouds and their role in precipitation processes, and ice-water interactions) is reviewed by Hobbs [66]
and a more comprehensive account is given in his monograph [25]. Work
on ice formation in the atmosphere has followed two distinct lines,
hailstone growth and aircraft icing. Recent investigations of hailstone
growth have provided also a general understanding of ice accretion
processes.

With the introduction of jet-powered aircraft flying higher (30,000
ft) above the weather conditions that cause icing and faster, research
in aircraft icing declined since icing problems occur at altitudes
below 20,000 ft (6100 m) and at airspeeds below 300 m/h (134 m/s),

primarily in clouds ranging from about 0 to -20°C in temperature [67].
The requirements for all-weather flying for light aircraft since 1960
due to the development of electronic systems and helicopter de-icing
systems have brought about a renewed interest on aircraft icing which
may occur in lower altitudes and air speeds. Recent aircraft icing
research at NASA is described in [68].

The increasing use of both aircraft and industrial gas turbine
engines for stationary powerplant applications in subfreezing environments has resulted in operational difficulties and occasional
catastrophic engine failures caused by ice formations on components
within the engine intake system. Over the years this icing problem has
been the object of considerable study in relation to aircraft gas
turbine applications. Chappell [108] reviews the atmospheric phenomena
which can give rise to icing conditions.

Atmospheric icing also occurs on transmission lines (power or
communication), radar towers and other structures in cold regions.
Recent literature can be found in the proceedings of International
Workshops on Atmospheric Icing of Structures (1982, 1984, 1986, 1988)
and other sources dealing with the physics of ice accretion, simulation
and modelling, design-oriented research, meteorological measurements
and damage observation, and ice load measurements and design practices.
Recent computer modelling of atmospheric ice accretion is reported by
Ackley and Templeton (1979) [67], Lozowski, Stallabrass and Hearty
(1979) [69], Oleskiw and Lozowski (1980) [70], Lozowski and Oleskiw
(1981) [71], Lozowski and Oleskiw (1983) [72], and Lozowski,
Stallabrass and Hearty (1984) [73]. Recent literature on icing on
structures (types of ice accretion, frost, rime, glaze, spray ice,
conditions governing type of accreted ice, accretion rates, spray
icing, structural design factors, techniques for minimizing structural
icing, data collection needs) was well reviewed by Minsk [74]. Icing
and snow accretion on electric wires was discussed by Kuroiwa [75, 76]
among many others. Atmospheric icing of transmission lines is reviewed
by Henry [112].

Marine Icing

Shipboard icing has been known for several centuries and noted in
the scientific literature for more than 100 years (Nature, 1881). Ship
icing research is relatively recent and the initial stimulus was
provided by the British Shipbuilding Research Association by sponsoring
cold room model tests in 1955 (Polar Record, 1958), following the loss
of the British trawlers Lorella and Roderigo. Ship icing is caused
primarily by sea spray. Canadian research on trawler icing at the
National Research Council started in 1968 after the loss of three
Canadian fishing vessels in preceding winters [77]. Japanese research
on ice accretion on ships started around 1960 after the loss of 24
fishing vessels and 375 crews [78]. Sutherby [79] reports icing
problems on ships based on the experiences of the Royal Navy ships
serving in the arctic waters for a period 1914 to 1949. The
exploration of oil in the Beaufort sea area using artificial islands
since around 1970 also presents serious atmospheric and marine icing
problems for superstructures. Icing problems also exist for coastal
and offshore structures.

Marine icing is reviewed by Fein and Freiberger (1965) [80], Lock (1972) [81] and Lozowski and Gates (1985) [82]. A comprehensive account of ice accretion on ships and marine structures is given by Shellard (1974) [83], Minsk (1977) [84], Stallabrass (1975, 1980) [77, 85], Horjen (1981) [86], Itagaki (1977) [87], and Makkonen (1984) [88]. These reviews and reports also present comprehensive lists of references. These reviews generally discuss the physics of the icing processes (drop deposition, heat balance, icing due to fog, precipitation particles, water vapor in air), factors causing the icing of ships and stationary offshore structures, meteorological conditions, geographical factors, theoretical methods for calculating the intensity of atmospheric icing (ice accretion process), test results in icing wind tunnels, physical modelling and ice prevention or controlling methods.

Atmospheric icing can occur on upper parts of a ship that are not reached by spray droplets. On stationary structures spray icing is usually limited to the bottom 10 m. The most common mechanism of atmospheric icing at sea is the formation of evaporation fog. Marine icing is a potential danger whenever air temperatures are below the freezing point of water and the sea temperature is 6°C or lower. The sea spray accumulation rate on such objects as cylinders or flat plates depends on water source temperature, air temperature, wind speed, liquid water content, droplet size, and other factors such as ship size and configuration, angle between ship course and water heading, and ship speed. The icing of vessels can result from the following causes: (a) supercooled fog, (b) freezing rain or freezing drizzle, (c) falling snow, in particular wet snow, (d) freezing sea spray (vessel and/or wind generated). Of these, the freezing of sea spray on a vessel is the most dangerous cause of icing.

A series of five Japanese studies are presented on the ice accumulation on ships (1963-1974) [89-93]. The related studies on ice formation in wind tunnels are also of interest [94-96]. Makkonen (1985) [97] developed an icing model which can be used in estimating the relative icing rates on objects of various sizes. His model also includes the effect of salinity so that it can be applied to icing of sea sprays on offshore structures. Lozowski, Gates, and Makkonen (1986) [98] report a rather extensive review on the existing empirical and numerical simulation icing models and discuss comparison of model predictions with results from wind tunnel studies. They examine the effects of droplet temperature, sponginess of the ice accreted, the heat transfer coefficient, and salinity. Gates et al. (1986 [99] discuss the significance of sponge ice (ice which traps unfrozen water in an ice matrix) for the modelling of marine ice accretion. Jurgensen (1986) [100] gives a very general discussion on the mechanisms of atmospheric and sea spray icings and countermeasures for ice removal by passive and active methods. Stallabrass's work [85, 101-105] on ship icing was summarized in [77, 85].

The equations of motion of particles in a stream of air appear to have been first developed by Albrecht [106] and later applied by Taylor [107] to droplets causing icing of aircraft. Langmuir [65] presented droplet trajectory results in graphical form for simple objects such as spheres, cylinders, and flat plates [77]. The rate of icing in the wet growth conditions typical of ship icing and icing in freezing precipitation depends on the rate at which the heat liberated in the freezing process is transferred to the environment.

A theoretical model for the heat transfer from the front half of a rough cylinder, based on boundary-layer theory, is described by Makkonen and Stallabrass (1985) [109]. The trajectories and temperature histories of small water droplets in icing wind tunnels is discussed by Gates, Lam and Lozowski (1988) [110]. Only the small droplets (10 μm diameter) remain in mechanical and thermodynamical equilibrium with the air stream and the larger droplets are sorted out due to gravity and inertia effects. Heat and mass transfer from freely falling drops at low temperatures was discussed by Zarling [111]. Because of recent industrial activities relating to oil and gas in the offshore region of the arctic seas, one expects further advances in the understanding of physical processes and the physical modelling of the marine icing problem in the future.

A recent icing technology bibliography (1987) [451] lists over 2,000 references under 26 different topics: Meteorology of Icing Clouds; Meteorological Instruments; Propeller Icing; Induction System Icing; Gas Turbine Engine and Inlet Icing Studies; Wing Icing; Windshield Icing; Ice Adhesion and Mechanical Properties; Heat Transfer; Helicopter Climatic Tests and Icing; Helicopter Rotor Blade Icing; Engine Snow Ingestion and Snow Measurements; Droplet Trajectories and Impingement; Ice Accretion Modelling; Icing Test Facilities and Icing Simulation; Aircraft Ice Formation; Runway Icing; Microwave Sensing and Ice Protection Systems; Iced Airfoil Performance; Land and Sea Ice Studies; Fluid and Two-Phase Flow Dynamics; Liquid Evaporation and Ice Crystal Formation Studies; Electrical modelling; Random Icing and Miscellaneous.

2.6 Ice Formation in Rivers and Lakes
Icing Phenomena

The hydrologic cycle is the central focus of hydrology dealing with the waters of the earth, their occurrence, circulation, distribution, and chemical and physical properties. From ancient times to about AD 1400, the concept of the hydro-logic cycle was speculated upon by the philosophers and biblical scholars. During the Renaissance, hydrology became an observational science and a correct understanding of the concept of the hydrologic cycle was achieved. Ice exists in various forms in the hydrologic cycle. Water is the most abundant substance on earth and is the principal constituent of all living things. Water is a key factor in air-conditioning the earth for human existence and in influencing the development of civilization. The kinds of precipitation (natural phenomena) may be classified as follows [113]: dew, hoarfrost, fog drip, rime, drizzle, freezing drizzle, rain, snow, snowflakes, snow grains, snow pellets, sleet, ice pellets, and hails.

In cold regions of the world, ice forms on rivers and lakes every year and remains for various durations. Ice in rivers and lakes is well discussed in [114, 115]. River and lake ice is of great practical and economic importance. A history of ice engineering is discussed briefly in [116]. Ice engineering had its origins in the 19th century and scientific studies on river ice formation were begun early in the 20th century (by Altberg in the U.S.S.R., Barnes in Canada, and Devik in Norway) [116]. Barnes' book [117] on Ice Engineering is considered to be a classic and its bibliography is a valuable source for the literature up to 1920, and particularly in the 19th century. Many interesting observations of ice formation phenomena, which are still under investigation today, are presented, and it provides a good general introduction to the subject. The literature on ice formation, growth and decay in rivers and lakes is very extensive and can be found in recent monographs [116-123] and reviews [64, 124-130]. Recent literature on river-ice problems can be found in the proceedings of the symposia on ice problems sponsored by the International Association for Hydraulic Research (IAHR) since 1970. Discussion here will be limited to a brief literature review of some of the heat transfer aspects of ice formation phenomena in rivers and lakes.

Frazil ice, anchor ice, and aufeis are typical ice forms associated with moving water in cold climates. Ice begins to form when water is cooled to 0°C and continues to lose heat to the surrounding. It is necessary to cool the water somewhat below the equilibrium freezing point (supercooling) before ice will nucleate [131]. Frazil-ice crystals characteristically are small discs measuring up to around 4 mm in diameter and 25-100 μm in thickness. Anchor ice also forms in supercooled water, probably by deposition and adhesion of frazil-ice crystals to the river bottom and obstacles in the river flow. Aufeis growth is caused by the freezing of an overflow of river water on a continuous ice cover [132].

The heat transfer phenomena involving river ice starting from the initial cooling of an open surface flow, the onset of ice formation in the form of frazil ice, the formation of an initial ice cover on a river due to the freezing together of small crystals forming at the surface of the turbulent water flow, the formation of pancake ice, the formation of floes, the actual formation of the initial ice cover resulting from floe accumulation, the transient growth of ice cover, the formation of ice ripples on underside of river ice cover, and the melting and decay of ice cover leading to the break up process and ice jams are extremely complicated and intriguing natural phenomena. Thermal regimes and energy budgets for various ice formation processes in rivers are discussed by Ashton [128]. The thermal energy transfer at the water's surface due to solar or short wave radiation, long wave radiation, convection and evaporation-condensation determines the thermal state of river. The heat transfer mechanisms depend on the stage of river ice formation and heat transfer analysis involving ice formation is often based on the assumption of steady or quasi-steady states. Heavy snow deposits can often greatly complicate the estimation of the rate of ice thickness growth.

The heat transfer processes for ice formation, growth and melting in lakes differ somewhat from those for rivers, but the basic features remain the same [128-130]. The thermal regimes of lakes depend on cooling period through maximum density at 4°C, wind mixing, stratification, inflow, throughflow, outflow, and evolution through the winter [129]. At the initiation of ice formation, two processes can take place - frazil ice growth in the center of a stream or lake, and shore ice growth along the borders of a stream or lake. Ice formation in lakes and rivers is discussed by Devik [136] and method for predicting river and lake ice formation is described by Bilello [137].

Heat Transfer

Before the onset of ice nucleation in water, supercooling is required so that the latent heat can be carried away. Ice nucleation has been studied for over 250 years. Supercooling phenomenon of water before freezing was noted by Fahrenheit (1724) and Gay-Lussac (1836). Monti (1891) showed that boiled water could be cooled to -7°C without freezing. Dorsey (1948) observed that laboratory water samples nucleate spontaneously between -3.4 and -21.1°C. Measurements in turbulent water reveal a supercooling of about -0.05°C. Devik [136, 138] and Carstens [139] discuss growth of ice in supercooled water.

The growth of an ice cover depends on air temperature and velocity (atmospheric conditions), water temperature and velocity, heat exchange (thermal radiation and convection) at the air-ice or the air-snow interface, turbulent heat exchange at the ice-water interface and possibly convection at the river bottom. Thermal regimes of lakes and rivers are discussed in [115, 116, 118, 119, 123, 126, 128, 129, 130, 140-162]. The turbulent flow heat transfer to an ice cover causes the formation of a wavy relief pattern (ice ripples) on the underside of the ice cover. The ripples are oriented with crests and troughs transverse to the main flow. The ice ripple phenomenon is of special interest and is apparently related to convective instability at the ice-water interface. The wave formation and heat transfer relating to ice ripples are discussed in [144, 162]. The deterioration of ice covers is related to melting heat transfer phenomena.

Frazil Ice

In rivers, only small degrees of supercooling are needed to form ice, as turbulence quickly carries away the heat of freezing, losing it to the atmosphere. If cooling is rapid and turbulent water is super-cooled a few hundredths to a tenth of 1°C, small discs of ice, called frazil ice, begin to form. Many excellent review articles on frazil ice formation are available dealing with nucleation mechanisms (homo-geneous and heterogeneous nucleation, a mass exchange mechanism between the surface water and the cold air, secondary nucleation), hydrodynamic and thermal conditions (turbulence and small supercooling), fundamental properties (concentration, permeability, shear strength, sintering, adhesive property), growth processes (crystal multiplication), heat transfer (nonequilibrium processes), kinetics of crystal growth, field observations, experimental investigations, and physical modelling. The main objective here is a brief literature review.

In 1887, Henshaw [163] discussed the nature of frazil ice and the prevention of its action in causing flood. In Chapter 5 of Barnes' classic book [117] on frazil ice, 11 references from 1856 to 1924 were listed. In Canada, the work of Barnes [117] represents the first major effort to present detailed summaries of available information on frazil ice formation and occurrence. Williams (1959) [164] summarizes the existing information on frazil ice and presents a selected biblio-graphy. Recently major review articles on frazil ice formation are presented by Carstens (1966, 1970) [139, 165], Michel (1963, 1971, 1978) [166, 118, 120], Osterkamp (1978) [167], Ashton (1978, 1980, 1983, 1986) [126, 128, 168, 169], Gerard (1979) [127], Larsen (1978) [153], Martin (1981) [170], and Daly (1987) [171]. Daly (1984) [172] presents a model for frazil ice formation with a comprehensive litera-ture review. Ettema, Karim, and Kennedy (1984) [173] present an exten-sive review of frazil ice formation, experimental results on the influences of turbulence and water temperature on frazil ice formation, and an analytical model, in which the rate of frazil ice formation is related to temperature rise of a turbulent volume of water from the release of latent heat.   Recent papers by Hanley and Michel [174], Hanley [175], Muller and Calkins [176], Gosink and Osterkamp [177], Osterkamp et al. [178], Forest [179], Osterkamp and Gosink [180], Andres [181], Ettema, Kennedy and Karim [182], Hanley and Tsang [183] also provide considerable insight on frazil ice phenomena.

From the viewpoint of heat transfer, the winter thermal regime of ice covered waters (heat transfer from water to air, solar short wave radiation and diffuse radiation, net long wave radiation, evaporation, condensation, conduction, snow or ice melting) and frazil ice formation phenomena are of special interest as natural phenomena.   State-of-the-art reviews on frazil ice are well covered in the literature.   Because of the practical importance of river and lake ice problems, one may expect considerable progress in clarifying unsolved problems and mechanisms in the future.

3.   RECENT DEVELOPMENTS

3.1 Ground Freezing

Artificial freezing of ground as a contruction technique for supporting poor soils has been used in practice and known in science for over a hundred years, but it is only in the last few decades that the method has gained industrial maturity and widespread applications. During this period of applications, many significant advances have been made in ground freezing technology. The time of the pioneers is over and the present problems are accurate predictions of freezing time and strength, heaving forces, the effects of freezing on soil properties and cost per unit frozen volume. Accuracy in predicting refrigeration requirements and placing freezing pipes is also of great importance. The ground freezing technique is truly interdisciplinary in nature involving soil mechanics, refrigeration, heat and mass transfer and numerical solutions of flow and strength problems. The subject is also closely related to permafrost and frost protection problems, which are fields that have experienced considerable attention and progress over the past twenty-five years or so. The attention here will be directed towards a brief review of literature relating to heat and mass transfer problems only.

Recent advances in ground freezing are well documented in the proceedings of the International Symposium on Ground Freezing (1st, Bochum, West Germany, 1978; 2nd, Trondheim, Norway, 1980; 3rd. Hanover, New Hampshire, U.S.A., 1982; 4th Sapporo, Japan, 1985; 5th, Nottingham, England, 1988) where the state-of-the-art reports can be found [184-186].   The ground freezing problems are discussed in recent monographs [33, 187-199] and physics of soils including heat transfer and thermal properties of soils are treated in [200-202], for example. Research work on frozen ground was basically motivated by frost heaving problems relating to cold storage, railways, paved highways, freezing of ground around construction sites using artificial freezing method for excavating tunnels in soft ground, freezing of ground surrounding the underground storage of cold liquid (LNG, -160°C, LPG, -42°C), low-temperature heat storage from cold atmosphere using heat pipes. The ground freezing research can be classified into the following four main topics: (1) thermal properties and processes, (2) frost action, (3) mechanical properties, and (4) engineering designs and case histories.

Kersten's work [203, 204] on thermal properties of soils is most well-known, but Farouki's recent monographs [205, 206] are considered to be the definitive reference on ground thermal properties. The freezing of water in porous particulates involves complex processes. Ice nucleation, the movement of water to enlarging ice crystals and the rejection and segregation of solutes and mineral particles (mechanism of frost heaving) appear to have received the most attention. The state-of-the art reviews on frost heave phenomena and its mechanism are reported by Miller (1980) [207], Loch (1981) [208], O'Neill (1983) [209], Anderson (1967) [210], Anderson and Morgenstern (1973) [211], Hoekstra (1969) [212], Nixon (1987) [213] and Anderson et al. (1984) [214]. It has been shown that when water freezes in porous, particulate, mineral matter, the ice crystals remain separated from mineral surfaces by an unfrozen, fluid-like interfacial layer of water. The fluidity of this water layer persists as low as -10°C, apparently diminishing as the temperature is lowered. Between -10°C and -40°C the mobility of the unfrozen interfacial water, while still evident, is very low [215].

The major factors governing frost heave are: (1) type of soil, (2) freezing rate, (3) water content and movement, and (4) overburden pressure. Frost heave phenomenon depends on (1) the driving force for the movement of water towards the frozen surface, (2) detailed ice growth from supercooled water, and (3) discrete ice lens growth. The existing frost heave theories are basically based on (1) freezing of supercooled water, (2) curvature and properties of soil particle surface, (3) electrical properties of soil particle surface, and (4) movement and behaviour of unfrozen water [216].

Gilpin (1980) [217] observed that over the years theories of the frost heave mechanism have been proposed by Everett (1961) (capillary approach), Jackson, Uhlmann, and Chalmers (1966), Takagi (1970), Miller (1978), Derjaginn and Churaev (1978). He notes that a number of distinctive features of the frost heave process that must be incorporated by any complete model. These include: (1) an explanation of the driving force by which water is drawn to the freezing zone, (2) a mechanism by which the water can force its way into the gap between a soil particle and an ice lens to cause growth of the ice lens, and (3) an explanation for the development of the banded ice lens structure that is normally observed in a frozen soil sample. The frost heave model proposed by Miller (1978) is perhaps the most complete one, but is considered to be very complex [217]. Gilpin proposed a model which exhibits all of the characteristics associated with the frost heave considering a basic physical understanding of the behaviour of water in a thin layer near a solid substrate [217-219].

Gilpin (1982) [220] notes that there are many approaches that can be taken in frost heave modelling. One can start with experimental data and generalize it by relating heave rate to frost penetration rate and/or heat fluxes (Horiguchi, 1979; Konrad, 1980; Konrad and Morgenstern, 1980; Nixon, Ellwood, and Slusarchuk, 1982; Outcalt, 1980; Ueda and Penner, 1978) [220]. This approach has the advantage that the resulting equations can be used directly in existing ground thermal regime models to give predictions under field conditions. An alternative approach is to use thermodynamic arguments combined with equations of energy and mass flow to construct a detailed physical model of the heaving process (Gilpin, 1980; Guyman, Hromadka and Berg, 1980; Hopke, 1980; Miller, 1978; Takagi, 1980; Sheppard, Kay and Loch, 1978). This approach has the advantage of providing insight into the mechanics of frost heave, but the resulting model is usually very cumbersome to use [220]. A simplified model which agrees with some of the empirical correlations was developed by Gilpin [220]. Gilpin's work is noted here only to show one approach to the very complex and difficult frost heave problems which are still under investigation. Ground freezing and frost heave is also well reviewed by Nixon (1987) [213]. Japanese literature on ground freezing and frost heave can be found in [192, 193, 221].

The earlier works on ground freezing by Taber (1929, 1930) [222, 223], Beskow (1935) [224], Berggren (1943) [225], Aldrich and Paynter (1953) [226] and Aldrich (1956) [227] are well-known. The coupled heat and mass transfer problems for frost heave phenomena were studied by the following investigators: Uhlmann and Jackson (1966) [228], Chalmers and Jackson (1970) [229], Harlan (1973) [230], Jackson, Uhlamann and Chalmers (1966) [231], Hoekstra (1966, 1969) [232, 233], Philip and De Vries (1957) [234], Jame and Norum (1976) [235], Fukuda, Orhum and Luthin (1980) [236], Kay and Groenevelt (1974) [237], Taylor and Luthin (1978) [238], Jame and Norum (1980) [239], Sheppard, Kay and Loch (1978) [240], Guymon and Luthin (1974) [241], Biermans, Dijkema and De Vries (1978) [242], Taylor and Luthin (1976) [243], Fukuda and Nakagawa (1985) [244], Kay and Perfect (1988) [452], and Ladanyi and Shen (1989) [453].

Numerical analysis of ground freezing and frost heave is reviewed in [193]. Takashi (1982) presents comprehensive reviews on frost heave force and mechanism [193] and mathematical analysis of frost heave mechanism [245]. Frost heaving in artificial ground freezing is reviewed by Ohrai and Yamamoto [246] and Ohrai [193]. Konrad [247] discusses the physics of freezing soils and an engineering frost heave approach. Lunardini [248] presents a review on conduction with freezing and thawing. Khakimov [249] presents theory and practice of artificial freezing of soils.

3.2 Permafrost

Early investigations on permafrost were motivated by construction problems (building foundations, dams and dykes, pipelines, roads, railways and airfields, utilities, excavation and handling of frozen ground) [31, 250] and recent research is necessitated by the construction on arctic land and offshore oil and gas pipelines [251]. The investigations on permafrost are truly interdisciplinary in nature, including geography, geology, geophysics, physics, chemistry, geo-technique, and civil and mechanical engineering. The recent literature dealing with all aspects of permafrost problems can be found in the proceedings of the International Conference on Permafrost (1st 1963, Purdue Univ., U.S.A.; 2nd 1973, Yakutsk, USSR; 3rd 1978, Edmonton, Canada; 4th 1984, Fairbanks, Alaska; 5th 1988, Trondheim, Norway), and the Canadian Permafrost Conferences (1st, 1963; 4th, 1982). The following monographs on permafrost are available. Brown (1970) [252]; Andersland and Anderson (1978) [189]; Lunardini (1981) [190]; Johnston (1981) [253]; Krzewinski and Tart (1985) [196]; Phukan (1985) [254];

Church and Slaymaker (1985) [255]; Harris (1986) [256]; CRREL (1980) [257]; Anderson and Williams (1985) [195].

Permafrost occurs extensively in northern regions of Europe, Asia, and North America, and in the southern mountainous regions of these land masses. Investigations on the characteristics and origin of permafrost were started well before 1900. By the 1950's it was known that polar permafrost could be divided into two distinctive zones, namely, continuous and discontinuous. The forms of ice in the ground can be classified as: (1) pore ice, (2) segregated ice, (3) foliated or ice-wedge ice, (4) pingo ice, and (5) buried ice. The thermal state of permafrost is sensitive to changes in thermal conditions at the surface causing melting of ice with subsequent thaw settlement.

A review of permafrost was presented by Terzaghi (1952) [258], Black (1954) [259], Lachenbruch (1962, 1968, 1970) [260-262], Ferrians (1969) [263], French (1979) [264], and Gold and Lachenbruch (1973) [265]. Ground temperature studies in North America in the 1950's and 1960's provide simplified methods for calculating the depth of freeze and thaw in soils with primary emphasis on the design of roads and runways and comprehension of the geothermal regime in polar regions [196]. The results of studies were summarized in a design manual [266]. Over the same period, Lachenbruch presented a series of practical studies relating to heat transfer aspects of permafrost problems [267-272].

Ground temperatures are controlled by heat flow at the air-ground interface and geothermal heat flux. The combined effects of radiative, convective and conductive heat transfer process at the air-ground interface controls the surface temperatures. Heat and mass transfer phenomena at the interface are very complex. However, the thermal boundary condition at the ground surface is required for ground thermal regime analysis, and the n-factor approach which correlates various ground surface temperatures to the air temperature was used in the 1960's [196]. Further advances in permafrost technology were motivated by the construction of a hot oil pipeline in ice-rich permafrost regions. Excellent review articles and literature on heat and mass transfer aspects of permafrost engineering can be found in the mono-graphs mentioned earlier [189, 190, 195, 196, 253, 255]. It is of interest to note here the following recent review articles on soil freezing, ice formation and thaw in [196, 255]: T.A. Hammer et al., Ground Temperatures; C.E. Heuer et al., Passive Techniques for Ground Temperature Control; J.F. Nixon, Active Freezing Techniques; D.C. Esch, Thawing Techniques for Frozen Ground; O.T. Farouki, Ground Thermal Properties; V.J. Lunardini, Review of Analytical Methods for Ground Thermal Regime Calculations; J.F. Nixon, Case Histories of Ground Temperature Effects [196]; L.W. Gold, The Ice Factor in Frozen Ground; M.W. Smith, Models of Soil Freezing; S.I. Outcalt, A Step Function Model of Ice Segregation [255]. All aspects and issues of permafrost research including ground freezing and pipelines are addressed in [273]. Crawford and Johnston [274] presented an overview of construction on permafrost.

3.3 Arctic Oil and Gas Pipelines
Problems

Today the Trans-Alaska Pipeline System (12 pump stations and 151 mainline valves), in relative harmony with the harsh and fragile arctic environment, is transporting crude oil from Prudhoe Bay to Valdez through 48-inch (122 cm) diameter, and 800 miles (1287 km) long pipeline at the design rate of 1.2 million barrels (190 million litres) per day (ultimate design throughput 2 million barrels per day). The Prudhoe Bay oil field contains known reserves of 9.8 billion barrels of crude oil and 26 trillion cubic feet of natural gas in one producing geological formation. The Alaska pipeline represents a monumental and landmark engineering feat in the arctic regions and is the most expensive privately financed construction project in history. The construction costing almost $8 billion was completed in 1977 and took almost ten years to complete.

The crude oil enters the pipeline at approximately 60°C (140°F) and may alter the thermal regime of critical permafrost areas if the pipe is in a buried mode. The pipeline construction modes utilize elevated construction, burial, and special burial using refrigeration systems at three locations totaling approximately 6.43 km in potentially thaw unstable permafrost to accommodate such requirements as free passage of big game animals. The design, construction and operation of the Trans-Alaska Pipeline are described in [275-284] among many other sources. For the elevated construction, about 62,000 of the 78,000 vertical support members placed in thaw unstable permafrost had heat-transfer devices (heat pipes) installed to dissipate more heat from the surrounding ground during winter than is introduced by conduction down the steel pipe during the summer months [278]. It is of interest to note that both passive and active systems were used in the Trans-Alaska Oil Pipeline to protect the permafrost.

Transport of gas and oil by pipelines in the permafrost areas and the related problems have received considerable attention since the 1960's. Changes in the ground thermal regime will occur as a result of the construction and operation of a buried pipeline. Major effects will be introduced by clearing, ditching, pipe laying and back filling during construction and, when operating, the pipeline will act as a heat source or sink depending on whether the line is operated at temperatures above or below 0°C, resulting in further disturbances to the ground thermal regime [277]. The gas pipeline in the arctic region will be refrigerated at the compressor station to protect permafrost and increase the throughput. The effects of buried hot lines and cold lines on permafrost are characteristically different in both the discontinuous and continuous zones. Buried hot lines will cause a large thaw bulb to develop in frozen soil. Thawing of ice-rich permafrost around and below the line can result not only in significant differential settlement, but also in loss of strength of the soil. The problem may be particularly critical with respect to stability if the water released upon thawing is unable to drain at a satisfactory rate or if the line is on a slope such that downslope movement of the soil and pipe may occur [277].

A chilled gas pipeline that is operated at temperatures below 0°C in unfrozen areas of the discontinuous permafrost zone or adjacent to large water bodies in the continuous zone, may be subject to potential danger for differential frost heave, particularly if the soil is fine grained, because a large frozen zone (bulb) will be formed around the line. This frozen zone will also act as a barrier to surface and subsurface movement of water and thus may create additional problems such as development of icings or ponding of water. Where burial is not feasible the pipeline will be insulated and carried above ground on piles or in a berm placed on the ground surface [277].

Recent Investigations

The potential effects of burying the Trans-Alaska Pipeline in permafrost terrain were recognized by Lachenbruch (1970) [272] well before its actual construction predicting changes in the ground thermal regime occurring as a result of the construction and operation of a buried pipeline. Gold et al. (1972) [285] reported the thermal effects in permafrost caused by buried hot and cold pipelines. The proceedings of the Canadian Northern Pipeline Conference (1972) [251] contain information on the technology required for the construction of pipelines over permafrost and the conference was apparently held in response to the needs after the Prudhoe Bay, Alaska discovery of oil and gas reserves and the associated increase in exploration in northern Canada. Problems and possibilities relating to Arctic oil and gas pipelines are also discussed in [286] (1975). The results of full scale field tests of hot pipeline were reported initially in [287, 288] (1972). Peyton (1971) [289] discusses thaw settlement, thaw liquefaction, thaw plug stability, erosion due to a hot oil pipeline in permafrost and presents results of thermal regime analysis.

At the Second International Conference on Permafrost (Yakutsk, 1973) [290], the following technical reviews were presented by specialists: L.W. Gold and H. Lachenbruch, Thermal Conditions in Permafrost - A Review of North American Literature; D.M. Anderson and N.R. Morgenstern, Physics, Chemistry, and Mechanics of Frozen Ground; K.A. Linell and G.H. Johnston, Engineering Design and Construction in Permafrost Regions.

At the Fourth International Conference on Permafrost (Fairbanks, Alaska, 1983) [290], panel sessions on (1) Frost Heave and Ice Segregation, (2) Pipelines in Northern Regions, and (3) Subsea Permafrost were held, and state-of-the-art reviews can be found in the reports of Panel Sessions [291]. The following subjects are covered in Pipelines in Northern Regions: O.J. Ferrians, Pipelines in the Northern USSR; A.C. Mathews, Hot-Oil and Chilled-Gas Pipeline Interaction with Permafrost; H.O. Jahns, Pipeline Thermal Considerations; M.C. Metz, Pipeline Workpads in Alaska; E.R. Johnson, Performance of the Trans-Alaska Oil Pipeline.

The most serious permafrost-related engineering problems for hot oil pipelines are caused by the thawing (thaw plug) of ice-rich permafrost, which results in a loss of bearing strength and a change in volume of ice-rich soil. A chilled (below 0°C) natural gas pipeline in the discontinuous permafrost region, on the other hand, may result in frost heaving (freeze plug) due to the freezing of pore water in soils surrounding the pipe and to the freezing of additional water attracted to the freezing front. At present the literature on Arctic oil and gas pipelines is fairly extensive and recent technical papers [292-315] are noted here for reference. Recent pipelines and frost heave problems are also reviewed in [316]. In recent years problems relating to oil and gas pipelines in offshore permafrost have also received attention.

3.4 Recent Applications of Heat Pipes and Thermosyphons in Cold Regions Passive Techniques [454]

Constructions in permafrost regions are faced with unique problems that arise from disturbances of the thermal regime in the ground. Heat pipes and thermosyphons (both open and closed, and both single- and two-phase) are used in cold regions to remove heat from soil and ensure the integrity of foundations and piles. A recent review by Heur, Long and Zarling [317] contains a comprehensive summary of the applications of passive techniques for ground temperature control. Five general classes of passive techniques are considered: natural convection devices, thermal barriers, thermal sinks, elevated construction, and ground surface modifications. Natural convection devices include both closed thermosyphons and open air convection pipes. A bibliographical summary of recent applications (mostly after 1975) is also given [317]. Thermosyphon including system description, uses, and applications are well reviewed.

The application of heat pipes on the Trans-Alaska Pipeline is reviewed by Heur [318]. The applications of heat pipes and thermosyphons in cold regions can also be found in [302, 304, 317-348, 454]. Because of the unique characteristics such as low temperature difference between heat source and sink, no moving parts, and requiring no external power, heat pipes and thermosyphons of various designs are particularly suitable for applications in cold regions. Other applications of heat pipes and thermosyphons in cold regions include: (1) ship de-icing system, (2) roadway snow melting, (3) low temperature heat storage [349], (4) artificial permafrost storage using heat pipes [350, 351], (5) closed-loop two-phase thermosyphon systems for solar collector [352], and low grade waste heat recovery [353], and (6) thermosiphon power generation system (under development) [354].

Active Techniques.

Air circulation ducts have been used for several decades to transmit heat from the pad beneath an insulated structure to the ambient air. When the structure is heated, the heat transfer may cause thawing in the frozen subsoils. Thus the use of insulation alone beneath a continuously heated structure on permafrost is not sufficient, and heat removal devices such as ventilating ducts and refrigeration equipment are required at least on a seasonal basis. Nixon presents a review on active freezing techniques [355] for both pads and piles, and on case histories of ground temperature effects [355].

3.5 Thermophysical Properties of Ice, Snow, and Sea Ice

Fukusako (1990) [464] reviewed recent data and information on the thermophysical properties of ice, snow and sea ice. These properties include thermal conductivity, specific heat, density, thermal diffusivity, latent heat of fusion, thermal expansion, and absorption coefficient. The available data are shown graphically for convenience in conjunction with the recommended correlation equations. Sixty-six related references are given. Thermal conductivity of a dendritic ice layer is discussed in [465].

4. FREEZING AND MELTING HEAT TRANSFER (General)

The literature on physical problems dealing with freezing of water or melting of ice with or without forced flow is very extensive. The following review papers provide an excellent overview of recent developments: Bankoff (1964) [356], Muehlbauer and Sunderland (1965) [357], Mori and Araki (1976) [358], Saitoh (1980) [359], Epstein and Cheung (1984, 1984) [360, 362], Viskanta (1983) [361], Lunardini (1981, 1985, 1987, 1988) [190, 363-365], Fukusako and Seki (1987) [366, 367], and Fukusako (1989) [368, 369]. The following bibliographies are given to show the scope of recent developments since the classical solutions of Stefan (1891) and Neumann (1860's):

Exact Solutions in Plane Geometries

Rubinsky and Shitzer (1978) [370], Crank (1978) [371], Carslaw and Jaeger (1959) [7], Cho and Sunderland (1981) [372], Tao (1978) [373], Tao (1979) [374], Lunardini (1988) [365], Lozano and Reemsten (1981) [376], Westphal (1967) [377], Evans, Isaacson and MacDonald (1950) [378].

Exact Solutions in Cylindrical and Spherical Geometries

Carslaw and Jaeger (1959) [7], Ozisik and Uzzell (1979) [379], Kreith and Romie (1955) [380], Rubinsky and Shitzer (1978) [370], Lunardini (1981) [190].

Approximate Solution Methods

(1) Quasi-Static Approximation

Lunardini (1981, 1988) [190, 365], Seban (1971) [382], London and Seban (1943) [383], Carslaw and Jaeger (1959) [7], Citron (1960, 1962) [384, 385], Lock (1969) [386], Pedroso and Domoto (1973) [387, 388], Gupta (1973) [389], Foss and Fan (1972, 1974) [390, 291], Riley, Smith and Poots (1974) [392], Hung and Shih (1975) [393], Jiji and Weinbaum (1978) [394], Hill and Kucera (1983) [395], Gutman (1986) [396].

(2) Heat Balance Integral Method

Goodman (1955) [397], Noble (1975) [298], Bell (1978) [399], Bell and Abbas (1985) [400], Lunardini and Varotta (1981) [401], Lunardini (1980-1985) [402-406], Cho and Sunderland (1981) [407], Yuen (1980) [408], Altman (1961) [409], Tien and Yen (1975) [410], Bell (1979) [411], Goodling and Khader (1975) [412], Poots (1962) [413].

Ice Melting in Forced Flow

Yen and Tien (1963, 1964) [414, 415], Pozvonkov et al. (1970) [416], Tien and Yen (1965) [417], Merk (1954) [418], Lunardini et al. (1986) [419], Lunardini (1986) [420], Yen and Galea (1969) [421], Griffin (1973, 1974, 1975) [422-424], Wilson and Sarma (1977) [425], Yen (1976) [426], Gilpin (1973-74 [427], Tien and Yen (1974) [428], Szekeley et al. (1984) [429].

The Melting in Natural Convection Including the Effect of Density Anomaly

(1) In a Confined Region

Tien and Yen (1966) [430], Yen et al. (1966) [431], Yen (1968, 1980) [432, 433], Rieger and Beer (1986) [434].

(2) In an Unconfined Region

Tkachev (1953) [435], Dumore et al. (1953) [436], Vanier and Tien (1970) [437], Bendell and Beghart (1976) [438], Ho and Chen (1986) [439].

The above references are noted here to show the general scope of recent developments and many references were not cited. It is not practical to conduct a comprehensive survey on freezing and melting heat transfer in cold regions within the scope of this work. Recent developments are well surveyed in [359-369]. It may be of some interest to mention some recent work in order to gain some historical perspectives. Since the classical works of Stefan (1891) and Neumann (1860's), the ice formation problem in pipes without main flow was solved by London and Seban [383] in 1943. In 1963, Yen and Tien [414] studied the modified Leveque problem for laminar heat transfer over a melting plate. In 1968, Zerkle and Sunderland [440] extended the classical Graetz problem to ice formation in pipe. In 1969, an analytical solution for solidification of a moving warm liquid onto an isothermal cold wall was given by Savino and Siegel [441]. The dendritic ice formation phenomena in water pipes without mainflow was observed independently by Kanayama [442] in 1967 and Gilpin [443] in 1977. The ice ripple phenomena (ice-band structure) inside a pipe at or near transition Reynolds numbers similar to those for ripples on underside of river ice covers [144] were observed first by Gilpin [444] in 1979. This work apparently marked the beginning of recent investigations [445, 446] on ice formation in water pipes and recent works are reviewed by Hirata [447]. The above brief review reveals that freezing and melting heat transfer studies are of rather recent developments. The growth and decay of ice was reviewed by Lock [448] in 1974. Numerical solution methods for freezing and melting heat transfer problems are viewed by Saitoh [359, 449] and Katayama et al. [450]. Further references on cold regions heat transfer problems can be found in [455-463]. Because of the extensive nature of recent literature on freezing and melting heat transfer, it is clear that one may never finish a review, but merely abandon it at an appropriate point. It is also clear that some important studies have no doubt been overlooked.

5. CONCLUDING REMARKS

Considerable progress in all fields of cold regions heat transfer has been made during the past 50 years. Cold regions heat transfer (ice formation problems in cold regions) has clearly emerged as an important interdisciplinary research in engineering. This review attempts to identify the subjects of cold regions heat transfer, but the following subjects were not mentioned: frost formation, solidification and melting of solutions, snow layer melting, ice formation in oceans, two-phase flow and heat transfer of ice and water, heat transfer between atmosphere and earth, freezing and melting problems in biology, relationship with refrigeration engineering, and other problems in cold climates (man and clothes). The physical phenomena relating to ice formation problems in engineering are numerous. The literature reporting scientific and engineering investigations on ice can be found in a section of the Journal of Glaciology published since 1947. The scope of this review is rather limited and the listing of references is incomplete. This review should be regarded as a starting point for a more comprehensive overview based on various ice forms and engineering applications [368, 369].

ACKNOWLEDGEMENTS

The preparation of this article was supported by an operating grant from the Natural Sciences and Engineering Research Council of Canada. The authors are indebted to Mrs. B. Bloedorn for her skillful typing of the manuscript. The authors wish to thank Dr. V.J. Lunardini for his valuable comments on the original manuscript.

REFERENCES

1. Layton, E.T. and Lienhard, J.H., 1988, "History of Heat Transfer - Essays in Honor of the 50th Anniversary of the ASME Heat Transfer Division", ASME, New York.
2. Cheng, K.C. and Fujii, T., 1988, "Review and Some Observations of the Historical Development of Heat Transfer from Newton to Eckert, 1700-1960, An Annotated Bibliography", E.T. Layton and J.H. Lienhard, eds., History of Heat Transfer, ASME, pp. 213-260.
3. Cheng, K.C., 1988, "Historical Development of Convective Heat Transfer from Newton to Eckert (1700 to 1960)", Procs. 3rd Int. Symp. on Transport Phenomena in Thermal Control, Taipei, Aug. 14-18, 1988, ed., G.J. Hwang, to be published by the Hemisphere Pub. Corp., New York.
4. Fourier, J.B., 1822, Theorie Analytique de la Chaleur, Paris, English Translation by A. Freeman in 1878, Dover Publications, New York (1955).
5. Carslaw, H.S., 1906, Fourier's Series and Integrals and the Mathematical Theory of the Conduction of Heat, London.
6. Carslaw, H.S., 1921, Introduction to the Mathematical Theory of the Conduction of Heat in Solids, MacMillan and Co., London.
7. Carslaw, H.S. and Jaeger, J.C., 1959, Conduction of Heat in Solids, 2nd ed., Oxford at the Clarendon Press.
8. Stefan, J., 1889, "Uber die Theorie der Eisbildung, insbesondere uber die Eisbildung im Palarmeere" , Wien, Akad. Mat. naturw., Vol. 98, 11a, pp. 965-983, also Ann. Phys. u. Chem. (Wiedemann) N.F. Vol. 42 (1891), pp. 269-286.
9. Stefan, J., 1889, "Uber einige Probleme der Theorie der Warmeleitung", Sitzber. Wien, Akad. Mat. naturw., Vol. 98, 11a, pp. 473-484.
10. Stefan, J., 1889, "Uber die Diffusion von Sauren und Basen gegen einander" Sitzber. Wien, Akad. Mat. naturw., Vol. 98, 11a, pp. 616-634.
11. Stefan, J., 1889, "Uber die Verdampfung und die Auflosung als Vorgange der Diffusion", Stizber. Wien, Akad. Mat. naturw., Vol. 98, pp. 1418-1442.

12. Rubinstein, L.I., 1971, The Stefan Problem (Translated from the Russian by A.D. Solomon), Amer. Math. Soc.
13. Datzeff, A., 1970, Sur le Probleme Lineaire de Stefan, Memoires De Sciences Physiques, Gauthier-Villars, Paris, 1970.
14. Yamaguchi, M. and Nogi, T., 1977, Stefan Problem (in Japanese), Sangyotosho, Tokyo.
15. Luikov, A.V., 1968, Analytical Heat Diffusion Theory, Academic Press, pp. 443-459.
16. Ozisik, M.N., 1980, Heat Conduction, John Wiley & Sons, pp. 397-438.
17. McKie, D. and Heathcote, N.H. de V., 1975, The Discovery of Specific and Latent Heats, Edward Arnold & Co., London.
18. Truesdell, C., 1953, "Notes on the History of the General Equations of Hydrodynamics", The American Mathematical Monthly, Vol. 60, pp. 445-458.
19. Bullard, E.C., 1965, "Historical Introduction to Terrestrial Heat Flow", Terrestrial Heat Flow, W.H.K. Lee, ed., American Geophysical Union, pp. 1-6.
20. Chang, J.H., 1958, Ground Temperature, Vol. 1, Blue Hill Meteorological Observatory, Harvard University.
21. Nakaya, U., 1954, Snow Crystals, Natural and Artificial, Harvard Univ. Press, Cambridge, pp. 1-4.
22. Dorsey, N.E., 1940, 1968, Properties of Ordinary Water-Substance, In All Its Phases: Water-Vapor, Water, and all the Ices, Hafner Publishing Co., New York.
23. Pounder, E.R., 1965, The Physics of Ice, Pergamon Press.
24. Fletcher, N.H., 1970, The Chemical Physics of Ice, Cambridge at the Univ. Press.
25. Hobbs, P.V., 1974, Ice Physics, Clarendon Press, Oxford.
26. Mason, B.J., 1971, The Physics of Clouds, 2nd ed., Clarendon Press, Oxford.
27. Williams, P.J., 1979, Pipelines and Permafrost: Physical Geography and Development in the Circumpolar North, Longman, London and New York.
28. Williams, P.J., 1986, Pipelines and Permafrost: Science in a Cold Climate, The Carleton Univ. Press, Canada.
29. Harris, S.A., 1986, The Permafrost Environment, Barnes & Noble Books, Tatowa, New Jersey.
30. Péwé, T.L., 1976, "Permafrost: Challenge of the Arctic", 1976 Yearbook of Science and the Future, Encyclopaedia Britanica, Inc., Chicago, pp. 91-103.
31. Muller, S.W., 1943, Permafrost or Permanently Frozen Ground and Related Engineering Problems, Strategic Studies No. 62, United States Army, Also 1947, J.W. Edwards, Inc., Ann Arbor, MI.
32. Beskow, G., 1935, 1947, Soil Freezing and Frost Heaving with Special Application to Roads and Railroads, Technological Institute, Northwestern Univ., Evanston, IL, pp. 1-3.
33. Tsytovich, N.A., 1975, The Mechanics of Frozen Ground, McGraw-Hill, New York.
34. ASME, 1957, Seventy-Seven Year Index, Technical Papers, 1880-1956, p. 250.
35. Scott, M., 1930, "Ice Formation on Aircraft and Its Prevention", J. of the Franklin Institute, Vol. 210, pp. 537-586.
36. Simpson, G., 1942, Ice Accretion on Aircraft, Meteorological Office, Professional Notes No. 82, H.M. Stationery Office, London.
37. Hardy, J.K., 1946, Protection of Aircraft Against Ice, R.A.E. Report, S.M.E. 3380.

38. Rodert, L.A., 1946, "Some Suggested Specifications for Thermal Ice-Prevention System for Aircraft", Trans. ASME, Vol. 68, pp. 781-789.

39. Schaefer, V.J., 1947, "Heat Requirements for Instruments and Airfoils During Ice Storms on Mt. Washington", Trans. ASME, Vol. 69, pp. 843-846.

40. Droege, W.C., 1947, "Instrumentation for Flight Testing of Thermal Anti-Icing Systems", Trans. ASME, Vol. 69, pp. 695-698.

41. Brock, G.W., 1947, "Liquid-Water Content and Droplet Size in Clouds of the Atmosphere", Trans. ASME, Vol. 69, pp. 769-770.

42. Tribus, M., 1947, "A Review of Some German Developments in Airplane Anti-Icing", Trans. ASME, Vol. 69, pp. 505-507.

43. Guibert, A.G., 1947, "Thermal Anti-Icing Survey on Mt. Washington", Trans. ASME, Vol. 69, pp. 829-832.

44. Tribus, M., Young, G.B.W. and Boelter, L.M.K., 1948, "Analysis of Heat Transfer Over a Small Cylinder in Icing Conditions on Mount Washington", Trans. ASME, Vol. 70, pp. 971-976.

45. Tribus, M., Young, G.B.W., and Boelter, L.M.K., 1948, "Limitations and Mathematical Basis for Predicing Aircraft Icing Characteristics from Scale-Model Studies", Trans. ASME, Vol. 70, pp. 977-982.

46. Tribus, M., 1951, "Intermittent Heating for Aircraft Ice Protection with Application to Propellers and Jet Engines", Trans. ASME, Vol. 73, pp. 1117-1130.

47. Weiner, F., 1951, "Further Remarks on Inter-mittent Heating for Aircraft Ice Protection", Trans. ASME, Vol. 73, pp. 1131-1137.

48. Hauger, H.H., 1954, "Intermittent Heating of Airfoils for Ice Protection, Utilizing Hot Air", Trans. ASME, Vol. 76, pp. 287-298.

49. Tribus, M., 1946, "Development and Application of Heated Wings", SAE Journal, Vol. 54, No. 6, pp. 261-269, 309.

50. Stickley, A.R., 1937-38, "Some Remarks on the Physical Aspects of the Aircraft Icing Problems", J. of the Aeronautical Sciences, Vol. 5, pp. 442-446.

51. Geer. W.C., 1939, "An Analysis of the Problem of Ice on Airplanes", J. of the Aeronautical Sciences, Vol. 6, pp. 451-459.

52. Boeke, F.L. and Paselk, R.A., 1946, "Icing Problems and the Thermal Anti-Icing System", J. of the Aeronautical Sciences, Vol. 13, pp. 485-497.

53. Tribus, M. and Guibert, A., 1952, "Impingement of Spherical Water Droplets on a Wedge at Supersonic Speeds in Air", J. of the Aeronautical Sciences, Vol. 19, pp. 391-394, 403.

54. Messinger, E.L., 1953, "Equilibrium Temperature of an Unheated Icing Surface as a Function of Air Speed", J. of the Aeronautical Sciences, Vol. 20, pp. 29-42.

55. Hardy, J.K., 1947-1951, "The Physics of the Deposition Process", Symp. on the Deposition of Ice on Exposed Surfaces, J. of Glaciology, Vol. 1, pp. 536-550.

56. Spencer, K.T., 1947-1951, "Aircraft Icing", J. of Glaciology, Vol. 1, pp. 68-69.

57. Brun, E.A., ed., 1957, Icing Problems and Recommended Solutions, AGARDograph 16, Chap. 1, pp. 7-69.

58. Comite d'Etude du Givrage, 1939, Rappport du 19 Mai, 1938, Publ. Sc. et Tech. du Min. de l'Air, B.S.T. 85.

59. Brun, E.A., 1943, Synthese des Connaissances sur le Givrage, G.R.A., R.T. 7.

60. Tribus, M., 1952, Modern Icing Technology, Inst. Univ. of Michigan.

61. Stallabrass, J.R., 1957, "Some Aspects of Helicopter Icing", Canadian Aeronautical Journal, Vol. 3, No. 8, pp. 273-283.
62. Stallabrass, J.R., 1958, "Canadian Research in the Field of Helicopter Icing", J. of the Helicopter Assoc. of Great Britian, Vol. 12, No. 4, pp. 169-196.
63. Stallabrass, J.R., 1959, Flight Tests of an Experimental Helicopter Rotor Blade Electrical De-Icer, NRC Aeronautical Report LR-263.
64. Gold, L.W., 1987, "Fifty Years of Progress in Ice Engineering", J. of Glaciology, Special Issue, p. 78-85.
65. Langmuir, I. and Blodgett, K.B., 1945, A Mathematical Investigation of Water Droplet Trajectories, General Electric Co., Report RL-224, RL-225, (US Army Air Forces Tech. Rep. No. 5418, 1946); Also, The Collected Works of Irving Langmuir, Vol. 10, Atmospheric Phenomena, Pergamon Press, pp. 335-393.
66. Hobbs, P.V., 1973, "Ice in the Atmosphere: A Review of the Present Position", Physics and Chemistry of Ice, ed., E. Whalley, S.J. Jones and L.W. Gold, Royal Society of Canada, Ottawa, pp. 308-319.
67. Ackley, S.F. and Templeton, 1979, Computer Modelling of Atmospheric Ice Accretion, CRREL Report 79-4.
68. Reinmann, J.J., Shaw, R.J. and Olsen, W.A., 1983, "Aircraft Icing Research at NASA", Procs. of First Int. Workshop, Atmospheric Icing of Structures, CRREL Special Report 83-17.
69. Lozowski, E.P., Stallabrass, J.R. and Hearty, P.F., 1979, The Icing of an Unheated Nonrotating Cylinder in Liquid Water Droplet-Ice Crystal Clouds, National Research Council of Canada, Report LTR-LT-96, 109 pp.
70. Oleskiw, M.M. and Lozowski, E.P., 1980, "Helicopter Rotor Blade Icing: A Numerical Simulation", Procs. of the 8th Int. Conf. on Cloud Physics, Clermont-Ferrand, France.
71. Lozowski, E.P. and Oleskiw, M.M., 1981, "Computer Simulation of Airfoil Icing without Run-Back", AIAA 19th Aerospace Sciences Meeting, St. Louis, AIAA-81-0402.
72. Lozowski, E.P. and Oleskiw, M.M., 1983, Computer Modelling of Time-Dependent Rime Icing in the Atmosphere, CRREL Report 83-2.
73. Lozowski, E.P., Stallabrass, J.R. and Hearty, P.F., 1984, "The Icing of an Unheated, Nonrotating Cylinder, Part I, A Simulation Model", J. Climate Appld. Meteorology, Vol. 22, pp. 2053-2062.
74. Minsk, L.D., 1980, Icing on Structures, CRREL Report 80-31.
75. Kuroiwa, D., 1965, Icing and Snow Accretion on Electric Wires, CRREL Research Report 123, AD 611750.
76. Kuroiwa, D., 1958, Icing and Snow Accretion, Monograph Series of the Research Institute of Applied Electricity, Hokkaido University, No. 6, pp. 1-30.
77. Stallabrass, J.R., 1980, Trawler Icing, A Compilation of Work Done at N.R.C., National Research Council, Canada, Mechanical Engineering Report, MD-56, 103 pp.
78. Fukusako, S., 1977, "Accretion of Ice on Ships", J. of the Japan Soc. of Mech. Engrs., Vol. 80, No. 709, pp. 1301-1305.
79. Sutherby, F.S., 1951, "Icing Problems on Ships", J. of Glaciology, Vol. 1, pp. 546-548.
80. Fein, N. and Freiberger, A., 1965, "A Survey of the Literature on Shipboard Ice Formation", Naval Engineers Journal, pp. 849-855.

81. Lock, G.S.H., 1972, Some Aspects of Ice Formation with Special Reference to the Marine Environment, Paper presented to the North-East Coast Institution of Engineers and Shipbuilders, Newcastle-Upon-Tyne, NECIES Trans., Vol. 88, No. 6, pp. 175-184.

82. Lozowski, E.P. and Gates, E.M., 1985, "An Overview of Marine Icing Research",Procs. Fourth Int. Offshore Mechanics and Arctic Engineering Symp., Vol. 2, pp. 6-15.

83. Shellard, H.C., 1974, The Meteorological Aspects of Ice Accretion on Ships, World Meteorological Organization, Marine Science Affairs, Report 10 (WMO397).

84. Minsk, L.D., 1977, Ice Accumulation on Ocean Structures, CRREL Report 77-17.

85. Stallabrass, J.R., 1975, "Icing of Fishing Vessels in Canadian Waters", Quarterly Bulletin of the Div. of Mech. Engrg. and the National Aeronautical Establishment, Jan. 1 to Mar. 31, pp. 25-43.

86. Horjen, I., 1981, Ice Accretions on Ships and Marine Structures, Norwegian Hydrodynamic Laboratories, Report No. 81-02.

87. Itagaki, K., 1977, Icing on Ships and Stationary Structures Under Maritime Conditions, CRREL Special Report 77-27.

88. Makkonen, L., 1984, Atmospheric Icing on Sea Structures, CRREL Monograph 84-2.

89 Tabata, T., Iwata, S. and Ono, N., 1963, "Studies on the Ice Accumulation on Ships I", Low Temperature Science, Ser. A., Vol. 21, Hokkaido univ., pp. 173-221, N.R.C. Canada, Technical Translation TT-1318, 1968, Also Defence Research Board, Canada, Translation No. T93J, 1967.

90. Ono, N., 1964, "Studies on the Ice Accumulation on Ships II, On the Conditions for the Formation of Ice and the Rate of Icing", Low Temperatures Science, Hokkaido Univ., Ser. A., Vol. 22, pp. 171-181, Defense Research Board, Canada, Translation T94J.

91. Tabata, T., 1969, "Studies on the Ice Accumulation on Ships II, Relation Between the Rate of Ice Accumulation and Air, Sea Conditions", Low Temperature Science, Hokkaido Univ., Ser. A., Vol. 27, pp. 339-349.

92. Ono, N., 1974, "Studies on the Ice Accumulation on Ships IV, Statistical Analysis of Ship-Icing Conditions", Low Temperature Science, Hokkaido Univ., Ser. A, Vol. 32, pp. 235-242.

93. Tabata, T., 1968, Research on Prevention of Ship Icing, Report to Hokkaido Prefectural Government, Defense Research Board, Canada, Translation No. T95J.

94. Takano, T., 1950, "Studies on Ice Formation with a Wind Channel: I, Construction and Efficiency of the Wind Tunnel; II, Ice Formation on Wing Models in the Wind Channel; III, Difference of Material, IV, Further Experiments on the Ice Formation on Wind Models in the Wing Channel", Low Temperature Science, Hokkaido Univ., Vol. 5, pp. 1-60 (in Japanese).

95. Oguchi, H., 1951, "Physical Investigation on Icing: I, General Classification of Ice Forms by the Character of their Microscopic Structures; II, On Meteorological Conditions of Icing; III, Density of Ice Formation; IV, Scale Effect of Ice Formation", Low Temperature Science, Hokkaido Univ., Vol. 6, pp. 95-146.

96. Iwata, S., 1975, "Ice Accumulation on Ships", Arctic Oil and Gas, Problems and Possibilities, Fondation Francaise D'etudes Nordiques, Paris, Vol. 1, pp. 363-386.

97. Makkonen, L., 1985, "Icing Rates on Cylinder Structure", Proc. Int. Workshop on Offshore Winds and Icing, Halifax, NS, eds., T.A. Agnew and V.R. Swail, pp. 140-151.

98. Lozowski, E.P., Gates, E.M. and Makkonen, L., 1986, "Towards the Estimation of the Icing Hazard for Mobile Offshore Drilling Units", Proc. 5th Int. Symp. OMAE, Vol. 4, pp. 175-182.

99. Gates, E.M., Marten, R., Lozowski, E.P. and Makkonen, L., 1986, "Marine Icing and Spongy Ice", Proc. 8th IAHR Symp. on Ice, Iowa City, Vol. 2 pp. 153-163.

100. Jurgensen, T.S., 1986, "Consequences of Sea Spray Icing on Marine Units and Brief Survey of Current Research Activities", Proc. Int. Symp. Offshore and Navigation Conference and Exhibition, Helsinki, pp. 447-470.

101. Stallabrass, J.R. and Hearty, P.F., 1967, The Icing of Cylinders in Conditions of Simulated Freezing Sea Spray, National Research Council of Canada, Mechanical Engineering Report MD-50.

102. Stallabrass, J.R., 1970, Method for the Alleviation of Ship Icing, National Research Council of Canada, Mechanical Engineering Report MD-51.

103. Stallabrass, J.R., 1971, "Meteorological and Oceanographic Aspects of Trawler Icing Off the Canadian East Coast", The Marine Observer, London, Vol. 41, pp. 107-121.

104. Stallabrass, J.R., 1973, Trawler Icing Off Eastern Canada, National Research Council Canada, Report No. LTR-LT-50, 1973.

105. Stallabrass, J.R., 1979, Icing of Fishing Vessels: An Analysis of Reports from Canadian East Coast Waters, National Research Council of Canada, Report No. LTR-LT-98.

106. Albrecht, F., 1931, "Theoretische Untersuchungen uber die Ablagerung von Staub Stromender Luft und Ihre Anwendung auf die Theorie der Staubfilter", Phys. Zeitschr., Vol. 32, pp. 48-56.

107. Taylor, G.I., 1940, Notes on Possible Equipment and Technique for Experiments on Icing of Aircraft, Ministry of Aircraft Production, Aeronautical Research Committee, R. & M. No. 2024.

108. Chappell, M.S., 1972, Stationary Gas Turbine Icing Problems: The Icing Environment, Report No. DME/NAE (4), Quarterly Bulletin, National Research Council Canada.

109. Makkonen, L. and Stallabrass, J.R., 1985, "The Effect of Roughness on the Rate of Ice Accretion on a Cylinder", Annals of Glaciology, Vol. 6, pp. 142-145.

110. Gates, E.M., Lam, W. and Lozowski, E.P., 1988, "Spray Evaluation in Icing Wind Tunnels", Cold Regions Science and Technology, Vol. 15, pp. 65-74.

111. Zarling, J.P., Heat and Mass Transfer from Freely Falling Drops at Low Temperatures, CRREL Report 80-18.

112. Henry, K., 1987, Atmospheric Icing of Transmission Lines, Cold Regions Technical Digest, No. 87-2, CRREL.

113. Encyclopaedia Britanica (Macropaedia), 1974, Hydrologic Sciences, Vol. 9, pp. 102-125.

114. Encyclopaedia Britanica (Macropaedia), 1974, Ice in Rivers and Lakes, Vol. 9, pp. 165-170.

115. Ashton, G.D., 1979, "River Ice", American Scientist, Vol. 67, pp. 38-45.

116. Ashton, G.D., ed., 1986, River and Lake Ice Engineering, Water Resources Publications, Littleton, CO.

117. Barnes, H.T., 1928, Ice Engineering, Renouf Publishing Co., Montreal, Canada.

118. Michel, B., 1971, Winter Regime of Rivers and Lakes, United States Army Cold Regions Research and Engineering Laboratory (CRREL) Science and Engineering Monograph II - Bla, 131 p.

119. Pivovarov, A.A., 1973, Thermal Conditions in Freezing Lakes and Rivers, John Wiley & Sons, New York.
120. Michel, B., 1978, Ice Mechanics, Les Presses De L'Université Laval, Quebec, Canada.
121. Carey, K.L., 1970, Icings, CRREL Science and Engineering Monograph III-D3.
122. Chow, V.T., ed., 1964, Handbook of Applied Hydrology, McGraw-Hill.
123. Starosolszky, O., 1970, Ice in Hydraulic Engineering, Institut for Vassbygging, Univ. of Trondheim, Norway, Report No. 70-1.
124. Altberg, W.J., 1936, Twenty Years of Work in the Domain of Underwater Ice Formation (1915-1935), Proc. Int. Union Geol. Geophys., Int. Assoc. Sci. Hydrol., pp. 373-407.
125. Kennedy, J.F., et al., 1974, "River-Ice Problems: A State-of-the-Art Survey and Assessment of Research Needs", J. of Hydraulics Div., ASCE, Vol. 100, pp. 1-15.
126. Ashton, G.D., 1978, "River Ice", Ann. Rev. Fluid Mech., Vol. 10, pp. 369-392.
127. Gerard, R., 1979, "River Ice in Hydrotechnical Engineering; A Review of Related Topics", Proc. Canadian Hydrology Symp., National Research Council of Canada, Ottawa.
128. Ashton, G.D., 1980, "Fresh Water Ice Growth, Motion and Decay", Dynamics of Snow and Ice Masses, Academic Press, pp. 261-304.
129. Ashton, G.D., 1982, "Theory of Thermal Control and Prevention of ice in Rivers and Lakes", Advances in Hydroscience, Vol. 13, pp. 131-185.
130. Carlson, R.F., 1981, "Ice Formation on Rivers and Lakes", the Northern Engineer, Vol. 13, No. 4, pp. 4-9.
131. Burgi, P.H. and Johnson, P.L., 1971, Ice Formation - A Review of the Literature and Bureau of Reclamation Experience, REC-ERC-71-8, Engineering and Research Center, Bureau of Reclamation, Denver, CO, 27 p.
132. Hays, R.B., 1974, Design and Operation of Shallow River Diversions in Cold Regions, REC-ERC-74-19, Engineering and Research Center, Bureau of Reclamation, Denver, CO, 30 p.
133. Acres Consultants Ltd., 1971, Review of Current Ice Technology and Evaluation of Research Priorities, Inland Waters Branch, Dept. of Environment, Ottawa, Report No. 17, 299 p.
134. Chalmers, B., 1961, The Growth of Ice in Supercooled Water, Edgar Marburg Lecture, Amer. Soc. for Testing and Materials, pp. 9
135. Committee on Glaciology, Polar Research Board, 1983, Snow and Ice Research, An Assessment, National Academy Press, Washington, D.C., pp. 73-78.
136. Devik, O., 1944, "Ice Formation in Lakes and Rivers", The Geographical Journal, Vol. CIII, No. 5, pp. 193-203.
137. Bilello, M.A., 1964, "Method for Predicting River and Lake Ice Formation", J. of Applied Meteorology, Vol. 3, pp. 38-44.
138. Devik, O., 1949, "Freezing Water and Supercooling, Anchor Ice and Frazil Ice", J. of Glaciology, Vol. 1, pp. 307-309.
139. Carstens, T., 1966, "Experiments with Supercooling and Ice Formation in Open Water", Geofys. Publ., Vol. 26, pp. 1-18.
140. Williams, G.P., ed., 1975, Thermal Regime of River Ice, Proc. Research Seminar, Tech. Memorandum No. 114, National Research Council.
141. Baines, W.D., 1961, "On the Transfer of Heat from a River to an Ice Sheet", Trans. of the E.I.C., Vol. 5, pp. 27-32.

142. Raphael, J.M., 1962, "Prediction of Temperature in Rivers and Reservoirs", J. of Power Division, ASCE, Vol. 88, pp. 157-181.
143. Ashton, G.D., 1971, The Formation of Ice Ripples on the Underside of River Ice Covers, Ph.D. Thesis, Univ. of Iowa, Univ. Microfilms, Ann Arbor, MI, 147 p.
144. Ashton, G.D. and Kennedy, J.F., 1972, "Ripples on Underside of River Ice Covers", J. of Hydraulics Div., ASCE, Vol. 98, pp. 1603-1624.
145. Ashton, G.D. and Kennedy, J.R., 1970, "Temperature and Flow Conditions During the Formation of River Ice", Ice Symposium, Reykjavik, IAHR, Paper 2.4, 12 p.
146. Ashton, G.D., 1972, "Turbulent Heat Transfer to Wavy Boundaries", Proc. Heat Transfer Fluid Mech. Inst., pp. 200-213.
147. Ashton, G.D., 1972, "Field Implications of Ice Ripples", Int. Assoc. Hydraul. Res., pp. 123-129.
148. Ashton, G.D.,. 1973, "Heat Transfer to River Ice Covers", Proc. Eastern Snow Conf., pp. 125-135.
149. Hsu, K., 1973, Spectral Evaluation of Ice Ripples, Ph.D. Thesis, Univ. of Iowa, Univ. Microfilms, Ann Arbor, MI.
150. Paily, P.P., Macagno, E.O. and Kennedy, J.F., 1974, "Winter-Regime Thermal Response of Heated Streams", J. of Hydraulics Div., ASCE, Vol. 100, pp. 531-551.
151. Viaud, P.R., 1975, "Theoretical and Experimental Study of Stationary Profiles of a Water-Ice Mobile Solidification Interface", I. Prigogine and S.A. Rice, eds., Advances in Chemical Physics, Vol. 32, Wiley, pp. 163-205.
152. Tatinclaux, J.-C. and Kennedy, J.F., 1977, "Ripple Formation at Ice-Flow Interfaces: Potential Effects on Iceberg Transport", A.A. Husseiny, ed., Iceberg Utilization, pp. 276-282.
153. Larsen, P., 1978, "Thermal Regime of Ice Covered Waters", IAHR Symp. on Ice Problems, Lulea, pp. 95-117.
154. Haynes, F.D. and Ashton, G.D., 1979, Turbulent Heat Transfer in Large Aspect Channels, CRREL Report 79-13.
155. Calkins, D.J., 1979, Accelerated Ice Growth in Rivers, CRREL Report 79-14.
156. O'Neill, K. and Ashton, G.D., 1981, Bottom Heat Transfer to Water Bodies in Winter, CRREL Special Report 81-18.
157. Chiang, L.A. and Shen, H.T., 1983, "Numerical Simulation of Thermal-Ice Conditions", Proc. Eastern Snow Conf., Vol. 28, pp. 24-36.
158. Natousek, V., 1984, "Types of Ice Run and Conditions for Their Formation", IAHR Ice Symposium, Hamburg, pp. 315-327.
159. Calkins, D.J., 1984, "Ice Cover Melting in a Shallow River", Can. J. Civil Eng., Vol. 11, pp. 255-265.
160. Ashton, G.D., "Deterioration of Floating Ice Covers", J. of Energy Resources Technology, Trans. ASME, Vol. 107, pp. 177-182.
161. Shen, H.T. and Lal, A.M.W., 1986, "Growth and Decay of River Ice Covers", Cold Regions Hydrology Symp., Am. Water Resources Assoc., pp. 583-591.
162. Gilpin, R.R., Hirata, T. and Cheng, K.C., 1980, "Wave Formation and Heat Transfer at an Ice-Water Interface in the Presence of a Turbulent Flow", J. Fluid Mech., Vol. 99, pp. 619-640.
163. Henshaw, G.H., 1887, "Frazil Ice: On Its Nature and the Prevention of its Action in Causing Floods", Trans. Can. Soc. Civ. Eng., Vol. 1, pp. 1-23.

164. Williams, G.P., 1959, "Frazil Ice, A Review of its Properties, With a Selected Bibliography", The Engineering Journal, Canada, Vol. 42, pp. 55-60.

165. Carstens, T., 1970, "Heat Exchanges and Frazil Formation", Int. Assoc. Hydraul. Res., Paper 2.11, 17 p.

166. Michel, B., 1963, "Theory of Formation and Deposit of Frazil Ice", Proc. Eastern Snow Conf., Vol. 8, pp. 129-149.

167. Osterkamp, T.E., 1978, "Frazil Ice Formation: A Review", J. of Hydraulics Div., ASCE, Vol. 104, pp. 1239-1255.

168. Ashton, G.D., 1983, "Frazil Ice", R.E. Meyer, ed., Theory of Dispersed Multiphase Flow, Academic Press, New York, Pp. 271-289.

169. Ashton, G.D., ed., 1986, River and Lake Ice Engineering, Water Resources Publications, Littleton, CO.

170. Martin, S., 1981, "Frazil Ice in Rivers and Oceans", Ann. Rev. Fluid Mech., Vol. 13, pp. 379-397.

171. Daly, S.F., 1987, "The Evolution of Frazil Ice in Rivers and Streams: Research and Control", K.C.Cheng, V.J. Lunardini and N. Seki, eds., 1987 Int. Symp. on Cold Regions Heat Transfer, ASME, pp. 11-16.

172. Daly, S.F., 1984, Frazil Ice Dynamics, CRREL Monograph 84-1.

173. Ettema, R., Karim, M.F. and Kennedy, J.F., 1984, Frazil Ice Formation, CRREL Report 84-18.

174. Hanley, T.O'D. and Michel, B., 1977, "Laboratory Formation of Border Ice and Frazil Slush", Can. J. Civ. Engg., Vol. 4, pp. 153-160.

175. Hanley, T.O'D., 1978, "Frazil Nucleation Mechanisms", J. of Glaciology, Vol. 21, pp. 581-587.

176. Muller, A. and Calkins, D.J., 1978, "Frazil Ice Formation in Turbulent Flow", IAHR Symp. on Ice Problems, Lylea, pp. 219-234.

177. Gosink, J.P. and Osterkamp, T.E., 1983, "Measurements and Analyses of Velocity Profiles and Frazil Ice-Crystal Rise Velocities During Periods of Frazil-Ice Formation in Rivers", Annals of Glaciology, Vol. 4, pp. 79-84.

178. Osterkamp, et al., 1983, "Water Temperature Measurements in Turbulent Streams During Periods of Frazil-Ice Formation", Annals of Glaciology, Vol. 4, pp. 209-215.

179. Forest, T.W., 1986, "Thermodynamic Stability of Frazil Ice Crystals", Proc. 5th Int. Offshore Mechanics and Arctic Engg. Symp., Vol. 5, ASME, pp. 266-270.

180. Osterkamp, T.E. and Gosink, J.P., 1982, "Selected Aspects of Frazil Ice Formation and Ice Cover Development in Turbulent Streams", Proc. Workshop on Hydraulics of Ice-Covered Rivers, D.D. Andres and R. Gerard, eds., National Research Council of Canada, pp. 131-147.

181. Andres, D.D., 1982,, "Nucleation and Frazil Production During the Supercooling Period", Proc. Workshop on Hydraulics of Ice-Covered Rivers, NRC, Canada, pp. 148-175.

182. Ettema, R., Kennedy, J.F. and Karim, F., 1982, "Influence of Turbulence and Temperature on Frazil Ice Formation", Proc. Workshop on Hydraulics of Ice-Covered Rivers, NRC, Canada, pp. 176-195.

183. Hanley, T. O'D. and Tsang, G., 1982, "Formation and Properties of Frazil in Saline Water", Proc. Workshop on Hydraulics of Ice-Covered Rivers, NRC, Canada, pp. 196-220.

184. Jessberger, H.L., ed., 1979, Ground Freezing, Elsevier Scientific Pub. Co., Amsterdam, Also in Engineering Geology, Vol. 13, Nos. 1-4.

185. Frivik, P.E., ed., 1981, Proc. 2nd Int. Symp. on Ground Freezing, Engineering Geology, Vol. 18, pp. 1-410.
186. Kinosita, S. and Fukuda, M., eds., 1985, Ground Freezing, A.A. Balkema, Rotterdam.
187. Jumikis, A.R., 1966, Thermal Soil Mechanics, Rutgers Univ. Press, New Brunswick, New Jersey.
188. Jumikis, A.R., 1977, Thermal Geotechnics, Rutgers Univ. Press, New Brunswick, New Jersey.
189. Andersland, O.B. and Anderson, D.M., 1978, Geotechnical Engineering for Cold Regions, McGraw-Hill.
190. Lunardini, V.J., 1981, Heat Transfer in Cold Regions, Van Nostrand Reinhold Co.
191. Higashi, A., 1981, Fundamentals of Cold Regions Engineering Science, Kokon-Shoin Co., Tokyo (in Japanese).
192. Kinosita, S., ed., 1982, Ground Freezing - Its Control and Application, Japan Soc. of Soil Mechanics and Foundation Engineering, Tokyo (in Japanese).
193. Kinosita, S., ed, 1982, Physics of Frozen Ground, Morikita Pub. Co., Tokyo (in Japanese).
194. Berg, R.L. and Wright, E.A., eds., 1984, Frost Action and Its Control, Technical Council on Cold Regions Engineering Monograph, ASCE, New York.
195. Anderson, D.M. and Williams, P.J., eds., 1985, Freezing and Thawing of Soil-Water Systems, Technical Council on Cold Regions Engineering Monograph, ASCE.
196. Krzewinski, T.G. and Tart, R.G., eds., 1985, Thermal Design Considerations in Frozen Ground Engineering, Technical Council on Cold Regions Engineering Monograph, ASCE.
197. Johnson, E.G., ed., 1988, Embankment Design and Construction in Cold Regions, Technical Council on Cold Regions Engineering Monograph, ASCE.
198. Lunardini, V.J., 1988, Heat Conduction with Freezing or Thawing, CRREL Monograph 88-1.
199. Hillel, D., 1980, Applications of Soil Physics, Academic Press.
200. Nerpin, S.V., 1970, Physics of the Soil, Israel Program for Scientific Translations.
201. Van Wijk, W.R., ed., 1963, Physics of Plant Environment, North-Holland Pub. Co., Amsterdam.
202. De Vries, D.A. and Afgan, N.H., 1975, Heat and Mass Transfer in the Biosphere, Part 1, Transfer Processes in the Plant Environment, John Wiley & Sons.
203. Kersten, M.S., 1949, Thermal Properties of Soils, Univ. of Minnesota Experiment Station, Bulletin No. 28.
204. Kersten, M.S., 1963, "Thermal Properties of Frozen Ground", Proc. 1st Int. Conf. on Permafrost, National Academy of Sciences, Washington, D.C., pp. 301-305.
205. Farouki, O.T., 1981, Thermal Properties of Soils, CRREL Monograph 81-1.
206. Farouki, O.T., 1986, Thermal Properties of Soils, Trans. Tech. Publications, Germany.
207. Miller, R.D., 1980, Freezing Phenomena in Soils, Applications of Soil Physics, D. Hillel, ed., Academic Press, pp. 254-299.
208. Loch, J.P.G., 1981, "State-of-the-Art Report - Frost Action in Soils", Engineering Geology, Vol. 18, pp. 213-224.
209. O'Neill, K., 1983, "The Physics of Mathematical Frost Heave Models: A Review", Cold Regions Science and Technology, Vol. 6, pp. 275-291.

210. Anderson, D.M., 1967, "The Interface Between Ice and Silicate Surfaces", J. of Colloid and Interface Science, Vol. 25, pp. 174-191.
211. Anderson, D.M. and Morgenstern, N.R., 1973, "Physics, Chemistry, and Mechanics of Frozen Ground: A Review", Permafrost: The North American Contribution to the Second International Conf., Yakutsk, Siberia, National Academy of Sciences, pp. 257-288.
212. Hoekstra, P., 1969, The Physics and Chemistry of Frozen Soils, Highway Research Board, Special Report 103, pp. 78-90.
213. Nixon, J.F., 1987, "Ground Freezing and Frost Heave", Proc. 1987 Int. Symp. on Cold Regions Heat Transfer, K.C. Cheng, V.J. Lunardini and N. Seki eds., ASME, pp. 1-10.
214. Anderson, D.M., Williams, P.J., Guymon, G.L. and Kane, D.L., 1984, "Principles of Soil Freezing and Frost Heaving", Frost Action and Its Control, ASCE, pp. 1-21.
215. Anderson, D.M. and Tice, A.R., "Thawing of Frozen Clays", Freezing and Thawing of Soil-Water Systems, D.M. Anderson and D.J. Williams, eds., ASCE, pp. 1-9.
216. Kinosita, S., 1981, "Mechanism of Frost Heaving", Refrigeration, Vol. 56, No. 649, pp. 969-976, (in Japanese).
217. Gilpin, R.R., 1980, "A Model for the Prediction of Ice Lensing and Frost Heave in Soils", Water Resources Research, Vol. 16, No. 5, pp. 918-930.
218. Gilpin, R.R., 1979, "A Model of the 'Liquid-Like' Layer Between Ice and a Substrate with Applications to Wire Regelation and Particle Migration", J. Colloid Interface Sci., Vol. 68, pp. 235-251.
219. Gilpin, R.R., 1980, "Theoretical Studies of Particle Engulfment", J. Colloid Interface Sci., Vol. 74, pp. 44-63.
220. Gilpin, R.R., 1982, "A Frost Heave Interface Condition for Use in Numerical Modelling", Procs. 4th Can. Permafrost Conf., The Roger J.E. Brown Memorial Volume, H.M. French, ed., National Research Council Canada, pp. 459-465.
221. Kinosita, S., 1988, "A Review of Researches on Frozen Ground in Japan", Int. Symp. on Cold Regions Development, Y. Chen et al., eds., Harbin, P.R.C., Vol. 1, pp. 24-34.
222. Taber, S., 1929, "Frost Heaving", J. Geology, Vol. 37, pp. 428-461.
223. Taber, S., 1930, The Mechanics of Frost Heaving, J. Geology, Vol. 38, pp. 303-317.
224. Beskow, G., 1935, Soil Freezing and Frost Heaving with Special Application to Roads and Railroads, Swedish Geological Society, No. 375, pp. 1-144.
225. Berggren, W.P., 1943, "Prediction of Temperature Distribution in Frozen Soils", Trans. American Geophysical Union, III, pp. 71-77.
226. Aldrich, H.P. and Paynter, H.M., 1953, Analytical Studies of Freezing and Thawing of Soils, U.S. Army Engineer Div., New England Arctic Construction and Frost Effects Laboratory Technical Report 42.
227. Aldrich, H.P., 1956, Frost Penetration Below Highway and Airfield Pavement, Bull. 135, Highway Res. Board, pp. 124-149.
228. Uhlmann, D.R. and Jackson, K.A. 1966, "Frost Heave in Soils, The Influence of Particles on Solidification", Physics of Snow and Ice, Sapporo Conf., Part 2, pp. 1361-1373.
229. Chalmers, B. and Jackson, K.A., 1970, Experimental and Theoretical Studies of the Mechanism of Frost Heaving, CRREL Research Report 199.

230. Harlan, R.L., 1973, "Analysis of Coupled Heat-Fluid Transport in Partially Frozen Soil", Water Resources Research, Vol. 9, pp. 1314-1323.

231. Jackson, K.A., Uhlmann, D.R. and Chalmers, B., 1966, "Frost Heave in Soils", J. Appl. Physics, Vol. 37, pp. 848-852.

232. Hoekstra, P. 1966, "Moisture Movement in Soils Under Temperature Gradient with Cold Side Temperature Below Freezing", Water Resources Research, Vol. 2, pp. 241-250.

233. Hoekstra, P., 1969, "Water Movement and Freezing Pressures", Soil Sci. Soc. Amer. Proc., Vol. 33, pp. 512-518.

234. Philip, J.R. and DeVries, D.A., 1957, "Moisture Movement in Porous Materials Under Temperature Gradients", Trans. Amer. Geophys. Union, Vol. 38, pp. 222-232.

235. Jame, Y.W. and Norum, D.I., 1976, "Heat and Mass Transfer in Freezing Unsaturated Soil in a Closed System", Proc. 2nd Conf. on Soil Water Problems in Cold Regions, Edmonton, Canada, pp. 1-18.

236. Fukuda, M., Orhum, D. and Luthin, J.N., 1980, "Experimental Studies of Coupled Heat and Moisture Transfer in Soils During Freezing", Cold Regions Science and Technology, Vol. 3, pp. 223-232.

237. Kay, B.D. and Groenevelt, P.H., 1974, "On the Interaction of Water and Heat Transport in Frozen and Unfrozen Soils, Basic Theory; I, The Vapor Phase; II, The Liquid Phase", Soil Sci., Amer., Vol. 38, pp. 395-404.

238. Taylor, G.S. and Luthin, J.N., 1978, "A Model for Coupled Heat and Moisture Transfer During Soil Freezing", Canadian Geotech. J., Vol. 15, pp. 548-555.

239. Jame, Y.W. and Norum, D.I., 1980, "Heat and Mass Transfer in Freezing Unsaturated Porous Medium", Water Resources Research, Vol. 16, pp. 814-819.

240. Sheppard, M.I., Kay, B.D., and Loch, J.P.G., 1978, "Development and Testing of a Computer Model for Heat and Mass Flow in Freezing Soil", Proc. 3rd Int. Permafrost Conf., pp. 76-80.

241. Guymon, G.L. and Luthin, J.N., 1974, "A Coupled Heat and Moisture Transport Model for Arctic Soils", Water Resources Res., Vol. 10, No. 5, pp. 995-1001.

242. Biermans, M.B.G.M., Dijkema, K.M. and DeVries, D.A., 1978, "Water Movement in Porous Media Towards an Ice Front", J. Hydrology, pp. 137-148.

243. Taylor, G.S. and Luthin, J.N., 1976, "Numerical Results of Coupled Heat-Mass Flow During Freezing and Thawing", 2nd Conf. on Soil-Water Problems in Cold Regions, Edmonton, Canada, pp. 155-172.

244. Fukuda, M., and Nakagawa, S., 1985, "Numerical Analysis of Frost Heaving Based Upon the Coupled Heat and Water Flow Model", Ground Freezing, S. Kinosita and M. Fukuda, eds., pp. 109-117.

245. Takashi, T., 1982, Mathematical and Physical Analysis of Frost Heave Mechanism, Seiken Co., Ltd., Osaka, Japan, Vol. 1, p. 123, Vol. 2, p. 94 (in Japanese).

246. Ohrai, T. and Yamamoto, H., 1989, Frost Heaving in Artificial Ground Freezing, Freezing and Melting Heat Transfer in Engineering - Selected Topics on Ice-Water Systems, and Welding and Casting Processes, K.C. Cheng and N. Seki, eds., Hemisphere Pub. Corp.

247. Konrad, J.-M., 1989, The Physics of Freezing Soils and an Engineering Frost Heave Approach, Freezing and Melting Heat Transfer in Engineering, K.C. Cheng and N. Seki, eds., Hemisphere Publ. Corp.

248. Lunardini, V.J., 1989, Conduction with Freezing and Melting,

Freezing and Melting Heat Transfer in Engineering, K.C. Cheng and N. Seki, eds., Hemisphere Publ. Corp.

249. Khakimov, K.R., 1966, Artificial Freezing of Soils, Theory and Practice, Israel Program for Scientific Translations, Jerusalem.

250. United States Navy, 1955, Arctic Engineering, Technical Publication, Navdocks TP-PW-11, Bureau of Yards and Docks, Washington, D.C.

251. Legget, R.F. and MacFarlane, I.C., eds., 1972, Procs. of the Canadian Northern Pipeline Research Conference, National Research Council Canada, Technical Memorandum No. 104.

252. Brown, R.J.E., 1970, Permafrost in Canada, Its Influence on Northern Development, Univ. of Toronto Press.

253. Johnston, G.H.J., ed., 1981, Permafrost, Engineering Design and Construction, John Wiley & Sons.

254. Phukan, A., 1985, Frozen Ground Engineering, Prentice-Hall.

255. Church, M. and Slaymaker, O., 1985, Field and Theory, Lectures in Geocryology, Univ. of British Columbia Press, Vancouver.

256. Harris, S.A., 1986, The Permafrost Environment, Barnes & Noble Books, New Jersey.

257. CRREL, 1980, Building Under Cold Climates and on Permafrost, Collection of Papers from a US-Soviet Joint Seminar, Leningrad, USSR, CRREL Special Report 80-40.

258. Terzaghi, K., 1952, "Permafrost", Boston Soc. Civil Engrs. J., Vol. 39, pp. 1-50.

259. Black, R.F., 1954, "Permafrost - A Review", Geol. Soc. Am. Bull., Vol. 65, pp. 839-856.

260. Lachenbruch, A.H., et al., 1962, "Temperatures in Permafrost", Temperature, Its Measurement and Control in Science and Industry, Vol. 3, Pt. 1, pp. 791-803, Reinhold Pub. Corp.

261. Lachenbruch, A.H. 1968, "Permafrost", Encyclopedia of Geomorphology, R.W. Fairbridge, ed., Reinhold Book Co., New York, pp. 833-838.

262. Lachenbruch, A.H., 1970, "Thermal Considerations in Permafrost", Proc. Geological Seminar on the North Slope of Alaska, Los Angeles, Amer. Assoc. Petroleum Geologists, Pacific Section, Al-Rio. J1-2 and J2-5.

263. Ferrians, O.J. et al., 1969, Permafrost and Related Engineering Problems in Alaska, Prof. Pap. US Geol. Surv. 678.

264. French, H.M. 1979, "Permafrost and Ground Ice", Man and Environmental Processes, A Physical Geography Perspective, K.J. Gregory and D.E. Walling, eds., Dawson Westview Press, Kent, England, pp. 144-162.

265. Gold, L.W. and Lachenbruch, A.H., 1973, Thermal Conditions in Permafrost: A Review of North American Literature, 2nd Int. Conf. on Permafrost, Yakutsk, NAS, Washington, D.C., pp. 3-25.

266. Anon., 1966, "Calculation Methods for Determination of Depths of Freeze and Thaw in Soils", U.S. Dept. of the Army Technical Manual TM5-852-6; Also U.S. Dept. of the Air Force Manual AFM 88-19, Chap. 6.

267. Lachenbruch, A.H., 1957, "Three-Dimensional Heat Conduction in Permafrost Beneath Heated Buildings", U.S. Geol. Surv. Bull., 1052-B, pp. 51-69.

268. Lachenbruch, A.H., 1957, "Thermal Effects of the Ocean on Permafrost", Geol. Soc. Am. Bull., Vol. 68, pp. 1515-1530.

269. Lachenbruch, A.H., 1959, Periodic Heat Flow in a Stratified Medium with Application to Permafrost Problems, U.S. Geol. Surv. Bull., 1083-A, 36 p.

270. Lachenbruch, A.H., 1962, Mechanics of Thermal Contraction Cracks and Ice-Wedge Polygons in Permafrost, Geol. Soc. Am. Spec. Paper, Vol. 70, 69 p.

271. Lachenbruch, A.H. and Marshall, B.V., 1969, "Heat Flow in the Arctic", Arctic, Vol. 22, pp. 300-311.

272. Lachenbruch, A.H., 1970, Some Estimates of the Thermal Effects of a Heated Pipeline in Permafrost, U.S. Geol. Surv. Circular 632, 23 p.

273. Polar Research Board, 1983, Permafrost Research: An Assessment of Future Needs, National Academy Press, Washington, D.C.

274. Crawford, C.B. and Johnston, 1971, "Construction on Permafrost", Canadian Geotechnical Journal, Vol. 8, pp. 236-251.

275. Pietsch, D.D., et al., 1975, "The Trans-Alaska Pipeline", Arctic Oil and Gas Problems and Possibilities", Mouton & Co. and Ecole Pratique des Hautes Etudes, Paris, Vol. 1, pp. 302-317.

276. Sharpe, T.A., 1975, "Problems of Ice and the Effect of Low Temperatures on Production Installations on Land", Arctic Oil and Gas Problems and Possibilities", Mouton & Co., Paris, pp. 268-286.

277. Hnatiuk, J., 1975, "Environmental Conditions Influencing Arctic Decisions and Design Criteria", Arctic Oil and Gas Problems and Possibilities, Vol. 1, pp. 11-35.

278. Turner, M.J., 1978, "Lessons Learned from the Trans-Alaska Pipeline Project", An Overview of the Alaska Highway Gas Pipeline: The World's Largest Project, ASCE, pp. 51-94.

279. Williams, P.J., 1979, "Pipelines and Permafrost - Physical Geography and Development in the Circumpolar North", Longman, London.

280. Williams, P.J., 1986, "Pipelines and Permafrost, Science in a Cold Climate", The Carleton Univ. Press, Canada.

281. Butler, B.R., 1975, "Special Problems of Design and Maintenance in Arctic Oil Operations", Arctic Oil and Gas Problems and Possibilities, Vol. 1, pp. 432-456.

282. Brew, D.A. and Gryc, G., 1975, "The Analysis of Impact of Oil and Gas Pipeline Systems on the Alaskan Arctic Environment", Arctic Oil and Gas Problems and Possibilities, Vol. 2, pp. 538-560.

283. Roscow, J.P., 1977, "800 Miles to Valdez: The Building of the Alaska Pipeline", Prentice-Hall.

284. Liguori, A., Maple, J.A. and Heuer, C.E., 1979, "The Design and Construction of the Alyeska Pipeline", Procs. 3rd Int. Conf. on Permafrost, National Research Council Canada, Vol. 2, pp. 151-157.

285. Gold. L.W. et al., 1972, "Thermal Effects in Permafrost", Procs. Canadian Northern Pipeline Research Conf., NRC Canada, Tech. Memorandum Technique 104, pp. 25-45.

286. Fondation Franciase d'Etudes Nordiques, 1975, Arctic Oil and Gas Problems and Possibilities, Mouton & Co. and Ecole Pratique des Hautes Etudes, Vols. 1 and 2.

287. Rowley, R.K. et al., 1972, "Performance of a 48-in Warm Oil Pipeline Supported on Permafrost", 25th Can. Geotech. Conf., Ottawa; Can. Geotech. J., Vol. 10, 1973, pp. 282-303.

288. Slusarchuk, W.A., Watson, G.H. and Speer, T.L., 1972, "Instrumentation Around a Warm Oil Pipeline Buried in Permafrost", 25th Can. Geotech. Conf., Ottawa; Can. Geotech. J., Vol. 10, 1973, pp. 227-245.

289. Peyton, H.R., 1971, "Arctic Pipelines", Procs. 8th World Petroleum Congress, Vol. 6, pp. 145-157.

290. North American Contribution, 1973, Second International Conference on Permafrost, National Academy of Sciences, Washington, D.C.

291. Univ. of Alaska and National Academy of Science, 1983, Procs. Fourth Int. Conf. on Permafrost, Vol. 2, National Academy Press, Washington, D.C.

292. Sykes, J.F., Lennox, W.C. and Unny, T.E., 1974, "Two-Dimensional Heated Pipeline in Permafrost", J. of Geotechincal Eng. Div., ASCE, Vol. 100, No. GT11, pp. 1203-1214.

293. Heise, H., 1973, "Arctic Gas Pipeline Study Yields Engineering Knowledge", Canadian Petroleum, Vol. 14, pp. 36-43.

294. Hwang, C.T., 1976, "Predictions and Observations on the Behaviour of a Warm Gas Pipeline on Permafrost", Can. Geotech. J., Vol. 13, pp. 452-480.

295. Hwang, C.T. and Yip, F.C., 1977, "Advances in Frost Heave Prediction and Mitigative Methods for Pipeline Application", ASME Winter Annual Meetings, ASME Paper.

296. Hwang, C.T., 1977, "On Quasi-Static Solutions for Buried Pipes in Permafrost", Can. Geotech. J., Vol. 14, pp. 180-192.

297. Hwang, C.T., 1980, "Thermal Design for Insulated Pipes", Can. Geotech. J., Vol. 17, pp. 613-622.

298. Morgenstern, N.R. and Nixon, J.F., 1975, "An Analysis of the Performance of a Warm-Oil Pipeline in Permafrost, Inuvik, N.W.T.", Can. Geotech. J., Vol. 12, pp. 199-208.

299. Konrad, J.-M. and Morgenstern, N.R., 1984, "Frost Heave Prediction of Chilled Pipelines Buried in Unfrozen Soils", Can. Geotech. J., Vol. 21, pp. 100-115.

300. Thornton, D.E., 1976, "Steady-State and Quasi-Static Thermal Results for Bare and Insulated Pipes in Permafrost", Can. Geotech. J., Vol. 13, pp. 161-171.

301. Nixon, J.F., Morgenstern, N.R. and Reesor, S.N., 1983, "Frost-Heave - Pipeline Interaction Using Continuum Mechanics", Can. Geotech. J., Vol. 20, pp. 251-261.

302. Wheeler, J.A., 1973, "Simulation of Heat Transfer from a Warm Pipeline Buried in Permafrost", 74th National Meeting of AIChE, New Orleans, LA, AIChE Paper 27b.

303. Wheeler,J.A., 1978, "Permafrost Thermal Design for the Trans-Alaska Pipeline", Moving Boundary Problems, D.G. Wilson et al., eds., Academic Press, pp. 267-284.

304. Jahns, H.O. et al., 1973, "Permafrost Protection for Pipelines", 2nd Int. Conf. on Permafrost, North American Contribution, pp. 673-684.

305. Speer, T.L., Watson, G.H. and Rowley, R.K., 1973, "Effects of Ground-Ice Variability and Resulting Thaw Settlements on Buried Warm-Oil Pipelines", 2nd Int. Conf. on Permafrost, North American Contribution, pp. 746-752.

306. Watson, G.H. et al., 1973, "Performance of a Warm-Oil Pipeline Buried in Permafrost", 2nd Int. Conf. on Permafrost, pp. 759-766.

307. Slusarchuk, W.A. et al., 1978, "Field Test Results of a Chilled Pipeline Buried in Unfrozen Ground, Procs. 3rd Int. Conf. on Permafrost, Vol. 1, pp. 877-883.

308. Carlson, L.E. and Butterwick, D.E., 1983, "Testing Pipeline Techniques in Warm Permafrost", Procs. 4th Int. Conf. on Permafrost, pp. 97-102.

309. Hanna, A.J. et al., 1983, "Alaska Highway Gas Pipeline Project (Yukon) Section Thaw Settlement Design Approach", Procs. 4th Int. Conf. on Permafrost, pp. 439-444.

310. Jahns, H.O. and Heur, C.E., 1983, "Frost Heave Mitigation and Permafrost Protection for a Buried Chilled-Gas Pipeline", Procs. 4th Int. Conf. on Permafrost, pp. 531-536.

311. Carlson, L.E. et al., 1982, "Field Test Results of Operating a Chilled, Buried Pipeline in Unfrozen Ground", 4th Can. Permafrost Conf., pp. 475-480.

312. Carlson, L.E. and Nixon, J.F., 1988, "Subsoil Investigation of Ice Lensing at the Calgary, Canada, Frost Heave Test Facility", Can. Geotech. J., Vol. 25, pp. 307-319.

313. Nixon, J.F., 1987, "Thermally Induced Heave Beneath Chilled Pipelines in Frozen Ground", Can. Geotech. J., Vol. 24, pp. 260-266.

314. Stanley, J.M. and Cronin, J.E., 1983, "Investigations and Implications of Subsurface Conditions Beneath the Trans Alaska Pipeline in Atigun Pass", Procs. 4th Int. Conf. on Permafrost, pp. 1188-1193.

315. Thomas, H.P. and Ferrell, J.E., 1983, "Thermokarst Features Associated with Buried Sections of the Trans-Alaska Pipeline", Procs. 4th Int. Conf. on Permafrost, pp. 1245-1250.

316. Dallimore, S.R. and Williams, P.J. 1985, "Pipelines and Frost Heave", Proc. of a Seminar at Caen, France, Carleton Univ., Ottawa, Canada.

317. Heur, C.E., Long, E.L. and Zarling, J.P., 1985, "Passive Techniques for Ground Temperature Control", Thermal Design Considerations in Frozen Ground Engineering, T.G. Krzewinski and R.G. Tart, eds., ASCE, pp. 72-154.

318. Heur, C.E., 1979, The Application of Heat Pipes on the Trans-Alaska Pipeline, CRREL Special Report 79-26, 27 p.

319. Waters, E.D., 1974, "Heat Pipes to Stabilize Pilings on Elevated Alaska Pipeline Sections", Pipeline and Gas Journal, Vol. 201, August, pp. 46-58.

320. Galate, J.W., Power, L.D. and Wheeler, J.A., 1974, "Further Evaluation of Thermal Piles for an Elevated Pipeline", EXXON Production Research Co. Report EPR 17 PS 74.

321. Glover, L.W., 1975, "Cryo-Anchor - A Design Approach for Arctic Foundations", MDAC Paper WD 2593, McDonnell Douglas Astronautics Co., West Hungtinton Beach, CA.

322. Long, E.L.,. 1963, "The Long Thermopile", Proc. Int. Conf. on Permafrost, National Academy of Sciences, Washington, D.C., pp. 487-491.

323. Long, E.L., 1973, "Designing Friction Piles for Increased Stability at Lower Installed Cost in Permafrost", Procs. 2nd Int. Conf. on Permafrost, Yakutsk, USSR, pp. 693-699.

324. Reed, R.E., 1966, Refrigeration of a Pipe Pile by Air Circulation, CRREL Technical Report 156.

325. Reid, R.L. and Tennant, J.S., 1974, "The Modeling of a Thermosyphon Type Permafrost Protection Device", AIAA paper No. 74-739, ASME Paper No. 74-HT-46; Also ASME J. of Heat Transfer, Vol. 97, 1975, pp. 382-286.

326. Waters, E.D., 1973, "Stabilization of Soil and Structure by Passive Heat Transfer Device", 74th National AIChE Meeting, New Orleans.

327. Weissman, T., 1971, "New System to Beat Permafrost", Oilweek, Vol. 22, No. 29, pp. 23-24.

328. Galate, J.W., 1976, "Passive Refrigeration for Arctic Pile Supports", ASME J. of Engineering for Industry, Vol. 98, pp. 695-700.

329. Haynes, F.D. and Zarling, J.P., 1982, "A Comparative Study of Thermosyphons Used for Freezing Soil", ASME Paper 82-WA/HT-40.
330. Zarling, J.P., Connor, B. and Goering, J., 1983, "Air Duct Systems for Roadway Stabilization Over Permafrost Areas", Procs. 4th Int. Conf. on Permafrost, National Academy Press, pp. 1463-1468.
331. Reid, R.L. and Evans, A.L., 1983, "Investigation of the Air Convection Pile as a Permafrost Protection Device", Procs. 4th Int. Conf. on Permafrost, pp. 1048-1053.
332. Evans. A.L. and Reid, R.L., 1982, "Heat Transfer in an Air Thermosyphon Permafrost Protection Device", ASME J. of Energy Resources Technology, Vol. 104, pp. 205-210.
333. Reid, R.L. and Hudgins, E.H. and Onufer, J.S., 1982, "Frost and Ice Formation in the Air Convection Pile Permafrost Protection Device", ASME J. of Energy Resources Technology, pp. 199-204.
334. Zarling, J.P. and Haynes, F.D., 1984, "Performance of a Thermosyphon with an Inclined Evaporator and Vertical Condenser", Procs. 3rd Int. Offshore Mechanics and Arctic Engineering (OMAE) Symp., New Orleans, Vol. 3, pp. 64-68.
335. Zarling, J.P. and Haynes, F.D., 1985, "Laboratory Tests and Analysis of Thermosyphons with Inclined Evaporator Sections", Procs. 4th Int. OMAE Symp., Dallas, Vol. 2, pp. 31-37.
336. Haynes, F.D. and Zarling, J.P., 1986, "Heat Transfer Characteristics of Thermosyphons with Inclined Evaporator Sections", Procs. 5th Int. OMAE Symp., Vol. 4, pp. 285-292.
337. Hayley, D.W., 1982, "Application of Heat Pipes to Design of Shallow Foundations on Permafrost", 4th Can. Permafrost Conf., pp. 535-544.
338. Womick, O. and LeGoullon, R.B., 1975, "Settling a Problem of Settling", The Northern Engineer, Vol. 7, pp. 4-10.
339. Babb, A.L. et al., 1971, "The Thermo-Tube: A Natural Convection Heat Transfer Device for Stabilization of Arctic Soils in Oil Producing Regions", Soc. of Petroleum Engineers SPE 3618.
340. Babb, A.L. et al, 1978, "A Natural Circulation Self-Refrigerated Pile for the Direct Support of Buildings in Permafrost Regions", AIChE Symp. Series, No. 174, Vol. 74, pp. 223-234.
341. Zarling, J.P., Haynes, F.D. and Daly, S.F., 1988, "On the Application of Thermosyphons in Cold Regions", Procs. 7th Int. OMAE Conf., Houston, pp. 281-286.
342. Haynes, F.D. and Zarling, J.P., 1988, "Thermosyphons and Foundation Design in Cold Regions", Cold Regions Science and Technology, Vol. 15, pp. 251-259.
343. Den Hartog, S.L., 1988, "A Thermosyphon for Horizontal Applications", Cold Regions Science & Technology, Vol. 15, pp. 319-321.
344. Davison, D.M. and Lo, R.C., 1982, "Preservation of Permafrost for a Fuel Storage Tank", 4th Can. Permafrost Conf., pp. 545-554.
345. Barthelemy, J.L., 1980, "Performance of a Natural Convection Heat Exchange System for Subgrade Cooling of Permafrost", Building Under Cold Climates and on Permafrost, CRREL Special Report 80-40, pp. 235-261.
346. Cady, E.C., 1980, "Cryo-Anchor - Proven Performance for Arctic Foundation Stabilization", Building Under Cold Climates and on Permafrost, CRREL Special Report 80-40, pp. 262-278.
347. Pearson, S.W., 1979, "Thermal Performance Verification of Thermal Vertical Support Members for the Trans-Alaska Pipeline", ASME J. of Energy Resources Technology, Vol. 101, pp. 225-231.

348. Waters, E.D., Johnson, C.L. and Wheeler, J.A., 1975, "The Application of Heat Pipes to the Trans-Alaska Pipeline", 10th Intersociety Energy Conversion and Engrg. Conf. (IECEC '75 Record), pp. 1496-1501.

349. Ryokai, K. et al., 1987, "Low Temperature Storage by Means of Heat Pipe", Cold Regions Technology Conf., Japan, pp. 22-26.

350. Matsuoka, K. et al., 1984, "Experimental Study on Ground Freezing Utilized Cold Air for Heat Sink Using a Heat Pipe", Trans. Japanese Assoc. of Refrigeration, Vol. 1, No. 1, pp. 43-52 (in Japanese).

351. Sawada, S., 1988, "Creation of an Artificial Permafrost", J. of Japanese Soc. of Snow and Ice, Vol. 50, No. 1, pp. 9-15 (in Japanese).

352. Cheng, K.C., Fukuda, K. and Lee, C.A., 1987, "Some Operating Characteristics of a Closed-Loop Two-Phase Thermosyphon Flat-Plate Solar Collector System Using Freons R-11 and R-113 by Indoor Test", Procs. ASME-JSME-JSES Solar Energy Conf., Vol. 2, pp. 1035-1045.

353. Cheng, K.C., Takuma, M. and Kamiya, Y., 1987, "Operating and Heat Transfer Characteristics of an Air-to-Air Heat Exchanger Using a Closed-Loop Two-Phase Thermosyphon System", 3rd Int. Symp. on Transport Phenomena in Thermal Control, Taipei, Taiwan, R.O.C., G.J. Hwang, ed., pp. 13-25, to be published by Hemisphere Pub. Corp.

354. Nakatani,H. and Tusima, K., 1988, "Thermo-Siphon Power Generation System", Int. Symp. on Cold Region Development, Harbin, China, Vol. 4, pp. 272-278.

355. Nixon, J.F., 1985, "Active Freezing Techniques", Thermal Design Considerations in Frozen Ground Engineering, T.G. Krzewinski and R.G. Tart, eds., ASCE, pp. 155-171; "Case Histories of Ground Temperature Effects", pp. 258-274.

356. Bankoff, S.G., 1964, "Heat Conduction or Diffusion with Change of Phase", Advances in Chemical Engineering, Vol. 5, Academic Press, pp. 75-150.

357. Muehlbauer, J.C. and Sunderland, J.E., 1965, "Heat Conduction with Freezing or Melting", Applied Mechanics Review, Vol. 18, pp. 951-959.

358. Mori, A. and Araki, K., 1976, "Methods for Analysis of the Moving Boundary-Surface Problem", Int. Chemical Engineering, Vol. 16, pp. 734-744.

359. Saitoh, T., 1980, "Recent Developments of Solution Methods for the Freezing Problem (I), (II)", Refrigeration, Vol. 55, No. 636, pp. 875-883, No. 637, pp. 1005-1015 (in Japanese).

360. Epstein, M. and Cheung, F.B., 1983, "Complex Freezing-Melting Interfaces in Fluid Flow", Ann. Rev. Fluid Mech., Vol. 15, pp. 293-319.

361. Viskanta, R., 1983, "Phase-Change Heat Transfer", Solar Heat Storage: Latent Heat Materials, Vol. 1, CRC Press Inc., Boca Raton, FL, pp. 153-222.

362. Cheung, F.B. and Epstein, M., 1984, "Solidification and Melting in Fluid Flow", Advances in Transport Processes, Vol. 3, pp. 35-117.

363. Lunardini, V.J., 1985, "Review of Analytical Methods for Ground Thermal Regime Calculations", Thermal Design Considerations in Frozen Ground Engineering, T.G. Krzewinski and R.G. Tart, eds., ASCE, pp. 204-257.

364. Lunardini, V.J., 1987, "Some Analytical Methods for Conduction Heat Transfer with Freezing/Thawing", 1987 Int. Symp. on Cold Regions Heat Transfer, K.C. Cheng, V.J. Lunardini and N. Seki, eds., ASME, pp. 55-64.

365. Lunardini, V.J., 1988, Heat Conduction with Freezing or Thawing, CRREL Monograph 88-1, 329 p.

366. Fukusako, S. and Seki, N., 1987, "Freezing and Melting Characteristics in Internal Flow", 1987 Int. Symp. on Cold Regions Heat Transfer, ASME, pp. 25-38.

367. Fukusako, S. and Seki, N., 1987, "Fundamental Aspects of Analytical and Numerical Methods on Freezing and Melting Heat-Transfer Problems", Ann. Rev. Numerical Fluid Mechanics and Heat Transfer, Hemisphere Pub. Corp., Vol. 1, pp. 351-402.

368. Fukusako, S., 1989, "Freezing and Melting Heat Transfer Problems in Cold Climates", J. of Heat Transfer Soc. of Japan, Vol. 28, No. 108, pp. 97-126 (in Japanese).

369. Fukusako, S., 1989, "An Overview of Freezing Problems in Cold Regions", JSME Hokkaido Branch Short Course (Jan. 27, 1989), Kitami, Japan (in Japanese).

370. Rubinsky, B. and Shitzer, A., 1978, "Analytical Solution to the Heat Equation Involving a Moving Boundary with Application to the Change of Phase Problem (The Inverse Stefan Problem)", ASME J. Heat Transfer, Vol. 100, pp. 300-304.

371. Crank, J., 1975, The Mathematics of Diffusion, Clarendon Press, Oxford.

372. Cho. S.H. and Sunderland, J.E., 1974, "Phase Change Problem with Temperature Dependent Thermal Conductivity", ASME J. Heat Transfer, Vol. 96, pp. 214-217.

373. Tao, L.N., 1978, "The Stefan Problem with Arbitrary Initial and Boundary Conditions", Quart. Appld. Math., Vol. 36, No. 3, pp. 223-233.

374. Tao, L.C., 1979, "On the Boundary Problem with Arbitrary Initial and Flux Conditions", J. Appl. Math. Phys. (ZAMP), Vol. 30, pp. 416-426.

375. Lunardini, V.J., 1987, Heat Conduction with Freezing or Thawing, CRREL Monograph 87-1.

376. Lozano , C.J. and Reemsten, R., 1981, "On a Stefan Problem with an Emerging Free Boundary", Numerical Heat Transfer, Vol. 4, pp. 239-245.

377. Westphal, K.O., 1967, "Series Solution of Freezing Problem with the Fixed Surface Radiating into a Medium of Arbitrary Varying Temperature", Int. J. Heat Mass Transfer, Vol. 10, pp. 195-205.

378. Evans, G.W., Isaacson, E. and MacDonald, J.K.L., 1950, "Stefan-Like Problems", Quart. J. of Applied Math., Vol. 8, pp. 312-319.

379. Ozisik, M.N. and Uzzell, J.C., 1979, "Exact Solution for Freezing in Cylindrical Symmetry with Extended Freezing Temperature Range", ASME J. Heat Transfer, Vol. 101, pp. 331-334.

380. Kreith, F. and Romie, F.E., 1955, "A Study of the Thermal Diffusion Equation with Boundary Conditions Corresponding to Solidification or Melting of Materials Initially at the Fusion Temperature", Proc. Phys. Soc., Sect. B, Vol. 68, pp. 277-291.

381. Frank, F.C., 1950, "Radially Symmetric Phase Growth Controlled by Diffusion", Proc. Royal Soc. A., Vol. 201, pp. 586-599.

382. Seban, R.A., 1971, "A Comment on the Periodic Freezing and Melting of Water", Int. J. Heat Mass Transfer, Vol. 14, pp. 1862-1864.

383. London, A.L. and Seban, R.A., 1943, "Rate of Ice Formation", Trans. ASME, Vol. 65, pp. 771-778.

384. Citron, S.J., 1960, "Heat Conduction in Melting Slab", J. Aerospace Sciences, Vol. 27, pp. 219-228.

385. Citron, S.J., 1962, "Conduction of Heat in a Melting Slab", Proc. 4th U.S. National Congress of Applied Mechanics, pp. 1221-1227.

386. Lock, G.S.H., 1969, "On the Use of Asymptotic Solution to Plane Ice-Water Problems", J. Glaciology, Vol. 8, No. 53, pp. 285-300.

387. Pedroso, R.I. and Domoto, G.A., 1973, "Perturbation Solutions for Spherical Solidification of Saturated Liquids", ASME J. Heat Transfer, Vol. 95, pp. 42-46.

388. Pedroso, R.I. and Domoto, G.A., 1973, "Exact Solution by Perturbation Method for Planar Solidification of a Saturated Liquid with Convection at the Wall", Int. J. Heat Mass Transfer, Vol. 16, p. 1816.

389. Gupta, J.P., 1973, "An Approximate Method for Calculating the Freezing Outside Spheres and Cylinders", Chem. Eng. Sciences, Vol. 28, pp. 1629-1633.

390. Foss, S.D. and Fan, S.S.T., 1972, "Approximate Solution to the Freezing of the Ice-Water System", J. Water Resources Res., Vol. 8, No. 4, pp. 1083-1086.

391. Foss, S.D. and Fan. S.S.T., 1974, "Approximate Solution to the Freezing of the Ice-Water System with Constant Heat Flux in the Water Phase", J. Water Resources Res., Vol. 10, No. 3, pp. 511-513.

392. Riley, D.S., Smith, F.I. and Poots, G., 1974, "The Inward Solidification of Spheres and Circular Cylinders", Int. J. Heat Mass Transfer, Vol. 17, pp. 1507-1516.

393. Hung, C.L. and Shih, Y.P., 175, "Perturbation Solution of Planar Diffusion-Controlled Moving Boundary Problems", Int. J. Heat Mass Transfer, Vol. 18, pp. 689-695.

394. Jiji, L.M. and Weinbaum, S., 1978, "Perturbation Solutions for Melting or Freezing in Annular Regions Initially not at the Fusion Temperature", Int. J. Heat Mass Transfer, Vol. 21, pp. 581-592.

395. Hill, J.M. and Kucera, A., 1983, "Freezing a Saturated Liquid Inside a Sphere", Int. J. Heat Mass Transfer, Vol. 26, pp. 1631-1637.

396. Gutman, L.N., 1986, "On the Problem of Heat Transfer in Phase-Change Materials for Small Stefan Numbers", Int. J. Heat Mass Transfer, Vol. 29, pp. 921-926.

397. Goodman, T.R., 1955, "The Heat Balance Integral and Its Application to Problems Involving a Change of Phase", Trans. ASME, Vol. 80, pp. 335-342.

398. Noble, B., 1975, "Heat Balance Methods in Melting Problems, Moving Boundary Problems in Heat Flow and Fission", J.R. Ockendon and W.R. Hodgkins, eds., Clarendon, Oxford, pp. 208-209.

399. Bell, G.E., 1978, "A Refinement of the Heat Balance Integral Method Applied to a Melting Problem", Int. J. Heat Mass Transfer, Vol. 21, pp. 1357-1362.

400. Bell, G.E. and Abbas, S.K., 1985, "Convergence Properties of the Heat Balance Integral Method", Numerical Heat Transfer, Vol. 8, pp. 373-382.

401. Lunardini, V.J. and Varotta, R., 1981, "Approximate Solution to Neumann Problem for Soil Systems", ASME J. Energy Resources Tech., Vol. 103, No. 1, pp. 76-81.

402. Lunardini, V.J., 1982, "Freezing of Soils with Surface Convection", Proc. 3rd Int. Symp. on Ground Freezing, CRREL, Hanover, NH, pp. 205-212.

403. Lunardini, V.J., 1983, "Approximate Solution to Conduction Freezing with Density Variation", ASME J. Energy Resources Tech., Vol. 105, pp. 43-45.

404. Lunardini, V.J., 1983, "Freezing and Thawing: Heat Balance Integral Approximations", ASME J. Energy Resources Tech., Vol. 105, pp. 30-37.

405. Lunardini, V.J., 1985, "Freezing of Soil with Phase Change Occurring Over a Finite Temperature Difference", Proc. 4th Int. OMAE Symp., pp. 38-46.

406. Lunardini, V.J., 1980, Phase Change Around a Circular Pipe, CRREL Report 80-27; Also ASME J. Heat Transfer, Vol. 103, 1981, pp. 598-600.

407. Cho, S.H. and Sunderland, J.E., 1981, "Approximate Temperature Distribution for Phase Change of a Semi-Infinite Body", ASME J. Heat Transfer, Vol. 103, pp. 401-403.

408. Yuen, W.W., 1980, "Application of the Heat Balance Integral to Melting Problem with Initial Supercooling", Int. J. Heat Mass Transfer, Vol. 23, pp. 1157-1160.

409. Altman, M., 1961, "Some Aspects of Melting Solution for a Semi-Infinite Slab", Chem. Eng. Symp. Ser., Vol. 57, pp. 16-23.

410. Tien, C. and Yen, Y.C., 1975, "An Approximate Analysis of Melting and Freezing of a Drill Hole Through an Ice Sheet in Antarctica", J. Glaciology, Vol. 14, pp. 421-432.

411. Bell, G.E., 1979, "Solidification of a Liquid About a Cylindrical Pipe", Int. J. Heat Mass Transfer, Vol. 22, pp. 1681-1686.

412. Goodling, J.S. and Khader, M.S., 1975, "Results of Numerical Solution for Outward Solidification with Flux Boundary Conditions", ASME J. Heat Transfer, Vol. 97, pp. 307-309.

413. Poots, G., 1962, "On the Application of Integral Method to the Solution of Problems Involving the Solidification of Liquids Initially at Fusion Temperature", Int. J. Heat Mass Transfer, Vol. 5, pp. 525-531.

414. Yen, Y.C. and Tien, C., 1963, "Laminar Heat Transfer Over a Melting Plate, The Modified Leveque Problem", J. Geophys. Res., Vol. 68, No. 12, pp. 3673-3678.

415. Tien, C. and Yen, Y.C. 1964, "An Additional Note to the Modified Leveque Problem", J. Geophys. Res., Vol. 69, No. 8, pp. 1672-1673.

416. Pozvonkov, F.M., Shurgalskii, E.F. and Akselrod, L.S., 1970, "Heat Transfer at a Melting Flat Surface Under Conditions of Forced Convection and Laminar Boundary Layer", Int. J. Heat Mass Transfer, Vol. 13, pp. 957-962.

417. Tien, C. and Yen, Y.C., 1965, "The Effect of Melting on Forced Convection Heat Transfer", J. Applied Meteorology, Vol. 4, No. 4, pp. 523-527.

418. Merk, H.J., 1954, "The Influence of Melting and Anomalous Expansion on the Thermal Convection in Laminar Boundary Layers", Appl. Sci. Res. III (A4), pp. 435-452.

419. Lunardini, V.J., Zission, J.R. and Yen, Y.C., 1986, "Experimental Determination of Heat Transfer Coefficients in Water Flowing Over a Horizontal Ice Sheet", CRREL Report 86-03.

420. Lunardini, V.J., 1986, "Free and Forced Convection Heat Transfer in Water Over a Melting Horizontal Ice Sheet", Proc. 5th Int. OMAE Symp., Vol. 4, pp. 227-236.

421. Yen, Y.C. and Galea, F., 1969, "Onset of Convection in a Water Layer Formed Continuously by Melting Ice", Physics of Fluids, Vol. 12, pp. 509-516.

422. Griffin, O.M., 1973, "Heat, Mass and Momentum Transfer During the Melting of Glacial Ice in Sea Water", ASME J. Heat Transfer, Vol. 95, pp. 317-323.

423. Griffin, O.M., 1974, "An integral Method of Solution for Combined Heat and Mass Transfer Problems with Phase Transformation", Proc. 5th Int. Heat Transfer Conf., Vol. 1, pp. 211-215.

424. Griffin, O.M., 1975, "A Note Concerning the Transport Process Near Melting of Glacial Ice in Sea Water", ASME J. Heat Transfer, Vol. 97, pp. 624-626.

425. Wilson, N.W. and Sarma, T.S., 1977, "Prediction of Heat, Mass and Momentum Transfer During Laminar Forced Convection Melting of Ice in Saline Water", Proc. Symp. on Sea Ice Process and Models, Vol. 2, pp. 167-176.

426. Yen, Y.C., 1976, "Heat Transfer Between a Free Water Jet and an Ice Block Held Normal to It", Letters in Heat Mass Transfer, Vol. 3, pp. 299-308.

427. Gilpin, R.R., 1973-74, "The Ablation of Ice by a Water Jet", Trans. Can. Soc. Mech. Engrs., Vol. 2, No. 2, pp. 91-96.

428. Tien, C. and Yen, Y.C., 1974, "Heat Transfer Analysis of Air Bubble System", Proc. 5th Int. Heat Transfer Conf., Vol. 5, pp. 139-143.

429. Szekeley, J., Grevet, H.H. and El-Kaddah, N., 1984, "Melting Rates in Turbulent Recirculating Flow Systems", Int. J. Heat Mass Transfer, Vol. 27, pp. 1116-1121.

430. Tien, C. and Yen, Y.C., 1966, "Approximate Solution of a Melting Problem with Natural Convection", Chem. Eng. Prog. Symp. Ser., Vol. 62, No. 64, pp. 166-172.

431. Yen, Y.C., Tien, C. and Sander, G., 1966, "An Experimental Study of a Melting Problem with Natural Convection", Proc. 3rd Int. Heat Transfer Conf., pp. 159-166.

432. Yen, Y.C., 1968, "Onset of Convection in a Layer of Water Formed by Melting Ice from Below", Physics of Fluids, Vol. 11, pp. 1263-1270.

433. Yen, Y.C., 1980, "Free Convection Heat Transfer Characteristics in a Melt Layer", ASME J. Heat Transfer, Vol. 102, pp. 550-556.

434. Rieger, H. and Beer, H., 1986, "The Melting Process of Ice Inside a Horizontal Cylinder: Effect of Density Anomaly", ASME J. Heat Transfer, Vol. 108, pp. 166-173.

435. Tkachev, A.G., 1953, "Heat Exchange in Melting and Freezing of Ice", Problems of Heat Transfer During a Change of State: A Collection of Articles, Moscow, Translated from a Publication of the state Power Press, AEC-TR-34-5, pp. 169-178.

436. Dumore, J.M., Merk, H.J. and Prins, J.A., 1953, "Heat Transfer from Water to Ice by Thermal Convection", Nature, Vol. 172, pp. 460-461.

437. Vanier, C.R. and Tien, C., 1970, "Free Convection Melting of Ice Spheres", AIChE J., Vol. 16, pp. 76-82.

438. Bendell, M.S. and Gebhart, B., 1976, "Heat Transfer from Ice Melting in Ambient Water Near its Density Extremum", Int. J. Heat Mass Transfer, Vol. 19, pp. 1081-1087.

439. Ho, C.J. and Chen, S., 1986, "Numerical Simulation of Melting of Ice Around a Horizontal Cylinder", Int. J. Heat Mass Transfer, Vol. 29, pp. 1359-1369.

440. Zerkle, R.D. and Sunderland, J.E., 1968, "The Effect of Liquid Solidification in a Tube Upon the Laminar-Flow Heat Transfer and Pressure Drop", ASME J. Heat Transfer, Vol. 90, pp. 183-190.

441. Savino, J.M. and Siegel, R., 1969, "An Analytical Solution for Solidification of a Moving Warm Liquid onto an Isothermal Cold Wall", Int. J. Heat Mass Transfer, Vol. 22, pp. 1719-1723.

442. Kanayama, K., 1967, "The Rate of Ice Formation at Freezing of Water Filled in Steel Pipes", Kitami Inst. of Tech. Research Report, Vol. 2, No. 1, pp. 1-10 (in Japanese).

443. Gilpin, R.R., 1977, "The Effects of Dendritic Ice Formation in Water Pipes", Int. J. Heat Mass Transfer, Vol. 20, pp. 693-699.

444. Gilpin, R.R., 1979, "The Morphology of Ice Structure in a Pipe at or Near Transition Reynold Numbers", Heat Transfer, AIChE Symp. Series 189, Vol. 75, pp. 89-94.

445. Gilpin, R.R., 1981, "Ice Formation in Pipe Containing Flow in the Transition and Turbulent Regimes", ASME J. Heat Transfer, Vol. 103, pp. 363-368.

446. Gilpin, R.R., 1981, "Modes of Ice Formation and Flow Blockage that Occur While Filling a Cold Pipe", Cold Regions Science & Technology, Vol. 5, pp. 163-171.

447. Hirata, T., 1987, "Recent Advances in the Study of Formation of Ice-Band Structure in Water-Flow Pipe", 1977 Int. Symp. on Cold Regions Heat Transfer, ASME, pp. 39-45.

448. Lock, G.S.H., 1974, "The Growth and Decay of Ice", Heat Transfer, 1974, Vol. 6, pp. 12-27.

449. Saitoh, T., 1986, Computer-Aided Heat Transfer, Yokendo, Tokyo (in Japanese).

450. Katayama, K. et al., 1974, "Numerical Methods in Unsteady Heat Conduction", Progress in Heat Transfer Engineering, Vol. 3, Yokendo, pp. 111-209 (in Japanese).

451. SAE, 1987, Icing Technology Bibliography, Aerospace Information Report, Air 4015, Society of Automotive Engineers, Inc., Warrendale, PA, 150 p.

452. Kay, B.D. and Perfect, E., 1988, "State of the Art: Heat and Mass Transfer in Freezing Soils", Ground Freezing 88, R.H. Jones and J.T. Holden, eds., A.A. Balkema/Rotterdam/Brookfield, pp. 3-21.

453. Ladanyi, B. and Shen, M., 1989, "Mechanics of Freezing and Thawing in Soils", Frost in Geotechnical Engineering, Vol. 1, H. Rathmayer, ed., VTT Symposium 94, Technical Research Centre of Finland, Espoo, pp. 73-103.

454. Cheng, K.C. and Zarling, J.P., 1991, "Applications of Heat Pipes and Thermosyphons in Cold Regions", Procs. 7th Int. Heat Pipe Conference, Hemisphere Pub. Corp., New York, 1991.

455. Cheng, K.C., Lunardini, V.J. and Seki, N., eds., 1987 Internaitonal Symposium on Cold Regions Heat Transfer, Procs., ASME, 270 p.

456. Seki, N., Cheng, K.C.and Lunardini, F.J., eds., 1989 International Symposium on Cold Regions Heat Transfer, Procs., Hokkaido University, Sapporo, Japan, 314 p.

457. Fukusako, S., 1990, "Recent Advances in Study of Water-Freezing and Ice-melting Problems", Trans. of the Japanese Association of Refrigeration, Vol. 7, No. 1, pp. 1-32 (in Japanese).

458. Fukusako, S., Tago, M. and Mitsuyu, T., 1986, "Thermal Conductivities of Ethylene Glycol- and Propylene Glycol-Water Antifreezes", 7th Japan Symposium on Thermophysical Properties, pp. 1-4.

459. Tago, M. and Fukusako, S., 1988, "Freezing Heat-Transfer Characteristics in Return Bend with a Rectangular Cross-Section", National heat Transfer Conf., Vol. 3-HTD-Vol. 96, ASME, H.R. Jacobs, ed., pp. 215-224.

460. Fukusako, S. et al., 1989, "Frosting Heat Transfer from a Bundle of Horizontal Tubes Immersed in Aggregative Fluidized Bed", HTD-Vol. 114, Heat Transfer with Phase Change, ASME, pp. 29-37.
461. Fukusako, S. and Yamada, M., 1989, "Freezing Characteristics of Ethylene Glycol Solution", Wärme- und Stoffübertragung, Vol. 24, pp. 303-309.
462. Fukusako, S., Horibe, A. and Tago, M., 1989, "Ice Accretion Characteristics Along a Circular Cylinder Immersed in a Cold Air Stream with Seawater Spray", Experimental Thermal and Fluid Science, Vol. 2, pp. 81-90.
463. Fukusako, S., Takahashi, M. and Sawaoka, M., 1989, "Characteristics of the Freezing Heat Transfer of Layered Air-Water Flow in a Circular Tube", JSME International Journal, vol. 32, Series II, pp. 91-97.
464. Fukusako, S., 1990, "Thermophysical Properties of Ice, Snow, and Sea Ice", Int. J. of Thermophysics, Plenum Pub. Corp., Vol. 11, No. 2, pp. 353-372.
465. Fukusako, S., Yamada, M. and Tago, M., 1989, "Thermal Conductivity of a Dendritic Ice Layer", Int. J. of Thermophysics, Vol. 10, No. 1, pp. 269-278.

FREEZING AND MELTING HEAT TRANSFER

Chapter 3

Conduction with Freezing and Thawing

VIRGIL J. LUNARDINI
U.S. Army Cold Regions Research and Engineering Laboratory,
Hanover, New Hampshire, 03755-1290, USA

ABSTRACT

Conduction of heat transfer with solidification is a subset of the mathematical theory called Stefan problems or moving boundary problems. The exact solutions available are examined in some detail to yield insight into useful techniques, but approximate methods tend to be more useful for practical engineering problems. The concepts involved in the heat balance integral method, the quasi-static method, and perturbation methods are noted. Graphs are presented to aid in the application of theory to practical problems, especially those dealing with soil systems. Numerical methods and problems with significant convective aspects have not been examined nor has an attempt been made to do more than survey the literature of conduction heat transfer with phase change.

CONTENTS

Melting Temperature Range
Subcooled Liquid — Frazil Ice
Solidification in Contact with a Cold Wall
Thaw Consolidation of Melted Medium
Freeze of a Flowing Fluid

4. CONVECTION AT FREE SURFACE

Single Phase Problems
    Exact Solution for Semi-infinite Medium
    Analogue Solution
    Quasi-steady Approximation
    Heat Balance Integral Approximation
    Constant Heat Flux From Liquid Region
    Freeze of a Finite Slab
Two Phase Problems
    Heat Balance Integral Approximation
    Insulated Semi-infinite Region

5. SPECIFIED SURFACE HEAT FLUX

Exact Solution For Semi-infinite Region
Approximate Solutions, Single Phase, Semi-infinite Region
    Heat Balance Integral
Two Phase Problem
Ablation with Complete Removal of Melt
    Constant Surface Heat Flux
    Variable Surface Heat Flux
    Finite Thickness Slab

6. THAW BENEATH INSULATED STRUCTURES

Semi-Infinite Strip (Road)
Rectangular Building

NOMENCLATURE

A	$A' - 1 + \rho_{iw}$
A'	thaw strain of soil
b	half length of structure
B	$2k_{21}S_T + \alpha_{21}$
B_o	$h\,D/k$
$c,\ c_p$	specific heat, (specific heat at constant pressure)
c_{12}	c_1/c_2
C	ρc — heat capacity or volumetric specific heat
d	insulation or wall thickness
D_a	amplitude of ambient temperature above T_m
g	gravitational acceleration
g_1	$\dfrac{\alpha_{12}\,(1+\sigma)}{(\delta/X)(\delta/X+2)\sigma+2}+1$
G	temperature gradient
h	surface coefficient of heat transfer
h_i	specific enthalpy of ith component in porous body
h_r	radiation coefficient

j_i	mass flux rate of ith component of porous body
k	thermal conductivity
k_{ij}	k_i/k_j
$\bar{\bar{k}}$	thermal conductivity tensor
ℓ	latent heat of fusion
L	$\rho\ell$ - volumetric latent heat
p	pressure
p_s	saturation pressure of water vapor
P	period of transient surface temperature
$q*$	$q/h(T_f - T_a)$
q_g, q_r	generated and radiant energy per unit volume, per unit time
q_w	constant heat flux rate from liquid
q_w*	$q_w c_1/(h\,\ell)$
Q	specified surface heat flux
Q_o	constant surface heat flux
S	X/X_s nondimensional phase change depth
S_T	$\dfrac{c_1}{\ell}(T_s - T_f)$, Stefan number
S_{Ta}	$\dfrac{c_1}{\ell}(T_a - T_o)$
S_{Te}	$\dfrac{c_1}{\ell}(T_a - T_f)$
S_{Tm}	$\dfrac{C_3(T_f - T_s)}{\gamma_d\,\ell\,\Delta\,\xi}$
t	time
t_o	time when phase change starts
T	temperature
T_a, T_∞	ambient temperature
T_f, T_o, T_s	freezing, initial, and surface temperature
T_m	lowest temperature for freeze, mushy zone
T_p	temperature of bottom of structure
$\widehat{T}_p$	temperature of insulation - ground interface
x,y,z	Cartesian coordinates
X	phase change depth
X_o	initial phase change depth
z	$\dfrac{x - X}{D}$, Landau transformation

Greek

α	thermal diffusivity
α_{ij}	α_i/α_j
α_4	$\dfrac{\alpha_2}{1 + \sigma_{32}\,\phi_m/S_{Tm}}$
β	$\delta/(2\sqrt{\alpha t})$ thermal penetration depth parameter

γ	$X/(2\sqrt{\alpha t})$ phase change depth parameter
δ	thermal disturbance depth or radius
Δ	$Q\delta/(\rho\alpha_1\ell)$
θ	$(T - T_f)/(T_a - T_f)$ dimensionless temperature
θ_c	$\frac{\pi}{2}(T_f - \bar{T}_s)$
μ	viscosity
ν	h/h'
ξ	ratio of unfrozen water to soil solid mass
ξ_o, ξ_f, ξ_s	values of ξ at T_f, T_m, T_s
ξ_B	value of z_o/a at center of rectangle
ξ_1	$\left(\dfrac{2\pi\rho\ell}{k\,\theta_c\,P}\right)^{1/2} X$
ρ	density
ρ_{12}	ρ_1/ρ_2
σ	hX/k_1 dimensionless phase change depth
τ	$\dfrac{\alpha t}{X_o^2}$
τ_1	$\alpha(t-t_m)/D^2$
ϕ	$(T_o - T_f)/(T_f - T_s)$
ψ_1	$[T_o + T_i(z,t_m) - T(x,t)]/\dfrac{\partial T_i(1-S,t_m)}{\partial t}$, dimensionless temperature
ω	$\rho\ell\,k\,\Omega/h^2(T_f - T_m)$
Ω	frequency of ambient temperature

Subscripts

f	frozen
i	insulation
ℓ	liquid
s	solid
t	thawed
v	vapor
w	water or wall
1,2,3	regions of material

1. BASIC EQUATIONS

Solid-liquid phase change occurs quite regularly in engineering problems dealing with permafrost, seasonally frozen ground, solar energy, the freezing of food or biological material, and metallurgy. During a melting or freezing process, the system will be divided into regions separated by a phase change region which is at the fusion temperature. For soil systems, including permafrost, the thermal properties of the frozen and unfrozen regions are different, but are not strong functions of temperature for each individual phase.

Before discussing any physical laws it is necessary to define the thermodynamic system being considered. In most cases it is possible to consider a material as a continuum. Thus we can consider the properties of a substance at a "point" by considering a finite volume of material, large enough to contain sufficient atoms for an average property to have meaning, yet small enough so that the concept of a mathematical point is valid. A porous material differs from a continuous material in that its structure is non-homogeneous when viewed from the usual macroscopic level. The material consists of a framework of solids enclosing numerous voids which can contain fluids or other solids. A soil system has a mineral skeleton whose voids contain air, water, water-vapor, ice, hydrocarbons, or various solutions. The concepts of solid and liquid relate to the thermodynamic state of the liquid contained in the pores. When porous systems are considered the basic equations for a continuum will be considered valid, with the average properties of the porous material used.

General Energy Equation for a Continuum

The general energy equation for the temperature of a body at any point and time may be written (Lunardini, 1981a)

$$\text{div } (\bar{\bar{k}} \text{ grad } T) + q_r + q_g = \rho c_p \frac{DT}{Dt} + \frac{T}{\rho} \left(\frac{\partial \rho}{\partial T}\right)_p \frac{Dp}{Dt} - \mu \phi \tag{1}$$

where ϕ is the frictional dissipation function, often negligible for flow through permeable materials. If the material is also isotropic with constant thermal conductivity

$$k \nabla^2 T + q_r + q_g = \rho c_p \frac{DT}{Dt} \tag{2}$$

where ∇^2 is the nabla operator and D/Dt is the substantive derivative. For a motionless solid with all the velocities zero, the dissipation term and the pressure term may be dropped. Assume the material is orthotropic with the radiation term treated as a boundary condition. Then eq. 1 becomes the general conduction equation

$$\frac{\partial}{\partial x}\left(k_x \frac{\partial T}{\partial x}\right) + \frac{\partial}{\partial y}\left(k_y \frac{\partial T}{\partial y}\right) + \frac{\partial}{\partial z}\left(k_z \frac{\partial T}{\partial z}\right) + q_g = \rho c \frac{\partial T}{\partial t} \tag{3}$$

Energy Balance at the Phase Change Interface

At the interface between the phases, energy will be released or absorbed as the material freezes or thaws respectively. The conservation of energy applied to the mass which undergoes phase change, is

$$-k_1 \frac{\partial T_1}{\partial x} + k_2 \frac{\partial T_2}{\partial x} = \pm \rho \ell \frac{dX}{dt} \quad ; \quad x = X \tag{4}$$

where the upper sign is for melting and the lower sign is for freezing. The derivation assumes that the interface motion is in the positive direction of the space coordinate, otherwise the signs for the latent heat are reversed. The phase change introduces a basic non-linearity into the boundary conditions of the problem.

Heat and Mass Flow in Porous Materials

A porous medium is a matrix of solids with voids between the solid particles. The void space can contain water in the vapor, liquid, and solid phases and non-condensable gases such as air; these four constituents and the solid matrix itself make up the porous medium. Conservation of mass and energy must hold for each of the constituents. The conservation of mass requires

$$\frac{\partial \rho_o \theta}{\partial t} = -\text{div}\left(\sum_{i=1}^{4} j_i \right) \tag{5}$$

where subscripts 0-4 denote solid, vapor, liquid, ice, and air components and θ is the moisture content. The flux of water vapor is governed by Fick's law and capillary water transport is based on the Darcy equation. The total flux of moisture can be written as

$$\frac{j}{\rho_o} = -D_\theta \nabla \theta - D_T \nabla T \tag{6}$$

where the diffusion coefficients are

$$D_{\theta V} = \frac{\varepsilon_\theta D_a}{\rho_o RT} \ P_s \left(\frac{\partial \eta}{\partial \theta}\right)_T \qquad\qquad D_{\theta \ell} = \frac{K_\ell}{\rho_o g} \left(\frac{\partial \psi}{\partial \theta}\right)_T$$

$$D_{TV} = \frac{\varepsilon_T D_a}{\rho_o RT} \ \eta \left(\frac{\partial p_s}{\partial T}\right)_\theta \qquad\qquad D_{T\ell} = \frac{K_\ell}{\rho_o g} \left(\frac{\partial \psi}{\partial T}\right)_\theta$$

$$D_\theta = D_{\theta V} + D_{\theta \ell} \qquad\qquad D_T = D_{TV} + D_{T\ell}$$

ε_θ, ε_T are diffusion correction factors, K_ℓ is the hydraulic conductivity, η is the relative humidity, R the gas constant, D_a the diffusion coefficient of water vapor in air, and ψ the moisture potential. The overall conservation of mass is then

$$\frac{\partial \theta}{\partial t} = D_\theta \ \nabla^2 \theta + D_T \ \nabla^2 T \tag{7}$$

Conduction and enthalpy flows will be the only significant energy flux terms.

$$\rho_o C \frac{\partial T}{\partial t} = \text{div} \ (k \ \nabla T) - \sum_{i=1}^{3} h_i I_i - \sum_{i=1}^{3} c_{pi} \ j_i \ \nabla T \tag{8}$$

where I is a source/sink. The convective heat transfer term, in Eq (8), is often ignored if the mass flux includes only diffusion. This is not acceptable if water fluxes such as filtration or groundwater are present, Luikov (1964, 1975). If the medium remains above the solidification temperature of the moisture then no ice can be present. Using the relations developed, the energy equation with no vapor flux is

$$\rho_o C \frac{\partial T}{\partial t} = \text{div} \ (k\nabla T) + c_{p2} \ \rho_o \left[D_{\theta \ell} \ \nabla \theta + D_{T\ell} \ \nabla T \right] \nabla T \tag{9}$$

We shall assume negligible vapor flux and convective heat transfer for a frozen medium,

$$\rho_o C \frac{\partial T}{\partial t} = \text{div}(k\nabla T) + \rho_o \ell \left[\frac{\partial \theta_3}{\partial t}\right] \qquad (10)$$

If melting/freezing occurs but there is no liquid flux

$$\rho_o C \frac{\partial T}{\partial t} = \text{div}(k\nabla T) - \rho_o \ell \frac{\partial \theta_2}{\partial t} \qquad (11)$$

Conduction in Porous Media

Notwithstanding the mass and heat flux relations just derived, pure conduction may be an excellent assumption for porous media, Porkhayev (1959), Martynov (1959). Consider the total heat flux that might occur in a porous medium.

$$q_{TOT} = -\left(k + \rho_o \left[h_v D_{TV} + h_w D_{T\ell}\right] + h + h_r\right)\nabla T - \rho_o \left(h_v D_{\theta V} + h_w D_{\theta\ell}\right)\nabla\theta \qquad (12)$$

If the second term on the right hand side of eq. 12 is negligible then pure conduction is an excellent assumption, with the thermal conductivity and other properties of the medium altered to account for the heat transfer and moisture effects. Clearly, such an assumption is not possible if a significant heat flow due to bulk water movement occurs. Figure 1 is a qualitative attempt to outline the important heat flow regimes in a soil system, Johansen (1975). Note that for nearly all practical cases, pure conduction will predominate or can be corrected with the use of an effective thermal conductivity. The figure is strictly valid only for a thawed soil. If a freeze interface exists, then the flow of soil moisture can be significantly increased and convection may need to be considered

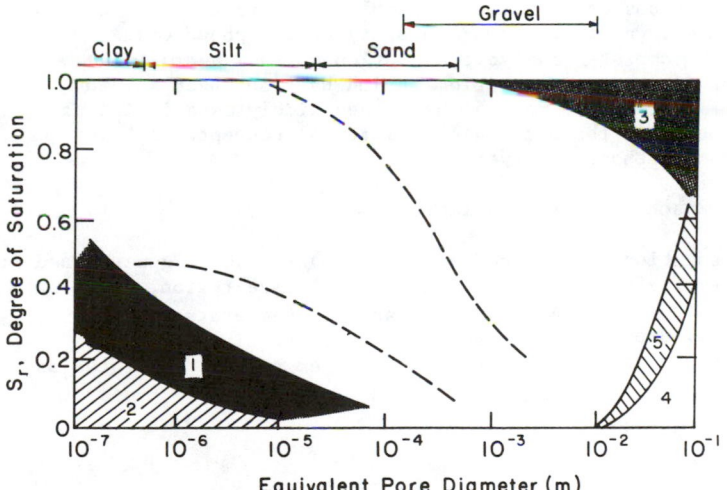

FIGURE 1. Influence of heat transfer mechanisms in soil (after Johansen 1975). 1. Thermal redistribution of moisture. 2. Vapor diffusion. 3. Free convection in water. 4. Free convection in air. 5. Radiation.

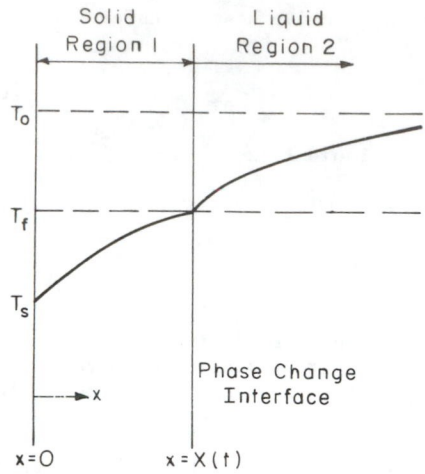

FIGURE 2. Temperature distribution in partially frozen medium.

explicitly. For many applications, it is reasonable to treat a soil system as a continuum and to evaluate the temperature field on the basis of conduction with the effects of moisture incorporated into the soil properties.

2. EXACT AND APPROXIMATE SOLUTION METHODS

Problems which can be formulated in terms of pure conduction will be considered and only solutions for plane geometries will be examined. Ideally, exact solutions of engineering problems of freezing and thawing are sought. However, due to the non-linearity of the phase change system, there are very few complete, analytic, solutions. Thus approximate solutions will be discussed for problems which have not been solved exactly. Several approximate methods have been widely used including quasi-steady methods and the heat balance integral concept. A brief discussion of these methods is given.

Neumann Problem, Exact Similarity Solution

The first exact solution method is due to Neumann (c 1860), generalized in Carslaw and Jaeger (1959). Initially, a semi-infinite region, shown in Figure 2, is at a constant temperature T_0 and the temperature of the surface is suddenly dropped to T_s and held constant. The medium is initially in a liquid state, $T_0 > T_f$. The problem can be formulated as

$$\frac{\partial^2 T_1}{\partial x^2} = \frac{1}{\alpha_1} \frac{\partial T_1}{\partial t} \tag{13}$$

$$\frac{\partial^2 T_2}{\partial x^2} = \frac{1}{\alpha_2} \frac{\partial T_2}{\partial t} \tag{14}$$

$$\lim_{x \to \infty} T_2 = T_o \tag{15}$$

$$T_1 (0, t) = T_s \tag{16a}$$

$$T_1(X,t) = T_2(X,t) = T_f \tag{16b}$$

The energy balance at the phase change interface is

$$k_1 \frac{\partial T_1}{\partial x} - k_2 \frac{\partial T_2}{\partial x} = \rho \ell \frac{dX}{dt} \; ; \; x = X \tag{17}$$

A solution to this problem is obtainable by using the well known similarity transformation

$$\eta = x/2\sqrt{\alpha_1 t} \tag{18}$$

Equation 18 will transform Eqs 13 and 14 into ordinary differential equations which have error function solutions. The error function has been numerically evaluated and tabulated, Carslaw and Jaeger (1959). Thus a formal solution to the conduction equation is available, if the similarity transformation is valid. This will be the case if the differential equations and all of the boundary and initial conditions can be expressed in terms of the single, independent variable η. However this similarity solution is only valid for problems where the phase change interface moves proportional to $\sqrt{t}$. This precludes the use of the similarity transformation for many interesting problems such as convection boundary conditions, variable initial temperature, and variable surface temperature.

The phase change interface is

$$X = 2\gamma \sqrt{\alpha_1 t} \tag{19}$$

where γ is a constant found from

$$\frac{\exp(-\gamma^2)}{\mathrm{erf} \, \gamma} - \frac{k_{21} \sqrt{\alpha_{12}} \, \phi \, e^{-\alpha_{12}\gamma^2}}{\mathrm{erfc} \, (\gamma \sqrt{\alpha_{12}})} = \frac{\gamma \sqrt{\pi}}{S_T} \tag{20}$$

The Stefan number, S_T, is the ratio of the sensible to the latent heat. The temperatures are now given by

$$T_1 = T_s + \frac{(T_f - T_s)}{\mathrm{erf} \, \gamma} \, \mathrm{erf} \, \frac{x}{2\sqrt{\alpha_1 t}} \tag{21}$$

$$T_2 = T_o - \frac{(T_o - T_f)}{\mathrm{erfc} \, (\gamma \sqrt{\alpha_{12}})} \, \mathrm{erfc} \, \frac{x}{2\sqrt{\alpha_2 t}} \tag{22}$$

The solution is also valid for the thaw case if the meaning of the property subscripts is interchanged, i.e., region 1 is now thawed. If the medium is initially at the phase change temperature only one phase will be present. This special case of the Neumann problem is also often referred to as the

Stefan solution, following the original work of Stefan (1891). However, it should be noted that problems of conduction with phase change are generally classed, mathematically, as Stefan Problems or moving boundary problems. If the initial temperature of the liquid is T_f then eq (20) is

$$\gamma e^{\gamma^2} \text{ erf } \gamma = \frac{S_T}{\sqrt{\pi}} \tag{23}$$

Now, when γ is small, the error function may be approximated as erf $\gamma \approx 2\gamma/\sqrt{\pi}$ and the phase change depth is

$$X = \sqrt{2 S_T \alpha t} \tag{24}$$

Equation 24 was presented by Stefan (1891) who used the quasi-steady approximation to be discussed next. The Stefan solution is not an exact solution in the sense of the Neumann solution. For rocks or metals $S_T \approx 2.0$ but for water this value can be very small. Therefore, eq. 24 is an acceptable approximation for water and for many soil systems.

Quasi-static Approximations and Perturbation Methods

The Neumann problem has shown that error functions can lead to an exact solution, suitable for certain boundary conditions. However, the method cannot be applied in general. Similarity solutions will not exist for finite domains, two phases present initially, non-uniform initial temperatures, and boundary temperatures which are arbitrary functions of time. Thus there are very few other exact solutions existing. This has prompted considerable interest in approximate methods that can yield solutions acceptable for engineering design. Aside from the usual numerical procedures, several analytical methods have been of great value including: the quasi-static approximation, the heat balance integral method, and variational methods.

If the phase change interface moves relatively slowly, then an assumption can be made that the moving interface will not exert a major influence upon the temperature field during short time periods. The quasi-stationary assumption neglects the moving interface in evaluating the temperature field. The assumption can handle initial conditions but is not generally valid if the temperature ahead of the interface is non-uniform or is changing due to the interface motion. The problem reduces to one of transient conduction with no phase change. The actual phase change is then solved through the interface boundary condition. The quasi-steady approximation further simplifies the quasi-stationary problem by dropping the transient term in the energy equation. The method cannot satisfy the initial conditions or sensible heat, however the problem is mathematically so simple that the idea has been used extensively.

Quasi-stationary approximation. The melting system will be examined again, for a Neumann problem with variable density, in order to illustrate the quasi-stationary idea. Neglect the temperature variations in the solid region and examine only the liquid equations. The equations are

$$\frac{\partial^2 \theta}{\partial x_1^2} = \frac{\partial \theta}{\partial \tau} + \left(\frac{\rho_2}{\rho} - 1\right) S_T \frac{\partial \theta}{\partial x_1} \left(\frac{\partial \theta}{\partial x_1}\right)_\xi \tag{25}$$

$$\theta(x_1, 0) = 0$$

$$\theta(o, \tau) = \phi_s$$

$$\theta(\xi, \tau) = \phi$$

$$\frac{d\xi}{d\tau} = -S_T \left(\frac{\partial \theta}{\partial x_1}\right)_\xi \qquad (26)$$

$$\xi(0) = 1$$

where $\xi = X/X_o$, $\theta = (T - T_o)/(T_s - T_f)$, $x_1 = x/X_o$. The quasi-stationary approximation, which tends to be valid if $S_T \ll 1$, reduces eq. 25 to

$$\frac{\partial^2 \theta}{\partial x_1^2} = \frac{\partial \theta}{\partial \tau} \qquad (27)$$

After solving eq. 27, the interface location is evaluated with eq. 26. Due to the limitations already mentioned, the method is best suited to single phase problems. Duda and Vrentas (1969a) showed that the quasi-stationary solution is the first term of an asymptotic series. They expanded the temperature and interface position, with the Stefan number as the perturbation parameter. The procedure is called a surface-volume perturbation since the differential equation and the boundary conditions both contain non-linearities, Van Dyke (1964), Cole (1968). This leads to a system of equations for which the zeroth order equation is essentially the quasi-stationary approximation. The method reduces the phase change problem to one of transient heat conduction with no phase change.

Quasi-steady approximation. The quasi-stationary method can be further simplified if the unsteady terms in the diffusion equation are also neglected. To justify this a new characteristic time will be used.

$$\tau_1 = \frac{\alpha t}{X_o^2} S_T \qquad (28)$$

The characteristic time is now long compared to the diffusion time X_o^2/α, (if $S_T < 1.0$) and is suited to long time movement of the interface. Jiji and Weinbaum (1978) used two time domains, the quasi-stationary for initial growth and eq 28 for later growth. In this way a two-phase problem could be handled. The non-dimensional equations are

$$\frac{\partial^2 \theta}{\partial x_1^2} = S_T \frac{\partial \theta}{\partial \tau_1} - (\rho_2/\rho - 1)S_T \frac{d\xi}{d\tau_1} \frac{\partial \theta}{\partial x_1} \qquad (29)$$

$$\frac{d\xi}{d\tau_1} = -\left(\frac{\partial \theta}{\partial x_1}\right)_\xi$$

The boundary conditions remain the same as those of eq. 25. For small Stefan numbers the diffusion equation reduces to

$$\frac{\partial^2 \theta}{\partial x_1^2} = 0 \qquad (30)$$

Thus, no transient term need be considered and the solution is extremely simple. Solutions are far easier to obtain, compared to the quasi-stationary equations, but the validity of the solution is more limited since the initial conditions cannot be met and the sensible heat is not accounted for. Nevertheless this concept is very widely used for freezing and thawing problems.

The quasi-steady method can also be examined from the viewpoint of perturbation solutions. The zeroth solution is the quasi-steady approximation which Pedroso and Domoto (1973a) demonstrated for a spherical system. Lock (1969) derived the zeroth and first order systems for the solidification of a semi-infinite medium. Duda and Vrentas (1969a,b) discussed the usefulness of perturbation methods for phase change problems. Pedroso and Domoto (1973a,b,c) noted the difficulty of using perturbation methods for inward, spherical, solidification. Jiji (1970) used perturbation for curvilinear solidification. Huang and Shih (1975) introduced a useful concept for perturbation methods by replacing the time variable τ by $\xi(\tau)$, the phase change interface position. This transformation is acceptable if ξ is a monotonic function of τ, a common relation for many practical problems. The temperature and the rate of change of the phase change interface are expanded as asymptotic series with the Stefan number as the parameter. These expansions allow many problems to be solved in a particularly simple fashion, Seeniraj and Bose (1982). An advantage of this procedure is that it is no longer necessary to expand the unknown functions about the initial value of the phase change depth when considering the conditions at the phase change interface. This greatly simplifies the formulation of the equations, but the solutions are not necessarily simpler.

Let us return to the Stefan problem. The solution was based on the assumption that the Stefan number is small which is equivalent to the quasi-steady approximation. The quasi-steady version of this problem reduces to

$$\frac{\partial^2 T_1}{\partial x^2} = 0 \qquad (31)$$

$$T_1(0,t) = T_s$$

$$T_1(X,t) = T_f$$

$$k_1 \frac{\partial T_1(X,t)}{\partial x} = \rho_1 \ell \frac{dX}{dt}$$

The solution to this system is

$$T_1 = [T_f - T_s(t)]\frac{x}{X} + T_s(t) \qquad (32)$$

$$X = \sqrt{\frac{2k_1}{\rho_1 \ell} \int_o^t [T_f - T_s(t')]dt'} \qquad (33)$$

The phase change depth X given here is identical to that of eq. 24. Thus, Stefan was one of the first to use the quasi-steady method.

A major limitation of the quasi-steady approximation is the failure to account for the sensible heat during the phase change. The total surface energy flow, during the time that a layer of thickness X freezes, is simply the latent heat. The method does not take into account any sensible heat, due to the assumption of a Stefan number of zero, although the temperature of the frozen layer does decrease with time.

Heat Balance Integral Method

An approximate method which has been used with good results for solidification problems involves the concept of the temperature penetration depth. The integral method introduced by Goodman (1958) is based on the same concepts as the momentum integral method of the boundary layer in fluid mechanics, Pohlhausen (1921), von Karman (1921). Consider the semi-infinite solid shown in Figure 3. At a time t, after the surface temperature has dropped to T_s, the temperature in the solid will be disturbed to a depth $\delta(t)$. Beyond this depth, the temperature of the solid remains at the initial temperature T_o and no energy is transferred. The penetration distance δ is analogous to the boundary layer thickness in fluid mechanics and the basic equations are satisfied on average over the volume of thickness $\delta(t)$, rather than at each point. The conduction equation can be integrated spatially over the distance $\delta(t)$. Then, using Leibniz's rule for a general function, the heat balance integral equation is

$$\frac{d}{dt} \int_o^\delta T(x,t)dx + \alpha \frac{\partial T(0,t)}{\partial x} - T_o \frac{d\delta}{dt} = 0 \tag{34}$$

This equation is valid if there is no phase change. The solution now involves choosing an approximation to $T(x,t)$ which satisfies the boundary and initial conditions. Equation 34 will then yield a differential equation for $\delta(t)$. In general, the accuracy of the approximate solution can be improved by using higher order temperature profiles, however, the algebraic work also increases. The convergence of the method for a given situation is unpredictable, as has been discussed by Langford (1973).

Consider the case of phase change where the properties of the frozen region differ from those of the thawed region. There will then be two integral equations as follows

$$\frac{d}{dt} \int_o^X T_1(x,t)dx - T_f \frac{dX}{dt} - \alpha_1 \left[\frac{\partial T_1(X,t)}{\partial x} - \frac{\partial T_1(0,t)}{\partial x} \right] = 0 \tag{35}$$

$$\frac{d}{dt} \int_X^\delta T_2(x,t)dx - T_o \frac{d\delta}{dt} + T_f \frac{dX}{dt} + \alpha_2 \frac{\partial T_2(X,t)}{\partial x} = 0 \tag{36}$$

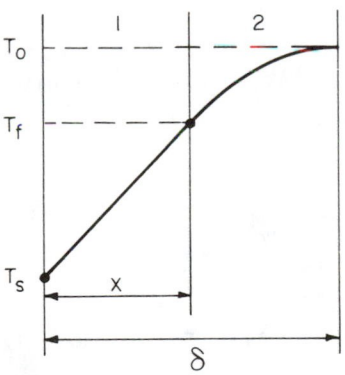

FIGURE 3. Heat balance integral geometry.

where $T_1(X,t) = T_2(X,t) = T_f$; $T_2(\delta,t) = T_o$.

The solution of a general problem with superheat or subcooling (the initial temperature not at the freezing temperature) will involve two coupled parameters X and δ and is usually difficult. However, if the initial temperature is T_f then the problem reduces to only one differential equation. The derivation of the method and some applications have been described by Goodman (1958, 1964).

The choice of the proper approximation to use for the assumed temperature profile can be eased with a refinement to the heat balance integral method suggested by Noble (1975) and carried out by Bell (1978). Instead of following only the phase change and initial temperature penetration depths, any number of isotherms can be followed by writing the heat balance integral for an arbitrary number of regions. The accuracy increases with an increasing number of subdivisions, but the computational work also increases. Bell (1978) showed that the error can be reduced from 6.5% with one region to 1.2% for two sub-divisions. Bell (1979, 1982) also noted that the approximate solution asymptotically approached the exact solution as the number of intervals increased. This procedure can reduce the errors for the heat balance integral method, but it also tends to negate the simplicity of the method. For more than two subdivisions or for more complicated boundary conditions it is likely that numerical solution of the set of simultaneous differential equation will be required. Bell and Abbas (1985) used a sub-division of the penetration depth and solved for the temperatures at each sub-division. They proved that the solution converges to the exact value, at least for a simple problem, (without phase change).

A simple, but useful, approximate solution to the Neumann Problem can be obtained by the use of the heat balance integral method. The system of equations for the freeze problem, is eqs. 35 and 36 with the following boundary and initial conditions.

$$T_1(o,t) = T_s \tag{37a}$$

$$T_1(X,t) = T_f \tag{37b}$$

$$T_2(\delta,t) = T_o \tag{38a}$$

$$\frac{\partial T_2(\delta,t)}{\partial x} = 0 \tag{38b}$$

$$T_2(X,t) = T_f \tag{39}$$

$$k_1\frac{\partial T_1(X,t)}{\partial x} - k_2\frac{\partial T_2(X,t)}{\partial x} = \rho_1\ell\frac{dX}{dt} \tag{40}$$

$$-k_1\left[\frac{\partial T_1(X,t)}{\partial x}\right]^2 + k_2\frac{\partial T_1(X,t)}{\partial x}\frac{\partial T_2(X,t)}{\partial x} = \rho_1\ell\alpha_1\frac{\partial^2 T_1(X,t)}{\partial x^2} \tag{41}$$

Assume that the solutions for the phase change interface, X, and the thermal penetration depth, δ, are

$$X = 2\gamma \sqrt{\alpha_1 t} \tag{42}$$

$$\delta = 2\beta \sqrt{\alpha_2 t} \tag{43}$$

A linear approximation for the temperature in region 1 and a quadratic approximation for T_2 are assumed leading to the following differential equation for X.

$$\left(\frac{1}{2} + \frac{1}{S_T}\right) \frac{dX}{dt} = \frac{\alpha_1}{X} \left(1 - \frac{2k_{21}\phi}{\delta/X-1}\right) \tag{44}$$

The solution is straightfoward and can be written as

$$\gamma^2 = \frac{-b_1 - \sqrt{b_1^2 - 4a\,S_T^2}}{2a} \tag{45}$$

with

$$a = \left(S_T + 2 + \frac{2k_{21}\,\phi S_T}{\alpha_{21}}\right)(S_T + 2)$$

$$b_1 = -2\,S_T\left(S_T + 2 + \frac{k_{21}\,\phi\,S_T}{\alpha_{21}}\right) - \frac{4}{3}\,\frac{(k_{21}\,\phi\,S_T)^2}{\alpha_{21}}\ .$$

As often occurs with the heat balance integral method, a more complicated approximation for T_1 does not significantly improve the accuracy (Lunardini, 1988). The approximate value for γ can be compared to exact values, eq. 20, given by Carslaw and Jaeger (1959), for water-ice systems, shown in Table 1. The approximate equation yields good results for small values of S_T. A typical soil is also compared in Table 2. The heat balance integral gives a very good approximate equation for γ over a wide range of S_T, ϕ, and property values. Equation 45 should be within 5% for all practical cases for soil systems.

TABLE 1. Accuracy of approximate γ-values for ice-water system, $k_{21} = 0.2717$, $\alpha_{21} = 0.1252$ (from Lunardini and Varotta 1981).

S_T	ϕ	Exact	eq 45	Error %, eq 45	Quadratic temperature	Error %
0.0063	0	0.056*	0.0559	−0.2	---	---
0.0063	5	0.049*	0.0486	−0.8	---	---
0.0314	0	0.124*	0.1243	+0.3	---	---
0.0314	1	0.115*	0.1156	0.5	---	---
0.1	0	0.2200	0.2181	−0.9	0.2232	1.5
0.1	5	0.1176	0.1190	1.2	0.1200	2.0
1	0	0.6201	0.5774	−6.9	0.6600	6.4
1	5	0.1449	0.1479	2.1	0.1497	3.3

*Values from Carslaw and Jaeger (1959).

TABLE 2. Accuracy of approximate γ-values, $k_{21} = 0.51$, $\alpha_{21} = 0.3355$ (from Lunardini and Varotta 1981).

S_T	ϕ	Exact	eq 45	Error %, eq 45	Quadratic temperature	Error %
				γ		
0.0058	0	0.0538	0.0538	0	0.0539	0.2
0.0058	5	0.0468	0.0465	0	0.0465	0
0.03	0	0.1219	0.1215	-0.30	0.1225	0.5
0.03	5	0.0860	0.0861	0.1	0.0865	0.6
0.1	0	0.2200	0.2181	-0.9	0.2232	1.5
0.1	5	0.1168	0.1174	0.5	0.1185	1.5
1	0	0.6201	0.5774	-6.9	0.6600	6.4
1	5	0.1460	0.1477	1.2	0.1496	2.5

Nixon and McRoberts (1973) found a semi-empirical relation for γ, with $\phi = 0$, as

$$\gamma = \left(1 - \frac{S_T}{8}\right) \sqrt{\frac{S_T}{2}} \tag{46}$$

In the limit as $\phi \to 0$, eq 45 reduces to

$$\gamma = \sqrt{\frac{S_T}{2 + S_T}} \tag{47}$$

The quasi-steady Stefan solution has been shown to be

$$\gamma = \sqrt{\frac{S_T}{2}} \tag{48}$$

All of the above equations are quite close, and accurate, for small S_T values, but eq. 47 gives the best overall accuracy.

3. TEMPERATURE BOUNDARY CONDITIONS

Modified Berggren Equation

The Neumann solution is widely used for soil freezing estimates but special names have been given to it which can be confusing. Berggren (1943) was one of the first to actually apply the Neumann solution to soil phase change problems. Aldrich and Paynter (1953) later used the Stefan form of the solution to arrive at the modified Berggren equation. Equation 19 can be changed to the Stefan form as

$$X = \lambda \sqrt{(2k_f/L)(T_f - T_s)t} \tag{49}$$

where λ can be determined from the exact solution. Stefan (1891) original-ly solved a similar problem for the growth of sea ice when the sensible heat to latent heat ratio was small and the water was at the freezing temperature. Equation 49 reduces to the Stefan equation, eq. 24, when λ equals one hence the name "Stefan form." The surface temperature of a soil system does not normally remain constant during the freeze season and the surface index I_s is often used in place of the mean surface temperature.

$$X = \lambda \sqrt{(2k_f/L)I_s} \tag{50}$$

The surface index is defined as

$$I_s = \int_o^\varepsilon \left[T_f - T_s(t') \right] dt' = (T_f - \overline{T}_s)\varepsilon \tag{51}$$

where ε is the length of the freeze season. Unfortunately, the surface index is rarely available for a location, however, the air temperature index I_f is usually tabulated and I_s can be replaced with the n-factor, defined as

$$n = I_s/I_f \text{ or } I_s/I_t \tag{52}$$

A procedure for obtaining a value of n at a given site is given by Lunardini (1978). Finally, the modified Berggren equation is written as

$$X = \lambda \sqrt{2k_f I_f n)/L}. \tag{53}$$

Berg and Aitken (1973) have shown that the modified Berggren equation gives good results for seasonal phase change depths. The coefficient λ can be found from eq. 20 with appropriate soil thermal properties and tempera-tures. Aldrich and Paynter (1953) assumed that the frozen and thawed properties were equal and published a widely used graph for λ (see Sanger 1969). Actually, the graph is only valid when the water content of a soil is zero. Nixon and McRoberts (1973) presented a graph of λ valid only for $\sqrt{\alpha_{ft}} = 1.43$. Lunardini (1980) solved eq. 20 numerically to find values of λ for soil systems. The geometric mean for the thermal conductivity of a soil mixture was used as recommended by Gold and Lachenbruch (1973). For soil systems, the thermal conductivity of the solids and gases will not vary significantly as phase change occurs (Kersten 1949) and there will be only a small error if it is assumed that the frozen state contains only ice with no unfrozen water. The specific heats of different soil solids and ice are all similar in magnitude (Lunardini, 1971). Figures 4 to 6 give the values of λ. Notice that when the volumetric water content $x_\ell = 0$ the λ values are the same for freezing or thawing, which is the Aldrich and Paynter (1953) case.

Constant Phase Change Rate

Stefan (1891) gives a solution for the case of a constant heat flux at the phase change interface assuring that the interface moves at constant velocity and an exact solution can be found using the similarity transfor-mation approach. Consider a semi-infinite solid initially at the fusion temperature. The solution follows as

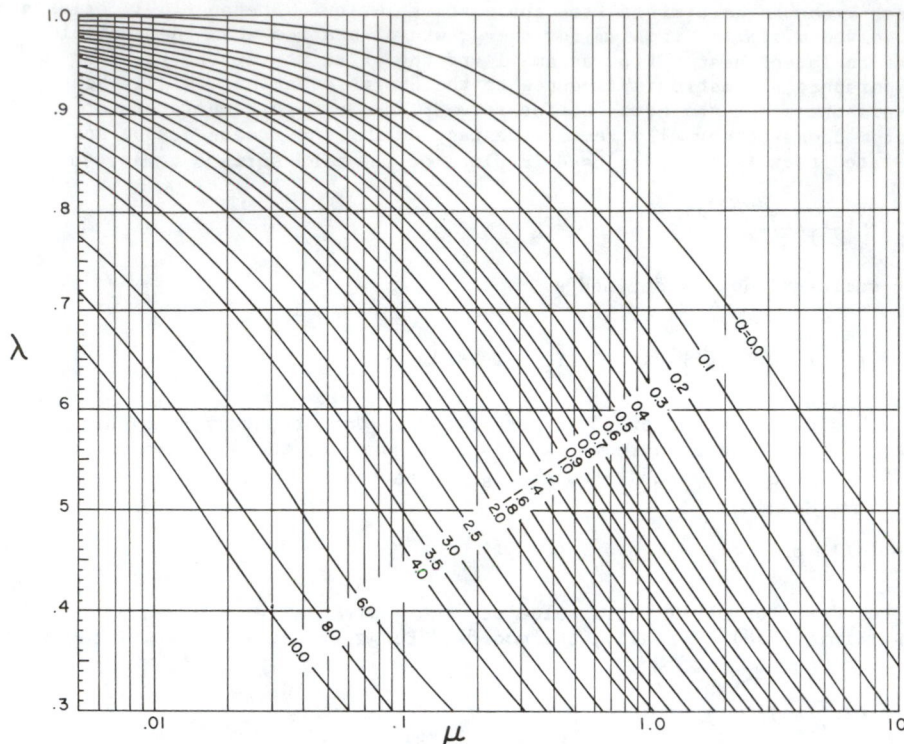

FIGURE 4. Neumann equation parameter, freeze/thaw, $\alpha \equiv \phi$, $\mu = S_T/2$, $x\ell = 0.0$.

$$T_1(x,t) = T_f + \frac{\ell}{c_1}[1 - e^{\frac{-V}{\alpha_1}(x-X)}] \qquad (54)$$

$$X = Vt \qquad (55)$$

The required surface temperature is

$$T_1(o,t) = T_f + \frac{\ell}{c_1}(1 - e^{\frac{-V^2 t}{\alpha_1}}) \qquad (56)$$

As noted by Stefan (1891), the surface temperature of an ice layer may drop rapidly to a constant value. During this transient phase the ice layer will grow linearly with time and later grow as the square root of time. This kind of problem is an inverse problem wherein applied boundary conditions must be found to obtain a given phase change interface motion. Rubinsky and Shitzer (1978) give a general solution as follows.

$$T_1(x,t) = T_f + \sum_{n=o}^{\infty} a_n(X) \frac{(x - X)^n}{n!} \qquad (57)$$

$$a_o = 0 \qquad (57a)$$

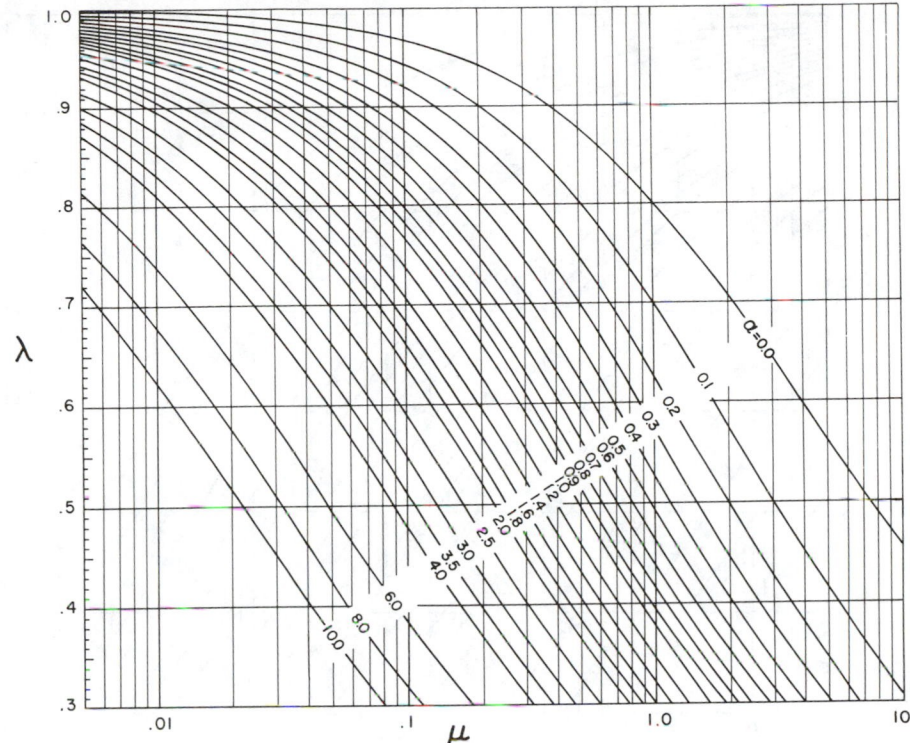

FIGURE 5. Neumann equation parameter, freeze, $x_\ell = 1.0$.

$$a_1 = \frac{\partial T_1(X,t)}{\partial x} \qquad (57b)$$

$$a_n = \frac{dX/dt}{\alpha}\left[\frac{da_{n-2}}{dX} - a_{n-1}\right] \qquad (57c)$$

The temperature gradient at the interface is obtained from the usual interface energy relation.

$$k_1 \frac{\partial T_1(X,t)}{\partial x} - Q_i = \rho\ell\frac{dX}{dt} \qquad (58)$$

where Q_i — heat flux supplied to the interface from the original phase. The problem does not have significant practical value since it requires that a variable surface temperature be imposed upon the solid to maintain melting at a constant rate.

Neumann Problem with Variable Properties

<u>Variable density</u>. If the density of the solid and liquid phases differ, as is usual, a solution can be found using the similarity technique. This problem actually involves convection in the frozen phase, due to the motion of the solid, but it can be formulated as a conduction problem. The problem can be posed according to Figure 7. Reference frame x_1 is attached to

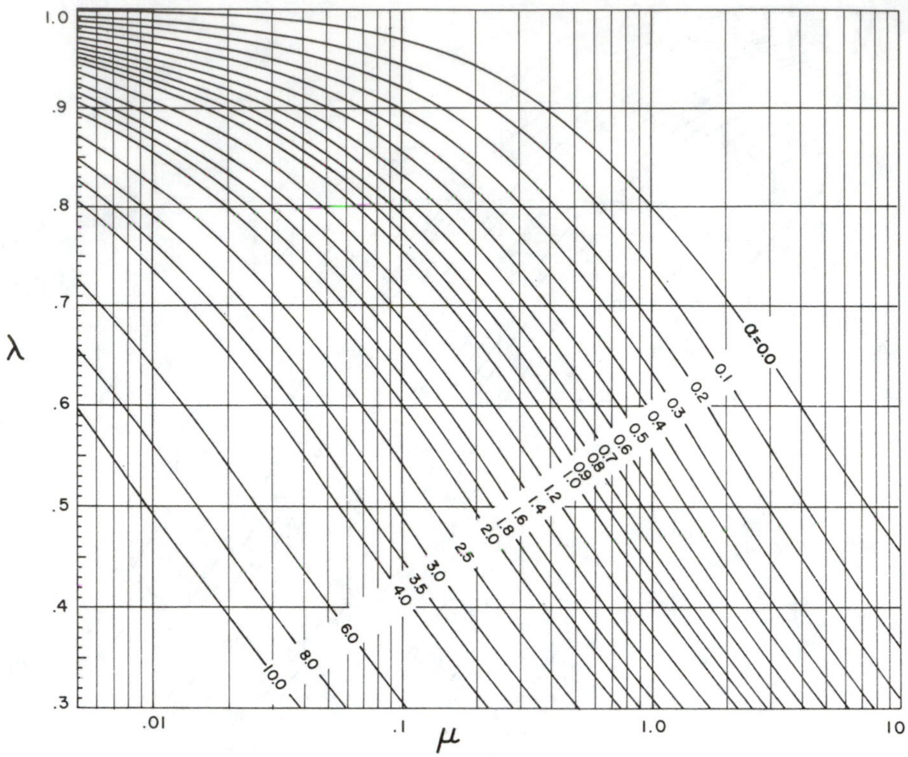

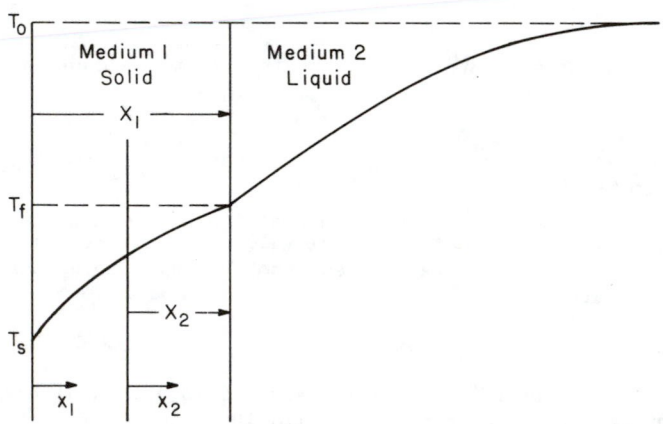

FIGURE 6. Neumann equation parameter, thaw, $x_{\ell} = 1.0$.

FIGURE 7. Freezing in a medium with variable density.

the free surface of medium 1 and moves as medium 1 expands, assuming that $\rho_1/\rho_2 < 1$, while reference frame x_2 is fixed at the location of the original free surface. The thickness of the solified material between the original free surface location and the phase change interface is X_2. The solution to this problem has been given by Crank (1975), and Carslaw and Jaeger (1959) with the interface given by

$$X_1 = 2\gamma\sqrt{\alpha_1 t} \tag{59}$$

The parameter γ is obtained from the following transcendental equation

$$\frac{k_{21}\sqrt{\alpha_{12}}\phi\exp(-\rho_{12}^2\,\alpha_{12}\gamma^2)}{\mathrm{erfc}(\gamma\sqrt{\alpha_{12}\rho_{12}})} - \frac{e^{-\gamma}}{\mathrm{erf}\gamma} + \frac{\gamma\sqrt{\pi}}{S_T} = 0 \tag{60}$$

The solution of equation (60) requires a numerical procedure, and the number of parameters makes a graphical presentation of the solution impractical. An approximate solution to the problem has been found, with the heat balance integral method, which yields an expression for γ that does not require numerical solution, Lunardini (1983). The approximate relation for γ is

$$\gamma^2 = \frac{b_1 - \sqrt{b_1^2 - 4aS_T^2}}{2a} \tag{61}$$

$$b_1 = 2S_T\left[2 + S_T + \frac{\phi k_{21}\rho_{12}S_T}{\alpha_{21}}\right] + \frac{4}{3}\frac{(\phi k_{21}S_T)^2}{\alpha_{21}} \tag{62}$$

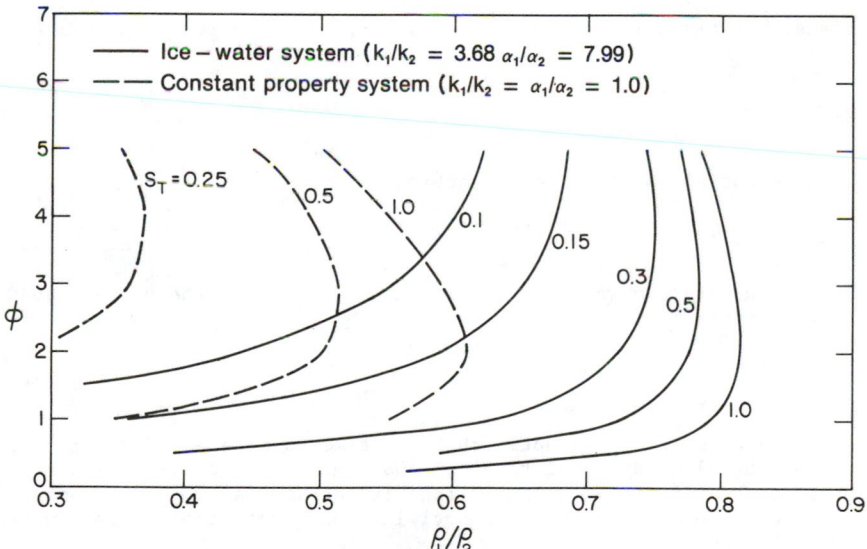

FIGURE 8. Range of validity of constant density solution, six percent accuracy.

$$a = (2 + S_T) \left[2 + S_T + 2S_T \frac{\phi k_{21} \rho_{12}}{\alpha_{21}} \right] \tag{63}$$

This approximate solution will be in error by less than 3% for most practical problems. The effect of the density variation, on the Neumann solution, will be small, unless the density ratio is quite small, as can be seen from Figure 8. The equations are valid for the thaw case except that region 1 is now thawed and the total thaw depth is given by $\rho_{12} X_1$.

Variable thermal conductivity. Cho and Sunderland (1974) have extended the Neumann Problem when the thermal conductivity varies linearly with temperature. The effect of variable conductivity on the phase change rate is small unless the rate of change of the thermal conductivity is large and the Stefan number is large. Pedroso and Domoto (1973) have presented a perturbation solution for the Stefan problem if the thermal properties are variable. General integral solutions are given for arbitrary thermal property functions. The perturbation technique used is not applicable to two phase problems.

Variable latent heat. For many materials the latent heat is fixed or varies weakly with the thermodynamic state. For soil systems, however, the latent heat is directly proportional to the water content. We will ignore the fact that all the water in a soil system need not change phase. For simplicity, based on data for clay soils, Lock (1969) assumed that the water content decays exponentially from the value at the soil surface. For a freezing system, initially at the freezing temperature, the zeroth-order, quasi-steady equation for the phase change interface is

$$\frac{d\xi}{d\tau^*} = \frac{1}{\xi f(\xi)} \tag{64}$$

where $\xi^2 = 4X^2 \rho \ell_o / [k \ P(T_f - \overline{T}_s)$, $\tau^* = \frac{2\pi t}{P}$, ℓ_o is the latent heat at at surface. If the water content is constant, then the Stefan solution results

$$\tau^* = \frac{\xi^2}{2} \tag{65}$$

With an exponential water content function

$$\tau^* = 1 - (1+\xi)e^{-\xi} \tag{66}$$

Finally, if the average value of the latent heat between the surface and the freezing depth is used

$$\tau^* = \frac{\xi}{2}(1-e^{-\xi}) \tag{67}$$

From these results it can be noted that the time to freeze a layer $\xi = 1$, for an exponential decay, will be about 50% that of the constant latent heat solution and 85% that for an average latent heat assumption. Some caution should therefore be used in applying the constant or average water content solutions.

Variable initial temperature. An extension of the similarity method was used by Tao (1978) to obtain an exact solution with arbitrary surface and

initial conditions. Unfortunately this exact solution is such that numerical computations are extremely difficult due to transient functions which
require an increasing number of series terms as time increases. The solution is perhaps best used to verify the accuracy of approximate and numerical solutions or for short time solutions. A modification of the heat
balance integral technique utilizing a single integration over an entire
nonconstant property volume has yielded accurate solutions, Yuen (1980),
Lunardini (1981, 1982, 1983a). This has been used for a modified Neumann
problem with a linear initial temperature distribution. Such an initial
temperature is common for soil systems with a geothermal temperature gradient, Figure 9. At zero time the surface temperature drops to T_S and
freezing commences. The integration of the energy equations over the
region $0 \leq x \leq X + \delta$, detailed by Lunardini (1981), is

$$\frac{d}{dt} \left[\rho_1 c_1 \int_o^X T_1(x,t)dx + \rho_2 c_2 \int_X^{X+\delta} T_2(x,t)dx - \rho_1 \ell X + (\rho_2 c_2 - \rho_1 c_1)T_f X \right.$$

$$\left. - \rho_2 c_2 (X + \delta) \left\{ T_o + \frac{G}{2} (X + \delta) \right\} \right] = -k_1 \frac{\partial T_1(0,t)}{\partial x} + k_2 G \qquad (68)$$

Quadratic temperature profiles which satisfy the boundary conditions are
used and the energy integral equation, can be written nondimensionally as

$$\tau = \alpha_1 \left(\frac{G}{T_o - T_f}\right)^2 = \int_o^\sigma \frac{\frac{1}{3} + \frac{1}{S_T} + c_{21}\phi \left[1 + \sigma + \frac{b + \sigma b'}{3}\right] + \frac{1}{6g_1}\left(1 - \frac{\sigma g_1'}{g_1}\right)}{\frac{1}{\sigma}\left(2 - \frac{1}{g_1}\right) - k_{21}\phi} \, d\sigma \qquad (69)$$

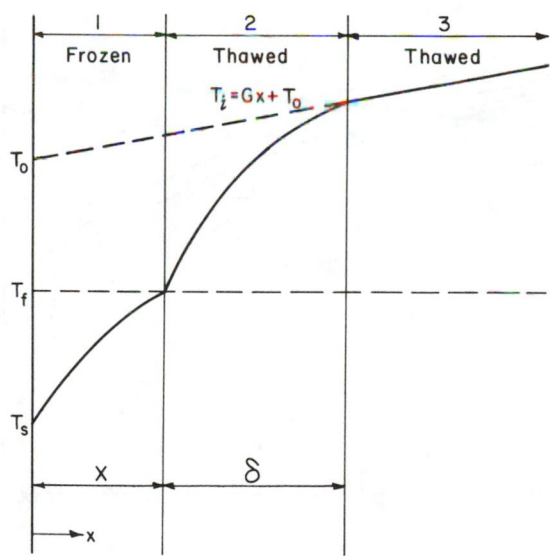

FIGURE 9. Freezing of a semi-infinite region with linear initial
temperature.

where $\sigma = GX/(T_o - T_f)$, $b = \delta/X$. The derivatives of b and g_1 can be found by solving simultaneously the following equations:

$$b' = \frac{db}{d\sigma} = \frac{\dfrac{k_{21}\phi\alpha_{21}}{2b} - \dfrac{b}{2}\left[\dfrac{4(g_1 - 1)}{S_T} - \dfrac{1}{g_1^2}\right]g_1'}{\dfrac{2(g_1 - 1)^2}{S_T} - 1 + \dfrac{1}{g_1}} \tag{70}$$

$$g_1' = \frac{\alpha_{21}}{b[\sigma(b + 2) + 2]}\left[1 - \frac{1 + \sigma}{b[\sigma(b + 2) + 2]}\left\{2b'[\sigma(b + 1) + 1]\right.\right.$$

$$\left.\left. + b(b + 2)\right\}\right] \tag{71}$$

Unlike the Neumann solution, the frozen zone for the general case reaches a steady-state value given by

$$T_{1_\infty} = \frac{(T_o - T_f)x}{\phi X_\infty} + T_s \tag{72}$$

$$\sigma_\infty = \frac{1}{k_{21}\phi} \tag{73}$$

Equation 69 was solved numerically (Lunardini, 1984) with the results presented in Figure 10. It is possible to present the results for soil

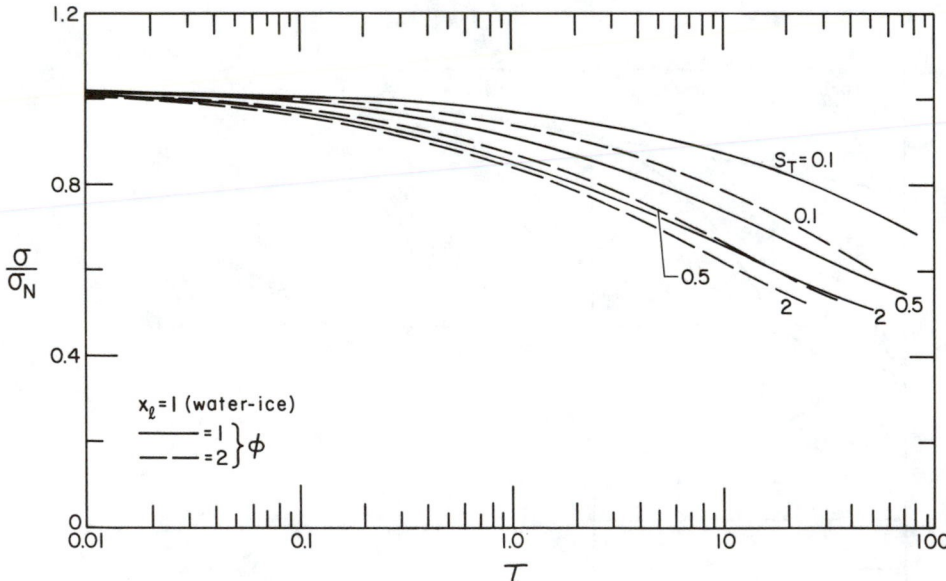

FIGURE 10. Ratio of freeze depth to that of Neumann solution, ice-water system, $k_{ft} = 3.98$, $\alpha_{ft} = 7.87$.

systems, quite efficiently, since the property ratios can be described as functions of the soil water content, x_ℓ, Lunardini and Varotta (1981). The effect of the geothermal gradient on the accuracy of the constant temperature solution is of interest for design in cold climate regions. For soil systems the Neumann solution will be quite acceptable even if the time is measured in years since τ is still quite small, but if G is large, then it will be necessary to use eq. 69 with appropriate property ratios. The effect of the initial temperature gradient can be compensated for by using a temperature parameter ϕ_N with the Neumann solution

$$\phi_N = \phi + \frac{1}{2k_{21}} \tag{74}$$

<u>Sinusoidal surface temperature.</u> It is of some interest to examine the relation between the Neumann problem and one with sinusoidal surface temperature. This variable surface temperature problem cannot be solved exactly, however an acceptable quasi-steady solution can be found. Consider a system which is initially at the fusion value and then undergoes a sinusoidal surface temperature variation. The first order quasi-steady solution will be obtained. The solution will start with freezing an initially thawed medium and then thawing the frozen system. If freeze starts at zero time the solutions are

$$\frac{T - T_f}{\theta_c} = \left(\frac{T_s(\tau^*) - T_f}{\theta_c}\right)\left(1 - \frac{y}{2\sin\frac{\tau^*}{2}}\right) +$$

$$\frac{S_{Ta}}{2} y (y - 2 \sin \frac{\tau^*}{2}) [2\sin^2 \frac{\tau^*}{2} - 1 - \frac{1}{6} \sin \frac{\tau^*}{2}(y + 2 \sin \frac{\tau^*}{2})] \tag{75}$$

$$\xi_1 = 2 \sin \frac{\tau^*}{2} - \frac{S_{Ta}}{3} [\frac{2}{3} + 2 \cos \frac{\tau^*}{2} - \frac{8}{3} \cos^3 \frac{\tau^*}{2}] \tag{76}$$

where $y^2 = 4x^2 \rho\ell/[k P (T_f - \overline{T}_s)]$, $S_{Ta} = \frac{c \theta_c}{\ell}$, $\tau^* = 2\pi t/P$.

These equations can be used for $0 < \tau^* < \pi$ with the properties of the frozen and thawed material. Lock et al. (1969) gives a similar solution for the phase change

$$\xi_1 = 2 \sin \frac{\tau^*}{2} - \frac{S_{Ta}}{3} \sin \frac{\tau^*}{2} \sin\tau^* \tag{77}$$

Seban (1971) noted that the even simpler Stefan equation gave acceptable results and also that the effect of convection due to density variation, in water, is small but may be important for some systems, see also Yen (1968). The same procedure can also be used for other surface temperature variations. If the surface temperature is not symmetric about the fusion temperature the results are more complicated, but follow in the same way.

The Neumann solution with a constant surface temperature equivalent to the sinusoidal temperature is

$$\xi_1 = 2 \sqrt{\frac{\tau^*}{\pi + S_{Ta}}} \tag{78}$$

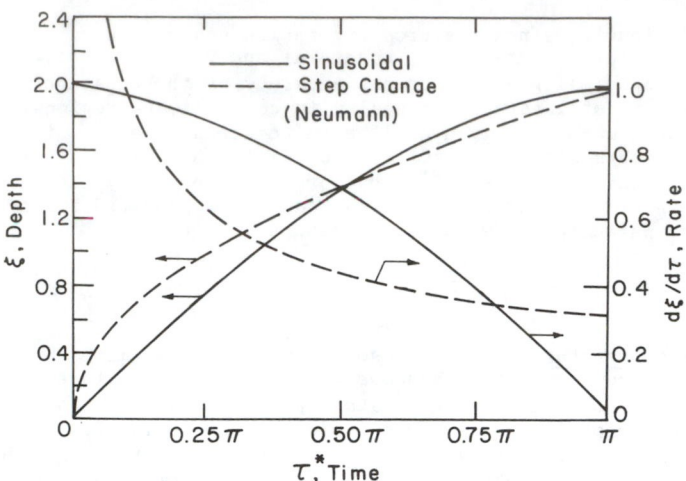

FIGURE 11. Freezing of semi-infinite medium with sinusoidal and step-change surface temperatures.

Figure 11 shows the freeze depths and the freezing rates for $S_{Ta} = 0.1$. The total freeze depths, for the two surface temperature cases, are within 1% but the freeze depths and especially the freezing rates differ considerably at intermediate times. Thus the Neumann solution will model a variable surface temperature if the total phase change depth is desired.

Melting Temperature Range

For soils, rocks, metal alloys, etc., melting will occur over a temperature range. Cho and Sunderland (1969) give an exact solution for the case of a binary eutectic mixture for a semi-infinite region and a finite slab. Tien and Geiger (1967) give an approximate heat balance integral solution when the mixture is initially at the liquidus temperature. A completely solidified region, an initially liquid region, and a region with both solid and liquid phases formed by isothermal planes at the liquidus and solidus temperatures exist. Equilibrium freezing occurs if the element is completely frozen just as the solid front reaches it. Normal non-equilibrium freezing is such that the element still has a liquid fraction which then freezes isothermally at the solidus temperature before the solid front moves on. The solid fraction distribution, in the freezing zone, was assumed linear with distance but Tien and Geiger (1967) have shown that this assumption is not important for the phase change process.

Soils. The variation of unfrozen water with temperature causes the soil system to freeze or thaw over a finite temperature range (Lunardini, 1981a). At any temperature below the normal freezing point, there will be an equilibrium state of unfrozen water, ice, and soil solids. There may be a residual amount of bound water which will remain unfrozen even at very low temperature, denoted by ξ_f. It will be assumed that for $T < T_m$, unfrozen water may exist but no phase change will occur. The region $T_m \leq T \leq T_f$ is called the zone of phase change. The form of the ξ function for soils can be expressed in various functional relations. The simplest relation is a linear one

$$\xi = \xi_o + \frac{\Delta\xi}{\Delta T_m} \; (T - T_f) \tag{79}$$

where $\Delta T_m = T_f - T_m$, $\Delta\xi = \xi_o - \xi_f$. A function which more closely approximates the soil water data is an exponential form

$$\xi = a_1 \; e^{\; b_1 T} + d_1 \tag{80}$$

Another functional relation, which can closely model the data, and is easy to manipulate analytically is a quadratic form

$$\xi = \xi_o + \frac{2 \; \Delta\xi}{\Delta T_m} \; (T - T_f) + \frac{\Delta\xi}{\Delta T_m^{\;2}} \; (T - T_f)^2 \tag{81}$$

The problem is one of conduction with a distributed energy source within the mushy zone. The energy equation then becomes (Lunardini, 1987)

$$\frac{\partial}{\partial x} \left(k \; \frac{\partial T}{\partial x} \right) - \ell \; \gamma_d \; \frac{\partial\xi}{\partial t} = \frac{\partial CT}{\partial t} \tag{82}$$

where γ_d is the dry unit weight. The thermal conductivity and the specific heat are functions of the unfrozen water and may be represented by

$$k = k_f + (k_t - k_f) \; \frac{\xi}{\xi_o} \tag{83}$$

$$C = C_f + (C_t - C_f) \; \frac{\xi}{\xi_o} \tag{84}$$

If ξ varies linearly with temperature then an exact solution may be found. The most general case will be a problem with 3 regions as shown in Figure 12. The solution to these equations follows from similarity.

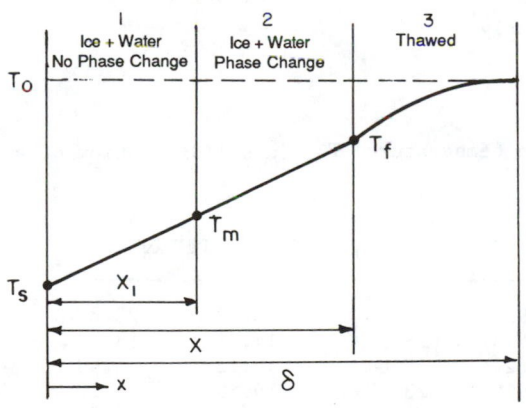

FIGURE 12.   Geometry for solidification with a phase change zone.

$$X = 2 \gamma \sqrt{\alpha_4 \, t} \tag{85}$$

$$X_1 = 2\psi \sqrt{\alpha_1 t} \tag{85a}$$

$$\frac{T_1 - T_s}{T_m - T_s} = \frac{\text{erf } (x/2\sqrt{\alpha_1 \, t})}{\text{erf } \psi} \tag{86}$$

$$\frac{T_2 - T_f}{\Delta T_m} = \frac{\text{erf } (x/2\sqrt{\alpha_4 t}) - \text{erf } \gamma}{\text{erf } \gamma - \text{erf } (\sqrt{\alpha_{14}} \, \psi)} \tag{87}$$

$$\frac{T_3 - T_o}{T_o - T_f} = \frac{-\text{erfc } (x/2\sqrt{\alpha_3 \, t})}{\text{erfc } (\sqrt{\alpha_{43}} \, \gamma)} \tag{88}$$

$$\frac{(T_m - T_s) \, e^{-\gamma^2 \, (1 - \alpha_{14})}}{\Delta T_m} = \frac{k_{21} \sqrt{\alpha_{14}} \, \text{erf } \gamma}{\text{erf } \gamma - \text{erf } (\sqrt{\alpha_{14}} \, \psi)} \tag{89}$$

$$\frac{\Delta T_m \, k_{23}}{T_o - T_f} \sqrt{\alpha_{34}} \, e^{-\gamma^2 \, (1 - \alpha_{43})} = \frac{\text{erf } \gamma - \text{erf } (\sqrt{\alpha_{14}} \, \psi)}{\text{erfc } (\sqrt{\alpha_{43}} \, \psi)} \tag{90}$$

Lunardini (1985) compared the solution to the Neumann solution for specific cases. Table 3 shows the effect of changing the width of the phase change zone by varying T_m. The Neumann solution has a temperature which always exceeds the gradual case and is significantly different within the zone of phase change. An approximation to the solution may be obtained with the heat balance integral method. Using

$$\delta - X = B \, X \tag{91}$$

the solution for γ follows directly as

TABLE 3. Effect of phase change temperature, T_m, on solidification of a soil (from Lunardini 1985).

Case	T_m	ψ	γ	X^* (cm)	X_1^* (cm)	$\Delta X = X - X_1$
1	-4	0.0617	1.395	33.33	8.13	25.2
2	-2	0.1135	1.6614	28.34	14.95	13.39
3	-1	0.1376	2.062	25.0	18.12	6.88
4	$-.5$	0.14922	2.6965	23.27	19.65	3.52
5	$-.1$	0.1571	5.058	21.41	20.69	0.72
Neumann	0	0.1606	----	21.15	21.15	0

*For t = 24 hours.

$$3 \alpha_{43} (B - 2 k_{32} \phi) (1 + \frac{B}{3}) - 1 - \frac{\phi \, k_{32}}{B} = 0 \tag{92}$$

$$\gamma^2 = \frac{1}{B(1 + \frac{B}{3}) \, \alpha_{43}} \tag{93}$$

The heat balance integral solution is within 5% of the exact solution, accuracy typical of the heat balance integral method.

If the unfrozen water has the exponential form discussed earlier the heat balance integral solution is

$$-\alpha_{23} \left(1 - \frac{2 \, k_{32} \, \phi}{B}\right) B \, (\frac{B}{3} + 1) + \frac{k_{32} \, \phi}{3B} + \frac{1}{3} =$$

$$\frac{1}{C_{23} \, S_{Tm} \Delta\xi} \left[\frac{\xi_s - \xi_o}{\ln(\xi_s - d_1)/(\xi_o - d_1)} + d_1 - \xi_o \right] \tag{94}$$

with γ given by Eq. 93. The exponential solution is 21% less than the linear case. Thus the form of the unfrozen water content function will be significant.

A quadratic relation can represent the unfrozen water content with acceptable accuracy and the heat balance integral solutions are

$$2 \, C_{32} \, \phi \, (1 - \frac{1}{P_1}) \, (3 + \frac{2 \, k_{32}}{P_1} \phi) + \frac{P_1}{2} + 1$$

$$+ \frac{C_{32} \, \phi}{S_{Tm}} \left[2 + P_1 - \frac{\phi}{5} (\frac{P_1}{2} + \frac{3}{2} \, P_1 + 3) \right] = 0 \tag{95}$$

where $P_1 = 2 k_{32} \, \phi/B$.

$$\gamma^2 = \frac{P_1 \, \alpha_{34}}{2 \, k_{32} \, \phi \, (\frac{2 \, k_{32} \, \phi}{3 \, P_1} + 1)} \tag{96}$$

The solutions may be compared as shown in Table 4. Since the Neumann solution assumes a step function for the water content, with all of the unbound water frozen, the phase change interface lags the other solutions by a significant amount. Nakano and Brown [1971] solved a version of this problem numerically. Their results showed a significant effect of a freezing zone on the temperature profile but the magnitudes of the differences did not follow those shown here.

Subcooled Liquid – Frazil Ice

A problem of some practical interest relates to the freezing of a liquid, initially below its fusion temperature, as shown in Figure 13. The liquid is in a metastable state and phase change results in the release of latent heat which warms the liquid to its normal fusion point. The solid phase remains at the freezing temperature. This can be related physically to the formation of frazil ice from water which supercools due to turbulence. The solution follows directly as

TABLE 4. Effect of ξ function, $T_0 = 4°C$, $T_s = T_m = -4°C$, $T_f = 0°C$, $k_2 = 0.00703$, $C = 0.165$, $\xi_0 = 0.2$, $\Delta\xi = 0.1218$, $k_{32} = 0.82219$, $C_{32} = 1.0$ (Lunardini 1985)

Solution	γ	Difference from linear ξ (%)	X (t = 24 hrs) (cm)
Exact, linear ξ	1.2365	− 4.8	29.54
H.B.I. − linear ξ	1.2994	---	31.04
H.B.I. − exponential ξ	1.0246	−21.1	24.48
H.B.I − quadratic ξ	1.1561	−11.0	27.62
Neumann (ξ, step function)	0.7846	−39.6	18.74

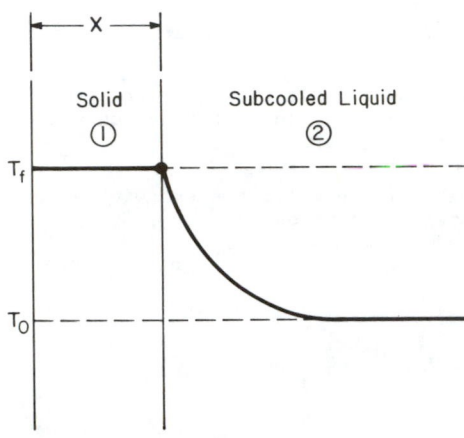

FIGURE 13. Geometry for subcooled liquid.

$$T_2 = T_0 + \frac{(T_f - T_0)\ \text{erfc}\ \left(\frac{x}{2\sqrt{\alpha_2 t}}\right)}{\text{erfc}\ \gamma} \tag{97}$$

$$X = 2\gamma\ \sqrt{\alpha_2 t} \tag{98}$$

$$\gamma\,\text{erfc}\ (\gamma)\ e^{\gamma^2} = \frac{c_2\ (T_f - T_0)}{\ell\ \sqrt{\pi}} \tag{99}$$

Solidification in Contact with Cold Wall

The freezing of liquid in contact with a cold wall is of importance in casting metals and in the intrusion of magmas during geological processes. At time $t = 0$, a semi-infinite liquid at T_0 is brought into contact with a semi-infinite cold wall at T_c. A solid phase forms instantly and grows with time, while the temperature between the cold wall, and solid phase remains constant, see Figure 14. The density of the solid and liquid

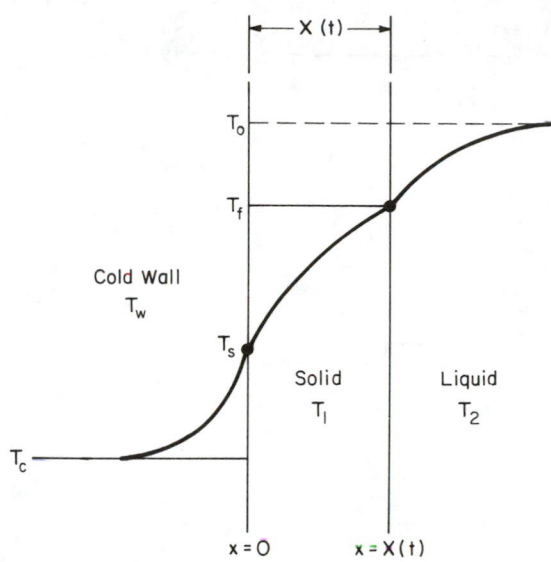

FIGURE 14. Freezing against a cold wall.

phases are assumed equal. The solution follows from the usual similarity transformation.

$$\frac{T_2 - T_f}{T_o - T_f} = \frac{1 - \text{erfc}\left(\frac{x}{2\sqrt{\alpha_2 t}}\right)}{\text{erfc}\left(\gamma \sqrt{\alpha_{12}}\right)} \tag{100}$$

$$\frac{T_w - T_c}{T_s - T_c} = \text{erfc}\left(\frac{-x}{2\sqrt{\alpha_w t}}\right) \tag{101}$$

$$\frac{T_1 - T_s}{T_f - T_s} = \frac{\text{erf}\left(\frac{x}{2\sqrt{\alpha_1 t}}\right)}{\text{erf}\,\gamma} \tag{102}$$

$$X = 2\gamma\sqrt{\alpha_1 t} \tag{103}$$

$$\frac{e^{-\gamma^2}}{\text{erf}\,\gamma} - \frac{k_{21}\sqrt{\alpha_{12}}\,\phi e^{-\gamma^2 \alpha_{12}}}{\text{erfc}\left(\gamma\sqrt{\alpha_{12}}\right)} = \frac{\gamma\sqrt{\pi}}{S_T} \tag{104}$$

The interface temperature between the cold wall and the frozen liquid is

$$T_s = \frac{T_f + T_c\, k_{wl}\sqrt{\alpha_{lw}}\,\text{erf}\,\gamma}{1 + k_{wl}\sqrt{\alpha_{lw}}\,\text{erf}\,\gamma} \tag{105}$$

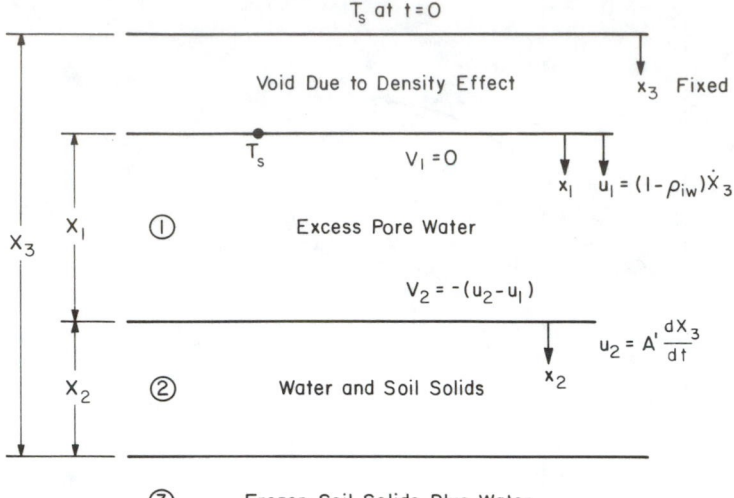

FIGURE 15. Geometry for thaw settlement.

Since the interface temperature is constant the solution will be identical to the Neumann solution with the same surface temperature.

Thaw with Consolidation of Melted Medium

An interesting problem arises due to the fact that some porous media, such as soils, can become more compact after thawing. Some of the water in the frozen soil is forced out as the solid particles settle and the thawed soil becomes denser. Consider a frozen soil system, saturated with water, as shown in Figure 15. The initial temperature is T_O and the surface temperature suddenly drops to T_s. As thaw progresses, pore water will be forced out of the thawed material due to the thaw strain which proceeds such that the interface between the pore water and the thawed region moves at the velocity

$$u_2 = A' \frac{dX_3}{dt} \tag{106}$$

A velocity is induced in the pore water due to the density difference of ice and water. The water surface moves at the velocity

$$u_1 = (1 - \rho_{iw}) \frac{dX_3}{dt} \tag{107}$$

where ρ_{iw} — ratio of density of ice to density of water. The solution to the above set of equations is

$$T_1 = T_s + P_4 (T_f - T_s) \operatorname{erf} \frac{x_1}{2\sqrt{\alpha_1 t}} \tag{108}$$

$$T_2 = T_s + P_3 \ (1 + 1/P_1)(T_f - T_s) \ \text{erf} \ \left(\frac{x_2}{2\sqrt{\alpha_2 t}} + K\gamma\right) \tag{109}$$

$$T_3 = T_o + \frac{(T_f - T_o) \ \text{erfc} \ \left(\frac{x_3}{2\sqrt{\alpha_3 t}}\right)}{\text{erfc} \ (\gamma\sqrt{\alpha_{23}})} \tag{110}$$

$$X_3 = 2\gamma \ \sqrt{\alpha_2 t} \tag{111}$$

where

$$P_1 = k_{21}\sqrt{\alpha_{12}} \ e^{\gamma^2(A^2\alpha_{21} - K^2)} \ \text{erf} \ (A \ \gamma\sqrt{\alpha_{21}}) - \text{erf} \ K\gamma$$

$$P_2 = \text{erf} \ [\gamma(K + 1 - A')]$$

$$P_3 = \frac{P_1}{P_1 + P_2}$$

$$P_4 = \frac{k_{21}\sqrt{\alpha_{12}} \ e^{\gamma^2(A^2\alpha_{21} - K^2)}}{P_1 + P_2}$$

The value of γ is obtained from

$$\frac{e^{-\gamma^2(K + 1 - A')^2}}{P_1 + P_2} - \frac{k_{32}\sqrt{\alpha_{23}} \ \phi \ e^{-\gamma^2\alpha_{23}}}{\text{erfc} \ (\gamma\sqrt{\alpha_{23}})} = \frac{\sqrt{\pi} \ \gamma \ \rho_{32}}{S_T} \tag{112}$$

where $K = \frac{\rho_w c_w}{\rho_2 c_2} A$. Nixon (1975) examined the case of a soil initially at the fusion temperature and the excess pore water vanished instantaneously as it was formed. Physically, the vanishing of the water layer is equivalent to an infinite thermal diffusivity for region 1. The equations then reduce to

$$T_2 = T_s + \frac{(T_f - T_s)\left[\text{erf} \ \left(\frac{x_2}{2\sqrt{\alpha_2 t}} + K\gamma\right) - \text{erf} \ K\gamma\right]}{\text{erf} \ [\gamma(K + 1 - A')] - \text{erf} \ K\gamma} \tag{113}$$

$$\frac{e^{-\gamma^2(K + 1 - A')^2}}{\text{erf} \ [\gamma(K + 1 - A')] - \text{erf} \ K\gamma} - \frac{k_{32}\sqrt{23} \ \phi \ e^{-\gamma^2\alpha_{23}}}{\text{erfc} \ \gamma\sqrt{\alpha_{23}}} = \frac{\sqrt{\pi} \ \gamma \ \rho_{32}}{S_T} \tag{114}$$

The equation for the frozen zone does not change. If $A' = 0$ and $\rho_{iw} = 1.0$, the above equations reduce to the familiar Neumann solution. Table 5 lists some calculated values for the thaw parameter. The thaw depth is

TABLE 5. Effect of convection and conductive resistance on thaw of strained soil, $S_T = 1.0$, $\phi = 0.0$.

A'	ρ_{iw}	K	Excess water layer	γ	γ (Nixon 1975)	γ (Neumann Solution)
0.35	1.0	0.5	Absent	0.7384	0.5771	0.6203
0.35	1.0	0.5	Present	0.6078	----	0.6203
0.35	0.92	0.386	Present	0.6413	----	0.6203
0.50	1.0	0.714	Absent	0.8276	0.5618	0.6203
0.75	1.0	1.071	Absent	1.4119	0.5386	0.6203
0.75	0.92	0.957	Present	0.6388	----	0.6203
0.90	1.0	1.286	Absent	1.7821	0.5248	0.6203
0.90	0.92	1.171	Present	0.6450	----	0.6203
1.0	1.0	1.429	Absent	----	0.5152	0.6203

Property ratios unity except as noted

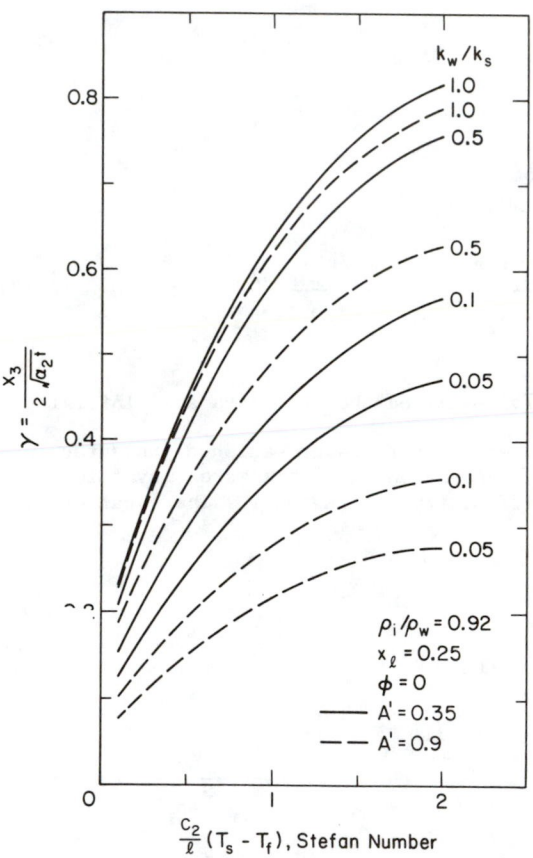

FIGURE 16. Effect of Stefan number on thaw rate with settlement.

controlled by the thermal resistance of the thawed layer and the excess water rather than the convection. As the soil consolidates the conductive resistance to heat flow decreases and this effect is augmented if the excess water layer is neglected. With no water layer, the thaw rate exceeds the Neumann case by 18%. For the same case with a water layer the thaw lags the Neumann case by 2%. This shows the actual minor effect of the convection. Figure 16 shows solutions for typical soil systems where x_ℓ is the volumetric water fraction of the media.

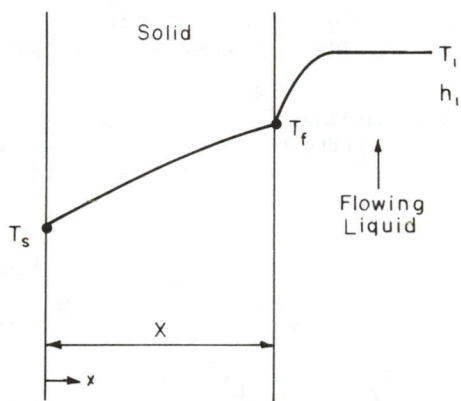

FIGURE 17. Freezing with convective heat flow from liquid region.

Freeze of a Flowing Fluid

The freezing of a river with convection from the water to the ice or the solidification of a fluid flowing in contact with a cold wall at constant temperature can be estimated as follows. As shown in Figure 17 the fluid contacts a cold surface at constant temperature T_s. Energy flows from a warm fluid at constant temperature T_1 and constant convective coefficient h_1. A perturbation method can be used with the time variable replaced by the interface position. Expand the temperature and the phase change inter-face speed in asymptotic series with the Stefan number as a parameter. Seeniraj and Bose (1982) developed this method and noted that the interface position S compared very well to analytical results of Savino and Siegel (1969) and numerical results of Beaubouef and Chapman (1967). Lapadula and Mueller (1966) used Biot's method to obtain the following result.

$$h_1 (T_1-T_f)t/(\rho\ell\ X_s) = - \frac{2\ S_T^2 + 10\ S_T + 15}{5(3 + S_T)} [S + \ln\ (1-S)] \tag{115}$$

This agrees well with the perturbation solution if $S_T < 1.0$.

4. CONVECTION AT FREE SURFACE

The thermal boundary condition at the free surface of a body can be specified in terms of temperature or energy flow. The most widely encountered boundary condition for a solid is that for which energy flows between the solid and an ambient fluid. The heat flux at the surface can be specified as

$$-k \ \frac{\partial T(0,t)}{\partial x} = h \ [T_a(t) - T(0,t)] \tag{116}$$

where h is the surface conductance.

Single Phase Problems

Problems with only a single phase (solid or liquid) which experiences temperature changes are much simpler mathematically than two-phase situations.

Exact Solution for Semi-infinite Medium. Consider a semi-infinite solid, initially at the fusion temperature, which starts to melt due to convection or a heat flux imposed on the free surface. Referring to Figure 18, the problem can be formulated as

$$\frac{\partial^2 \theta}{\partial y^2} = \frac{\partial \theta}{\partial \tau} \tag{117}$$

$$\theta(\sigma,\tau) = 0 \tag{118}$$

$$\frac{\partial \theta}{\partial y} (\sigma,\tau) = - \ \frac{1}{S_{Te}} \ \frac{d\sigma}{d\tau} \tag{119}$$

$$\frac{\partial \theta}{\partial y} (0,\tau) = -1 + \varepsilon \ \theta(0,\tau) \tag{120}$$

where $y = xh/k$, $\tau = \alpha_1 t \ h^2/k^2$. For a constant surface heat flux ($\varepsilon = 0$) and $S_{Te} = .2$, Lozano and Reemsten (1981) obtained the following expression for the first four terms of the interface position

$$\sigma = 0.2\tau - 0.004\tau^2 + 0.0002667\tau^3 - 0.272 \ x \ 10^{-4}\tau^4 \tag{121}$$

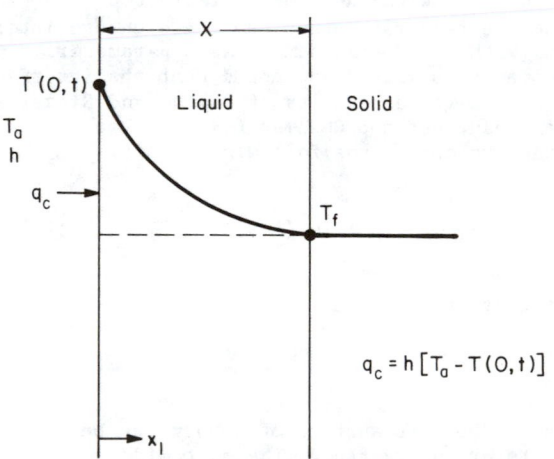

FIGURE 18. Melt of a semi-infinite medium with surface convection.

In general

$$\sigma = S_{Te} \, \tau \left[1 - \frac{1}{2} S_{Te}^{\,2} \, \tau + \frac{5}{6} S_{Te}^{\,4} \, \tau^2 - \frac{17}{8} S_{Te}^{\,6} \, \tau^3 + \frac{827}{120} S_{Te}^{\,8} \, \tau^4 + \ldots \right]$$

(122)

The solution presented here is exact but it is perhaps best used to check approximate methods, due to the time-consuming computations necessary if many terms are needed in eq. 122. We shall use the results of this section to compare approximate solutions. Westphal (1967) considered freezing of a semi-infinite medium with convection, letting the ambient temperature be a function of time. He obtained an exact solution using infinite series for the case when the ambient temperature is constant and water is the medium. Table 6 shows the interface vs. time. The results are in close agreement (less than .7%) with the solution given by eq. 122 for small values of time. This is expected since the Stefan number is small and the surface heat flux is relatively constant at early times. The solution method is closely allied to the concept used by Tao (1978) to solve the extended Neumann problem.

Analogue solution. There are no exact solutions for this problem when the initial temperature is different from the fusion temperature and the ambient temperature can vary. Kreith and Romie (1955) used an electrical analogue to obtain a solution, for the freeze case, when the initial temperature was at the fusion value. The depth of freeze for this problem agrees within 10% with the Neumann solution ($T_o = T_f$, $T_s = T_a$) if X h/k is greater than one. The heat flow to the surface encounters two thermal resistances: a conductive resistance that increases with time as the frozen layer grows and a convective resistance that is constant. After a certain time, the relative effect of the surface resistance is approaching zero and the solidification proceeds essentially as in the constant surface temperature case.

TABLE 6. Phase change depth vs. time, convection with constant ambient temperature, $S_{Te} = 2/35$.

	Time ($2h/k\sqrt{\alpha t}$)		
$\sigma = \frac{h}{k} X$	Exact (Westphal 1967)	Quasi-steady (eq 125)	Heat balance integral (eq 127)
0.003565	0.5	0.500	0.501
0.01418	1.0	1.000	1.002
0.03161	1.5	1.499	1.503
0.05562	2.0	2.000	2.004
0.08543	2.5	2.497	2.505
0.1208	3.0	2.994	3.005
0.1611	3.5	3.491	3.504
0.2056	4.0	3.984	4.004
0.2534	4.5	4.471	4.503
0.3028	5.0	4.940	5.002

Quasi-steady approximation. The freeze problem reduces to a simple case for the quasi-steady approximation. The solution is

$$T = (T_f - T_a) (x + \frac{k_1}{h})/(X + k_1/h) + T_a(t) \tag{123}$$

If the surface coefficient h is a function of time, then the equation for the phase change interface is

$$\frac{dp_1^2}{dt} = \frac{2k_1}{p_1 \ell} (T_f - T_a) - 2p_1 \frac{k_1}{h^2} \frac{dh}{dt} \tag{124}$$

where $p_1 = X + k_1/h$. With $T_a(t)$ constant, Eq. (124) was first reported by London and Seban (1943). The equation was given in its present form by Foss and Fan (1972) and was used for surface temperature calculations during freezing and thawing by Lunardini (1978, 1978a). If T_a is constant, then

$$(X + \frac{k_1}{h})^2 = \frac{2k_1}{\rho_1 \ell} (T_f - T_a)t + (\frac{k_1}{h})^2 = 2\alpha_1 S_{Te} t + (\frac{k_1}{h})^2 \tag{125}$$

and the surface temperature is

$$T_s = T_a(t) + \frac{T_f - T_a(t)}{p_1} \frac{k_1}{h} \tag{126}$$

Equations 125 and 126 agree very well with the analog solutions of Kreith and Romie (1955), particularly for small Stefan numbers as is expected. Comparison with the exact solution of Table 6 confirms the accuracy of this approximation for small Stefan numbers.

Heat balance integral approximation. Goodman (1958) solved this problem using the integral method with quadratic temperature approximations.

$$S_{Te}\tau = h^2 \alpha_1 t/k_1^2 = \frac{1}{12\beta} \{[(1+2\beta) + (2+\beta)\sigma][1 + \beta\sigma(2+\sigma)]^{\frac{1}{2}}$$

$$- \frac{2(\beta-1)}{\sqrt{\beta}} \ln \frac{[1+\beta\sigma(2+\sigma)]^{\frac{1}{2}} + [(1+\sigma)\beta]^{\frac{1}{2}}}{1 + \sqrt{\beta}} - 4\beta(\beta-1)\ln \frac{-1+\beta(2+\sigma) + [1+\beta\sigma(2+\sigma)]^{\frac{1}{2}}}{2\beta}$$

$$+ (\beta^2+5\beta) \frac{\sigma^2}{2} + 2(\beta^2+4\beta-2)\sigma - (1+2\beta)\} \tag{127}$$

where $\beta = 1 + 2S_{Te}$. The surface temperature is given by

$$\frac{T_f - T(0,t)}{T_f - T_a} = \frac{(\beta-1)\sigma^2 + 2(\beta-2)\sigma-2 + 2[1 + \beta\sigma(2+\sigma)]^{\frac{1}{2}}}{(\beta-1)(2+\sigma)^2} \tag{128}$$

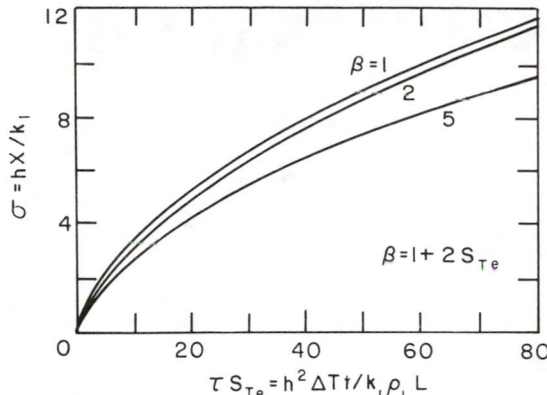

FIGURE 19. Thickness of melt versus time, aerodynamic heating, eq 127.

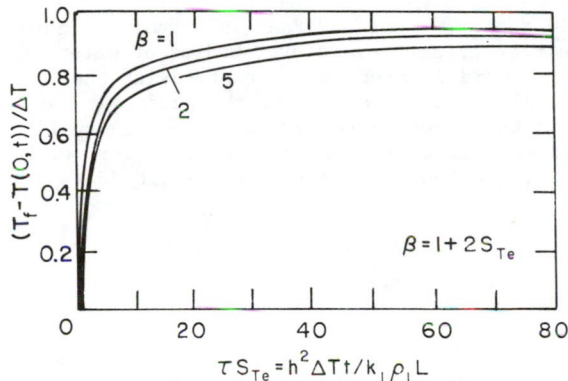

FIGURE 20. Surface temperature versus time, aerodynamic heating, eq. 128.

These equations are plotted as Figures 19 and 20 and the integral approximation is quite close to the analog solution. The heat balance integral approximation is compared to the exact solution of Westphal (1967) in Table 6. If the Stefan number is zero ($\beta = 1$), then the integral solution for the phase-change interface and the surface temperature are

$$S_{Te}\tau = \frac{\sigma^2}{2} + \sigma \tag{129}$$

$$\frac{T_f - T(0,t)}{T_f - T_a} = \frac{\sigma}{1+\sigma} \tag{130}$$

These are identical to Eqs. (125) and (126) of the quasi-steady approximation. Cho and Sunderland (1981) presented an approximate method of solving this problem for the single phase case. They assumed that the temperature profile was of the same form as for the case of the non-melting problem. Their results agree very well with eq. 126.

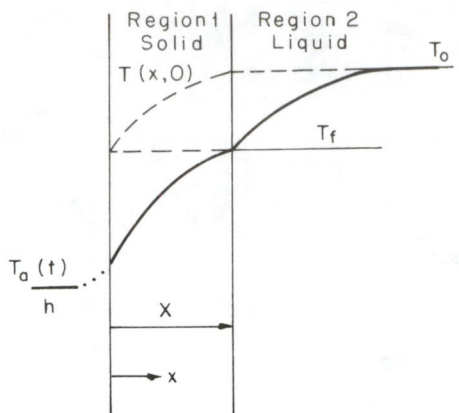

FIGURE 21. Freezing with surface heat flow.

Constant heat flux from liquid region. Foss and Fan (1974) solved the problem of the freezing or melting of ice layers over bodies of water if the heat flux from the melting liquid is assumed constant, using the quasi-steady method. The initial air temperature and the surface temperature of the water are at the freezing temperature, see Figure 21. At $t = 0$ the air temperature drops below freezing and may then vary with time. At the solid-liquid interface the heat flux from the liquid is assumed to be constant, therefore,

$$k_2 \frac{\partial T_2(X,t)}{\partial x} = q_w \tag{131}$$

$$T_1 = \frac{(T_f + \frac{h\,T_a\,X}{k_1})}{1 + \frac{h\,X}{k_1}} (1 + \frac{h\,x}{k_1}) - \frac{h}{k_1} T_a x \tag{132}$$

The solution depends upon the functional form of the time variation of the ambient temperature. For a constant ambient temperature the solution is

$$q_w^{*2} S_{Te} \tau = \ln \left[\frac{1 - q_w^*}{1 - q_w^{*(1+\sigma)}} \right] - \sigma\, q_w^* \tag{133}$$

The steady-state value of X is

$$X_{max} = k_1 \left[\frac{T_f - T_a}{q_w} - \frac{1}{h} \right] \tag{134}$$

$$\sigma_{max} = \frac{S_{Te}}{q_w^*} - 1 \tag{135}$$

If the ambient temperature varies with time, a numerical solution will normally be necessary. Foss and Fan (1974) presented a solution for a

104

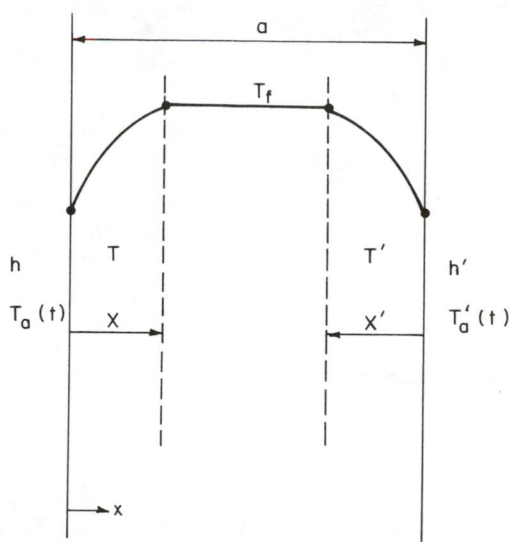

FIGURE 22. Freezing of a finite slab with convective cooling.

particular yearly sinusoid, with daily fluctuations included, and a
particular value of q_w^*, see Lunardini (1988).

Freeze of a finite slab. The freeze (or thaw) of a finite slab of material
is of interest for energy storage problems. Consider the slab – shown in
Figure 22 – which is initially at the freezing point and is suddenly
immersed in an ambient fluid with a variable temperature. The mathematical
description of the problem is the same for both sides of the slab but the
ambient conditions can differ on each side. Gutman (1986) considered a
perturbation solution of this problem. The time for the slab to solidify
completely, τ_f, can be found from

$$\tau_f = \tau_{fo} + S_T \, \tau_{f1} + p \, \tau_{f2} \tag{136}$$

where $S_T = \dfrac{c(T_f - T_m)}{\ell}$, T_m = mean ambient temperature, $\tau = \alpha S_T h^2 t/k^2$,
$p = D/(T_f - T_m)$.

$$\tau_{fo} = \frac{1}{2(1-\eta)} \left[(1+\eta)A^2 - (1-\nu^2)(1-\eta) - 2A^2\sqrt{\eta} + \frac{(1-\eta)(\nu^2-\eta)}{A^2} \right] \tag{137}$$

$$\tau_{f1} = -\psi[\sigma_1(\tau_{fo}) + \nu \, \eta \, \sigma_1(\tau_{fo} \, \eta/\nu^2)] \tag{138}$$

$$\tau_{f2} = \frac{\psi(1-\cos\omega\tau_{fo})}{\omega(1+\sigma_o)} \tag{139}$$

where

$$A = B_o + \nu + 1$$

105

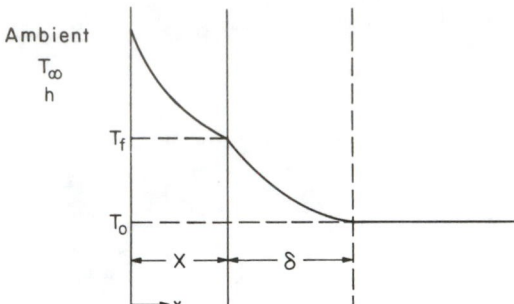

FIGURE 23. Surface convection for a semi-infinite body.

$$\psi = [(1+\sigma_o)^{-1/2} + \eta \ (\nu^2+2\eta \ \tau_{fo})^{-1/2}]^{-1}$$

$$\eta = \frac{T_f - T_m'}{T_f - T_m}$$

$$\sigma_o = \sqrt{1+2\tau} - 1$$

T_m' = mean ambient temperature.

For the case of a constant ambient temperature, the solution compares very well with the exact results in Table 6. For a slab insulated at x = D (also the case of a semi-infinite medium), the solidification time is

$$\tau_f = \frac{B_o \ (B_o+2)}{2} + \frac{B_o^2 \ (B_o+3)}{6(B_o+1)} \ S_T + (1-\cos\omega\tau_{fo})D_a\omega/(T_f - T_m) \tag{140}$$

Two Phase Problems

There is no exact solution for the case when the medium is initially at a different value than the fusion temperature, but Lunardini (1981, 1982) has obtained an approximate solution. The geometry of the problem is shown in Figure 23. Using the heat balance integral, the overall energy balance for the volume of interest is

$$\frac{d}{dt} \ [\rho_1 c_1 \int_o^X T_1(x,t)dx + \rho_2 c_2 \int_X^{X+\delta} T_2(x,t)dx + \rho_2 c_2 \theta_2 + \rho_1 \ell X$$

$$+ (\rho_2 c_2 - \rho_1 c_1)T_f X - \rho_2 c_2 T_0(X+\delta)] = -k_1 \ \frac{\partial T_1(0,t)}{dx} \tag{141}$$

The energy balance equation leads to

$$2\tau = \int_0^\sigma Q \ d\sigma' \tag{142}$$

$$PQ = (2\phi + \alpha_{21}\sigma)\sigma \; S_{Te} + (1 + C_{21}S_{Tm})g + 2[\sigma(1 + C_{21}S_{Tm}) + \tfrac{1}{3} C_{21}S_{Tm}\phi](P + \alpha_{21})$$

$$+ \; S_{Te} \; \sigma^2 + \tfrac{1}{3} C_{21}S_{Tm}g + 2[\sigma(1 + C_{21}S_{Tm}) + \tfrac{1}{3}(C_{21}S_{Tm}\phi](\sigma + 1)$$

$$- \; \frac{4\phi(P + \alpha_{21})}{2\phi - b(\sigma + 1)} \left\{ \frac{S_{Te}\sigma^2(\phi + \tfrac{1}{3}\alpha_{21}\sigma)}{2\phi(\sigma + 1) + \alpha_{21}\sigma(\sigma + 2)} + c_{21} \; S_{Tm}(\sigma + \tfrac{1}{3}\phi) + \sigma \right\} \tag{143}$$

where

$$P = \phi + \alpha_{21}\sigma, \quad g = 2[P(1 - \sigma) - \sigma\phi],$$

$$b = (2k_{21}S_{Tm} + \alpha_{21})/S_{Te},$$

$$S_{Tm} = c_1 \, (T_f - T_o)/\ell$$

$$\phi = hX/k_1 = \sqrt{\frac{b(\sigma + 1)}{2} + \frac{b^2(\sigma + 1)^2}{4} + \alpha_{21} \; \sigma(\tfrac{\sigma}{2} + 1)b}$$

It can be shown that when $S_{Tm} = S_{Te} = 0$, eq. 142 leads to the quasi-steady solution, eq. 125. The numerical solution to eq. (142), when $S_{Tm} = 0$, is identical to the single-phase heat balance integral solution given earlier. The surface temperature is

$$\frac{T_1(0,t) - T_f}{T_a - T_f} = \frac{\sigma(2\phi + \alpha_{21}\sigma)}{\sigma(2\phi + \alpha_{21}\sigma) + 2(\phi + \alpha_{21}\sigma)} \tag{144}$$

The nondimensional surface heat transfer rate is

$$q^* = \frac{(\phi + \alpha_{21}\sigma)}{\sigma(\phi + \tfrac{1}{2}\alpha_{21}\sigma) + (\phi + \alpha_{21}\sigma)} \tag{145}$$

Equation 142 was solved by numerical quadrature (Lunardini 1983a). Figure 24 is a plot of the solution for some values of Stefan number and S_{Tm}, with property ratios given as functions of the volumetric water content for soil systems.

Insulated semi-infinite region. Figure 24 can also be used for the case of a slab insulated with a layer of material when the insulation temperature is T_∞. The conductive resistance of the insulation must equal the convective resistance of the air layer. Then

$$\frac{d}{k_1} = \frac{1}{h} \tag{146}$$

The dimensionless phase change depth is given by

$$\sigma_c = \frac{k_1}{dk_1} X_c \tag{147}$$

The graph can then be used by assuming that the insulation layer has no latent heat and phase change starts at $t = t_o$ when the temperature of the insulation/slab interface reaches T_f. The single-phase solution, with $S_{Te} = 0$, can be rewritten as

$$X_c = \sqrt{k_{1i}^2 d^2 + \frac{2k_1(T_a - T_f)(t - t_o)}{\rho_1 \ell}} - k_{1i}d. \tag{148}$$

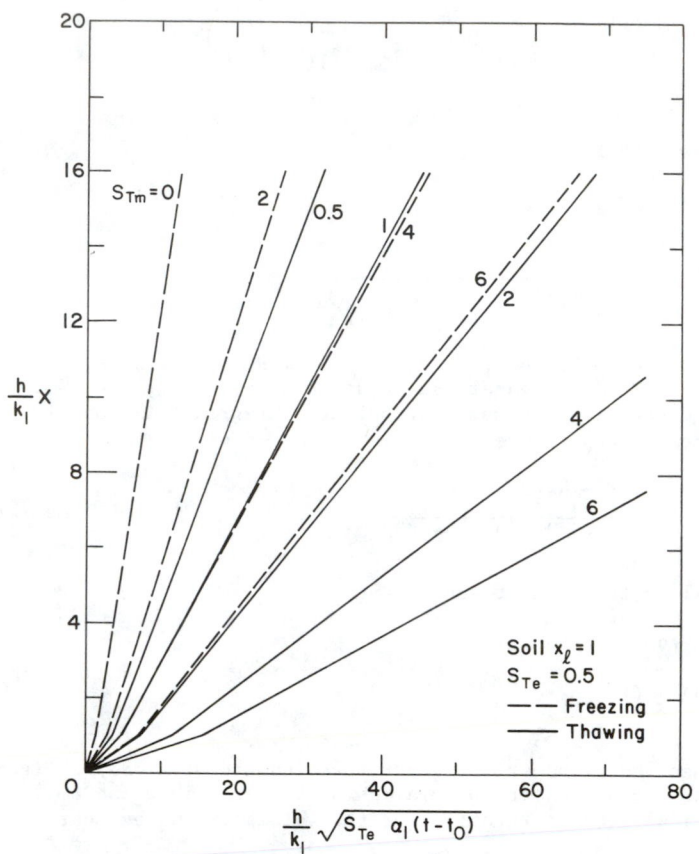

FIGURE 24. Surface convection for soil, volumetric water content = 1.0.

5. SPECIFIED SURFACE HEAT FLUX

Exact Solution for the Semi-infinite Region

Tao (1979) has found an exact solution to the freeze problem shown in
Figure 25. A liquid, initially at a temperature $V(x)$, has a heat flux $Q(t)$
imposed upon its free surface. Solidification starts when the surface
temperature of the liquid reaches the melting point such that $Q(t) > 0$ and
$V(0) = T_f$. Tao (1979) gives a numerical example for the special case
when $V(x) = T_f$ and the heat flux is a constant, Q_o. The first four
terms of the phase change interface can be written as

$$\xi = \tau - \frac{1}{2}\tau^2 + \frac{5}{6}\tau^3 - \frac{17}{8}\tau^4 + \dots \tag{149}$$

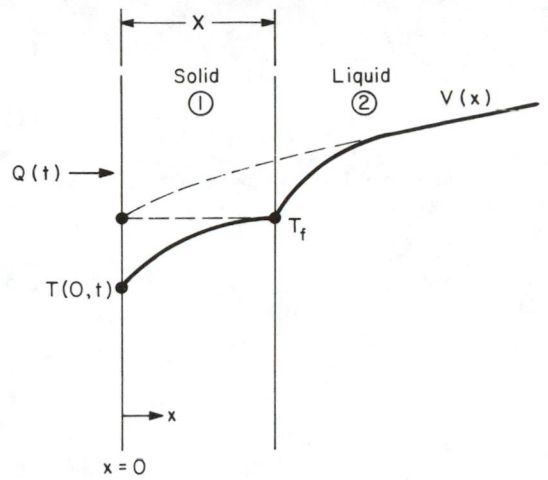

FIGURE 25. Melt of a semi-infinite medium with surface heat flux.

where $\xi = XQ/(\rho\alpha_1 \ell)$, $\tau = \dfrac{Q}{\rho^2 \alpha_1 \ell^2} \displaystyle\int_0^t Q(t)dt$. Evans et al. (1950) also
presented an exact solution for this problem by assuming Taylor Series
expansions for $X(t)$ about $t = 0$. The solution is a series

$$\xi = \tau - \frac{1}{2}\tau^2 + \frac{5}{6}\tau^3 - \frac{17}{8}\tau^4 + \frac{827}{120}\tau^5 \ldots \tag{150}$$

The equations are only valid for values of $\tau < 0.4$, unless many more terms
are included. This will not limit the use of Eq. (150) too significantly
if the latent heat is large relative to the heat flux. The exact solution
of Lozano and Reemsten (1981), Eq. (121), has precisely the same form as
Eq. (149). All three solutions are exact, although they all appear
distinctly different, and they reduce to the identical series for the phase
change interface.

Approximate Solutions, Single Phase, Semi-infinite Region

Kreith and Romie (1955) present an analogue solution for this case with the
initial temperature at the fusion temperature. The solutions, in graphical
form were given for the freeze depth and the surface temperature.

Heat balance integral method. The single phase problem is formulated as

$$\frac{\partial^2 T_1}{\partial x^2} = \frac{1}{\alpha_1}\frac{\partial T_1}{\partial t} \tag{151}$$

$$T_1(X,t) = T_f \tag{152}$$

$$T_1(x,0) = T_f \tag{153}$$

$$X(0) = 0 \tag{154}$$

$$k_1 \frac{\partial T_1}{\partial x} (X,t) = \rho_1 \ell \frac{dX}{dt} \tag{155}$$

$$\frac{\partial T_1 (0,t)}{\partial x} = Q(t)/k_1 \tag{156}$$

A heat-balance integral solution for the problem has been obtained by Goodman (1958). The heat balance integral is

$$\frac{d}{dt} \left[\int_o^X T(x,t)dx - (T_f + \frac{\alpha_1 \rho_1 \ell}{k_1})X \right] = -\alpha_1 Q/k_1 \tag{157}$$

The solutions are

$$\tau = \frac{\xi}{6} (5 + \xi + \sqrt{1 + 4\xi}) \tag{158}$$

$$\frac{4c_1}{\ell} (T_f - T(0,t) = 2\xi - 1 + \sqrt{1 + 4\xi} \tag{159}$$

When the heat flux is constant, the results agree very well with the analogue solution of Kreith and Romie (1955), and also with the exact solution of Evans et al. (1950) for small values of time. The quasi-steady solution to this problem is extremely simple but valid only for short times.

$$\tau = \xi \tag{160}$$

Gutman (1986) used a perturbation method to obtain the following solution for a constant heat flux.

$$\xi = \tau - \frac{1}{2} \tau^2 + \frac{5}{6} \tau^3 \tag{161}$$

$$\frac{c}{\ell} [T_f - T(\zeta,\tau)] = \tau - \zeta + \frac{1}{2} \zeta^2 - \tau^2 - \tau (\zeta^2 - \frac{7}{3} \tau^2) \tag{162}$$

where $\zeta = xh/k$. The solution clearly corresponds to the first three terms of Eq. (149). Cho and Sunderland (1981) obtained an approximate heat flux solution using an analogy to the non-freezing solution. The results agree with Eq. (158), but the generality of the method is unknown.

Two Phase Problem

The two phase problem has been solved by Lunardini (1982) using HBI, with collocation. The geometry for a melt problem is shown in Figure 26. For semi-infinite solids the following temperature approximations can be used

$$T_1 = T_f + \frac{Q}{k_1} \left[\frac{\alpha_{21} X}{\delta + \alpha_{21} X} - 1 \right] (x - X) + \frac{Q \alpha_{21} (x-X)^2}{2k_1 (\delta + \alpha_{21} X)} \tag{163}$$

$$T_2 = T_f - 2 \frac{(T_f - T_0)}{\delta} (x - X) + \frac{(T_f - T_0)}{\delta^2} (x - X)^2 \tag{164}$$

The surface temperature will increase from T_0 to the fusion value T_f when melting begins and the phase change solution can then be obtained. The surface boundary condition is

$$-k_1 \frac{\partial T_1(0,t)}{\partial x} = Q(t) \tag{165}$$

The collocation method allows a simple relation to be derived between δ and X, which is

$$\Delta = \frac{B}{2} + \sqrt{\frac{B^2}{4} + \alpha_{21} B \xi} \tag{166}$$

The phase change depth ξ is

$$\frac{\alpha_{21}}{6} \xi^3 + \xi^2 \left[\frac{\Delta}{2} + \alpha_{21} (1 + c_{21} S_T)\right] + \xi \left[(1 + c_{21} S_T) \Delta + \frac{1}{3} k_{21} S_T(\Delta - B)\right]$$

$$+ \frac{1}{3} c_{21} S_T \Delta(\Delta - B) = \tau (\Delta + \alpha_{21} \xi) \tag{167}$$

where $S_T = c_1 (T_f - T_o)/\ell$. For the single phase case, when $S_T = 0$ this reduces to Eq. (158). The surface temperature (for $t > t_0$) is given by

$$\frac{T_1(o,t) - T_o}{T_f - T_o} = 1 + \left[\frac{\alpha_{21} \xi^2 + 2\xi \Delta}{2(\Delta + \alpha_{21} \xi)}\right] \frac{1}{S_T} \tag{168}$$

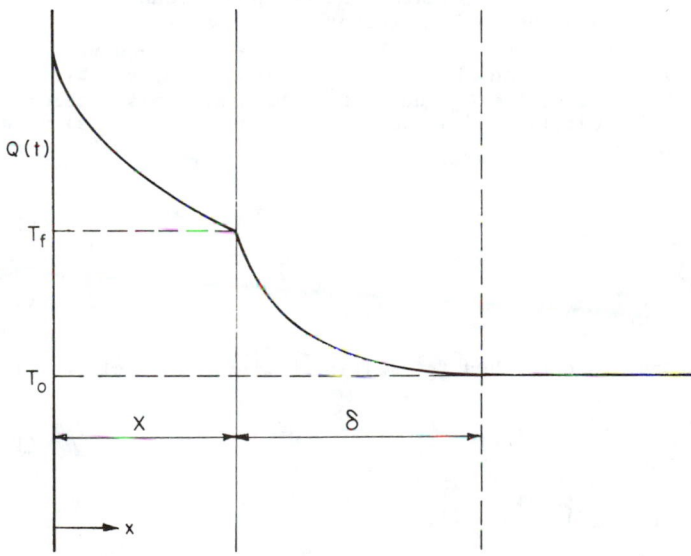

FIGURE 26. Specified surface heat flux for a semi-infinite medium.

Ablation with Complete Removal of Melt

Phase change problems for which the melting (or vaporizing) material is removed from the system might be useful for ice melting from vertical surfaces where the water can run off the surface due to gravity.

Constant surface heat flux. For this problem, there are two time domains to consider: the time before the surface temperature reaches the fusion

111

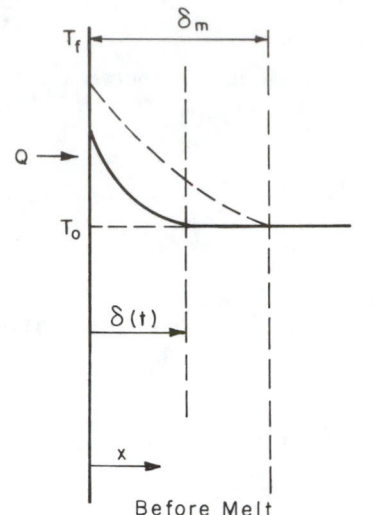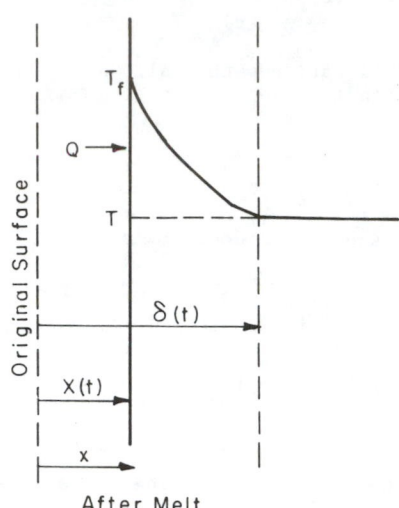

FIGURE 27. Complete removal of melt from surface.

value, during which no phase change occurs, and the phase change with removal of melt during which the surface temperature remains at T_f. These are shown in Figure 27. Initially, the solid is at T_o and at t = 0, a heat flux, Q is applied. The thermal penetration depth is $\delta(t)$; when the surface temperature is T_f, $\delta = \delta_m$ and melting begins. This problem has been solved exactly, Carslaw and Jaeger (1959), but the integral method approximation is given by

$$\delta_m = \frac{2k(T_f - T_o)}{Q} \tag{169}$$

$$t_m = \frac{2}{3} \frac{k^2(T_f - T_o)^2}{\alpha Q^2} \tag{170}$$

Once the surface temperature reaches T_f, melting begins. The solution is

$$\tau = -\frac{1}{3S_T}\left\{\phi - 2 + \frac{2(1 + S_T)}{S_T} \ln[S_T(1 - \frac{\phi}{2}) + 1]\right\} \tag{171}$$

$$\sigma = -\frac{1}{3}\left[\phi - 2 + \frac{2}{S_T}\ell n\left(S_T(1 - \frac{\phi}{2}) + 1\right)\right] \tag{172}$$

where $\phi = Q(\delta - X)/[k_1(T_f - T_o)]$, $\sigma = XQ/[k_1(T_f - T_o)]$.

These are the parametric equations for σ as a function of τ. Goodman (1958) plotted these equations for some values of S_T, as shown on Figure 28. Landau (1950) solved this problem numerically and the results agree quite well with Equations (171) and (172), but as S_T become large the values tend to diverge for small times. A steady state solution for σ_∞ can be easily found.

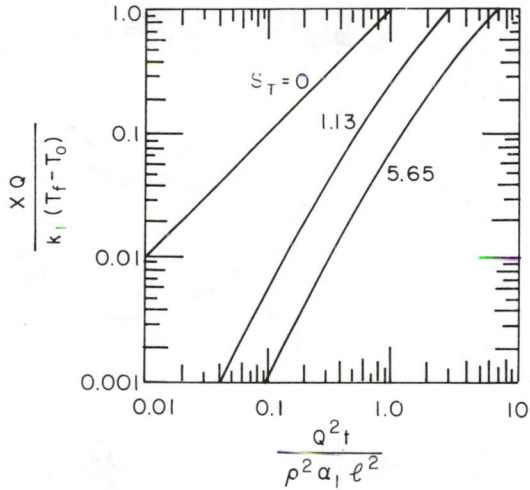

FIGURE 28. Melt-line location vs. time with complete melt removal, eq. 172, (adapted from Goodman 1958).

$$\sigma_\infty = \frac{1}{1 + S_T} \tag{173}$$

For large values of time

$$\sigma = \frac{S_T}{1 + S_T} \tau \tag{174}$$

This equation is identical to the exact solution of Landau (1950).

Variable surface heat flux. The procedure detailed in the previous section was used for the case of variable surface heat flux, Altman (1961). The surface temperature is

$$T(0,t) = T_o + \frac{Q\delta}{4k} \tag{175}$$

The equation for δ and the time when the surface temperature is T_f are

$$\delta(t) = \sqrt{20\alpha} \ [\frac{1}{Q} \int_o^t Q \ dt]^{\frac{1}{2}} \tag{176}$$

$$\delta_m = \frac{4k(T_f - T_o)}{Q(t_m)} \tag{177}$$

$$\int_o^{t_m} Q \ dt = \frac{4k^2(T_f - T_o)^2}{5\alpha \ Q(t_m)} \tag{178}$$

113

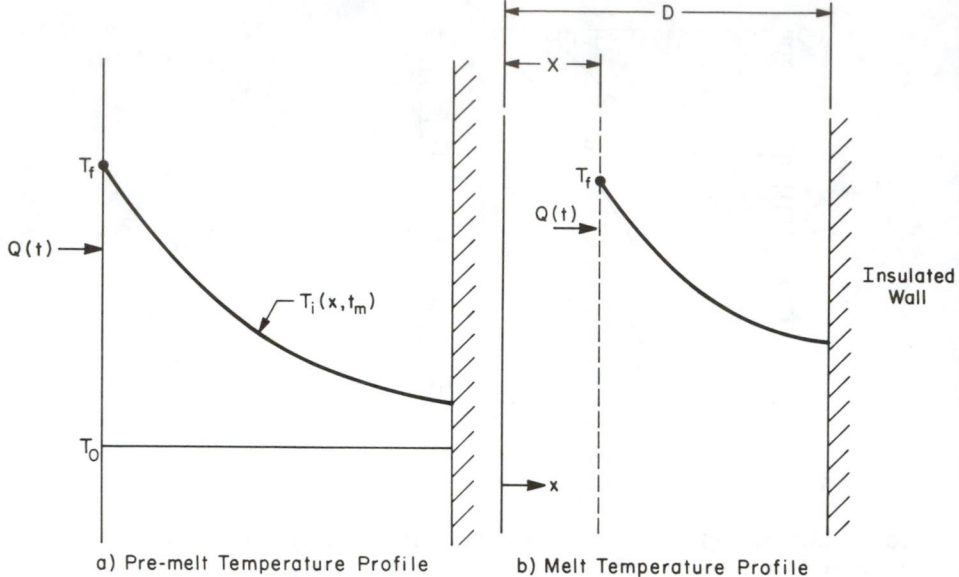

a) Pre-melt Temperature Profile   b) Melt Temperature Profile

FIGURE 29. Ablation of finite thickness slab.

δ_m and t_m can be evaluated if the transient function $Q(t)$ is specified. The equation for the phase change depth can be solved numerically, given $Q(t)$,

$$X = \frac{1}{\rho \ell} \int_o^t [Q(t) - \frac{4k(T_f - T_o)}{(\delta - X)}] \, dt \tag{179}$$

For $Q(t)$ constant, the solution is

$$\tau = - \frac{1}{5S_T} [\phi - 4 + 4(1 + \frac{1}{S_T}) \ell n \ (1 + S_T(1 - \frac{\phi}{4}))] \tag{180}$$

$$\sigma = - \frac{1}{5} [\phi - 4 + \frac{4}{S_T} \ell n \ (1 + S_T(1 - \frac{\phi}{4}))] \tag{181}$$

<u>Slab with finite thickness</u>. An approximate method has been given by Citron (1960), for ablation of a slab with finite thickness. The slab is originally at T_o and is insulated at $x = D$, while a constant heat flux Q_o impinges on the free surface as shown in Figure 29. Melting begins when the free surface temperature reaches the fusion value T_f and continues until the entire slab has melted. An approximate solution method is facilitated by immobilizing the phase change interface using a Landau Transformation. The first order solution is

$$\psi_1 \ (z, \tau_1) = A + z + \frac{A}{1-p} (p \ e^{\lambda_1 z} - e^{\lambda_2 z}) \tag{182}$$

$$\lambda_{1,2} = - \frac{\dot{S}}{2} \pm \sqrt{\frac{\dot{S}^2}{4} + \frac{\dot{S}}{S}} \tag{183}$$

where $S = X/D$, $A = S - 1 + \dfrac{1}{S}$, $p = \lambda_2/\lambda_1 \, e^{(\lambda_2 - \lambda_1)(1 - S)}$. The equation for the phase change interface is

$$\frac{\ell k}{c_1 Q_o D} \dot{S} = S \left\{ 1 - \frac{\lambda_2 A \left(1 - \dfrac{\lambda_1}{\lambda_2} p \right)}{1 - p} \right\} \tag{184}$$

The phase change interface equation can be solved easily by numerical techniques. A starting solution must be used at $\tau_1 = 0$ since the equation is singular there. The starting solution is

$$S = \frac{c_1 Q_o D}{\ell k} \left(\frac{2 \tau_1}{3} \right)^{\frac{3}{2}} \tag{185}$$

Citron (1960) showed that the first order solution, for the complete melt time, is about 9% high for $\dfrac{c_1 Q_o D}{\ell k} = 5.52$ and is essentially exact for $\dfrac{c_1 Q_o D}{\ell k} = 56.4$

6. THAW BENEATH INSULATED STRUCTURES

The effect of heated structures on the underlying or surrounding medium is important for the melting of permanently frozen ground (permafrost) or the freezing of thawed soils which can lead to frost heave complications. Lachenbruch (1957a,b, 1959) and Jumikis (1978) applied linear conduction theory to the effect of heating on permafrost. Applications of the quasi-steady method have included uninsulated buried pipes (Hwang (1977), Thornton (1976), Porkhayev (1963)), insulated buried pipes (Lunardini (1981b), Seshadri and Krishnayya (1980)), and three dimensional structures, Porkhayev (1970). Widely used calculated results are those of Porkhayev (1970). Lunardini (1982a, 1983b) derived new quasi-steady relations for insulated geometries including semi-infinite strips (roads, sewers), rectangular buildings, circular storage tanks, and buried pipes. Consider an infinite strip as shown in Figure 30. Initially the temperature of the

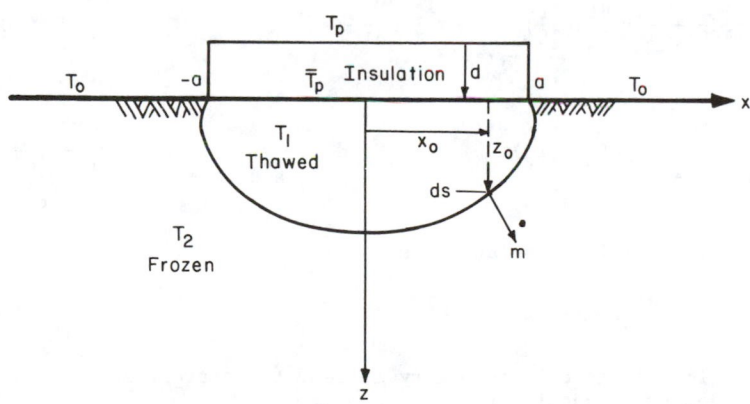

FIGURE 30.   Insulated semi-infinite strip.

semi-infinite space is uniform at T_o and the insulated surface of the strip jumps to T_p at time zero. The temperature beneath the strip then starts to vary with time. The properties of the material differ for regions below and above the fusion temperature T_f. The energy equations cannot be solved exactly, but if the Stefan number, S_T, is small they reduce to the steady state case; thus it is not necessary to solve the transient conduction equation but only the much simpler Laplace equation. The accuracy of the quasi-steady method depends upon the magnitude of the Stefan number; for systems with a large latent heat relative to the sensible heat it can be expected that the quasi-steady approximation will be reasonably good. This covers soil systems and many phase change materials used for latent heat storage.

General Quasi-Steady Relations

General equations have been derived which are valid for a class of important phase change problems. The solutions to the quasi-steady energy equations are

$$\phi_1 = \frac{f - f_o}{1 - f_o} = \frac{T_1 - T_f}{\bar{T}_p - T_f} \tag{186}$$

$$\phi_2 = \frac{f}{f_o} = \frac{T_2 - T_o}{T_f - T_o} \tag{187}$$

The function f is the solution to the equivalent steady state equations and f_o is the value of the function on the phase change surface where the temperature is at the fusion value. Once f_o has been determined, the total solution will be known. The heat flow through the insulation is equated to the heat flow through the thawed soil at $\xi = z/a = 0$, Seshradi and Krishnayya (1980), Lunardini (1981b, 1983b). Another approach assumes that the insulation is accounted for by considering an excess layer of soil, with a thermal resistance equal to that of the insulation, applied only to the thawed zone temperature relations, Porkhayev (1963, 1970).

The general interface equation

$$\left(\frac{1}{f_o - 1 + 2\alpha' \frac{\partial f(o,o)}{\partial \xi}} + \frac{\beta'}{f_o} \right) \left(\frac{\partial f}{\partial \xi} \right)_{\zeta_o, \xi_o} = \frac{2}{\pi} \frac{d\xi_o}{d\tau} \tag{188}$$

where

$$\xi_o = z_o/a \ , \ \tau = \frac{2k_1(T_p - T_f)t}{\pi a^2 L} \ , \ \alpha' = \frac{k_1 \ d}{2 \ a \ k_i} \ , \ \text{and } \beta' = \frac{k_{21} (T_f - T_o)}{(T_p - T_f)}$$

The steady-state, or limiting solution is

$$f_{o\infty} = \frac{\beta'}{1+\beta'} [1 - 2\alpha' \frac{\partial f(o,o)}{\partial \xi}] \tag{189}$$

To evaluate Eq. (188) it is only necessary to find the appropriate, steady-state, geometric function f.

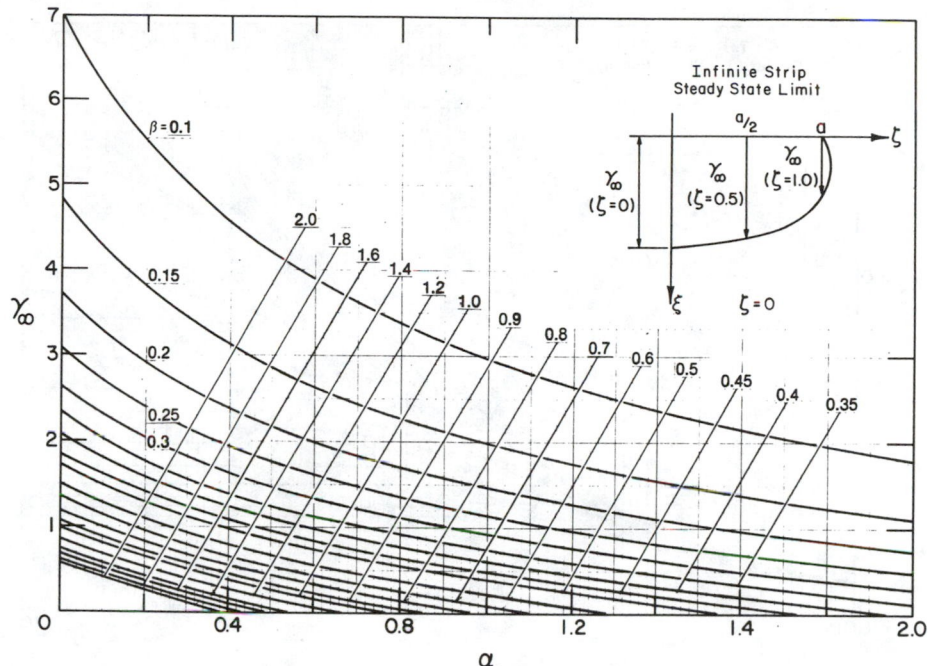

FIGURE 31. Steady state limit phase change beneath infinite strip, $\zeta=0$.

Semi-infinite Strip

The semi-infinite strip can represent roads, shallow rivers, or rectangular buildings with a large aspect ratio. The limiting interface beneath the center, at $\tau = \infty$, is

$$\xi_{o\infty} = \cot \frac{\beta'}{1 + \beta'} (\pi + 4\alpha') + \sqrt{\cot^2 \left[\frac{\beta'}{1+\beta'} (\pi + 4\alpha') \right] - (\zeta^2 - 1)} \qquad (190)$$

and is given in Figure 31. The interface position can be evaluated exactly if $\beta' = 0$. For this case

$$\tau = (1 + \frac{4\alpha'}{\pi})(\frac{\gamma^3}{3} + \gamma) = \frac{2}{\pi}\{(\frac{\gamma^3}{3} - \gamma) \, \text{ctn}^{-1}\gamma + \frac{1}{3} \ln(1 + \gamma^2) + \frac{1}{6} \gamma^2\} \qquad (191)$$

where γ is the value of ξ_o at the center of the strip. The temperatures, in the thawed and frozen zones, can be found with Eqs. (186) and (187). The transient behavior can be evaluated from

$$\gamma = \gamma(\alpha' = 0) - K_i \qquad (192)$$

and Figures 32-34.

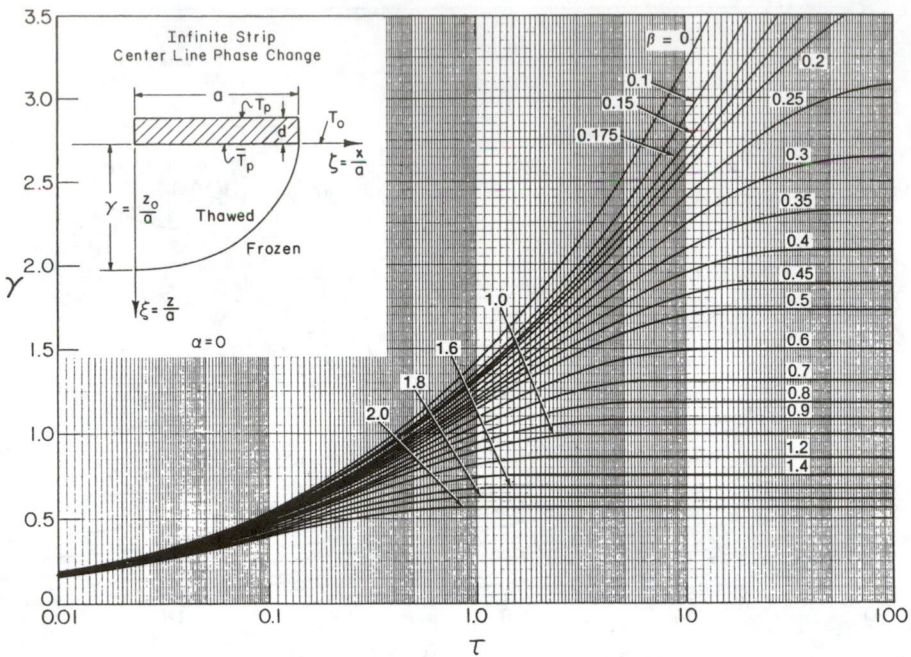

FIGURE 32. Thaw beneath center of infinite strip, $\alpha' = 0.0$.

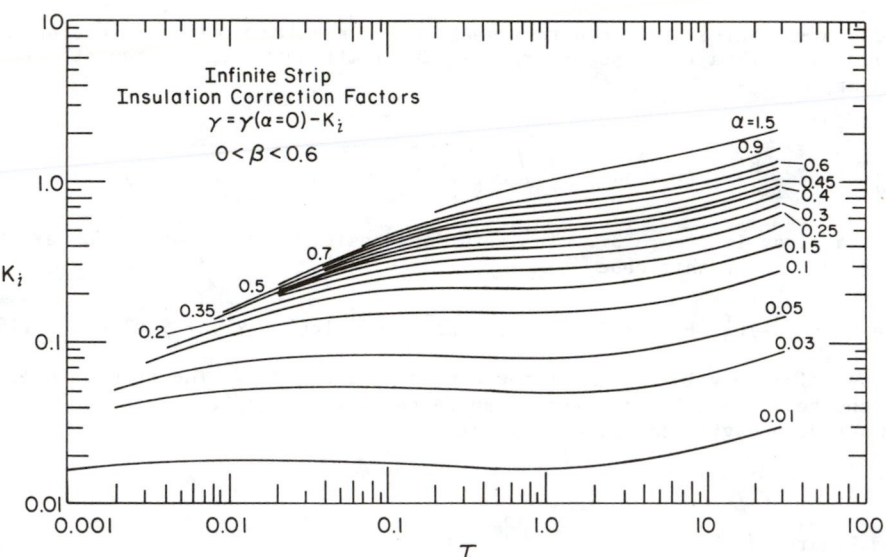

FIGURE 33. Thaw beneath center of insulated infinite strip, $0 < \beta' < 0.6$.

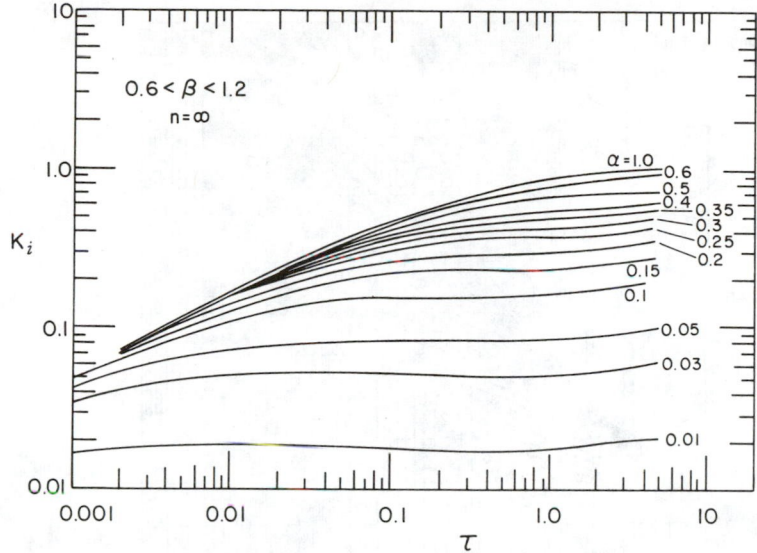

FIGURE 34. Thaw beneath center of insulated infinite strip, $.6 < \beta' < 1.2$.

Rectangular Building

The steady state solution is given by Lunardini (1982a) and is

$$\xi_{B\infty}^2 = \left(\frac{1+n^2}{2}\right)^2 + n^2 \cot^2\left[\frac{\beta'}{1+\beta'}\left(\frac{\pi}{2} + \frac{2\alpha'\sqrt{1+n^2}}{n}\right)\right] - \frac{(1+n^2)}{2} \tag{193}$$

where n = b/a is the aspect ratio. Numerical quadrature leads to the plots given by Figures 35–41 for n = 1 and 2. The correction factor is used with

$$\xi_B = \xi_B(\alpha' = 0) - K_{in} \tag{194}$$

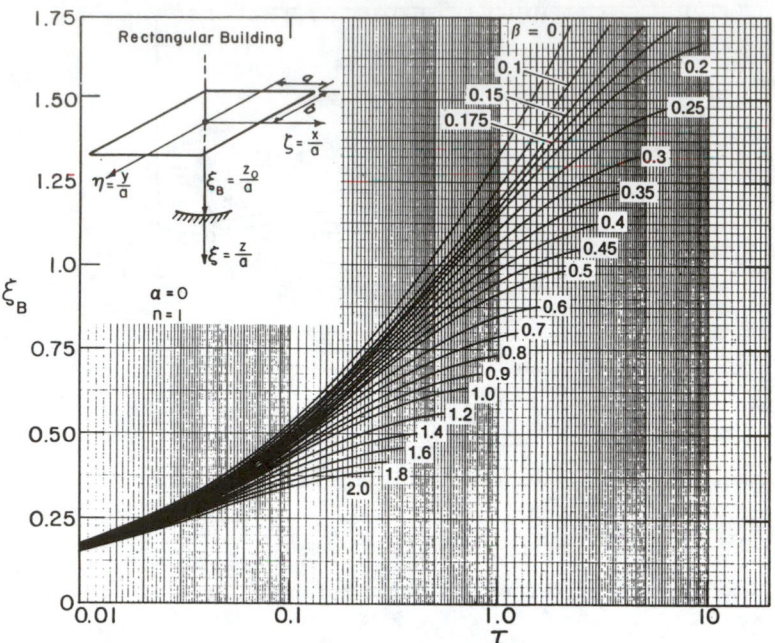

FIGURE 35.   Phase change beneath center of uninsulated rectangle, n=1.

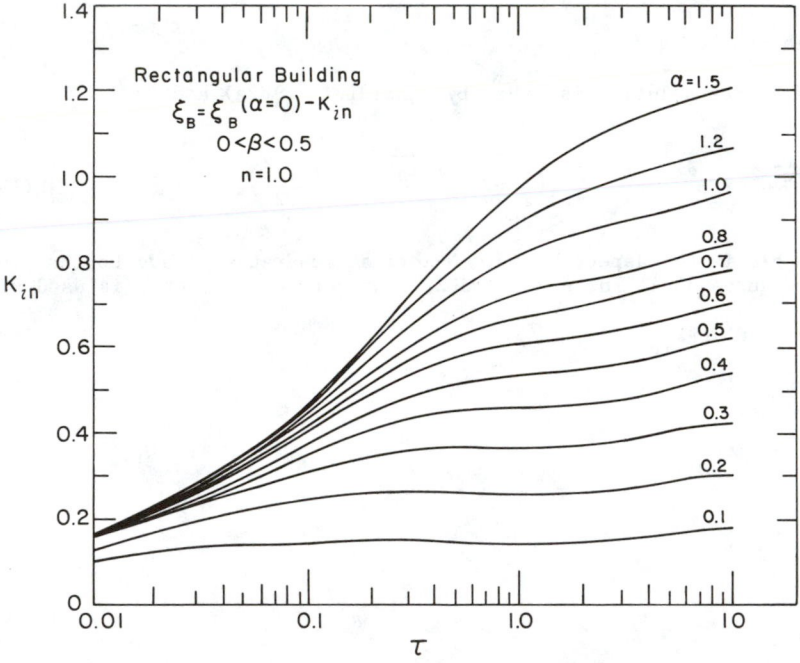

FIGURE 36.   Insulation factor K_{in}, n = 1, $0 < \beta' < 0.5$.

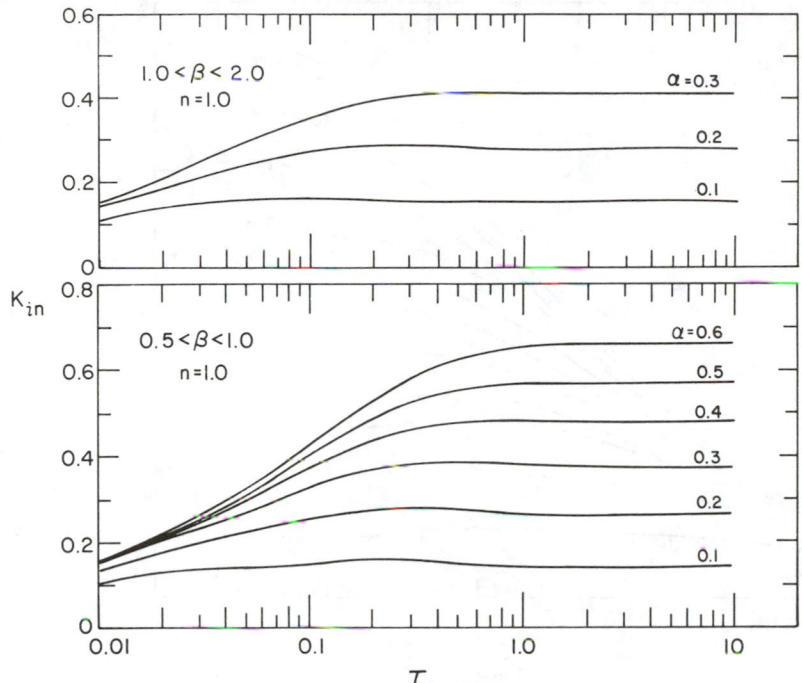

FIGURE 37. Insulation factor K_{in}, $0.5 < \beta' < 2.0$.

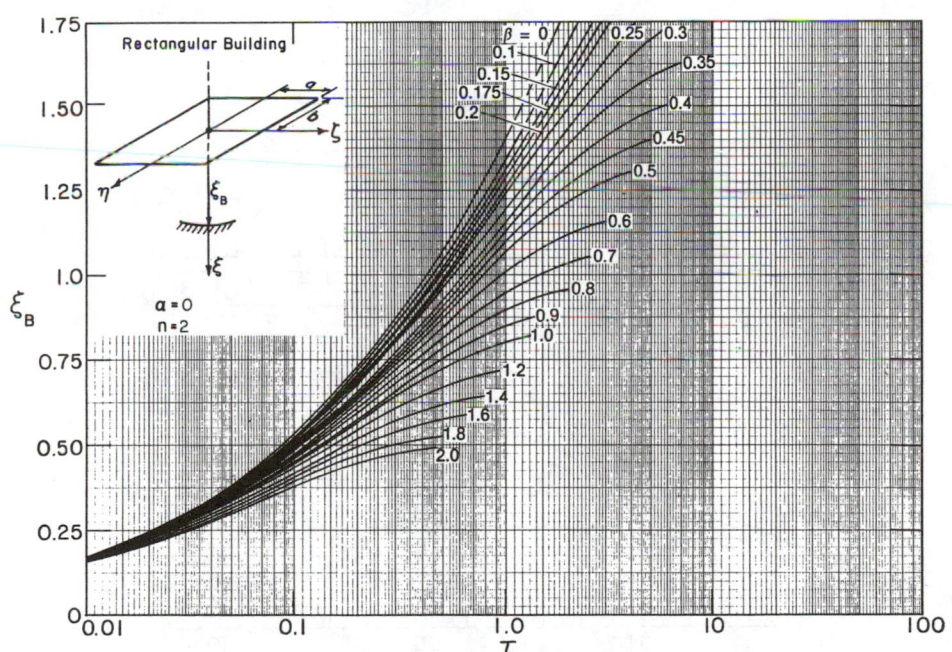

FIGURE 38. Phase change beneath center of rectangle, n = 2.

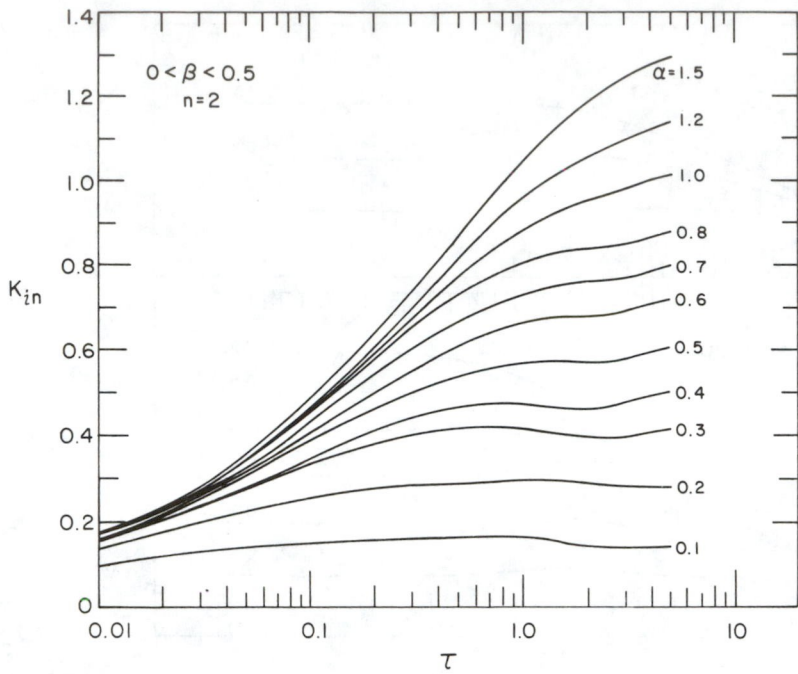

FIGURE 39. Insulation factor for rectangular building, n=2, 0<β'<0.5.

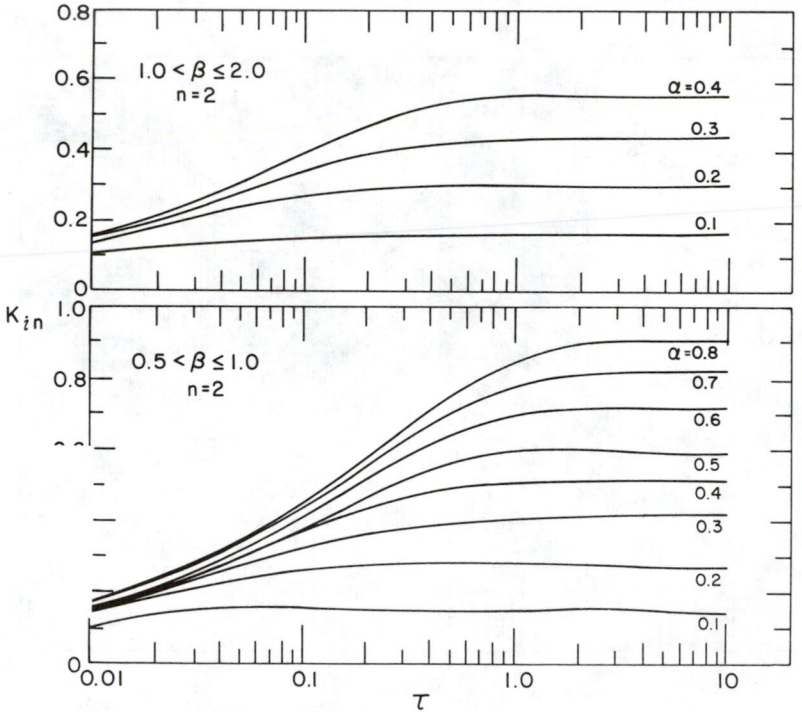

FIGURE 40. Insulation factor for rectangular building, n=2, 0.5<β'<2.0.

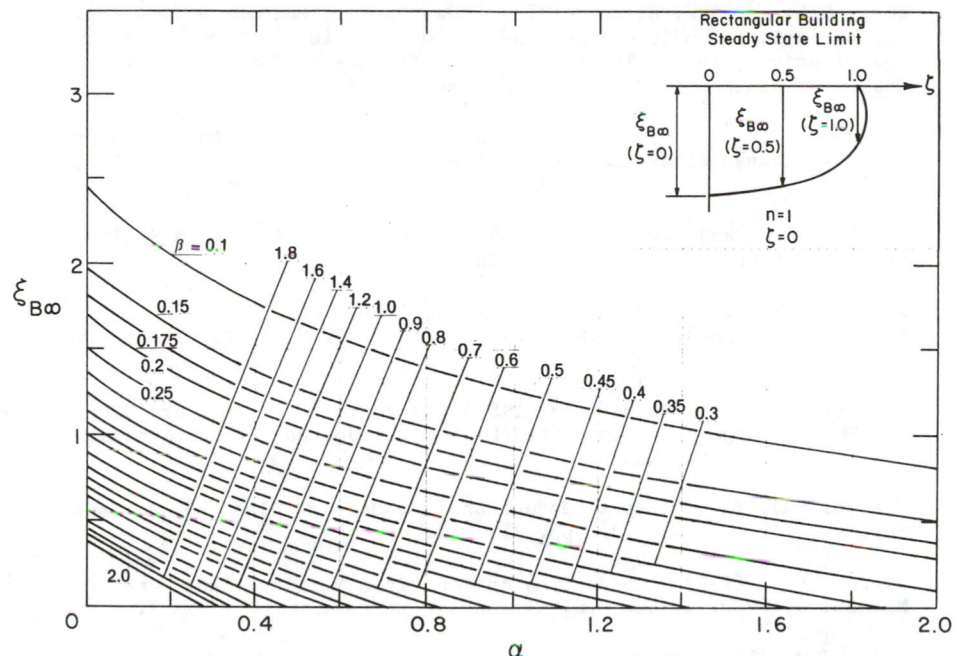

FIGURE 41. Steady state limit of thaw beneath rectangle.

REFERENCES

Aldrich, H.P. and H.M. Paynter (1953) "Frost Investigations. Fiscal year 1953, first interim report. Analytical Studies of Freezing and Thawing of Soils." U.S. Army Arctic Construction and Frost Effects Laboratory (ACFEL) Technical Report 42.

Altman, M. (1961) "Some Aspects of the Melting Solution for a Semi-infinite Slab," Chem. Engr. Progress Symposium Series, Vol. 57, 16-23, Buffalo. Andersland, O.B. and D.M. Anderson (1978) "Geotechnical engineering for cold regions." New York: McGraw-Hill.

Beaubouef, R.T. and A.J. Chapman (1967) "Freezing of Fluids in Forced Flow," Int. J. Heat Mass Transfer. Vol. 10, pp. 1581-1588.

Bell, G.E. (1978) "A Refinement of the Heat Balance Integral Method Applied to a Melting Problem," Int. J. Heat Mass Transfer, Vol. 21, 1357-1362.

Bell, G.E. (1979) "Solidification of a Liquid about a Cylindrical Pipe," Int. J. Heat Mas Transfer, Vol. 22, pp. 1681-1686.

Bell, G.E. (1982) "The Prediction of Frost Penetration," Int. J. Numer. Anal. Methods Geomech., Vol. 6, pp. 287-290.

Bell, G.E. and S.K. Abbas (1985) "Convergence Properties of the Heat Balance Integral Method," Numerical Heat Transfer, Vol. 8, pp. 373-382.

Berg, R.L. and G.W. Aitken (1973) "Some Passive Methods of Controlling Geocryological Conditions in Roadway Construction." North American Contributions, Second International Conference on Permafrost, Yakutsk, USSR. Washington, D.C.: National Academy of Sciences, p. 581-586.

Berggren, W.P. (1943) "Prediction of Temperature Distribution in Frozen Soils." Transactions, American Geophysical Union, Vol. 24, Part 3, p. 71-77.

Carslaw, H.W. and J.C. Jaeger (1959) "Conduction of Heat in Solids." Oxford: Clarendon Press, 2nd edition.

Cho, S.H. and Sunderland, J.E., (1974), "Phase Change Problems with Temperature Dependent Thermal Conductivity," J. Heat Transfer, V. 96, 214-217.

Cho, S.H. and J.E. Sunderland (1981) "Approximate Temperature Distribution for Phase Change of a Semi-Infinite Body." Journal of Heat Transfer, Vol. 103, No. 2, pp. 401-403.

Citron, S.J. (1960) "Heat Conduction in a Melting Slab," J. Aerospace Science, Vol. 27, No. (3), pp. 219-228.

Citron, S.J. (1962) "Conduction of Heat in a Melting Slab," Applied Mechanics, Proc. of 4th U.S. National Congress, pp. 1221-1227.

Cole, J.D. (1968) "Perturbation Methods in Applied Mathematics." Blaisdell, New York.

Crank, J. (1975), "The Mathematics of Diffusion," Clarendon Press, Oxford.

Duda, J.L. and J.S. Vrentas (1969a) "Perturbation Solutions of Diffusion - Controlled Moving Boundary Problems," Chem. Engr. Sci., Vol. 24, 461-470.

Duda, J.L. and J.S. Vrentas (1969b) "Mathematical Analysis of Bubble Dissolution," A.I.Ch.E.Jl 15, 351.

Evans, G.W., Isaacson, E. and MacDonald, J.K.L. (1950) "Stefan-Like Problems," Quarterly of Applied Math., Vol. 8, No. 3, 312-319.

Farouki, O.T., (1981), "Thermal Properties of Soils," USACRREL Monograph 81-1, Hanover, N.H.

Foss, S.D. and Fan, S.S.T. (1972) "Approximate Solution to the Freezing of the Ice-Water System," J. Water Resources Research, Vol. 8, No. 4, 1083-1086.

Foss, S.D. and Fan, S.S.T. (1974) "Approximate Solution to the Freezing of the Ice-Water System with Constant Heat Flux in the Water Phase," J. Water Res. Research, Vol. 10, No. 3, 511-513.

Gold, L.W. and A.H. Lachenbruch (1973) "Thermal Conditions in Permafrost-A Review of North American Literature." North American Contributions, Second International Conference on Permafrost, Yakutsk, USSR. Washington, D.C.: National Academy of Sciences, p. 3-25.

Goodman, T.R. (1958) "The Heat-Balance Integral and Its Application to Problems Involving a Change of Phase." American Society of Mechanical Engineers Transactions, Vol. 80, p. 335-342.

Goodman, T.R. (1964) "Application of Integral Methods to Transient Nonlinear Heat Transfer," Advances in Heat Transfer, Vol. 1, T.F. Irvine and J.P. Hartnett, Eds., 52-122, Academic Press, New York.

Gutman, L.N. (1986) "On the Problem of Heat Transfer in Phase Change Materials for Small Stefan Numbers," Int. J. Heat Mass Transfer, Vol. 29, No. 6, pp. 921-926.

Hwang, C.T. (1977) "On Quasi-Static Solutions For Buried Pipes in Permafrost." Canadian Geotechnical Journal, v. 14, p. 180-192.

Hwang, C.T., Murray, D.W. and Brooker, E.W., (1972) "A Thermal Analysis for Structures on Permafrost." Can. Geotech. J., 9(2), pp. 33-46.

Jiji, L.M. (1970) "On the Application of Perturbation to Free-Boundary Problems in Radial Systems." J. Franklin Inst. 289, p. 282.

Jiji, L.M. and S. Weinbaum (1978) "Perturbation Solutions for Melting or Freezing in Annular Regions Initially Not at the Fusion Temperature." International Journal of Heat and Mass Transfer, 21, p. 581-592.

Johansen, O., (1975), "Thermal Conductivity of Soils," USACRREL, TL 637, (1977) Hanover, N.H.

Jumikis, A.R., (1978), "Graphs for Disturbance Temperature Distribution in Permafrost Under Heated Rectangular Structures," Proceedings Third International Conference on Permafrost, Vol. I, p. 589-598, NRC, Ottawa, Canada.

Kersten, M.S. (1949), "Thermal Properties of Soils." University of Minnesota Experiment Station Bulletin No. 28.

Kreith, F. and F.E. Romie (1955) A Study of the Thermal Diffusion Equation With Boundary Conditions Corresponding to Solidification or Melting of Materials Initially at the Fusion Temperature. Proceedings of Physical Society, Section B, Vol. 68, p. 277-291.

Lachenbruch, A.H., (1957a), "Three Dimensional Heat Conduction in Permafrost Beneath Heated Buildings," U.S. Geological Survey Bulletin 1052-B, U.S. Gov. Printing Office, Washington, D.C.

Lachenbruch, A.H., (1957b), "Thermal Effects of the Ocean on Permafrost," Bull. Geological Society of America 68, p. 1515-1530.

Lachenbruch, A.H., (1959), "Periodic Heat Flow in a Stratified Medium with Applications to Permafrost Problems," U.S. Geological Survey Bulletin 1083-A.

Landau, H.G. (1950) "Heat Conduction in a Melting Solid," Quart. Appl. Math, Vol. 8, 81-94.

Langford, D. (1973) The Heat Balance Integral method, Int. J. Heat Mass Transfer, Vol. 16, pp. 2424-2428.

Lapadula, C. and W.K. Mueller (1966) "Heat Conduction with Solidification and a Convective Boundary Condition at the Freezing Front," Int. J. Heat Mass Transfer, Vol. 9, pp. 702-704.

Lock, G.S.H. (1969) "On the Use of Asymptotic Solutions to Plane Ice-Water Problems," J. Glaciology, Vol. 8, No. 53, 285-300.

Lock, G.S.H., Gunderson, J.R., Quon, D., and Donnelly, J.K., (1969), "A Study of One-Dimensional Ice Formation with Particular Reference to Periodic Growth and Decay," Int. J. Heat Mass Transfer, Vol. 12, 1343-1352.

London, A.L. and Seban, R.A. (1943) "Rate of Ice Formation," Trans. ASME, Vol. 65, No. 7, 771-779.

Lozano, C.J. and R. Reemsten (1981) "On a Stefan Problem With an Emerging Free Boundary. Numerical Heat Transfer, Vol. 4, p. 239-245.

Luikov, A.V. (1964), "Heat and Mass Transfer in Capillary-Porous Bodies," Advances in Heat Transfer, Vol. I, T.F. Irving, J.P. Hartnett, Eds., pp. 123-184.

Luikov, A.V. (1975), "Systems of Differential Equations of Heat and Mass Transfer in Capillary-Porous Bodies (review)," Int. J. Heat Mass Transfer, Vol. 18, pp. 1-14.

Lunardini, V.J. (1971) "Presentation of Some Thermal Properties of Soil Systems." Engineering Institute of Canada Meeting, Paper 71-CSME-38. Quebec City, Canadian Society of Mechanical Engineers.

Lunardini, V.J. (1978) "Theory of n-factors and Correlation of Data." Proceedings, Third International Conference on Permafrost, Edmonton, Alberta, Vol. 1, p. 40-46.

Lunardini, V.J. (1978a) "A Correlation of N-Factor Data," Vol. 1, Proceedings Applied Techniques for Cold Environments, pp. 233-244, ASCE Anchorage, Alaska, May 17-18.

Lunardini, V.J. (1980) "The Neumann Solution Applied to Soil Systems," USA CRREL Report 80-22, Hanover, NH.

Lunardini, V.J., (1981) "Application of the Heat Balance Integral to Conduction Phase Change Problems," U.S. CRREL Report 81-25, Hanover, NH.

Lunardini, V.J., (1981a) "Heat Transfer in Cold Climates," Van Nostrand Reinhold, New York.

Lunardini, V.J. (1981b) "Phase Change Around Insulated Buried Pipes: Quasi Steady Method," Journal of Energy Resources Technology, v. 103, no. 3, p. 201-207.

Lunardini, V.J., (1982) "Freezing of Soil with Surface Convection," Proceedings Third International Symposium on Ground Freezing, pp. 205-213.

Lunardini, V.J. (1982a), "Conduction Phase Change Beneath Insulated Heated or Cooled Structures," USACRREL Report 82-22.

Lunardini, V.J. (1983), "Approximate Solution to Conduction Freezing with Density Variation," J. Energy Res. Tech., Vol. 105, pp. 43-45.

Lunardini, V.J. (1983a), "Freezing and Thawing: Heat Balance Integral Approximations," J. Energy Resources Tech., Vol. 105(1), pp. 30-37.

Lunardini, V.J. (1983b), "Thawing Beneath Structures on Permafrost," Proceedings of 4th International Conference on Permafrost, July 1983, Fairbanks, Alaska, pp. 750-756.

Lunardini, V.J. (1984) "Freezing of a Semi-Infinite Medium with Initial Temperature Gradient," J. Energy Resources Technology, Vol. 106, pp. 103-106.

Lunardini, V.J., (1985), "Freezing of Soil with Phase Change Occurring over a Finite Temperature Difference," Proc. 4th International Offshore Mechanics and Arctic Engineering Symposium, pp. 38-46, ASME.

Lunardini, V.J. (1988) "Heat Conduction with Freezing or Thawing," USACRREL Monograph 88-1.

Lunardini, V., and Varotta, R. (1981) "Approximate Solution to Neumann Problem for Soil Systems," ASME Journal of Energy Resources Technology, Vol. 103, No. 1, pp. 76-81.

Martynov, G.A. (1959), "Principles of Geocryology," Part 1, Chapter VI, Academy of Sciences, USSR, NRC TT-1065, (1963) Ottawa, Canada.

Nakano, Y. and J. Brown (1971) "Effect of a Freezing Zone of Finite Width on the Thermal Regime of Soils," J. Water Resources Research, Vol. 7, No. 5, pp. 1226-1233.

Neumann, F. (ca. 1860) Lectures given in the 1860's. cf. Riemann-Weber, Die partiellen Differentialgleichungen der mathematischen Physik. 5th edition, 1912, Vol. 2, p. 121.

Nixon, J.F. (1975), "The Role of Convective Heat Transport in the Thawing of Frozen Soil." Canadian Geotechnical J., Vol. 12, pp. 425-429.

Nixon, J.F. and E.L. McRoberts (1973) "A Study of Some Factors Affecting the Thawing of Frozen Soils." Canadian Geotechnical Journal, Vol. 10, p. 439-452.

Noble, B. (1975) "Heat Balance Methods in Melting Problems," in J.R. Ockendon and W.R. Hodgkins (eds.), Moving Boundary Problems in Heat Flow and Diffusion, pp. 208-209, Clarendon, Oxford.

Pedroso, R.I. and Domoto G.A. (1973) "Planar Solidification with Fixed Wall Temperature and Variable Thermal Properties," J. Heat Transfer, Vol. 95, No. 4, p. 533-535.

Pedroso, R.I. and G.A. Domoto (1973a) "Perturbation Solutions for Spherical Solidification of Saturated Liquids," J. Heat Transfer, Vol. 95, No. 1, 42-46.

Pedroso, R.I. and G.A. Domoto (1973b) "Exact Solution by Perturbation Method for Planar Solidification of a Saturated Liquid With Convection at the Wall." Int. J. Heat Mass Transfer Vol. 16, p. 1816.

Pedroso, R.I. and G.A. Domoto (1973c) "Inward Spherical Solidification Solution By the Method of Strained Coordinates," Int. J. Heat Mass Transfer Vol. 16, p. 1037.

Pohlhausen, K. (1921) "Zur näherungsweisen Integration der Differentialgleichung der laminaren Reibungsschicht." ZAMM 1, 252-268.

Porkhayev, G.V., (1959), "Principles of Geocryology," Part II, Chapter IV, Academy of Sciences USSR, NRC TT-1249, (1966) Ottawa, Canada.

Porkhaev, G.V., (1963), "Temperature Fields in Foundations," Proceedings Permafrost Int. Conference, National Academy of Science, Washington, D.C. Pub. No. 1287, pp. 285-291.

Porkhaev, G.V., (1970), "Thermal Interaction Between Buildings, Structures, and Perennially Frozen Ground," Nauka Publisher, Moscow.

Rubinsky, B. and A. Shitzer (1978) "Analytic Solutions to the Heat Equation Involving a Moving Boundary with Applications to the Change of Phase Problem (the Inverse Stefan Problem)," J. Heat Transfer, Vol. 100, pp 300-304.

Sanger, F.J. (1969) "Foundations of Structures in Cold Regions." Cold Regions Research and Engineering Laboratory, Hanover, New Hampshire, Monograph III-C4.

Savino, J.M. and R. Siegel (1967) "Experimental and Analytical Study of the Transient Solification of a Warm Liquid Flowing Over a Chilled Flat Plate." NASA TN D-4015.

Seban, R.A. (1971) "A Comment on the Periodic Freezing and Melting of Water," Int. J. Heat Mass Transfer, Vol. 14, 1862-1864.

Seeniraj, R.V. and T.K. Bose (1982) "Planar Solification of a Warm Flowing Liquid Under Different Boundary Conditions," Wärme-und Stoffübertragung 16, pp. 105-111.

Seshadri, R., and Krishnayya, A.V.G., (1980), "Quasi-steady Approach for Thermal Analysis of Insulated Structures," Int. J. Heat Mass Transfer, Vol. 23, pp. 111-121.

Stefan, J. (1891) "Uber die Theorie des Eisbildung, insbesonder uber die Eisbildung im Polarmere." Ann. Phys. u Chem., Neue Folge, Bd. 42, Ht. 2, p. 269-286.

Tao, L.N. (1978) "The Stefan Problem with Arbitrary Initial and Boundary Conditions," Quart. Applied Math, Vol. 36(3), pp 223-233.

Tao, L.N. (1979) "On Free Boundary Problems with Arbitrary Initial and Flux Conditions," J. Applied Math and Physics (ZAMP), Vol. 30, pp. 416-426.

Thornton, D.C. (1976) "Steady State and Quasi-Static Thermal Results for Bare and Insulated Pipes in Permafrost," Canadian Geotechnical Journal, v. 13, no. 2, p. 161-170.

Tien, R.H. and Geiger, G.E. (1967) "A Heat Transfer Analysis of the Solidification of a Binary Eutectic System," J. Heat Transfer, Trans. ASME, Vol. 89, 230-234.

Van Dyke, M. (1964) "Perturbation Methods in Fluid Mechanics." Academic Press, New York.

Von Karman, Th. (1946) "Uber laminare und turbulente Reibung." ZAMM 1, 233-252 (1921); NACA TM 1092.

Westphal, K.O. (1967) "Series Solution of Freezing Problem with the Fixed Surface Radiating into a Medium of Arbitrary Varying Temperature," Int. J. Heat Mass Transfer, V10, pp. 195-205.

Yen, Y.C. (1968) "On the Effect of Density Variation on Natural Convection in a Melted Water Layer," Chem. Engr. Prog. Symp. Ser., No 29, A.I.Ch.E., 245.

Yuen, W.W. (1980) "Application of the Heat Balance Integral to Melting Problems with Initial Subcooling," International Journal of Heat Mass Transfer, Vol. 23, pp. 1157-1160.

Chapter 4

Numerical and Analytical Aspects in Freezing and Melting

TAKEO SAITOH
Department of Mechanical Engineering II, Tohoku University,
Sendai 980, Japan

ABSTRACT

Recent developments of numerical and analytical methods on multidimensional freezing and melting problem have been reviewed in this chapter. Elucidation of these problems is of great importance from the view point of the cost-effective thermal energy storage systems with a high coefficient of performance(COP). A particular attention was focused on the efficient numerical methods which have been developed recently. Among these are the enthalpy method(EM), the boundary fixing method(BFM), and the growth ring method(GRM). A detailed description was given to the GRM, which is a powerful method applicable to the general freezing and melting problems in arbitrary geometries. Some illustrative numerical examples were shown for the typical two-dimensional freezing problem.

CONTENTS

NOMENCLATURE

a	:	thermal diffusivity
A	:	area
B	:	shape function
C, C_p	:	specific heat
D	:	diffusion coefficient or reference length
erf	:	error function
F	:	freezing front position
Gr	:	Grashof number
i	:	enthalpy
k	:	thermal conductivity
L	:	latent heat
m	:	flow rate
n	:	normal direction
$\mathbf{n}$	:	unit vector
p	:	Peclet number
Pr	:	Prandtl number
r	:	radial coordinate
R	:	freezing front position
Ra	:	Rayleigh number
Re	:	Reynolds number
S	:	surface
Ste	:	Stefan number ($Ste = C_p \Delta T / L$)
t	:	time
Δt	:	time step
T	:	temperature

T_s	: initial temperature
T_i^*	: dimensionless initial temperature
u, v	: velocity
U	: temperature
V	: temperature or volume
$\mathbf{V}$	: velocity vector
x, y, z	: coordinates
Δx	: mesh length
x^*	: $x/(4at)^{1/2}$
Y	: concentration
y^*	: $y/(4at)^{1/2}$

Greek symbols

α	: thermal diffusivity
β	: Ste^{-1}
δ	: thickness
θ, Θ	: temperature
λ	: constant in stationary solution[see ref.(26)] or angle
ρ	: density
σ	: angle
ϕ	: angle
ψ	: stream function
Ω	: vorticity

Subscripts and Superscripts

f : freezing front or fuel

i : initial condition

in : inlet

l : liquid

o : oxidant

s : solid

1. INTRODUCTION

Heat transfer with freezing and melting is of practical importance in many areas. For example, in engineering ice making, in the freezing and thawing of foods, in icing of electrical power plants, in aerodynamic heating of spacecraft, during the casting of metals and eutectic alloys, in various chemical processes, in many geophysical problems such as the thawing of soil, in bubble growth[1] and collapse in boiling and cavitation, and also in diffusion flame problems and absorption problems with rapid chemical reactions.

Particular attention has recently been focused on the cost- effective thermal energy storage (TES) systems employing the latent heat of fusion energy storage concept[2] . The development of such TES subsystems is of considerable importance in the utilization of natural energies like solar thermal, wind, sky radiation, oceanic, and geothermal energies, and also in energy conservation, waste heat recovery, and as a constant temperature reservoir for the heat pump.

Mathematically, these problems belong to the so-called moving (or free) boundary problems (MBP's), which are characterized by the possession of a moving interface dividing the relevant field into two regions[3] . Such problems become perfectly non-linear because the positions of the moving interfaces are neither fixed in space nor known a priori. Due to this non-linearity, the analytical solutions can be found only in limited situations, for example in Neumann's or Stefan's solution for a one-dimensional problem.

Typical phenomena categorized as MBPs are listed in Table 1 [4] ' [5] . Some of these problems are schematically shown in Fig.1. The joint conditions at the moving interface are also classified in the Table. Note that there is no latent heat term in the joint condition for two problems: the Burke-Schumann diffusion flame[6] and the gas absorption with a rapid chemical reaction. As seen in the Table, phase change heat transfer has many applications in various fields.

Since the works of Neumann and Stefan over a century

Table 1 Principal phenomena belonging to moving
boundary problems

Phenomena	Remarks
1.Freezing and melting of water	
2.Evaporation and condensation of a droplet	Involves phase change and has joint condition of the form:
3.Solidification and melting of metal	
4.Ablation of spacecraft	$\dot{R} = A\dfrac{\partial T_1}{\partial x} - B\dfrac{\partial T_2}{\partial x}$
5.Growth and collapse of bubbles in a superheated liquid	
6.Burke-Schumann diffusion flame	with joint condition of the form:
7.Gas absorption with rapid chemical reaction	$\dfrac{\partial Y_1}{\partial x} = A\dfrac{\partial Y_2}{\partial x}$
8.Oil flow problem	
9.Poroelastic problem	
10.Diffusion of oxygen and lactic acid in tissues	
11.Sublimation, recrystallization, cavitation, etc.	

ago, these phase-change problems have been studied by many researchers in various fields. In earlier times, analytical solutions for one-dimensional problems were obtained by Lightfoot[7], Burke-Schumann[6], Friedman[8], Kolodner[9], Dressel[10,11], Landau[12], Danckwerz[13], Miranker[14], Miranker and Keller[15], Boley[16,17], Sherman[18,19], and so on. It is emphasized that an exact solution is available only in limited situations, namely for one-dimensional planar geometry.

Concerning the analytical methods, review papers have been issued by Boley[20], Bankoff[21], Muelbauer and Sunderland[22], Moriyama and Araki[23], and Lunardini[24].

A thorough numerical method for one-dimensional MBP's was first presented by Murray and Landis[25], though the origin of their method can be traced to the earlier work of Landau, whose method was known as Landau transformation. Their first method was the variable space network method, in

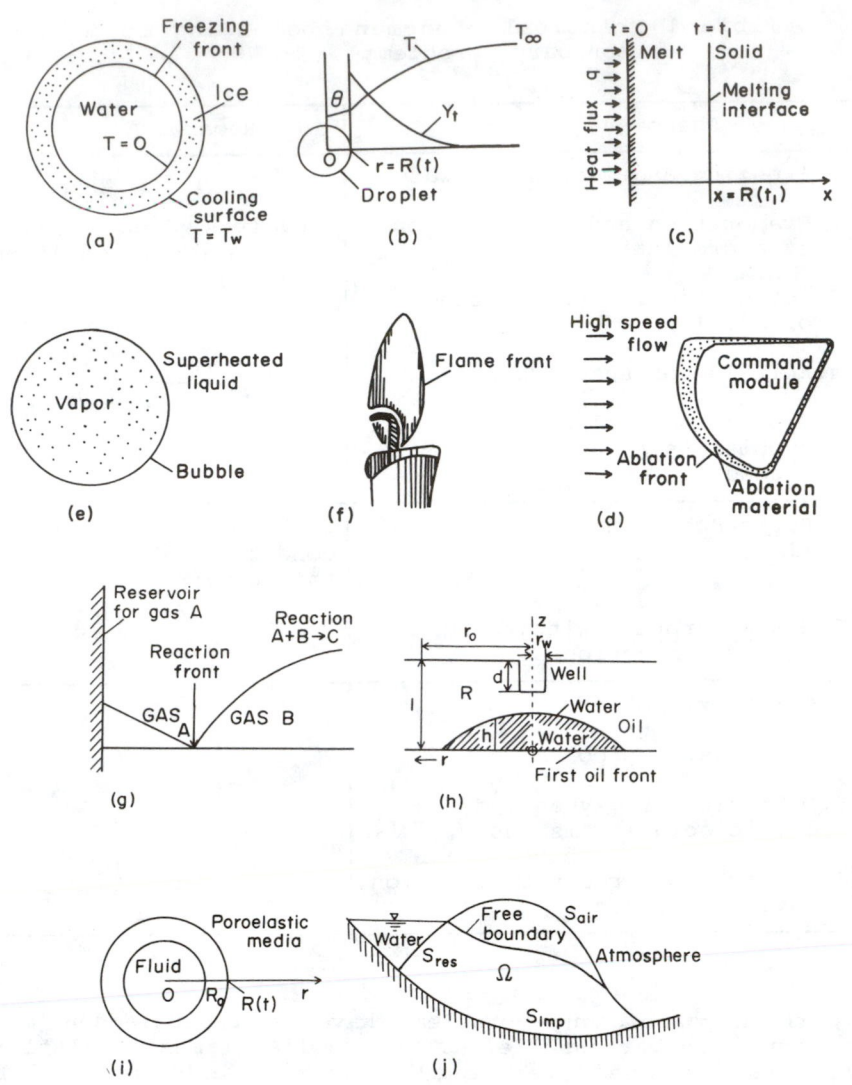

FIGURE 1   Schematic description of phenomena belonging to MBP's.

which the total number of grids in both phases is always kept constant, just like the pleats of an accordion. The second method was the fixed space network method, which utilized the grids fixed in space before the moving interface is pursued. This method has been widely used for solving one-dimensional MBP since its presentation over 15 years age.

Currently, however, the Murray-Landis method is applied only to one-dimensional problems.

In this article, special attention will be focused on

the numerical and analytical methods for solving multi-dimensional moving boundary problems, which have recently become important. With the great advancement in supercomputers in recent years, many complicated problems with multi-dimensional geometry, involving both heat conduction in the solid phase and natural convection in the liquid phase, have been successfully investigated.

One of the difficulties encountered in solving the multi-dimensional MBPs was that the phase-change problem itself is inherently transient. As a consequence, the Fourier number at complete freezing or melting reaches a high value, for example 5 - 10, thereby necessitating a tremendous iteration to finish the computation up to complete freezing or melting. Another difficulty encountered was in taking account of the multi-dimensional configuration, i.e., two- or three-dimensional shapes of both the system boundary and the moving interface. Tracking the multi-dimensional moving boundary becomes much more difficult than in a one-dimensional case. Many interesting and effective numerical and analytical methods for the multi-dimensional problem have recently been presented by many investigators. These methods may be broadly categorized into seven groups: (1) Analytical method, (2) Fixed space network method (FSNM) (3) Single-phase method (SPM) including the enthalpy method, (4) Conformal mapping method (CMM), (5) Boundary fixing method (BFM), (6) Growth ring method (GRM), and (7) Finite element method (FEM). Among others, there are also the Isotherm migration method (IMM) and the Boundary element method (BEM).

2. RECENT ANALYTICAL AND NUMERICAL METHODS FOR MULTI-DIMENSIONAL FREEZING PROBLEMS

In this section, typical numerical and analytical methods for solving multi-dimensional freezing and melting problems will be introduced. The advantages and special features of each method are described along with the disadvantages.

2.1 Analytical Method

Rathjen and Jiji[26] have extended the so-called Lightfoot method, which was presented in 1929 by Lightfoot[7], to solve the one-dimensional moving boundary problem (Neumann's problem). The solution included as the superposition of the solutions of two problems; one was the solution of an ordinary heat conduction problem with a constant initial temperature, and the other was that of the moving heat source problem.

The solution by Rathjen and Jiji is characterized by the variable $x/t^{1/2}$, $y/t^{1/2}$. Here x, y, and t mean spatial coordinates and time, respectively. In extending Lightfoot's method to the two-dimensional case, it was not necessary to assume an equality of the conductivities k and specific heats c as was the case in Lightfoot's method, but only that the ratio k/c is the same for both the solid and the liquid phases.

If $U(x,y,t)$ is a solution of an ordinary heat conduction problem and $V(x,y,t)$ is a solution of the heat conduction problem with a moving line heat source, the temperature in the solid is given as

$$T(x,y,t) = U(x,y,t) + V(x,y,t) \tag{1}$$

The former solution $U(x,y,t)$ can be easily obtained as

$$U(x,y,t) = -1 + (1 + T_i^*)\,\mathrm{erf}\left(\frac{x}{\sqrt{4\alpha t}}\right)\,\mathrm{erf}\left(\frac{y}{\sqrt{4\alpha t}}\right) \tag{2}$$

According to Rathjen and Jiji, $V(x,y,t)$ is represented by the following equation:

$$V(x,y,t) = \frac{\beta}{4\pi\alpha}\int_0^t\int_0^\infty \frac{\partial s(x',t')}{\partial t'}$$

$$\Big[G(x',t;x;t) \times G(s(x',t'),t';y;t)$$

$$+ G(s(x',t'),t';x;t) \times G(x',t';y;t)\Big]\,dx'\,\frac{dt'}{(t-t')} \tag{3}$$

Here, the function G has been defined as

$$G(x',t';x;t) = \exp\left[-\frac{(x-x')^2}{4\alpha(t-t')}\right] - \exp\left[-\frac{(x+x')^2}{4\alpha(t-t')}\right] \tag{4}$$

Rathjen and Jiji obtained the solution of the two-dimensional freezing in a corner portion of a square prism via the non-linear integro-differential equation with a singular kernel, and its solution was compared with the finite difference solution. A fairly good agreement between the two was accomplished.

Fig.2 shows a comparison of analytical and finite-difference solutions when the Stefan number, Ste, and the dimensional initial temperature, T_i, are 4.0 and 0.3, respectively.

Later, a similar approach was undertaken by Budhia and Kreith[27] for the freezing and melting problems in a wedge.

A similar method has also been presented by Boley[28] in which a simultaneous integro-differential equation was solved, using the moving front and the heat flux at the boundary as unknown variables.

Although these analytical methods are often quite useful to verify the validity of various numerical methods, it should be clearly noted that finding the solution itself is frequently very difficult because of the singularity of the kernel and because of the simultaneousness of the equations[29].

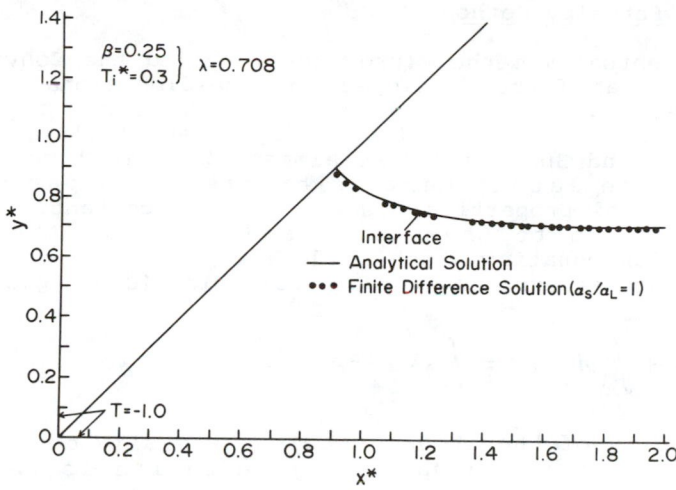

FIGURE 2   Comparison of analytical and finite-difference solutions [Rathjen and Jiji [26]].

2.2 Fixed Space Network Method

The most primitive method of investigating the multi-dimensional MBP is the so-called fixed space network method, in which grids are always fixed in space and the moving boundary is searched for. This method can be regarded as an extension of the Murray- Landis II method proposed for one-dimensional MBP's.

Springer and Olson [30] were the first to apply this method to the two-dimensional freezing and melting problem in an axisymmetrical configuration.

Later, Lazaridis [31] presented a thorough formulation for multi-dimensional MBP's. However, this method requires quite complicated handling of the moving boundary, i.e., the determination of the interface, temperature distribution including its gradient near the moving interface, and so on.

2.3 Single-Phase Method

Since the most prominent feature of the moving boundary problem is that the boundaries exist in the domain considered, which are neither fixed in space nor known a priori, an ideal method to use for this particular problem may be one in which the moving interface can be left out of consideration completely.

In this section, two typical methods, the enthalpy method and the Burke-Schumann method, both of which make use of the above concept, will be described.

2.3.1 Enthalpy Method

The enthalpy method (previously called the Conventional method) was first developed for one-dimensional MBP's by Eyres, et al.[32] and Baxter[33]. However, a thorough formulation was presented by Katayama and Hattori[34], and Shamsunder and Sparrow[35] at almost the same time. In this method, the latent heat is treated as a part of the thermophysical properties: namely, as heat content.

According to Shamsunder and Sparrow, the energy conservation equation which holds true in the domain of the phase change is expressed by the following (see Fig.3):

$$\frac{d}{dt} \int_V \rho i \, dV + \int_A \rho i \, \vec{v} \cdot \vec{dA} = \int_A k \nabla T \cdot \vec{dA} \tag{5}$$

The validity of the above equation is easily confirmed by the fact that the usual energy equation is reducible directly from the above equation.

Katayama and Hattori have introduced an apparent heat content method employing the concept of the Dilac delta function for the freezing and melting problem with a discrete temperature. They defined a temperature function for heat content as shown in Fig.4.

For example, the finite difference formulation for the one-dimensional problem is represented as

$$\int_{T_{i,j}}^{T_{i,j+1}} c\rho^* \, dT = \frac{\Delta t}{(\Delta x)^2} \left[\int_{T_{i,j}}^{T_{i-1,j}} k \, dT + \int_{T_{i,j}}^{T_{i+1,j}} k \, dT \right] \tag{6}$$

Here, the relationship among c, ρ^*, and L is given by

$$\int_{T_1}^{T_2} c\rho^* \, dT = L\rho + \int_{T_1}^{T_2} c\rho \, dT \tag{7}$$

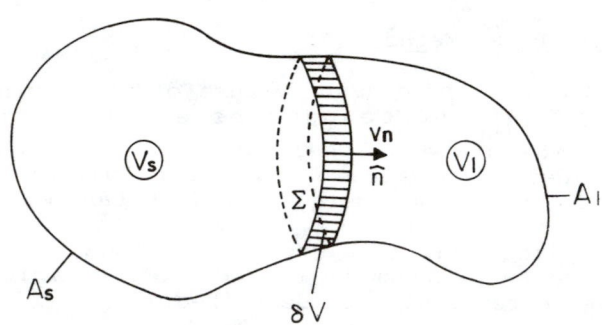

FIGURE 3[35]   Control volume for derivation of interface condition.

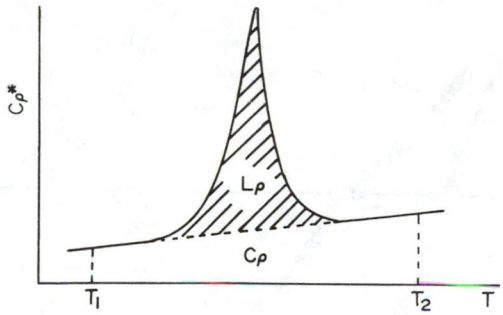

FIGURE 4   Temperature function [34].

For the problem with a discrete freezing temperature, the following equation is used.

$$\lim_{\epsilon \to 0} \int_{T_f - \epsilon}^{T_f + \epsilon} c\rho^* \, dT = L\rho \tag{8}$$

It is assumed that the latent heat is liberated in the small temperature range 2ϵ. The temperature function is then determined so that the total heat content coincides with the latent heat L.

The numerical result obtained by Katayama and Hattori for the Neumann problem is shown in Fig.5. Another result for the one-dimensional melting problem by Shamsunder is shown in Fig.6, in which discrete and dispersed models are compared. It must be noted here that there exists a peculiar fluctuation of the obtained solution. This is apparently due to the introduction of latent heat into the thermophysical properties. Owing to this numerical oscillation of both the temperature and the shape of the moving interface, this method has posed some difficulties in obtaining an accurate temperature gradient at the interface, which is often important in two-dimensional problems accompanying natural convection effects in the liquid phase. However, this method is applicable to dispersed latent heat types of freezing problems, such as fish, domestic animal meat, glucose solutions, and eutectic alloys. Further, this method is inherently able to handle the two- and three-dimensional problems.

Crowley[36] has also studied the enthalpy method for two-dimensional freezing in a prism.

2.3.2 Burke-Schumann Method

In 1928, Burke and Schumann[37] proposed a theory for a cylindrical diffusion flame, as shown in Fig.7. The essential part consists of two concentric tubes. Combustible gas moves upward through the inner tube, and comes into contact with the oxidant which flows through the outer tube.

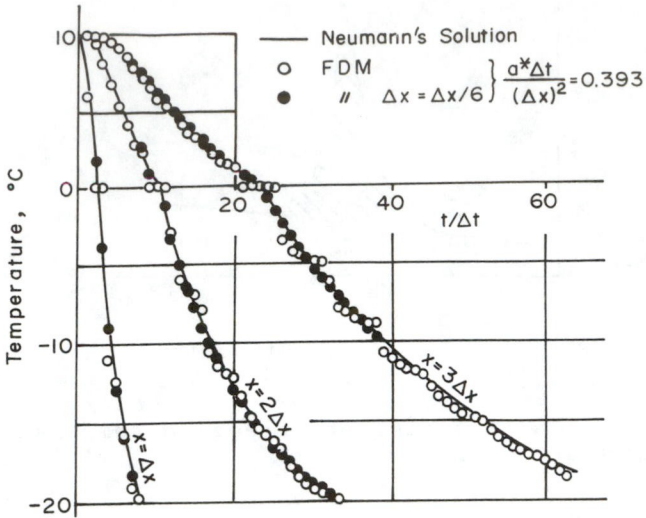

FIGURE 5   Comparison between the computed results and the Neumann's solution [Katayama and Hattori[34]].

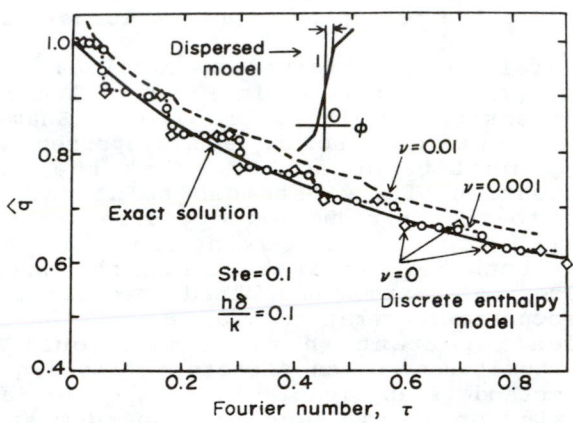

FIGURE 6   Comparison between the discrete enthalpy model and the dispersed model.

The following assumptions were made for the analysis. (1) The velocities of fuel and oxidant are the same, (2) the diffusion coefficients of the two streams are the same, (3) there is no axial diffusion, and (4) properties are constant.
The governing equations for this problem under the above assumptions can be written as follows.

$$v\frac{\partial Y_f}{\partial y} = D_f\left(\frac{\partial^2 Y_f}{\partial r^2} + \frac{1}{r}\frac{\partial Y_f}{\partial r}\right) \tag{9}$$

$$v\frac{\partial Y_0}{\partial y} = D_0\left(\frac{\partial^2 Y_0}{\partial r^2} + \frac{1}{r}\frac{\partial Y_0}{\partial r}\right) \tag{10}$$

Boundary conditions are:

$$r = 0 \quad ; \qquad \frac{\partial Y_f}{\partial r} = 0 \tag{11}$$

$$r = R(y) \quad :$$
$$\left.\begin{array}{l} Y_f = Y_0 = 0 \\ -\rho D_0\frac{\partial Y_0}{\partial r} = i\cdot\rho D_f\frac{\partial Y_f}{\partial r} \end{array}\right\} \tag{12}$$

$$r = r_2 \quad ; \qquad \frac{\partial Y_0}{\partial r} = 0 \tag{13}$$

$$t = 0 \quad ; \qquad \left.\begin{array}{ll} Y_f = Y_{fi} & 0 \le r \le r_1 \\ Y_0 = Y_{0i} & r_1 < r \le r_2 \end{array}\right\} \tag{14}$$

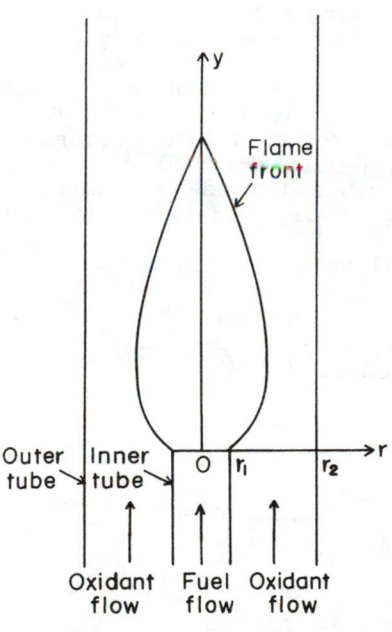

FIGURE 7   Burke–Schumann diffusion flame.

Equation (12) indicates the joint condition at the moving interface. Of these, the latter corresponds to the heat balance equation for the ordinary phase change problem, but note that there is no latent heat term, including latent heat L. The lack of a latent heat term makes it possible to adopt a different approach to this diffusion flame problem.

Burke-Schumann employed negative oxygen concentration, - Y_o/i, so that the concentration gradient of the oxygen was equal to that of the fuel at the flame front. Finally, the new concentration was defined as

$$Y = Y_f - Y_0 / i \tag{15}$$

Thus, the problem reduced to a one-phase problem only for Y. The moving interface R(y) could be obtained by setting Y=0 in the concentration equation.

$$\sum \frac{1}{\mu} \frac{J_1(\mu r_1) J_0(\mu R(y))}{\{J_0(\mu r_2)\}^2} \exp\left(-\frac{D \mu^2 R(y)}{v}\right) = \frac{r_2^2 Y_{0i}}{2 r_1 i Y_i} - \frac{r_1}{2} \tag{16}$$

Here, J_0 and J_1 represent Bessel function of the first kind, and μ were the positive root of equation $J_1(\mu r_2)=0$. An extension of their method to a phase change problem with a latent heat term has been done by Saitoh [4],[5] (Equi-gradient Matching Method).

2.4 Conformal Mapping Method

A conformal mapping method was presented by Kroeger and Ostrach[38] for the continuous casting problem for metals and alloys.
A typical continuous casting problem is schematically shown in Fig.8. Liquid metal, usually 50-100 K above its freezing temperature, was introduced continuously at the top of a bottomless mold. They clarified the various qualitative aspects of the combined forced and natural convection flow field in the liquid pool, together with heat conduction in the solid.

Non-dimensional governing equations are as follows:

$$\left(\frac{Re}{\sqrt{Gr}} + \frac{\partial \psi}{\partial y}\right) \frac{\partial \Omega}{\partial x} - \frac{\partial \psi}{\partial x} \frac{\partial \Omega}{\partial y} = \frac{\partial \Theta}{\partial y} + \frac{1}{\sqrt{Gr}} \nabla^2 \Omega \tag{17}$$

$$\nabla^2 \psi = -\Omega \tag{18}$$

$$\left(\frac{Re}{\sqrt{Gr}} + \frac{\partial \psi}{\partial y}\right) \frac{\partial \Theta}{\partial x} - \frac{\partial \psi}{\partial x} \frac{\partial \Theta}{\partial y} = \frac{1}{P_r \sqrt{Gr}} \nabla^2 \Theta \tag{19}$$

Here,

$$u = \frac{\partial \psi}{\partial y} \quad ; \quad v = -\frac{\partial \psi}{\partial x} \qquad (20) \qquad \text{and} \qquad \Omega = \frac{\partial v}{\partial x} - \frac{\partial u}{\partial y} = -\nabla^2 \psi \qquad (21)$$

Boundary conditions are:

$$\left.\begin{array}{l} x = 0; \quad 0 < y < 1 \quad : \quad \psi = \dfrac{\partial \psi}{\partial x} = 0; \quad \Theta = 1 \\[2ex] 0 < x < x_p; \quad y = 0: \quad \psi = \dfrac{\partial^2 \psi}{\partial y^2} = 0; \quad \dfrac{\partial \Theta}{\partial y} = 0 \\[2ex] \qquad\qquad y = \hat{y}(x): \quad \psi = \dfrac{\partial \psi}{\partial n} = 0; \quad \Theta = 0. \end{array}\right\} \qquad (22)$$

Solid region:

$$\frac{\partial \theta}{\partial x} = \frac{1}{P}\left(\frac{\partial^2 \theta}{\partial x^2} + \frac{\partial^2 \theta}{\partial y^2}\right) \qquad (23)$$

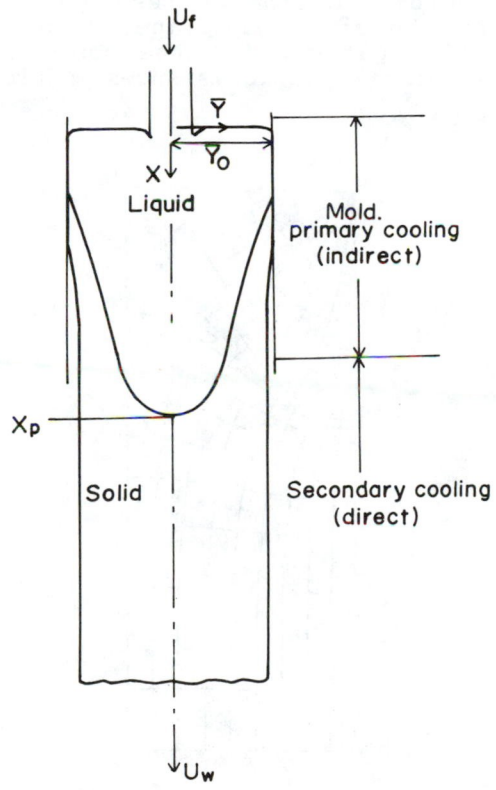

FIGURE 8   A schematic of continuous casting[38] .

145

with boundary conditions:

$$\left.\begin{array}{l} y = \hat{y}(x): \quad \theta = 1 \\ 0 < x < \infty; \quad y = 1 \quad : \quad \theta = 0 \\ x_p < x < \infty; \quad y = 0: \quad \dfrac{\partial \theta}{\partial y} = 0 \\ x \to \infty; \quad 0 < y < 1 \; ; \quad \theta = 0. \end{array}\right\} \qquad (24)$$

A mapping procedure is only briefly described here. The complete slab of the liquid and solid domain can be transformed into the half plane u>0 using conformal mapping

$$\omega = \sinh \frac{\pi z}{2} = u + iv \qquad (25)$$

The moving interface C is mapped into contour C'. Then, Theodorsen's integral equation is used to perform the conformal mapping of the curve C' in w-plane and the unit circle C''.

In Fig.9, coordinate line correspondence between the mapped plane and the physical plane is designated. An example of the flow pattern for the combined forced and natural convection in the liquid is shown in Fig.10.

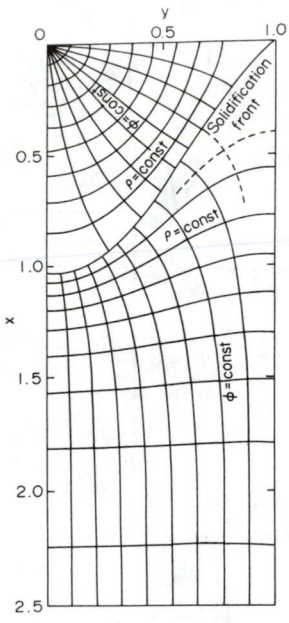

FIGURE 9   Coordinate lines for mapped and physical planes [38]

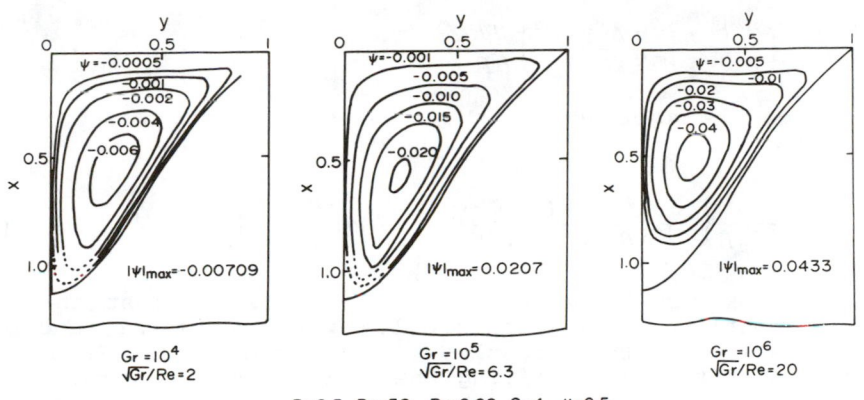

FIGURE 10　　Natural convection streamlines for combined forced and natural convection in the liquid phase [38].

Although the conformal mapping method is quite an interesting one, capable of being applied to problems dealing with arbitrary shape, the use of such a transformation also adds considerable complexity to the solution procedures; namely, it requires the solution of additional programs for mapping. To date, an extension to the problem involving a non-steady moving boundary has been unresolved.

2.5 Isotherm Migration Method(IMM)

This method was first presented by Rose[39], Chernous'ko[40], and Dix and Cizek[41] for the one-dimensional problem. Later, an extension to the two-dimensional problem was presented by Crank and Gupta[42], and Crank and Crowley[43]. Most existing methods for the phase change problem utilize the so-called Eulerian formulation, in which temperature and spatial distance are taken as dependent and independent variables, respectively. By contrast, IMM is based on the Lagrangian formulation, with temperature being an independent variable and distance being a dependent variable. Since Gibbs' free energy is maintained at a constant during the phase change process, and since temperature is often kept constant on the boundary of the system and on the moving interface, it may be quite effective to adopt temperature as an independent variable.

The heat conduction equation is represented in Cartesian coordinates as

$$\frac{\partial T}{\partial t} = \frac{\partial^2 T}{\partial x^2} + \frac{\partial^2 T}{\partial y^2} \tag{26}$$

Now, changing the dependent variable T(x,y,t) into y(T,x,t), the above equation is transformed into

$$\frac{\partial y}{\partial t} = -\left\{\frac{\partial^2 T}{\partial x^2} - \frac{\partial^2 y}{\partial T^2}\left(\frac{\partial y}{\partial T}\right)^{-2}\right\}\frac{\partial y}{\partial T} \tag{27}$$

The heat balance equation at the moving interface is also changed:

$$\frac{\partial y}{\partial t} = \frac{1}{\rho C_p L}\left\{1 + \left(\frac{\partial y}{\partial x}\right)^2\right\}\left[k_s\left(\frac{\partial y}{\partial t}\right)^{-1}\bigg|_{-} - k_l\left(\frac{\partial y}{\partial T}\right)^{-1}\bigg|_{+}\right] \tag{28}$$

Crank and Gupta have applied the IMM against the two-dimensional solidification problem of molten metal, and made a comparison of their results with those of Allen, Severn and Lazaridis, thereby verifying the accuracy of their results.

2.6 Boundary Fixing Method (BFM)

An application of the coordinate transformation method to the multi-dimensional MBP was proposed by Saitoh [29], [44], [45]. This method can be regarded as an extension of the Landau transformation method [12] to the multi-dimensional case, which was presented qualitatively in his paper of 1950 for the one-dimensional problem. The most remarkable feature of the multi-dimensional MBP is that the moving interface exists in an arbitrarily shaped domain. Therefore, the general numerical method should be one which can consider the arbitrariness of the shapes of both the moving front and the domain boundary.

The boundary fixing method considers the arbitrary geometry of both the moving interface and the domain boundary via a change of an independent variable. BFM was also studied by Duda et al.[46], almost concurrently with the author's study.

2.6.1 Formulation of the Problem

For the purpose of illustration, governing equations are presented for the two-dimensional freezing problem, since the extension to three-dimensions is accomplished in a similar manner. A schematic description and coordinate system for two-dimensional freezing in an arbitrary domain are shown in Fig.11. The liquid is initially contained in a domain surrounded by a boundary represented by the function $B^+(\phi^+)$. The temperature at the boundary is given arbitrarily by the function $T_w(\phi^+, t^+)$. At the time $t^+ = t^+$, the corresponding freezing front position is assumed to be at $r^+ = F^+(\phi^+, t^+)$. The problem is then to find the freezing front $F^+(\phi^+, t^+)$ and the temperature distribution $T^+(r^+, \phi^+, t^+)$ in the solid.

For simplification, it was postulated that the temperature of the liquid is equal to the freezing point, and that the properties are constant.

The basic equation and the boundary conditions for this problem under the above assumptions are described below, using the cylindrical polar coordinate system.

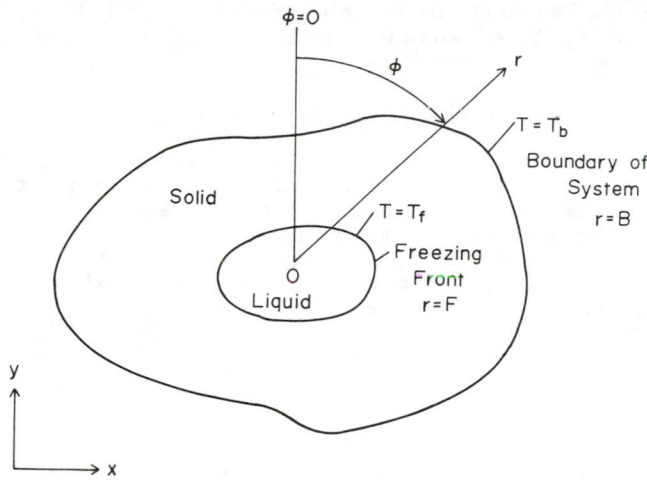

FIGURE 11 Model for the two-dimensional freezing and coordinate system.

$$\frac{\partial T}{\partial t} = \nabla^2 T \tag{29}$$

Here,

$$\nabla^2 \equiv \frac{1}{r}\frac{\partial}{\partial r}\left(r\frac{\partial}{\partial r}\right) + \frac{1}{r^2}\frac{\partial^2}{\partial \phi^2} \tag{30}$$

Boundary conditions:

$$\phi = 0 \quad \text{and} \quad 1 \quad ; \quad \frac{\partial T}{\partial \phi} = 0 \tag{31}$$

$$r = B(\phi) \quad ; \quad T = T_\omega(\phi, t) \tag{32}$$

$$r = F(\phi, t) \quad ; \quad T = 0 \tag{33}$$

$$\frac{\partial F}{\partial t} = Ste\left[1 + \left(\frac{1}{\phi_0 F}\frac{\partial F}{\partial \phi}\right)^2\right]\frac{\partial T}{\partial r} \tag{34}$$

The governing equations have been normalized by virtue of the following dimensionless variables.

$$r = \frac{r^+}{D}, \quad t = \frac{at^+}{D^2}, \quad \phi = \frac{\phi^+}{\phi_0}, \quad T = \frac{T^+}{T_0^+},$$

$$F = \frac{F^+}{D}, \quad B = \frac{B^+}{D}, \quad T_W = \frac{T_W^+}{T_0^+} \tag{35}$$

In the above equations, the joint condition (33) is transformed in the radial direction(e.g. see Rathjen and Jiji [26]).

2.6.2 Solution by the Boundary Fixing Method

The principal difficulties in the analysis of the multi-dimensional freezing problem arise from the fact that the position of the moving boundary is not known a priori, and that its shape is multi-dimensional.

In order to overcome these difficulties, the next independent variable which takes a constant value at both the moving and the fixed boundaries is adopted.

$$\eta = \frac{r - F(\phi,t)}{B(\phi) - F(\phi,t)} \tag{36}$$

The governing equation is transformed by this variable into:

$$\frac{\partial T}{\partial t} = \left[\frac{1}{(B-F)^2} + \frac{1}{S}\left(\frac{\partial \eta}{\partial \phi}\right)^2\right]\frac{\partial^2 T}{\partial \eta^2}$$
$$+ \left[\frac{1}{r(B-F)} - \frac{\partial \eta}{\partial t} + \frac{1}{S}\frac{\partial^2 \eta}{\partial \phi^2}\right]\frac{\partial T}{\partial \eta} + \frac{2}{S}\frac{\partial \eta}{\partial \phi}\frac{\partial^2 T}{\partial \phi \partial \eta} + \frac{1}{S}\frac{\partial^2 T}{\partial \phi^2} \tag{37}$$

Boundary conditions:

$$\phi = 0 \quad \text{and} \quad 1 \; ; \quad \frac{\partial T}{\partial \phi} = 0 \tag{38}$$

$$\eta = 0 \; ; \quad T = 0 \tag{39}$$

$$\frac{\partial F}{\partial t} = Ste\frac{\partial T}{\partial \eta}\frac{1}{B-F}\left[1 + \left(\frac{1}{\phi_0 F}\frac{\partial F}{\partial \phi}\right)^2\right] \tag{40}$$

$$\eta = 1 \; ; \quad T = T_W(\phi,t) \tag{41}$$

Here, $r = \eta(B-F) + F$, $\quad S = (\phi_0 r)^2$.

In the above equations, $\partial \eta/\partial t$, $\partial \eta/\partial \phi$, **and** $\partial^2 \eta/\partial \phi^2$ are given by the following equations:

$$\frac{\partial \eta}{\partial t} = -\frac{\partial F}{\partial t}\frac{1-\eta}{B-F}$$
$$\frac{\partial \eta}{\partial \phi} = -\frac{1}{B-F}\left[\frac{\partial F}{\partial \phi} + \eta\left(\frac{dB}{d\phi} - \frac{\partial F}{\partial \phi}\right)\right]$$
$$\frac{\partial^2 \eta}{\partial \phi^2} = -\frac{1}{B-F}\left[\frac{\partial^2 F}{\partial \phi^2} + 2\frac{\partial \eta}{\partial \phi}\left(\frac{dB}{d\phi} - \frac{\partial F}{\partial \phi}\right) + \eta\left(\frac{d^2 B}{d\phi^2} - \frac{\partial^2 F}{\partial \phi^2}\right)\right] \tag{42}$$

These functions can be readily obtained only if the moving and fixed boundaries are prescribed.

2.6.3 Computed Results

Computed results for an ellipse with a short diameter of 0.08 m, an aspect ratio of 3/2, and a cooling rate of 3.333 x 10^{-3} K/s is shown in Fig. 12. The cooling rate C is defined as

$$C_1 = -\frac{T_W(t)}{t} \tag{43}$$

In this case, the temperature at the fixed boundary decreases linearly with time. The initial freezing front position, F_0 can be calculated by

$$F_0(\phi, t_0) = B(\phi) - \delta \left[1 + \left(\frac{1}{B} \frac{dB}{d\phi} \right)^2 \right]^{1/2} \tag{44}$$

where, δ designates the initial thickness of the frozen layer. The boundary function $B(\phi)$ is expressed by the following equation:

$$B(\phi) = \{1 - 5/9 \sin^2 \phi\}^{-1/2} \tag{45}$$

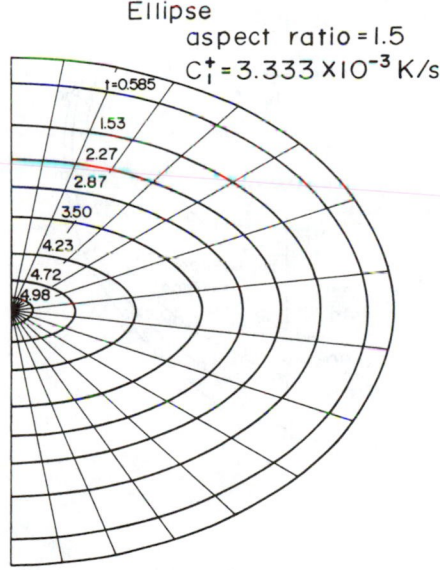

FIGURE 12   Computed results via BFM for an ellipse.

It is assumed that the temperature distribution within the initial frozen layer is linear.

Next, as an example of the case in which natural convection in the liquid is combined with heat conduction in the solid, the two-dimensional freezing around a horizontal circular cylinder in quiescent water was solved by the BFM. For the analysis, it was assumed that (1) Flow is laminar, (2) Properties are constant, (3) Boussinesq approximation is valid.

For brevity, only computed results will be shown. Fig.13 shows a comparison between numerical and experimental freezing front contours at different times. The water temperature is 6.9°C, the diameter of the cylinder is 0.04 m, and the cylinder surface temperature is -12.4°C. Experimental data obtained by Saitoh and Hirose [47] are shown by the circle. The agreement between the two is excellent except near the vicinity of the downstream edge, where separation may occur in the real flow. This was not considered in the analysis.

As shown in the above, the boundary fixing method is a very simple tool for solving multi-dimensional MBP's. For this reason, this method has been widely used to date.[48-63]

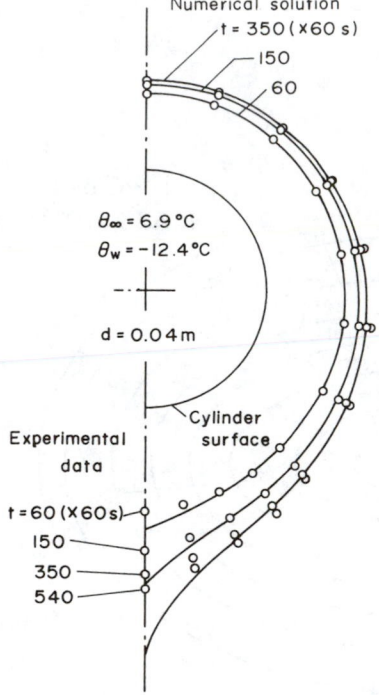

FIGURE 13   Comparison between the BFM results and the experimental results.   The solid lines indicate computed results and the circles the experimental results.

However, even this method and other similar methods using coordinate transformation have at least two major disadvantages. First of all, the moving and fixed boundaries should be smooth, i.e., the derivatives of shape functions should exist up to the second order. Secondly, this method fails to be applied to so-called multiply-connected domains shown in Fig.14.

2.7 Growth Ring Method (GRM)

Although two well-used methods, i.e.,the Boundary fixing method and the Enthalpy method, have many advantages over other existing methods, these methods still have severe disadvantages. Namely, BFM largely depends on the smoothness of the boundary shape function, and with the Enthalpy method it is impossible to include a natural convection effect in the liquid phase.

An alternative method was recently developed by the author [64], [65]. This method is based on the multi-lateral element (MEM) method, in which the representative point is placed on the barycenter of a triangular or quadrilateral element. Note that it is not the circumcenter but the barycenter where the representative point is positioned. The existing triangular element method (TEM) has used the circumcenter. The barycenter method makes it possible to consider the arbitrary geometry of both the fixed and the moving boundaries, thereby eliminating difficulties encountered with, for example, BFM. Another advantage of this method is that the usual explicit finite difference can be incorporated in the formulation. This considerably reduces the troublesome numerical procedures that are necessary prior to computation. New freezing front positions are calculated by using temperature distribution in the vicinity of the freezing front (see Fig.15). Then, the adjacent element is enlarged by the same amount as the newly frozen area thus obtained. This procedure is continued in a step-by-step manner until the size of the frozen area reaches nearly the same element size as that of the old element, as shown in Fig.16.
Since the freezing front produced in this method closely resembles the growth rings of trees, it is called the "Growth Ring Method (GRM)".

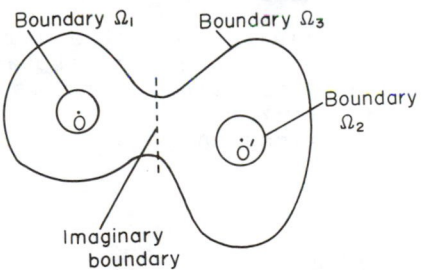

FIGURE 14   An example of doubly-connected region.

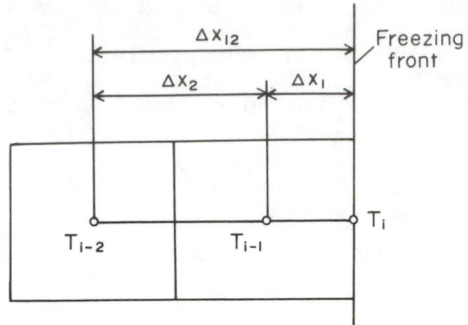

FIGURE 15   Grid points near moving interface.

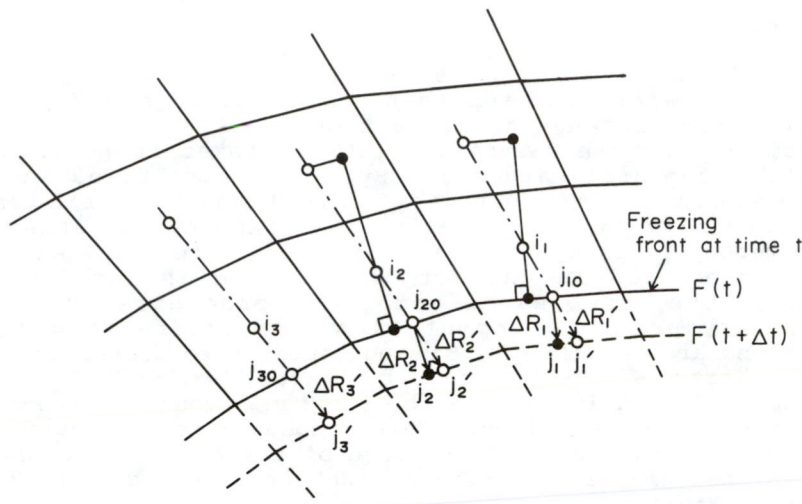

FIGURE 16   Handling the freezing front.

2.7.1 Governing Equations

For the sake of illustration, the formulation will be shown for two-dimensional freezing in a square prism in which water initially at the freezing point is contained (two-dimensional Stefan problem).

The governing equations and the boundary conditions are listed below.

$$\frac{\partial T}{\partial t} = \nabla^2 T \tag{46}$$

$$\text{on} \quad B \quad : \quad T = T_W(t) \tag{47}$$

on $\quad F \quad : \quad T = 0$ $\qquad$ (48)

$$\rho L\mathbf{V} \cdot \mathbf{n} = k\frac{\partial T}{\partial n}$$ (49)

$$\nabla^2 \equiv \frac{\partial^2}{\partial x^2} + \frac{\partial^2}{\partial y^2}$$ (50)

2.7.2 Formulation by Multilateral Element Method

By integrating the control volume shown in Fig.17, and by applying Gauss' theorem and Green's theorem, the above equation (46) is rewritten as

$$\iiint_V \left(\frac{\partial T}{\partial t} - \frac{\partial^2 T}{\partial x^2} - \frac{\partial^2 T}{\partial y^2}\right) dV$$ (51)

$$= \frac{\partial T}{\partial t}dV - \iint_S \frac{\partial T}{\partial n}dS = 0$$ (52)

The first attempt to solve the above equation by employing triangular elements was done by MacNeal[66],[67], as early as 1949. This method has been called the triangular element finite difference method (TEM), and has been used by many researchers[68] - [75]. In this method, the representative point is chosen at the circumcenter of the element so that the line connecting the circumcenters is normal to the side of the element. However, the triangular element method has the following disadvantages over other existing methods such as FEM and BEM.

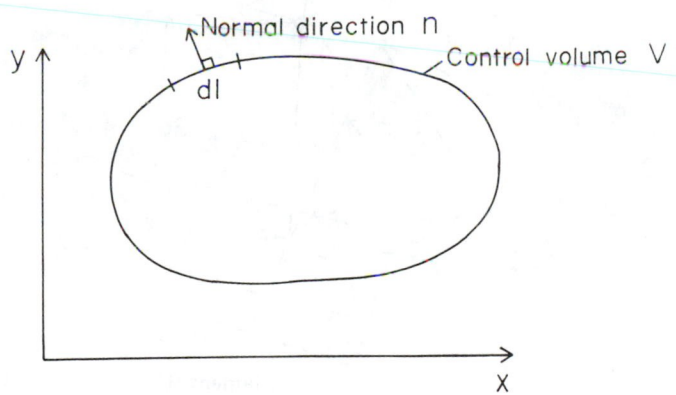

FIGURE 17   Control volume and normal direction.

(i) The triangle should be an acute triangle if the representative point comes inside the element.

(ii) Extension to three-dimensions is impossible.

(iii) Application to the quadrilateral element is not easy.

For the above reasons, at this point we must use the barycenter TEM, in which the barycenter is selected as a representative point.

The finite difference expression of equation (46) based on the barycenter method can be written as follows (see Fig.18 for quadrilateral element used).

$$\frac{T_i^{(n+1)} - T_i^{(n)}}{\Delta t} = \sum_{j=1}^{4} \frac{T_j^{(n)} - T_i^{(n)}}{dx_j \sin \sigma_j} \frac{dl_j}{dS_i} \tag{53}$$

Solving with respect to $u_i^{(n+1)}$, it follows that

$$T_i^{(n+1)} = T_i^{(n)} + \Delta t \sum_{j=1}^{4} C_{ij}(T_j^{(n)} - T_i^{(n)}) \tag{54}$$

Here,

$$C_{ij} = \frac{dl_j}{dS_i \, dx_j \, \sin \sigma_j} \tag{55}$$

The stability limit is

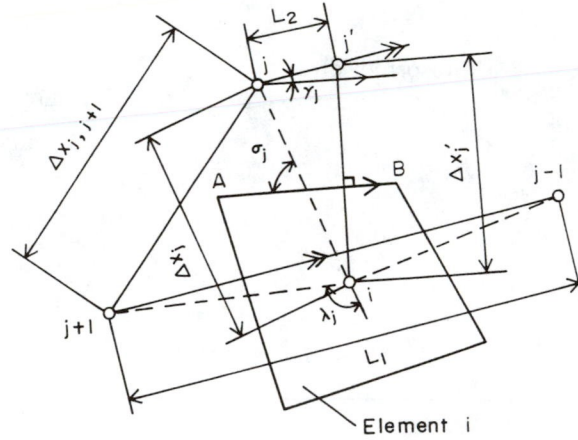

FIGURE 18   Quadrilateral element and temperature correction.

$$\Delta t \leq \left[\sum_{j=1}^{4} C_{ij} \right]^{-1} \tag{56}$$

A similar formulation is possible for the usual triangular element shown in Fig.19.

2.7.2.1 Temperature Correction

If the barycenter were used as a representative point of the element, the temperature gradient between the adjacent two points (shown in Fig.18) should be corrected, since the line connecting the two points is usually not normal to the side AB.

One of the correction methods is to use Taylor's expansion, and as a result the corrected temperature T_j is expressed as,

$$T'_j = T_i + (T_j - T_i)\left(\sin^2 \sigma - \frac{\sin \sigma \cos \sigma}{\tan \lambda} \right) - (T_{j+1} - T_i)\frac{dx_j \sin \sigma \cos \sigma}{d\lambda_{j+1} \sin \lambda} \tag{57}$$

The temperature at the (n+1)th time level is then given by

$$T_i^{(n+1)} = T_i^{(n)} + \Delta t \sum_{j=1}^{4} C'_{ij}(T_j^{'(n)} - T_i^{(n)}) \tag{58}$$

Here, $C_{ij}' = C_{ij}$.

2.7.3 Handling the Moving Boundary

The most important procedure for this method is the handling of the moving interface. Given that the freezing front at time t is line A_0B_0 as shown in Fig.19, on which

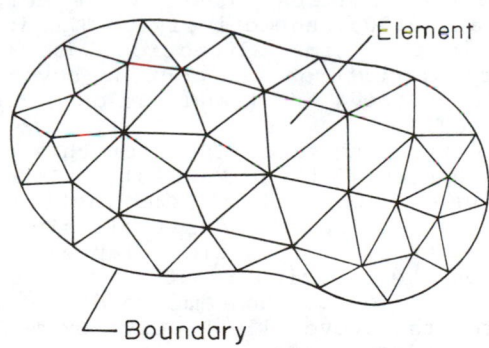

FIGURE 19   Grid generation using triangular elements.

temperature is constant (i.e., freezing point), the freezing front increment dR during time   t is determined in the following manner.

First, an imaginary point j' is defined as an intersection of line jj', which is parallel with line C_0D_0 and line ij', and perpendicular to line A_0B_0. The temperature $T_{j'}$ at point j' is obtained in the same way as explained in section 2.7.2.1.

The freezing front increment dR is then calculated using

$$dR = Ste\frac{\partial T}{\partial x}\Delta t \tag{59}$$

Here,

$$\frac{\partial T}{\partial x}\bigg|_i = \frac{h_1 + h_{12}}{h_1 h_{12}}T_i - \frac{h_{12}}{h_1 h_{12}}T_{i-1} + \frac{h_1}{h_2 h_{12}}T_{i-2} \tag{60}$$

At the subsequent time level, i.e., $t=t_0 + 2$ t, the procedure is repeated, except that point i' on the boundary of the element i is moved to i'' on the new moving front A_1B_1. Note that the new boundary condition $T=T_f$ is imposed on point i'' without changing the size of element i. Timewise iteration is continued until the thickness of the frozen layer is approximately equal to the size of a typical element.

2.7.4 Numerical Example

As an example, the numerical results for the two-dimensional freezing of water in a regular prism initially at its fusion temperature, has been taken up as a standard 2-D freezing problem [36], [45], [76].

The computed results by the GRM are shown in Fig.20, in which freezing front contours   and generated elements (quadrilateral element in this case) are simultaneously designated. The side length of the regular square, Stefan number Ste, and the time step used in the computation, were 0.079 m, 0.49, and 0.003,respectively. The initial freezing front position in the normal direction was set to be 0.92. Therefore, every circumferential contour beyond this initial line denotes the freezing front and the quadrilateral element the mesh generated.

The computer running time (CPU) for this case was some 100s on NEAC 2200/ACOS 1000(30 MFLOPS) at Tohoku University. Note that the use of the SOR scheme instead of the usual explicit scheme is particularly desirable since the mesh size in this example gets smaller as time elapses.

Fig.21 shows the freezing front histories in the normal and diagonal directions for the same condition as in Fig.20.

As seen from the above illustrative examples, the Growth Ring Method is a quite simple,   but very powerful method applicable to arbitrary multi-dimensional domains. The GRM can also handle problems involving natural convection in the liquid phase.

It is advantageous that the mesh may **be** regenerated halfway through the computation in order to reduce the number of elements and resultant computer time.

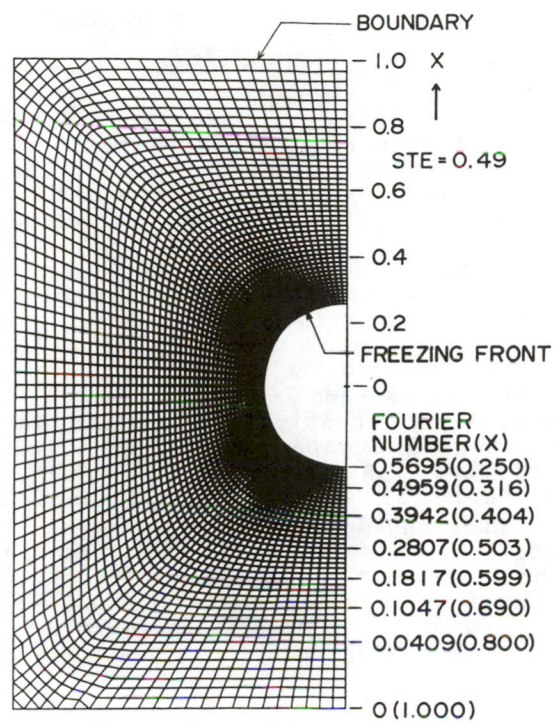

FIGURE 20   Numerical solution via the GRM for two-dimensional Stefan problem.

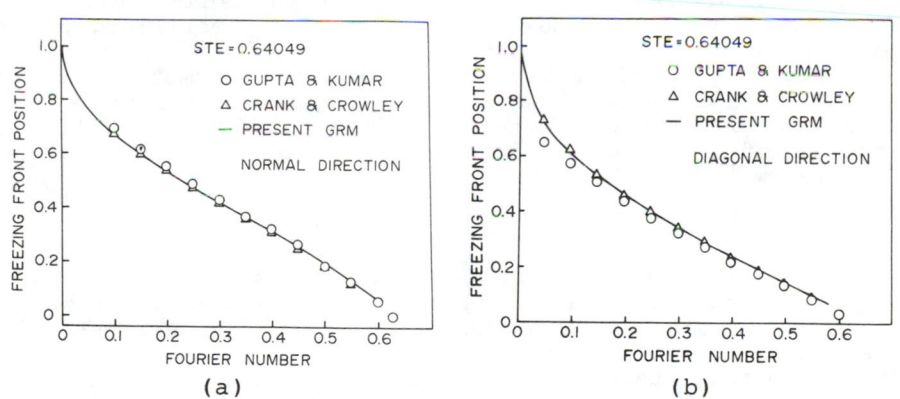

FIGURE 21   Time variation of freezing front position in the normal, (a) and the diagonal(b) directions.

3. APPLICATION TO LATENT HEAT THERMAL ENERGY STORAGE

The need to develop efficient, cost-effective thermal energy storage (TES) subsystems has been raised in recent years in many industrial countries like the United States, Canada, Japan and European countries, covering such areas as electric utilities, and residential/commercial and industrial applications.

The purposes of installing a TES system are summarized as follows:

(i) High energy density — Most natural energy sources including solar, thermal, wind, geothermal, oceanic, etc., or industrial waste heat, have features of diluteness and intermittency. Therefore, in order to use these energies efficiently, it is very important to collect and store heat before utilization.

(ii) Load leveling — Energy storage devices could provide a means of meeting peak demands for electrical generation without introducing superfluous facilities. The energy is stored during off-peak hours and distributed during periods when more energy is needed.

(iii) Efficient utilization — Energy storage devices could reduce the capacity of heating and cooling facilities, and also reduce the rating capacity of electrical equipment. Increasing the energy storage density and reducing the capacity of facilities will make it possible to use residential or building space more effectively.

In the following section, the heat transfer and design aspects in the LHTES system will be mentioned. Because of space restriction, particular attention will be focused on (i) natural convection (buoyancy) effects on phase-change heat transfer, and (ii) simulation techniques for designing the LHTES unit.

3.1 Natural Convection Effects on Phase-Change Heat Transfer

The effects of natural convection have been found to be of the first order in many melting and solidification problems which may appear in engineering, metallurgy, and geophysics.

The effects of natural convection on the motion of the solid-liquid interface have been reviewed by Viskanta[51]. It is not our purpose here to completely describe all the aspects of the natural convection effect, so only a few examples will be introduced.

3.1.1 Melting Around an Embedded Heat Source

The time variation of the size and shape of the melt region around a horizontal cylinder embedded in solid paraffin was clarified by White, Bathelt, Leidenfrost, and Viskanta[79].

In the initial stages, heat conduction is a dominant mode of heat transfer, and the melting front is nearly concentric. However, as natural convection develops, its shape becomes asymmetrical with the axis of the cylinder. Then, a strong thermal plume conveys hot liquid to the upper liquid region, and melting occurs mainly in the upper portion. Fig.22 depicts experimentally obtained melt contours from a uniformly heated cylinder [80].

3.1.2 Comparison of Experiment and Numerical Computation by BFM

Numerical analysis via BFM was performed for the above mentioned heat transfer melting problem with the natural convection effect [81], [82].
A schematic diagram and coordinate system are shown in Fig.23.
The governing equations using an independent variable

$$\eta = \frac{r - B(\phi)}{F(\phi,t) - B(\phi)} \tag{61}$$

are described as follows:

$$\frac{\partial \Omega}{\partial t} + \frac{1}{\pi r} \frac{\partial \eta}{\partial r} \frac{\partial(\psi, \Omega)}{\partial(\eta, \phi)} = Pr \nabla^2 \Omega - \frac{\partial \Omega}{\partial \eta} \frac{\partial \eta}{\partial t}$$
$$- Pr \cdot Ra \left[\sin \pi\phi \frac{\partial T}{\partial \eta} \frac{\partial \eta}{\partial r} + \frac{\cos \pi\phi}{\pi r} \left(\frac{\partial T}{\partial \eta} \frac{\partial \eta}{\partial \phi} + \frac{\partial T}{\partial \phi} \right) \right] \tag{62}$$

$$\frac{\partial T}{\partial t} + \frac{1}{\pi r} \frac{\partial \eta}{\partial r} \frac{\partial(\psi, T)}{\partial(\eta, \phi)} = \nabla^2 T - \frac{\partial T}{\partial \eta} \frac{\partial \eta}{\partial t} \tag{63}$$

$$\Omega = \nabla^2 \psi \tag{64}$$

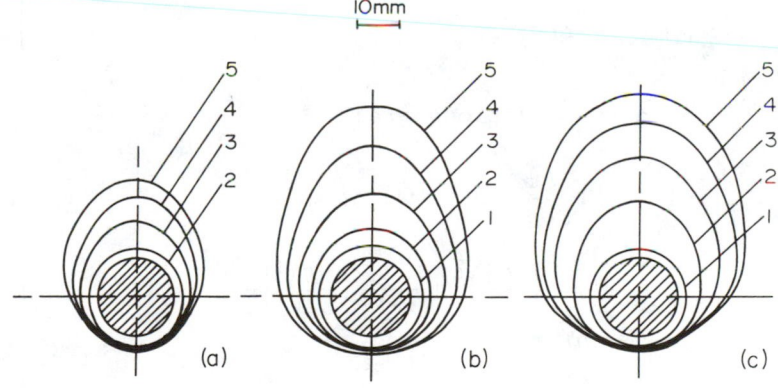

FIGURE 22 Experimentally determined melt contours during melting from a uniformly heated cylinder [79]. (a) Ste=0.461, (b) Ste=0.775, and (c) Ste=0.996.

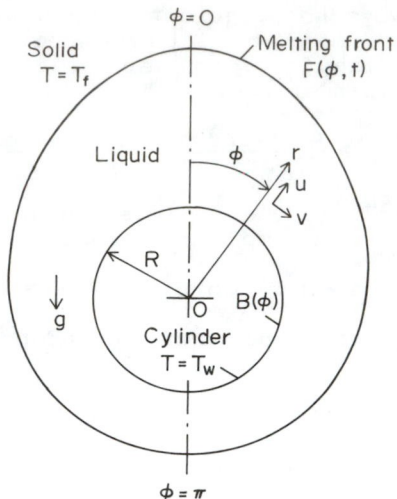

FIGURE 23   Schematic diagram and coordinate system of natural convection melting around a horizontal cylinder.

where,

$$\nabla^2 \equiv \left[\left(\frac{\partial \eta}{\partial r}\right)^2 + \frac{1}{\pi^2 r^2}\left(\frac{\partial \eta}{\partial \phi}\right)^2\right]\frac{\partial^2}{\partial \eta^2} + \left(\frac{1}{r}\frac{\partial \eta}{\partial r} + \frac{1}{\pi^2 r^2}\frac{\partial^2 \eta}{\partial \phi^2}\right)\frac{\partial}{\partial \eta}$$

$$+ \frac{2}{\pi^2 r^2}\frac{\partial \eta}{\partial \phi}\frac{\partial^2}{\partial \eta \partial \phi} + \frac{1}{\pi^2 r^2}\frac{\partial^2}{\partial \phi^2}$$

$$\frac{\partial \eta}{\partial r} = \frac{1}{F-B}, \quad \frac{\partial \eta}{\partial \phi} = \frac{-1}{F-B}\{B_\phi + \eta(F_\phi - B_\phi)\}, \quad r = \eta(F-B) + B$$

$$\frac{\partial^2 \eta}{\partial \phi^2} = \frac{-1}{F-B}\{B_{\phi\phi} + 2\frac{\partial \eta}{\partial \phi}(F_\phi - B_\phi) + \eta(F_{\phi\phi} - B_{\phi\phi})\}$$

$$\left.\right\} \quad (65)$$

Boundary and joint conditions are;

$$\eta = 0; \qquad T = 1, \qquad \psi = \frac{\partial \psi}{\partial \eta} = 0, \qquad \Omega = \Omega_\omega|_{\eta=0},$$

$$\Omega_\omega = \left[\left(\frac{\partial \eta}{\partial r}\right)^2 + \frac{1}{\pi^2 F^2}\left(\frac{\partial \eta}{\partial \phi}\right)^2\right]\frac{\partial^2 \psi}{\partial \eta^2}$$

$$\eta = 1; \qquad \psi = \frac{\partial \psi}{\partial \eta} = 0, \quad \Omega = \Omega_\omega|_{\eta=1}$$

$$\phi = 0, 1; \qquad \left.\frac{\partial T}{\partial \phi}\right|_+ = \left.\frac{\partial T}{\partial \phi}\right|_-, \quad \psi = \Omega = 0$$

$$\left.\right\} \quad (66)$$

$$\eta = 1; \quad T = 0, \quad \frac{\partial F}{\partial t} = -Ste\left[1 + \left(\frac{F_\phi}{\pi F}\right)^2\right]\frac{1}{F - B}\frac{\partial T}{\partial \eta}\bigg|_{\eta=1} \tag{67}$$

A comparison of the numerical results with the experimental results by Bathelt and Viskanta[80] is shown in Fig.24. The values of the Prandtl number, Rayleigh number, and Stefan number used are Pr=51.1, Ra=75300, and Ste=0.099, which correspond to the case of d=1.59 cm, and T_w. Comparing the numerical results with the experimental ones, it is seen that a high level agreement is achieved both qualitatively and quantitatively.

The readers who have further interest in this subject are encouraged to examine references[81]-[98],[121]-[123].

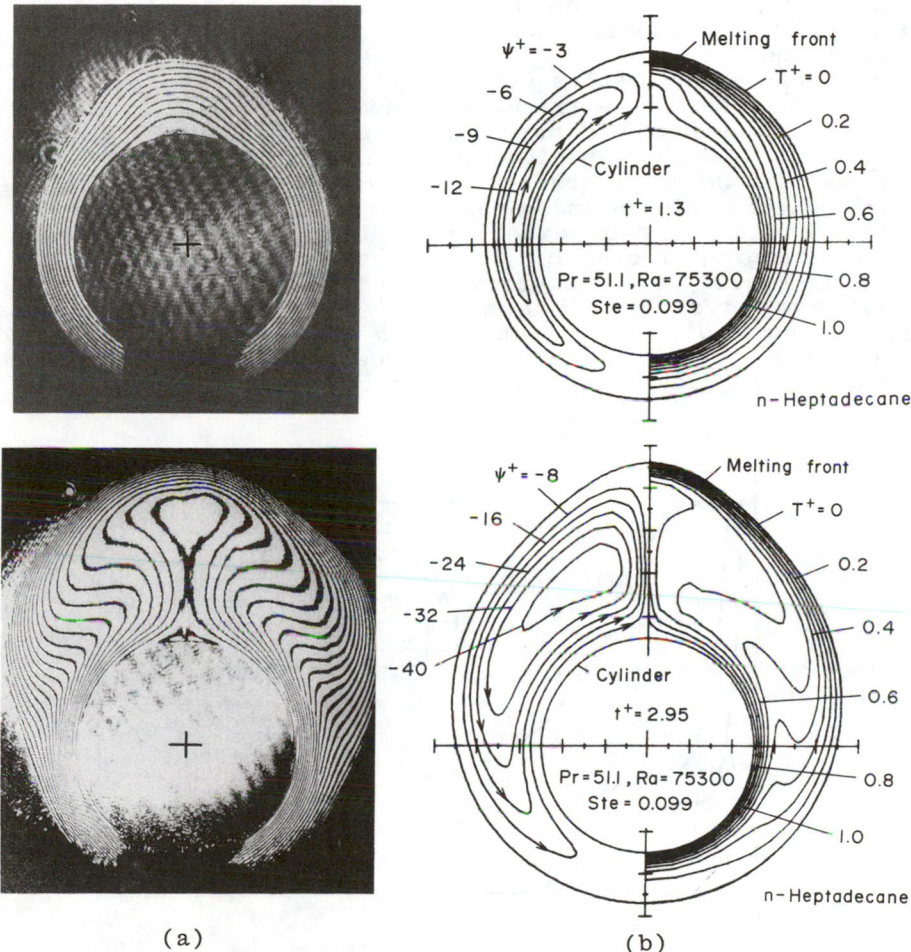

(a) (b)

FIGURE 24   Typical time sequences of streamlines and isotherms in the melt region, and melting front contour for Pr=51.1, Ra=75300, and Ste=0.099. (a) Experimental data by Bathelt and Viskanta[80], (b) Numerical result.

3.2 Simulation of a Latent Heat Thermal Energy Storage System

In the following, the results of a computer simulation of a latent heat TES system using spherical capsules will be presented. The heat pump assisted LHTES system seems to be a quite promising means for developing heat storage units in many utility devices. The readers who have further interest in these systems are recommended to examine references [99]-[117].

Approximate simulations for the spherical capsule bed were carried out by Green and Vliet[118], and Wood, et al.[119].

However, a thorough simulation considering both heat transfer between the working fluid and the capsule, and a solidification process inside each capsule was conducted by Saitoh and Hirose[105].

Here, only a few results will be presented.

Fig.25 shows the transient response of a LHTES unit with varying spherical capsule diameters. The PCM used is $Na_2HPO_4 \cdot 12H_2O$. The capsule diameter is one of the important parameters which determine the thermal performance of the LHTES unit. Fig.26 indicates a comparison of the simulation results with that of experiments for the melting process (i.e. charging process). A moderate agreement was attained.

In Fig.27, a schematic diagram for a combined LHTES and heat pump system is presented. This system will be a promising tool in the future for many space heating and cooling systems since the total coefficient of performance (TCOP) is as high as 4.2 on the average.

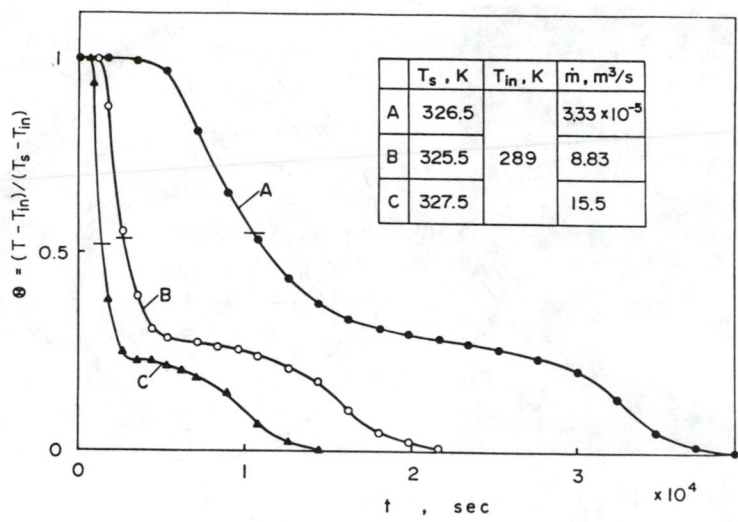

	T_s , K	T_{in} , K	$\dot{m}$, m^3/s
A	326.5		3.33×10^{-5}
B	325.5	289	8.83
C	327.5		15.5

FIGURE 25   Variation of the experimental outlet temperature history curve with flow rate (discharging process).

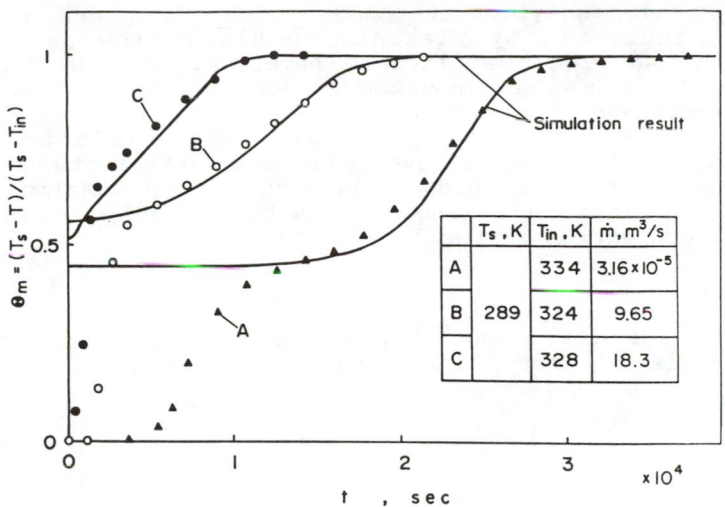

FIGURE 26   Typical variation of the experimental thermal response curve with flow rate (charging process).

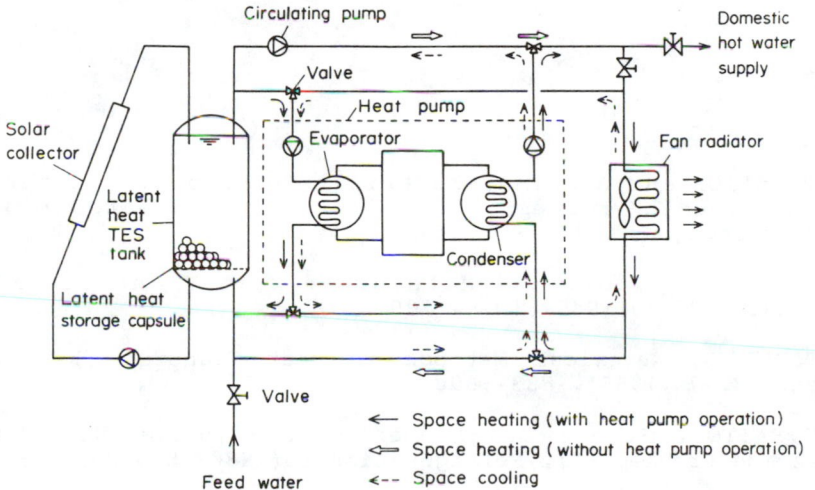

FIGURE 27   A schematic diagram of Latent heat thermal energy storage/ Heat pump system.

4. CONCLUDING REMARKS

Recent developments of the numerical and analytical methods for the two- and three-dimensional freezing and melting problems have been discussed in this article.   The growing need to facilitate the optimal design of latent heat thermal energy storage and other engineering systems using phase change phenomena will continue to require powerful numerical and analytical methods.

As indicated by Viskanta in his review paper[120], powerful and widely applicable numerical methods are still lacking, though a few methods presented recently seem to be efficient among existing ones.

In closing, it is important that efforts should continue to be made in developing time-efficient and highly accurate numerical methods to deal with more complicated and multi-dimensional freezing and melting problems, including natural convection in the liquid phase.

ACKNOWLEDGMENTS

The author extends his sincere thanks to Professor Raymond Viskanta for his many valuable comments.

REFERENCES

1.T.SAITOH and A.SHIMA, Numerical Solution for Spherical Bubble Growth Problem in a Uniformly Ultraheated Liquid, J.Mech.Engng Sci.,19(1977),1439-1454.

2.T.SAITOH, Numerical Methods in Heat Transfer (XIX), Sci.Mach.,37-10(1985),1275-1280.

3.T.SAITOH, Numerical Methods in Heat Transfer (XV), Sci. Mach., 37-7 (1985), 883-889.

4.T.SAITOH, Recent Developments of Solution Methods in Freezing Problems(I), Refrigeration,55(1980),875-883.

5.T.SAITOH, Recent Developments of Solution Methods in Freezing Problems(II), Refrigeration,55(1980),1005-1015.

6.T.SAITOH, A Numerical Method for Diffusion Flame Problem Treated as a Moving Boundary Problem, Memoirs of Sagami Inst.Tech.,6(1972),1-14.

7.N.M.H.Lightfoot, The Effect of Latent Heat on the Solidification of Steel Ingots, Third Rep. of the Committee on Heterogeneity of Steel Ingots, J.Iron Steel Inst.,1(1929),364.

8.A.Friedman, Remarks on Stefan-type Free Boundary Problem for Parabolic Equations, J.Math.Mech.,9(1960),887.

9.I.I.Kolodner, Free Boundary Problem for the Heat Equation with Applications to Problems of Change of Phase, Communs.Pure Appl.Math.,9(1956),1.

10.F.G.Dressel, The Fundamental Solution of Parabolic Equation I, Duke Math.J.,7(1940),186.

11.F.G.Dressel, ibid. II,Duke Math.J.,13(1946),61.

12.H.G.Landau, Heat Conduction in a Melting Solid, Q. Appl. Math.,8(1950),81.

13.P.V.Danckwerz, Unsteady-state Diffusion or Heat Conduction with Moving Boundary, Trans.Farady Soc.,46(1950),701.

14.W.L.Miranker, A Free Boundary Value Problem for the Heat Equation, Quart.Appl.Math.,16(1958),121.

15.W.L.Miranker and J.B.Keller, The Stefan Problem for a Nonlinear Equation, J.Math.Mech.,9(1960),67.

16.B.A.Boley, A Method of Heat Conduction Analysis of Melting and Solidification Problems, J.Math.Phys.,40(1961),300.

17.B.A.Boley, Uniqueness in Melting Slab with Space and Time Dependent Heating, Q.Appl.Math.,27(1970),481.

18.B.Sherman, A Free Boundary Problem for the Heat Equation with Heat Input at a Melting Interface, Q.Appl.Math.,23(1966),337.

19.B.Sherman, Some Generalizations in the Theory of Free Boundary Problems for the Heat Equation, Rocketdyne Res.Rep.,68-7(1968).

20.B.A.Boley, Survey of Recent Developments in the Fields of Heat Conduction in Solids and Thermo-Elasticity, Nucl.Eng.Des.,18 (1972),377.

21.S.G.Bankoff, Heat Conduction or Diffusion with Change of Phase, Advances in Chem.Eng.,5(1964), Acad.Press,N.Y.,75-150.

22.J.C.Muehlbauer and J.E.Sunderland, Heat Conduction with Freezing or Melting, Appl.Mech.Rev.,18-12(1965),951.

23.A.Moriyama and K.Araki, Methods for Analysis of the Moving Boundary Surface Problem, Int.Chem.Eng.,16(1976),734.

24.V.J.Lunardini, "Heat Transfer in Cold Climate",D van Nostrand, Reinhold,N.Y.(1981).

25.W.D.Murray and F.Landis, Numerical and Machine Solution of Transient Heat Conduction Problems Involving Melting or Freezing, Trans.ASME,81(1959),106.

26.K.A.Rathjen and Jiji, Heat Conduction with Melting or Freezing in a Corner, J.Heat Transfer, Trans.ASME, 93(1971), 101.

27.H.Budhia and F.Kreith, Heat Transfer with Melting in a Wedge, Int.J.Heat Mass Transfer, 16(1973),195.

28.B.A.Boley, Upper and Lower Bounds for the Solution of a Melting Problem, Q.Appl.Math.,21(1963),1.

29.T.Saitoh, A Study on Free Boundary Problem(Part IV, Second Method and Application to Freezing Problem), Proc.11th Heat Transfer Symp.Japan(1974),333-336.

30.G.S.Springer and D.R.Olson, Method of Solution of Axisymmetric Solidification and Melting Problems,ASME Paper 62-WA-246(1962),1.

31.A.Lazaridis, A Numerical Solution of the Multidimensional Solidification(or Melting) Problem, Int.J.Heat Mass Transfer,13(1970),1945.

32.N.R.Eyers et al.,The Calculation of Variable Heat Flow in Solids, Phil.Trans.Royal Soc.London(A), 240(1946),1.

33.D.C.Baxter, The Fusion of Slabs and Cylinders, J.Heat Transfer, Trans.ASME,84(1962),317.

34.K.Katayama and M.Hattori, A Study of Heat Conduction with Freezing(1st Report,Numerical Method of Stefan Problem), Trans. JSME,40-33(1974),1404.

35.N.Shamsunder and E.M.Sparrow, Analysis of Multidimensional Conduction Phase Change via the Enthalpy Model, J.Heat Transfer, Trans.ASME,97(1975),333.

36.A.B.Crowley, Numerical Solution of Stefan Problems, Int.J.Heat Mass Transfer,21(1978),215.

37.S.P.Burke and T.E.W.Schumann, Diffusion Flames,Ind.Engng Chem.,20(1928),998.

38.P.G.Kroeger and S.Ostrach, The Solution of a Two-dimensional Freezing Problem Including Convection Effects in the Liquid Region, Int.J.Heat Mass Transfer, 17(1974),1191.

39.M.E.Rose, On the Melting of a Slab, SIAM J.Appl.Math., 15-3(1967),495.

40.F.L.Chernous'ko, Solution of Nonlinear Heat Conduction Problems in Media with Phase Changes, Int.Chem.Eng.,10-1(1970),42.

41.R.C.Dix and J.Cizek, The Isotherm Migration Method for Transient Heat Conduction Analysis, Proc. 4th Int.Heat Transfer Conf., Paris vol.1,ASME,N.Y.(1971).

42.J.Crank and R.S.Gupta, Isotherm Migration Method in Two Dimensions, Int.J.Heat Mass Transfer,18(1975),1101.

43.J.Crank and A.B.Crowley, Isothermal Migration along Orthogonal Flow Lines in Two Dimensions, Int.J.Heat Mass Transfer, 21(1978),393.

44.T.Saitoh, Unpublished report, Sagami Inst.Tech.(1972).

45.T.Saitoh, Numerical Method for Multi-dimensional Freezing Problems in Arbitrary Domains, J.Heat Transfer, Trans.ASME,100(1978),294.

46.J.L.Duda,M.F.Malone,R.H.Notter, and J.S.Vrentas, Analysis of Two-dimensional Diffusion-controlled Moving Boundary Problems, Int.J.Heat Mass Transfer, 18(1975),1975.

47.T.SAITOH and K.HIROSE, Numerical Method for for the Two-Dimensional Freezing Problem around a Horizontal Cylinder Encompassing a Density Inversion, Trans.JSME,46(1980),971-980. also see: Bull.JSME,24(1981),147-152.

48.J.Crank, "Free and Moving Boundary Problems", Oxford Univ.Press(1984).

49.N.Shamsunder, Comparison of numerical Methods for Diffusion Problems with Moving Boundaries, in "Moving Boundary Problems(D.G.Wilson,A.D.Solomon, and P.T.Boggs ed.),Academic Press(1978).

50.R.Viskanta, "Solar Heat Storage: Latent Heat Materials",vol.1(G.A.Lane ed.),CRC Press,Florida(1983).

51.R.Viskanta,"Alternative Energy Sources III", vol.1 (T.N. Veziroglu ed.), Hemisphere Publ.Corp.,Washington (1983).

52.I.Onaka, "Computer-Aided Heat Transfer and Solidification Analysis",Maruzen(1985).

53.T.Maekawa and I.Tanasawa, A Study on Natural Convection in a Horizontal Fluid Layer Induced by Differences of Density and Surface Tension, Trans.JSME,51-465(1985).

54.T.Maekawa and I.Tanasawa, Instability Problem of Combined Natural Convection Driven by Density and Surface Tension Differences in a Horizontal Fluid Layer, Trans.JSME,51-465(1985).

55.I.Tanasawa, A Study of Natural Convection with Surface Tension Marangoni Effect, Rep.of Japanese Sci. Foundation (1985).

56.C.J.Ho and R.Viskanta, Heat Transfer During Inward Melting in a Horizontal Tube, Int.J.Heat Mass Transfer,27(1984),705.

57.K.Mitachi, K.Aoki, K.Kitamura, and M.Furuuchi, Natural Convection Heat Transfer in a Horizontal Tube with

Solidification and Heat Generation, Trans.JSME,51-472(1985),3996.

58.Y.Nagano, G.Son, and M.Hishida, Solidification of High Prandtl Number Fluid in a Horizontal Tube, Trans.JSME,51-467(1985).

59.K.Yamagami and M.Kurashige, Infiltration of a Fluid into a Dry Poro-Elastic body, J.Appl.Mech.,48 (1981),259.

60.N.Ramachandran,J.P.Gupta, and Y.Jaluria, Two-Dimensional Solidification with Natural Convection in the Melt and Convective and Radiative Boundary Conditions, Num.Heat Transfer, 4(1981),469.

61.R.S.Gupta and K.S.Kumar, Int.J.Heat Mass Transfer, 28(1985),1355.

62.T.Saitoh and N.Sasaki, Two-Dimensional Analysis of Combustion of Fuel Droplet or Sphere under Conditions of Natural Convection, Trans.JSME,50-453(1984),1397.

63.C.F.Hsu, E.M.Sparrow, and S.V.Patankar, Numerical Solution of Moving Boundary Problems by Boundary Immobilization and Control-volume Based Finite Difference Scheme, Int.J.Heat Mass Transfer, 24(1981),1335.

64.T.Saitoh, Numerical Method for Multidimensional Freezing Problem, Trans.JAR,3(1986),to appear.

65.T.Saitoh and K.Kato, Numerical Analysis of Multi-dimensional Freezing Problems via Growth Ring Method, Proc. 23rd Heat Transfer Symp.Japan(1986).

66.R.H.MacNeal, The Solution of Partial Differential Equations by means of Electrical Networks, Ph.D. Thesis, California Inst.Tech.(1949).

67.R.H.MacNeal, An Asymmetrical Finite Difference Networks, Q.Appl.Math.,11(1953/54),295.

68.A.M.Winslow, Numerical Solution of the Quasilinear Poisson Equation in a Nonuniform Triangle Mesh, J.Comp.Phys.,2(1967), 149.

69.G.M.Dusinberre, Triangular Grids for Heat Flow Studies, ASNE J.(1960),61.

70.I.Ohnaka, Classification of Numerical Methods for Transient Heat Transfer Problem and Improved Inner Nodal Point Method, Iron and Steel, 12(1979),1737.

71.C.Kleinstreuer and J.T.Holdman, A Triangular Finite Element Mesh Generator for Fluid Dynamic Systems of Arbitrary Geometry, Int.J.Num.Meth.Eng.,15(1980),1325.

72.J.R.Freeman, TRIDIF, A Triangular Mesh Diffusion Code,

J.Comp. Phys.,41(1981),142.

73.M.L.Hodgdon and J.R.Freeman, IEEE Trans. Magnetics, 2 (1982), 510.

74.T.Saitoh, "Heat Transfer Handbook",JSME(1986).

75.Y.Kawashima and K.Nishimoto, Numerical Method for Heat Conduction Problem by means of Dual Difference Method, Trans.JSME,50-455(1984),1727.

76.T.SAITOH, An Experimental Study of the Cylindrical and Two-Dimensional Freezing of Water with Varying Wall Temperature, Tecnol.Rep., Tohoku Univ.,41(1976),61-72.

77.T.SAITOH, Numerical Methods in Heat Transfer (XVI), Sci.Mach.,37-8(1985),967-973.

78.T.SAITOH, Numerical Methods in Heat Transfer (XX), Sci.Mach.,3711(1985),1397-1399.

79.A.G.Bathelt,R.Viskanta, and W.Leidenfrost, An Experimental Investigation of Natural Convection in the Melted Region around a Heated Horizontal Cylinder, J.Fluid Mech.,90(1979),227.

80.A.G.Bathelt and R.Viskanta, Heat Transfer at the Solid-Liquid Interface during Melting from a Horizontal Cylinder, Int.J.Heat Mass Transfer, 25(1982),369.

81.T.SAITOH and K.HIROSE, Latent Heat Thermal Energy Storage Around a Horizontal Cylinder, Technol.Rep.,Tohoku Univ.,49-1(1984),17-27.

82.T.SAITOH, Solar Energy Utilization and Heat Transfer Techniques, Proc.Symp. on Heat Transfer Engineering, Soc.Chem.Engrs of Japan(1981),1-4.

83.W.N.Hale,Jr. and R.Viskanta, Solid-Liquid Phase Change Heat Transfer and Interface Motion in Materials Cooled or Heated from Above or Below, Int.J.Heat Mass Transfer, 23(1980), 283.

84.C.Gau and R.Viskanta, Flow Visualization during Solid-Liquid Phase Change Heat Transfer I: Freezing in a Rectangular Cavity, Int.Comm.Heat Mass Transfer,10(1983),173.

85.P.D.Van Buren and R.Viskanta, Interferometric Observation of Natural Convection during Freezing from a Vertical Flat Plate,J. Heat Transfer,Trans.ASME,102(1980),375.

86.R.D.White,A.G.Bathelt,W.Leidenfrost, and R.Viskanta, Study of Heat Transfer and Melting from a Cylinder Embedded in a Phase Change Material, ASME Paper No.77-HT-42(1977).

87.E.M.Sparrow,R.R.Schmidt, and J.W.Ramsey, Experiments on the Role of Natural Convection in the Melting of Solids,

J.Heat Transfer,Trans.ASME,100(1978),11.

88.A.G.Bathelt and R.Viskanta, Heat Transfer at the Solid-Liquid Interface during Melting from Horizontal Cylinder, Int.J.Heat Mass Transfer, 23(1980),1443.

89.C.Gau,R.Viskanta, and C.J.Ho, Flow Visualization during Solidification Phase Change Heat Transfer II: Melting in a Rectangular Cavity, Int.Comm.Heat Mass Transfer,10(1983),183.

90.C.J.Ho and R.Viskanta, Heat Transfer during Melting from an Isothermal Vertical Wall,J.Heat Transfer, Trans. ASME, 106 (1984), 2 .

91.C.Gau and R.Viskanta, Inward Solidification of a Superheated Liquid in a Cooled Horizontal Tube, Wärme und Stoffübertragung,17(1982),39.

92.C.J.Ho and R.Viskanta, Experimental Study of Solidification Heat Transfer in an Open Rectangular Cavity, J.Heat Transfer,Trans.ASME,105(1983),671.

93.C.J.Ho and R.Viskanta, Heat Transfer inside a Horizontal Tube, Int.J.Heat Mass Transfer,27(1984),705.

94.E.M.Sparrow,S.V.Patankar, and S.Ramadhyani, Analysis of Melting in the Presence of Natural Convection in the Melt Region, J.Heat Transfer,Trans.ASME,99(1977),520.

95.T.SAITOH, Natural Convection Heat Transfer from a Horizontal Ice Cylinder, Appl.Sci.Res.,32(1976),429-451.

96.T.SAITOH, An Experimental Study for Two-Dimensional Freezing Around a Horizontal Cylinder Passing Through a Maximum Density Point, Refrigeration,53(1978),891-896.

97.T.SAITOH and K.HIROSE, A Study for Two-Dimensional Freezing in a Horizontal Circular Cylinder Passing Through a Maximum Density Point, Refrigeration,54(1979),845-852.

98.T.SAITOH and K.HIROSE, Thermal Instability of Natural Convection Flow over a Horizontal Ice Cylinder Encompassing a Maximum Density Point,J.Heat Transfer, Trans. ASME, 102 (1980), 261-267.

99.T.SAITOH, Heat Transfer Problems with Boundary Moving, HTSJ News,20-76(1981),22-30.

100. T.SAITOH, Heat Transfer Problem in Solar Energy Utilization, Refrigeration,57(1982),369-384.

101. T.SAITOH and K.HIROSE, High Rayleigh Number Solutions to Problems of Latent Heat Thermal Energy Storage in a Horizontal Cylinder Capsule, J.Heat Transfer, Trans. ASME, 104 (1982), 545-553.

102. T.SAITOH, Recent Advances in Solution Methodologies of

Solidification Problem, Proc. of the 136th Conference, Japan Soc. for the Prom.of Sci.(1983),2-5.

103.T.SAITOH, Optimum Design for Latent Heat Thermal Storage Systems, Refrigeration,58(1983),749-756.

104.T.SAITOH and K.HIROSE, Performance of Latent Heat Thermal Energy Storage Reservoir of Spherical Capsule Type, Refrigeration,58-672(1983),933-940.

105.T.SAITOH and K.HIROSE, High-Performance Phase Change Thermal Energy Storage Using Spherical Capsules, Proc.1984 National Heat Transfer Conf.(ASME Paper No.84-HT-9),Niagara Falls,N.Y.,(1984).

106.T.SAITOH and K.HIROSE, Experimental Performance of Spherical Capsule-type Latent Heat Thermal Energy Storage System, Refrigeration,59(1984),519-525.

107.T.SAITOH and K.HIROSE, High-Rayleigh Number Melting Heat Transfer in a Spherical Capsule, Trans.Japanese Assoc.Refrig.,1(1984),157-162.

108.T.SAITOH, Numerical Methods in Heat Transfer (XVII), Sci. Mach.,37-9(1985),1085-1087.

109.T.SAITOH and K.HIROSE< On the Phase-Change Thermal Energy Storage/Heat Pump System, Trans.JSME,51-462(1985),705-711.

110.T.SAITOH and K.HIROSE, Performance Simulation of Spherical Capsule Latent Heat Thermal Energy Storage Unit, Trans.JSME,51-466(1985),1867-1873.

111.T.SAITOH and K.HIROSE, Spherical Capsule -Type Latent Heat Thermal Energy Storage/Heat Pump System with Application to Solar System, Proc.Int.Symp. on Thermal Application of Solar Energy, Int.Solar Energy Soc.,Hakone(1985),319-323.

112.T.SAITOH and K.HIROSE, Spherical Capsule-type Latent Heat Thermal Energy Storage for Solar Energy, Trans. JSME,51 (1985),3060-3067.

113.T.SAITOH and K.HIROSE, Solar-Assisted Latent Heat Thermal Energy Storage/Heat Pump System using Spherical Capsules, Rep. Toyoda Phys.Chem.Res.Inst.,38(1985),45-53.

114.T.SAITOH and K.HIROSE, Latent Heat Thermal Energy Storage System with Heat Pump for Elimination of Supercooling Problem, Res.Rep. of Iwatani Foundation,8(1985),9-13.

115.R.M.Goldstein and J.W.Ramsey, Heat Transfer to a Melting Solid with Application to Thermal Energy Storage Systems, Heat Transfer Studies: A Festschrift for E.R.G. Eckert, ed., J.P. Hartnett et al.,McGraw-Hill,N.Y.(1979),189.

116.A.G.Bathelt,R.Viskanta, and Leidenfrost, Latent Heat-of-Fusion Energy Storage: Experiments on Heat Transfer from

Cylinders during Melting,J.Heat Transfer, Trans. ASME, 101 (1979), 732.

117.K.Katayama,A.Saito,Y.Utaka,A.Saito,H.Matsui,H.Maekawa, and A.Z.A.Saitullah, Heat Transfer Characteristics of the Latent Heat Thermal Energy Storage Capsule, Solar Energy,27(1981),91.

118.T.F.Green and G.C.Vliet, Transient Response of a Latent Heat Storage Unit,: An Analytical and Experimental Investigation,ASME J.Solar Energy Eng.,103(1981),275.

119.R.J.Wood,S.D.Gladwell,P.W.O'Callaghan, and S.D.Probert, Low Temperature Thermal Energy Storage using Packed Beds of Encapsulated Phase-Change Materials, Int.Conf.Energy Storage, Brighton(1981),145.

120.R.Viskanta, Natural Convection in Melting and Solidification, 5th NATO Advanced Study Inst. on Natural Convection: Fundamentals and Applications(1984).

121.R.R.Gilpin, Cooling of a Horizontal Cylinder of Water Through its Maximum Density Point at $4°C$, Int.J.Heat Mass Transfer, 18(1975),1307.

122.K.C.Cheng, M.Takeuchi, and R.R.Gilpin, Transient Natural Convection in Horizontal Pipes with Maximum Density Effects and Supercooling, Num.Heat Transfer, 1(1978),101.

123.R.R.Gilpin, T.Hirata, and K.C.Cheng, Longitudinal Vortices in a Horizontal Boundary Layer in Water Including the Effects of the Density Maximum at $4°C$, AIAA-ASME Thermophysics and Heat Transfer Conf.,Palo Alto,CA.(1978).

ICE FORMATION

Chapter 5

Order of Magnitude Analysis on Liquid Solidification in Pipe Flows

G. J. HWANG and R. S. CHEN
Department of Power Mechanical Engineering, National Tsing Hua University,
Hsinchu, Taiwan 30043, ROC

ABSTRACT

An order of magnitude analysis on liquid solidification in
internal laminar forced flows is presented in this article.
General formulations of heat conduction equations in solid
shells, energy balance equations on liquid-solid interfaces,
and continuity, momentum, and energy equations for liquid
phases are investigated using the dimensionless parametric
approach. This article discusses the ranges of parameters
and the related limiting cases, and also summarizes both the
theoretical and experimental results of solidification-free
zones, liquid-solid interface profiles, heat transfer rates,
and pressure drops.

CONTENTS

NOMENCLATURE

a	dimensionless radius of liquid-solid interface, δ/R_i
Bi	Biot number, hR_i/k or hL/k
C	specific heat
D	diameter

Gr_m	Grashof number, as given in ref. [16]
Gz_m	Graetz number, as given in ref. [16]
h	heat transfer coefficient
k	thermal conductivity
L	latent heat or half distance between parallel plates
$\dot{m}$	mass transfer rate
O( )	order of magnitude
Nu_m	Nusselt number, as given in ref. [16]
P	pressure
p	dimensionless pressure
p^*	dimensionless pressure drop, $(P_0-P)/(\rho W_{m0}^2/2)$
Pe	Peclet number, $W_{m0}R_i/\alpha$
Pr	Prandtl number, ν/α
Pr_m	Prandtl number, as given in ref. [16]
q	heat transfer rate for a tube of length Z
q^*	dimensionless heat transfer rate, $q/(T_0-T_f)\dot{m}C$
R,ϕ,Z	cylindrical coordinates
r,ϕ,z	dimensionless cylindrical coordinates
Ra	Rayleigh number
R_i	inner radius of tube
Re_D	Reynolds number, $W_{m0}D/\nu$
Ste	Stefan number, $C_s(T_0-T_\infty)/L$
T	temperature
t	dimensionless time
U,V,W	velocity components in R, ϕ and Z directions
u,w	dimensionless velocity components in r, z directions
u^*,w^*	dimensionless velocity components, au and a^2w, respectively

Greek letters

α	thermal diffusivity
β	angle or thermal expansion coefficient
δ	radius of liquid-solid interface
δ_s	thickness of solid shell
ϵ	superheat ratio, $(T_0-T_f)/(T_f-T_\infty)$, or small value
θ	dimensionless temperature, $(T-T_\infty)/(T_0-T_\infty)$
λ	product of conductivity and superheat ratios, $k(T_0-T_f)/[k_s(T_f-T_\infty)]$
μ	dynamic viscosity
ν	kinematic viscosity
ξ,η	dimensionless coordinates, z and $r/a(z,t)$, respectively
ρ	density
τ	time
ϕ	dimensionless temperature, $(T-T_f)/(T_0-T_f)$, or angular position
ϕ_s	dimensionless temperature, $(T-T_\infty)/(T_f-T_\infty)$

Subscripts

B	bulk
D	diameter
f	freezing condition
ℓ	liquid phase
m	mean value
0	condition at Z=0
s	solid phase
w	wall
∞	ambient

1 INTRODUCTION

The phenomenon of liquid freezing in forced flows is promoted and utilized in many processes such as the casting of metals, desalination of water and storage of cooling fluids. On the other hand, solidification is not always desirable. Ice formation in water mains, freezing of liquid metals in heat exchangers, and phase changes in hydraulic systems, for example, may have harmful effects. This article presents an order of magnitude analysis on liquid freezing in internal laminar forced flows. Papers on liquid solidification in laminar pipe flows are listed in Table 1 and reviewed as follows:

Only a limited amount of research has been done on the subject of liquid solidification inside ducts. This is due to the difficulties arising from the influence of the liquid-solid interface on the flow and heat transfer characteristics of the liquid phase.

One of the earliest attempts to analyze the effect of liquid solidification in a tube upon laminar flow heat transfer and pressure drop was made by Zerkle and Sunderland [1]. They used a parabolic axial velocity profile to obtain heat transfer coefficients and conducted an experiment using water as a working fluid in a circular tube 1.5 in. in diameter. A discrepancy of 140% at z=0.03, based on the theoretical value, was observed between their theoretical and experimental heat transfer rates. They concluded that the difference might be caused by the effects of free convection, and an empirical equation for these effects was suggested. In contrast to the parabolic form used by Zerkle and Sunderland [1], Özisik and Mulligan [2] employed a rectangular velocity profile in an analysis of the transient solidification problem in a tube with a Dirichlet thermal boundary condition. The predicted values of heat transfer coefficients were greater than the theoretical ones of Zerkle and Sunderland [1], but coincidently the results [2] agreed with the experimental data in reference [1].

Using the available flow and heat transfer data, Des Ruisseaux and Zerkle [3] developed a theoretical technique for predicting the conditions under which a hydraulic system freezes shut. The theoretical predictions were then compared with their experimental results. Depew and Zenter [4] experimented with the laminar flow heat transfer and the pressure drop with freezing at the wall. A test section of 0.786 in. in diameter was used in their experiment. A 100% discrepancy due to the effects of free convection was observed at z=0.03.

In a convectively cooled circular tube, Lock, Freeborn, and Nyren [5] used quasi-steady state equations in both the liquid and solid phases to analyze the growth of an ice layer. Due to a finite Biot number, this analysis yielded the length of the ice-free zone and provided a description of the ice-water interface and its effect on pressure.

TABLE 1. Summary of papers on liquid solidification in pipe flows

Ref.	Yr.	Authors	Theo.	Exp.	Fluid	Section	Velocity	B.C.
1	68	Zerkle Sunderland	✓	✓	H_2O	O	D	Ⓣ
2	68	Özisik Mulligan	✓		H_2O	O	☐	Ⓣ
3	69	Des Ruisseaus Zerkle	✓	✓	H_2O	O	D	Ⓣ
4	69	Depew Zenter		✓	H_2O	O	D	Ⓣ
5	70	Lock Freeborn Nyren	✓		H_2O	O	D	Bi= finite
6	70	Bilenas Jiji	✓		H_2O	O	D	Ⓣ
7	71	Lock Nyren	✓		H_2O	O	zero	$h=h(\phi)$
8	73	Hwang Yih	✓		H_2O	O	D	Bi= finite
9	75	Ou Cheng	✓		Pr>>1	☐	$W_f(X,Y)$ Ra≠0	Bi= finite
10	76	Hwang Sheu	✓	✓	H_2O	O	W(R,Z)	Ⓣ
11	76	Mulligan Jones		✓	H_2O	O	D	Ⓣ
12	77	Liu Hwang		✓	H_2O	O	D	Ⓣ
13	77	Cheng Wong	✓		H_2O	=	D	Bi= finite
14	80	Lee Hwang	✓		Pe=1∿∞	O	D	Bi= finite
15	81	Sadeghipour Özisik Mulligan	✓		liquid metal	O	▯	Bi= finite
16	89	Lee Hwang	✓		Pe=1∿30	O	D	Ⓣ

O: circular   ☐: square   = : parallel plates   ▯: slug flow
D: fully developed flow   Ⓣ: constant wall temperature

Bilenas and Jiji [6] re-examined the problem in ref. [1] using a variational technique and the solution was expressed in terms of the short-time, asymptotic, and steady-state components. Two forms of the variational solution were presented. One had limited validity in the entrance region of the tube, and the other, while less general, was more accurate. Lock and Nyren [7] extended their work [5] to an ice formation problem in a long circular tube cooled by external convection. This analysis utilized a regular perturbation expansion for the temperature and the interface location. Hwang and Yih [8] studied the length of the ice-free zone presented by Lock et al. [5] by recalculating numerically the eigenvalues and eigenfunctions. Ou and Cheng [9] further extended the liquid solidification-free zone in a convectively-cooled horizontal square channel with the effects of free convection. It was concluded that the free convecion effects should be included in the analysis of ice formation in a convectively-cooled pipe or channel when the Rayleigh number is greater than 10^4.

Considering a gradual development of axial velocity from a uniform distribution to a fully developed one, Hwang and Sheu [10] presented theoretical and experimental investigations of liquid solidification in the combined hydrodynamic and thermal entrance flow in a circular tube with a uniform wall temperature. The theoretical solution assumed that a quasi-steady condition prevails; axial variations in the liquid-solid interface were gradual and the effects of the radial velocity component were considered significant. An experiment using water as a working medium was carried out to verify the theoretical results, and employed two double-pipe heat exchangers with small inner tube diameters to prevent the free convection effects. Reasonable agreement between their theoretical curves and experimental data was observed.

Experiments on heat transfer and pressure drop in a horizontal tube with internal solidification were performed by Mulligan and Jones [11]. The experiments involved a steady, hydrodynamically developed laminar flow with a steady-state frozen deposit at the inside wall, a constant and uniform wall temperature, and a Graetz number in the range of significant natural convection. It is shown that Oliver's correlation [16] of combined forced and free convection is applicable when the L/D is significantly greater than 50, and that the correlation is more accurate when corrected for the presence of a solid phase thickness. Using a 95 cm long tube for flow development before cooling, Liu and Hwang [12] conducted an investigation on the effect of liquid solidification in the thermal entrance region of a circular tube in the small inner section. Heat transfer data from the experiment [12] showed better agreement with the theoretical curve than the results [1,4,11].

To study a channel under a meteorological environment in the permafrost region, Cheng and Wong [13] made a theoretical analysis of liquid solidification in the thermal entrance region of a parallel-plate channel with a uniform external convection. This analysis was carried out by using the confluent hyper-geometric function in the solution for laminar flow and steady-state conditions. Using a finite Biot number, the length of the solidification-free zone was also found in this analysis. Lee and Hwang [14] examined the solidification of a low Peclet number fluid flow in the thermal entrance region of a round pipe. The velocity was assumed to be laminar and fully developed throughout the pipe and the fluid temperature was taken to be uniform at $Z=-\infty$. The pipe wall was adiabatic at $Z<0$ and cooled convectively at $Z\geq0$. The exact solutions in the liquid and solid phases were obtained by using the method of separating the variables and constructing two sets of orthonormal functions from the non-orthogonal eigenfunctions at $Z=\pm0$. The length of the ice-free zone was determined and the case corresponding to $Pe=\infty$ was in excellent agreement with the existing solution. Without the axial conduction effect, Sadeghipour, Özisik and Mulligan [15] studied the transient solidification of liquid metals in the thermal entry region of a circular tube. The effect of the Biot number and the difference between the freezing and the ambient temperature on the length of freeze-free zone were examined. Lee and Hwang [16] investigated the combined effects of axial conduction and solidification on heat transfer and pressure drop in pipe flows by using a modified Galerkin finite element method. The profile of the solid-liquid interface, the heat transfer rate and the pressure drop were presented for $Pe=1$, 3, 5, 10 and 30, and for the modified superheat ratio, $c=0.1$, 0.5, 5,0 and ∞.

2 ORDER OF MAGNITUDE ANALYSIS

2.1 Descriptions and Assumptions

Consider a laminar forced convection in the hydrodynamic entrance or fully developed flow in a duct. The duct cross section can be circular or rectangular with a finite or an infinite aspect ratio. When the duct wall is cooled convectively at $Z\geq0$ by using an external cold stream with a temperature below the freezing temperature of the fluid in the duct, solidification may occur at a certain distance Z_f along the duct. If the external convection is strong enough, the distance may be close to zero. The growth of the solid shell may be rapid or gradual depending on the magnitude of the external heat transfer rate. Fig. 1 illustrates the configuration and coordinate system in a pipe.

The basic assumptions employed in the analysis are:
1. The flow is laminar, incompressible, axisymmetric, and without viscous dissipation.
2. The thermal resistance of the duct wall is negligible.
3. The distribution of the heat transfer coefficient h from the duct wall to the external cold stream is peripherally uniform.
4. The physical properties are constant.

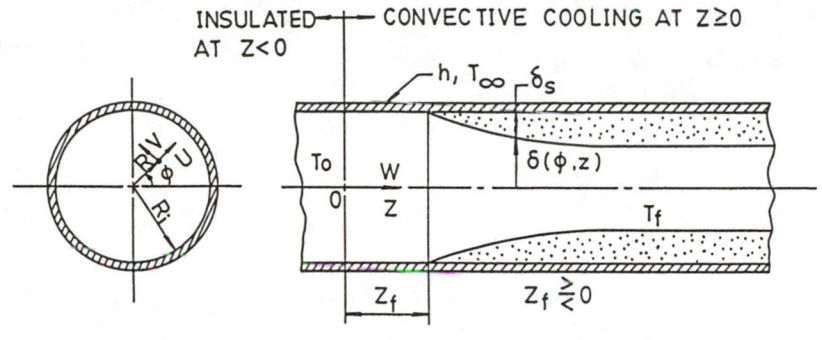

INSULATED—+— CONVECTIVE COOLING AT Z≥0
AT Z<0

SOLIDIFICATION – FREE LENGTH

FIGURE 1.   Configuration and coordinate system

Based on the above assumptions, the governing equations for both the liquid and solid phases are given in the following sections.

2.2 Liquid Phase

The transient energy equation for the liquid phase is

$$\frac{\partial T}{\partial \tau} + U\frac{\partial T}{\partial R} + W\frac{\partial T}{\partial Z} = \alpha\left(\frac{\partial^2 T}{\partial R^2} + \frac{1}{R}\frac{\partial T}{\partial R} + \frac{\partial^2 T}{\partial Z^2}\right) \tag{1}$$

for $0 \leq R \leq R_i$, $-\infty < Z \leq Z_f$, $\tau > 0$, and $0 \leq R \leq \delta$, $Z_f < Z < \infty$, $\tau > 0$ with the boundary and initial conditions

$$
\begin{array}{lll}
\dfrac{\partial T}{\partial R} = 0 & R=R_i, & -\infty < Z < 0, \quad \tau > 0 \\[2mm]
-k\dfrac{\partial T}{\partial R} = h(T_w - T_\infty) & R=R_i, & 0 \leq Z \leq Z_f, \quad \tau > 0 \\[2mm]
T = T_f & R=\delta, & Z_f < Z < \infty, \quad \tau > 0 \\[2mm]
\dfrac{\partial T}{\partial R} = 0 & R=0, & -\infty < Z < \infty, \quad \tau > 0 \\[2mm]
T = T_0 & 0 < R < R_i, & Z=-\infty, \quad \tau > 0 \\[2mm]
T = T_f & 0 < R < \delta, & Z=+\infty, \quad \tau > 0 \\[2mm]
T = T_0 & 0 < R < R_i, & -\infty < Z < Z_f, \quad \tau=0 \\[2mm]
 & 0 < R < \delta, & Z_f < Z < \infty, \quad \tau=0.
\end{array}
\tag{2}
$$

For a fully developed flow at $Z \leq Z_f$, the velocity components are given by

$$U = 0$$

$$W = 2 \; W_{m0}[1-(R/R_i)^2] \tag{3}$$

but for the accelerating flow in the converging section at $Z>Z_f$, the governing equations for the velocity components U and W are

$$\frac{\partial U}{\partial R} + \frac{u}{R} + \frac{\partial W}{\partial Z} = 0 \tag{4}$$

$$\frac{\partial U}{\partial \tau} + U\frac{\partial U}{\partial R} + W\frac{\partial U}{\partial Z} = -\frac{1}{\rho}\frac{\partial P}{\partial R} + \nu(\frac{\partial^2 U}{\partial R^2} + \frac{1}{R}\frac{\partial U}{\partial R} - \frac{U}{R^2} + \frac{\partial^2 U}{\partial Z^2}) \tag{5}$$

$$\frac{\partial W}{\partial \tau} + U\frac{\partial W}{\partial R} + W\frac{\partial W}{\partial Z} = -\frac{1}{\rho}\frac{\partial P}{\partial Z} + \nu(\frac{\partial^2 W}{\partial R^2} + \frac{1}{R}\frac{\partial W}{\partial R} + \frac{\partial^2 W}{\partial Z^2}) \tag{6}$$

for $0 \leq R \leq \delta$, $Z_f < Z < \infty$, and $\tau > 0$ with the boundary and initial conditions

$$U = \frac{\partial W}{\partial R} = 0 \qquad\qquad R=0, \; Z>Z_f, \; \tau>0$$

$$U = W = 0 \qquad\qquad R=\delta, \; Z>Z_f, \; \tau>0 \tag{7}$$

$$U = P-P_0 = 0 \text{ and}$$

$$W = 2W_{m0}[1-(R/R_i)^2] \qquad 0<R<\delta, \; Z=Z_f, \; \tau>0$$

$$U = W-2W_{m0}[1-(R/R_i)^2] = 0 \qquad 0<R<\delta, \; Z>_f, \; \tau=0.$$

Note that the boundary conditions for P in the R-direction and for U and W at $Z=\infty$ are not clearly defined.

2.3 Solid Phase

The energy equation in the solid phase is

$$\rho_s C_s \frac{\partial T}{\partial \tau} = k_s \; (\frac{\partial^2 T}{\partial R^2} + \frac{1}{R}\frac{\partial T}{\partial R} + \frac{\partial^2 T}{\partial Z^2}) \tag{8}$$

for $\delta \leq R \leq R_i$, $Z_f < Z < \infty$, and $\tau > 0$ with the boundary and initial conditions

$$T = T_f \qquad\qquad R=\delta, \; Z_f<Z<\infty, \; \tau>0$$

$$-k\frac{\partial T}{\partial R} = h(T_w-T_\infty) \qquad\qquad R=R_i, \; Z_f<Z<\infty, \; \tau>0 \tag{9}$$

$$T = T_f \qquad\qquad \delta<R<R_i, \; Z_f<Z<\infty, \; \tau=0.$$

Also note that the boundary conditions in the Z-direction cannot be given explicitly.

Equations (1) and (8) are coupled with the interface equation

$$\rho_s L \frac{\partial \delta}{\partial \tau} = k_s(\frac{\partial T}{\partial R})_s - k(\frac{\partial T}{\partial R})_\ell \tag{10}$$

for $R=\delta$, $Z>Z_f$ and $\tau>0$ with an initial condition $\delta(Z,0)=R_i$.

2.4   Characteristic Quantities and Transformations

To facilitate the order of magnitude analysis, characteristic quantities for each independent and dependent variable should be properly chosen. Some of the characteristic quantities can be determined from the posed boundary and initial conditions, but some must be obtained by the judgement of physical relationships among terms in the governing equation. In the present problem, the characteristic quantities of R, W, and T are determined from the posed conditions as:
1. The inner radius R_i of the pipe is chosen as the characteristic length in the radial direction.
2. The average axial velocity W_{m0} is a constant and is selected as the characteristic axial velocity.
3. The temperature difference T_0-T_∞ , is the largest temperature difference in the present system, and is taken as the characteristic quantity for the temperature difference.

The characteristic quantities of Z, τ, U, and P are obtained by the following physical relationships:
1. The characteristic quantity of Z is determined by considering

$$O\ (W\frac{\partial T}{\partial Z})\ =\ O\ [\alpha(\frac{\partial^2 T}{\partial R^2}\ +\ \frac{1}{R}\frac{\partial T}{\partial R})] \tag{11}$$

in energy equation (1) and the characteristic axial length is taken as $W_{m0}R_i^2/\alpha$ or $W_m\delta^2/\alpha$.
2. The characteristic quantity of τ is determined by considering

$$O\ (\rho_s L\frac{\partial\delta}{\partial\tau})\ =\ O\ [k_s(\frac{\partial T}{\partial R})_s] \tag{12}$$

in the interface equation (10) and the characteristic time is $R_i^2/ste\alpha_s$, where $Ste=C_s(T_0-T_\infty)/L$ is the Stefan number representing sensible to the latent heat ratio.
3. The characteristic radial velocity is established by using the continuity equation (4) and the value is α/R_i.
4. The characteristic quantity of P is determined by using the relation

$$O\ (W\frac{\partial W}{\partial Z})\ =\ O\ (\frac{1}{\rho}\frac{\partial P}{\partial Z}) \tag{13}$$

in the axial momentum equation (6) and the characteristic pressure difference is defined as ρW_{m0}^2.

Using the above characteristic quantities, the dimensionless transformations are:

$$r = R/R_i$$

$$a = \delta/R_i$$

$$z = Z/(W_{m0}R_i^2/\alpha) = Z/(W_m\delta^2/\alpha)$$

$$t = \tau/(R_i^2/Ste\alpha_s)$$

$$\theta = (T-T_\infty)/(T_0-T_\infty)$$

$$w = W/W_{m0}$$

$$u = U/(\alpha/R_i)$$

$$p = (P-P_0)/\rho W_{m0}^2$$

By substituting the transformations into equation (1), the dimensionless energy equations in the liquid phase are

$$[Ste(\frac{\alpha_s}{\alpha})]\frac{\partial\theta}{\partial t} + 2(1-r^2)\frac{\partial\theta}{\partial z} = \frac{\partial^2\theta}{\partial r^2} + \frac{1}{r}\frac{\partial\theta}{\partial r} + [\frac{1}{Pe^2}]\frac{\partial^2\theta}{\partial z^2} \tag{14}$$

for $0 \le r \le 1$, $-\infty < z \le z_f$, and $t > 0$, and

$$[Ste(\frac{\alpha_s}{\alpha})]\frac{\partial\theta}{\partial t} + u\frac{\partial\theta}{\partial r} + w\frac{\partial\theta}{\partial z} = \frac{\partial^2\theta}{\partial r^2} + \frac{1}{r}\frac{\partial\theta}{\partial r} + [\frac{1}{Pe^2}]\frac{\partial^2\theta}{\partial z^2} \tag{15}$$

for $0 \le r < a(z,t)$, $z_f < z < \infty$, and $t > 0$.
The dimensionless governing equations for velocities u and w are:

$$\frac{\partial u}{\partial r} + \frac{u}{r} + \frac{\partial w}{\partial z} = 0 \tag{16}$$

$$[\frac{Ste}{Pr}\frac{\alpha_s}{\alpha}]\frac{\partial u}{\partial t} + [\frac{1}{Pr}](u\frac{\partial u}{\partial r} + w\frac{\partial u}{\partial z})$$
$$= -[\frac{Pe^2}{Pr}]\frac{\partial p}{\partial r} + \frac{\partial^2 u}{\partial r^2} + \frac{1}{r}\frac{\partial u}{\partial r} - \frac{u}{r^2} + [\frac{1}{Pe^2}]\frac{\partial^2 u}{\partial z^2} \tag{17}$$

$$[\frac{Ste}{Pr}\frac{\alpha_s}{\alpha}]\frac{\partial w}{\partial t} + [\frac{1}{Pr}](u\frac{\partial w}{\partial r} + w\frac{\partial w}{\partial z})$$
$$= -[\frac{1}{Pr}]\frac{\partial p}{\partial z} + \frac{\partial^2 w}{\partial r^2} + \frac{1}{r}\frac{\partial w}{\partial r} + [\frac{1}{Pe^2}]\frac{\partial^2 w}{\partial z^2} \tag{18}$$

for $0 \le r \le a(z,t)$, $z_f < z < \infty$, and $t > 0$.
The dimensionless energy equation in the solid phase is

$$[Ste]\frac{\partial\theta}{\partial t} = \frac{\partial^2\theta}{\partial r^2} + \frac{1}{r}\frac{\partial\theta}{\partial r} + [\frac{1}{Pe^2}]\frac{\partial^2\theta}{\partial z^2} \tag{19}$$

for $a \le r \le 1$, $z_f \le z < \infty$, and $t > 0$,
and the dimensionless interface equation is

$$[\theta_f]\frac{\partial a}{\partial t} = (\frac{\partial\theta}{\partial r})_s - [\frac{k}{k_s}](\frac{\partial\theta}{\partial r})_\ell \tag{20}$$

2.5 Parameters and Limiting Cases

Table 2 summarizes the literature examining the dimensionless parameters in internal flows. Regardless of the value of the Prandtl number, most papers, except in references [14,16], considered the large Peclet number. The terms with the coefficient $1/Pe^2$ in equations (14)-(19) vanish. In references [14,16] the axial conduction was considered in equation (14), but the terms were neglected in equations (15)-(19). For the unsteady terms of equations (14)-(19), a parameter $Ste=C_s(T_f-T_\infty)/L$ indicating the ratio of sensible heat to latent heat appears. As was done in references

TABLE 2. Summary of dimensionless parameters

Ref.	a	z	t	Pe	Ra	λ	Bi	Ste
1	✓	✓	∞	∞	e	✓	∞	−
2	✓	✓	✓	∞	−	✓	∞	✓
3	✓	✓	∞	∞	e	✓	∞	−
4	−	✓	∞	∞	e	✓	∞	−
5	✓	✓	✓	∞	−	✓	✓	ε
6	✓	✓	✓	∞	−	✓	∞	✓
7	✓	−	✓	∞	−	0	✓	✓
8	−	✓	−	∞	−	✓	✓	−
9	−	✓	−	∞	✓	−	✓	−
10	✓	✓	✓	∞	e	✓	∞	ε
11	−	✓	∞	∞	e	✓	∞	−
12	−	✓	−	∞	e	✓	∞	−
13	✓	✓	∞	∞	−	✓	✓	−
14	✓	✓	∞	✓	−	✓	✓	−
15	✓	✓	✓	∞	−	✓	✓	✓
16	✓	✓	∞	✓	−	✓	∞	−

✓ : included   − : not included   e : experiment only

0 : zero value ∞ : infinite value ε : small value

[1,3,13,14] for the steady state condition, the unsteady terms in equations (14)-(20) can be omitted completely. If the Stefan number is small, a quasi-steady approximation can be employed to simulate the transient growth of the interface profile. With this approximation [5,10], all the transient terms in equations (14)-(19) are neglected except the transient term in equation (20). But in references [2,6,7,15], all the transient terms are kept without the assumption of the small Stefan number.

In the present analysis, a large Peclet number and a small Stefan number are assumed. During the liquid freezing process, the quasi-steady equations without the axial conduction and viscous terms are used. Energy equation (14) in the liquid phase for $0 \leq r \leq 1$, $0 \leq z \leq z_f$, and $t > 0$ can be approximated as

$$2(1-r^2)\frac{\partial \phi}{\partial z} = \frac{\partial^2 \theta}{\partial r^2} + \frac{1}{r}\frac{\partial \theta}{\partial r} \tag{21}$$

After the transformation $\phi=(\phi-\phi_\infty)/(1-\phi_\infty)$ where $\phi=(T-T_f)/(T_0-T_f)$, the quasi-steady energy equation without the axial conduction term can be obtained from equation (15)

$$u\frac{\partial \phi}{\partial r} + w\frac{\partial \phi}{\partial z} = \frac{\partial^2 \phi}{\partial r^2} + \frac{1}{r}\frac{\partial \phi}{\partial r} \tag{22}$$

for $0 \leq r \leq a(z,t)$, $z_f < z < \infty$, and $t > 0$.
Considering a large Peclet number, the pressure term in equation (17) is

$$\frac{\partial p}{\partial r} \approx \frac{Pr}{Pe^2}\left[-\frac{\partial^2 u}{\partial r^2} - \frac{1}{r}\frac{\partial u}{\partial r} + \cdots\cdots\right] \approx 0$$

thus the pressure is not a function of the radial position. Therefore, the continuity and momentum equations in the quasi-steady state are

$$\frac{\partial u}{\partial r} + \frac{u}{r} + \frac{\partial w}{\partial z} = 0 \tag{23}$$

$$\frac{1}{Pr}\left(u\frac{\partial w}{\partial r} + w\frac{\partial w}{\partial z}\right) = -\frac{1}{Pr}\frac{dp}{dz} + \frac{\partial^2 w}{\partial r^2} + \frac{1}{r}\frac{\partial w}{\partial r} \tag{24}$$

for $0 \leq r < a(z,t)$, $z_f < z < \infty$, and $t > 0$.
After the transformation $\theta=\phi_s\theta_f$ where $\phi_s=(T-T_\infty)/(T_f-T_\infty)$, the quasi-steady state energy equation in the solid phase without the axial conduction is

$$\frac{\partial^2 \phi_s}{\partial r^2} + \frac{1}{r}\frac{\partial \phi_s}{\partial r} = 0 \tag{25}$$

for $a \leq r < 1$, $z_f < z < \infty$, and $t > 0$.
Equations (22) and (25) are coupled with the unsteady interface equation

$$\frac{\partial a}{\partial t} = \frac{\partial \phi_s}{\partial r} + \lambda\frac{\partial \phi}{\partial r} \tag{26}$$

evaluated at the interface $r=a(z,t)$ where $\lambda =k(T_0-T_f)/k_s(T_f-T_\infty)$ is a product of the conductivity ratio and the superheat ratio. In Table 2 the product λ appears in most of the papers except for the references [7,9]. In reference [7], a zero value of λ was employed and in reference [9] only the solution in the solidification-free zone was discussed.

The solution of the above equations (21)-(26) for $a=1$ can be obtained without too much difficulty by using a numerical technique, but the solution within the boundaries involving a converging liquid-solid interface is quite difficult. Therefore a transformation was carried out in the literature [1,10]. By introducing the transformations in the coordinates $\eta=r/a(z,t)$ and $\xi=z$, and in the dependent variables $u^*=a(z,t)u(r,z)$ and $w^*=a^2(z,t)w(r,z)$, equations (22), (23), and (24) become

$$u^*\frac{\partial \phi}{\partial \eta} + w^*\frac{\partial \phi}{\partial \xi} - [\frac{1}{a}\frac{\partial a}{\partial z}] w^*\eta\frac{\partial \phi}{\partial \eta} = \frac{1}{\eta}\frac{\partial}{\partial \eta}(\eta\frac{\partial \phi}{\partial \eta}) \tag{27}$$

$$\frac{\partial u^*}{\partial \eta} + \frac{u^*}{\eta} + \frac{\partial w^*}{\partial \xi} - [\frac{1}{a}\frac{\partial a}{\partial z}](2w^* + \eta\frac{w^*}{\partial \eta}) = 0 \tag{28}$$

$$\frac{1}{Pr}\{u^*\frac{\partial w^*}{\partial \eta} + w^*\frac{\partial w^*}{\partial \xi} - [\frac{1}{a}\frac{\partial a}{\partial z}](2w^{*2} + w^*\eta\frac{\partial w^*}{\partial \eta})\}$$
$$= -\frac{a^4}{Pr}\frac{dp}{d\xi} + \frac{1}{\eta}\frac{\partial}{\partial \eta}(\eta\frac{\partial w^*}{\partial \eta}) \tag{29}$$

The coefficient $(\partial a/\partial z)/a$ is the percentage change of the liquid phase radius per unit of dimensionless axial length. Note that all the terms with a coefficient of $(\partial a/\partial z)/a$ signify the effect due to the axial interface variation. The terms $2w^*$ and $2w^{*2}$ in equations (28) and (29) come from the acceleration of the mean velocity along the axial direction. The terms $w^*\eta\partial\phi/\partial\eta$, $\eta\partial w^*/\partial\eta$ in equations (27)-(29) are due to the coordinate transformation. If no solidification occurs, all the terms with $(\partial a/\partial z)/a$ vanish, and equations (27)-(29) are reduced immediately to the well known equations for the convection problem in the combined hydrodynamic and thermal entrance region of a circular tube. In references [1-3,6-10], a gradual variation in the liquid-solid interface i.e. $(\partial a/\partial z)/a \approx 0$ was employed.

The appropriate boundary and initial conditions for equations (21), (25), and (26), and equations (27)-(29) without the terms with the coefficient $(\partial a/\partial z)/a$ can be obtained from equations (2), (7), and (9). The conditions are:

$$\frac{\partial \theta}{\partial r} = 0 \qquad\qquad r=0, \qquad 0<z\leq z_f$$

$$\frac{\partial \theta}{\partial r} = -Bi\theta_w \qquad\qquad r=1, \qquad 0<z\leq z_f$$

$$\theta = 1 \qquad\qquad 0\leq r\leq 1, \qquad z=0$$

$$\phi_s = 1 \qquad\qquad r=a, \qquad z_f<z<\infty$$

$$\frac{\partial \phi_s}{\partial r} = -Bi\phi_{sw} \qquad\qquad r=1, \qquad z_f<z<\infty \tag{30}$$

$$a = 1 \qquad\qquad z_f < z < \infty, \qquad t = 0$$

$$u^* = \frac{\partial w^*}{\partial \eta} = \frac{\partial \phi}{\partial \eta} = 0 \qquad \eta = 0, \qquad \xi_f < \xi < \infty$$

$$u^* = w^* = \phi = 0 \qquad \eta = 1, \qquad \xi_f < \xi < \infty$$

$$u^* = \theta - \theta(a\eta, \xi_f) = 0 \qquad 0 \le \eta \le 1, \qquad \xi = 0$$

$$w^* - 2(1 - \eta^2) = 0 \qquad 0 \le \eta \le 1, \qquad \xi = 0$$

Note that the value of the pressure term $-a^4 dp/d\xi$ is determined at each axial position by considering the continuity equation for the axial velocity

$$\int_0^1 w^* \eta \, d\eta = \frac{1}{2} \tag{31}$$

Note that a parameter Bi indicating the relative importance of the convective and conductive heat transfer appears in the thermal boundary conditions at r=1. If Bi=0, an insulated thermal boundary condition results and if Bi=∞, $\theta_w = \phi_{sw} = 0$ the wall temperature T_w equals the ambient temperature T_∞. Papers discussing the problems with a finite Biot number and a constant wall temperature, Bi=∞ are listed in Table 2.

3 RESULTS AND DISCUSSION

3.1 Solidification-free Zone

Due to a finite Biot number employed in references [5,8,9,13,14,15], the wall temperature drops gradually from the inlet temperature at Z=-∞ to the freezing temperature at Z=Z_f, where Z_f is the length of the solidification-free zone. The thickness of the solid shell increases at Z $\ge Z_f$. Fig. 2 shows the length of the solidification-free zone as compared to the superheat ratio $\varepsilon = (T_0 - T_f)/(T_f - T_w)$ for the case of Pe$\to\infty$ with the Biot number as a parameter for circular pipes [5,8] and parallel-plate channels [13]. Lock et al. [5] made a first attempt to analyze an ice formation in a convectively cooled pipe. Hwang and Yih [8] recomputed the problem by using a power series solution and a numerical technique. Cheng and Wong [12] also studied the liquid solidification in a convectively cooled parallel-plate channel. It can be seen from this figure that the length of the solidification-free zone in the circular tube is shorter than in the parallel-plates with the same superheat ratio and Biot number. This is due to a better convective cooling from the circular tube than the parallel-plate channel.

The studies of solidification-free zones for liquid metals or a low Peclet number fluid were also reported in [14,15]. Lee and Hwang [14] kept the axial conduction term and

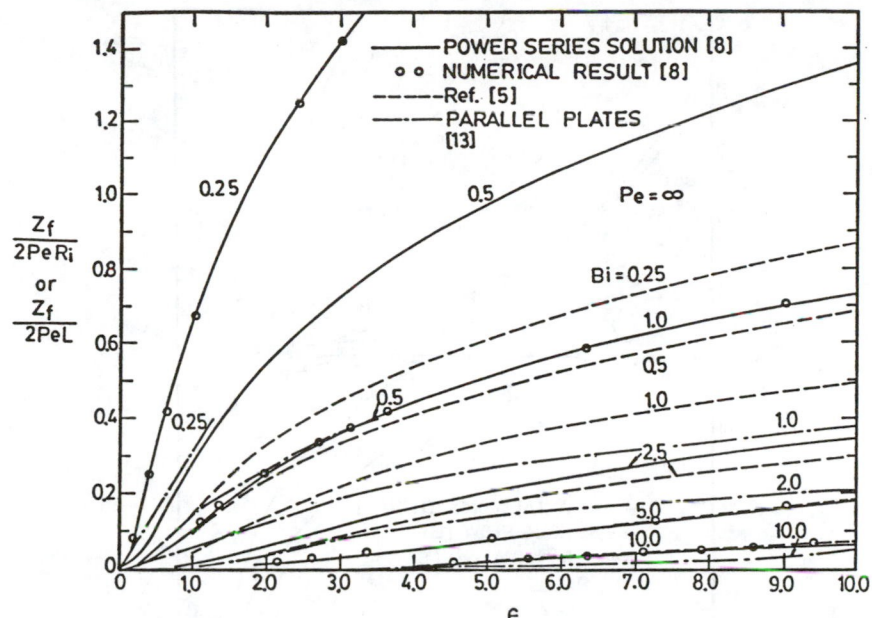

FIGURE 2. The length of solidification zone versus the
superheat ratio in steady state

examined the length of the solidification-free zone with the
Peclet number as a parameter.   Sadeghipour, Özisik, and
Mulligan [15] investigated the solidification of liquid
metals without the axial conduction term. The comparison of
these two results is shown in Fig. 3.   It can be seen that
negative values of the solidification-free length are found
for cases using a small Peclet number and a large Biot
number, for example Pe$\leq$1.5, and Bi$\geq$2 [14].   The curves
Bi=0.1 and 0.5 of reference [15] shows a reasonable trend in
comparison to the data from reference [14].   It can also be
noted that the results from reference [8] without the axial
conduction effect are in excellent agreement with the
results of large Pe from reference [14].

Figs. 2 and 3 show that there are three limiting cases for
zero length of the solidification-free zone:
1.   Pe$\to\infty$, $\varepsilon\to0$ (or $\theta_f\to1$, $\lambda\to0$)
2.   Pe$\to\infty$, Bi$\to\infty$
3.   Pe$\to0$, Bi$\to\infty$, $\theta_f\to0$ (or $\varepsilon\to\infty$)
The first two cases can be observed clearly in both Figs. 2
and 3.   The last case is seen in Fig. 3 only.   The liquid
solidifies at Z_f<0 for the case of Pe=0.5, Bi = 10 for the
values of θ_f=0.2-0.9 and the length $Z_f\to0$ as $\theta_f\to0$.

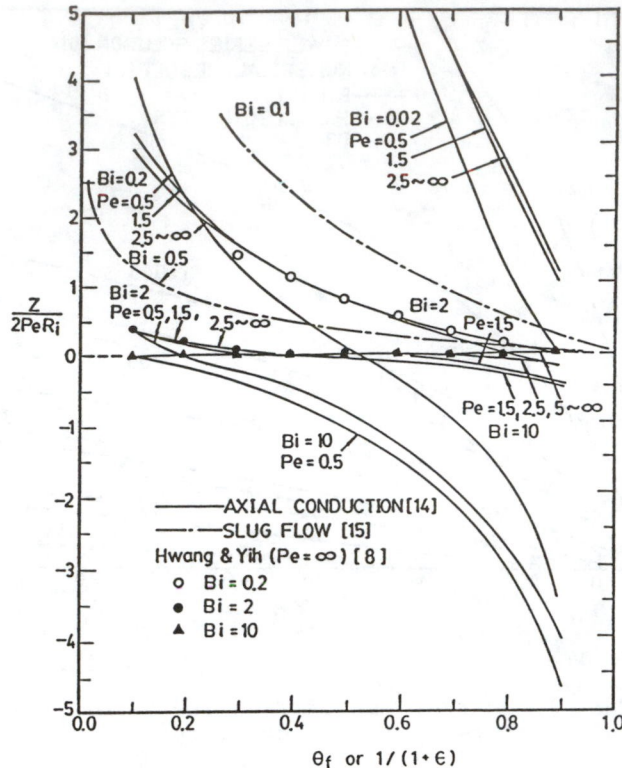

FIGURE 3.   The dimensionless length $(Z_f/2PeR_i)$ versus θ_f for various Pe and Bi in steady state

3.2 Liquid–Solid Interface

Both the steady and transient liquid–solid interface profiles with laminar forced convection are reviewed in this paper. Fig. 4 summarizes the steady state liquid–solid interface profiles [1,2,5,6,10] with laminar forced convection. The relationship between a^λ and Z/PeR_i was originally derived by Zerkle and Sunderland [1] as

$$a^\lambda = \exp\left\{-1/\left[-\frac{\partial\phi}{\partial\eta}(1,z)\right]\right\} \tag{32}$$

Özisik and Mulligan [2] used a slug flow resulting in a larger value in the radius of the liquid–solid interface in entrance region than the one obtained from the parabolic axial velocity [1]. The variational solution using a parabolic velocity distribution by Bilenas and Jiji [6]

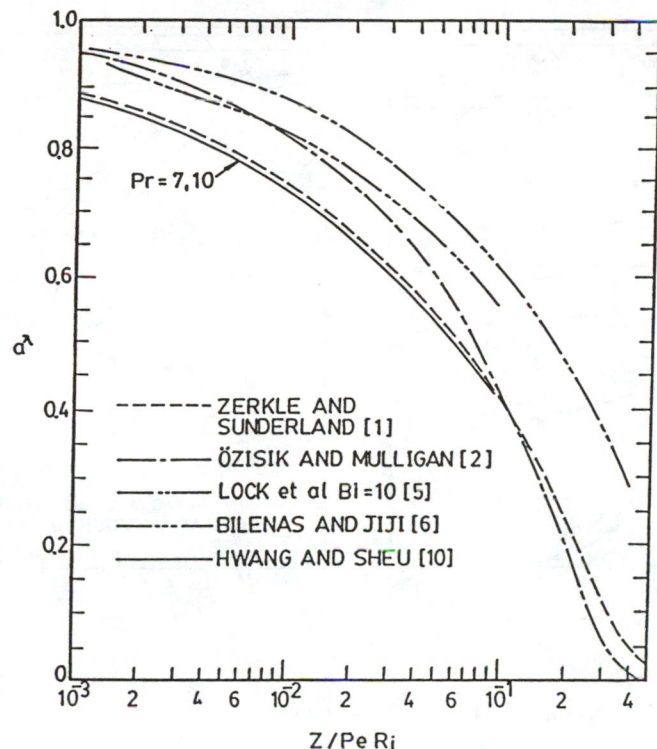

FIGURE 4. Steady state radius of liquid-solid interface versus Z/PeR_i

shows a discrepancy of 0.1 in the value of a^λ from the solution determined by Zerkle and Sunderland [1]. Hwang and Sheu [10] presented the solution in the combined hydrodynamic and thermal entrance regions with a developing axial velocity from a uniform velocity profile. Solution [10] coincides well with the solution using a parabolic velocity profile from reference [1]. This indicates the rapid growth of the momentum boundary layer for a liquid with a large Prandtl number, for example $Pr \geq 7$, in the hydrodynamic entrance region.

The transient growth of the solid shell was presented in references [5,6,10]. Due to different values selected in these references, a direct comparison is impossible. However, Figs. 5(a) and (b) show the transient variations in the liquid-solid interface for different parameters in their works. The trends of these curves are quite reasonable. Note that parameter λ cannot be combined with the dimensionless radius in the unsteady state as was done in the case of the steady state, shown previously in Fig. 4.

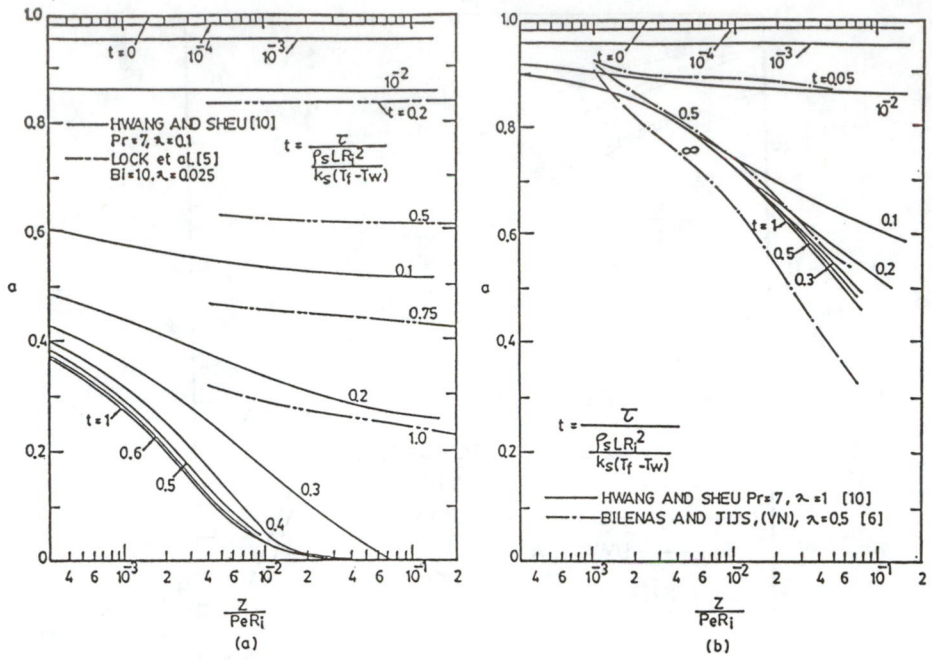

FIGURE 5. Transient interface growth

3.3 Heat Transfer Rates

Due to different flow conditions, there are two types of heat transfer characteristics:

1. Laminar forced convection with slug, parabolic or developing axial velocity as reported in references [1,2,5,6,8,10,12,13,14,15,16].
2. Laminar forced convection with natural convection effects as presented in references [1,3.4,9,11].

Fig. 6 shows variations in the dimensionless laminar heat transfer rate with the dimensionless axial position. In this figure, the rate of heat transfer from the liquid is independent of the tube wall temperature and the thickness of the phase shell. The theoretical result of Zerkle and Sunderland [1] is in excellent agreement with the result for a developing axial velocity [10] for Pr=7 and 10. This indicates that the result using the parabolic velocity profile represents a limiting case of developing flow [10] for large Prandtl number fluids. On the contrary, the slug flow result [2] cannot be realized in the combined hydrodynamic and thermal entrance regions for fluids of larger Prandtl numbers. The experimental data from various references [1,4,10,11] are also plotted for comparison. Due to free convection effects, the data [1,4,11] lie

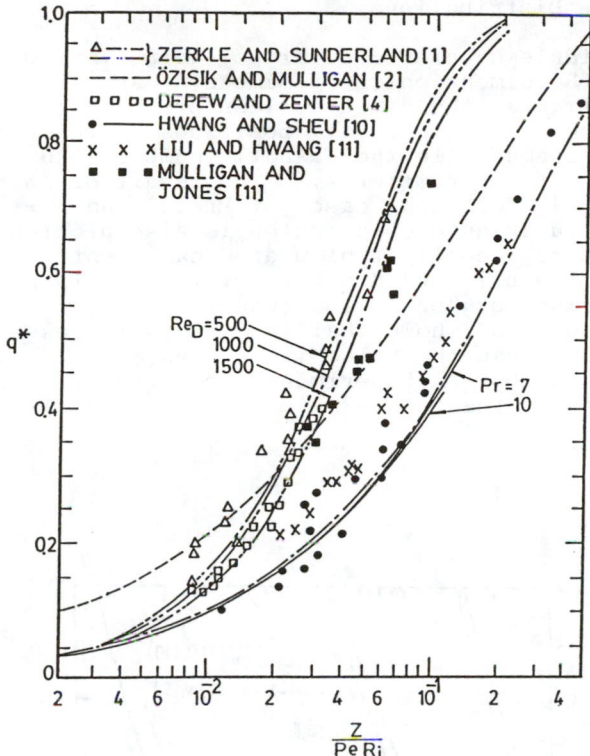

FIGURE 6. Heat transfer rates versus Z/PeR_i

considerably above the theoretical curves [1,10]. It is noted that the Grashof number is of the order of 10^6 in reference [1] and is of the order of 10^5 in references [4,11], but is only of the order of 10^4 in references [10,12].

A semi-empirical equation with natural convection effects

$$Nu_m(\frac{\mu_W}{\mu_B})^{0.14} = 1.75[Gz_m + 5.6\times10^{-4}(Gr_mPr_mZ/D)^{0.70}]^{1/3} \qquad (33)$$

given by Oliver [17] was employed in references [1,3,4] for the computation of heat transfer rate. The curves obtained from equation (22) which were presented in reference [1] are also plotted in Fig. 6.

3.4 Pressure Distributions

The dimensionless pressure drops with various Prandtl numbers and the dimensionless parameter λ are shown in Fig. 7. In this figure, the dimensionless pressure drop from the results of ref. [10], at all positions Z/PeR_i, increases with the increment in the Prandtl number and with the decreases in the parameter λ. The result of Zerkle and Sunderland [1] for the case of $Pr=10$ and $\lambda=1$ in the hydrodynamic fully developed region is also plotted in this figure. Due to the effect of hydrodynamic entrance length, the value of the pressure drop from ref. [10] is higher than the one obtained by Zerkle and Sunderland [1]. The curve from reference [5] shows that the dimensionless pressure drop is less than the value in reference [10] due to a finite Biot number. Experimental results of references [1,4,11] show a reasonable agreement with the theoretical curves [1,10].

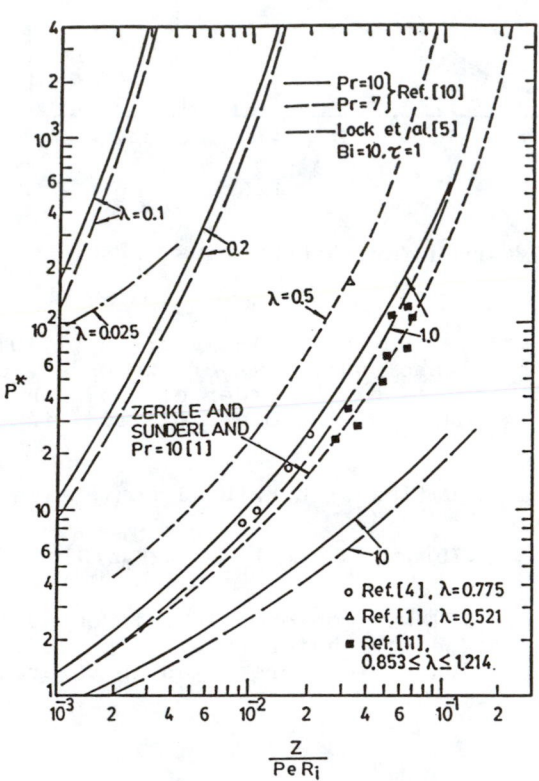

FIGURE 7. Pressure drops versus Z/PeR_i

4. CONCLUDING REMARKS

1. The solidification phenomenon is promoted and utilized in many processes such as the casting of metals, desalination of water and storage of cooling fluids. On the other hand, solidification is not always desirable. Ice formation in water mains, freezing of liquid metal in heat exchangers and phase changes in hydraulic systems, for example, may have harmful effects.

2. The dimensionless parameters a, z, t, Pe, λ, Bi, and Ste tabulated in Table 2 can be obtained by applying the order of magnitude analysis to the governing equations. Most of the papers [1-13,15] reviewed in the present article neglected the axial conduction term by considering a large Peclet number, and some of papers [5,10] employed a quasi-steady condition with a small Stefan number.

3. Due to different thermal boundary and flow conditions, the liquid flowing in a pipe solidifies at $Z=Z_f \geq 0$. For a uniform wall temperature less than the freezing temperature of the liquid, the solidification begins at $Z=0$. For convective cooling and $Pe \to \infty$, a length of solidification-free zone $Z_f > 0$ appears. For convective cooling and $Pe \to 0$, solidification may occur at $Z_f < 0$.

4. The order of magnitude analysis is a powerful technique for determining the simplified governing equations in a given complex system with proper parameters. It is still necessary to have an experimental verification of the theoretical solutions. In the present analysis, the dimensionless heat transfer rate and the pressure distribution are verified by the experimental data.

REFERENCES

1. Zerkle, R.D., and Sunderland, J.E., "The Effect of Liquid Solidification in a Tube Upon Laminar-Flow Heat Transfer and Pressure Drop," J. Heat Transfer, Vol. 90, 1968, pp. 183-190.

2. Özisik, M.N., and Mulligan, J.C., " Transient Freezing of Liquids in Forced Flow inside Circular Tubes," J. Heat Transfer, Vol. 91, 1969, pp. 385-390.

3. Des Ruisseaux, N., and Zerkle, R.D., "Freezing of Hydraulic System," Canadian J. Chemical Engineering, Vol. 47, 1969, pp. 233-237.

4. Depew, C.A., and Zenter, R.C., "Laminar Flow Heat Transfer and Pressure Drop with Freezing at the Wall," Int. J. Heat Mass Transfer, Vol.12, 1969, pp. 1710-1714.

5. Lock, G.S.H., Freeborn, R.D.J., and Nyren, R.H., "Analysis of Ice Formation in a Convectively-Cooled Pipe," Proc. Int. Heat Transfer Conf., 4th. Paris-Versailles Vol. 1 (Cu 2.9), 1970, pp. 1-11.

6. Bilenas, J.A., and Jiji, L.M., "Variational Solution of Axisymmetric Fluid Flow in Tubes with Surface Solidification," J. Franklin Institute, Vol. 289, 1970, pp. 265-279.

7. Lock, G.S.H., and Nyren, R.H., "Analysis of Fully-Developed Ice Formation in a Convectively-Cooled Circular Tube," Int. J. Heat Mass Transfer, Vol. 14, 1971, pp. 825-834.

8. Hwang, G.J., and Yih, I., "Correction on the Length of Ice-Free Zone in a Convectively-Cooled Pipe", Int. J. Heat Mass Transfer, Vol. 16, 1973, pp. 681-683.

9. Ou, J.W., and Cheng, K.C., "Buoyancy Effects on Heat Transfer and Liquid Solidification-Free Zone in a Convectively-Cooled Horizontal Square Channel," Appl. Sci. Res. Vol. 30, 1975, pp. 355-366.

10. Hwang, G.J., and Sheu, J.P., "Liquid Solidification in Combined Hydrodynamic and Thermal Entrance Region of a Circular Tube," Canadian J. Chemical Engineering, Vol. 54, 1976, pp. 66-71.

11. Mulligan, J.C., and Jones, D.D., "Experiments on Heat Transfer and Pressure Drop in a Horizontal Tube with Internal Solidification," Int. J. Heat Mass Transfer, Vol. 19, 1976, pp. 213-219.

12. Liu, H.L., and Hwang, G.J., "An Experiment on Liquid Solidification in Thermal Entrance Region of a Circular Tube," Letters in Heat Mass Transfer, Vol. 4, 1977, pp. 437-444.

13. Cheng, K.C., and Wong, S.L., "Liquid Solidification in a Convectively-Cooled Parallel-Plate Channel," Canadian J. Chemical Engineering, Vol. 55, 1977, pp. 149-155.

14. Lee, S.L., and Hwang, G.J., "Solidification of a Low Peclelt Number Fluid Flow in a Round Pipe with the Boundary Condition of the Third Kind," Canadian J. Chemical Engineering, Vol. 58, 1980, pp. 177-184.

15. Sadephipour, M.S., Özisik, M.N., and Mulligan, J.C., "Transient Solidification of Liquid Metals in the Thermal Entry Region of a Circular Tube," Nuclear Science and Engineering, Vol. 79, 1981, pp. 9-18.

16. Lee, S.L., and Hwang, G.J., "Liquid Solidification in Low Peclet Number Pipe Flows," Canadian J. Chemical Engineering, Vol. 67, 1989.

17. Oliver, D.R., "The Effect of Natural Convection on Viscous Flow Heat Transfer in Horizontal Tube," Chem. Engng. Sci. Vol. 17, 1962, pp. 335-350.

Chapter 6

On the Freeze Occlusion of Water

G. S. H. Lock
Department of Mechanical Engineering, University of Alberta,
Edmonton, Alberta, Canada T6G 2G8

ABSTRACT

This paper presents an exploration of an important class of
problems dealing with the restriction or stoppage of water flows by
means of freezing. Several representative problems are analysed with a
view to defining the conditions under which partial or complete
occlusion may take place. Specific reference is made to the velocity
field in the water, and to the prevailing temperature fields in both
ice and water. The role of viscous dissipation is also discussed.

In keeping with the spirit of Gilpin's work, the treatment is less
concerned with detailed solutions than with the essential structure of
the problems: analysis is based on first principles. Accordingly,
normalization and ordering are used to define the characteristic
conditions of occlusion. The discussion is limited to laminar flow
under quasi-steady conditions: the treatment is readily extended to a
wider range of circumstances.

CONTENTS

NOMENCLATURE

d   - tube diameter
g   - gap
h   - heat transfer coefficient
j   - heat flux density
k   - thermal conductivity
L   - overall length
p   - absolute pressure
Q   - volumetric flow rate per unit width
$\underset{\sim}{R}$ - radial displacement

T	- absolute temperature
t	- time
U,V	- velocities
u,v	- normalized velocities
X,Y	- Cartesian coordinates
x,y	- normalized coordinates
s,S	- distance along the ice-water interface

Subscripts

c	- centerline, characteristic
f	- free surface
i	- ice, ice-water interface
L	- $X = L$
m	- midplane
0	- $X = 0$
s	- substrate
w	- water
∞	- bulk

Superscript

$\circ$	- occlusion

Greek Symbols

α	- angle
Δ	- lateral width
δ	- boundary layer thickness
ϵ	- superheat ratio
π,Π	- pressure
θ,ϕ	- temperature
κ	- thermal diffusivity
λ	- latent heat of freezing
μ	- absolute viscosity
ν	- momentum diffusivity
ρ	- density
τ	- normalized time

Non-Dimensional Groups

Ec	- Eckert number
Gz	- Graetz number
Bi	- Biot number
Pe	- Peclet number
Pr	- Prandtl number
Re	- Reynolds number
Ste	- Stefan number

1. INTRODUCTION

The presence of sub-zero temperatures in the vicinity of flowing water often causes the water to freeze. Almost as often, the resulting ice becomes an impediment to the water flow which may eventually be completely occluded: the word occlude is taken to mean partial or complete stoppage of the flow system. The occlusion of water flow by freezing is a widespread phenomenon which has received fragmentary treatment limited either by the particular application or by conditions which did not approach a complete shut down. This paper will treat freeze occlusion as a single, though variable, process *sui generis*, and will attempt to present a unified view of the subject.

Freeze occlusion is perhaps most familiar in an environmental context when rivers and estuaries are covered in ice which, in extreme circumstances, may extend down to the bed. Likewise, the arrest of inclined water flows - from aufeis to the icicle to the waterfall - bear witness to the natural environment's enormous occlusive power. Perhaps less obvious, because they are usually less visible, are the occlusive effects attributable to such phenomena as hanging dams in a river, or dendritic crystals inside a buried water pipe.

Occlusion is not limited to free flowing water; it is also found in porous media and various particulate aggregates. Seepage and filtration of water through soil may be influenced by the ice growth produced, for example, by a chilled gas line or a cryopile. Similarly, the functioning of most plants and animals is greatly compromised when they are exposed directly to sub-zero temperatures which impede and ultimately arrest the flow of aqueous biofluids.

For the most part, the effect of occlusion is regarded as undesirable, even threatening, but there are some circumstances when occlusion is actually sought[*]. In either circumstance the process is the same and may therefore be modelled in the same way. This paper illustrates some models of occlusion which have been used or may be used in future work. No attempt will be made to develop complete analytic or numerical solutions to the various problems discussed, but the basis of the analysis will be detailed along with any implications or limitations.

In certain circumstances the process of partial occlusion is straight-forward: growing ice simply reduces the area through which the water flows, the flow field itself being otherwise unchanged. The shape and movement of the ice-water interface in such problems may not be a simple function of time, but the physical process is clear and explicit. In other circumstances, the impedance introduced by the growing ice alters the flow field in a fundamental way, thus coupling the water flow and heat flow systems. Obviously, such a coupling will complicate the process overall, and this is especially significant when occlusion eventually continues to completion.

Since freeze occlusion may occur in a wide variety of circum-stances, it is worthwhile making an attempt to divide the phenomenon into its principal types or categories. Useful divisions are illus-trated in Fig. 1 which, in the interests of simplicity and generality, shows a water flow (in the absence or presence of a porous medium) bounded by a solid substrate and a free surface. The three main sources for the nucleation and growth of ice crystals suggest the following divisions:

---

*For example, the isolation of a water valve for inspection without shutting down the system.

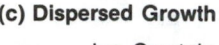

(a) Substrate Growth

Free Surface

Water Flow →

|‖‖‖‖‖‖‖‖‖‖‖‖‖| Ice
////////// Cold
Substrate

(b) Surface Growth

Cold Environment ↔

⌐ Initial Water Surface
|‖‖‖‖‖‖‖‖‖‖‖‖| Ice
Water Flow →

////////// Substrate

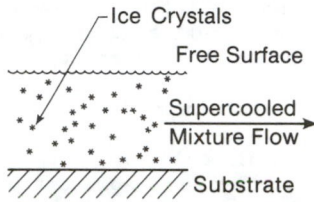

(c) Dispersed Growth

⌐Ice Crystals

Free Surface

Supercooled
Mixture Flow →

////////// Substrate

Fig. 1. Categories of occlusion

(a) <u>Substrate growth</u>, in which the substrate is cooled sufficiently to cause ice to form and grow on its surface. There may or may not be supercooling in the water, and the ice is assumed to remain attached to the substrate.

(b) <u>Free surface growth</u>, for which there must be a free water surface exposed to a suitably cold environment. Ice is assumed to nucleate on, and to grow at and from, the surface.

(c) <u>Dispersed growth</u>, resulting from the nucleation and subsequent growth of ice throughout the bulk of supercooled water.

It is obvious that the illustrations in Fig. 1 could be supplemented by other configurations and circumstances. Occlusion in an enclosed channel, for example, would be very like that shown in Fig. 1b if growth occurred only along the top; and if the free flowing water indicated in the figure is replaced by flow through a porous medium, as in ground water flow, yet further possibilities arise. Since there are too many particular variations to treat in a single, exploratory paper, the following discussion will be limited to type (a) in Fig. 1. The treatment, however, will attempt to deal with generalities as far as possible.

2. FREEZING OF A GRAVITY-DRIVEN FILM

The simplest model of occlusion for the water film depicted in Fig. 2 assumes quasi-steady, quasi-one-dimensional conditions* in which the interface equation

---

*The plane of the substrate lies normal to the page; Ste ≪ 1, and variations in ice and water film thickness are gradual.

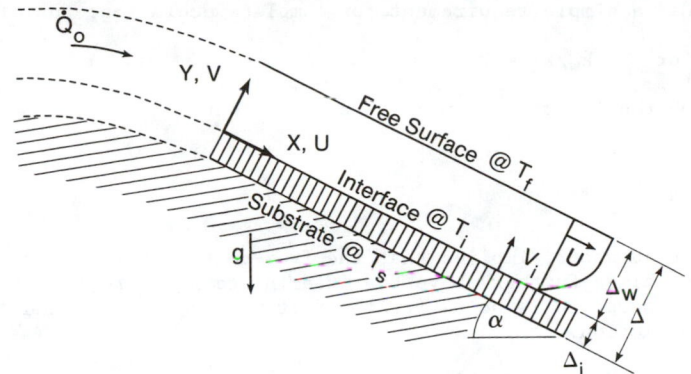

Fig. 2. Occlusion of a film: 1-D

$$\rho_i \lambda V_i = k_i \left(\frac{\partial T_i}{\partial y}\right)_i - k_w \left(\frac{\partial T_w}{\partial y}\right)_i \tag{1}$$

reduces to

$$\rho_i \lambda V_i = - \frac{k_i \theta_s}{\Delta_i} - h\theta_f$$

where $\theta = T - T_i$. Thus when growth stops

$$\theta_f = - \theta_s/Bi \tag{2}$$

where $Bi = h\Delta_i/k_i$. This may be an adequate treatment of partial occlusion, but it does suffer from several drawbacks among which are: information on the time elapsed, the length over which the freezing takes place, and the heat transfer coefficient in the water. Suitable values of h may not be available, especially as $\Delta_i \to \Delta$ and occlusion becomes complete.

Should heat transfer in the water be purely by conduction, Eq. (1) reduces to the no growth limit.

$$\theta_f = \frac{k_i}{k_w} \frac{\Delta_w}{\Delta_i} \theta_s \tag{3}$$

from which we may obtain

$$\frac{\Delta_i}{\Delta} = \frac{1}{1 - \epsilon \frac{k_w}{k_i}} \tag{4}$$

where $\epsilon = \theta_f/\theta_s$ is the superheat ratio.

This provides a simple requirement for complete occlusion: namely,

$\Delta/\Delta_i = 1$, or $- \epsilon k_w/k_i = 0$.

To avoid substantial occlusion it is evident that

$$\theta_f \geq - \frac{k_i}{k_w} \theta_s \tag{5}$$

which, for cold water, implies that the free surface temperature must be over four times further above the freezing temperature than the substrate temperature is below it; the fact that $k_w/k_i < 1$ tends to encourage occlusion.

A better model would incorporate the essential continuity requirement: namely, that the rate of film water depletion must equal the rate of ice growth, as expressed by

$$\rho_w \dot{Q}_o - \rho_i \int_o^X V_i \, dX = \rho_w \int_o^{\Delta_w} U \, dy \tag{6}$$

$$\text{or} \quad V_i = - \frac{\rho_w}{\rho_i} \frac{\partial}{\partial X} \int_o^{\Delta_w} U \, dY \tag{7}$$

As demonstrated later, the velocity in a slow laminar, gravity-driven film may be expressed by

$$u = \frac{\Delta_w^2 \, g \, \sin \alpha}{\nu} \, [(\frac{Y}{\Delta_w}) - \frac{1}{2} (\frac{Y}{\Delta_w})^2] \tag{8}$$

in which Δ_w is a slowly-varying function of X.

Thus, if density difference is ignored,

$$V_i = -g \frac{\sin \alpha}{3\nu} \frac{d}{dX} (\Delta_w^3)$$

which integrates formally over the length L to give

$$\Delta_{wL}^3 - \Delta_{wo}^3 = \frac{-3 \, \nu}{g \, \sin \alpha} \int_o^L V_i \, dX \tag{9}$$

An elementary solution of Eq. (9) may be obtained when the ice-water interface remains planar: that is if V_i is taken as constant. This leads to

$$\frac{\Delta_{wL}}{\Delta_{wo}} = [1 - \frac{3 \, \nu \, V_i \, L}{\Delta_{wo}^3 \, g \, \sin \alpha}]^{1/3} \tag{10}$$

from which we may estimate the corresponding length for complete occlusion as

$$L° = \frac{\Delta_{wo}^3 \, g \sin \alpha}{3 \, \nu \, V_i} \tag{11}$$

The above elementary relations reveal the important fact that the water film thickness under such conditions varies almost linearly* with distance from the leading edge; the occlusion length evidently varies as the cube of the upstream film thickness.

For a time-dependent treatment of the interface velocity we must return to the interface Eq. (1). Assuming for simplicity that sensible heat is not important in either the water or the ice, Eq. (1) may be used to generate the familiar Stefan solution.

$$V_i = \frac{d\Delta_i}{dt} = \left[\frac{k_i \, (T_i - T_s)}{2 \, \rho_i \lambda t}\right]^{1/2}$$

Substituting this into Eq. (9), and again taking V_i independent of X, we obtain

$$\frac{\Delta_{wx}}{\Delta_{wo}} = \left[1 - \frac{3\sqrt{2} \, \nu \kappa_i}{\Delta_{wo}^3 \, g \sin \alpha} \, Ste^{1/2} \, \eta\right]^{1/3} \tag{12}$$

where $\eta = X/2 \, (\kappa_i t)^{1/2}$ is a similarity variable. Again we notice the third power relation between film thickness** and distance from the leading edge: such a relation only applies if V_i is independent of X.

In quasi-steady analyses such as the above, the effect of time enters the results through the boundary conditions rather than the governing equations. One must therefore be cautious in interpreting transient effects reflected in the results. This is especially true in predicting movement of the toe of the film ($X=L°$). For $X < L°$, Eq. (12) will apply, but at $X = L°$ we find that

$$L° = \frac{\Delta_{wo}^3 \, g \sin \alpha}{3 \, \nu \, Ste^{1/2}} \, \left(\frac{2t}{\kappa_i}\right)^{1/2}$$

Now as $L°$ increases, the water spilling over to reach it does not necessarily flow over ice which has previously grown as assumed, i.e. according to the Stefan solution. The water may, for example, flow

---

*Using the first two terms of a binomial expansion:

$$\left(\frac{d\Delta_w}{dX}\right)_{X=0} = \frac{- \nu \, V_i}{\Delta_{wo}^2 \, g \sin \alpha}$$

**The water film gradient at the leading edge is now given by the time-dependent relation

$$\left(\frac{d\Delta_w}{dX}\right)_{X=0} = - \frac{\nu \, Ste^{1/2}}{\Delta_{wo}^3 \, g \sin \alpha} \, \left(\frac{V_i}{2t}\right)$$

directly on to the subcooled substrate, where $T_s \neq T_i$. Movement of the toe is intrinsically a dynamic phenomenon and should be treated by incorporating the transient terms in the governing equations. An exception may occur if V_i increased with time, the resulting retraction of the toe then being treated quasi-steadily according to the global form of Eq. (6)

$$L^\circ = \frac{\dot{Q}_o \, \rho_w}{\overline{V}_i \, \rho_i}$$

where $\overline{V}_i$ is the mean interface velocity integrated over L° at any particular time.

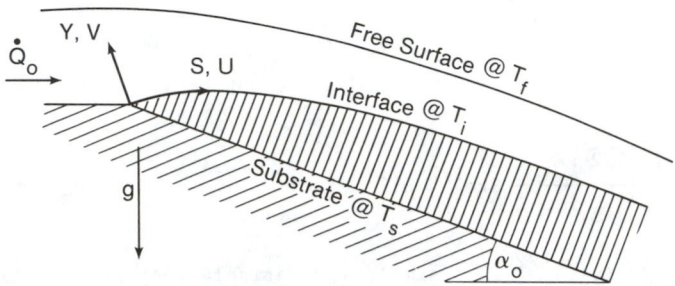

Fig. 3. Occlusion of a film: 2-D

A fuller analysis of the laminar, gravity-driven film requires a more realistic physical model (see Fig. 3) and a discussion of the governing equations[*]:

$$\frac{\partial U}{\partial X} + \frac{\partial V}{\partial Y} = 0$$

$$U\frac{\partial U}{\partial S} + V\frac{\partial U}{\partial Y} = -\frac{1}{\rho}\frac{\partial \Pi}{\partial S} + g \sin \alpha + \nu \left(\frac{\partial^2 U}{\partial Y^2} + \frac{\partial^2 U}{\partial S^2}\right)$$

$$U\frac{\partial V}{\partial S} + V\frac{\partial V}{\partial Y} = -\frac{1}{\rho}\frac{\partial \Pi}{\partial S} + g \cos \alpha + \nu \left(\frac{\partial^2 V}{\partial Y^2} + \frac{\partial^2 V}{\partial S^2}\right)$$

$$U\frac{\partial \theta}{\partial S} + V\frac{\partial \theta}{\partial Y} = \kappa \left(\frac{\partial^2 \theta}{\partial Y^2} + \frac{\partial^2 \theta}{\partial S^2}\right) + \frac{\nu}{C_p} \left(\frac{\partial U}{\partial Y}\right)^2$$

(13)

where $\Pi = P - P_{oi}$. After normalization, the continuity equation is satisfied automatically and the equations of motion may be rewritten

---

[*]S and Y in these equations form an orthogonal non-rectangular system of coordinates. If the film is thin, the effect of interface curvature may be ignored.

$$\left[\frac{Y_c^2 U_c}{\nu\, S_c}\right] \left(u\frac{\partial u}{\partial s} + v\frac{\partial u}{\partial y}\right) = -\left[\frac{\Pi_c Y_c}{\mu\, S_c U_c}\right]\frac{\partial \pi}{\partial s} + \left[\frac{Y_c^2\, g\, \sin\alpha_o}{\nu\, U_c}\right]\frac{\sin\alpha}{\sin\alpha_o}$$

$$+ \frac{\partial^2 u}{\partial y^2} + \left[\frac{Y_c}{X_c}\right]^2 \frac{\partial^2 u}{\partial s^2}$$

$$\left[\frac{U_c^2\, Y_c}{g\, S_c^2}\right]\left(u\frac{\partial v}{\partial s} + v\frac{\partial v}{\partial y}\right) = -\left[\frac{\Pi_c}{\rho g Y_c \cos\alpha_o}\right]\cos\alpha_o\,\frac{\partial \pi}{\partial y} - \cos\alpha \qquad (14)$$

$$+ \left[\frac{\nu\, U_c}{g Y_c S_c}\right]\left(\frac{\partial^2 v}{\partial y^2} + \left[\frac{Y_c}{X_c}\right]^2\frac{\partial^2 v}{\partial s^2}\right)$$

If the film is gravity driven, the implied balance between the gravitational and viscous forces reveals that the velocity scales are given by

$$U_c = \frac{\Delta_{wo}^2\, g\, \sin\alpha_o}{\nu} \qquad (15)$$

$$V_c = \frac{\Delta_{wo}^3\, g\, \sin\alpha_o}{\nu L}$$

which were tacitly assumed earlier in Eq. (8). Variations in static pressure are two-fold: hydrostatic water pressure variations across the film, and hydrostatic air pressure variations along the film. Since the latter are small in comparison to the former, we may define the pressure scale by:

$$\Pi_c = \rho_w \Delta_{wo}\, g\, \sin\alpha_o \qquad (16)$$

The equations of motion may thus be re-written:

$$\frac{Re_\Delta \Delta_{wo}}{L}\left(u\frac{\partial u}{\partial s} + v\frac{\partial u}{\partial y}\right) = -\left[\frac{\Delta_{wo}}{L}\cot\alpha_o\right]\frac{\partial \pi}{\partial s} + \frac{\sin\alpha}{\sin\alpha_o} + \frac{\partial^2 u}{\partial y^2} + \left[\frac{\Delta_{wo}}{L}\right]^2\frac{\partial^2 u}{\partial s^2}$$

$$Re_\Delta \left(\frac{\Delta_{wo}}{L}\right)^2 \sin\alpha_o\left(u\frac{\partial v}{\partial s} + v\frac{\partial v}{\partial y}\right) = -\cos\alpha_o\,\frac{\partial \pi}{\partial y} - \cos\alpha \qquad (17)$$

$$+ \frac{\Delta_{wo}}{L}\sin\alpha_o\left(\frac{\partial^2 v}{\partial y^2} + \left[\frac{\Delta_{wo}}{L}\right]^2\frac{\partial^2 v}{\partial s^2}\right)$$

where $Re_\Delta = \dfrac{U_c\,\Delta_{wo}}{\nu} = \dfrac{\Delta_{wo}^3\, g\, \sin\alpha_o}{\nu^2}$

A number of observations on the momentum Eqs. (17) are now possible. First, it is evident that lateral velocities are less than longitudinal velocities by a factor of Δ_{wo}/L. Second, longitudinal variations in pressure may be ignored if $\tan\alpha_o \gg \Delta_{wo}/L$, i.e. the substrate inclination is steep; they may not be important for shallow inclinations either if the film is slender enough. Third, the lateral viscous terms are much less than the longitudinal viscous terms. Fourth, the effect of inertia may be ignored if $Re_\Delta \ll L/\Delta_{wo}$.

Restrictions on the use of a parabolic velocity profile are thus evident: $\Delta_{wo}/L \ll Re_\Delta$, $\tan \alpha_o$ and 1. The third condition is commonly satisfied, and the second will apply for all but the very smallest inclinations permissible in a gravity-driven flow. The first condition may be re-stated as $\Delta_{wo}/L \ll \nu/\dot{Q}_o$, thus tying the relative importance of inertia to the upstream flow rate or velocity. Clearly, when $\dot{Q}_o = O(\nu L/\Delta_{wo})$ the effect of inertia may not be ignored, and the longitudinal equation of motion must then be written.

$$u\frac{\partial u}{\partial s} + v\frac{\partial u}{\partial y} = \frac{\sin\alpha}{\sin\alpha_o} + \frac{\partial^2 u}{\partial y^2}$$

which is of the boundary layer type in which the lateral length scale Y_c is no longer Δ_{wo}, but is given instead by

$$Y_c = (\frac{\nu L}{U_c})^{1/2}$$

or, in more familiar notation,

$$\delta(S) = (\frac{\nu^2 S}{g\sin\alpha_o})^{1/4} \tag{18}$$

$$U_c(S) = (gS\sin\alpha)^{1/2}$$

Since $\delta \leq \Delta_{wo}$, it is evident that this inertial region extends over the range

$$\frac{S}{\Delta_{wo}} \leq \frac{\Delta_{wo}^3 \, g\sin\alpha_o}{\nu^2} \tag{19}$$

Equations (18) reveal that the flow is not a Blasius type.

In order to calculate the temperature gradients within the water it is necessary to subject the energy equation to the same type of analysis. After normalization, we obtain

$$[\frac{U_c Y_c^2}{k S_c}] \, (u\frac{\partial\phi}{\partial s} + v\frac{\partial\phi}{\partial y}) = \frac{\partial^2\phi}{\partial y^2} + [\frac{y_c}{S_c}] \frac{\partial^2\phi}{\partial s^2} + [\frac{\mu U_c^2}{k \theta_f}] \, (\frac{\partial u}{\partial y})^2 \tag{20}$$

in which Y_c now refers to the thermal system. Inspection reveals that longitudinal conduction may be neglected if $\Delta_{wo} \ll L$, while neglect of viscous dissipation is valid when

$$\frac{\mu U_c^2}{k\theta_f} = PrEc \ll 1$$

The relative importance of advection may be gauged from the coefficient

$$\frac{U_c Y_c^2}{k S_c} = Gz_o = Pe_\delta \frac{\delta_*}{L} = Re_\Delta \frac{\Delta_{wo}}{L} Pr(\frac{\delta_*}{\Delta_{wo}})^2$$

where $\delta_* = Y_c$ in the thermal system. It thus appears that since $\delta < \Delta_{wo}$ and $\delta_* < \delta$ (for $Pr \geq 1$) the importance of inertia in the momentum equation implies the importance of advection in the energy equation; the converse, however, is not necessarily true. More precisely, the thermal boundary layer thickness when inertia is important is given by

$$\delta_*(S) = (\frac{\kappa^2 S}{g \sin \alpha_o})^{1/4} \tag{21}$$

whereas the corresponding relation when inertia is unimportant is

$$\delta_*(S) = (\frac{\nu \kappa S}{\Delta_{wo}^2 g \sin \alpha_o})^{1/2} \tag{22}$$

The inertialess advection region described by Eq. (22) thus extends up to

$$\frac{S}{\Delta_{wo}} \leq \frac{\Delta_{wo}^3 g \sin \alpha_o}{\nu \kappa}$$

whereas the inertial advection region described by Eq. (21) extends up to

$$\frac{S}{\Delta_{wo}} \leq \frac{\Delta_{wo}^3 g \sin \alpha_o}{\kappa^2}$$

By comparison with Eq. (19) it is evident that the advection region extends much further downstream than the inertial region when $Pr > 1$: the factor is as great as Pr^2 which, in cold water, is more than two orders of magnitude. It is therefore clear that the neglect of inertia implicit in the use of a parabolic velocity profile does not necessarily justify the neglect of advection.

The relative importance of viscous dissipation, on the other hand, implies that $PrEc = 0(1)$, which may be stated as

$$\left. \begin{array}{l} \theta_f = 0 \ [\frac{\mu}{k} (\frac{\dot{Q}_o}{\Delta_{wo}})^2 \ ] \\ \\ \theta_f = 0 \ [\frac{\mu}{k} (U_c)^2] \end{array} \right\} \tag{23}$$

For a laminar, gravity-driven film this yields

$$\theta_f = 0 \ [\frac{(\Delta_{wo}^2 \ \rho \ g \ \sin \alpha_o)^2}{\mu k}]$$

which, for a vertical surface in cold water, reduces to

$$\theta_f = 0 \ (10^{11} \ \Delta_{wo}^4)$$

as the maximum temperature difference at which viscous dissipation is important. This implies that for a film 1 mm thick, $\theta_f = 0(10^{-1} \ ^\circ C)$ whereas for a film 1 cm thick, $\theta_f = 0 \ (10^3 \ ^\circ C)$! Corresponding values for shallow inclinations are reduced substantially.

Whether or not the temperatures generated by viscous dissipation would be significant in the delay or prevention of occlusion depends not only upon the scale of any imposed temperature field, e.g. if the water is heated initially, but on the rate at which heat is conducted away through the ice. In the earlier discussion it was observed that growth ceases when $\theta_f = -\theta_s/Bi$, and therefore if this cessation is attributable to viscous dissipation it follows that

$$\Delta_{wo} = 0 \ ([\frac{(T_i - T_s)}{h\Delta_i \sin^2\alpha_o}]^{1/4} \ x \ 10^{-3})$$

for cold water. It is thus evident that the appropriate magnitude of Δ_{wo} is not likely to be more than the order of a few milimetres except for slender, near-horizontal ice layers.

A more precise treatment of occlusion demands a return to Eq. (7) in which the integration is carried out from the solution[*] of Eqs. (17). The interface velocity itself obviously depends upon the two coupled temperature fields. In normalized form,

$$v_i = - \ \frac{\rho_i \ \lambda \ \Delta_i \ V_i}{k_i \ \theta_s} = (\frac{\partial \phi_i}{\partial y})_i + \frac{k_w}{k_i} \frac{\theta_f}{\theta_s} \frac{\Delta_i}{\delta_*} (\frac{\partial \phi_w}{\partial y})_i$$

or, if Ste $\ll$ 1,

$$v_i = 1 + \frac{k_w}{k_i} \frac{\theta_f}{\theta_s} \frac{\Delta_i}{\delta_*} (\frac{\partial \phi_w}{\partial y})_i \quad\quad\quad (24)$$

bearing in mind that y in the water is measured normal to the interface, not the substrate. Equation (24) in turn depends upon the solution of the energy equation

---

[*]Strictly, movement of the interface creates a non-Newtonian frame of reference in the water, but the interface velocity is usually small enough to ignore in the calculation of fluid velocity.

$$Gz\ u\frac{\partial \phi_w}{\partial s} = \frac{\partial^2 \phi_w}{\partial y^2} + PrEc\ \left(\frac{\partial u}{\partial y}\right)^2 \tag{25}$$

for a slender, gravity-driven film. Equations (24) and (25) are evidently coupled not only through the water and ice temperature fields, but through the fact that the changing interface shape will change the water film thickness profile and hence the water temperature profile. For the special situation where sensible heat in the water may be neglected along with that in the ice, we obtain from Eq. (7)

$$f(\xi) = f(o) - \frac{\nu\ \kappa_i\ \rho_i\ Ste^{1/2}}{\Delta_{wo}^3\ \rho_w\ g\ \sin\ \alpha_o}\ \eta$$

where

$$\xi = \frac{\sin\alpha}{\sin\alpha_o}\ \left(\frac{\Delta_w}{\Delta_{wo}}\right)^3$$

and

$$\eta = \int_o^X \phi_s^{1/2}\ dX/2(\kappa_i t)^{1/2}$$

This is essentially the same result as given by Eq. (12).

3. FREEZING OF INTERNAL AND EXTERNAL FLOWS

3.1 Internal Flows: The Closed Channel

There are many situations where flowing water does not have a free surface but is surrounded completely by a substrate, e.g. pipes, closed channels, etc. Such internal flows are not essentially different from film flows and therefore they need not be treated here in as much detail. Figure 4 depicts a representative situation in which two horizontal, flat and parallel substrates initially bound the flow of water. When the substrate temperatures Ts_1, Ts_2 (here taken equal for simplicity) fall beneath the equilibrium freezing temperature ice will form as shown if the water enters the channel above the freezing temperature. The flow is driven by an overall pressure difference whose magnitude P_O-P_L will be strongly influenced by the obstructing effect of the ice. We may again begin our analysis by considering Eq. (1) applied to the quasi-steady, quasi-one-dimensional system shown in Fig. 4. When ice growth is arrested

$$\theta_m = -\ \frac{\theta_s}{Bi} \tag{26}$$

using Eq. (2). It is worth noting that heat transfer coefficients are available for a wide range of internal flows, thus making Eq. (26) a simple, rapid, though crude, method suitable for partial occlusion problems. Its limitations include the neglect of variations in temperature, velocity and pressure with respect to time and distance

downstream. In essence, Eq. (26) assumes or neglects the shape of the
ice water interface; it is only as valuable as our ability to predict
$\theta_m(X)$ permits. Viscous dissipation tends to maintain or increase θ_m
which otherwise decays with increasing distance along the channel.

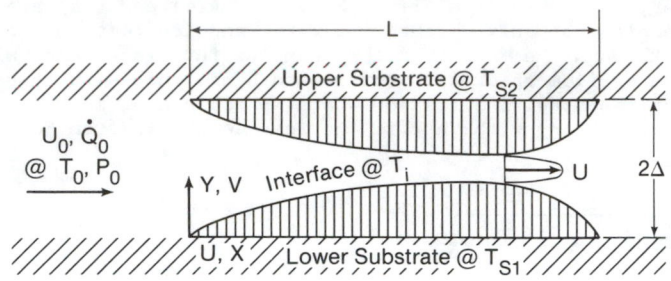

Fig. 4. Occlusion of internal flow

The impeding effect of the ice will be felt in an overall relation
of the form

$$\dot{Q} = \frac{P_o - P_L}{R} \tag{27}$$

where R is a hydraulic resistance which will, in general, increase with
time during a freezing process; it's rate of increase will also vary
with time, as will be discussed later. Equation (14) applied to the
closed channel of Fig. 4 reveals that for a pressure-driven flow

$$\frac{\Pi_c Y_c^2}{\mu X_c U_c} = 0(1)$$

$$\Pi_c = 0\left(\frac{\mu X_c U_c}{Y_c^2}\right) = 0\left(\frac{\mu X_c \dot{Q}}{Y_c^3}\right)$$

which may be written in the form of Eq. (27)

$$\dot{Q} = \frac{\Pi_c}{R} \tag{27a}$$

where $\Pi_c = P_o - P_L$ and $R = 0 \, (\mu L / \Delta_w^3)$. It is important to note that in
this expression Δ_w is a representative measure of water thickness
chosen to ensure that a proportionality is converted into an equality.
With this understanding, it is clear that as $\Delta_w \to o$, $R \to \infty$ and
occlusion tends to completion.

Despite the simplicity of the above hydraulic equation, it is not
obvious how flow rate or pressure drop, or indeed velocity, will in
fact vary with time. Provided $\rho \Delta_w^4 (P_o - P_L)/(\nu L)^2 \ll 1$, Eq. (14)
reduces to the Poiseuille problem with its familiar parabolic velocity

$$u = - \frac{1}{2} \frac{\partial \pi}{\partial x} (1 - y^2)$$

where y is measured from the mid-plane. From this, it may be shown that

$$U_m = - \frac{1}{2\mu} \frac{\partial \Pi}{\partial X} \Delta_w^2$$

consistent with the ordering above, but still assuming that Π is prescribed. Figure 5 shows a number of alternatives. In Fig. 5a, the water flow rate is held fixed so that the hydraulic gradient on either side of the occluding ice will also remain fixed; pressure drop across the growing ice increases with time. In Fig. 5b, the pressure drop[*] across the ice is held constant, even though its hydraulic resistance gradually increases; the flow rate obviously decreases with time. Figures 5c, 5d, and 5e are all variations of another constraint: namely, prescribing the overall pressure difference at specific points upstream and downstream of the ice. In Fig. 5c the immediate downstream pressure P_L is fixed, whereas in Fig. 5d the immediate upstream pressure P_o is fixed; in Fig. 5e the fixed static pressures are located well upstream P_u and downstream P_d. It is clear that two constraints must be specified if conditions are to be fully defined: two pressures, or a single pressure combined with either the flow rate or the occlusive pressure drop. Stated another way, the hydraulic characteristics of the growing ice are not sufficient to completely define its own role within the overall system of which it forms a dynamic part.

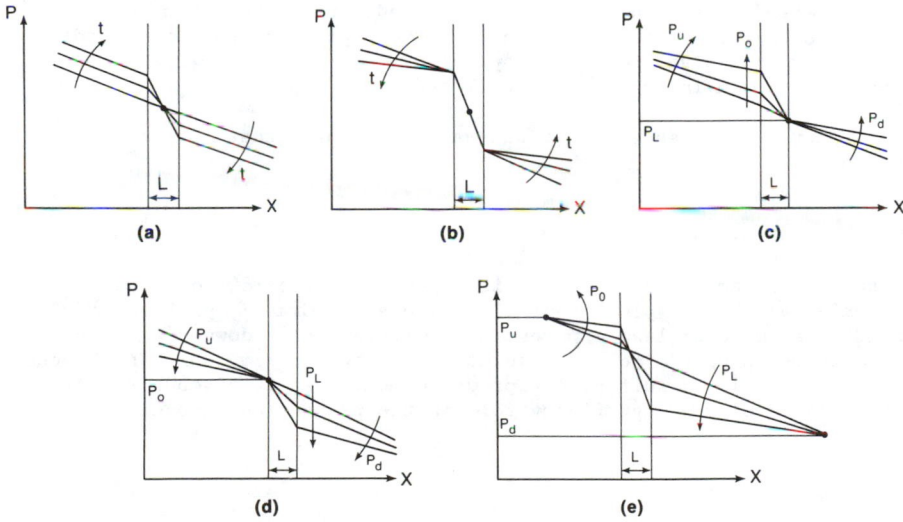

Fig. 5. Static pressure profiles

---

*For purposes of presentation, the pressure midway in the ice is also taken as constant.

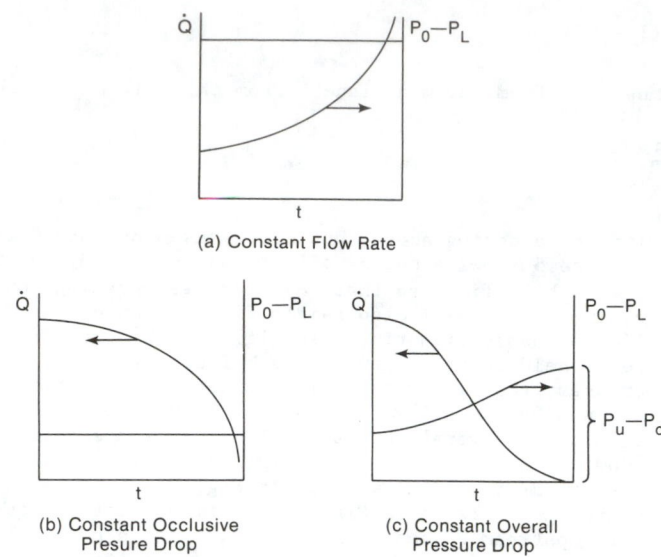

(a) Constant Flow Rate

(b) Constant Occlusive
Presure Drop

(c) Constant Overall
Pressure Drop

Fig. 6. Hydraulic characteristics

Corresponding flow rate and occlusive pressure drop curves are shown in Fig. 6. The constant flow rate and constant occlusive pressure drop conditions are illustrated in Fig. 6a and 6b, respectively. Figure 6c depicts the behaviour for a fixed overall pressure drop. It is clear from these curves that the constraints exert a significant effect when occlusion itself is significant. The flow rate and the fluid velocity are markedly different when total occlusion is approached in particular, the rapid closure corresponding to a fixed occlusive pressure drop is very different from the gradual closure accompanying a fixed overall pressure drop.

Under quasi-steady conditions we would expect that for no further growth

$$\theta_m(X) = - \frac{k_i \Delta_w}{2k_w \Delta_i} \theta_s$$

if the water temperature profile is taken as parabolic* for convenience. In a typical thermal entrance, or Graetz, problem θ_m would decay more or less exponentially with distance downstream, but the details require a complete solution of the problem, which is beyond our purpose. It is sufficient for us to note that if viscous dissipation is to play a significant role in the prevention of occlusion,

$$PrEc = \frac{\mu U_c^2}{k_w \theta_m} = 0(1)$$

and hence

---

*This is not strictly accurate, as Eq. (25) implies. However, it is only to be used here as an order of magnitude, for which it will suffice.

represents the appropriate velocity scale: substituting into the hydraulic equation we obtain

$$\left.\begin{array}{l} \dfrac{\Pi_c}{L} = 0 \; (\dfrac{-2\mu k_i \theta_s}{\Delta_i \Delta_w^3})^{1/2} \\[20pt] \dfrac{\Pi_c}{L} = 0 \; (\dfrac{-\mu k_i^2 \theta_s}{h\Delta_i \Delta_w^4})^{1/2} \end{array}\right\} \qquad (28)$$

if we begin with Eq. (26).

At first glance, the implication of $\Delta_w \to 0$ is readily understood, but not so the implication of $\Delta_i \to 0$. However, if we again make use of the hydraulic equation it is evident that:

$$\dot{Q}\Pi_c = 0(jL) \qquad (29)$$

where $j = -k_i\theta_s/\Delta_i$ is the density of the heat flux being removed through the ice. This is consistent with the overall thermodynamic requirement that the work dissipation rate must balance the heat loss rate under the assumed steady conditions. It is thus to be expected that the overall pressure gradient necessary to maintain such a balance would tend to infinity if either the water flow rate shrinks to zero ($\Delta_w \to 0$) or the heat withdrawal rate tends to infinity ($\Delta_i \to 0$).

The overall thermodynamic relation is obviously more convenient than Eqs. (28) which require a detailed knowledge of the system from point to point. Equation (29) merely requires the heat flow rate and waterflow rate to estimate the overall pressure gradient for a dissipative prevention of occlusion. On the other hand, the overall rates are based on average quantities which may differ substantially from local quantities. Equation (29) is therefore only a convenient beginning.

The structure of Eqs. (28) reveals the curious fact that the pressure gradient is not a monotonic function of either ice or water thickness; in fact it exhibits a minimum when $\Delta_w = 3\Delta/4$ or $4\Delta/5$, depending on which of Eqs. (28) are used. It thus appears that the minimum pressure gradient necessary for the prevention of occlusion is given by

$$(\dfrac{\Pi_c}{L})_{min} = 0 \; [\dfrac{5\mu k_i (T_i - T_s)}{\Delta^4}]^{1/2}$$

or

$$(\dfrac{\Pi_c}{L})_{min} = 0 \; [\dfrac{10\mu k_i^2 (T_i - T_s)}{h\Delta^5}]^{1/2}$$

Using the first of these

$$\left(\frac{\Pi_c}{L}\right)_{min} = 0 \left[\frac{(T_i - T_s)^{1/2}}{10\Delta^2}\right]$$

for cold water. For a substrate temperature of -10°C, we find that the minimum pressure gradient is of the order of 0.3 x 10^{-3} kPa/m, 0.3 x 10 kPa/m and 0.3 x 10^3 kPa/m for channel widths of 1 m, 1 cm and 1 mm, respectively. Above this minimum pressure gradient, the double value of Δ_w (or Δ_i) raises an interesting question. Unfortunately, the answer lies beyond this exploratory paper, which must also leave discussion of turbulent flow and banding phenomena for another occasion.

3.2 External Flows: The Row of Tubes

A third category of substrate growth problems covers situations in which the water flows externally but, unlike the finite film with its free surface at which there is a discontinuity in density and/or phase, the typical external flow extends away from the surface more or less indefinitely. Figure 7 depicts the example of the cross flow of water over a row of tubes whose surface temperatures are maintained below the freezing temperature. Ice grows in a series of annuli which, depending upon the bulk flow conditions (U_∞, T_∞), the cylinder diameters d, and the initial gap g_0, may or may not merge to form a continuous wall of ice thereby producing total occlusion. Precise behaviour obviously depends upon the details of fluid flow and heat transfer.

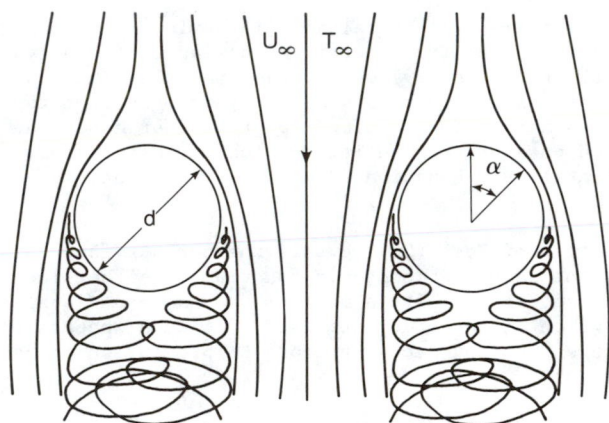

Fig. 7. Cylinders in cross flow: open spacing

For a sufficiently large initial gap, flow around any given tube in the row is not too different from that around a single cylinder. This is indicated in Fig. 7 for sub-critical conditions. However, even though the tubes may be positioned to maintain such a gap-diameter range, the gap-diameter ratio of ice surfaces cannot possibly be thus limited if the aim is total occlusion. Figure 8 shows the effect of a very small gap-diameter ratio in terms of the pressure distribution around a circular cylinder at various Reynolds numbers. As we would expect, the pressure variations are larger with the precise shape lying somewhere between the extremes of single cylinder behaviour, on the one hand, and behaviour characteristic of a row of parallel jets on the other.

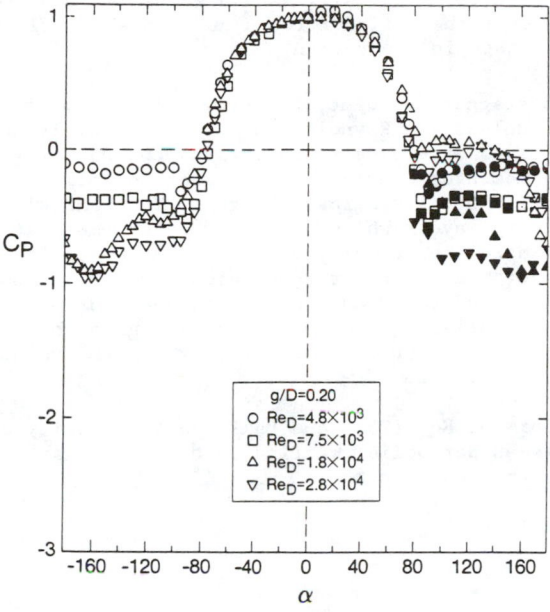

Fig. 8. Pressure coefficients around a cylinder situated in a closely-spaced row: g is gap and D is diameter

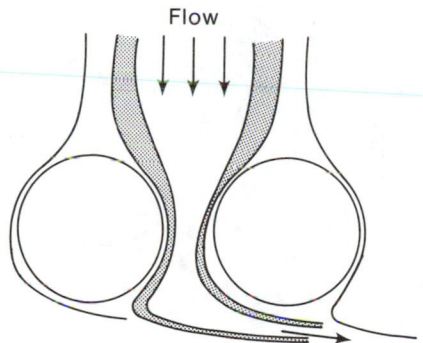

Fig. 9. Cylinders in a cross flow
with close spacing

Of particular interest is the fact that the pressure distribution is not generally symmetric about the cylinder, indicating that flow in the gap is not symmetric about the central plane. This may be seen, for example, by comparing the open triangle data on both sides of the diagram. The solid triangle data on the right hand side, which was taken from the adjacent cylinder, is symmetric with the open triangle data on the left side. This feature may lead to the lateral motion of the emergent fluid, producing a 'wake' flowing across the cylinder rears, at right angles to the bulk flow direction. This is depicted in Fig. 9. Obviously such behaviour will have a significant effect on the heat transfer rates over the cylinder rears and thereby influence the ice-water interface shape in the region.

At the forward stagnation point, the flow splits almost symmetrically[*] and, unless the Reynolds number is extremely low, will produce a boundary layer extending round towards the narrow gap. Eventually, the two momentum boundary layers meet to form a more or less parabolic velocity profile across the gap. The same will be true of the thermal boundary layers which, for cold water, meet at a point further downstream where the velocity profile is more nearly fully developed. The higher heat transfer rates associated with the higher velocities in the gap tend to impede ice growth and thus produce a gap region resembling a parallel channel. In such a region the convection problem takes on a form of a Graetz problem in which the velocity may be treated as fully-developed.

Applying the form of Eq. (25), and taking both velocity and temperature profiles as parabolic, we find that the mid-gap temperature varies according to the equation

$$\frac{d\theta_m}{dX} = -\left(\frac{15 \; \kappa}{2U_m \Delta_w^2}\right) \theta_m + \frac{15 \; \nu \; U_m}{6C_p \Delta_w^2}$$

which may be re-written as

$$\frac{d\theta_m}{dx} = -\beta_1 \theta_m + \beta_1 \beta_2 \tag{30}$$

$$\text{where} \quad \beta_1 = \frac{15 \; \kappa}{2U_m \Delta_w^2} \quad \left.\right]$$

$$\text{and} \qquad \beta_2 = \frac{1}{3} \, PrEc\theta_o \tag{31}$$

This has the solution

$$\theta_m(X) = (\theta_o - \beta_2)e^{-\beta_1 X} + \beta_2 \tag{32}$$

where X is measured downstream from the point of intersection of the thermal boundary layers, where

$$\theta_m(0) = T_\infty - T_i = \theta_o$$

---

[*]The forward stagnation point is actually offset by a small angle α_o.

Equation (32) immediately reveals the effect of viscous dissipation on water temperature. As $\beta_2 \to 0$, the water temperature becomes an exponentially decaying function of X. Equation (31) reveals that the rate of decay increases dramatically as the ice interfaces approach each other, the precise details depending on both the gap width and the gap velocity. The magnitude and variations of β_1 obviously depend upon the overall flow conditions, as the curves shown in Fig. 5 have illustrated. When the effect of viscous dissipation is small, it is evident from Eq. (32) that the heat flux from the water reduces with distance along the gap and thus reinforces the tendency for the gap to remain parallel. Under these conditions

$$\theta_m(X) \simeq \theta_o e^{-\beta_1 X}$$

and therefore at the no growth limit[*], when

$$\theta_m(X) = - \frac{k_i \Delta_w}{2k_w \Delta_i} \theta_s$$

it follows that

$$X = \frac{1}{\beta_1} \ln \left(\frac{-2 \, k_w \Delta_i \theta_o}{k_i \Delta_w \theta_s} \right)$$

from which we may estimate the gap length for various degrees of occlusion, or vice versa, provided that $-k_i/2k_w\epsilon \leq \Delta_i/\Delta_w < \infty$ and the hydraulic constraints are known. Alternatively, we may substitute a representative value of $X = \pi(g_o + d)/4$ into the expression and then determine the relation between the hydraulic conditions, the thermal conditions and the gap width.

When viscous dissipation is significant in Eq. (30) $\theta_m = O(\beta_2)$, and therefore when dissipation leads to no growth, $\beta_2 \to \theta_o$,

$$\frac{\mu U_c^2}{3k_w} = O(\theta_0) = \frac{k_i \Delta_w \theta_s}{2k_w \Delta_i}$$

from which the appropriate velocity scale may be related to either the bulk water temperature or the gap width and substrate temperature. In turn, the velocity may be related to the hydraulic requirement through expressions like Eqs. (27) or (27a), but again the hydraulic constraints must be specified in order to determine the appropriate curves from Fig. 5. If the tube row is being used to curtain off a river, for example, the conditions of fixed overall pressure drop would appear to be appropriate.

---

[*]For simplicity, the effect of substrate curvature is ignored. The logarithmic function may be substituted if greater accuracy is needed.

4. SOME GENERAL OBSERVATIONS

This paper has suggested three major categories of freeze occlusion, and has offered a discussion of one of these: namely, substrate growth. To a certain extent, the discussion may be applied to the categories of free surface growth and dispersed growth. In the former instance, the essential difference lies in the transformation of the free surface into a fixed surface, by nucleation, flocculation and aggregation: after this transformation, the problem is essentially one of internal flow. Dispersed growth, on the other hand, may be substantially different, not least because the assumption of stable thermodynamic equilibrium does not apply to ice nuclei growing in supercooled water. The flow medium is then a heterogeneous mixture in which the crystals will grow, so long as the liberated latent heat does not offset the sensible heat deficit. If the number density of the ice nuclei is high enough, the mixture may not be Newtonian, and in any event the agglomerative process may produce large masses which can substantially alter the flow and thereby generate total occlusion in a very different way.

Substrate growth is perhaps the easiest category of occlusion to discuss in depth. From the three examples discussed in Sections 2 and 3, it is evident that occlusion may be treated as a coupled ice conduction - water convection problem in which the interface is treated simply as a boundary of separation, changing in shape at such a low rate that it may be incorporated as a "fixed" surface. The validity of this assumption is obviously determined by the ratio of the interface velocity to a typical fluid velocity. In most situations, the interface velocity may be ignored in the convection problem.

Commonly, the low interface velocity implies that the ice conduction problem and the water convection problem may both be treated in quasi-steady form. In the former, the strict requirement of Ste $\ll 1$ is clear. In the latter, the corresponding requirement may be found by equating the water time scale ($t_c \gg \Delta_w^2/\nu$) with the ice time scale ($t_c = \rho_i \lambda \Delta_i^2/k_i \theta_s$). From this it is evident that Ste $\ll (\Delta_i/\Delta_w)^2$ Pr κ_w/κ_i or Ste $\ll (\Delta_i/\Delta_w)^2$ in cold water. This is no more stringent than Ste $\ll 1$ except for $\Delta_i < \Delta_w$, which only applies to conditions of secondary importance in an occlusion study.

It is apparent that the ice conduction problem may be treated as a classic Stefan problem in which the substrate temperature may be prescribed as a function of time and position, providing the variations do not invalidate quasi-steady or quasi-one-dimensional assumptions. The occlusion time scale, and the no growth limit, if any, both stem from the solution of the interface Eq. (1). This equation in turn contains the water heat flux at the ice-water interface and is thus dependent upon solutions to the energy and momentum equations for the water. The convection problem is typically a Graetz problem in which the velocity is prescribed in a form consistent with a balance between the viscous forces and the appropriate driving forces: typically, this implies a parabolic velocity profile except near a leading edge or point of entry, where inertia may be important. Such solutions provide the various types of information sought in an occlusion study: the occlusion length of an aufeis water film; the no growth limit of a partially frozen water pipe; or the occlusion time for a curtain of ice in a river.

Normalization generates the principal variables appropriate to an occlusion problem. An order of magnitude analysis describes their role. Many of the non-dimensional parameters discussed above are well known, and their appearance in occlusion studies simply confirms their widespread importance. Among these are: the conductivity ratio k_w/k_i, the superheat ratio $(T_f-T_i)/(T_i-T_s)$, and the familiar numbers of Reynolds, Biot, Stefan, etc. In determining the pressure gradient necessary to prevent occlusion (through viscous dissipation) the group $(\Delta_i/\Delta_w) \cdot \mu U_c^2/k_i(T_i-T_s)$ may also be useful: it measures the viscous dissipation rate in relation to the cooling rate.

As a final observation, it is worth noting that occlusion phenomena cannot in general be treated in isolation. Typically they occur in only one part of a much larger water system and must be treated as a dynamic element within it. For purposes of analysis it may be convenient to separate the occlusion region from the rest, but complete knowledge of both the effect and the behaviour of the ice growth demand that it be viewed within the context of a larger system.

ACKNOWLEDGEMENT

This work was written in 1985 while the author was a Visiting Scholar at the Scott Polar Research Institute, University of Cambridge. I am indebted to the Institute's director, Dr. David Drewry.

Chapter 7

Ice-Band Structure on the Freezing of Flowing Water in Pipe

TETSUO HIRATA
Department of Mechanical Engineering, Shinshu University,
Nagano 380, Japan

ABSTRACT

Ice formation phenomena in water flow in pipes were investigated under the conditions which produce the ice-band structure, having a cyclic pattern of contractions and expansions of the flow passage along the length of the pipe. The spacing of the ice-bands, the heat transfer coefficient in the contraction region, and the friction factor in the pipe were examined. The onset conditions for pipe freeze-off were obtained analytically by introducing a modified Reynolds number based on a total pressure.

CONTENTS

1. INTRODUCTION

The freezing of liquids in forced flow in a pipe has been studied for years by many researchers. Almost all of the problems are for the condition that ice thickness increases monotonously along the length of a pipe. A number of theoretical and experimental studies, such as the profile of an ice layer and the effects of solidification on heat transfer and pressure drop, have been reported for a transient freezing process as well as for a steady-state condition. It would seem that the problem of pipe freezing has been fully investigated.

At a solid-liquid interface, the interactions among the flow, the shape of the interface, and the heat transfer at the interface, exist. If the ice layer is very thin, the temperature gradient in the ice is large; therefore, the ice-water interface is stable against a disturbance of water flow. Conversely, when the ice layer is thick and exceeds a certain

223

value, the temperature gradient is smaller and the interface is unstable.

In reference to the unstable ice layer, Gilpin et al.[1] studied the ripple undulation formed on the ice surface of a cold plate in a water stream, and examined the morphology of the ice surface for freezing water flowing through a pipe. The ripple undulation phenomenon was also observed in the freezing water flowing through a pipe, and was named ice-band structure. Gilpin et al. concluded that the ice-band structure has significant effects on the pressure drop and on freeze-off conditions [2,3]. It was also revealed that the freeze-off conditions were obtained by introducing a modified Reynolds number based on a total pressure [4].

In the present paper, the problem of the freezing of flowing water in a pipe containing the ice-bands will be reviewed. The contents in part of this paper will be borrowed from the previous paper [4].

2. PIPE FREEZING PROCESS

The variation in the flow rate of water in a pipe containing a growing ice layer is shown in Fig.1 for various values of Re_D and θ, where Re_D is a pipe Reynolds number and θ is a cooling temperature ratio defined by

$$\theta = (T_f - T_w)/(T_\infty - T_f) \tag{1}$$

Larger values of θ, therefore, produce a thicker ice layer. In Fig.1, when the freeze-off does not occur (θ=7.1 and 7.8), the flow rate shows a sinuous variation during the first stage of ice growth and finally approaches a steady-state value. The cyclic variation of the flow rate is caused by a change in the friction factor in a pipe associated with the appearance of the ice-band structures (see Fig.3). In this case, more than 7 hours was needed to obtain the steady-state

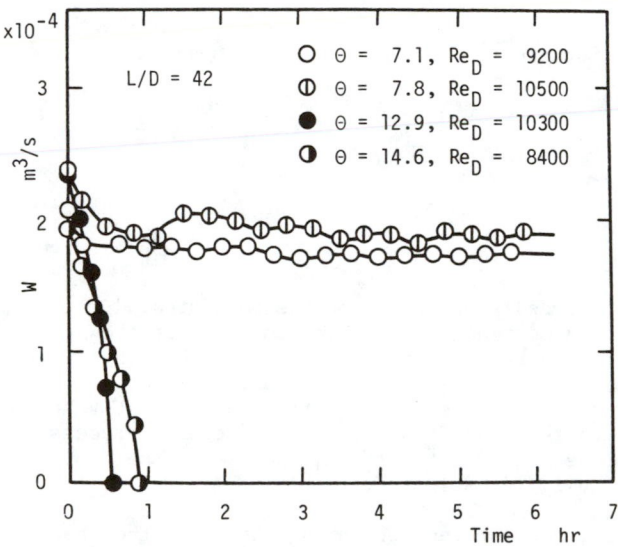

FIGURE 1. Variation of flow rate with time for several conditions of Reynolds number and temperature ratio [4].

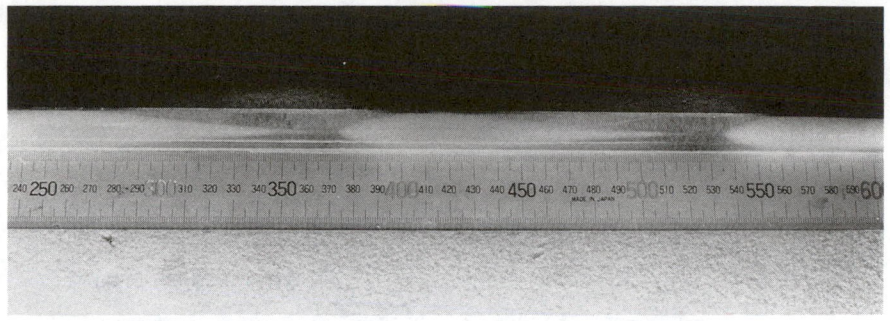

FIGURE 2.   Photograph showing freeze-off ice layer; D=19.9mm, L=697mm.

ice layer. Also shown in Fig.1 is the variation in the flow rate when
the freeze-off occurs. The flow rate decreases rapidly and the complete
blockage of the flow passage takes place in only one hour.

Fig.2 shows a photograph of the freeze-off ice layer. It can be observed
that the cyclic ice-band is formed along the whole length of the pipe,
and it is suggested that the freeze-off first occurs at the contraction
region in the ice-bands, and then develops to the freezing of pockets
of water trapped along the pipe. It should be noted that the ice-band
structure can be fully developed in a short period of time.

3.   FORMATION OF ICE-BAND STRUCTURE

When the ice-water interface is unstable, the transient freezing process
is dependent upon given temperatures and flow conditions to reach steady-
state ice-bands. Considering how the ice layer develops, the freezing
process may be classified into two major groups; one is the process for
a thinner ice layer and the other, for a thicker one.

3.1 Thinner Ice Layer

Fig.3 is a schematic representation of the transient freezing process
of growing and developing ice-bands for a thinner ice layer. Fig.3(a)
shows an ice growing during the first stage of freezing, where the ice
surface is smooth, that is, the thickness increases monotonously along
the length of a pipe. In the tapered flow passage at the entrance region,
the acceleration of flow becomes higher with increasing ice thickness;
therefore, a flow laminarization occurs when the flow is initially
turbulent [2]. Fig.3(b) shows the change in ice shape caused by a
transition of the flow. With increasing ice thickness, the flow becomes
highly accelerated and turbulent. The ice layer downstream of the
transition point, then, becomes thinner due to the enhancement of heat
transfer by turbulence. In the transition region, the point of inflection
in the velocity profile has been observed [5] and is typical of the
unstable nature of a flow in an adverse pressure gradient. One notes that
the profile shown in Fig.3(b) can be also observed in a steady-state
condition. In Gilpin et al's cold plate experiment, this was called the
smooth transition mode. In a 'step' transition mode, a separation of
the flow occurs as shown in Fig.3(c), where the flow passage has a step
expansion and the flow becomes turbulent in the downstream region. In

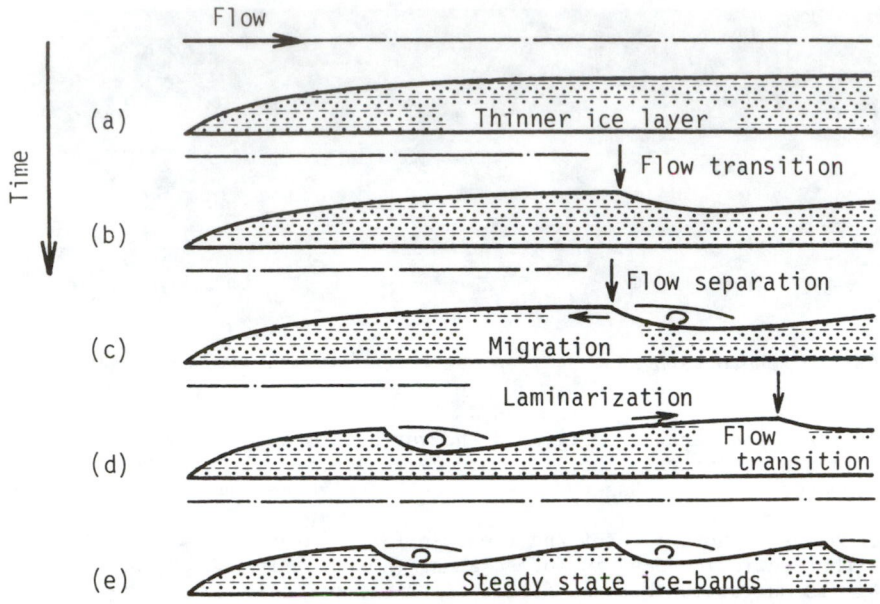

FIGURE 3. Schematic representation of transient development of ice-band structure for thinner ice layer.

the separated flow region, the heat transfer is greatly enhanced; therefore, the ice melts away and the separation point migrates upstream very slowly (several centimeters per hour). Fig.3(d) shows the appearance of another transition with increasing ice thickness downstream of the separation and the sequence of events shown in Figs.3(c) and (d) occurs. Ultimately the migration of the separation point stops at the location where a heat balance at the ice-water interface occurs and the steady-state ice-band structure is obtained as shown in Fig.3(e).

3.2 Thicker Ice Layer

In Fig.4 the transient process of attaining an ice-band structure is schematically shown for a thicker ice layer. In this case, a distinctive feature in the process is the simultaneous occurrence of the flow transition and the separation over the length of a pipe. In Fig.4(b), it is shown that the flow transition initially occurs at a certain distance from the inlet and, before it develops to the separation, a laminarization takes place downstream of the transition region; so, the subsequent transitions occur one after another. The ice-band structure is, therefore, formed simultaneously over the length of a pipe and reaches the steady-state in a shorter time than did the thinner ice layer. As previously shown in Fig.2, when the freeze-off occurs (which is equivalent to the thicker ice layer), the ice-band structure is completed in only one hour. If the ice-band, in that case, is formed by a manner of the thinner ice layer, it should take several hours to obtain the fully developed ice-bands. This implies that when the freeze-off occurs, the ice-bands are simultaneously formed as shown in Fig.4.

The process of attaining freeze-off is considered as follows [4]: The

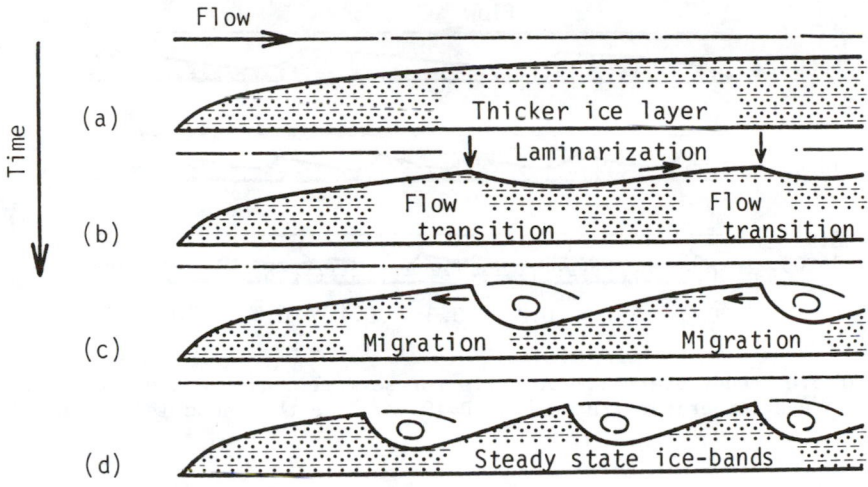

FIGURE 4. Schematic representation of transient development of ice-band structure for thicker ice layer.

formation of the ice-band results in an acceleration of the flow in the contraction region. Once the ice thickness exceeds a certain value, the flow undergoes a rapid acceleration and laminarization; then the heat transfer rate becomes smaller and the ice layer, thicker. These changes create a larger friction factor, a smaller flow rate and a further growth of ice; resulting in a quick freeze-off.

4. STEADY-STATE ICE-BAND STRUCTURE

4.1 Ice Profile

Typical profiles of the ice-band structure in a steady-state condition are shown in Fig.5. The spacing of each ice-band becomes shorter with increasing Reynolds number. For a larger Reynolds number, the heat transfer rate in the separated flow region increases, since the recirculating flow is stronger. The migration of the separation point in the upstream direction is quicker and therefore shorter ice-band spacing is obtained in the steady-state profile. In this case the ice surface downstream of the separation point formed a comparatively large expansion angle. Conversely, for a smaller Reynolds number, the ice-band spacing became long and the expansion angle small, caused by a lower heat transfer rate in the separated flow region. For the result of Re_D=15200, the regularity of the ice-band structure was broken in the region of x>0.2 and a three-dimensionalprofile with very small ripples was obtained. The same phenomena have been observed on the ice surface of a cold plate in a water stream for larger Reynolds numbers with strong turbulence [1].

The steady-state ice layer profiles in a pipe might be classified into three modes, depending on the conditions of Re_D and θ. The first mode is the smooth transition as shown in Fig.3(b); the second, the two-dimensional ice-band structure and the third, the three-dimensional ripples. However, the relationship between the modes and the conditions of Re_D and θ has not been clarified.

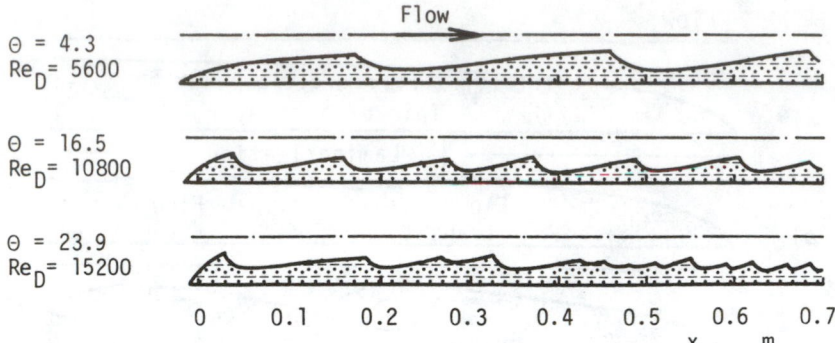

FIGURE 5. Steady-state ice profiles for several conditions of Reynolds numbers and temperature ratio [4], D=19.9mm (ice thickness is not in scale).

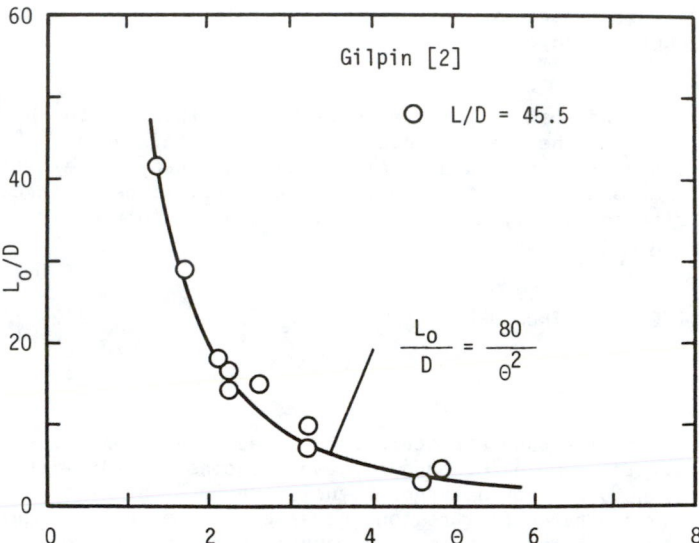

FIGURE 6. The normalized distance to the first separation point.

For two-dimensional ice-bands, the first separation point is important from two physical aspects. One is that the flow, which is initially laminar at the pipe inlet, becomes turbulent at the first separation point; the other, the location of the first separation point can indicate whether or not an ice-band structure exists in a freezing pipe. The latter is particularly important since ice-bands produce a high value of pressure drop compared to a smooth pipe (see Fig.9).
Gilpin [2] reported that the distance to the first separation point was correlated by a function of θ as shown in Fig.6.

$$L_o/D = 80/\theta^2 \qquad \text{for} \quad 1.4 \leq \theta \leq 4.9 \qquad (2)$$

It is evident that L_o becomes shorter with increasing ice thickness.

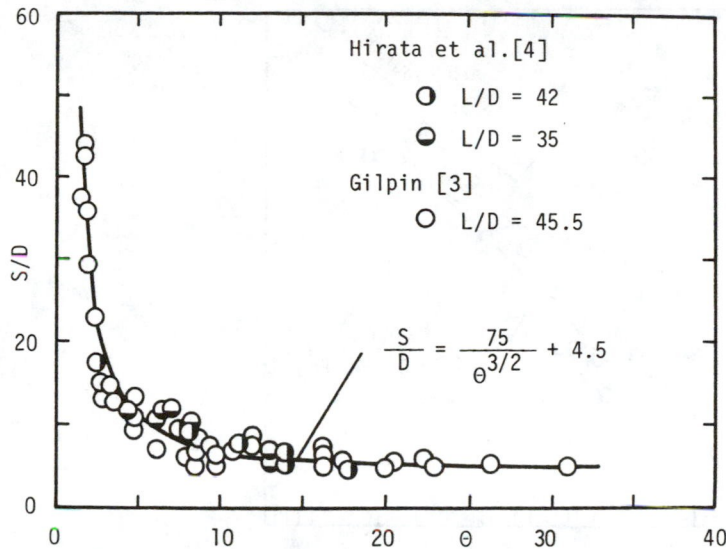

FIGURE 7. The normalized spacings between subsequent separation points.

This means that the first separation point moves upstream for a thicker ice layer. The L_o/D is represented as a function of θ; however, for a wider range of θ the Reynolds number might be involved.

In Fig.7 the ice-band spacings (the distances between each separation point) are plotted against θ. It can be seen that the spacing becomes longer with increasing θ. For smaller θ the ice is thinner; therefore, it needs to travel some distance downstream to recover the thickness of the ice layer which was thinned in the separated flow region. On the other hand, for larger θ, S takes a value nearly five times as large as the pipe diameter. The correlation equation is given by

$$\frac{S}{D} = \frac{75}{\theta^{3/2}} + 4.5 \qquad \text{for} \quad 1.6 \leq \theta \leq 31 \qquad (3)$$

In this case the Reynolds number is not involved. Gilpin [3] observed that the ratio, S/D, appears to be largely independent of Re_D, since the Reynolds number loses importance in the highly accelerating and decelerating flow through a fully developed ice-band structure.

4.2 Heat Transfer

The property of the ice-band structure of most interest may be the diameter of the narrowest water passage in the contraction region, since it is directly related to the friction factor in a pipe. The heat balance at the ice-water interface of a steady-state ice layer is described as follows: Assuming that the heat flow in the axial direction through the ice is low compared to the heat flow in the radial direction, we have

$$\pi dh(T_\infty - T_f) = 2\pi\lambda_i \frac{T_f - T_w}{\ln(D/d)} \qquad (4)$$

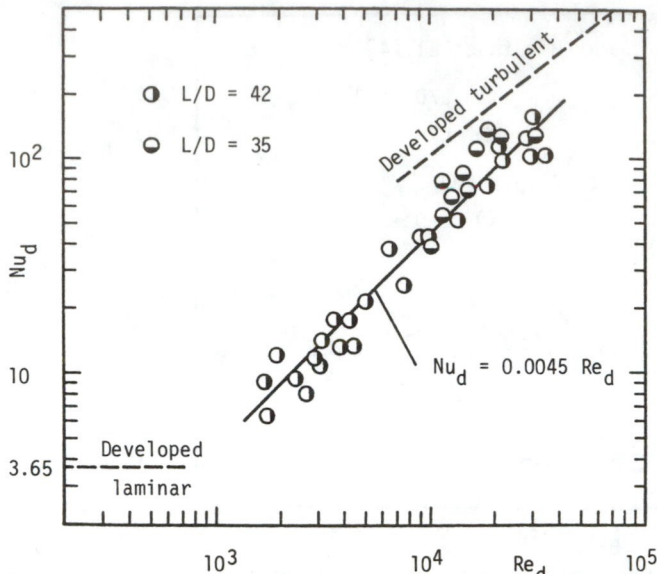

FIGURE 8. The Nusselt number at the contraction region of the ice-bands plotted against Reynolds number [4].

If the Nusselt number is defined as $Nu_d = hd/\lambda_w$, Eq.(4) can be rewritten as:

$$Nu_d = - \frac{2\lambda^*\theta}{\ln B} \qquad (5)$$

where λ^* is the thermal conductivity ratio of ice to water and B is the diameter ratio, d/D. From Eq.(5) it is realized that the logarithmic value of d is inversely proportional to Nu_d. In Fig.8, Nu_d is plotted against the Reynolds number, defined as $Re_d = vd/\nu$. The results for fully developed laminar and turbulent flows in a smooth pipe without ice are also shown. It is clear that the Nusselt numbers in the ice-bands show a smaller value than that of turbulent flow due to the laminarization of flow in the contraction region. A correlation for the Nusselt number was found to be

$$Nu_d = 0.0045 \, Re_d \qquad \text{for} \quad 1.6 \times 10^3 \le Re_d \le 3.0 \times 10^4 \qquad (6)$$

4.3 Pressure Drop

The flow in the ice-bands undergoes a series of accelerations and flow separations due to contraction and expansion; therefore, the pressure drop in a pipe containing cyclic two-dimensional ice-bands is caused mainly by two factors. One is due to viscous drag at the ice-water interface and the other is due to pressure loss occurring in the sudden downstream expansion of each separation point. The former is, in general, negligibly small compared to the latter and the pressure drop in the ice-band will be approximated by the factor of sudden expansion. In that case the

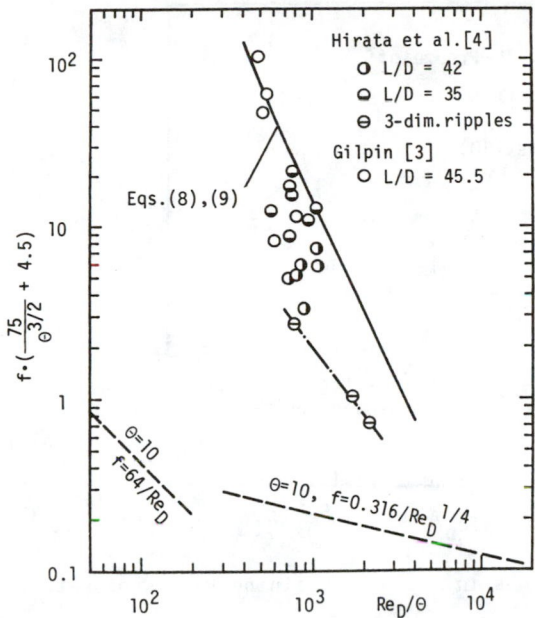

FIGURE 9. Friction factor for a pipe containing a steady-state ice-band structure.

friction factor will be

$$f = \frac{\rho v^2}{2} \, n(1 - B^2)^2 \, / \, \frac{\rho V^2 L}{2D} \tag{7}$$

where n is the number of ice-bands in a pipe, L/S. Introducing Eq.(3), we have

$$f(\frac{75}{\theta^{3/2}} + 4.5) = \frac{(1 - B^2)^2}{B^4} \tag{8}$$

Making the substitution $Re_D = B \, Re_d$ into Eqs.(5) and (6) yields

$$\frac{Re_D}{\theta} = - \frac{2\lambda^*}{0.0045} \frac{B}{\ln B} \tag{9}$$

The friction factor can be expressed in terms of Re_D and θ by eliminating B from Eqs.(8) and (9).

In Fig.9, the predicted value coincides well with experimental data in the region of smaller Re_D/θ but overestimates at larger Re_D/θ. This is concerned with the expansion angle of the ice-band in the separated flow region. In the case of a thicker ice layer, the expansion angle is large, therefore, the flow characteristic related to the friction factor becomes closer to that given by Eq.(7). For a thinner ice layer the actual value of f becomes smaller than that given by Eq.(7), since the expansion angle

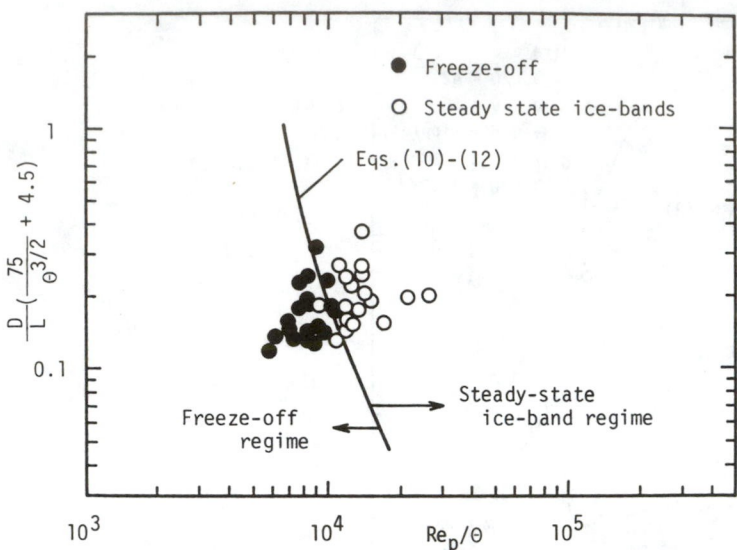

FIGURE 10. Freeze-off conditions in a pipe containing a flow of water.

becomes smaller. It can be said, however, that the predicted results give an upper bound value of the friction factor.

The results for three-dimensional ripples are also shown in Fig.9 and it is indicated that the value of f is smaller because of a larger flow passage with small ripples as shown in Fig.5. In any case, it should be noted that the pressure drop in a pipe containing ice-bands is 10-100 times as large as that for a smooth pipe.

5. FREEZE-OFF CONDITIONS

The freeze-off of a pipe causes a flow stop and, sometimes, a breakage of a pipe due to the volume expansion resulting from solidification. The freeze-off conditions are derived by considering whether or not the thickness of the ice ultimately exceeds the steady-state value expressed by Eq.(6). In consequence, those conditions are given by Eqs.(10)-(12) as follows [4]:

$$\text{Re}_p = - \frac{2\lambda^*\theta}{0.0045} \frac{\sqrt{1 + n(1 - B^2)^2}}{B \ln B} \tag{10}$$

$$\theta = [\frac{1}{75} (\frac{L/D}{n} - 4.5)]^{-2/3} \qquad \text{for} \quad 1.6 \leq \theta \leq 31 \tag{11}$$

$$n = - \frac{1 + \ln B}{(1 - B^2)^2 + (1 - B^4)\ln B} \qquad \text{for} \quad n > 0 \tag{12}$$

where Re_p is a modified Reynolds number defined by a total pressure P_o at the inlet of the pipe and is given as follows:

$$Re_p = \frac{D}{\nu} \sqrt{\frac{2P_o}{\rho}} \tag{13}$$

The onset of the freeze-off can be calculated from Eqs.(10)-(12) by eliminating n and B. In Fig.10 the predicted value for freeze-off is compared with experimental data. The solid and the open points indicate a freeze-off and a steady-state ice-band, respectively. It can be seen that the freeze-off occurs easily for smaller Re_p and larger θ.

It should be noted that the predicted results can be used only for a straight pipe. The effect of the pressure drop in a whole piping system, including valves and bends, on the freeze-off conditions is given in reference [6].

6. CONCLUDING REMARKS

The phenomena of ice-band structure in a pipe containing a flow of freezing water are reviewed. The study for this freezing problem was started only recently and many problems remain unknown.

In this connection, the author wishes to acknowledge his continuing interest in pipe freezing problems after having an opportunity of working with the late Professor R.R. Gilpin on the formation of ice ripples on a cold flat plate [1, 5].

NOMENCLATURE

B	diameter ratio, d/D
d	minimum diameter at contraction of an ice-band
D	diameter of pipe
f	friction factor of a pipe containing ice-bands
h	heat transfer coefficient at contraction of an ice-band
L_o	distance from inlet of a pipe to the first separation point
L	length of pipe
Nu_d	Nusselt number, hd/λ_w
n	number of ice-bands in a pipe, L/S
P_o	total pressure at inlet of a pipe measured on the basis of outlet value
Re_d, Re_D	Reynolds numbers, vd/ν, VD/ν
Re_p	Reynolds number based on P_o, defined in Eq.(13)
S	spacing between ice-bands
T_f, T_w, T_∞	freezing, pipe wall and water temperatures
v	mean velocity of water at contraction of an ice-band
V	mean velocity of water in a pipe without ice
W	water flow rate
x	distance along pipe
ρ	density of water
θ	cooling temperature ratio, defined in Eq.(1)
λ_i, λ_w	ice and water thermal conductivities
λ^*	thermal conductivity ratio, λ_i/λ_w
ν	kinematic viscosity of water

REFERENCES

1. Gilpin, R. R., Hirata, T., and Cheng, K. C., Wave Formation and Heat Transfer at an Ice-Water Interface in the Presence of a Turbulent

Flow, *J. Fluid Mech.*, vol. 99, part 3, pp. 619-640, 1980.

2. Gilpin, R. R., The Morphology of Ice Structure in a Pipe at or Near Transition Reynolds Numbers, *A.I.Ch.E. Symp. Ser.*, vol. 75, pp. 89-94, 1979.

3. Gilpin, R. R., Ice Formation in a Pipe Containing Flows in the Transition and Turbulent Regimes, *J. Heat Transfer*, vol. 103, no. 2, pp. 363-368, 1981.

4. Hirata, T., and Ishihara, M., Freeze-off Conditions of a Pipe Containing a Flow of Water, *Int. J. Heat Mass Transfer*, vol. 28, no. 2, pp. 331-337, 1985.

5. Hirata, T., Gilpin, R. R., and Cheng, K. C., The Steady State Ice Layer Profile on a Constant Temperature Plate in a Forced Convection Flow - Part II, *Int. J. Heat Mass Transfer*, vol. 22, no. 10, pp. 1435-1443, 1979.

6. Hirata, T., Effects of Friction Losses in Water-Flow Pipe Systems on the Freeze-off Conditions, *Int. J. Heat Mass Transfer*, vol. 29, no. 6, pp. 949-951, 1986.

Chapter 8

Freezing Fractures in Water Pipes

HIDEO INABA
Department of Mechanical Engineering, Okayama University,
Okayama 700, Japan

ABSTRACT

A review work concerning the freezing behavior and rupture mechanisms of water pipes with a quiescent water has been performed. An attempt is first made to demonstrate the case of water supply pipe accidents by freezing in cold climate regions. Subsequently, the freezing behavior of the quiescent water in domestic water supply pipes and the pipe rupture mechanism by freezing are explained by the results of both laboratory and field experiments. In addition, this review work refers to the usage of pipe freeze-off, where artificial freezing is used as a temporary measure to plug a pipe during a repair operation.

TABLE OF CONTENTS

NOMENCLATURE

Bi   Biot number
C   constant, defined in Eq.(2) and Eq.(5) or specific heat
D,d   pipe diameter
D_1 inner diameter of pipe
D_2 outer diameter of pipe
E_* modulus of elasticity
Fo   nondimensional time, $\alpha t_c/d^2$

G	shear modulus
H	cooling rate, defined in Eq.(2)
h	heat transfer coefficient
L	latent heat
m	Poisson's number
P	pressure
P_1	water pressure
P_c	critical pressure, defined in Eq.(5)
$p*$	non-dimensional pressure
R	radius ratio, r_2/r_1 or wall thickness
Re	Reynolds number
r_1	inner radius
r_2	outer radius or inner radius
r_3	outer radius
S	displacement of ice-water interface
T	temperature
T_m, T_f	freezing temperature or ice nucleating temperature
$T*$	nondimensional temperature
t	time
t_c	critical time
U	mean flow velocity
U_s	displacement of pipe
V	specific volume
W	consumption rate of liquid nitrogen
X	ice thickness
α	thermal diffusivity
ϵ	strain
ϵ_θ	circumferential strain
ϵ_z	axial strain
γ	Poisson's ratio
λ	thermal conductivity
σ_s	tensile stress
μ	chemical potential
μ^i	chemical potential of solid phase
μ^ℓ	chemical potential of liquid phase
ν	kinematic viscosity
ρ	density
τ	shear stress

1. INTRODUCTION

The utility delivery systems (water pipes and containers) in cold regions
are often exposed to subfreezing temperatures. Freeze-off and breakage of
water pipe lines are a common problem. The goal of the utility engineers
is to eliminate or minimize the problems of pipe freezing. The freezing
and rupture of a water pipe are complex phenomena involving the physical
phenomena of supercooling, pressure melting and the dynamic characteristics
of water pipes. The objective of this article is to review the freezing
behavior and rupture mechanisms of water pipes with a quiescent water,
since the understanding of the pipe freezing mechanism will improve pipe
design. In addition, this article also refers to the usage of pipe freeze-
off, where artificial freezing is used as a temporary measure to plug a
pipe during a repair operation.

2. FREEZING PROBLEM OF WATER PIPE LINES IN COLD REGIONS

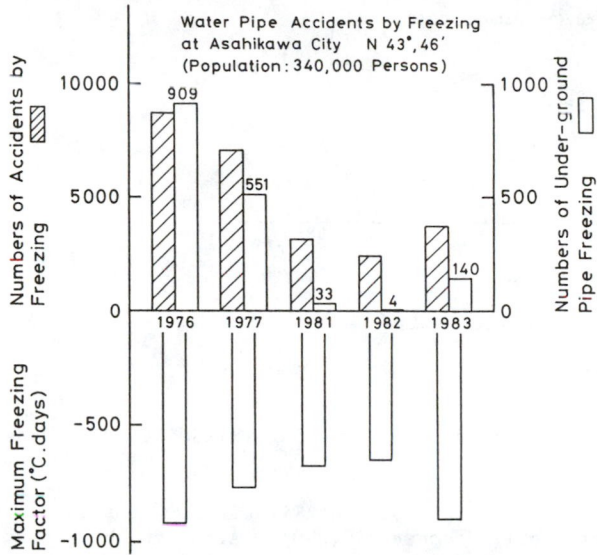

FIGURE 1. Water pipe accidents by freezing (Asahikawa city in Japan).

In cold climates one often experiences freezing problems, namely, freeze-
off and rupture of water pipe lines. The most important concern in designing a utility system in cold regions is to minimize or eliminate the
possibility of pipe damage due to freezing. The rupture of water pipes
occurs in the following stages: (1) volume expansion of water by freezing
in a water pipe, (2) external force of bending and shearing to underground
water pipes by frost heaving of frozen moist soil (non-homogeneous frost
heave and irregular melting of frozen soil) and (3) aging and brittleness
of buried water pipes exposed to subzero temperature. Often the accessories (flange joint, valve, etc.) of a water pipe line which has a large
heat loss may cause the distortion and breakage in severe temperatures.
The rupture of pipe lines may occur when the subfreezing condition prevails
around the pipe lines. The accumulated degree-day (Freezing factor,-°C·day)
as a parameter of coldness is about -750°C·days under normal winter conditions in Hokkaido, which is the northernmost island in Japan. The accumulated degree-day over -850°C·days pertains to severe weather conditions.
In the northern part of Hokkaido (North latitude, 43°45'), the water mains
buried 1 - 1.5 m underground are often damaged by freezing, since the water
temperature in the underground pipes drops to a minimum temperature of 1 -
3°C in winter and the soil freezes to a depth of 0.8 - 1.2 m.   However,
the majority of pipe freezing problems occur at water supply pipes near
the main, which are exposed to the severe subfreezing ambient condition.
   Statistical data of the pipe freezing problems in Asahikawa city
(population 340,000, domestic water supplied houses; 100,000 [1]) are shown
in Fig.1 in the form of a relationship between rate of pipe freezing accidents and years. It is seen that the incidence of pipe freezing increases
abruptly when the accumulated degree-day is over -850°C·days. Many underground water pipes were frozen off due to the severe weather conditions of
the winter of 1976-1977. Considerable time and money was spent in restoring service under difficult working conditions. Figure 2 presents the
ratio of the numbers of pipe freezing accidents to total water pipe acci-

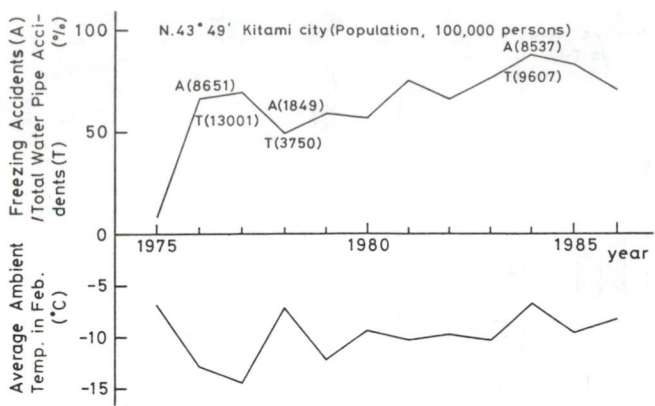

FIGURE 2. Rate of water pipe accidents by freezing (Kitami city in Japan).

dents per year at Kitami city (population 100,000 and domestic water supplied houses 40,000). Although the water supply facilities have been improved in recent years, pipe freezing troubles still represent about 70 % of the total water pipe accidents which need restoring service.

There are two freezing categories of water pipe: with water quiescent and with water flowing. Pipe freezing with quiescent water occurs due to a supercooling process. When quiescent tap water in a pipe is cooled down, ice normally does not nucleate until the water temperature has been super-cooled to 4 - 6°C below freezing point. The growth of ice in the super-cooled water results in the formation of dendritic ice, which consists of thin feathery crystals interspersed in the water. It is well known that during a cold night the hot water pipes in a house are more likely to burst by freezing than the cold water pipes. This phenomenon is commonly explained by dendritic ice growth, which occurs in supercooling water. Because heated water destroys potential nucleation sites, ice does not nucleate in the hot water pipe until the water has been cooled to -6 - -7 °C. As a result, a large amount of dendritic ice grows in the hot water pipe and the pressure required to break through the dendritic structure is greater than that required for the cold water pipe. Therefore, it is conceivable that there might be sufficient line pressure to start the flow in the cold pipe, but not in the hot pipe. As the annular growth of solid ice begins from pipe wall, the dendritic ice becomes more solid. An in-crease of water pressure by a volumetric change may cause pressure melting, depending on the ambient subfreezing temperatures. The freezing process in a water pipe with flowing water differs from that in the water pipe with quiescent water only in freezing rate, due to the addition of heat via the water flow. Pipe rupture in this case may occur with a water pressure rise caused by the freezing of the remaining water, which is trap-ped in the ice wall. Non-homogeneous heat loss in the water pipe line may be caused near pipe accessories with a large heat transfer surface area (fitting joint, valve, curved portion of the pipes, etc.) by the use of thermally non-homogeneous pipe materials, and because of the difference of insulating material thickness in the water flow direction. Even with a straight pipe consisting of thermally homogeneous material exposed to a uniform cold environment, there appears an ice band structure which pro-duces a flow passage with a cyclic variation in cross-section along the length of the pipe [2]. As a result, the freeze-off of a pipe containing the ice band structure may occur first at each neck in the band, resulting in a series of water pockets trapped along the freezing pipe. Subsequently, freezing of the water pockets results in the breakage of the pipe. The freeze-up of a water pipe with flowing water is a function of the ice growth process, and also depends on the pressure-discharge characteristics of the water supply. That is, the decrease in discharge caused by the unstable flow behavior due to an increase in ice growth results in an increased pressure loss. As a result, the freeze-up may be enhanced by a lack of discharge at the pump.

The possibility that a pipe will rupture also depends on the dynamic characteristics of the pipe material, pipe line structure and environmental conditions. The following measures may be considered to prevent the water pipe line from freezing and rupturing:
(1) The most effective way to prevent ice formation in a water pipe is to ensure that the water pipe is not exposed to the subfreezing environment. However, it is difficult to adopt this measure because of the high cost of burying the water pipe deeper into the ground and using thicker thermal insulating materials.
(2) To prevent freezing in a water pipe, the water could be heated to above zero temperature using an electric resistance heater, geothermal energy, etc. and by the circulation of the hot water.
(3) Another way to prevent freezing is to minimize ice growth in the pipe by diminishing the heat loss from the water pipe lines using thermal insulating materials.

When the ice growth proceeds, pipe rupture may be prevented by the following measures:
(1) Absorption of the volumetric change of freezing by using flexible pipe (artificial rubber pipe, new chemical products of water pipe).
(2) Installation of a small amount of damping space (introduction of compressible gas into a small compressible tube placed in a selected pipe section).
(3) Coating the inside of the common water pipe with elastic materials (rubber, etc.).
(4) Installation of pressure-relief devices into selected pipe section.

Another freezing problem which can arise occurs where a pipe line has to be started up while emptying water during severe subfreezing weather [3]. It is probable that the leading slug of water in the pipe will be cooled to the freezing point before it reaches the pipe exit. Downstream of this point, ice formation could result in a pipe blockage. In recent years, the technique of plugging a pipe with artificial ice, instead of using a valve, has been developed [4]. That is, freeze-off of the pipe is used as a temporary valve during a repair operation. This technique is economical because it diminishes repair time and allows the repair operation to take place in a limited section of the damaged water pipe.

3. FREEZING BEHAVIOR OF A WATER PIPE WITH QUIESCENT WATER

3.1 The Time Freezing History For Various Stages Of The Freezing Process

Figure 3 shows schematically the water and pipe wall temperatures and ice formation for various stages of the freezing process. When the water pipe of initial temperature T_i is exposed to the subfreezing environment temperature T_∞, wall temperature (A) as well as the water temperature adjacent to the top wall of the pipe decreases abruptly. A natural convection process occurs. The density difference of water due to variation in

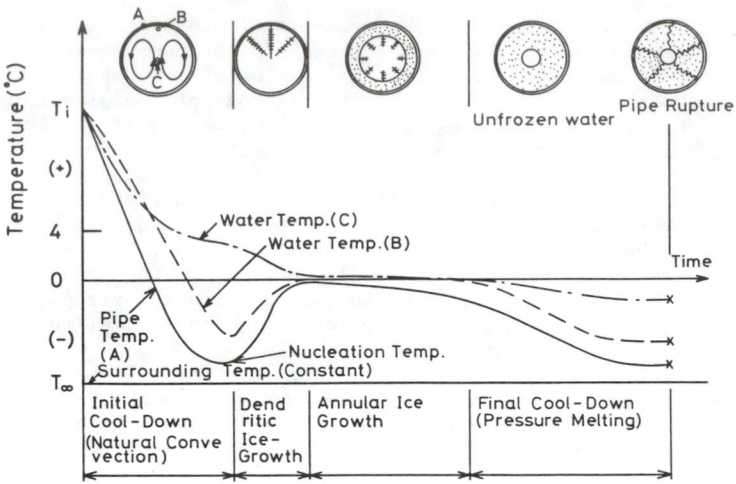

FIGURE 3. The stages of the freezing process in a water pipe with quiescent water.

temperature cause two eddies to form. The different time history of water temperature (C) at the center line from the other temperatures corresponds to the inversion of the convective patterns due to the maximum density of water at about 3.98°C for 1 atmospheric pressure prior to freezing. Subsequently, the water is supercooled significantly to -4 $\approx$ -7°C before ice nucleates. Ice nucleation occurs suddenly at the top. The dendritic ice spreads from the nucleation site down to the center of the water pipe and releases heat from fusion into the supercooled water. During the process of dendritic ice growth, the water temperature returns to a uniform 0°C. After dendritic ice growth is complete, the normally solid annulus of ice begins to grow inward from the pipe wall. Eventually, the water temperature in the cold region inside the annular ice is decreased by pressure melting, which occurs with a water pressure rise due to volumetric change in the freezing process. As the freezing proceeds, the first annular ice ruptures and subsequently, cracks on the inner surface of the pipe result in a pipe rupture.

3.2 Supercooling Phenomenon In A Water Pipe

Supercooled water plays an important role in ice formation in the water pipe, since ice nucleation and the quantity of the above-mentioned dendritic ice that forms in the supercooled water are controlled by the cooling rate and the amount of supercooling $\Delta T(=T_m$(freezing temperature) - T (supercooled water temperature)). The solid phase, liquid phase and vapor phase become stable in terms of the stability theory of thermo-chemical energy. However, during the freezing process, the stagnant water in the pipe will undergo substantial supercooling which remains in a quiescent condition before ice nucleation occurs. This behavior of supercooling can be explained in terms of Gibb's free energy theory. Free energy for ice nucleation in the water is caused by the mutual interaction between a free energy decrease of water (function of supercooling rate and

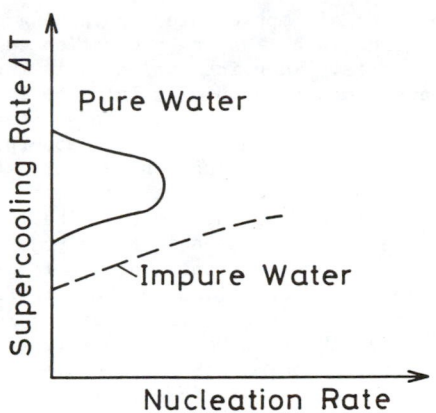

FIGURE 4. The relationship between supercooling rate and nucleation rate.

geometrical dimension of ice particles) as an ice formation factor and a surface energy increase (function of surface tension and geometrical dimension of ice particles) as a deterrent factor of ice nucleation.
When the free energy overcomes the surface energy, ice nucleation occurs in the water pipe. The free energy of supercooled water becomes a thermodynamic motive force of ice formation and is proportional to a chemical potential difference $\Delta\mu$ between the liquid phase (water) and the solid phase (ice). Using the supercooling rate ΔT, the $\Delta\mu$ is expressed as follows:

$$\Delta\mu = \mu^\ell - \mu^i = \frac{L\Delta T}{T_m} \tag{1}$$

where L = latent heat of water, T_m = freezing temperature μ^ℓ and μ^i = chemical potential of the liquid and solid phases, respectively.
Figure 4 represents the relationship between ice nucleation rate and supercooing rate [5]. The spontaneous nucleation rate of pure water, where ice nucleates without the aid of other solid particles or the wall surface, is increased with an increase of chemical potential difference $\Delta\mu$, which represents the supercooling rate in Eq.(1). Since the viscosity of water increases abruptly if the supercooling rate exceeds a certain limit, the ice nucleation rate decreases rapidly with a suppression of water movement.

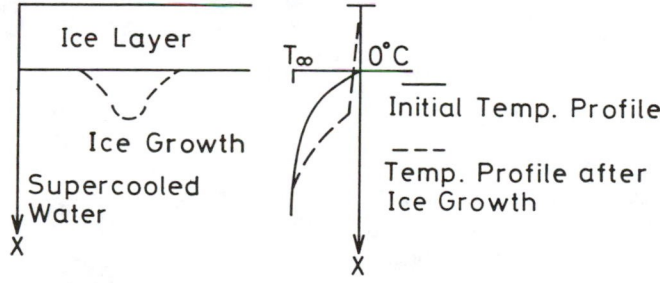

FIGURE 5. Ice growth process during supercooling process.

On the other hand, the ice nucleation rate of impure water containing dust particles and insoluble impurities, which can act as ice nucleation centers, becomes larger with a low supercooling rate. When ice nucleates in super-cooled water, the water medium is transformed very quickly into the ice crystal phase, since the interface of water-ice is covered with a water phase. Therefore, the ice nucleation speed would depend on the absorption rate of the latent heat of fusion by the thermal conduction heat transfer in the water pipe. That is, the ice growth rate can be related to the amount of supercooling and the cooling rate of the water pipe. The ice growth from the flat ice-water interface is illustrated in Fig.5 [6]. When a part of the water-ice interface stretches out into the supercooled water, due to hydrodynamic or thermodynamic fluctuations on the interface, the temperature gradient of the supercooled water near the stretched inter-face becomes larger. As a result, ice growth at the stretched interface could be prompted by a large temperature gradient in the supercooled water. Consequently, the surface energy increase on the interface during this process suppresses the ice crystal growth. When the supercooling energy as unstable factor overcomes the surface energy as a stable factor, dendritic ice crystals grow rapidly in the supercooled water medium, according to the cooling rate and the supercooling rate. Measurements [6] made on tap water show nucleation temperatures in the range -5 ~ -6°C and -6 ~ -7°C for water from the cold water tap and the hot water tap, respectively. Fresh water obtained from lakes or streams has nucleation temperatures of -3 ~ -5°C. The effectiveness of dust particles and insoluble impurities as nucleation sites is destroyed by contact with hot water or by prolonged contact with water. Therefore, the hot water during the cooling process is supercooled significantly, typically reaching -6 ~ -7°C. The increase in the amount of dendritic ice with an increased supercooling rate (hot water pipe), might cause an ice blockage of the water pipe if the water pressure is not large enough to break through the dendritic structure.

Figure 6 [7] shows the effect of the initial supercooling in the pipe at the time of ice nucleation in relation to the total volume fractions of ice and the pressure gradient required to start the flow. Data were obtained using a copper tube with an inside diameter of 12.6 mm.

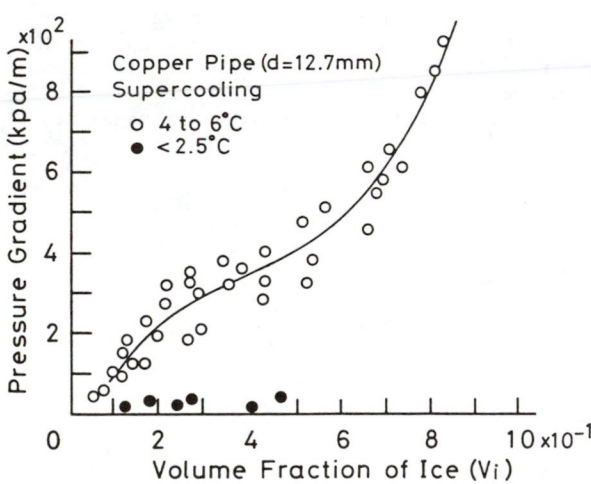

FIGURE 6. Pressure gradient vs. volume fraction of ice for various supercooling conditions.

For supercooling of 2.5°C or less, no significant start-up pressure gradient is measured under any circumstances. It can be seen that when only 7 - 20% of the pipe volume is frozen in a pipe that is supercooled to 4 - 6 °C, a pressure gradient of 50 - 200 kPa/m is required to start the flow. The effect of pipe diameter is shown in Fig.7 in the range of supercooling from 4 - 6°C. The data shows that the pressure gradient required to start the flow in a large pipe is smaller than that required in a small pipe. The amount of remaining water at the bottom might be increased with an increase in pipe diameter, since the dendritic ice grows from the top of the water pipe toward the pipe center. The critical condition for a dendriric ice closure of the cross section of the test copper pipe is demonstrated in Fig.8 for various cooling rates H (°C/min) and pipe diameters D (mm) under the nucleation temperature of -5°C. Symbols in this figure indicate the extent of dendritic ice growth. Two critical curves are drawn with the following expression for the cooling rate H.

$$H = C(\frac{-2 - T_m}{D})^{5/4} \tag{2}$$

where D = pipe diameter, T_m = ice nucleation temperature, C = 21 - 29 (constant) for copper pipe.
At a high rate of cooling the dendritic ice growth in a pipe does not block its entire cross section. On the contrary, at a low cooling rate such as in an underground water pipe, the dendritic ice growth might block the entire pipe cross section, becoming an obstacle to restarting a flow in the pipe. Figure 9 shows the time history of water pressure rise in a copper pipe with an inside diameter of 5 mm [8]. It is interesting to note that the water pressure in the pipe rises abruptly to 4 MPa within 6 seconds. This abrupt increase of water pressure corresponds to the appearance of dendritic ice growth. After 6 seconds, the water pressure increases gradually; its small pressure change with time corresponds to annular ice growth inward from the pipe wall.

3.3 Pressure Melting In The Water Pipe

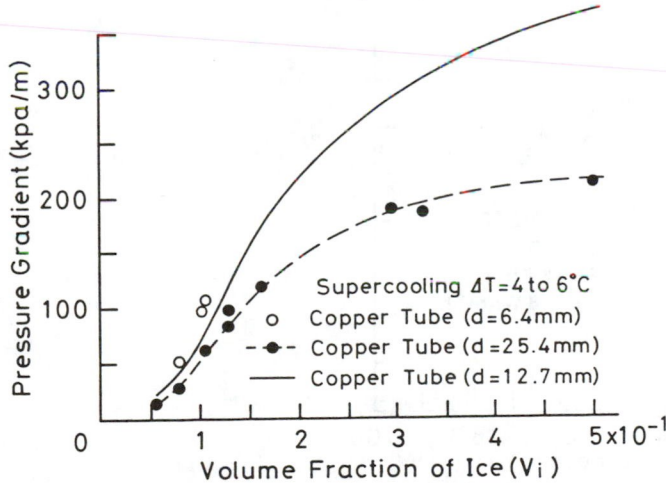

FIGURE 7. Pressure gradient vs. volume fraction of ice for various pipe diameters.

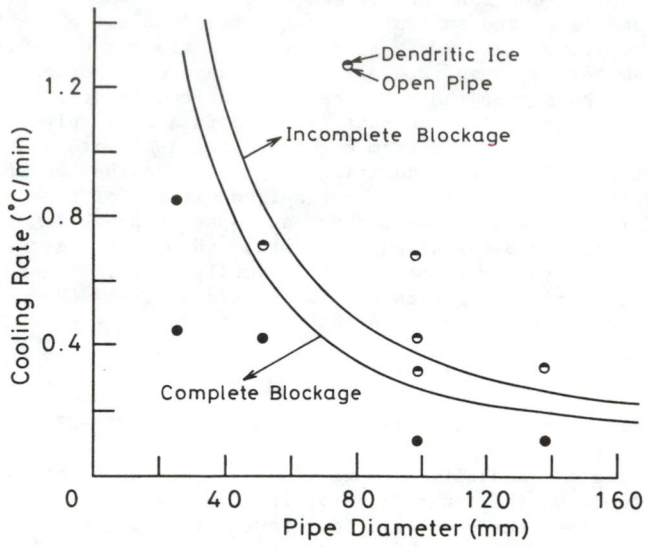

FIGURE 8. The effect of cooling rate on ice blockage in a pipe.

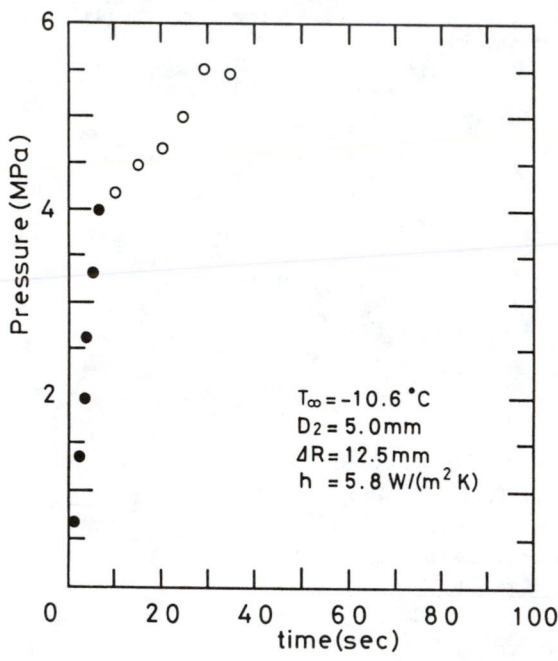

FIGURE 9. Water pressure rise with time in a frozen pipe.

As shown in Fig.3, the pressure rise occurs in the frozen pipe with the volumetric change of water in freezing and deformation of pipe materials. The drop in temperature with an increase in pressure can be estimated by using the Clausius-Clapeyron correlation as follow.

$$\frac{dT_m}{dP} = \frac{(V_w - V_i)\ T_m}{L} \tag{3}$$

where T_m = freezing temperature, V_w, V_i = specific volume of water and ice, respectively, and L = latent heat of fusion.
The rate of temperature drop to the rise in pressure is dT_m/dP = – 0.075 K/MPa for T_m = 273.15 K, L = 333.6 kJ/kg, V_w = 10^{-3} m^3/kg and V_i = 1.0905 x 10^{-3} m^3/kg. Thus the temperature drop is about 0.0075 K per 1 atmospheric pressure. On the other hand, the temperature drop caused by the solution of air into the water is about 0.0025 K per 1 atmospheric pressure. Therefore, a temperature drop of about 0.01 K can be brought by 1 atmospheric pressure. It needs about 100 atm. pressure (= 10 MPa) to reduce the freezing point by 1°C. Since the critical strength of a steel water pipe is about 250 Mpa, the freezing temperature at this pressure falls to about –25°C. It is possible for unfrozen water to remain in the pipe if the pipe is not cooled down below –25°C.

4.   RUPTURE BEHAVIOR OF A WATER PIPE BY FREEZING

4.1 Dynamic Behavior Of A Water Pipe By Freezing

The time histories of water pressure, strain on the water pipe and the various freezing stages in a water pipe exposed to a subfreezing environment are shown in Fig.10. These dynamic changes in the water pipe correspond with the freezing behaviors shown in Fig.3. The appearance of dendritic ice growth after the water is subcooled causes a rapid pressure

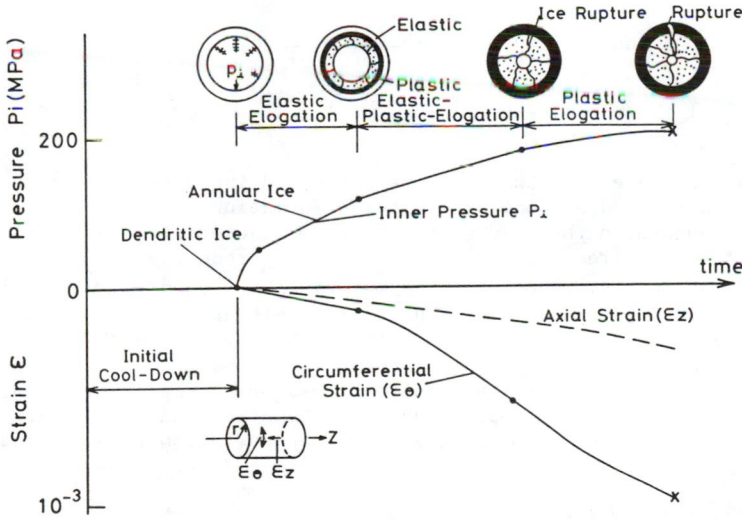

FIGURE 10.   The time history of pressure rise and strain of frozen water pipe.

rise P_i in the water pipe. Subsequently, the annular ice growth inward from the cooled pipe wall brings about a gradual pressure increase due to the volumetric change of the freezing process. The elastic deformation of the pipe from an increase in water pressure produces circumferential strain ε_θ and axial strain ε_z (in the case of a closed pipe). Plastic deformation from the inner surface to the outer surface of the pipe wall, after it deforms over the yield point at a certain time, occurs with an increase in pressure from freezing. This deformation produces a rapid increase of circumferential strain. The annular ice layer formed inside the pipe wall is broken at a certain pressure. Then the remaining water sinks into cracks in the ice layer and the freezing of this water produces a new annular ice layer. Eventually, the plastic deformation spreads into the entire pipe wall. The amount of circumferential strain is controlled by a processing hardness of pipe material. Due to the subfreezing environment, unfrozen water appears due to the temperature drop caused by a rise in pressure. The unfrozen water temperature can be determined by the pressure using the Clausius–Clapeyron theory. As the freezing proceeds, the water pipe ruptures according to the strength of the pipe material. If the plastic deformation occurs once in the pipe wall, the water pipe might be likely to rupture due to fatigue, caused by a cyclic deformation of freezing and melting. The critical water pressure P_{ic} that starts the yield of pipe material can be estimated in terms of Mises' yield theory [9] as follows;

$$P_{ic} = \frac{\sigma_s}{\sqrt{3}} \frac{R^2 - 1}{R^2} \tag{4}$$

where σ_s = tensile stress of the material, R = radius ratio r_2/r_1, r_1 = inner radius and r_2 = outer radius.
After the plastic deformation has been established in the entire pipe wall, the pipe rupture takes place when the critical pressure P_{ic} in the following equation (the known characteristics of strain hardening by shearing stress) reaches a maximum value in consideration of the strain hardening and dimension change of the pipe.

$$P_{ic} = 2 \int_{r_1}^{r_2} \tau(C) \frac{dr}{r} \quad \text{and} \quad P_{ic} = \int_{C_1}^{C_2} \tau(C) \frac{1}{\exp(C) - 1} \, dC \tag{5}$$

$$C = - \log_e\left(\frac{r_2^{*2}}{r^2} - \left(\frac{r_2^2}{r^2} - 1 \right) \exp(\varepsilon_z) \right) + \varepsilon_z$$

where r_2^* = initial outer radius, r_2, r_1 = outer and inner radius after deformation, and r = radial coodinate after deformation.
The critical pressure in the case of the known characteristics of strain hardening by tensile stress can be calculated by introducing $\tau = \sigma_s/\sqrt{3}$ and $C = \sqrt{3}\,\varepsilon$ into the Eq.(5). The following materials are used for water pipe at the present time; strainless (working tensile stress σ_s = 700 MPa), steel (σ_s = 300 – 400 MPa), ductile cast iron (σ_s = 400 – 700 MPa), copper (σ_s = 300 – 350 MPa), hard vinyl chloride (σ_s = 50 – 60 MPa) and hard rubber (σ_s = 3 – 5 MPa). Besides the tensile stress of the materials used, another cause of pipe rupture depends on the elongation characteristics and temperature-dependence of hardening (brittleness in low temperature).

4.2 Dynamic Characteristics Of Ice

The dynamic behavior of ice growth starting from the pipe wall during the

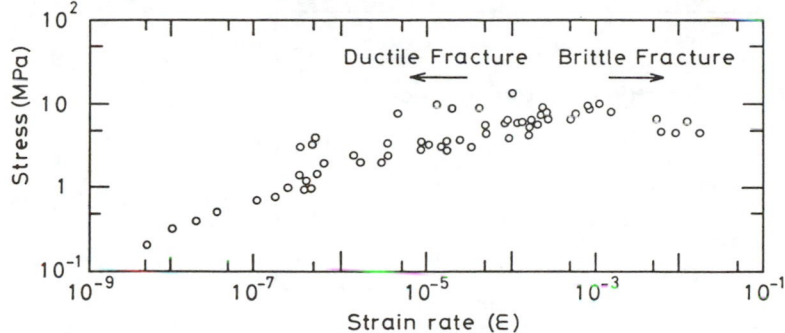

FIGURE 11. Variation of compressive stress on ice with strain rate.

freezing process plays an important role in estimating the pipe rupture. The ice as a visco-elastic substance has both elastic and plastic characteristics. The pure ice at − 5°C has modulus of elasticity E = 8.5 − 9.9 GPa, shear modulus G = 3.4 − 3.8 GPa and Poisson's ratio γ = 0.31 − 0.36). Figure 11 shows the relationship between the critical failure stress in compression and strain rate. Both ductile and brittle fractures appear at the boundary of strain rate at 10^{-4} (s^{-1})[11]. It is interesting that the critical compressive stress becomes 10 MPa, while the critical tensile stress becomes 1 MPa, which is 10 % of the compressive stress. However, the ice behaves like an elastic substance within a very short time (less 100 seconds) when the acting pressure is below 1 MPa. As a result, the dynamic characteristics of ice pertains to plastic deformation.

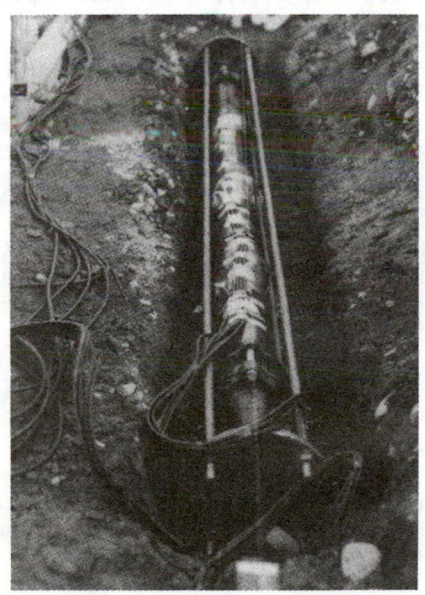

FIGURE 12. The whole view of test pipe buried in underground.

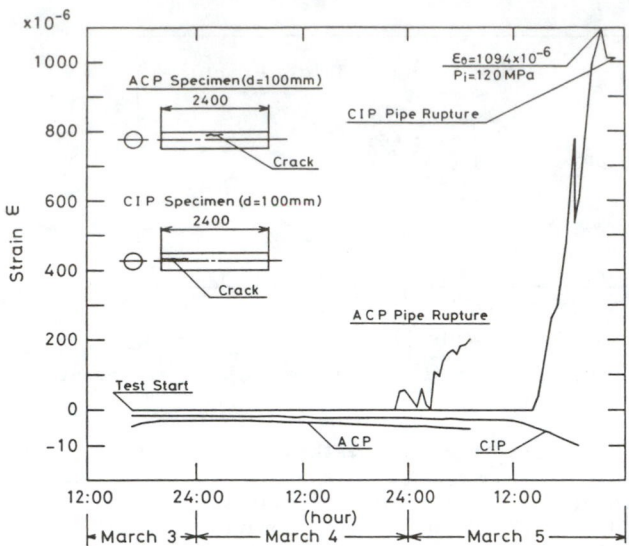

FIGURE 13. The time history of circumferential strain and pipe wall temperature.

4.3 Field Experiment On Pipe Rupture By Freezing

The mechanism of pipe rupture was investigated in a field test [1]. Four kinds of water pipe (pipe diameter of 100 mm, the length of 3 m) used for 25 - 30 years (asbestos cement pipe ACP of σ_s = 22.5 MPa, E = 26 GPa, cast iron pipe CIP of σ_s = 200 MPa, E = 110 GPa, ductile cast iron pipe DIC of σ_s = 400 MPa, E = 160 GPa, and vinyl chloride pipe VC of σ_s = 55 MPa, E = 3 GPa) were buried in the ground at a depth of 1 m, as shown in Fig.12. Some strain guages were set on these test water pipes to measure the strain on the pipes. The time histories of circumferential strain and pipe wall temperature are shown in Fig.13. Asbestos cement pipe (ACP) ruptured with a long crack in the axial direction after 38 hours. The strain on the pipe took place in 29.5 hours as shown in Fig.14(a) and then the pipe rupture occurred within another 8.5 hours. The pressure of 5.2 MPa was estimated from the amount of the circumferential strain (200 x 10^{-6}) put on the pipe wall, which was kept at the wall temperature of -4.6 - -5.1°C. The rupture of ACP (σ_s = 22 MPa) might have been caused by the aging effect of the material strength for long term usage. It took about 45 hours to cause strain on the pipe wall and 8 hours later the pipe rupture occurred, as shown in Fig.14(b). The circumferential strain increased abruptly due to the increase of ice formation with a decrease in the pipe wall temperature. Pipe stress of 120 MPa, estimated from the amount of measured strain of 1.09 x 10^{-6}, is not consistent with the working stress of 230 Mpa for CIP pipe. The difference between the measured and working stresses might be explained by the fact that the strain gauge set on the pipe wall surface was not matched with the part of the pipe that ruptured. The ductile cast iron can absorb the pipe circumferential elongation of about 4.5 %, caused by a volumetric change of 9 % from the water to ice phase. On the other hand, the vinyl chloride pipe should rupture due to its small circumferential elongation limit of about 1.8 %, as estimated by the dynamic properties of σ_s = 55 MPa, E = 3 GPa.

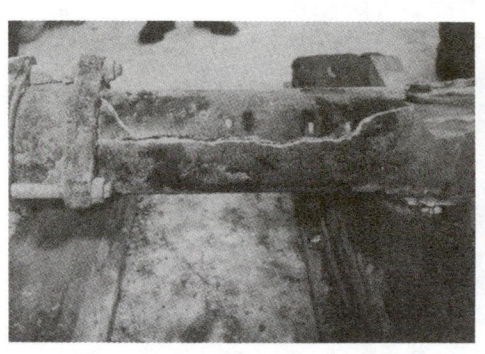

(a)

(b)

FIGURE 14. (a) Photograph of ACP pipe rupture (b) Photograph of CIP pipe rupture.

5. FREEZING BEHAVIOR OF A WATER PIPE BY PRESSURE MELTING

5.1 Theoretical Model Of Water Pipe Freezing Under Pressure Melting

As noted earier, when a water pipe filled with quiescent water is exposed to the subfreezing environment, the water pressure increase due to the volumetric expansion of phase change would cause a drop of water temperature in accordance with the Clausius-Clapeyron equation. Therefore, we must consider the drop of water temperature by pressure and the ice layer formed inside the pipe wall. Figure 15 shows the theoretical model for pipe freezing when an ice thickness of X forms between the water phase and the inner side of the pipe wall (inner radius r_2, outer radius r_3 and wall thickness $R = r_3 - r_2$). If the temperature of the remaining water is the same as that of the water-ice interface, T_1, during the freezing process, the following energy balance equation can be obtained at the ice-water interface.

$$L\rho_i \frac{dS}{dt} = \lambda_i \frac{dT}{dr} - \frac{1}{2} \rho_w C_w \frac{dT_1}{dt} dS \qquad (6)$$

where, L = latent heat of fusion, λ = thermal conductivity, ρ = density, C = specific heat, t = time, T = temperature, T_1 = freezing temperature, P_1, P_2 = pressure shown in Fig.15 and subscript i and w corresponds to the ice and water phase. The displacement U_s of the pipe can be expressed using a formula based on the elasticity theory [9].

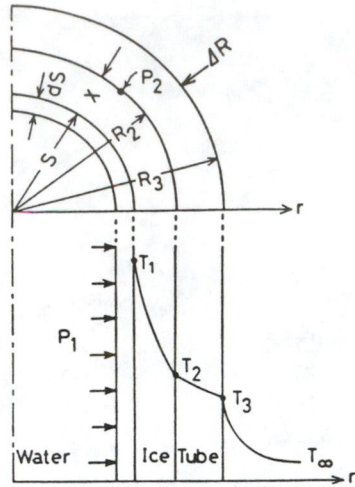

FIGURE 15. The theoretical model of pipe freezing.

$$U_s = \left[\frac{1 - \frac{1}{m_i}(\frac{S}{R_2}) \ P_1 - P_2}{E_i \ (1 - (\frac{S}{R_2})^2)} \right]s + \left[\frac{1 + \frac{1}{m_i}(P_1 - P_2)}{E_i \ (1 - (\frac{S}{R_2})^2)} \right]s \tag{7}$$

where E = modulus of elasticity, and m_i = Poisson's number.
The pressure P_2 between the ice layer and the inside wall of the pipe can be estimated by double cylinder model as follows;

$$P_2 = \frac{2 \frac{P_1}{E_i} \frac{S^2}{(R_2^2 - S^2)}}{\left[\frac{m_i - 1}{m_i E_i} - \frac{m_t - 1}{m_t E_t} \right] + 2\left[\frac{1}{E_i} \frac{S^2}{(R_2^2 - S^2)} + \frac{1}{E_t} \frac{R_3^2}{(R_3^2 - R_2^2)} \right]} \tag{8}$$

where the subscript t means the water pipe.

5.2 Freezing Behavior Of Steel Water Pipe

The annular ice growth from the steel pipe wall was calculated by using the following physical properties of steel pipe; λ_t = 46.5 W/(m K), E_t = 206 GPa, m_t = 3.45 and ice λ_i = 2.2 W/(m K), E_i = 9.8 GPa, specific volume V_i = 1.08 x 10^{-3} m3/kg, and m_i = 2.94. The steel pipe was exposed to a subfreezing temperature T_∞ = -5°C and the free convection of the heat transfer coefficient was h = 5.8 W/(m2K). In Fig.16, the time history of the ice growth X is presented under both calculation conditions: with volumetric change ($\sigma_w = \sigma_i$) and without it ($\sigma_w > \sigma_i$)[12]. The results of the calculations, taking into consideration of the volumetric change of phase change, show that the ice growth in the pipe decreases gradually with time and will leave unfrozen water in the water pipe. Figures 17 (a), (b) and (c) show the parameter effects of subfreezing temperature T_∞, the outer diameter of the pipe D_2 and pipe thickness R on the water pressure P_1 in the pipe. The water pressure P_1 approaches a certain constant value

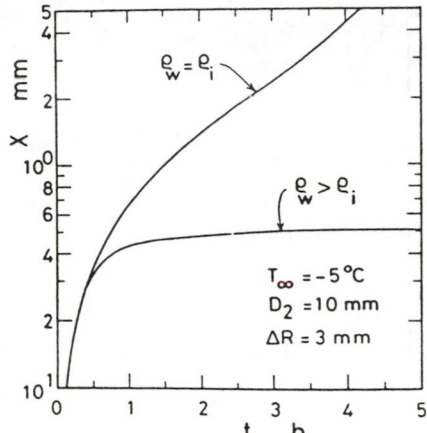

FIGURE 16. A freezing limit.

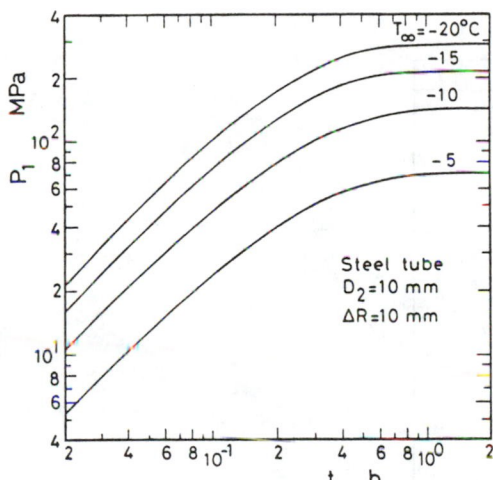

FIGURE 17(a). Effect of subfreezing temperature on freezing limit.

determined by the subfreezing temperature T_∞. With the assumption of a strength limit of $\sigma_s = 245$ MPa for the steel pipe, the steel pipe of $D_2 = 10$ mm and wall thickness $R = 10$ mm might be fractured below a subfreezing temperature of $T_\infty = -15°C$. The following function of the relationship of maximum ice thickness X_{max} to pressure $(p_1)_{max}$ can be predicted as the freezing limit [12].

$$X_{max} = 0.01\ D_2^{1.03}\ |T_\infty|$$

$$(p_1)_{max} = 14.5\ |T_\infty| \qquad (9)$$

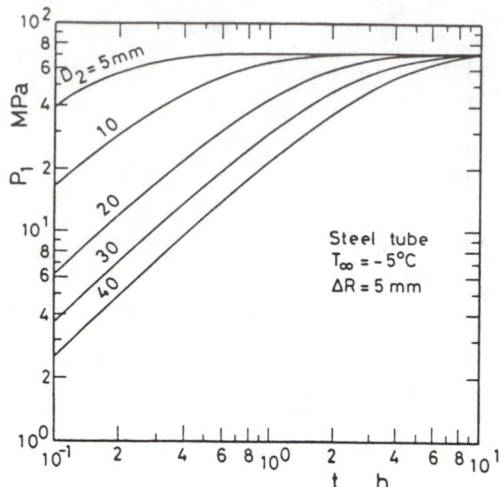

FIGURE 17(b). Effect of pipe wall thickness on freezing limit.

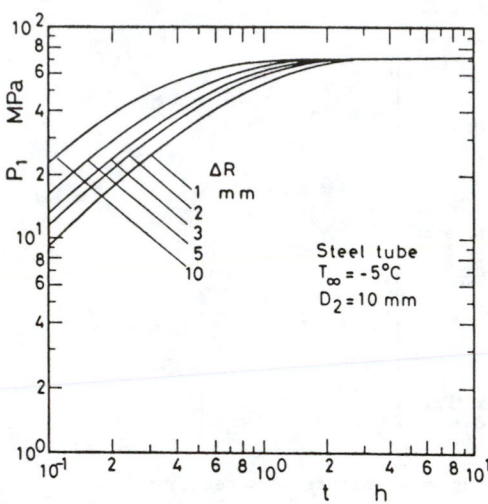

FIGURE 17(c). Effect of inner diameter of pipe on freezing limit.

in the ranges of $D_2 = 10 - 60$ mm, $\Delta R = 3 - 10$ mm and $T_\infty = 0 - 20°C$.

5.3 Freezing Behavior Of Water Pipes For Various Conditions

In Figs. 18 – 20, freezing behaviors for steel pipe ($E_t = 206$ GPa, $m_t = 3.45$, limit strength at fracture $P_c = 245$ MPa), Cast iron pipe ($E_t = 206$ GPa, $m_t = 2.95$, $P_c = 245$ MPa) and ductile cast iron pipe ($E_t = 150$ GPa,

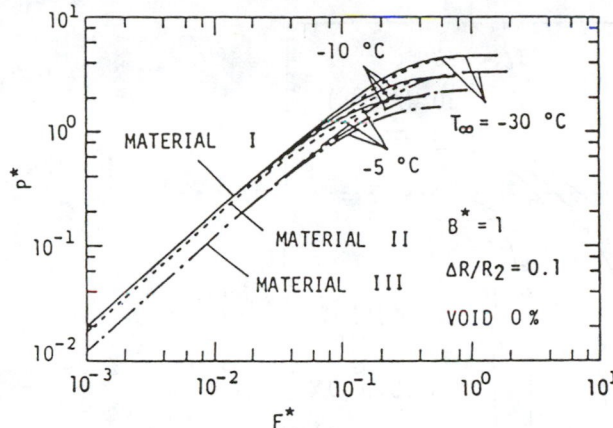

FIGURE 18. Influence of pipe material properties on non-dimensional pressure.

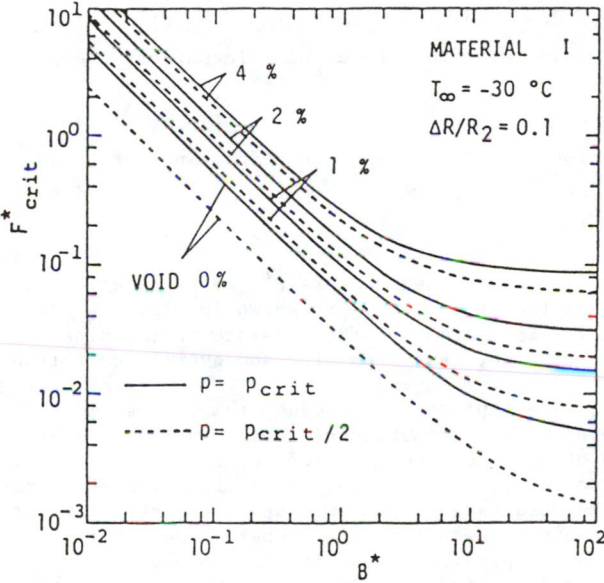

FIGURE 19. The relationship between the critical non-dimensional time and Biot number.

$m_t = 3.7$, $P_c = 290$ MPa) are shown for various environmental conditions. In these figures, the maximum water pressure P_1 in the pipe is non-dimensionalized as $P^* = P_1/P_c$ and the Fourier number is expressed as $F^* = (\alpha_i t/R_2^2) \times (C_i(T_1 - T_\infty)/L)$, where α_i = thermal diffusivity of ice and C_i = specific heat of ice. The heat transfer coefficient h on the outer surface

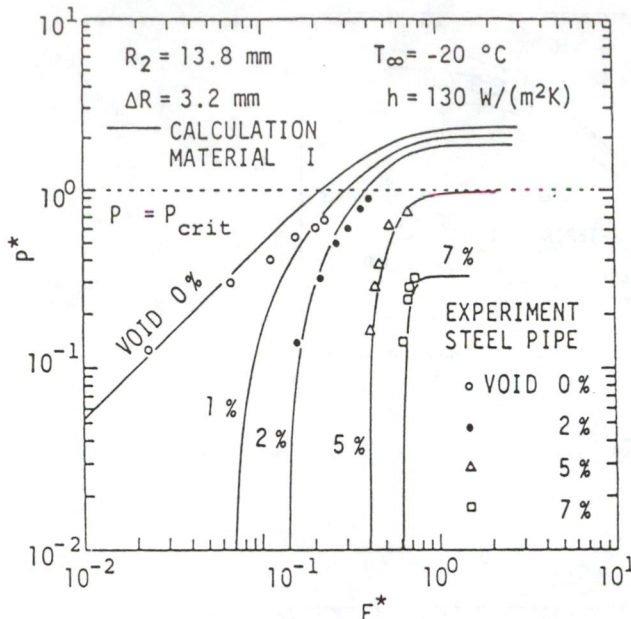

FIGURE 20. The effect of void ratio in pipe on non-dimensional pressure vs. non-dimensional time.

of the pipe is non-dimensionalized as a function of Biot number Bi as

$$B^* = Bi/\log_e(1 + \Delta R/R_2) \tag{10}$$

where $Bi = hR_3/\lambda_t$.

The non-dimensional time history of the pressure P^* for different pipe materials and environmental temperature T_∞, are shown in Fig.18 [13]. As the modulus of elasticity E_t increases, the pressure in the pipe increases, in the order: steel pipe, cast iron pipe and ductile cast iron pipe.

As mentioned earlier, pipe rupture is caused by the volumetric expansion of the phase change of water. Compressible space in the pipe absorbs the volumetric expansion of phase change. Therefore, it may be necessary to delay the critical time until the pipe rupture occurs. Figure 19 represents the effect of the void ratio of the air space to the entire steel pipe volume in a relationship between the Biot number B^* and critical Fourier number F^*_{crit} when the pipe ruptures with $T_\infty = -30°C$ and $\Delta R/R_2 = 0.1$. In this figure, solid lines indicate $P = P_c$ and dotted lines indicate $P = P_c/2$. It can be seen that the amount of time it takes the pipe to rupture increases with an increase in void ratio and with a decrease in Biot number. Figure 20 shows the relationship between non-dimensional pressure P^* and Fourier number F^* in the steel pipe for various void ratios. It should be noted that for $T_\infty = -20°C$, $h = 130 \ W/(m^2K)$, $R_2 = 13.8$ mm and $\Delta R = 3.2$ mm, the pipe does not rupture, since the value of P^* is smaller than 1 when the void ratio is over 5 %. As a result, it is possible to design a pipe line that will not rupture from freezing with the aid of a

void placed in the pipe. One should also pay attention to the water hammer
phenomenon, which will break the pipe when a compressible gas is used as
a void.

6. USAGE OF PIPE FREEZE-OFF

In recent years, utility engineers have used pipe freeze-up and the pipe
pressure test instead of a flow stopping valve in order to restore service.
Using the pipe freeze up method allows the possibility of working in a
smaller area near the damaged pipe and shortens the amount of time needed
to restore service. Figure 21 represents the relationship between the
consumption rate of liquid nitrogen W, which created an ice thickness of
10 mm in the pipe filled with quiescent water, and steel pipe diameter d.
The solid line in Fig.21 represents the experimental correlation, W =
$0.038\ d^{1.79}$[14]. The critical time, t_c, when the entire cross section of
the pipe is blocked by ice is shown in Fig.22 for various pipe diameters.
In Fig.22, the solid line indicates the curve of the obtained experimental
correlation, $t_c = 0.3\ d^{1.89}$. Figure 23 shows the relationship between non-
dimensional time Fo* ($= \alpha t_c/d^2$) for pipe freeze-off conditions with flowing water, and the Reynolds number Re (= Ud/ν) for various cooling parameters T^* (= $(T_m - T_w)/(T_\infty - T_m)$)[15]. The results in Fig.23 are for
steel pipe with a diameter of 12.7 mm and a length of 500 mm. The symbols
are : α = thermal diffusivity of water, U = mean velocity of water flow,
ν = kinematic viscosity, T_m, T_w, T_∞ = freezing,   wall and main flow
temperature, respectively. It can be seen that the critical time until
pipe freeze-off increases with an increase in Reynolds number and with a
decrease in cooling parameter T^*.

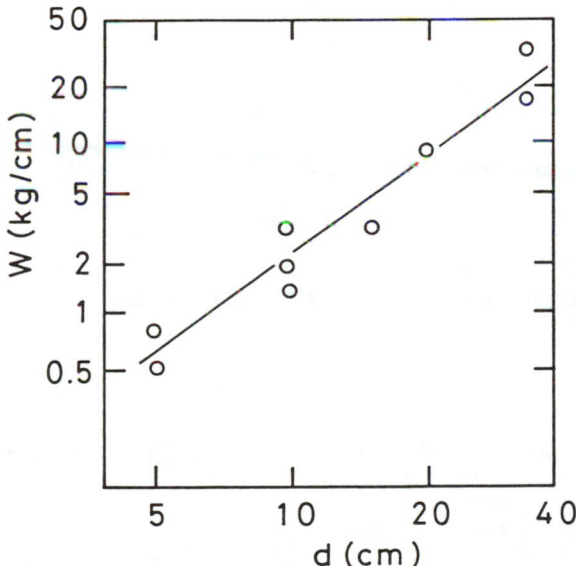

FIGURE 21. Variation of liquid nitrogen consumption with pipe diameter.

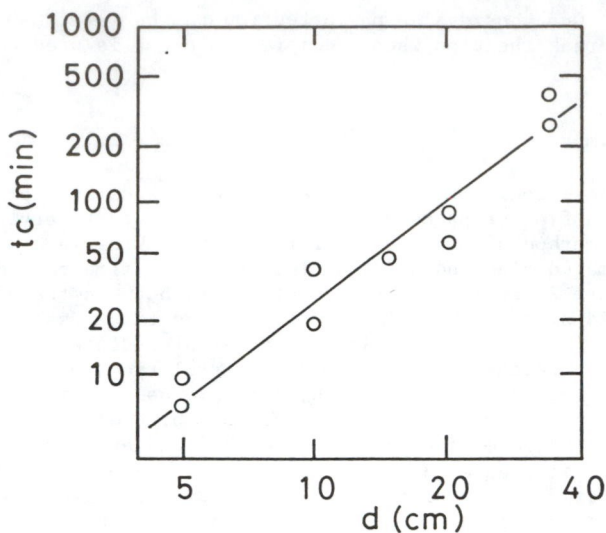

FIGURE 22. Variation of critical pipe freeze-off with pipe diameter.

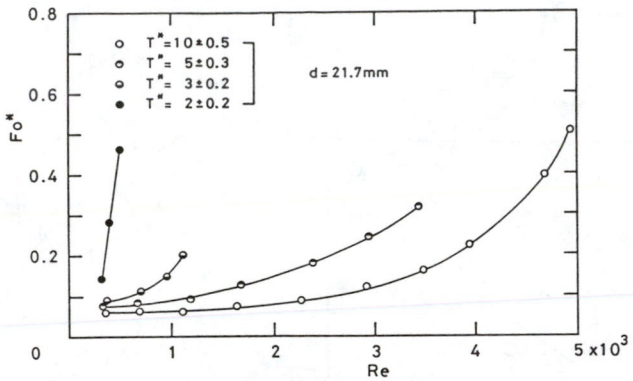

FIGURE 23. The relationship between critical Fourier number and Reynolds number.

7. CONCLUDING REMARKS

A review has been conducted on the fundamental aspects of pipe freezing and pipe rupture mechanisms in a water pipe with quiescent water. The present results on water pipe freezing are useful for utility engineers in designing a water pipe line system in cold regions.

The author wishes to acknowledge the permission by Asahikawa Water Supply Bureau, Hokkaido, Japan to publish the results of the field work.

REFERENCES

1. Asahikawa Water Supply Bureau, Field Test on Water Pipe Freezing, Asahikawa Water Supply Bureau Report, March 1982.

2. Gilpin, R.R., Ice formation in a Pipe Containing Flows in The Transition and Turbulent Regimes, *J. Heat Transfer. Trans. of ASME*, vol.103, no. 2, pp.363-368, 1981.

3. Gilpin, R.R., Modes of Ice Formation and Flow Blockage That Occurs While Filling a Cold Pipe, *Cold regions Science and Technology*, vol.5, no. 2, pp.163-171, 1981.

4. Sampson, P. and Gibson, R.D., A Mathematical Model of Nozzle Blockage by Freezing - Turbulent Flow, *Int. J. Heat Mass Transfer*, vol.25, no.1, pp.119-126, 1982.

5. Maeno, N. and Fukuda, M., *The Structure and Properties of Snow and Ice*, Kokon-Shoin Pub., 1986.

6. Gilpin, R.R., The Effect of Cooling Rate on the Formation of Dendritic Ice in a Pipe with No Main Flow, *J. Heat Transfer, Trans. of ASME*, vol.99, no.3, pp.419-424, 1977.

7. Gilpin, R.R., The Effect of Dendritic Ice Formation in Water Pipe, *Int. J. Heat Mass Transfer*, vol.20, no.5, pp.693-699, 1977.

8. Horiuchi, K. and Maeno, N., Pressure Behavior in Small Tube by Freezing, *Proc. Japan Snow and Ice Association*, pp.66-70, 1980.

9. Japan Society of Mechanical Engineers, *Handbook of Mechanical Engineering - Part 4 Material Strength*, JSME Pub., 1968.

10. Gold, L.W., Engineering Properties of Fresh-Water Ice, *J. Glaciology*, vol.19, no.81, pp.197-223, 1977.

11. Gold, L.W., The Process of Failure of Columnar Ice, *Phil. Mag.*, vol.26, no.2, pp.311-328, 1972.

12. Sugawara, M., Seki, N. and Kimoto, K., Freezing Limit of Water in a Closed Circular Tube, *Wärme- und Stoffübertragung*, vol.17, pp.187-192, 1983.

13. Oiwake, S., Saito, H., Inaba, H. and Tokura, I., Study on Dimensionless Criterion of Fracture of Closed Pipe Due to Freezing of Water, *Wärme- und Stoffübertragung*, vol.20, pp.323-328, 1986.

14. Takeuchi, K. and Nakajima, S., Freeze-up in a Large Diameter Water Pipe, *Proc. 30th Japan Water Supply Meeting*, pp.87-89, 1979.

15. Inaba, H., Fukuda, T., Saito, H. and Tokura. I., A study of Freezing Behavior in a Pipe with Water Flow, *Trans. of The Japan Association of Refrigeration*, vol.4, no.2, pp.71-79, 1987.

CONVECTIVE HEAT TRANSFER IN ICE-WATER SYSTEMS AND CONVECTIVE INSTABILITY

Chapter 9

Onset of Convection and Heat Transfer Characteristics in Ice-Water Systems

YIN-CHAO YEN
U.S. Army Cold Regions Research and Engineering Laboratory,
Hanover, New Hampshire 03755, USA

ABSTRACT

This review discusses the problems associated with the anomalous
temperature-density relations of water. It deals with the subjects
of onset of convection, the temperature structure and natural convective heat transfer and the laminar forced convective heat transfer in
the water/ice system. The onset of convection in a water/ice system
was found to be dependent on thermal boundary conditions, not a
constant value as in the classical fluids. This system also exhibits
a unique temperature distribution in the melt layer immediately after
the critical Rayleigh number is exceeded and soon after it estab-
lishes a more or less constant temperature region which expands to
about two-thirds of the melt layer depth. The constant temperature
is approximately 3.2°C for water layer formed from melting above but
varies for melt layers formed from below. The heat flux across the
water/ice interface was found to be a weak power function and to
increase linearly with temperature for melt layer formed from above
and below, respectively. Both theoretical and experimental melting
studies of ice spheres, cylinders and vertical plates show a minimum
heat flux at the inversion temperature ranged from 5.1° to 5.6°C.
For the case of laminar forced convection melting heat transfer, the
presence of an interfacial velocity reduces heat transfer in comparison with the case without phase change.

CONTENTS

NOMENCLATURE

a	wave number, also as defined in Eq. (59)
a_o	defined as $\int_0^\infty \exp(-\delta^3)d\delta$
a_N	defined as $\int_0^\infty \exp\left[-\delta^3 + \frac{\beta\delta}{a_N}\right]d\delta$
A	defined as $(T_1-T_{max})/(T_1-T_2)$
c_p	specific heat at constant pressure
d	thickness, distance to 4°C front and unstable layer depth
f	a factor defined in Eq. (30)
F_o	Fournier number defined as $\alpha t/\ell^2$
g	gravitational constant
Gr	Grashof number defined as $g\beta_\infty\theta_m\ell^3/\nu^2$
h	heat transfer coefficient
H	total layer depth
k	thermal conductivity
k_f	Kutateladze number $(= 1/\beta)$
ℓ	characteristic length
L	latent heat of fusion
m	a factor defined in Eq. (22)
n	a factor defined in Eq. (23)
N	a factor defined in Eq. (24)
Nu	Nusselt number
P_1, P_2 and P_3	defined in Eqs. (19), (20), and (21)
Pr	Prandtl number
Q	defined in Eq. (8)
R	radius
Ra	Rayleigh number defined in Eq. (3)
S	shape factor defined in Eq. (17)
St	Stefan number defined as $C_p(T_m-T_\infty)/L$
t	time
T	temperature
ΔT	$T_\infty - T_m$ also $T_1 - T_2$
v	velocity
Ta	Taylor number defined as $\frac{4\Omega_1^2}{\nu^2} R_1^4 \frac{(1-\mu)(1-\mu/\eta^2)}{(1-\eta^2)^2}$

Greek Letters

α	thermal diffusivity
β	coefficient of volumetric expansion, also $C_p\Delta T/L$
$\beta_1, \beta_2, \beta_3$	constants in Eq. (26)
β_∞	factor defined in Eq. (25)

γ_1, γ_2	coefficient in Eq. (2)
λ	defined as H/d
λ_1, λ_2	parameters defined in Eqs. (4) and (5)
μ	viscosity, ratio of angular velocity (Ω_2/Ω_1)
ν	kinematic viscosity
ρ	density
η	R_1/R_2, y/δ, y/δ_t similarity variable defined in Eq. (55)
Ω	angular velocity
δ	momentum boundary thickness
δ_t	thermal boundary thickness
δ_t^{**}	defined in Eq. (60)
∂^{**}	defined in Eq. (61)
$\psi(\beta)$	defined as a_o/a_N
$\theta(\beta)$	defined as $(a_o/a_N)^{4/3}$
Φ	defined as $k\Delta T/\rho L$
θ_T	defined as h/h_o
θ_m	defined as $T_m - T_\infty$
τ_w	shear stress
ω	defined in Eq. (54)

Subscripts

c	critical
i	ice
iv	inversion
m	melting
md	average diameter
mH	average height
max	maximum
o	initial, without melting
p	plate
w	wall
∞	bulk
1,2	lower and upper boundary, or inner and outer

1. INTRODUCTION

The phenomenon of convection in a horizontal water layer formed either
by melting from below or above are of special interest because the
system divides itself into two regions with an unstable zone lying under
a stable layer. This is caused by the nonlinear temperature-density
relations of water at its maximum density. Figure 1 schematically shows
the temperature and density profiles of a water layer continuously
formed by melting ice from below. It can be seen from curve A that the

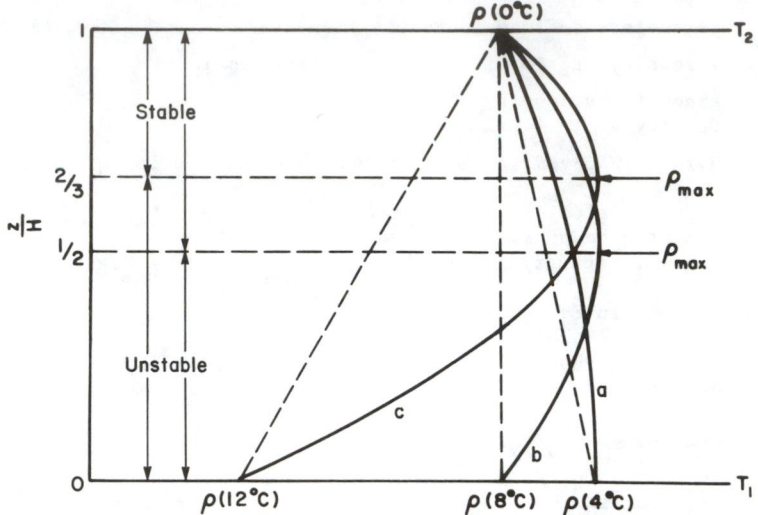

FIGURE 1. Schematic representation of density profile in a conduction state for the upper boundary $T_2 = 0°C$ (equivalent to the case of melting from below).

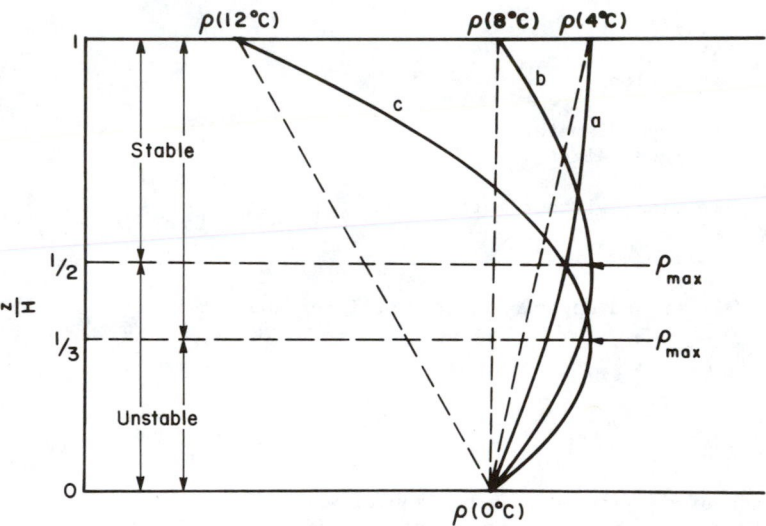

FIGURE 2. Schematic representation of density profile in a conduction state for the lower boundary $T_1 = 0°C$ (equivalent t the case of melting from above).

water layer is stabley stratified as long as $T_1 < 4°C$, and unstabley stratified for $T_1 > 4°C$ (the upper boundary is maintained at the phase transition temperature, i.e. $T_2 = 0°C$). It is distinctly shown that the water layer between the ice-water interface and the 4°C isotherm is always stable. However, the relative height (i.e. the ratio of z/H where z is the distance from the warm plate to the point where ρ_{max} exists, and H is the total layer depth) of the unstable region increased as T_1 increased. Figure 2 is a simple inversion of Figure 1 and shows the case of ice melting from above. In this case, T_1 is at the melting temperature (i.e. $T_1 = 0°C$) and $T_2 > 0°C$. In contrast to Figure 1, the relative height of the unstable layer decreased with increasing T_2.

Therefore, for water with its maximum density near 4°C, even for a constant temperature gradient, the density profile is a nonlinear function of the vertical ordinate z. These situations are quite different from the classic Rayleigh problem, in which a constant temperature gradient and a linear temperature-density relation of the fluid are assumed. In the classical problem, a fluid layer of a constant depth rigidly confined from top and bottom, but of infinite horizontal dimensions, is either cooled from above or heated from below. Convection occurs if the temperature gradient exceeds a certain critical value. The criterion is usually expressed in terms of a Rayleigh number with a critical value of ≈ 1708, which has been verified by Benard [1] who observed the regular cell structure for Ra > 1708. This classical problem has been studied intensively since the early works of Rayleigh [2] and Jeffreys [3]. A comprehensive analysis and summary of these studies are compiled in the book by Chandrasekhar[4]. The aim of this study is to review and analyze the stability problem caused by the density anomaly of water and its effect on the heat transfer characteristics in ice-water systems due to the evolution of the interfacial velocity resulting from the density difference between ice and water.

2. ANALYTICAL STUDIES ON THE ONSET OF CONVECTION IN A HORIZONTAL WATER, LAYER

The influence of the density anomaly on water at the onset of convection has been reported by quite a few investigators. Veronis [5] was first to investigate the phenomenon of penetrative convection. By representing the water density near 4°C as the quadratic expression

$$\rho = \rho_{max} [1-\gamma(T-T_{max})^2 \tag{1}$$

where ρ_{max} is water density corresponding to T_{max} ($\approx 3.98°C$) and $\gamma \approx 7.68\times110^{-6}$ ($°C^{-2}$) incorporating the linear stability theory, Veronis summarized the computed critical Rayleigh numbers Ra_c for both free (TABLE 1) and rigid (TABLE 2) boundaries. In both tables, λ is the ratio of the total layer depth H to the 4°C layer depth 'd' in the conductive state (i.e. $\lambda = H/d$), 'a' is the wave number, and Ra_c is the critical Rayleigh number. In TABLE 2, μ is defined as the ratio of rotation rates between the outer and inner cylinder $((1-\mu) = \lambda)$, and Ta_c is the critical Taylor number. The values of a^2/λ^2 can be compared with

TABLE 1. Critical Rayleigh Number and Corresponding Wave Numbers for Various Values of λ in Rigid-Free Boundary Problems [5].

λ	a^2	$\dfrac{Ra_c}{\pi^4}$	$\dfrac{a^2}{\lambda^2}$	$\dfrac{Ra_c}{2\lambda^4\pi^4}$
0	0.500	6.75	--	--
2.0	0.500	13.43	0.5	6.75
1.2	0.505	16.69	0.351	4.03
1.4	0.519	21.92	0.265	2.81
1.6	0.535	31.37	0.209	2.40
1.8	0.630	50.86	0.194	2.42
1.9	0.750	67.07	0.205	2.58
2.0	0.930	87.18	0.233	2.73
2.5	1.64	222.2	0.262	2.84
3.0	2.29	547.9	0.254	2.83
3.5	3.10	854.7	0.253	2.84

TABLE 2. Critical Rayleigh Numbers and Corresponding Wave Numbers for Various Values λ in Rigid-Rigid Boundaries [5].

$(1-\mu)=\lambda$	a	Ta_c	$\dfrac{a^2}{2(1-\mu)^4\pi^4}$	$\dfrac{Ta_c}{2(1-\mu)^4\pi^4}$
0	3.12	1.708×10^3	--	--
1	3.12	3.39×10^3	0.986	17.40
1.25	3.13	4.462×10^3	0.635	9.383
1.5	3.20	6.417×10^3	0.461	6.506
1.6	3.24	7.688×10^3	0.415	6.022
1.8	3.49	1.182×10^4	0.386	5.779
1.9	3.70	1.494×10^4	0.384	5.885
2.0	4.00	1.868×10^4	0.405	5.993
2.5	5.06	5.619×10^4	0.415	6.069
3.0	6.10	9.558×10^4	0.419	6.057
3.5	7.10	1.771×10^5	0.417	6.058

values of $a^2/(1-\mu)^2\pi^2$ and $Ra_c/2\lambda^4\pi^4$ with $Ta_c/2(1-\mu)^4\pi^4$. There is a quantitative difference between the results of the two cases due to the different boundary conditions. However, the qualitative behavior is the same in both cases. In both Tables, the unstable layer 'd' is used in defining the Rayleigh number. For the specific cases of $T_2 = 4°C$ and 8°C, i.e. $\lambda = 1$ and 2, Veronis reported Ra_c values of 13.43 π^4 and 87.18

π^4, respectively. These tabulated values of Ra_c clearly demonstrate that the onset of convection is dependent upon the thermal boundary conditions imposed.

This work was extended by Sun et al. [6] by adding a cubic term to the temperature-density relation

$$\rho = \rho_{max} [1 - \gamma_1 (T - T_{max})^2 - \gamma_2 (T - T_{max})^3] \tag{2}$$

which is valid up to 30°C. Following a procedure commonly employed in linear stability analysis, a modified Rayleigh number

$$Ra = \frac{2\gamma_1 A g(\Delta T)^2 H^3 \left[1 + \frac{3}{2} \frac{\gamma_2}{\gamma_1} A \Delta T\right]}{\alpha \nu} \tag{3}$$

is derived in which $A = (T_1 - T_{max})/(T_1 - T_2)$, H is the total melt layer depth, and ΔT is the overall temperature difference (i.e. $\Delta T = T_1 - T_2$). They found that the Ra values were functions of two parameters defined as follows

$$\lambda_1 = \left(-\frac{1}{A}\right) \frac{1 + 3 \frac{\gamma_2}{\gamma_1} A \Delta T}{1 + \frac{3}{2} \frac{\gamma_2}{\gamma_1} A \Delta T} \tag{4}$$

and

$$\lambda_1 = \left(-\frac{1}{A^2}\right) \frac{\frac{3}{2} \frac{\gamma_2}{\gamma_1} A \Delta T}{1 + \frac{3}{2} \frac{\gamma_2}{\gamma_1} A \Delta T} \tag{5}$$

Ra_c values were computed for ranges $-4.25 < \lambda_1 < -0.5$ and $-1.4 < \lambda_2 < 1.6$. Numerical values of $(Ra_c)_{\lambda_2 = 0}$ (i.e. for the case of the parabolic temperature-density relation, $\gamma_2 = 0$, where $\lambda_1 = (-1/A)$ as function of λ_1 were reported and shown in Figure 3 for both rigid-free and rigid-rigid boundaries. The Ra_c values for cubic and parabolic temperature-density relations were expressed in a ratio as $(Ra_c)_{\lambda_1}/(Ra_c)_{\lambda_2 = 0}$ and

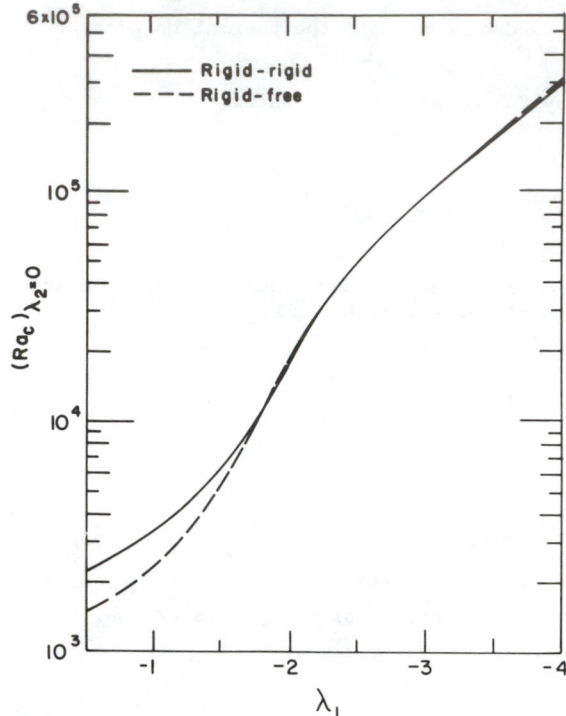

FIGURE 3. $(Ra_c)_{\lambda_2 = 0}$ as a function of λ_1 [6].

was shown verse λ_2 using λ_1 as the parameter for rigid-free (Fig. 4) and rigid-rigid boundary conditions (Fig. 5). These results are valid as long as the melt layer contains a density extremum within the boundaries.

Merker et al. [7] studied this problem with the representation of the temperature-density of water approximated by three different polynominals having 2, 3 and 5 terms. Linear stability analysis was used and the resulting perturbation equations were solved with the aid of Galerkin's method. Figure 6 shows the general diagram of stability for fluids having a density extremum in which the values of β_1 and β_2 are the coefficients of thermal expansion evaluated at T_1 and T_2, respectively, and the values of Ra_1 and Ra_2 are the corresponding Rayleigh numbers defined as $Ra = g\beta\Delta TH^3/\alpha\nu$ in which H is the total layer depth, and $\Delta T = T_1 - T_2$, g is the gravitational acceleration, α is the thermal diffusivity and ν is the kinematic viscosity. Since Ra_1 is always defined as positive if the layer is unstably stratified, whereas Ra_2 changes sign, Ra_1 was used by Merker et al. to describe the layer stability. Figure 7 shows the critical Rayleigh number variation with T_1 in the case of a constant wall temperature (i.e. T_w = constant) with the density-temperature relation represented by a polynominal of the

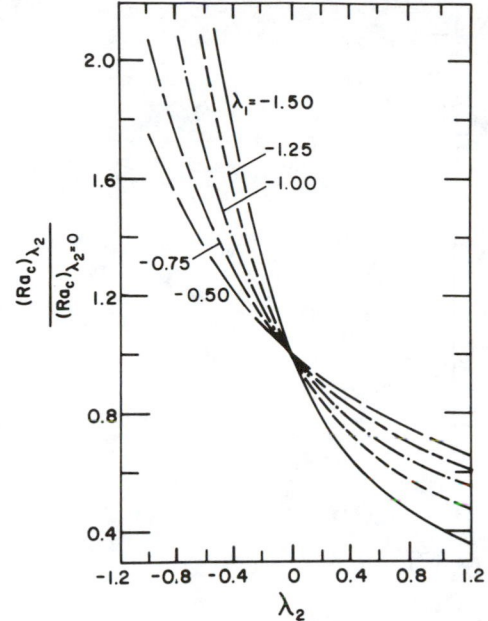

FIGURE 4. $\dfrac{(Ra_c)_{\lambda_2}}{(Ra_c)_{\lambda_2=0}}$ vs λ_2 with λ_1 as a parameter for rigid-free cases [6].

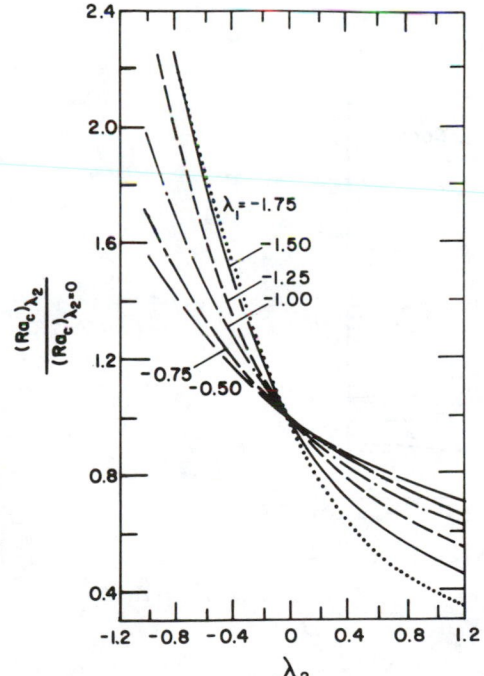

FIGURE 5. $\dfrac{(Ra_c)_{\lambda_2}}{(Ra_c)_{\lambda_2=0}}$ vs λ_2 with λ_1 as parameter for rigid-rigid cases [6].

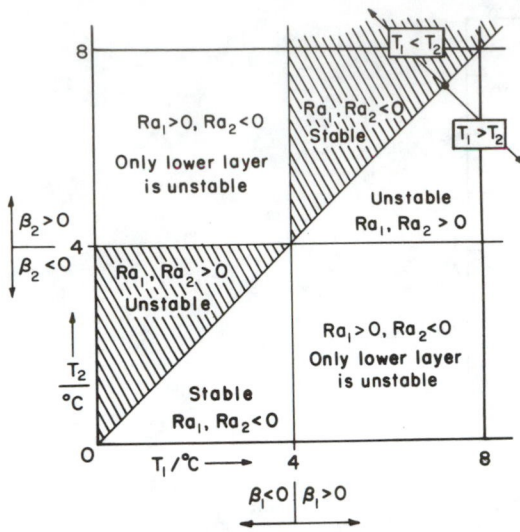

FIGURE 6. Principal stability diagram [7].

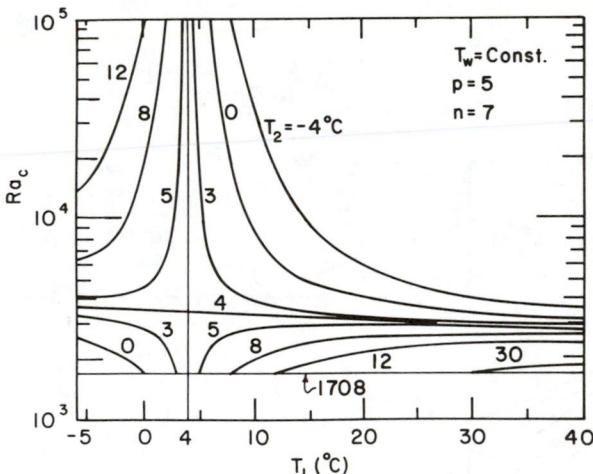

FIGURE 7. Critical Rayleigh number Ra_c as a function of T_1 with T_2 as the parameter for T_w = constant [7].

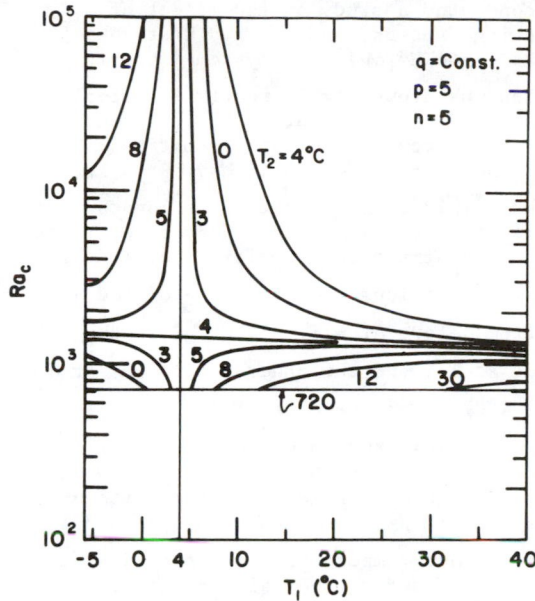

FIGURE 8. Critical Rayleigh number Ra_c as a function of T_1 with T_2 as the parameter for q_w = constant [7].

order p=5 and n=7 (number of Galerkin terms). Figure 8 shows the case of a constant wall heat flux density (i.e. q_w= constant). In both figures, it can be noted that in the region below the isotherm T_2= 4°C, the water layer has a density profile with no maximum value, i.e. the layer is completely unstably stratified. However, the non-linearity of the density profile is slight and the effect on the critical Rayleigh number is not large. The Ra_c are in the range of 1708 to ≈ 3600 for T_w= constant and 720 to ≈ 1600 for q_w= constant. On the other hand, for the region above the isotherm T_2= 4°C, the layer has a density profile containing a maximum value between the two boundaries and the layer is only partially unstably stratified. The non-linearity of the density profile is strong and the effect on Ra_c is great. The Ra_c values are significantly larger than those obtained from the classical problem. For a constant wall temperature and rigid boundary conditions, this work should be comparable with the work of Sun et al. [6] in which a cubic density-temperature relation was used and the entire layer depth (as in this work) was used in defining the critical Rayleigh number.

3. EXPERIMENTAL STUDIES ON THE ONSET OF CONVECTION IN A CIRCULAR-HORIZONTAL MELT LAYER

Yen [8] conducted the first experimental study, which focused solely on the determination of the onset of convection in a water layer formed by melting ice. Boger et al. [9] conducted a study on the effect of buoyancy on the melting and freezing process, and claimed that the critical Rayleigh number, Ra_c ($Ra = g\beta\Delta TH^3/\alpha\nu$) in a system involving phase change and density inversion was around 1700, nearly identical to the value reported by Globe et al. [10] and Schmidt et al. [11] who heated the horizontal layers of water from below, but did not test in the region of density inversion. However, in order to have $Ra_c \approx 1700$, as in the classical case of normal fluid (i.e. density is a monotonic function of temperature). Values of H, β, ΔT, α and ν used to evaluate Ra_c were found to be dependent on the direction of melting (upward or downward) and the thermal boundary conditions. In the case of ice at the top (equivalent to melting from below, $\Delta T = T_1 - 4°C$), H is the total layer depth, β is evaluated at T_1, and the other properties of α and ν are evaluated at $1/2$ ($T_1 + 4°C$). For the case of ice at the bottom (equivalent to melting from above), they proposed to use $\Delta T = 4°C$. H is the portion of water layer depth and was evaluated by equating the heat transported through the convective layer to the heat conducted through the ice, i.e. $H = - 4 d_i k_e/k_i T_i$, in which d_i, k_i and T_i are thickness, thermal conductivity, and temperature (maintained at the end of the sample) of ice, k_e is the effective thermal conductivity of water of the convective layer, β is evaluated at $0°C$ and α and ν are evaluated at the mean temperature of the buoyant region, i.e. $2°C$. In Yen's work, a rather large, cylindrical bubble-free ice sample (12.7 cm in diameter and 25 cm in height) was used. At the beginning of the experiment, there was only one solid phase. A constant temperature boundary was applied at the lower end which results in melting from below. Due to the rapid transition from the conductive to the convective state of the melt layer, the upper limit of the temperature was restrained. For T_1 ranging from 7.72° to 25.5°C and using the criteria recommended by Boger et al. in evaluating Ra_c, he found, on the contrary, that the Ra_c values were a strong function of T_1 and could be represented by

$$Ra_c = 14,200 \exp (-6.64 \times 10^{-2} T_1) \qquad (6)$$

The initial temperature of ice T_o was found to have no significant effect on the Ra_c values. Figure 9 shows the variation of Ra_c with the lower boundary temperature.

It should be emphasized that the study of a melt layer is quite distinct from the classical problem of a horizontal layer of fluid of invariant depth. For a water layer formed by melting, the layer depth not only grows with time, but a simultaneous addition of water is necessary to make up the volume shrinkage due to the phase transition (an intricate device has to be found in order to conduct studies of this nature). Water layers can be formed either by melting from below or above. For the case of melting from below (i.e. $T_1 > 4°C$, $T_2 = 0°C$, for the special case of $T_1 = 4°C$, the entire water layer will be always

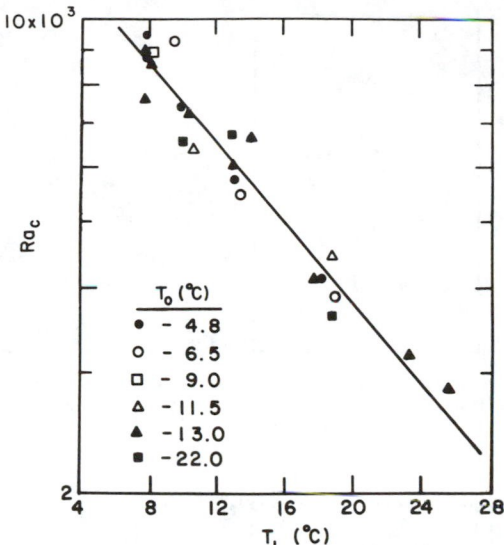

FIGURE 9. Ra_c as a function of T_1 [8].

stable), there will be an unstably stratified region extending from the lower boundary up to the 4°C isotherm and a stable region above the 4°C isotherm which extends to the upper boundary (the interface of the water-ice). On the other hand, for the case of melting from above (for $T_2 > 4°C$, $T_1 = 0°C$), there will be a stable layer adjacent to the upper boundary and an unstable region between the 4°C isotherm and the water-ice interface, while the $T_2 = 4°C$, the entire melt layer will be un-stably stratified. Figure 10 shows schematically the interdependence of the stable and unstable regions with the thermal boundary conditions. In a follow-up study by Yen et al. [12] on melting from below and above, they reported that the critical layer depth, H_c (at which the heat transfer mode changes from conduction to convection), could be deter-mined by either locating the inflection point on the melting front versus time or by locating the sudden jump of the temperature gradient in the stable region. Figure 11 shows the variation of the temperature gradient as a function of time and the upper boundary temperature for cases of melting from above. It can be seen that the higher the upper boundary temperature, the less steep the temperature gradient jump at the onset of convection. Figure 12 shows the comparison of Ra_c values for both melting from below and above. Experimental Ra_c values were determined using appropriate values of the critical layer depth H_c, A = $(T_1 - T_{max})/T_1$, $\Delta T = T_1$ and A = T_{max}/T_2, and $\Delta T = -T_2$ respectively for melting from below and above. Theoretical Ra_c values were determined from the graph developed by Sun et al. [6] (Fig. 5 in this paper) after evaluating parameters λ_1 and λ_2 from Eqs. (4) and (5). Figure 13 demon-strates another method of comparing Ra_c values, but it uses T_1 or T_2

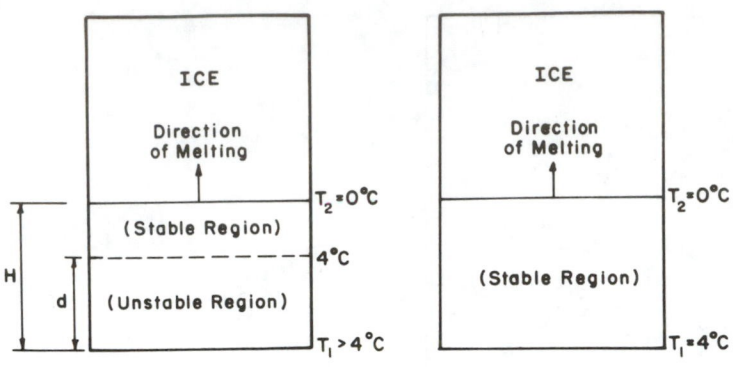

a) Melting from Below

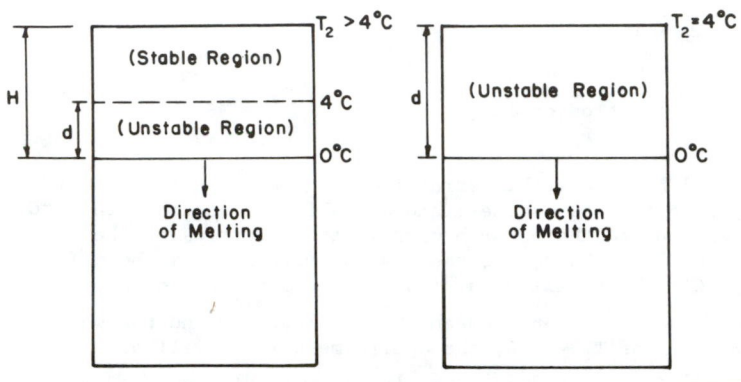

b) Melting from Above

FIGURE 10. Schematic representation of the interdependence of the stable and unstable regions with the thermal boundary conditions.

explicitly as a variable. This figure covered T_1 from 7.72° to 25.2°C and T_2 from 6.3° to 13.09°C. The corresponding ranges of λ_1 and λ_2 are λ_1 from -1.970 to -0.830, λ_2 from -0.311 to 0.390 for melting from below, and λ_1 from -3.444 to 1.618, and λ_2 from 0.112 to 0.508 for melting from above.

The experimental results of Yen [8] and Yen and Galea [12] were also compared with the analytical work of Veronis [5] and the experimental work of Legros et al. [13]. Veronis considered the case equivalent to melting from above while Legros et al. studied the case equivalent to melting from below. In both cases the layer was invariant, while in the studies of Yen [8] and Yen and Galea [12], a water layer was formed (initially at zero depth) continuously as a result of the phase transition. The results of Veronis were plotted in terms of $Ta_c/2(1-\mu)^4\pi^4$ versus λ, in which Ta_c was the critical Taylor number. the work of

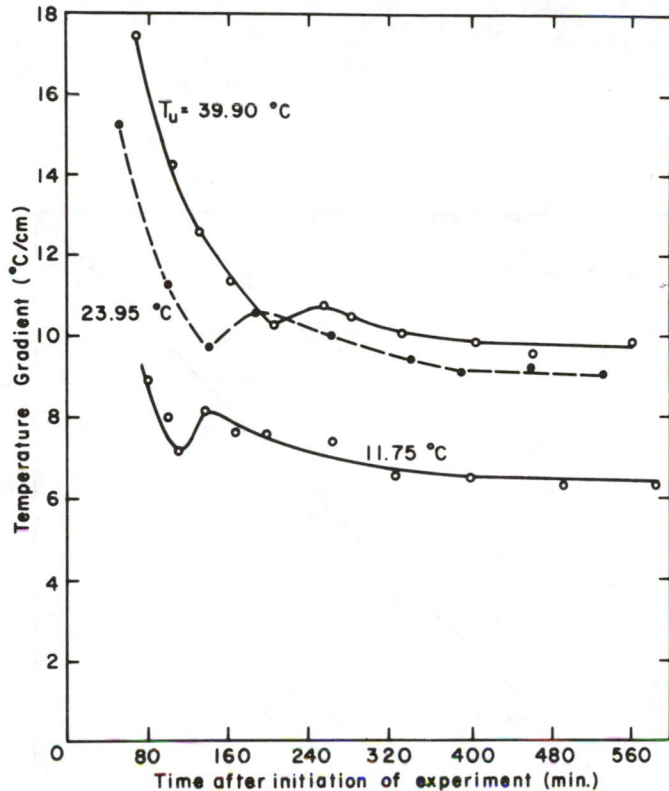

FIGURE 11. Variation of temperature gradient in the stable region of the water layer (near the upper boundary).

Legros et al. was expressed in terms of $Ra_c/2\lambda^4\pi^4$, in which Ra was defined as $Ra = g\gamma(\Delta T)^2 d_c^2/\alpha\nu$, where γ was the temperature coefficient as defined in Eq. (1), and ΔT was the temperature difference across the unstable layer. In the cases of melting from above and below, $\Delta T = 4°C$ and $\Delta T = T_1-4°C$ respectively. The values of d_c were determined as follows: for melting from above $d_c = (4/T_2)H_c$ and for melting from below, $d_c = [(T_1-4)/T_1]H_c$. Figure 14 shows that there is remarkable agreement among these studies.

Some unique features were observed in the studies reported by Yen [8] and Yen and Galea [12] during the formation and continuous deepening of the water layer. Prior to the onset of convection, the water-ice interface had a **planar** surface, and the prevailing heat transfer mode was conduction. Cells began to form as soon as the Ra_c value was attained. Initially, these cells possesed regular patterns and were hemispherical and higher in the center. They had a circular cross-

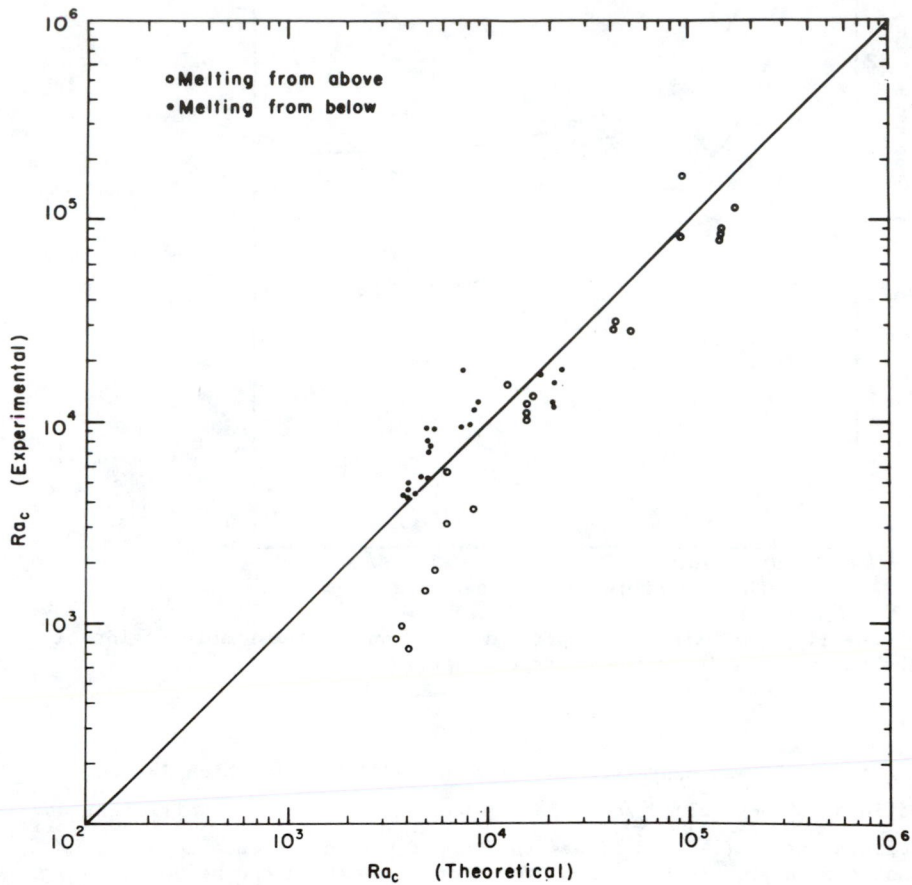

FIGURE 12.   Comparison between experimental and theoretical Ra_c [12].

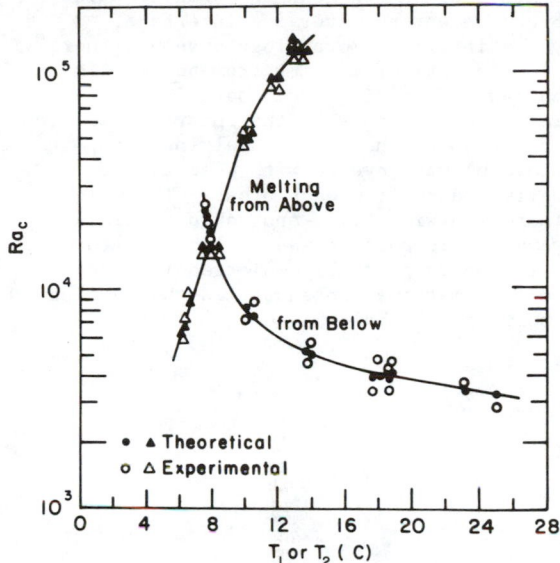

FIGURE 13. Comparison between experimental and theoretical Ra_c using T_1 or T_2 as the variable [6].

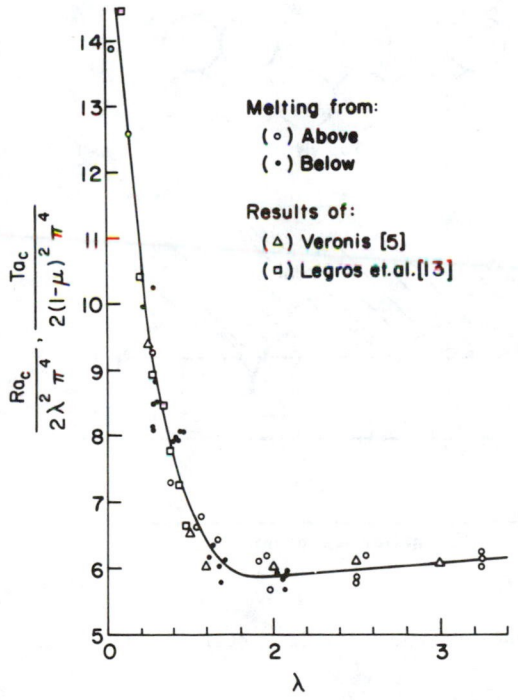

FIGURE 14. A comparison of $\dfrac{Ta_c}{2(1-\mu)^4\pi^4}$ and $\dfrac{Ra_c}{2\lambda^4 \pi^4}$ vs λ [18].

section and were different from the well known hexagonal Bénard cells and were evenly distributed over the entire water-ice interface. Figure 15a shows schematically the interface morphology observed shortly after the onset of convection in the case of melting from below. It is interesting to note that in the case of melting from above, a stable region formed over the unstable region adjacent to the interface as soon as a melted layer formed. In contrast to the case of melting from below, the interface (see Figure 15b) was covered with a series of circular concentric ridges equally spaced from each other. The height between the crest and trough increased with time, and could attain a value as large as 5.0 mm. However, this pattern was replaced by that of small inverted hemispherical cells which gradually enlarged in size and finally formed an irregular and unsymmetrical interface as the melting progressed (as in the case of melting from below).

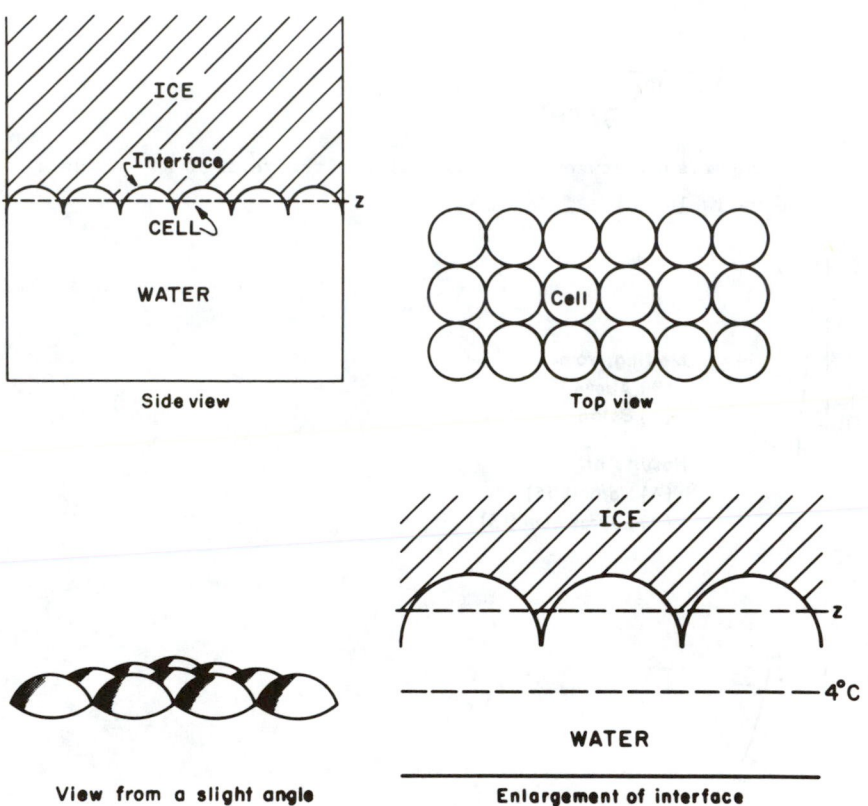

a) Melting from below.

FIGURE 15. Schematic of water-ice interface shortly after the onset of convection [12].

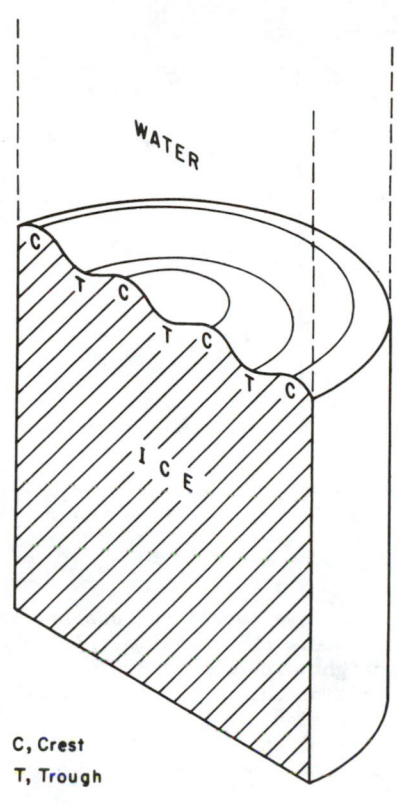

C, Crest
T, Trough

b) Melting from above.

FIGURE 15 (cont'd).

4. TEMPERATURE STRUCTURE AND HEAT TRANSFER IN A HORIZONTAL WATER LAYER

Townsend [14] was the first to have made a detailed study on the temperature structure and natural convective heat transfer in water over an ice surface. The experimental tank consisted of a 30 cm x 30 cm bottom made of dural 0.95 cm thick, and sides made of perspex of the same thickness. Distilled water was used to fill the tank to a depth of 15 cm and the tank bottom was cooled by a shallow dish filled with liquid nitrogen the dish was not in direct contact with the bottom, however it should be noted that the purpose of bottom cooling was only to maintain the lower boundary at 0°C. The free water surface temperature was electrically maintained through another dural plate suspended about 5 mm above the water surface. The most striking feature of the steady temperature distribution was the creation of a region of constant mean temperature. Townsend indicated that after a constant mean temperature region was established, the only observed changes were in the upward extension of its boundary. Figure 16 shows the temperature distributions during the approach to equilibrium. Figure 17a shows the equilibrium distribution of the mean temperature. It can be seen from the figure that the constant temperature region has a mean temperature of 3.2°C, significantly below the temperature of maximum density (i.e.≈

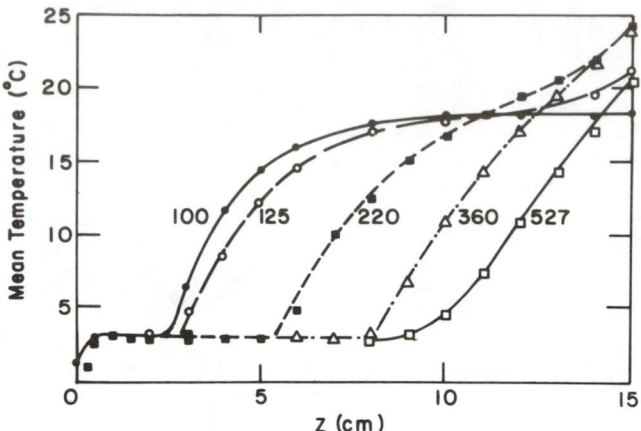

FIGURE 16. Temperature distribution during the approach to equilibrium (the numbers by the curves are the elapsed times in minutes from the application of cooling) [14].

4°C). Townsend also noted the peculiarities of "overshoot" regions at each end where the gradient of the mean temperature changes sign (Fig. 17a). The mean temperature increased with height and reached 3.6°C before decreasing to 3.2°C. At the upper limit of the region it fell to 2.9°C before increasing rapidly in the stable region above. The upper overshoot was postulated as being a consequence of the slowing-down of the lower parts of the rising columns of cold water as their tops entered the stable layer.

However, Townsend stated that the lower-overshoot cannot be interpreted in a similar manner, as the descending columns of hotter water were not observed, and indicated that this may be attributed to the subsidence of water with a temperature between 3.2° to 4°C. Townsend also noted large fluctuations in temperature just within the region of stable stratification, but they decreased very rapidly with height and were very small when the local temperature exceeded 10°C (Fig. 17b).

Townsend evaluated the downward heat flux q (i.e. the heat extracted through the tank bottom) by fixing $z = z_r$ as a reference plane in the stable region and assuming horizontal homogeneity as follows:

$$q = - \rho C_p \frac{d}{dt} \int_o^{z_r} (T - T_r) \, dz + k \left(\frac{\partial T}{\partial z}\right)_{z=z_r} \tag{7}$$

where the first term on the right represents the rate of change of heat content below the fixed plane z_r, T is the local mean temperature, T_r is the reference temperature taken to be 3.2°C. ρ, C_p and k are the density, heat capacity and thermal conductivity of the water, and z is the distance measured vertically upward from the ice surface. With the measurement of T in the stable and unstable regions as functions of time, the two terms on the right of Eq. (7) can be evaluated. The heat

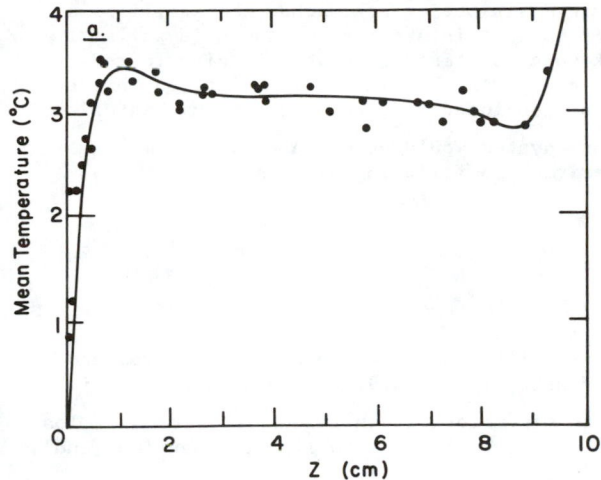

a) Unstable and "constant temperature" regions,

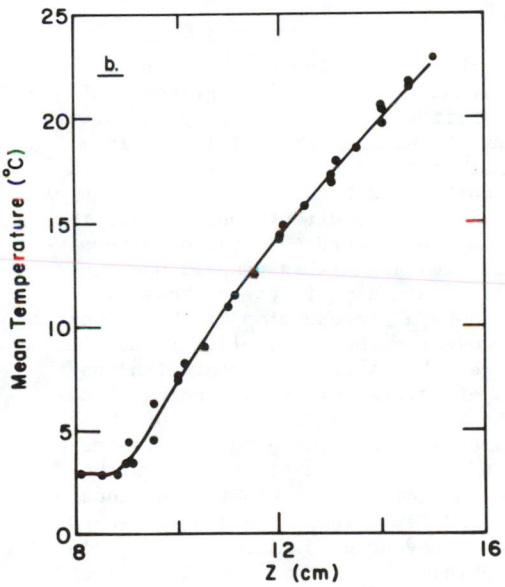

b) Stable and disturbed regions.

FIGURE 17. Mean temperature distribution for thermal equilibrium corrected to a surface temperature of 23°C [14].

flux was reported to initially be 34 mw cm^{-2}, and it decreased to about 26 mw cm^{-2} as the temperature distribution approached the equilibrium. He attributed this reduction to heat intake through the side walls of the test tank and the thick cold constant temperature region formed. However, since there was no change in the temperature distribution and the pattern of convection, he concluded that the equilibrium heat flux in a horizontally homogeneous system would be close to 34 mw cm^{-2}. Using this flux Townsend derived the following expression:

$$Q = \frac{q}{\rho C_p} = 0.156 \ (\Delta T)^{5/3} \ (\gamma \beta g)^{1/3} \ (\alpha/\nu)^{1/3} \tag{8}$$

in which ΔT is the temperature difference between the tank bottom and the temperature of maximum density $T_{max} = 3.97°C$, and γ is defined in Eq. (1). Equation (8) is the dimensionally analogous form for heat loss from a smooth heated plane involving fluids with linear expansion, and is expressed as

$$Q = 0.193 \ (\Delta T)^{4/3} \ (\alpha \beta g)^{1/3} \ (\alpha/\nu)^{1/3} \tag{9}$$

in which ΔT is the temperature difference between the surface and the temperature far above the plane.

Myrup et al. [15] also studied this convection problem with ice, which they found valuable for understanding and modeling geophysical and astrophysical convection systems. Their experimental set-up was similar to that of Townsend. In their study the water was initially at room temperature. The tank bottom was maintained at a temperature slightly higher than 0°C by a water and ice bath beneath the bottom. The top of the tank was covered with an aluminum plate in direct contact with the water surface and conditioned to room temperature. They made extensive measurements on mean temperature profiles as well as temperature fluctuations at every level. Similar temperature profiles to those of Townsend were reported. Figure 18 shows a reproduction of the temperature record at a single height 2.8 cm above the lower plate (bottom). It is evident that distinctly different kinds of temperature fluctuations occurred. About 160 to $\approx$ 220 minutes after the beginning of the experiment, there were some periodic warm fluctuations, some of which were warmer than 4°C. This phenomenon was attributed to the effects of gravity waves bobbing up and down in the stable layer just above the convective layer. From 220 minutes on, the temperature fluctuations distinctly changed. The irregular cold deviations were due to penetrations of the convective plumes of the buoyant cold fluid arising from the lower boundary layer. Photographic work was conducted to delineate the formation and subsequent development of the three distinct layers. The boundary layer nearest the lower plate was occupied by the densest concentration of the dye. The dark upper layer was fluid warmer than 4°C, which the dye-laden convective currents did not penetrate. Between these two layers was the convective layer. However, no attempt was made to evaluate heat fluxes in this study.

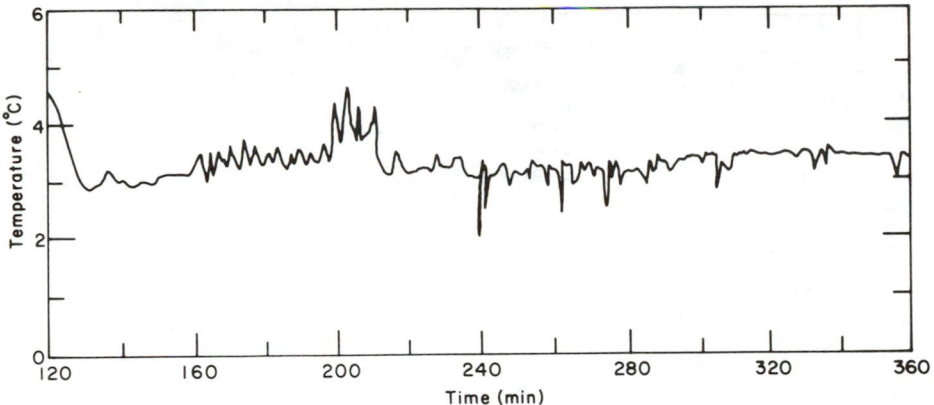

FIGURE 18. Temperature at 2.8 cm above the lower boundary maintained near 0°C [15].

5. HEAT TRANSFER AND TEMPERATURE STRUCTURE IN A CIRCULAR-HORIZONTAL MELT LAYER

Along with the study of the determination of the onset of convection in a melt layer, studies which measured the mean temperature of a growing melt layer were made intermittently by Yen [8] and Yen and Galea [12]. The ice-water system under those studies was drastically different from the work reported by Townsend [14] and Myrup et al. [15] in that, at the onset of the experiment, only solid ice was present. The depth of the melt layer began at zero and continuously increased in depth as melting progressed either from above or below. Thermistors with 0.02 s response time were used to intermittently measure the temperature profile of the growing layer. To minimize possible disturbances of the temperature field during the insertion of the thermistors into the layer, the thermistors were installed at different radii and distributed over a circular area. Typical mean temperature profiles are shown in Figures 19a, b and c for melting from below and in Figures 20a, b and c for melting from above. Each data point represents four thermistor readings from the same layer depth. Note that for both melting cases, the thermistor readings contained a random component associated with the fluctuating temperature fields in the convection region, demonstrating that the convective motion was unsteady. Figure 19a shows the temperature profile just before the onset of convection for T_1 at 4.02°, 6.65°,

7.46°, 8.11° and 9.84°C respectively, with an initial ice temperature T_o

at -11.5°C. It clearly shows that for 4.02°C at 418 minutes after the beginning of the experiment, the temperature profile remained linear (theoretically the melt layer will remain stable indefinitely as long as $T_1 \approx 4$°C) as expected. However, for T_1 at 9.84°C, the convection com-

menced a mere 36 minutes after the experiment started. This demonstrates the dependence of layer stability on the imposed T_1. At the

onset of convection, the temperature curves all deviated from the linear, and were concaved in an upward direction. In all these figures, the number attached to each curve is the elapsed time after the initiation of the experiment. A close time spacing was chosen in order to

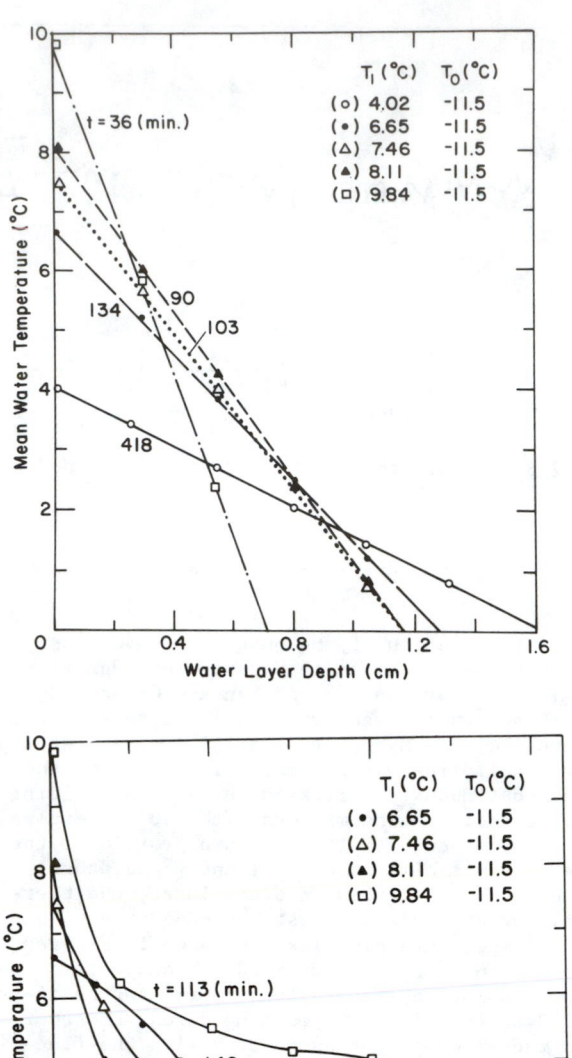

a) Prior to the onset of convection.

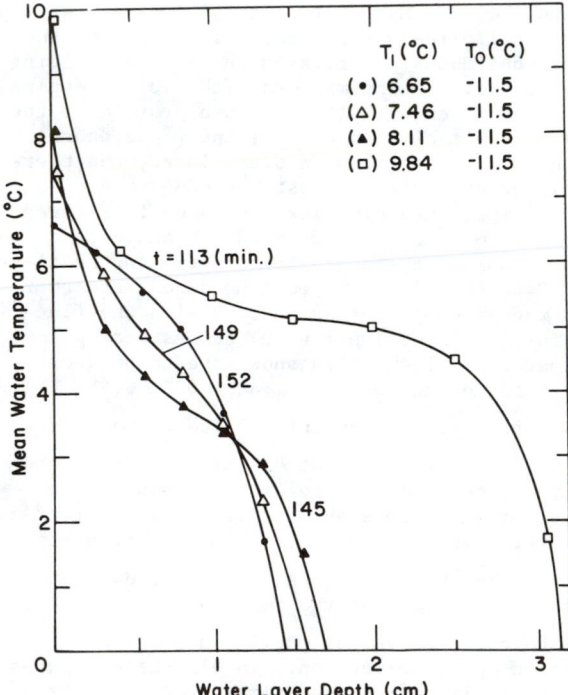

b) Development of the convective layer.

FIGURE 19. Mean temperature profiles for melting from below [16].

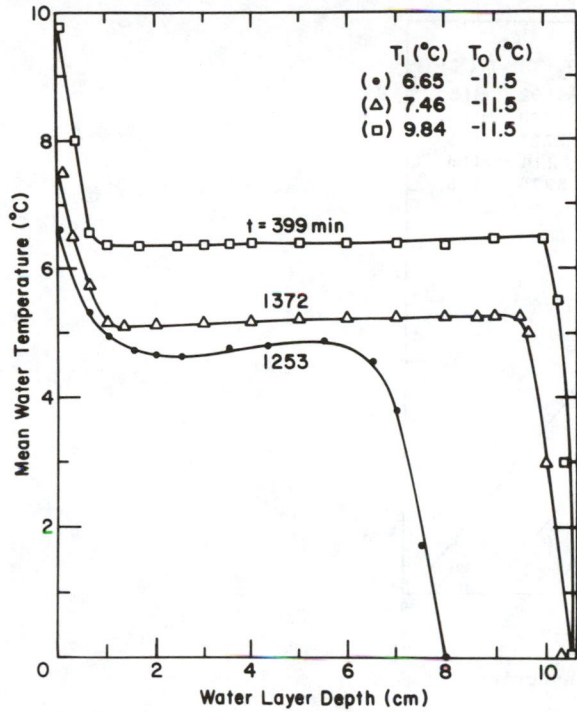

c) Pseudo-steady state.

FIGURE 19 (cont'd).

show the significance of the effect of T_1 on the onset of convection and the intensity of the convective mixing of the unstable region. Figure 19c shows the pseudo-steady state temperature distribution of the melt layer. This consisted of a nearly constant temperature zone occupying most of the total and well developed layer depth. However, it was noted that temperature was dependent on the T_1 imposed, i.e. the higher the value of T_1, the higher the temperature of the constant temperature zone. The four thermistor readings at each level exhibited considerable fluctuations, especially at the intersection between the convective and the upper conductive layer adjacent to the water-ice interface.

Figures 20a, b and c show some typical mean temperature profiles of the melt layer formed by melting from above. Contrary to the previous cases, the layer is most stable when T_2 is the highest, since the driving force to create the convective motion results from the density differences between 0° and ≈4°C. Figure 20a shows the mean temperature profiles just prior to the onset of convective motion for T_2 ranging from 11.92° to 39.75°C. The melt layer became progressively more stable as T_2 increased. Figure 20b distinctly shows that after almost the same time lapse, for T_2 = 39.75°C and at t = 238 minutes, the temperature was just beginning to deviate from the linear distribution, but on the other hand, for T_2 = 11.95°C and at t = 256 minutes, a constant temperature

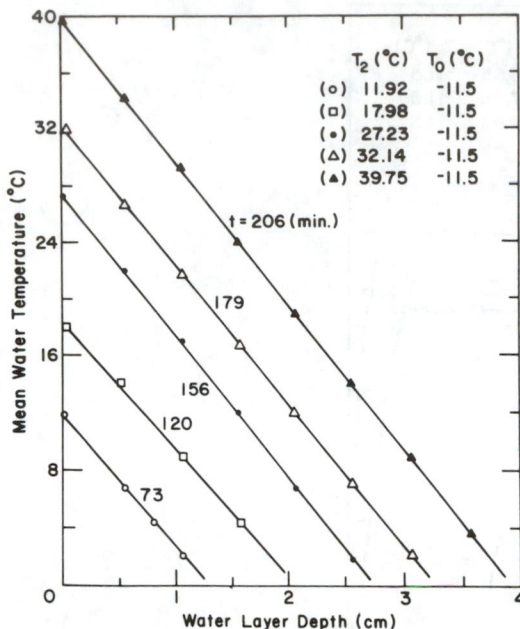

a) Prior to the onset of convection.

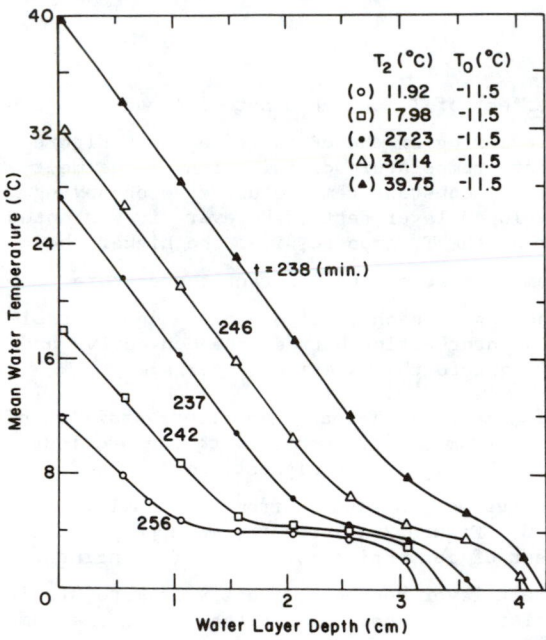

b) Development of the convective layer.

FIGURE 20. Mean temperature profile for melting from above [16].

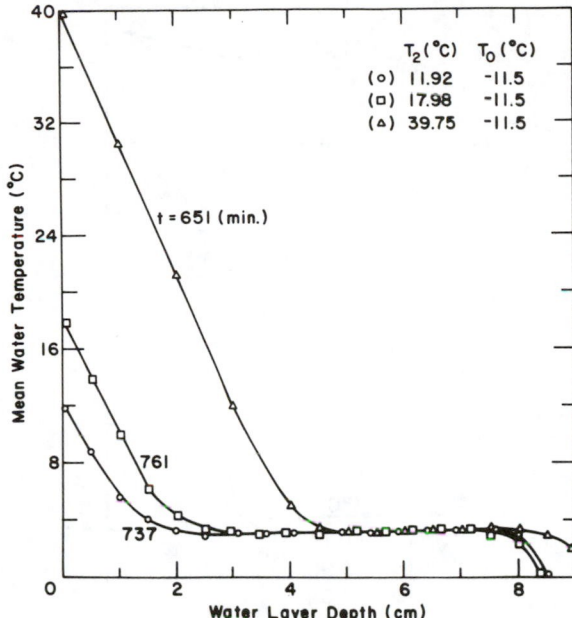

c) Pseudo-steady state.

FIGURE 20 (cont'd).

region had already been well developed. Figure 20c clearly shows the establishment of the well developed constant temperature zone as shown in Figure 19c, but contrary to the case of melting from below, the constant temperature zone had a temperature of $\approx 3.2°C$ regardless of what value of T_2 was imposed on the upper boundary. Slight temperature reversals can be observed both in Figures 19c and 20c. Overshoots were noted in all temperature traverses (regardless of whether the melting was from above or below) in the pseudo-steady conditions, substantiating the findings of Towsend [14].

Typical mean temperature data have been made dimensionless by the use of molecular scales of velocity, length and temperature for convection in water over ice by Townsend [14]

$$\omega_o = (\alpha\gamma gQ^2)^{1/5} \ , \ Z_o = \alpha/\omega_o \quad \theta_o = Q/\omega_o \tag{10}$$

or by the convection scale defined as

$$\omega_x = (\gamma gZ_cQ^2)^{1/4} \ , \ Z_x, \quad \theta_x = Q/\omega_x \tag{11}$$

in which the convection scale length scale Z_x is taken as the height of

287

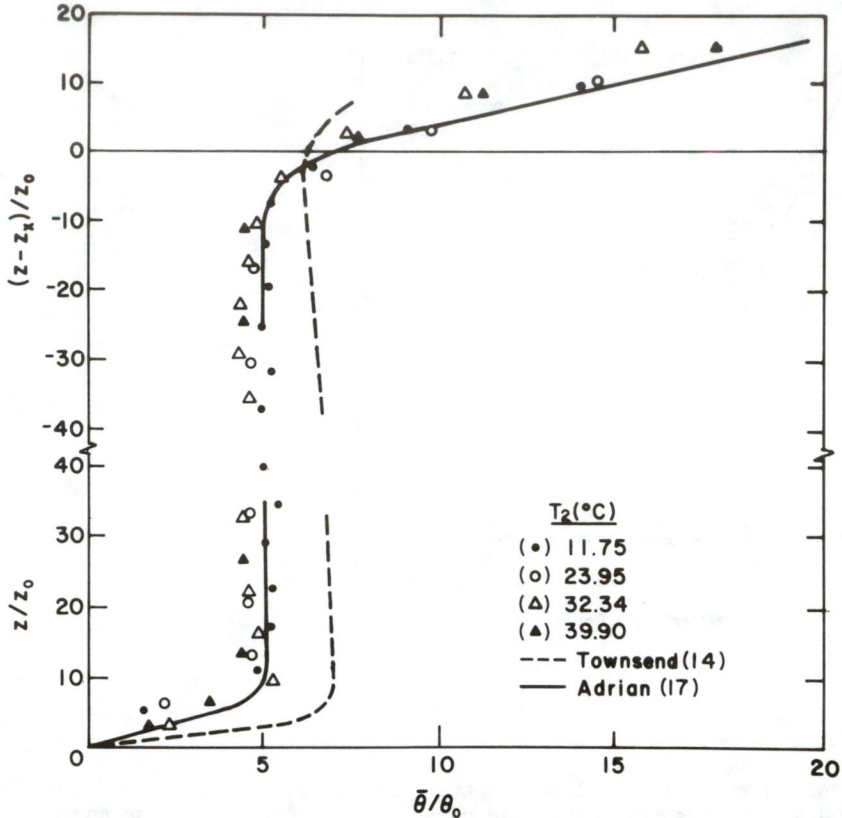

FIGURE 21. Non-dimensional mean temperature profile for melting from above [18].

the convection layer, as indicated by Adrian [17], and designated as the height at which the mean temperature reached 3.98°C.

The molecular scales are based on the assumption that in certain regions of the flow, the convection-layer depth is not dynamically significant, but molecular diffusion is important. Hence, the molecular scales are appropriate for boundary layer regions such as the conduction layer and the interfacial layer. On the other hand, the convection scales are intended to apply to the core of the convection layer, where the length scale of the convection motion is on the order of the layer depth. Figure 21 shows the typical non-dimensional mean temperature profile for the case of melting from above, including the work of Townsend [14] and Adrian [17] on convection in water over ice (a thin layer at maximum). It should be pointed out that in Townsend's and Adrian's work, the water layer did not result from melting. Rather, an invariant and rather deep water layer (≈ 15 cm) was used. The results of Adrian agree remarkably well with those of Yen [16].

Figure 22 shows the non-dimensional mean temperature profile for the case of melting from below. The shape of the curve is the opposite

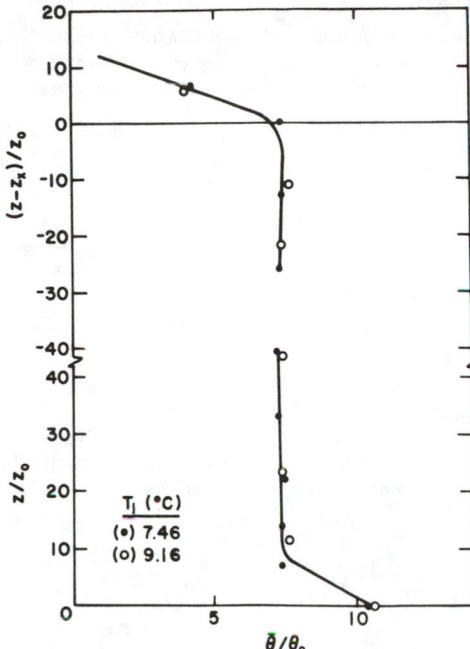

FIGURE 22. Non-dimensional mean temperature profile for melting from below [18].

of the curve shown in Figure 21. It should be pointed out that though the temperature of the convection layer, as indicated in Figure 19c, depends on the value of T_1, the non-dimensional mean temperature

(derived from the limited available data) fell more or less on a single curve as shown in Figure 21, with a somewhat higher ratio of $\bar{\theta}/\theta_o$ for

the convection layer. The only experimental work closely related to the case of melting from below was conducted by Legros et al. [13] in experiments with an upper boundary temperature less than 4°C. However, their work was aimed at the determination of the critical temperature difference across the layer and no attempt was made to measure the temperature profile within the layer.

Yen [18] also derived heat flux expressions for the melt layer formed by melting from below and above. For experiments in which the melting rates were determined, the total upward or downward heat flux for melting from below or above was evaluated by

$$q = \rho \, \frac{dz}{dt} \, [L + c_i \, (T_m - T_o)] + \frac{\rho c_p}{2t} \, [z(t)(T + T_m)] - k_i \, \frac{dT_i}{dz} \, \bigg|_{z=H_c} \qquad (12)$$

The term $\rho c_p/2t[z(t)(T + T_m)]$ represents the mean sensible heat content variation of the entire layer at a depth of $z(t)$. In the case of melting ice, $T_m = 0°C$ and T stands for either T_1 or T_2. The contribution of this term to the overall heat flux equation was found to be much more significant in the case for melting from above. For those experiments in which the mean layer temperature was measured as the melting progressed, the heat flux was approximated by

$$q = \frac{\Sigma \frac{k}{2} \left[\left(\frac{dT}{dz}\right)_2 + \left(\frac{dT}{dz}\right)_1 \right] (t_2 - t_1)}{\Delta t} \tag{13}$$

in which $(dT/dz)_2$ and $(dT/dz)_1$ are the mean temperature gradients at the stable region near the upper boundary for melting from above and from the lower warm plate for melting from below. Subscripts 1 and 2 indicate the beginning and end of each period, and Δt is the total time period. For evaluating a value of q, at least a dozen or more of these periods (with varying durations) were used. The heat flux from above was found to be a weak function of T_2 and can be represented by

$$q = 177 \ (T_2)^{0.303} \tag{14}$$

with an average $q \approx 474 \ w/m^2$. The above expression is valid for T_2 ranging from 11.75° to 39.9°C. For melting from below, Yen [18] reported

$$q = 1900 + 315 \ (T_1) \tag{15}$$

For T_1 ranging from 7.7° to 25.5°C, the heat flux was found to be a linear function of T_1. The higher the value of T_1 and thus the higher the temperature in the convective zone, the greater the convective motion in the unstable region. This consequently reduced the stable layer thickness adjacent to the ice and the thermal boundary. The works of Townsend [14] and Adrian [17] are similar to the case of melting from above. However, in their investigations, a rather deep invariant water layer was used and no phase transition occurred at the bottom of the tank. They both reported a nearly identical heat flux of approximately 340 w/m^2 independent of the initial water temperature. The discrepancy in reported q values, (i.e. in Yen's work the melt layer was formed as melting progressed and in Townsend's and Adrian's work, there was only a liquid phase to begin with), were assumed to be due to the real difference in the transfer processes involved in the experimental system.

6. HEAT TRANSFER STUDIES IN NON-PLANAR GEOMETRIES

The earliest and most comprehensive theoretical study of melting free convective heat transfer, other than the horizontal geometries such as spheres and cylinders, was reported by Merk [19]. Employing a third-order density-temperature polynomal of water density and with the application of the Von Karman-Pohlhausen integral method for the case of $Pr \gg 1$, he successfully solved the boundary layer equation and developed a general Nusselt number ratio as

$$\frac{Nu}{Nu_o} = \left[\left(P_1 + \frac{6175}{13671} P_2\ m + \frac{8471}{31031} P_3\ n\right) \frac{(1 + \frac{S}{2})^3}{1 - St}\right]^{1/4} \qquad (16)$$

where S is the shape factor of the temperature profile and is connected to the Stefan number $St = C_p(T_m - T_\infty)/L$ as

$$S = 2 + \frac{3}{St} - \frac{3}{St}\left(1 - \frac{4}{3} St\right) \qquad (17)$$

or

$$St = \frac{6S}{(S+2)^2} \qquad (18)$$

Where T_∞ is the bulk water temperature, the values of P_1, P_2 and P_3 are expressed as

$$P_1 = 1 - \frac{89}{217} S + \frac{19}{434} S^2 \qquad (19)$$

$$P_2 = 1 - \frac{765}{1343} S + \frac{303}{2686} S^2 - \frac{19}{2686} S^3 \qquad (20)$$

and

$$P_3 = 1 - \frac{6475}{8471} S + \frac{3537}{16942} S^2 - \frac{199}{8471} S^3 + \frac{19}{16942} S^4 \qquad (21)$$

The values of m and n are defined respectively as

$$m = (\beta_2 + 3\ \beta_3\ T_\infty)\theta_m/N\ \beta_\infty \qquad (22)$$

$$n = \beta_3\ \theta_m^2/(N\ \beta_\infty) \qquad (23)$$

where $\theta_m = T_m - T_\infty$, N and β_∞ are defined as

$$N = 1 + \beta_1 T_\infty + \beta_2 T_\infty + \beta_2 T_\infty^2 + \beta_3 T_\infty^3 \qquad (24)$$

and

$$\beta_\infty = (\beta_1 + 2\beta_2 T_\infty + 3 \beta_3 T_\infty^2)/N \tag{25}$$

in which the β's are the coefficients in the formula for the specific volume of the water, i.e.

$$\frac{1}{\rho} - \frac{1}{\rho_0} (1 + \beta_1 T + \beta_2 T^2 + \beta_3 T^3) \tag{26}$$

and Nu_0 is the Nusselt number for the case of St = m = n = 0. Merk [19] reported that for large values of the Prandtl number, neither the shape of the body nor the position on the surface influence the ratio of Nu/Nu_0. Therefore, this ratio can be replaced by $\overline{Nu}/\overline{Nu}_0$, and can be expressed by the following only if the effect of melting is considered (i.e. for m = n = 0)

$$\frac{\overline{Nu}}{\overline{Nu}_0} = \left[\frac{1 - \frac{89}{217} S + \frac{19}{434} S^2}{1 - \frac{1}{2} S + \frac{1}{4} S^2} (1 + \frac{1}{2} S)^5 \right] \tag{27}$$

Figure 23 is a graphic representation of Eq (27), and clearly shows that for S < 0 or for melting $\overline{Nu} < \overline{Nu}_0$, while for solidification, i.e. S > 0, $\overline{Nu} > \overline{Nu}_0$. For the case with convective inversion but no melting, S = 0 and hence $p_1 = p_2 = p_3 = 1$. Equation (16) becomes

$$\frac{\overline{Nu}}{\overline{Nu}_0} = (1 + 0.4912 \, m + 0.2730 \, n)^{1/4} \tag{28}$$

By defining $\overline{Nu}_0$ for large values of the Prandtl number as

$$\overline{Nu}_0 = C(GrPr)^{1/4} = C \left(\frac{gL^3}{\alpha\nu} \beta_\infty \theta_m \right)^{1/4} \tag{29}$$

Equation (28) can be written as

$$f = \frac{\overline{Nu}}{\left[C \frac{g \, L^3}{\alpha\nu} \right]^{1/4}} = [\beta_\infty \theta_m (1 + 0.4912 \, m + 0.2730 \, n]^{1/4} \tag{30}$$

where f is a dimensionless number. for small values of T_∞, $\beta_\infty \approx \beta_1 + 2$

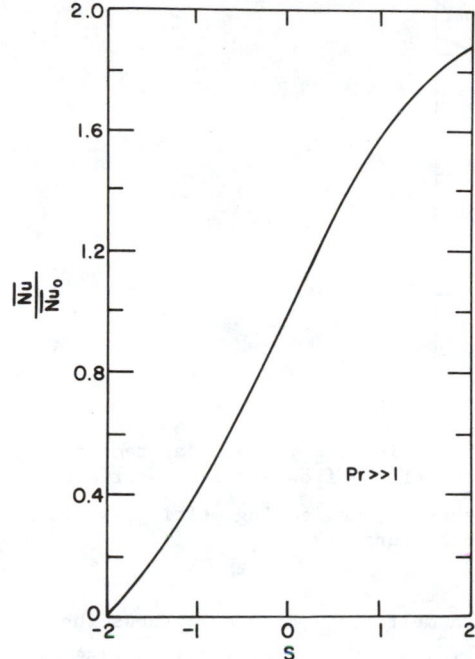

FIGURE 23. The influence of melting (S<0) and solidification (S>0) on heat transfer in thermal convection for large values of the Prandtl number [19].

$\beta_2 T_\infty$, $m \approx \beta_2 \theta_m / \beta_\infty$, and $n \approx 0$, Equation (30) can be simplified to

$$f = [1.509 \, \beta_2 \mid T_\infty - T_{iv} \mid \mid T_m - T_\infty \mid]^{1/4} \tag{31}$$

in which T_{iv} is the inversion temperature and is

$$T_{iv} = 0.663 \, \frac{\beta_1}{\beta_2} - 0.316 \, T_m \tag{32}$$

Using appropriate values of β_1 and β_2, and taking $T_m = 0°C$, Merk report-ed a value of $T_{iv} = 5.005°C$, indicating the inversion temperature for melting of ice in water is somewhat greater than $\approx 4°C$. The significance of the inversion temperature is clearly seen from Eq. (31) at $T_\infty = T_{iv}$; $\overline{Nu} = 0$ and the direction of the flow in the boundary layer along the surface of the body is inverted (for $T_\infty < T_{iv}$, the flow is upward and for $T_\infty > T_{iv}$ the flow is downward). Using the general Eq. (16), Merk derived the minimum Nusselt number at $T_{iv} = 5.30°C$ for $S = 0$ (no melt-

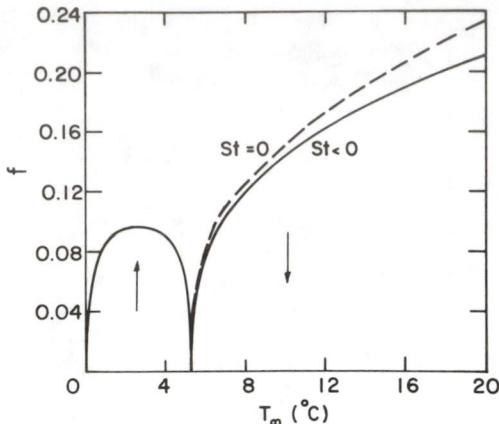

FIGURE 24. Convective inversion for melting ice in water at temperature T_∞. The arrows indicate the direction of the flow along the surface.
The dashed-dotted curve shows the behavior, neglecting melting, while in the full curve melting is taken into account [19].

ing) and T_{iv} = 5.31°C for S < 0 (with melting). Figure 24 shows the effect with melting and without melting on the convective inversion of the value of f (i.e. the value of $\overline{Nu}$). It is clearly shown that the effect of melting is only appreciable for $T_\infty > T_{iv}$ and may be neglected for $T_\infty < T_{iv}$. Experimental results of Dumore et al. [20] and analytical results from a non-melting vertical plate study by Ede [21] generally confirm Merk's findings.

Tkachev [22], using photographic techniques, reported a minimum Nusselt number for melting ice cylinders at 5.5°C and was the first to notice the peculiar nature of the maximum density boundary layer. He suggested that under certain conditions the boundary layer might be split by a predominantly upward motion immediately adjacent to the ice surface and by a region of downward motion outside this. Tkachev conducted melting experiments on spheres as well as vertical and horizontal cylinders. Using the same initial cylinder diameters but with various bulk water temperatures, he found that the coefficient of heat transfer is lowest for a water temperature of about 5.5°C. He correlated his data with the following dimensionless expressions as

$$Nu_{md} = 0.40 \ (GrPr)_{md}^{1/4} \qquad\qquad (33)$$

and

$$Nu_{md} = 0.104 \ (GrPr)_{md}^{1/4} \qquad\qquad (34)$$

for cylinders and for values in the range of $10^2 < GrPr < 10^7$ and $(GrPr_{md} > 10^7$, respectively. The corresponding equations for spheres are, respectively,

$$Nu_{md} = 0.54 \ (GrPr)_{md}^{1/4} \tag{35}$$

for laminar flow $[10^3 < (GrPr)_{md} < 10^7]$ and

$$Nu_{md} = 0.135 \ (GrPr)_{md}^{1/4} \tag{36}$$

for turbulent motion $[(GrPr)_{md} > 10^7]$. Subscript md represents that the physical properties, and the diameter of the sphere or cylinder was evaluated at the arithmetic mean temperature [i.e. $(T_m + T_\infty)/2]$ and diameter [i.e. $(d_o + d_f)/2]$ where d_o and d_f are the initial and final diameter at the end of the experiment. Based on his experimental data, Tkachev further presented an expression for determining the time required for complete melting as

$$StFo_{md} Nu_{md} = 0.305 \tag{37}$$

The occurrence of a split boundary layer flow was verified by the analytical and experimental work of Schecter and Isbin [23], in which an isothermal, vertical and non-melting plate was used. Figures 25 and 26 show the comparison of theoretical and experimental results for the unidirectional and inverted convections, respectively. The theoretical curves in Figures 25 and 26 are given by

$$Nu = 0.892 \left[\frac{\overline{Gr} \left(\frac{1}{3} + \frac{m}{5} + \frac{n}{7} \right)}{(0.952 + Pr)} \right]^{1/4} Pr^{1/2} \tag{38}$$

and

$$\overline{Nu} = 0.652 \left[\overline{Gr} \ Pr \left(\frac{1}{3} + \frac{m}{5} + \frac{n}{7} \right) \right]^{1/4} \tag{39}$$

where m and n are defined in Eqs. (22) and (23). They reported that a test of which type of region will prevail for given conditions of plate and bulk temperature can be stated, i.e. if

$$\frac{1}{3} + \frac{m}{5} + \frac{n}{7} \geq 0$$

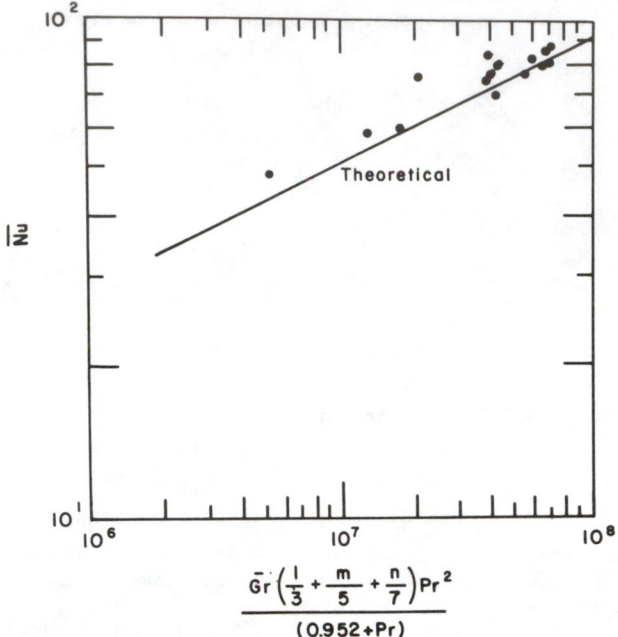

FIGURE 25. Comparison of theoretical (Eq. (37)) and experimental Nu for unidirectional convection [22].

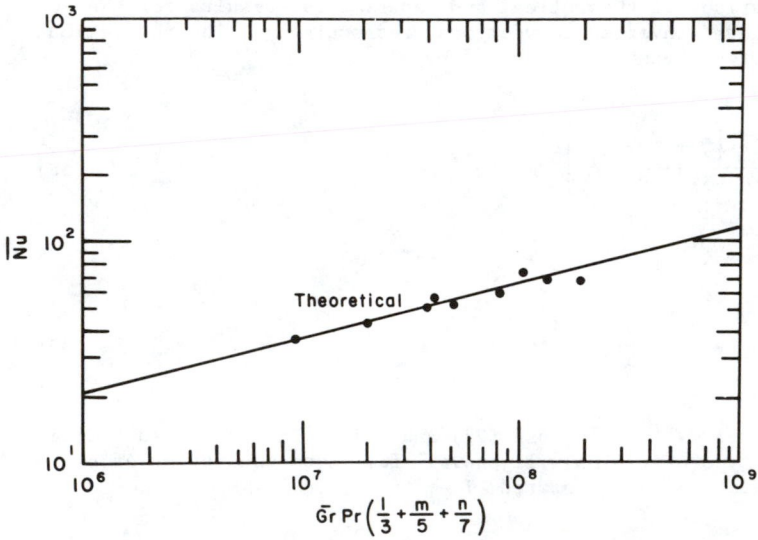

FIGURE 26. Comparison of theoretical (Eq. (38)) and experimental Nu for an inverted regime [22].

normal convection (unidirectional) and if

$$\frac{1}{3} + \frac{m}{5} + \frac{n}{7} < 0$$

inverted convection. They concluded that the heat transfer coefficient can be predicted for both regions with a deviation in the Nusselt number of $\pm$ 10% provided the absolute value of $[1/3 + m/5 + n/7] > 0.05$ by using equation (38) or (39), depending on the convective region. However, it should be noted that the boundary layer equations as approached either by the Von Karman-Pohlhausen integral method or by the similarity transformation method did not yield meaningful results under split-flow conditions.

To resolve this problem Vanier and Tien [24] used an accurate numerical method to solve the similarity equations for a semi-infinite vertical plate at a constant temperature T_w immersed in an indefinitely large volume of water at a bulk temperature T_∞. They reported that a new solution is necessary for every combination of T_w and T_∞ and by obtaining several hundred such solutions, the authors were able to map out temperature zones for each flow regime. The split boundary layer was found to be confined to two distinct triangular regions within which the similarity equations became quite intractable for $T_w = 0°C$ (in the case of melting ice, it is a special problem). They confirmed the findings of Merk [19], who stated that the melting heat transfer rates were closely similar to those for non-melting.

Vanier and Tien [25] conducted experimental work aimed at relating their numerical plate results to the more practical geometry of spheres (including the effect of changing body configuration). This was also partially motivated by a lack of detailed analysis and correlations of the experimental results on the melting of ice spheres and cylinders presented by Dumoré et al. [20] and Tkachev [22]. They presented their results in a least-squares fitted semi-empirical equation as

$$Nu = 2 + C(GrPr)^{1/4} \tag{40}$$

and found that for $T_\infty > 7°C$, it appeared that the results are not affected by the maximum density and the best value of C is 0.422 ± 0.0006 in the range of $0.7 \times 10^6 < (GrPr) < 2.4 \times 10^8$ (Fig. 27, curve b). However, for $T_\infty < 7°C$, nearly same value of C is found but with considerably more scattering (twice the standard deviation), indicating the need for at least incorporating another parameter such as T_∞ to adequately describe the heat transfer under these conditions. This was done by separating the low temperature data into positive and negative deviations. For $T_\infty < 3.8°C$, Nu values were higher than expected (curve a) while for $4.1° < T_\infty < 7.1°C$, Nu values were too low (curve c). to check the one-quarter power assumption in Eq. (40), a two-parameter fit was carried out resulting in

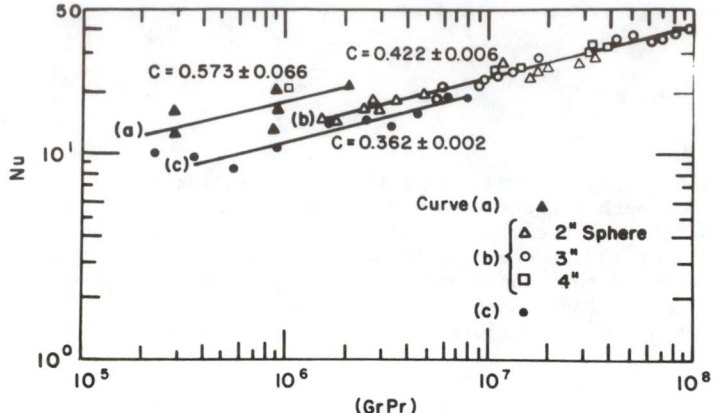

FIGURE 27. Nusselt number as a function of the Rayleigh number (GrPr) [25].

$$Nu = 2 + 0.437 \ (GrPr)^{0.248} \tag{41}$$

which provides an excellent verification. However, when the Grashof numbers were calculated using the physical properties values evaluated at the arithmetic mean temperature (i.e. $T_\infty/2$), the constant C was found to be 0.52 for $T_\infty > 14°C$. This showed a remarkable agreement with the results reported by Tkachev [22]. The effect of sphere diameter and maximum density on heat transfer can be seen in Figure 28. These curves are in general agreement with the flat plate results reported by Vanier and Tien [24] showing a sharp minimum between $5° < t_\infty < 6°C$. To ascertain the effect of sphere diameter, Vanier and Tien [25] proposed a correlation of the sphere results with those from a theoretical analysis of a melting plate by

$$\frac{\overline{Nu}_p}{\overline{Nu}} = C(\frac{L}{D})^{3/4} \tag{42}$$

where L and D are the characteristic height of the plate and the diameter of the sphere. The least square-fitted constant C was found to be 1.106 ± 0.144. The scaled-up experimental data are shown in Figure 29. This figure clearly indicates that a melting sphere behaves very similarly to a melting flat plate and that if all the transfer parameters are equal (including temperature, characteristic length, and surface area) about 11% more heat is transferred to the plate than to the sphere. This is due to the effect of curvature on the flow velocities and is in good agreement with the analytical results reported by Merk and Prins [26] for non-melting free convection systems without maximum density effects, i.e.

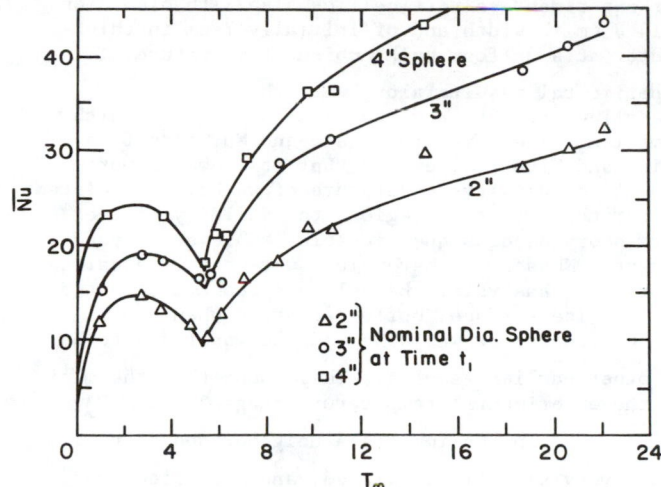

FIGURE 28. Nusselt number variation with bulk temperature for approximately constant sphere diameters [25].

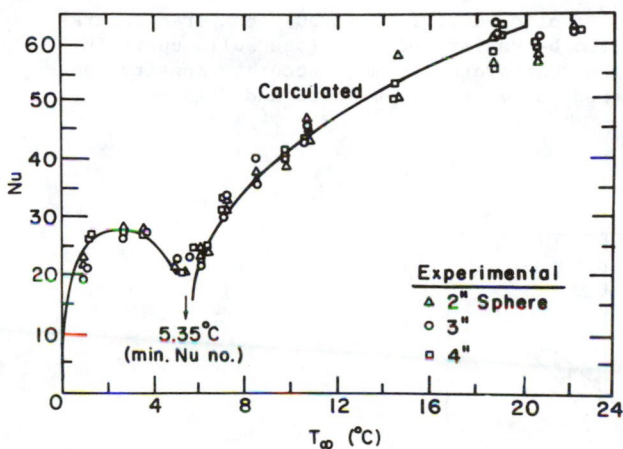

FIGURE 29. Correspondence between the melting of flat plates and spheres of ice [25].

$$\frac{\overline{Nu}_p}{\overline{Nu}} = 1.14 \; (\frac{L}{D})^{3/4} \; . \tag{43}$$

The minimum Nusselt number for spheres occurs at $T_\infty = 4.34 \pm 0.2°C$, as compared to the value of $5.31°C$ based on Mark's [19] theoretical results.

The most recent experimental work on heat transfer and ice melting in ambient water near its density extremum was reported by Bendell and

Gebhart [27]. In their experiment, a vertical ice slab with dimensions of 30.3 cm in height, 14.8 cm in width and of initially 3 cm in thickness was immersed in water at a uniform bulk ambient temperature, T_∞.

Figure 30 shows the experimental results along with the analytical results of Gebhart and Mullendorf [28] and the results predicted with the Boussinesq approximation. The work of Gebhart and Mollendorf is similar to that of Vanier and Tien [24] except that it showed a more. accurate representation of the density-temperature of water. As pointed out by Vanier and Tien for the inversion region, the validity of the simplest boundary layer theory becomes questionable. However, beyond that region the experimental Nusselt number values were nearly equal to those predicted by theoretical analysis. Bendell and Gebhart reported that for a melting vertical ice surface, upflow occurred when $T_\infty \le 5.6°C$. For $T_\infty \ge 5.6°C$, downflow was observed and was found to be in good agreement with other earlier results. They found that the minimum Nusselt number for the experimental temperature range $2.2°C \le T_\infty \le 25.2°C$ occurred at $T_\infty = 5.6°C$. In the immediate neighborhood of the flow direction inversion, very slow flows existed, and the effective Grashof number became very small. Therefore, the validity of the simplest boundary-layer theory becomes questionable and no theoretical results were found in the literature for this regime. In general, the theoretical results of Gebhart and Mollendorf, after multiplying a factor of $(0.102/0.303)^{0.75}$ (solid curve in Fig. 30), compared remarkably well with those reported by Vanier and Tien (see solid curve in Fig. 29) even though a rather elaborate and more accurate density-temperature of water was claimed to be used in Gebhart and Molendorf's study.

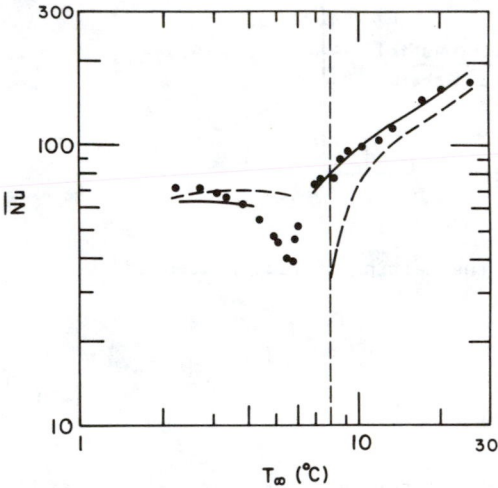

FIGURE 30. Variation of mean Nusselt numbers with ambient temperature T_∞ (0, points). The solid curve is from the analytical work of Gebhart and Mollendorf [28]. The dashed curve is the prediction with Boussinesq approximation [27].

7. FORCED CONVECTIVE HEAT TRANSFER OVER A MELTING SURFACE

In 1963 Yen and Tien [29] were the first to investigate the problem of laminar heat transfer over a melting plate. With the assumption of a linear velocity profile ($v_x = cy$) and with coordinates fixed on the melting surface, the energy equation was given as

$$cy \frac{\partial T}{\partial x} + v_{yo} \frac{\partial T}{\partial y} = \alpha \frac{\partial^2 T}{\partial y^2} \tag{44}$$

where v_{yo} is the interfacial velocity. Equation (44) was solved by the iteration process for the first approximation by letting $v_{yo} = 0$ and by defining $X = (c/9\alpha x)^{1/3} y$, the first approximate solution is

$$\frac{T - T_m}{T_\infty - T_m} = \frac{\int_o^X \exp(-\delta^3) d\delta}{a_o} \tag{45}$$

where δ is a dummy variable, and a_o is defined as $\int_o^\infty \exp(-\delta^3) d\delta$ and the final solution can be written as

$$\frac{T - T_m}{T_\infty - T_m} = \frac{\int_o^X \exp(-\delta^3 + \frac{\beta\delta}{a_N}) d\delta}{a_N} \tag{46}$$

where a_N is the limiting value of the sequences of $a_n = \int_o^\infty (\exp(-\delta^3) + \beta\delta/a_{n-1}) d\delta$ as $n \to \infty$. The melting rate is given by

$$v_{yo} = \frac{\Phi}{a_N} \left(\frac{c}{9\alpha x}\right)^{1/3} \tag{47}$$

in which $\Phi = \frac{K(T_\infty - T_m)}{\rho L}$. The Nusselt number is

$$Nu = \left(\frac{\ell}{a_n}\right) \left(\frac{c}{9\alpha x}\right)^{1/3} \tag{48}$$

The ratio of the Nusselt number with melting to that without melting, neglecting the factor $(c/9\alpha x)^{1/3}$, is

$$\psi(\beta) = \frac{Nu}{Nu_o} = \frac{a_o}{a_N} \tag{49}$$

If a reasonable estimation of $c = \tau_w/\mu$ is given where τ_w is the wall shear stress, the ratio of the Nusselt number becomes

$$\theta(\beta) = \frac{Nu}{Nu_o} = \left(\frac{a_o}{a_N}\right)\left(\frac{\tau_w}{\tau_{wo}}\right)^{1/3} \tag{50}$$

Since τ_w is proportional to the velocity gradient and assuming that the distribution of the velocity profile is comparable to that of the temperature profile, Eq. (50) becomes

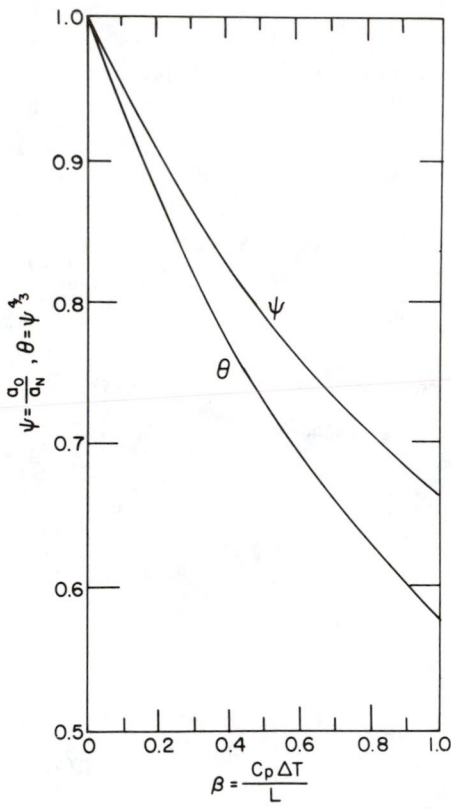

FIGURE 31. Relationship between ψ, θ and β [29].

TABLE 3. Numerical Values of a_N as a Function of β [29].

$\beta(=-St)$	a_N	$\beta(=-St)$	a_N
0.0	0.893	0.6	1.176
0.1	0.941	0.7	1.219
0.2	0.990	0.8	1.262
0.4	1.035	0.9	1.305
0.5	1.131	1.0	1.346

$$\theta(\beta) = \frac{Nu}{Nu_o} = \left(\frac{a_o}{a_N} \right)^{4/3} \tag{51}$$

Figure 31 shows the effect of $\beta(= - St)$ on the values of $\psi(\beta)$ and $\theta(\beta)$. TABLE 3 shows the values of a_N as a unique function of β. From the numerical values of a_N, it becomes obvious that the surface heat flux reduces the steepness of the temperature gradient, and consequently decreases the heat transfer coefficient. Figure 32 illustrates the effect of β on the temperature distribution. It clearly indicates that at the same value of X, the dimensionless temperature $T - T_m / T_\infty - T_m$ is lowered as β is increased from 0 to 1.

Although the assumption $v_x = cy$ simplified Yen and Tien's analysis, it is known that for fluids of high Prandtl numbers, the thickness of the thermal boundary layer is less than that of the velocity boundary layer, and in view of this Tien and Yen [30] performed an additional analysis of the same problem, but proposed that the portion of the velocity profile within the thermal boundary layer be defined as

$$v_x = c(x)y \tag{52}$$

The energy equation is solved as

$$T^+(\eta) = \frac{T - T_m}{T_\infty - T_m} = \frac{\int_o^\eta \exp\left[-\eta^3 + \frac{\omega}{a_N} \right] d\eta}{\int_o^\infty \exp\left[-\eta^3 + \frac{\omega}{a_N} \eta \right] d\eta} \tag{53}$$

in which

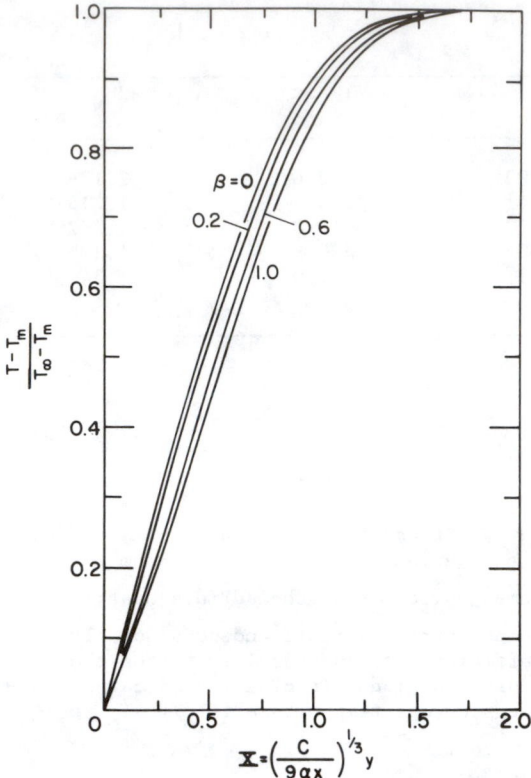

FIGURE 32. The effect of β on the temperature distribution [29].

$$\omega = \beta \left. \frac{dT^+}{d\eta} \right|_0 \tag{54}$$

and

$$\eta = y \, [c(x)]^{1/2} \, [9\alpha \int_0^X [c(x)]^{1/2} \, dx]^{-1/3} \tag{55}$$

The Nusselt number is

$$Nu = \left[\frac{1}{\int_0^\infty \exp\left(-\eta^3 + \frac{\omega}{a_N} \eta\right)} \right]^{1/3} \cdot \frac{[c(x)]^{1/2}}{9\alpha \int_0^X [c(x)]^{1/2} \, dx} \tag{56}$$

The ratio of Nusselt number is now

$$\frac{Nu}{Nu_o} = \frac{\int_0^\infty ex[-\eta^3]d\eta}{\int_0^\infty \exp(-\eta^3 + \frac{\omega}{a_N}\eta)d\eta} \cdot \sqrt{\frac{[c(x)]}{[c(x)_o]}} \cdot \left[\frac{\int_0^x [c(x)_o]^{1/2}dx}{\int_0^x [c(x)]^{1/2}dx}\right]^{1/3} \tag{57}$$

The first part on the right side of Eq. (57) is equal to $\psi(\beta) = a_o/a_N$. If it is assumed that the effect of melting on the temperature gradient is comparable to the effect on the velocity gradient, then Eq. (57) is equivalent to Eq. (51), indicating that results based on a rather restricted assumption are applicable to more realistic cases.

Pozvonkov et al. [31] also conducted an analytical study on heat transfer at a melting flat surface under the conditions of forced convection and laminar boundary layer. In their study, the boundary layer equations were solved using the assumptions of constant thermal-physical properties of the fluid, the independence of the ratio of momentum to thermal boundary layer thickness (δ/δ_t) in the direction of flow, and the fourth degree polynomal representation of the velocity and temperature distribution. The ratio of the local Nusselt number with melting to the local Nusselt number without melting is given as

$$\frac{Nu_x}{Nu_{xo}} = \sqrt{\left(\frac{a}{2}\frac{1}{1 + \frac{1}{k_f}}\right)} \sqrt{\left(\frac{\delta_t^{**}}{\delta_{to}^{**}}\right)} \tag{58}$$

in which k_f is the Kutateladze number $(=-1/st)$ 'a' is expressed in terms of k_f by

$$a = 3[\sqrt{(k_f^2 + \frac{4}{3}k_f)} - k_f] \tag{59}$$

and

$$\delta_t^{**} = \int_0^1 \left(1 - \frac{T}{T_\infty}\right)\frac{u}{u_\infty}d\eta \tag{60}$$

where T/T_∞ and u/u_∞ are the dimensionless thermal and velocity boundary layer distributions. The value of δ_{to}^{**} is for the case when there is no melting. To evaluate the values of δ_t^{**} and δ_{to}^{**}, the value of δ_t^{**} and δ_{to}^{**} ($\delta^{**} = \delta_o^{**}$) when there is no melting) has to be calculated from

$$\delta^{**} = \int\limits_{0}^{1} \left(1 - \frac{u}{u_{\infty}}\right) \frac{u}{u_{\infty}} \, d\eta \tag{61}$$

where $\eta = y/\delta_t$ or $\eta = y/\delta$. The value of δ_o^{**} can be found directly from Eq. (61), but the value of δ_{to}^{**} will contain the unknown boundary layer thickness ratio $\epsilon_o = \delta_{to}/\delta_o$. By retaining only the first power in ϵ_o in the expression δ_{to}^{**}, the first approximation of ϵ_o can be evaluated from

$$\frac{\delta_o}{\delta_{to}} = \sqrt{\frac{\delta_{to}^{**}}{\delta_o^{**}}} \sqrt{Pr} \tag{62}$$

For the case of no melting. Equation (62) is reduced from the general expression of

$$\frac{\delta}{\delta_t} = \sqrt{\frac{\delta_t^{**}}{\delta^{**}}} \frac{\sqrt{\frac{A}{a}}}{\sqrt{(1 + \frac{1}{k_f})}} \cdot \sqrt{Pr} \tag{63}$$

where

$$A = \frac{2}{1 + \dfrac{\lambda_o}{\epsilon \, Pr}} + \frac{6 \, \lambda_o}{\epsilon \, Pr} \tag{64}$$

$$a = 3 \, [\sqrt{(k_f^2 + \frac{4}{3} k_f)} - k_f]$$

in which $\lambda_o = a/6 \, k_f$ and $\epsilon = \delta_t/\delta$. Using a repeated application of Eq. (61) and the expression for δ_{to}^{**} derived from Eq. (60), a final value of

$$\delta_o/\delta_{to} \ (\text{or } \epsilon_o = \delta_{to}/\delta_o)$$

can be found. This final value of ϵ_o is used as a first approximation to calculate the values of δ_t^{**} and δ^{**} from Eqs (60) and (61). Equation (63) is then used to compute a new value of δ/δ_t (or $1/\epsilon$), and is used again in Eqs. (60) and (61) to compute a second set of values of δ_t^{**} and δ^{**}. A new value of δ/δ_t is computed from Eq (63) and is repeated until

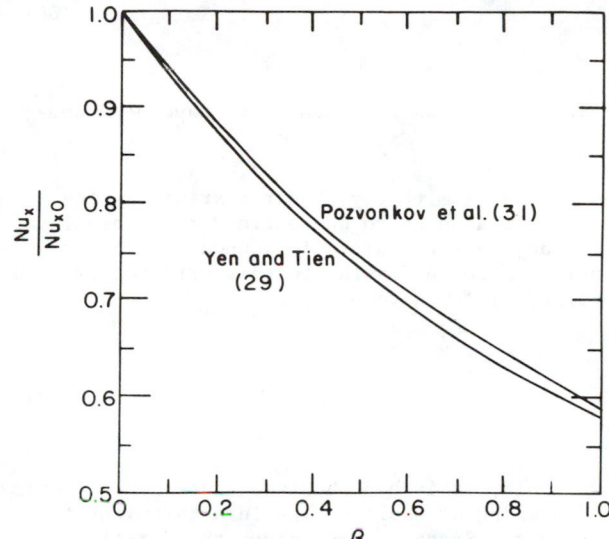

FIGURE 33. Comparison of Nu_x/Nu_{xo} vs. β [32].

this final value is nearly identical to the previously calculated one. The results from this rather complicated boundary layer integral analysis are compared with the much simplified work of Yen and Tien [29] and are shown in Figure 33 (Yen and Tien [30]).

The effect of melting on forced convection heat transfer between a melting body and a surrounding fluid was studied quantitatively from the point of view of the boundary layer theory, the film theory and the penetration theory by Tien and Yen [33].

For the boundary layer theory, exact solutions of the equations of motion, energy and continuity were obtained and the relationship between the ratio of $\theta_T (= h/h_o)$ of the heat transfer coefficient with melting to the heat transfer coefficient without melting was derived in terms of the parameter $\beta (= -S_t)$ as

$$\theta_T = \left(\frac{K \cdot Pr}{T^{+\prime}(0,Pr,0)} \right)/\beta \tag{65}$$

where T^+ is a dimensionless temperature and is identical to the one defined in Eq. (53), but with η defined as

$$\eta = \left(\frac{y}{2} \right) \left(\frac{V_\infty}{\nu x} \right)^{1/2} .$$

The parameter K is given as

$$K = \frac{\beta}{Pr} \, T^{+\prime} \, (0,Pr,K) \tag{66}$$

Numerical values of $(K \cdot Pr/T^{+\prime}(0,Pr,K)$ as function of θ_T have been compiled by Stewart [34].

In the analysis based on the film theory, the resistance to the interphase transport phenomena is assumed to be confined to a thin layer of stagnant film immediately adjacent to the solid boundary. The model is assumed to be one-dimensional (perpendicular to the surface) and in a steady state. The parameter θ_T is found to be

$$\theta_T = \frac{\ln(1+\beta)}{\beta} \tag{67}$$

In the penetration model analysis, it is hypothesized that the transport process is effected by the sweeping of small eddies in a turbulent field into contact with the interfaces. Since in most cases the duration of contact is rather short, the eddy may be treated as a semi-infinite solid and a transient state heat conduction equation can be used in describing the energy transfer process. A solution was attained, and the ratio θ_T is expressed as

$$\theta_T = \frac{1}{(1 - \mathrm{erf}\phi) \; \mathrm{erf}(\phi^2)} \tag{68}$$

in which ϕ is related to β by

$$(1 - \mathrm{erf}\phi) \; \sqrt{\pi} \; \phi \; \exp(\phi^2) = - \beta \tag{69}$$

Figure 34 shows the effect of β on the values of θ_T from the considerations of the boundary, film and penetration theories along with the work reported by Yen and Tien [29] and Merk [19]. Although the quantitative results differ among each other, they all show the same qualitative trend and indicate that the process of phase transition (melting) inhibits the heat transfer rate. The results based on the Leveque solution agree with those based on the boundary layer theory for small values of β, but deviate from each other as β increases. This is expected since the solution obtained by Yen and Tien, in a sense, is an asymtotic solution of that based on the boundary layer theory for small β. The difference in results between the boundary, film and penetration models makes it necessary to exercise caution when selecting among these results for use in practical applications.

8. DISCUSSION AND CONCLUSION

This review only covered the problems associated with the anomalous density-temperature relation of water. The discussion and conclusion

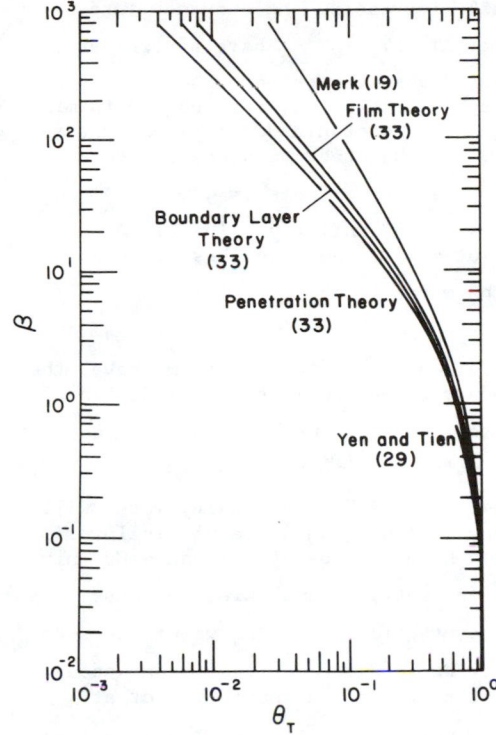

FIGURE 34. Comparison of various theories: effect of β on θ_T [33].

will be grouped into three sections: 1) the onset of convection; 2) the temperature structure and natural convective heat transfer; and 3) the laminar forced convective heat transfer.

8.1 Onset of Convection

The criterion for the onset of convection in a water layer containing a density extremum was found both experimentally (Yen [8], Yen and Galea [12] and analytically (Veronis [5], Sun et al. [6] and Merker et al. [7]) to not be a constant value as in classical Bénard problem, but rather to be dependent on the thermal boundary conditions (when the layer is formed by a phase transition, one of the thermal boundaries is at the ice melting point, i.e. $T_m = 0°C$). The experimental values compared very favorably to analytical values, as shown in Figure 12, in terms of Ra_c experimental versus Ra_c theoretical, and in Figure 13 as Ra_c versus T_1 and T_2 explicitly. It can be concluded that in the case of melting from the top, the higher the values of T_2, the greater the Ra_c becomes, or in other words, the further removed the temperature T_2 from 4°C, the less prone the layer is to the onset of convection. On the other hand, in the case of melting from below, as the temperature T_1 increases, Ra_c reduces exponentially and approaches the value ≈ 1708 asymptotically, as reported in the classical Bénard problem. This becomes evident by the fact that as T_1 increases, the buoyancy forces created by the temperature difference $\Delta T = T_1 - T_{max}$ have a stronger influence on the stability of the layer than the effect produced by the density extremum (i.e. $\approx 4°C$), and subsequently the continuously forming layer behaves like a normal fluid (i.e. a monotonic density-temperature relation), as in the Bénard problem. On the other hand, as T_1 decreases and approaches the temperature of the density extreme ($\approx 4°C$), Ra_c increases and approaches a limiting value of infinity. This is also expected, since if T_1 is maintained at $\approx 4°C$, the water has its highest density at the lower boundary and the water layer will always remain stable.

For the case of a water layer formed by melting ice from above, the trend of variation of Ra_c with boundary temperature is reversed. The higher the temperature T_2, the greater the Ra_c becomes. This can be explained by the fact that if T_2 is maintained in the range $0 < T_2 \leq 4°C$, the entire layer is unstable because the higher density water will lie above the less dense water, and will consequently result in lower Ra_c values. If T_2 is maintained at a higher temperature than $4°C$, only a fraction of the layer $(= 4H/T_2)$ is potentially unstable, and thus the layer is less prone to the onset of convection. The Ra_c value seems to increase as the effect of the density extreme becomes less pronounced. It is also interesting to note that the two Ra_c curves intersect at exactly $T_1 = T_2 = 8°C$. This clearly indicates that under these particular thermal conditions, these two systems are identical and have a unique Ra_c value regardless of how the water layer was formed.

8.2 Temperature Structure and Natural Convective Heat Transfer

The most striking phenomenon of the temperature distribution either in the constant water layer depth (Townsend [14], Myrup et al. [15]) or in the continuously growing layer of melting ice (Yen [16]), is the formation of a nearly constant temperature region which eventually expands to fill up to two-thirds of the entire layer depth. The only significant difference observed is that the temperature in the constant temperature region in a layer of water formed by melting ice from below is dependent on the boundary temperature T_1 (see Fig. 19c), but in the layer of water formed by melting from the top, the temperature has a fixed value ($\approx 3.2°C$) independent of the imposed upper boundary temperature T_2 (Fig. 17a) and Myrup et al. [15].

The heat flux for melting from the top is found to be a weak function of T_2 and can be expressed as $q = 177 \, (T_2)^{0.303}$, in which q is given in W/m^2. However, in melting from below, the heat flux is found to be strongly dependent on T_1 and can be approximated by $q = -1900 + 315 \, (T_1)$. this is because as T_1 is maintained at a higher value, there will be a stronger current of mixing in the constant temperature region, reducing the laminar layer thickness on the lower boundary and the upper water-ice interface and increasing the heat transfer rate. On the other hand, the effect of T_2 on heat transfer is much weaker because there is

a stable layer overlying the expanding unstable region. In addition, the buoyancy force is created merely by a temperature difference of 4°C (i.e. 4° - 0° = 4°C). The work of Townsend [14] and Adrian [17] are similar to work dealing with the melting of ice from the top and they reported a nearly identical heat flux of approximatley 340 W/m², somewhat lower than those obtained from the melting experiment. The discrepancy can probably be attributed to the unsteady nature of the melting experiment (i.e. the depth of the layer is intially at zero and deepens as the melting process advances).

The work of Merk [19], Tkachov [22], Vanier and Tien [24, 25], Schecter and Isbin [23], Bendell and Grebhart [27], Gebhart and Mollendorf [28] were conducted either theoreticlly or experimentally and were aimed at classifying the implication of the density extreme on the heat transfer characteristics of a melting system and a non-melting system. Tkachov was the first to suggest that under certain thermal conditions the boundary layer might be split, i.e. with a predominantly upward motion immediately adjacent to the ice surface and a region of downward motion outside this. Based on his experimental work on ice spheres, and vertical and horizontal cylinders, he reported that the heat transfer coefficient is lowest at T_∞ of about 5.5°C. Merk was the

first to take up this moving boundary problem analytically and solved it using an approximation method. For the case of no melting with a small value of T_∞, he reported the inversion temperature to be at ≈ 5.005°C.

Based on a rather complete calculation (i.e. without the limitation on T_∞), Merk reported the inversion temperature to be around 5.30°C for no

melting and 5.31°C with melting, and he further indicated that the effect of melting is only appreciable for $T_\infty > T_{iv}$ and may be neglected

for $T_\infty < T_{iv}$ (Fig. 24).

The analytical work of Vanier and Tien [24] also reported the existence of three types of flow regions, i.e. laminar upward, downward and dual flow (sometimes termed "inverted" or "split") as well as regions of no solution. These regions were determined by the specific combination of T_w and T_∞. Their calculated results were found to be

consistent with the observations of Tkachev, the analytical and experimental work of Schecter and Isbin, the analytical results of Merk, as well as the experimental results of Dumore et al. [20]. These analytical findings were verified with their experimental work on melting spheres similar to those reported by Tkachev and Dumore et al. A minimum heat flux occurred around $T_{iv} = 5.35 \pm 0.2$°C as compared with Merk's

5.31°C and Tkachev's 5.5°C. Their findings are also in good agreement with the most recent work of Bendell and Beghart [27]. Based on the melting of vertical ice plates, they reported an inversion temperature of 5.6°C.

Though Gebhart and Mollendorf [28] claimed to have used a more elaborate and accurate density-temperature representation, their results are comparable with those reported by Vanier and Tien [24] even though the latter used a less complex density-temperature relation.

8.3 Laminar Forced Convective Heat Transfer

Based on the limited analytical work on the forced convective heat transfer over a melting surface, it can be concluded that the interfacial velocity resulting from a phase transition tends to retard the heat transfer. Using a simplistic approach to this problem, Yen and Tien [29] found the ratio of the Nusselt number with melting to the Nusselt number without melting to be a strong function of $\beta (= C_p (T_\infty - T_m)/L = - St)$ (Fig. 31). This result was in good agreement with those reported by Pozvonkov et al. [31] from a rather complicated integral solution of the transport equations (Fig. 33). This reduction of heat transfer rate is attributed to the lowering of the temperature gradient and is equivalent to the case of mass injection through the laminar layer into the main stream. This evidence (Fig. 34) was further demonstrated by Tien and Yen [33] based on their quantitative analysis of this moving boundary problem with the classical boundary, film, and penetration theories.

REFERENCES

1. Bénard, H. Les Tourbillons Cellularies dans une Nappe Liquid, Rev. Gen. Sci. Pures Appl.,Vol. 11, pp. 1261-1271, 1900.

2. Rayleigh, Lord, On Convective Currents in a Horizontal Layer of Fluid when the Higher Temperature is on the Underside, Phil. Mag., Vol. 32, pp. 529-546, 1916.

3. Jeffreys, H., The Stability of a Layer of Fluid Heated from Below, Phil. Mag. Vol. 2, pp. 833-844, 1926.

4. Chandrasekhar, S., Hydrodynamic and Hydromagnetic Stability. Clarendon Press, Oxford, 1961.

5. Veronis, G., Penetrative Convection, Astrophys. J., Vol. 137, pp. 641-663, 1963.

6. Sun, Z.-S., Tien, C., and Yen, Y.-C., Thermal Instability of a Horizontal Layer of Liquid with a Maximum Density, A.I.Ch.E. JL., Vol. 15, pp. 910-915, 1969.

7. Merker, G.P., Wass, P., and Grigull, U., Onset of Convection in a Horizontal Water Layer with Maximum Density Effects, Int. J. Heat Mass Transfer, Vol. 22, pp. 505-515, 1979.

8. Yen, Y.-C., Onset of Convection in a Layer of Water Formed by Melting Ice from Below, Phys. Fluids, Vol. 11, pp. 1263-1270, 1968.

9. Boger, D.V., and Westwater, J.W., Effect of Buoyance on the Melting and Freezing Process, J. Heat Transfer, Vol. 69, pp. 81-89, 1967.

10. Globe, S., and Dropkin, D., Natural Convection Heat Transfer in Liquids Confined by Two Horizontal Plates and Heated from Below, J. Heat Transfer, Vol. 81, pp. 24-28, 1959.

11. Schmidt, E., and Silveston, P.L., Natural Convection in Horizontal Liquid Layers, Chem. Engr. Symp. Series, Vol. 55, no. 29, pp. 163-169, 1959.

12. Yen, Y.-C., and Galea, F., Onset of Convection in a Water Layer Formed Continuously by Melting Ice, Phys. Fluids, Vol. 12, pp. 509-516, 1969.

13. Legros, J.C., Longree, D., and Thomas, G., Bénard Problem in Water Near 4°C, Physica, Vol. 72, pp. 410-414, 1974.

14. Townsend, A.A., Natural Convection in Water Over an Ice Surface, Quart. J. Roy. Met. Soc., Vol. 90, pp. 248-259, 1964.

15. Myrup, L., Gross, D., Hoo, L.S., and Goddard, W., Upside Down Convection, Weather, Vol. 25, pp. 150-157, 1970.

16. Yen, Y.-C., In Frontiers in Hydrology, eds. W.H.C. Maxwell and L.R. Beard, pp. 305-325, Water Resources Publications, Littleton, Colorado, 1984.

17. Adrian, R.J., Turbulent Convection in Water Over Ice, J. Fluid Mech., Vol. 69, pp. 753-781, 1975.

18. Yen, Y.-C., Free Convection Heat Transfer Characteristics in a Melt Layer, J. of Heat Transfer, Vol. 102, pp. 550-556, 1980.

19. Merk, H.J., The Influence of Melting and Anomalous Expansion on the Thermal Convection in Laminar Boundary Layers, Appl. Sci. Res., Section A, Vol. 4, pp. 435-452, 1954.

20. Dumoré, J.M., Merk, H.J., and Prins, J.A., Heat Transfer from Water to Ice by Thermal Convection, Nature, Vol. 172, pp. 460-461, 1953.

21. Ede, A.J., The Influence of Anomalous Expansion on Natural Convection in Water, Appl. Sci. Res., Vol. 5, pp. 458-460, 1955.

22. Tkachev, A.G., In Problems of Heat Transfer During a Change of State: A Collection of Articles, eds. S.S. Kutateladze, pp. 169-178, AEC-Tr-3405, translated from a publication of the State Power Press, Moscow-Leningrad, 1953.

23. Schechter, R.S., and Isbin, H.S., Natural Convection Heat Transfer in Regions of Maximum Fluid Density, A.I.Ch.E. Jl., Vol. 4, pp. 81-89, 1958.

24. Vanier, C.R., and Tien, C., Effect of Maximum Density and Melting on Natural Convection Heat Transfer from a Vertical Plate, Chem. Engng. Prog. Sym. Ser., Vol. 64, pp. 240-254, 1968.

25. Vanier, C.R., and Tien, C., Free Convection Melting of Ice Spheres, A.I.Ch.E. Jl., Vol. 16, pp. 76082, 1970.

26. Merk, H.J., and Prines, J.A., Thermal Convection in Laminar Boundary Layers, III, Appl. Sci. Res., Section A, Vol. 4, pp. 207-223, 1954.

27. Bendell, M.S., and Gebhart, B., Heat Transfer and Ice Melting in Ambient Water Near its Density Extremum, Int. J. Heat Mass Transfer, Vol. 19, pp. 1081-1087, 1976.

28. Gebhart, B., and Mollendorf, J.C., Buoyancy-induced Flows in Water Under Conditions in which Density Extrema may Arise, J. Fluid Mech. Vol. 89, Part 4, pp. 673-707, 1978.

29. Yen, Y.-C. and Tien, C., Laminar Heat Transfer Over a Melting Plate, the Modified Leveque Problem, J. of Geophysical Res., Vol. 68, pp. 3673-3678, 1963.

30. Tien, C., and Yen, Y.-C., An Additional Note on the Modified Leveque Problem, J. of Geophysical Res., Vol. 69, pp. 1672-1673, 1964.

31. Pozvonkov, F.M., Shurgalskii, E.F., and Akselrod, L.S., Heat Transfer at a Melting Flat Surface Under Conditions of Forced Convection and Laminar Boundary Layers, Int. J. Heat Mass Transfer, Vol. 13, pp. 957-962, 1970.

32. Yen, Y.-C., and Tien, C., Heat Transfer at a Melting Flat Surface Under Conditions of Forced Convection and Laminar Boundary Layers, Int. J. Heat Mass Transfer, Vol. 14, pp. 1975-1976, 1971.

33. Tien, C., and Yen, Y.-C., The Effect of Melting on Forced Convection Heat Transfer, J. of Applied Meteorology, Vol. 4, pp. 523-527, 1965.

34. Stewart, W.E., Interaction of Heat, Mass and Momentum Transfer, D.Sc. Thesis, Mass. Inst. Tech., Cambridge, Mass., 1950.

Chapter 10

Convective Flow Generated by a Pipe in a Semi-Infinite Porous Medium Saturated with Water in the Neighborhood of 4°C

L. ROBILLARD and P. VASSEUR
Department of Mechanical Engineering, Case Postale 6079,
Succursale "A", Campus de l'Université de Montréal,
Montréal, Québec, Canada H3C 3A7

ABSTRACT

In relation with design requirements for cold regions, the problem of heat losses from buried pipes is considered with special emphasis on the particular effect of the density inversion that occurs in water at temperatures around 4°C. The problem is solved by both numerical and perturbation approaches.

CONTENTS

NOMENCLATURE

a	dimensionless distance between the flat plate and the locus $\alpha=\infty$
d	dimensionless buried depth, (Fig. 2)
g	gravitational acceleration, m s^{-2}
K	permeability of the porous medium, m^2
k	thermal conductivity of the saturated porous medium, J s^{-1}m^{-1}(°C)$^{-1}$
Nu	Nusselt number, (Eq. (26))
p'	pressure, Pa
Q	dimensionless heat flow by unit length of pipe
q	dimensionless local heat flow
r_p'	pipe radius, m
Ra	linear Rayleigh number, $(K(\rho c)_f r_p' g\beta_1 \Delta T/k\nu)$
Ra^*	nonlinear Rayleigh number, $(K(\rho c)_f r_p' g\beta_2 \Delta T^2/k\nu)$
s	salinity
T	dimensionless temperature
T_r'	reference temperature, °C

315

$\Delta T'$ temperature difference, $(T_w'-T_p')$, °C
x,y   cartesian coordinates, (Fig. 2)

Greek symbols

α, β bicylindrical coordinates
β_n volumetric coefficient of expansion, (Eq. (1)), $(°C)^{-n}$
γ inversion parameter, (Eq. (5))
θ angle between x axis and gravity, (Fig. 2)
μ dynamic viscosity, kg $m^{-1}s^{-1}$
ν kinematic viscosity, $m^2 s^{-1}$
ρ density, kg m^{-3}
$(\rho c)_f$ specific heat capacity of fluid, J $m^{-3}(°C)^{-1}$
$(\rho c)_p$ specific heat of saturated porous medium, J $m^{-3}(°C)^{-1}$
ψ dimensionless stream function

1 INTRODUCTION

One of the important factors affecting the rate of transfer by natural convection is the temperature-density relationship of the convective fluid. The importance of this factor is amplified when the heat is being transferred to a medium which has a maximum density such as water in the region of 4°C. The existence of a maximum density implies that the thermal expansion coefficient changes its sign. More important is the possibility of flow reversals due to the change in direction of the buoyancy force. Consequently, unusual flow patterns may be expected in areas of water exposed to near freezing temperatures. An understanding of the mechanisms of such flows is of practical interest due to the common occurrence of cold water in our environment and in many processes in technology. These problems arise in cryogenic heat exchanger operations, the freezing of rivers and lakes, the design of utility systems for northern climates,..etc.

Because of the increasing interest of the subject in industrial applications, considerable work has been done in the past. For recent literature on the effects of maximum density on free convection phenomena, one may cite the works by Joshi and Gebhart [1], El-Henawy et al. [2], Carey et al.[3,4], Vasseur and Robillard [5], Gebhart and Mollendorf [6] and Qureshi and Gebhart [7] who studied free convection about a vertical flat plate adjacent to a mass of cold water. Concerning the effects of maximum density on convective motion of enclosed fluids, experimental work on the cooling of quiescent water in a pipe has been performed by Gilpin [8,9] and Seki et al. [10] while a numerical solution to this problem has been obtained by Cheng and Takeuchi [11] and Robillard et al. [12]. Studies of natural convective heat transfer inside a rectangular cavity have been carried out experimentally by Lankford and Bejan [13] and Inaba and Fukuda [14] and analytically by Lankford and Bejan [13], Vasseur and Robillard [15,16], Robillard and Vasseur [17, 18, 19], Watson [20] and Desai and Forbes [21]. In all these studies it was found that the effect of density inversion on the flow patterns, temperature profiles and average Nusselt number was unexpectedly large.

Relatively few studies have been concerned with the problem of natural convection in a cold water-saturated porous medium. Of particular interest are the works of Sun et al. [22] and Yen [23], which focused on the onset of natural convection in a porous layer saturated with cold water and heated from below. Ramilson and Gebhart [24] examined the possible similarity solutions for vertical, buoyancy-induced flow in a porous medium saturated with water near 4°C. More recently, Gebhart et al. [25] obtained multiple steady-state solutions for natural convection in porous media saturated with cold pure or saline water. Also, theoretical results have been obtained for rectangular porous cavities (Poulikakos [26], Blake et al. [27], Altimir [28]) and horizontal porous annuli (Vasseur et al. [29]).

The purpose of the present work is to shed new light on the buoyancy-driven flow and heat transfer of cold water under the influence of density inversion. The problem considered here is that of a pipe buried in a saturated, semi-infinite, permeable medium. Both the cylinder and the medium surfaces are maintained at constant uniform temperatures. This geometry is of great practical interest, as demonstrated by the investigations devoted to the classical problem based on the linear Boussinesq approximation. A detailed survey of the literature is given by Bau [30]. To our knowledge, nothing has been done to study the effects of the density inversion on the thermal convection associated with pipes buried in a permeable medium saturated with cold water, in spite of their common occurrence in northern climates. The study of this problem is the main objective of this chapter.

In the following sections, we shall first review the various density state equations available in literature to represent the low temperature behavior of water in the region of the density extremum. A specific state equation with appropriate governing parameters, that keeps simplicity in handling the results and yet gives an exhaustive view of the problem considered, will be suggested. Then, the problem will be formulated in terms of the Darcy's and energy equations. Next, we shall study the flow pattern and heat transfer using numerical techniques and a perturbation method, respectively. Finally, essential results will be summarized and the main consequences of the presence of an extremum in the density temperature relationship on the convective flow and heat transfer will be discussed.

2 STATE EQUATION

For many liquids, the variation of the density with temperature can be approximated in the form of a polynomial, such that

$$\rho = \rho_r \left[1 - \sum_{n=1}^{N} \beta_n (T' - T_r')^n \right] \tag{1}$$

where ρ_r is the density of the liquid at the reference temperature T_r' and β_n are the coefficients of thermal expansion.

In most cases of free convection, a linear density temperature relationship is sufficient to predict the essential features of fluid motion and heat transfer. N is chosen equal to 1 and β_1 is a constant whose value depends on the fluid under consideration and on the range of temperatures involved. Thus Eq. (1) may be rewritten into the following dimensionless state equation

$$\frac{\rho - \rho_r}{\rho_r} = -(\beta_1 \Delta T') T \tag{2}$$

where $\Delta T'$ is a characteristic temperature difference and $T = (T' - T_r')/\Delta T'$ is a dimensionless temperature.

In the case of water at low temperature, Eq. (2) is inappropriate essentially because of the existence of a density extremum with temperature variation. This extremum occurs as a result of the decrease in hydrogen bonding between water molecules which causes the density of cold water to first increase with increasing temperature above the equilibrium melting while the thermal molecular motion opposes this effect. As a result, the density of pure water at atmospheric pressure reaches a maximum value at 3.98°C. Bismuth, antimony and gallium are other examples of pure liquids exhibiting a maximum in their density-temperature curves. With the existence of a maximum density, the use of Eq. (2) with an average value for β, as assumed in the Boussinesq approximation, becomes physically unrealistic since, for instance, there exists particular situations for which this average value may turn out to be zero.

For cold water, various approximations of the equation of state have been developed in the past. For instance, Moore and Weiss [31] have proposed a parabolic-type relationship of the form:

$$\rho = \rho_r [1 - \beta_1 (T' - T_r') - \beta_2 (T' - T_r')^2] \tag{3}$$

with $\beta_2 = 0.5\beta_1/(T_r' - T_m') = 8 \times 10^{-6} \, °C^{-2}$ and $T_m' = 3.98°C$. The density given by this equation agrees with the experimental value to 1 part per million (p.p.m.) in the temperature range 2-6°C, and to 5 p.p.m. in the temperature range 0-2°C and 6-8°C. The buoyancy force is then determined with an accuracy better than 4% over the range 0-8°C (Landolt-Bornstein [32]).

When considering the temperature range 0-20°C, third and fourth degree polynomials have been proposed by Vanier and Tien [33] and Fujii [34] respectively. Also, in Fujii [34], a fourth degree polynomial is given for the range -10 to 10°C which is of interest when dealing with supercooled water (see for instance Cheng et al. [35]). The error involved in Fujii's polynomials is less than one unit at the last digit of the tabulated data of Weast [36].

More recently, Gebhart and Mollendorf [37] have derived a new density state equation valid in the temperature range from phase equilibrium up to 20°C. It includes both pure and saline water and applies to 40 ppt (parts per thousand) salinity, s, and to 1000 bars pressure, p'. The agreement with modern density data over this whole region is about 10 ppm (RMS). The relation is

$$\rho(T', s, p') = \rho_m(s, p')[1 - \alpha(s, p')|T' - T_m'|^{q(s, p')}] \tag{4}$$

where $\rho_m(s, p')$ is the maximum density at s and p, $T_m'(s, p')$ is the corresponding inversion temperature and $q(s, p')$ is the exponent. Salinity is in parts per thousand, and pressure is in bars. The forms and values of q, α, ρ_m and T_m' are given in all detail by Gebhart and Mollendorf [37]. Eq. (4) with the appropriate choice of α, q, ρ_m and T_m to represent the density variation for pure water remains always a compromise and does not take into account, no more than Eq. (3), the skewness observed in experimental data, for the ranges below and above the density extremum.

Fig. 1 gives the dimensionless density difference $(\rho_m - \rho)/\rho_m$ for water in the neighborhood of 4°C as a function of temperature. A comparison is done between experimental data for pure water taken from Weast [36] and results from Eqs. (3) and (4). The particular values of α and q chosen in this figure provide the best fit of Eq. (4) within the range of temperature 0-20°C.

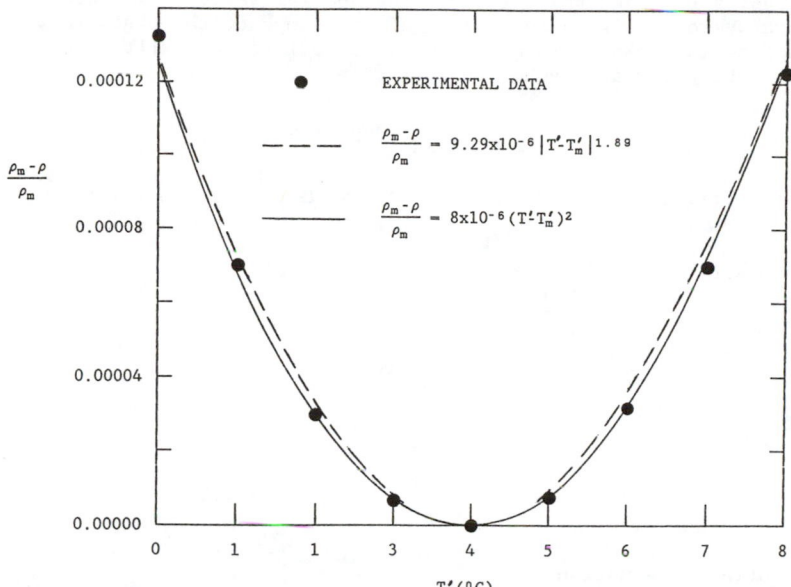

FIGURE 1. Density variation of water in the neighborhood of 4°C.

While the agreement between Eq. (3) and experimental data is perfect in the neighborhood of 4°C, discrepancies increase with $|T'-T_m'|$. It may be noticed in Fig. 1 that the experimental data exhibit a certain amount of skewness. Obviously this behavior cannot be reproduced either by the second order approximation Eq. (3) or by Eq. (4).

The objective of the present investigation is to predict in a general and exhaustive way the essential features of free convection where a density inversion is involved. For that purpose, the second order polynomial Eq. (3) is appropriate since it has the advantage of adding only one more parameter (β_2), by comparison to the more standard form, Eq.(2). With dimensional analysis, the consequence is that one additional governing parameter is introduced, in the form of a Rayleigh number (Robillard, Vasseur [19]) or in the form of an inversion parameter γ (Nguyen et al.[38]), defined as

$$\gamma = -2 \frac{(T_m' - T_r')}{\Delta T'} = \frac{\beta_1}{\beta_2 \; \Delta T'} \qquad (5)$$

Introducing equation (5) into equation (3) one obtains

$$\frac{\rho - \rho_r}{\rho_r} = - \beta_2 (\Delta T')^2 \; T(\gamma + T) \qquad (6)$$

The inversion parameter γ relates the temperature for maximum density T_m' to a reference temperature (a similar parameter has been used by Carey et al. [3] to study the effects of pure and saline water at a vertical isothermal surface). Assuming that $0 \leq T'-T_r' \leq \Delta T'$ is satisfied for the entire flow field, then T_m' lies within that flow field and there exists an inversion of density provided that $-2<\gamma<0$. The special value $\gamma = -1$ corresponds to the case where T_m' is in the middle of the range of the fluid temperature (complete inversion). When $\gamma \leq -2$ or $\gamma \geq 0$, no inversion will be observed and the flow behavior approaches that of a common fluid when $|\gamma| \gg 1$.

3 GOVERNING EQUATIONS FOR A POROUS MEDIUM

Neglecting inertial and Brinkman terms, the equations for a porous medium are as follows.

The Darcy-Oberbeck-Boussinesq equation is

$$\vec{V}' = -\frac{K}{\mu}(\rho \vec{g} - \nabla p') \tag{7}$$

the energy equation is

$$(\rho c)_p \frac{\partial T'}{\partial t'} + (\rho c)_f (\vec{V}'.\nabla)T' = k\nabla^2 T' \tag{8}$$

and the continuity equation is

$$\nabla.\vec{V}' = 0 \tag{9}$$

Symbols are defined in Nomenclature. The density difference $\rho - \rho_r$ is the key factor in generating convective flow. The gravity term $\rho_r g$ and the corresponding hydrostatic pressure can be cancelled out from Eq. (7) which becomes

$$\vec{V}' = -\frac{K}{\mu}\left[(\rho - \rho_r)\vec{g} - \nabla p_d'\right] \tag{7a}$$

where p_d' is the residual pressure.

By scaling time, length, temperature, velocity and pressure with $(\rho c)_p r_p'^2/k$, r_p', $\Delta T'$, $k/r_p'(\rho c)_f$ and $\mu k/K(\rho c)_f$ respectively, one obtains the governing equations in their nondimensional form

$$\vec{V} = \frac{K(\rho c)_f r_p'}{k\nu}(\frac{\rho - \rho_r}{\rho_r})\vec{g} - \nabla p_d \tag{10}$$

$$\frac{\partial T}{\partial t} + (\vec{V}.\nabla)T = \nabla^2 T \tag{11}$$

$$\nabla.\vec{V} = 0 \tag{12}$$

The link between the density difference in Eq. (10) and the temperature is done through the use of an appropriate equation of state, i.e., Eq. (2) for standard fluid behavior, or Eq. (6), for fluid having an extremum in temperature, such as water slightly above the freezing point. Eq. (10)

becomes:

First degree density-temperature relationship (linear state equation)

$$\vec{V} = -\, Ra\ \vec{e}_g T - \nabla p_d \tag{13}$$

with

$$Ra = \frac{K(\rho c)_f r_p'}{k\nu}\, g\beta_1 \Delta T \tag{14}$$

being the usual Rayleigh number for a porous medium and $\vec{e}_g$ a unit vector pointing in the direction of gravitational acceleration.

Second degree density-temperature relationship (non-linear state equation)

$$\vec{V} = -\, Ra^*\vec{e}_g T(\gamma+T) - \nabla p_d \tag{15}$$

with

$$Ra^* = \frac{K(\rho c)_f r_p'}{k\nu}\, g\beta_2 \Delta T^2 \tag{16}$$

being a special form of the Rayleigh number called the non-linear Rayleigh number. By comparing Eqs. (13) and (15), it is seen that the non-linear convection tends to the linear one when $|\gamma|$ becomes large, the product γRa^* remaining finite. The asymptotic limits of nonlinear convection are their linear counterparts, including reversed gravity.

A solution may be sought directly from governing Eqs. (11) and (12) together with (13) or (15). Those equations are expressed in terms of the primitive variables. An alternative exists to express velocities in terms of the stream function ψ and to eliminate the pressure term by taking the curl of Eqs. (13) or (15). We obtain

Linear state equation

$$\nabla^2 \psi = Ra\, \frac{\partial T}{\partial y} \tag{17}$$

Non-linear State Equation

$$\nabla^2 \psi = Ra^*\frac{\partial}{\partial y} T(\gamma+T) \tag{18}$$

y being the horizontal coordinate.

One must keep in mind however that this last procedure increases by one the order of the derivative of the Darcy equation and therefore increases by one the degree of freedom of the new set of governing equations. This is of particular importance when convective flow takes place in a doubly-connected region. A two-dimensional doubly-connected region is limited by two distinct boundaries that do not intersect. This is the case of Fig. 2. Physically, one must allow for the possibility of a net (circulating) flow between these two boundaries and therefore cannot prescribe a priori the same value of the stream function on the two boundaries. In fact, there exists an infinite set of solutions for (17) or (18), each solution corresponding to a particular value of the circulating flow. Among these solutions we must pick up the one that satisfies the original Darcy equations, (13) or (15). In other words, Eqs. (13) or (15) must provide the additional boundary condition required to fix the value of the net circulating flow.

4 APPLICATION TO THE PROBLEM OF A BURIED PIPE

Bau [30] has solved analytically and numerically the two-dimensional linear convection occurring in the case of a pipe buried in a porous medium limited by a flat plate of infinite extent (see Fig. 2). The pipe and the plate were maintained at constant but different temperatures. We attempt here to obtain solutions for the same geometry with the use of a nonlinear state equation. As mentioned earlier, a nonlinear state equation is involved in practice when the fluid is water in the neighborhood of 4°C.

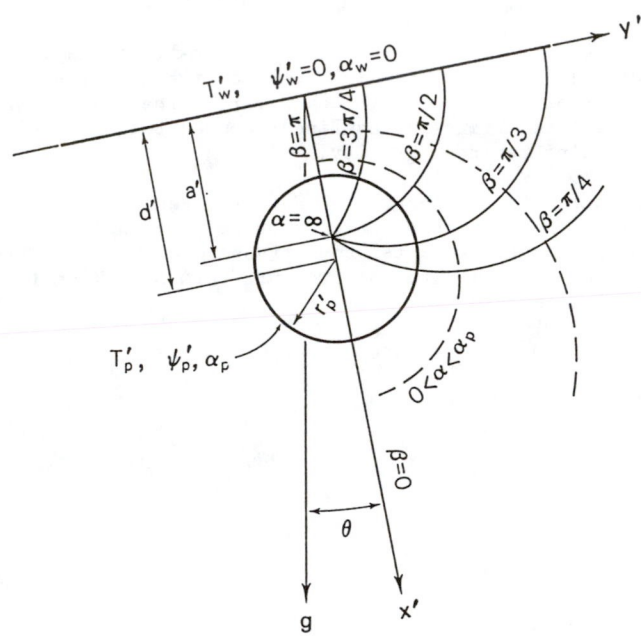

FIGURE 2. Flow geometry and coordinate system.

Following Bau's approach, a conformal mapping (bicylindrical coordinates)

$$x+iy = a \cot \frac{\alpha-i\beta}{2} \tag{19}$$

is used to transform the physical plane (Fig. 2) into a rectangular domain with coordinates α and β. r_p' is the length scale, $T_p'-T_w'=\Delta T'$ is the temperature scale and $a=a'/r_p'=\sinh\alpha_p$ is the non-dimensional distance between the flat plate and the locus $\alpha=\infty$.

The energy Eq. (11) and the stream function Eq. (18), expressed in terms of α and β, become

$$\frac{a^2}{(\cosh\alpha-\cos\beta)^2}\frac{\partial T}{\partial t} - \frac{\partial\psi}{\partial\beta}\frac{\partial T}{\partial\alpha} + \frac{\partial\psi}{\partial\alpha}\frac{\partial T}{\partial\beta} = \nabla^2_{\alpha\beta}T \tag{20}$$

and

$$\nabla_{\alpha\beta}\psi = aRa^* \mathscr{L}[T(\gamma+T)] \tag{21}$$

where

$$\nabla^2_{\alpha\beta}(\ ) = \frac{\partial^2}{\partial\alpha^2}(\ ) + \frac{\partial^2}{\partial\beta^2}(\ )$$

and

$$\mathscr{L}[\ ] = [H(\alpha,\beta)\sin\theta + G(\alpha,\beta)\cos\theta]\frac{\partial}{\partial\alpha}[\ ] + [H(\alpha,\beta)\cos\theta - G(\alpha,\beta)\sin\theta]\frac{\partial}{\partial\beta}[\ ]$$

The time dependent term has been retained in the energy equation. Although we search for steady state solutions, this term will be required in the numerical method to be used.

The coefficient $a^2/(\cosh\alpha-\cos\beta)^2$ of the time dependent term and the functions

$$H(\alpha,\beta) = \frac{1-\cosh\alpha\cos\beta}{(\cosh\alpha-\cos\beta)^2} \tag{22a}$$

and

$$G(\alpha,\beta) = \frac{\sinh\alpha\sin\beta}{(\cosh\alpha-\cos\beta)^2} \tag{22b}$$

appearing in the stream function equation are the result of coordinate transformation. Thus, the conformal mapping of Eq. (19) reduces the geometry to a rectangular domain with usual governing equations, except for the time dependent term of (20) and for the source term of (21) which is coordinate dependent through (22a) and (22b).

In general, for situations where θ is different from zero, the limits $\beta=0$ and $\beta=\pi$ of the rectangular region are not impermeable to mass or heat flux. However, the present investigation is restricted to the case $\theta=0$. Thus the hypothesis of symmetry is retained and the x axis of Fig. 2 is associated to an adiabatic streamline connecting two boundaries. Only the half domain (simply-connected domain) needs then to be considered for the solution and ψ may be set to zero on the plate and on the pipe. The boundary conditions are

$$\psi = 0 \qquad T = 0 \qquad \text{at } \alpha = 0$$

$$\psi = 0 \qquad T = 1 \qquad \text{at } \alpha = \alpha_p \qquad (23)$$

$$\psi = 0 \qquad \frac{\partial T}{\partial \beta} = 0 \qquad \text{at } \beta = 0, \pi$$

The local heat flux at the pipe $(\alpha=\alpha_p)$ or at the flat plate $(\alpha=0)$ can be obtained from the temperature field

$$q_{p,w} = \left(\frac{\cos\alpha - \cos\beta}{a}\right) \frac{\partial T}{\partial \alpha}\bigg|_{\alpha=\alpha_p, 0} \qquad (24)$$

The overall heat flow Q per unit length around the pipe is

$$Q = \int_0^{2\pi} \frac{\partial T}{\partial \alpha}\bigg|_{\alpha=\alpha_p} d\beta \qquad (25)$$

Q is an invariant under the coordinate transformation (19). It is equivalent to the heat flow integrated along the plate i.e., to the overall heat flow obtained by replacing α_1 by 0 in Eq. (25).

The Nusselt number is

$$Nu = \frac{Q}{Q_c} = \frac{\alpha_p}{\pi} Q \qquad (26)$$

Eqs. (20) and (21) together with boundary conditions (23) completely determine the solution of the problem in terms of the dimensionless depth or aspect ratio d, the non-linear Rayleigh number Ra" and the inversion parameter γ.

5 PERTURBATION SOLUTION

Following the treatment by Bau [30], a perturbation approach valid for low Ra*, with a pure conduction temperature field as a starting point, may be used to obtain approximations of the steady-flow and temperature fields.

The perturbation expansion is done in the usual, way, i.e., in terms of Ra*

$$T = \sum_{s=0}^{\infty} (a\, Ra^*)^s\, T_s$$

$$\psi = \sum_{s=0}^{\infty} (a\, Ra^*)^s\, \psi_s \qquad\qquad\qquad (27a,b,c)$$

$$Nu = \sum_{s=0}^{\infty} (a\, Ra^*)^s\, N_s$$

These expansions are introduced in Eqs. (21) and in the steady state form of Eq. (20).

The functions H and G are expressed in their corresponding Fourier series. The resulting equations at order s are

$$\nabla_{\alpha\beta}^2 \psi_s = 2\sum_{k=1}^{\infty} ke^{-k\alpha}\left[\sin k\beta\, \frac{\partial}{\partial\alpha} - \cos k\beta\, \frac{\partial}{\partial\beta}\right]\left[\gamma T_{s-1} + \sum_{j-1}^{s} T_{s-j}\, T_{j-1}\right] \qquad (28)$$

$$\nabla_{\alpha\beta}^2 T_s = \sum_{j=1}^{s} \frac{\partial\psi_j}{\partial\alpha}\, \frac{\partial T_{s-j}}{\partial\beta} - \frac{\partial\psi_j}{\partial\beta}\, \frac{\partial T_{s-j}}{\partial\alpha} \qquad (29)$$

$$N_s = \frac{\alpha_p}{2\pi} \int_0^{2\pi} \frac{\partial T_s}{\partial\alpha}\Big|_{\alpha=\alpha_p} d\beta \qquad (30)$$

The zero order solution can be easily shown to be given by

$$T_0 = \frac{\alpha}{\alpha_p} \qquad \psi_0 = 0 \qquad N_0 = 1 \qquad (31)$$

while for first order solution, Eqs. (28) and (29) give

$$\nabla_{\alpha\beta}^2 \psi_1 = -\frac{2}{\alpha_p} \sum_{k=1}^{\infty} (\gamma + \frac{2\alpha}{\alpha_p})\, ke^{-k\alpha}\sin k\beta \qquad (32)$$

$$\nabla_{\alpha\beta}^2 T_1 = -\frac{1}{\alpha_p}\, \frac{\partial\psi_1}{\partial\beta} \qquad (33)$$

with the appropriate boundary conditions (23). Solutions are

$$\psi_1 = \sum_{n=1}^{\infty} g_{1,n}^* \sin n\beta$$

$$T_1 = \sum_{n=1}^{\infty} f_{1,n}^* \cos n\beta \tag{34a,b}$$

where

$$g_{1,n}^* = \gamma \left[F_1 \, e^{-n\alpha_p} + F_2 \, e^{-n\alpha} \right] + F_3 \tag{35a}$$

and

$$f_{1,n}^* = \frac{F_1}{2} \left[\gamma \, e^{-n\alpha_p} + F_4 \right] \left[\coth n\alpha_p + F_2 \coth n\alpha \right]$$

$$\tag{35b}$$

$$+ \frac{1}{4} (\gamma + \frac{2}{n\alpha_p}) F_3 + \frac{1}{6} (F_2^3 \, e^{-n\alpha} + F_1 \, e^{-n\alpha_p})$$

with

$$F_1 = \frac{\sinh n\alpha}{\sinh n\alpha_p} \qquad F_3 = F_1 \, F_4 + F_5 \qquad F_5 = F_2 e^{-n\alpha} + F_1 \, e^{-n\alpha_p}$$

$$F_2 = - \frac{\alpha}{\alpha_p} \qquad F_4 = (1 + \frac{1}{n\alpha_p}) e^{-n\alpha_p}$$

The terms in Eqs. (35a,b) affected by the coefficient γ correspond to the ones established by Bau [1984] for the case of convection with a linear state equation. By using the slow converging terms of series (35a) the kind of transformation done by Bau, it may be shown that the first order stream function is zero at infinity.

Since all terms in the first order temperature are periodic in β, there is no net contribution to the total heat flow that can be obtained from Eq. (30) and

$$N_1 = 0 \tag{36}$$

A non-zero contribution to the Nusselt number requires the development of the second (and higher) order temperature terms of Eq. (27a)

6 NUMERICAL SOLUTION

The governing partial differential Eqs. (20) and (21) in their unsteady form, are converted into finite difference equations. Forward time and central differences are used to approximate the diffusive and convective terms. Overall, the method is second order accurate in space. The solution technique is to seek the steady-state as a limit of the unsteady problem.

The numerical rectangular domain α, β is divided into an uniform mesh. The number of grid points in the α and β directions are varied, depending upon

the aspect ratio of the cavity. Convergence with mesh size has been verified by employing coarser and finer grids on selected test problems. In
most of the calculations the number of grid points are 36x36. Trial calculations were necessary in order to optimize computation. Starting from initial data, stream function is advanced first by iteratively solving a difference approximation of Eq. (21) using successive overrelaxation. For
small Rayleigh numbers, an optimal overrelaxation parameter is used. As the
Rayleigh number increases, it is necessary to decrease the overrelaxation
in order to avoid divergence in the iterations. The iterative procedure is
stopped once the following convergence criteria has been satisfied

$$\left[\frac{\sum\limits_{i}\sum\limits_{j} \left| \psi_{i,j}^{n+1} - \psi_{i,j}^{n} \right|}{\sum\limits_{i}\sum\limits_{j} \left| \psi_{i,j}^{n+1} \right|} \right] \leq 10^{-4} \qquad (37)$$

where the superscripts n and (n+1) indicate the value of the nth and
$(n+1)^{th}$ iterations respectively and i and j indices denote grid locations
in the (α,β) coordinates. Further decrease of the convergence criteria
(10^{-4}) did not cause any significant change in the final results. The velocity field is obtained from the stream function and the temperature field
explicitly advanced using the alternating direction implicit method of
Peaceman and Rachford [39]. The entire procedure is then repeated.

During the iteration, the fields always evolved smoothly from arbitrarily
initial state to final, steady state. Iteration is terminated based on the
following criteria

$$\left[1 - \frac{\psi_{max}^{n}}{\psi_{max}^{n+1}} \right] \leq 10^{-4} \qquad (38)$$

where ψ_{max} is the maximum value of the stream function inside the cavity.

The number of iterations needed varies according to the Rayleigh number,
the aspect ratio and the overrelaxation parameter. The range typically is
from about 1200 to 1500 iterations. Typical values of the time steps range
from 10^{-3} to 10^{-4}. The very initial state is motion free and consists of
a conductive temperature gradient. As the Rayleigh number increases, the
steady state results of the previous run may serve as initial conditions
for the next one. The CPU time required for convergence was from 300 to
1200 seconds on an IBM 4381 computer.

To have an additional check on the accuracy of the results, an energy balance is used for the system. For this the heat transfer through each plane
α=constant is evaluated at each grid location $0 \leq \alpha \leq \alpha_p$ and compared with the
input at α=0. For most of the results reported here, the energy balance is
satisfied to within 2% and never exceeds 6%. Worst cases are when the flow
near the pipe is directed downward.

7 RESULTS AND DISCUSSION

As stated earlier, with increasing $|\gamma|$, the free convection arising from the parabolic type state equation (6) is expected to reproduce asymptotically the free convection based on a linear relationship between density and temperature (linear convection). Two distinct cases of linear convection are involved in the limit $|\gamma| \to \infty$: i) hot pipe/cold plate and ii) cold pipe hot plate. Both cases are discussed in the next section.

7.1 Fluid with a Linear State Equation

Figs. 3a and 3b show the flow fields with continuous lines and temperature fields with dotted lines, obtained numerically for an aspect ratio d=5 and for Rayleigh numbers Ra of 5 and -5 respectively. The heat fluxes on the flat wall, q_w and on the pipe surface, q_p are given on Figs. 4a and 4b, where the heavy line corresponds to the pure conduction. The Nusselt number Nu is given in Fig. 5 for aspect ratios d=3 and d=5. In Fig. 3a, the flow near the pipe is in the upward direction (Ra>0). This type of convection is obtained in the case of a relatively hot pipe with gravity in the downward direction. The flow field consists in two cells symmetrically located at a finite distance with respect to the vertical axis. The reversed buoyancy situation (Ra<0) is shown in Fig. 3b with the flow near the relatively cold pipe in the downward direction. The entire flow field is characterized by two symmetrical cells rotating in directions opposite to the ones in Fig. 3a and whose centers are now located at infinity ($\alpha, \beta \to 0$). For this situation, the velocities remain important far from the pipe, especially in the downward direction. It is seen from Fig. 4b that the corresponding heat flux on the flat wall, which was above the pure conduction curve for the normal buoyancy situation, is now reduced below the pure conduction curve in the neighborhood of y=0. Naturally, to insure a Nusselt number above unity, the heat flux becomes more important than the pure conduction curve at large values of y (y≥15). Thus in addition to velocities, heat fluxes remain important far from the origin of the x-y plane and consequently the mesh grid, based on constant increments $\Delta \alpha$ and $\Delta \beta$, is less appropriate to solve the reversed buoyancy situation. For example, at a

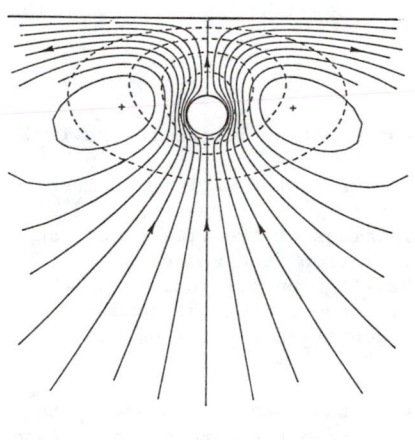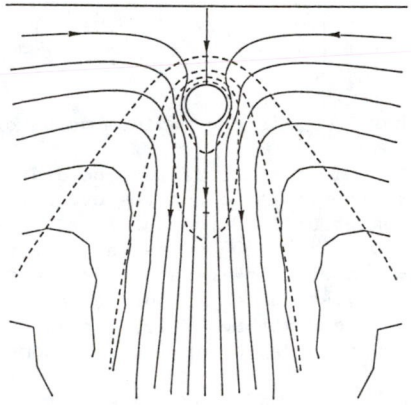

a) Ra=5; $\Delta \psi$=0.308   b) Ra=-5; $\Delta \psi$=1.000

FIGURE 3. Flows and temperature fields for convection with a linear state equation

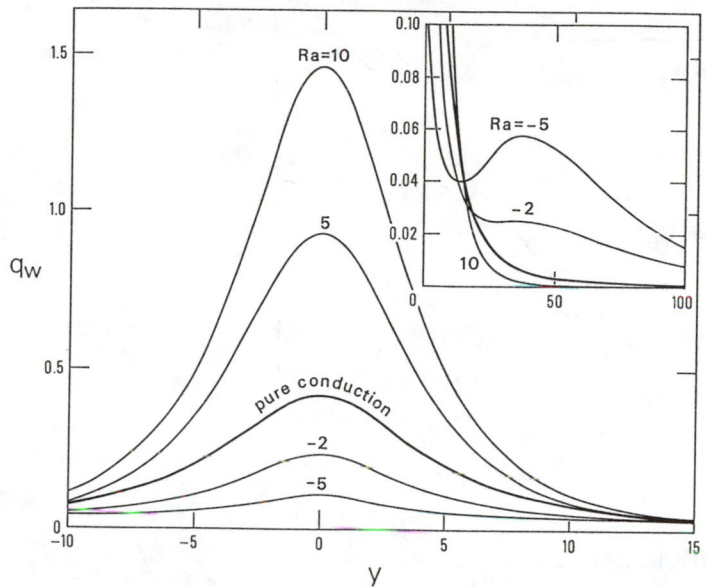

a) flat plate

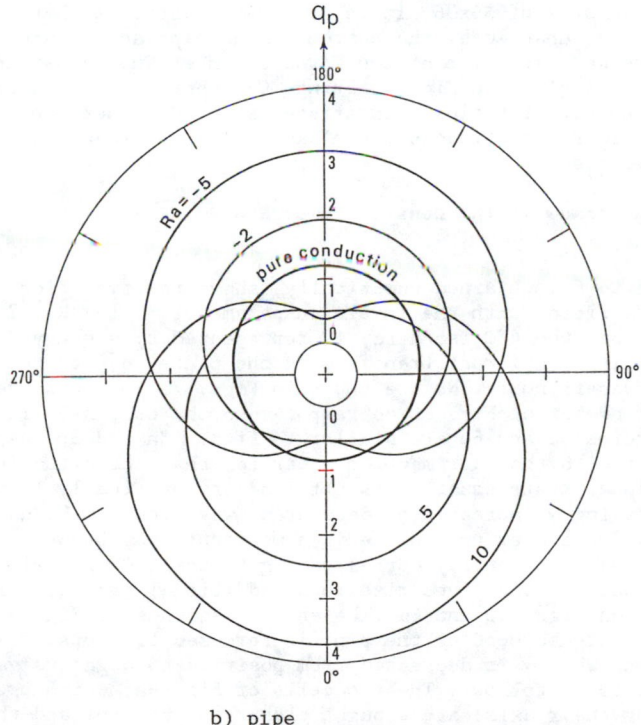

b) pipe

FIGURE 4. Local heat fluxes on the boundaries.

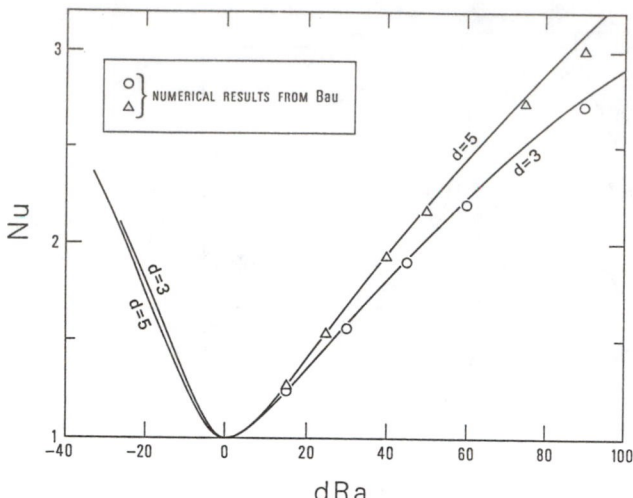

FIGURE 5. Nusselt number function of Ra.

Rayleigh number of -5, the Nusselt number, based on the heat flux evaluated at the flat wall, differs from the one evaluated on the pipe boundary by as much as 7% for a mesh size of 36x36. It is observed in Fig. 4b that the local heat transfer is enhanced at the bottom of the pipe and reduced at the top for positive Ra. The plot of the Nusselt Number Nu against the product of d and Ra, as given in Fig. 5, brings together the curves corresponding to different aspect ratios. Comparison may be done between the present numerical results (continuous lines) and those obtained by Bau [30] for positive Ra (symbols).

7.2 Fluid with an Extremum in the Density-Temperature Relationship

The set of Figs. 6a to 6e, obtained numerically, shows the evolution of the flow and temperature fields with the inversion parameter γ, for $Ra^*=2$ and d=5. In Figs. 6b to 6d, the 4°C isotherm, is represented by a heavy dashed line. The corresponding local heat transfers on the plate, q_w, on the cylinder, q_p and the Nusselt number Nu are shown in Figs. 7a,7b and 8a respectively. The Nusselt number of Fig. 8b corresponds to the case d=3. It may be noticed that Figures 6a and 6e are similar to Figs. 3a and 3b respectively, in agreement with the fact mentioned earlier that the situation corresponding to linear state equation is obtained asymptotically from the non-linear one by having γ increased or decreased away from -1. When -2<γ<0, the density field does not change in a monotonic way between the two boundaries. The maximum density corresponding to the 4°C line lies between the two boundaries and gives rise to an additional pair of cells that appear clearly on Figs. 6c and 6d. The entire sequence of Fig. 6 shows how the flow in the neighborhood of the pipe is reversed from upstream to downstream direction, when γ is decreased from positive to negative values. The flow field evolves as follows. The two cells of Fig. 6a, which produce an upward flow along the x axis, are brought closer to the pipe and their intensity decreases, as shown in Figs. 6b, 6c and 6d. When γ becomes nega-

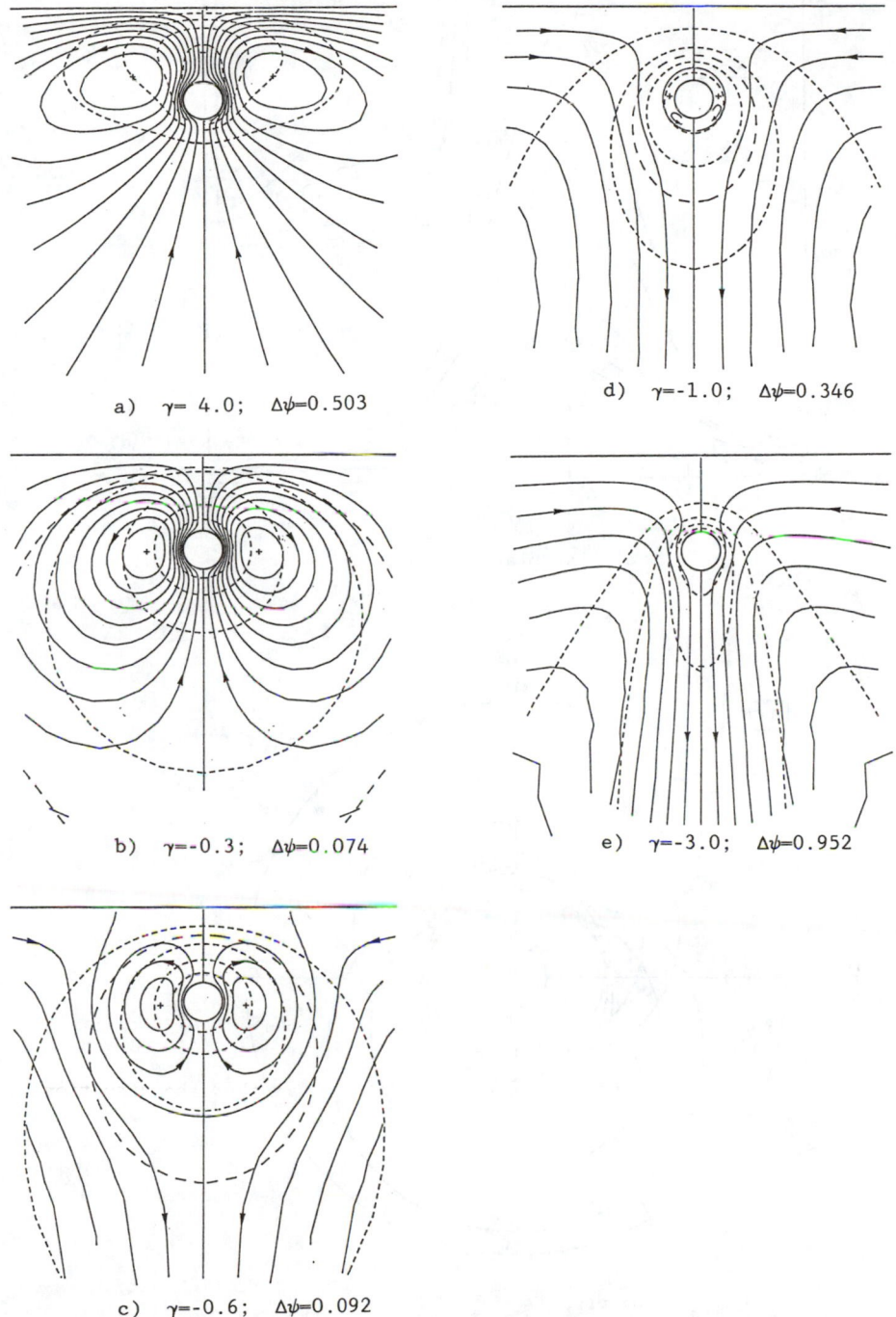

a) $\gamma= 4.0$; $\Delta\psi=0.503$

d) $\gamma=-1.0$; $\Delta\psi=0.346$

b) $\gamma=-0.3$; $\Delta\psi=0.074$

e) $\gamma=-3.0$; $\Delta\psi=0.952$

c) $\gamma=-0.6$; $\Delta\psi=0.092$

FIGURE 6. Flow and temperature fields for convection with a non-linear (second degree) state equation.

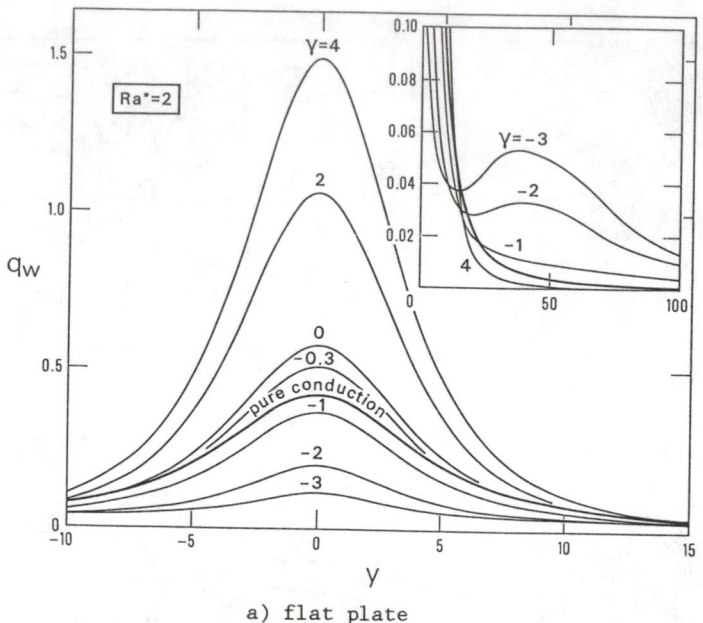

a) flat plate

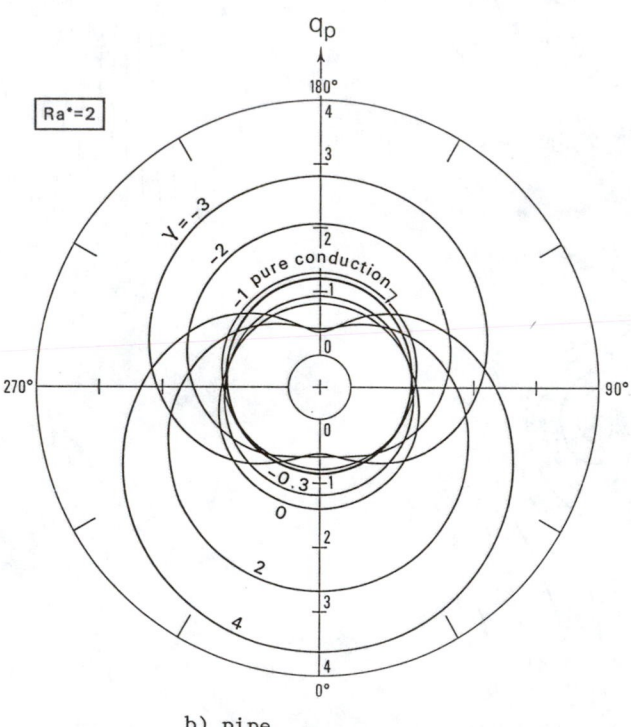

b) pipe

FIGURE 7. Local heat fluxes on the boundaries.

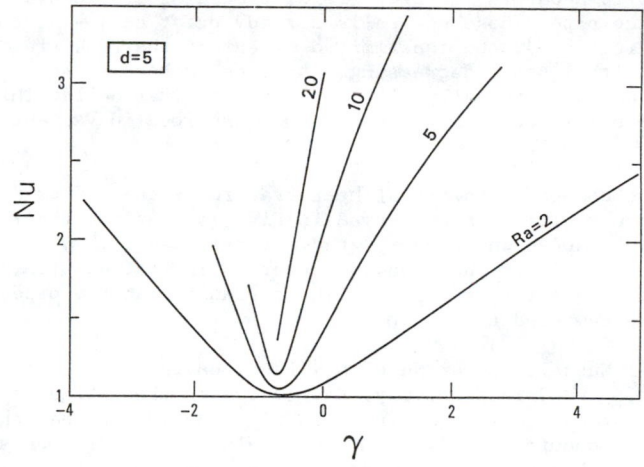

a)

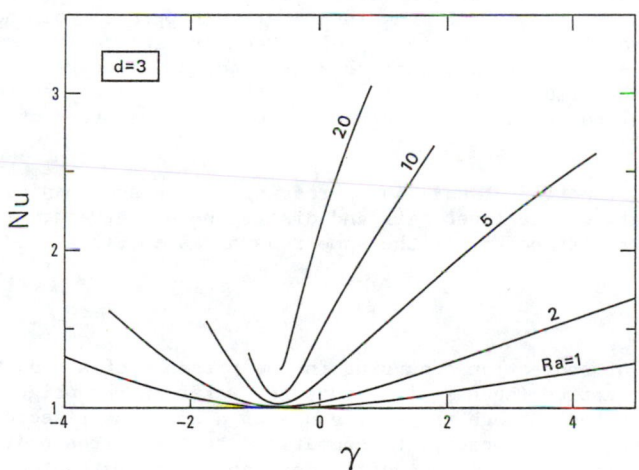

b)

FIGURE 8. Nusselt number variation with γ.

333

tive, two additional cells occur in the flow field, producing a downward flow, far away, below the pipe. Those new cells already exist at $\gamma=-.3$, but they do not appear in Fig. 6b, due to the limited extent of the flow field shown in that figure. With γ still decreasing, those cells becomes stronger (Figs. 6c, 6d, and 6e) and gradually eliminate the original cells. This behavior of the flow with γ is comparable to the one described in Vasseur et al. [29].

Figs. 7a and 7b show the change of the local heat transfer with γ, from normal to reversed buoyancy. It may be observed in Fig. 7a that the local heat transfer on the plate above the pipe is gradually reduced with γ increasing, in a way comparable to the behavior observed in Fig. 4a with Ra decreasing. Similarly the behavior of the local heat transfer at the pipe, given in Fig. 7b, may be compared to Fig 4b.

The change of the global Nusselt number Nu with γ is given in Figs. 8a and 8b for aspect ratios of 5 and 3 respectively. Each curve in those figures corresponds to a different Ra^*. It is observed in Figs. 8a and 8b that the minimum Nu for all curves occurs at $-.7<\gamma<-.6$. This value of γ corresponds to the complete inversion observed in the case of horizontal concentric pipes (Nguyen et al. [32]) and also in the case of a rectangular cavity with isothermal vertical boundaries (Robillard and Vasseur [17]). This minimum value of the heat transfer requires a given equilibrium between the intensities of the convective cells forming the flow field. Fig. 6c represents approximately that equilibrium.

A comparison is done in Figs. 9a and 9b between the numerical and analytical (first order) solution. In those figures, $|\psi|_{max}$ is the extremum value of the stream function within the inner cell. It is a measure of the intensity of flow within that cell. In Fig. 9a, $|\psi|_{max}$ is given as a function of γ while Ra* is maintained constant (Ra*=0.5). $|\psi|_{max}$ is seen to decrease monotonously to zero. Thus, at $\gamma\approx-1.05$, the inner cell disappears from the flow field. For low intensities of the inner cell, the correspondence between both numerical and analytical approaches is good but discrepancies become significant with growing intensity of that cell, owing mainly to the fact that the analytical curve is based exclusively on the first term of the series expansion (27b).

In Fig. 9b, $|\psi|_{max}$ is a function of Ra*. An increasing Ra* is accompanied by an increasing strength of the inner cell and discrepancies between analytical and numerical results occur for the same reasons as in Fig. 9a.

8 CONCLUSIONS

The essential features of convection occurring in the presence of a density extremum in the state equation (such as it occurs for water in the neighborhood of $4°C$) are revealed through the simple use of a non-linear second degree density-temperature relationship. By comparison to the corresponding problem with a linear state equation, one additional governing parameter is introduced through dimensional analysis. This new parameter may take the form given in Eq. 5. This form has been used in the literature and is called the inversion parameter γ.

The existence of a density extremum within the flow field between the solid boundaries, which occurs when the inversion parameter is between 0 and -1, results in doubling the number of convective cells within the flow field. These additional cells are part of the base flow. Their occurrence is pre-

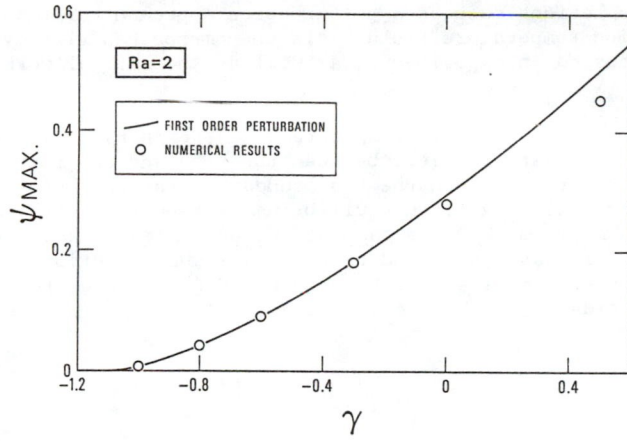

a) ψ vs γ: $Ra^* = 2$

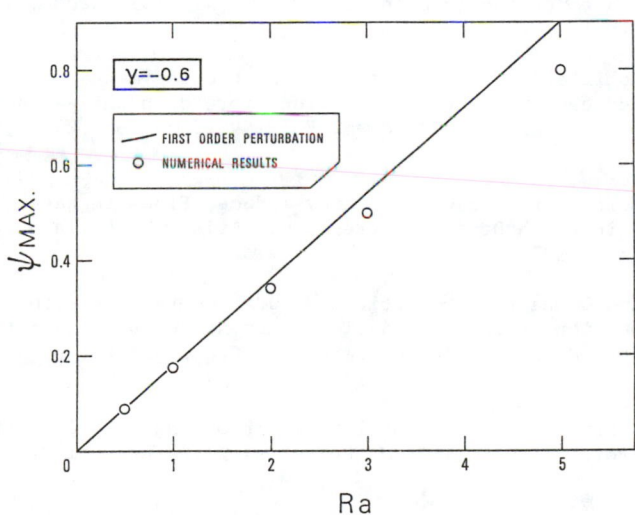

b) ψ vs Ra^*: $\gamma = -0.6$

FIGURE 9.   Comparison between numerical and perturbation approaches

dicted by analytical solutions such as a perturbation approach built up from the pure conduction temperature field. This phenomenon has already been observed and discussed in experimental as well as theoretical work in the past.

The occurrence of additional cells prevents direct heat exchange by convection between the solid boundaries, i.e., between the pipe and the flat plate. Thus the heat transfer between the two boundaries is greatly reduced. Moreover there exists a given equilibrium between the cells for which the reduction is maximum. For the case of a pipe buried in a semi-infinite porous medium, it has been found that the maximum reduction occurs when the inversion parameter γ is somewhere between -0.6 and -0.8, for the range of Ra^* and d considered in the present study.

REFERENCES

1. Joshi, Y. and Gebhart, B., Vertical Transient Natural Convection Flows in Cold Water, *Int.J. Heat Mass Transfer*, Vol. 27, No 9, pp. 1573-1582, 1984.

2. El-Henawy, I., Hassard, B., Gebhart, B. and Mollendorf, J. C., Numerically Computed Multiple Steady Stated of Vertical Buoyancy-Induced flows in Cold Pure Water, *J. Fluid Mech.*, Vol. 122, pp. 235-250, 1982.

3. Carey, V. P., Gebhart, B. and Mollendorf, J. C., Buoyancy Force Reversals in Vertical Convection Flows in Water, *J. Fluid Mech.*, Vol. 97, pp. 279-297, 1980.

4. Carey, V. P. and Gebhart, B., Visualisation of the Flow Adjacent to a Vertical Ice Surface Melting in Cold Pure Water, *J. Fluid Mech.*, Vol. 107, pp. 37-55, 1981.

5. Vasseur P. and Robillard, L., Etude théorique et expérimentale de la formation d'une couche limite par convection libre dans une eau à basse température, *The Canadian J. Chem. Eng.*, Vol. 59, pp. 24-34, 1981.

6. Gebhart B. and Mollendorf, J. C., Buoyancy-induced Flows in Water Under Conditions in which Density Extremes may Arise, *J. Fluid Mech.*, Vol 89, pp. 673-707, 1978.

7. Qureshi, Z. H. and Gebhart, B., Vertical Natural Convection with a Uniform Flux Condition in Pure and Saline Water at the Density Extremum, *Proc. 6th Int. Heat Transfer Conference*, NC-6, Toronto, pp. 217-222, 1978.

8. Gilpin, R. R., Cooling of a Horizontal Cylinder of Water Through its Maximum Density Point at 4°C, *Int. J. Heat Mass Transfer*, Vol. 18, pp. 1307-1315, 1975.

9. Gilpin, R. R., The Effects of Dentritic Ice Formation in Water Pipes, *Int. J. Heat Mass Transfer*, Vol. 20, pp. 693-699, 1977.

10. Seki, N., Fukusako, S. and Nakaoka, M., Experimental Study on Natural Convection Heat Transfer with Density Inversion Between Two Horizontal Cylinders, *J. Heat Transfer*, Vol. 97, pp. 556-561, 1975.

11. Cheng K. C. and Takeuchi, M., Transient Natural Convection of Water in a Horizontal Pipe with Constant Cooling Rate Through 4°C, *J. Heat Transfer*, Vol. 98, pp. 581-587, 1976.

12. Robillard, L., Nguyen, T. H., Vasseur, P. and Chandra Shekar, B., Convective Heat Loss of Water in a Circular Pipe Cooled at Constant Rate, *Wärme-und Stoffübertragung*, Vol. 19, pp. 181-185, 1985.

13. Lankford, K. E. and Bejan, A., Natural Convection In a Vertical Enclosure Filled with Water Near 4°C, *J. Heat Transfer*, Vol. 108, No 4, pp. 755-763, 1986.

14. Inaba, H. and Fukuda, T., An Experimental Study of Natural Convection in an Inclined Rectangular Cavity Filled With Water at its Density Extremum, *J. Heat Transfer*, Vol. 106, pp. 109-115, 1984.

15. Vasseur, P. and Robillard, L., Transient Natural Convection Heat Transfer in a Mass of Water Cooled Through 4°C, *Int. J. Heat Mass Transfer*, Vol. 23, pp. 1195-1205, 1980.

16. Vasseur, P. and Robillard, L., Natural Convection in a Rectangular Cavity with Wall Temperature Decreasing at a Uniform Rate, *Wärme-und-Stoffübertragung*, Vol. 105, pp. 199-207, 1982.

17. Robillard, L. and Vasseur, P., Effet du maximum de densité sur la convection libre de l'eau dans une Cavité Fermée, *Canadian J. Civil Eng.*, Vol. 6, pp. 481-493, 1979.

18. Robillard, L. and Vasseur, P., Transient Natural Convection Heat Transfer of Water with Maximum Density Effect and Supercooling, *J. Heat Transfer*, Vol. 103, pp. 528-534, 1981.

19. Robillard, L. and Vasseur, P., Convective Response of a Mass of Water Near 4°C to a Constant Cooling Rate Applied on its Boundaries, *J. Fluid Mech.*, Vol. 118, pp. 123-141, 1982.

20. Watson, A., The Effect of the Inversion Temperature on the Convection of Water in an Enclosed Rectangular Cavity, *Quart. J. Mech. Appl. Math.*, Vol. XXV, pp. 423-446., 1972.

21. Desai, V. S. and Forbes, R. E., Free Convection in Water in the Vicinity of Maximum Density, *Am. Soc. Mech. Eng.*, *Environmental and Geophysical Heat Transfer*, Vol. 93, pp. 41-47, 1971.

22. Sun, Z. C., Tien, C. and Yen, I. C., Onset of Convection in a Porous Medium Containing Liquid with a Density Maximum, *Proc. 4 th Int. Heat Transfer Conference*, Paris, Versailles, V.IV. NC-211, 1972.

23. Yen, I. C., Effects of Density Inversion on Free Convection Heat Transfer in Porous Layer heated from Below, *Int. J.Heat Mass Transfer*, Vol. 17, pp. 1349-1356, 1974.

24. Ramilson, J. M. and Gebhart, B., Buoyancy Induced Transport in Porous Media Saturated with Pure or Saline Water at Low Temperatures, *Int. J. Heat Mass Transfer*, Vol. 23, pp. 1521-1530, 1980.

25. Gebhart, B., Hassard, B., Hastings, S. P. and Kazarinoff, Multiple Steady State Solutions for Buoyancy Induced Transport in Porous Media Saturated with Cold Pure and Saline Water, *Numerical Heat Transfer*, Vol. 6, pp. 337-352, 1983.

26. Poulikakos, D., Maximum Density Effects on Natural Convection In a Porous Layer Differentially Heated in the Horizontal Direction, *Int. J. Heat Mass Transfer*, Vol. 27, pp. 2067-2075, 1984.

27. Blake, R. K., Bejan, A. and Poulikakos, D., Natural Convection Near $4°C$ in a Water Saturated Porous Layer Heated from Below, *Int. J. Heat Mass Transfer*, Vol. 27, pp. 2355-2364, 1984.

28. Altimir, I., Convection naturelle tridimensionnelle en milieu poreux saturé par un fluide présentant un maximum de densité, *Int. J. Heat Mass Transfer*, Vol. 27, pp. 1813-1824, 1984.

29. Vasseur, P., Robillard, L. and Chandra Shekar, B., Natural Convection of Water Within a Horizontal Cylindrical Annulus with Density Inversion Effects, *ASME J. Heat Transfer*, Vol. 106, pp. 117-123, 1983.

30. Bau, H., Convective Heat Losses From a Pipe Buried in a Semi-Infinite Porous Medium, *ASME 84-HT-77*, 1984.

31. Moore, D. R. and Weiss, N. O., Nonlinear Penetrative Convection, *J. Fluid Mech.*, Vol. 61, pp. 553-581, 1973.

32. Landolt-Bornstein, Zahlenwerte und Funktionen, Vol. 2, Springer, Berlin, pp. 36-37, 1971.

33. Vanier, C. R. and Tien, C., Effect of Maximum Density and Melting on Natural Convection Heat Transfer From a Vertical Plate, *Chem. Eng. Progr. Symp. Ser.*, No 82, pp. 240-252, 1968.

34. Fujii, T., Fundamentals of Free Convection Heat Transfer, Progr. in Heat *Transfer Eng*, Vol. 3, pp. 66-68, 1974.

35. Cheng, K. C., Takeuchi, M. and Gilpin, R. R., Transient Natural Convection in Horizontal Water Pipes with Maximum Density Effect and Supercooling, *Numerical Heat Transfer*, Vol. 1, pp. 101-115, 1978.

36. Weast, R. C., Handbook of Chemical and Physics, *The Chemical Rubber Company*, 1972.

37. Gebhart, B. and Mollendorf, J. C., A New Density Relation For Pure and Saline Water, *Deep-Sea Res.*, Vol. 24, pp. 831-848, 1977.

38. Nguyen, T. H., Vasseur, P. and Robillard, L., Natural Convection Between Horizontal Concentric Cylinder With Density Inversion of Water For Low Rayleigh Numbers, *Int. J. Heat Mass Transfer*, Vol. 25, pp. 1559-1568, 1982.

39. Peaceman, D. W. and Rachford, H. A., The Numerical solution of Parabolic and Elliptic Differential Equations, *J. Soc. Indust. Applied Mathematics*, Vol. 3, No 1, pp. 28-41, 1955.

ICE AND SNOW

Chapter 11

Heat Transfer Problems on Ice and Snow

S. FUKUSAKO AND N. SEKI
Department of Mechanical Engineering, Hokkaido University,
Sapporo 060, Japan

ABSTRACT

A brief review of recent developments in the research of the heat-
transfer problems on ice and snow has been carried out covering such
subjects as fundamental aspects of analytical and numerical methods
on freezing and melting problems, freezing of water without main
flow, freezing of water with main flow, melting of ice, and melting
of snow. An attempt was furthermore made to review freezing and
melting problems of a phase-change medium.

CONTENTS

NOMENCLATURE

A	non-dimensional temperature, defined in Eq.(9)
a	constant
a_ν	absorption coefficient
B	diameter ratio, d/D
b	constant
c	specific heat
D	inner diameter of tube
d	minimum diameter at contraction
Gr_H	Grashof number, $g\ (T_{wh}-T_f)H^3/\nu^2$
g	gravitational acceleration
H	thickness of melt water layer or height between parallel plates
h	thickness of unstable layer or heat-transfer coefficient
$h_{n,\phi}$	local heat-transfer coefficient around cylinder
L	latent heat of fusion or pipe length
L_f	latent heat of fusion
L_v	latent heat of vaporization
Ma	Marangoni number, defined in Eq.(6)
n	number of ice-bands
Nu	Nusselt number, defined in Eq.(7)
Pr	Prandtl number
P_{max}	maximum pressure
R	nondimensional temperature, $(T_m-T_\infty)/(T_w-T_\infty)$
ΔR	pipe wall thickness
Ra	Rayleigh number, defined in Eqs.(4) and (5)
Ra"	Rayleigh number, defined in Eq.(8)
Re_D	Reynolds number, vD/ν
Re_d	Reynolds number, vd/ν
Re_H	Reynolds number, $2HU\infty/\nu$
Re_p	Reynolds number, defined in Eq.(1)
Re_x	Reynolds number, $U\infty x/\nu$
r	radial coordinate
S_∞	salinity content at ambient saline water
T_f	fusion temperature
T_h	heating wall temperature
T_i	initial temperature
T_m	temperature at maximum density
T_v	vapor temperature
T_w	wall temperature
T_{wc}	cooled wall temperature
T_{wh}	heating wall temperature
T_o	initial temperature
T_2	heating wall or water-surface temperature
T	environmental temperature or main flow temperature
ΔT	Temperature difference
t	time
U	main flow velocity or mean velocity
v	mean velocity in tube
x	distance from leading edge

x_{max} maximum thickness of ice layer
x_{tr} distance from leading edge to ice-thickness transition

Greek Symbols

α heat-transfer coefficient
β coefficient of thermal expansion
δ_i thickness of ice layer
θ_c nondimensional cooling temperature, $(T_f-T_w)/(T_\infty-T_f)$
θ_h nondimensional heating temperature, $(T_f-T_{wh})/(T_\infty-T_f)$
κ thermal diffusivity
κ_ν extinction coefficient
λ_i thermal conductivity of ice
λ_w thermal conductivity of water
λ_∞ thermal conductivity of main water flow
ν kinematic viscosity or wave length of radiation beam
σ_0 surface tension force of water at $0^\circ C$
τ nondimensional time
Φ constant
ϕ degree

1 INTRODUCTION

The problems of water-freezing and ice-melting have received much attention. The growth or decay of an ice cover or on a lake and the freezing or melting of water in a pipe are a few of the practical phenomena which are commonly encountered in cold regions. The freezing of flowing water in a pipe subjected to extreme ambient temperatures causes a variety of detrimental effects, such as an increase in hydraulic pressure loss, a decrease of the flow rate, and occasionally damage to the pipe resulting from flow blockage caused by ice formation. In heavy snowfall regions, snow presents a number of problems encompassing, for instance, snow clearance and avalanches. "Road-heating" has recently been adopted as one of the technical methods for snow removal. Problems essentially associated with the transfer of thermal energy to or from fluids undergoing a change of phase are also encountered in a wide range of technologies, including such diverse applications as the freeze drying of food-stuff, the casting of metals, glass and thermoplastics, and the storage of thermal energy.

The objective of the present article is to review previous studies on heat-transfer problems of ice and snow layers, and also to demonstrate the need for additional research. Furthermore, attention is given to the freezing or melting problems of a phase-change medium other than water or ice, since it is closely related to the development of thermal energy storage technology.

2 MATHEMATICAL ASPECTS OF FREEZING OR MELTING PROBLEMS

2.1 Solutions within the Framework of Neumann's Solution

The problems of thermal conduction with freezing and melting have been studied extensively. Over 100 years ago, Stefan and Neumann considered the melting of a semi-infinite solid at a uniform initial temperature T_o, $T_o < T_f$, when the surface temperature is suddenly raised above the fusion temperature T_f. As discussed by Carslaw and Jaeger [1], exact solutions of the moving-boundary problem are possible only when the position of the phase interface remains fixed in $\xi = r/(\kappa t)^{1/2}$ - space, where r is a radial position coordinate in one, two, or three dimensions, κ is the thermal diffusivity, and t is the elapsed time. Within the framework of Neumann's solution, a number of modified or approximate solutions have been determined by Yang [2], Tien and Geiger [3], Cho and Sunderland [4], Komori and Hirai [5], Churchill and Gupta [6], and Cho and Sunderland [7].

2.2 Analytical Methods

So far, analytical methods for solving heat conduction problems with a moving boundary include the superposition of the solutions to two auxiliary problems: the use of the boundary conditions specified at the moving boundary, and the use of Green's function.

Budhia and Kreith [8] devised an analytical solution to predict the temperature and the motion of the interface in a liquid solidifying or melting in both a wedge and a square container. They superpositioned two solutions, one of which was the solution for the problem of heat conduction without a phase change, but used the same initial and boundary conditions as those in the actual problem. The other solution dealt with heat conduction with a phase change, but with the initial and boundary temperatures maintained at zero degrees. The latent heat due to the phase change was represented by a moving surface source along the interface between the solid and liquid.

Analytical solutions to the inverse Stefan problem, wherein the conditions at the moving rather than the fixed boundary are specified, were studied by Langford [9] and Bluman [10]. They placed the temporal location of the interface between the phase and calculated the required temperature and heat flux variations at the fixed boundary using the similarity solutions, respectively.

The use of Green's functions for solving melting or solidification problems was applied to one-dimensional problems in the Cartesian and cylindrical coordinate systems by Chuang and Szekely [11,12], but they included complicated surface conditions, such as the radiation effect discovered by Hassanein and Kulcinski [13].

2.3 Approximate Methods

2.3.1. Heat Balance Integral Method. The heat balance integral method originally proposed by Goodman [14] has been applied extensively to a variety of problems with phase changes. Some examples are the growth of a deposited layer on a cold surface in a gas stream by Libby and Chen [15], the unidimensional solidification of binary eutectic systems with a time-dependent surface temperature by Tien and Geiger [16], the problem of the fluidized-bed coating by Gutfinger and Chen [17], and the melting of a semi-infinite solid with temperature dependent thermal properties by Imber and Huang [18]. Recently, refinements of the method were proposed, such as subdividing the dependent variable temperature into equal intervals by Bell[19], carrying out a double space integration by El-Genk and Cronenberg [20], and the introduction of the collocation method by Yuen [21]. Furthermore, by using Biot's variational equations, Lapadula and Müeller [22] obtained a much simpler differential equation for solving the convection problem with a moving boundary. The heat balance integral method may be utilized more often in future, since the accuracy of the results are usually sufficient for most practical situations, provided that the assumed temperature profile is carefully chosen.

2.3.2. Series Expansion Method. Westphal [23] developed the series solution of one-dimensional freezing problems, based on the power series expansion in time originally proposed by Evans et al.[24]. The inverse Stefan-like problem was studied by Rubinsky and Shitzer [25] using an infinite series expansion in transformed coordinates for both Cartesian and spherical geometries.

2.2.3. Perturbation Method. Lock et al.[26] applied the perturbation method to the formation of an ice layer at the edge of a semi-infinite domain of water, using the Stefan number as a perturbation parameter. Lock and Nyren [27] also extended the perturbation method for ice formation problem by examining a long circular tube cooled by external convection . Huang and Shih [28] developed a new perturbation method for one-dimensional moving-boundary problems by using Landau's transformation and by applying the regular parameter perturbation technique. A similar perturbation method was used for predicting the degree of melting around a horizontal heating cylinder under a constant heat flux boundary condition by Prusa and Yao [29].

2.4 Numerical Method

2.4.1. Murray-Landis Method. Particularly excellent finite-difference methods were proposed by Murray and Landis [30] (Method I: Variable space network and Method II: Fixed space network). These have been extensively modified to solve freezing problems, such as freezing, outside a sphere by Teller and Churchill [31] and the multidimensional solidification problem by Lazaridis [32]. Refinements or extension of these methods have been developed by Seider and Churchill [33], Heitz and Westwater [34], Shoji [35], and Rao and Sastri [36].

2.4.2. Variable Time Step Method. The variable time-step method was proposed by Douglas and Gallie [37]. Space-direction was subdivided into a finite number of equal intervals and a time-step was determined so that the moving boundary traversed one space mesh during that time. This was used to solve the one-dimensional Stefan Problem under a general boundary condition by Gupta and Kumar [38].

2.4.3. Enthalpy Model and Effective Heat Capacity Formulation Method. Usually, when heat conduction problems involving a change of phase are solved using numerical methods, the major difficulty in generating a numerical solution is the representation of the discontinuity of the temperature gradient at the phase change boundary which could not be a priori estimated. One of the ways to overcome this difficulty is to use the enthalpy model by Katayama and Hattori [39] and Shamsundar and Sparrow [40]. The other is to use the apparent heat capacity formulation method by Bonacina et al.[41].

The enthalpy method was developed to solve the problem of

multidimensional conduction phase changes with density changes by Shamsundar and Sparrow [42]. A refinement of the method which eliminated the numerically induced oscillations and produced accurate predictions was studied by Voller and Cross [43]. An improvement of the apparent heat capacity formulation method was proposed by Goodrich [44] to compensate for the fact that the predicated phase change interface location usually advances in an nonphysical oscillatory fashion and this is accompanied by a distortion of the temperature profile in the region undergoing the phase change.

2.4.4. Isotherm Migration Method. The isotherm migration method by Dix and Cizek [45], proposed that since the moving boundary is essentially an isotherm, its movement can be traced directly as a part of the solution. That was recently extended to solve the two-dimensional and cylindrical solidification problems by Crank and Crowley [46] and Gupta and Kumar [47].

2.4.5. Landau Transformation Method. The novel idea of immobilizing the moving boundary by introducing a new variable to express the location of a particle as the ratio of its distance from the moving boundary to the instantaneous thickness of the original phase, was first devised by Landau [48]. This so-called Landau Transformation method was applied to one-dimensional freezing problems by Beaubouff and Chapman [49] and to two-dimensional or multi-dimensional freezing problems by Duda et al.[50], Saitoh [51], and Sparrow et al.[52]. In order to refine the method, Hsu et al.[53] devised a coordinate transformation system for their control volume formulation. Recently, the method was extended to the problem of heat generation with a one-dimensional phase change by Cheung et al.[54], for melting problems in a rectangular cavity by Ho and Viskanta [55] and Gadgil and Gobin [56], and in a horizontal tube by Ho and Viskanta [57]. More recently, Gupta and Kumar [58] analyzed the multi-dimensional freezing and melting problems on the basis of the coordinate transformation system which transforms only one space variable while others remain unchanged.

2.5 Other Methods

2.5.1. Finite–Element Method. For the numerical methods mentioned above, the fundamental equation will generally be reduced to finite–difference equations by using the rectangular mesh, which may make it difficult to analyze the problem within a more geometrically complex boundary. In this case, the finite element method appears to be helpful. Comini and Del–Giudice [59] formulated the finite element method for non-linear heat conduction coupled with distributed convection by including the enthalpy method. Bonnerot and Jamet [60] proposed a refinement for the finite element method by using space–time finite elements which allow one to move the nodes of the triangulation at each time-step in such a way that the moving boundary is always approximated by a polygonal curve whose vertices are nodes of the triangulation.

2.5.2. Conformal Transformation Method. The conformal transformation technique was used by Siegel [61] to estimate the shape of steady two dimensional freezing on a cold surface in a flowing, warm liquid. Sproston [62] extended the method to a flowing liquid within a cooled rectangular pipe. He applied the method in conjunction with the logarithmic hodograph plane in such a way that neither iteration nor any approximation technique was required. It seems, however, that the transformation technique may make it difficult to apply this method to any configuration except the simple one.

2.5.3. Similarity Rule Method.

The similarity rule method was used for the analysis of multi-dimensional solidification by Shamsundar and Srinivasen [63], who noted that this method could be applied to problems in which sensible heat contribution is much smaller than the latent heat contribution and the volume change based on phase change is ignored. Recently, this method was also extended to the solidification problem with changes in volume by Shamsundar [64].

2.5.4. Monte Carlo and Relaxation Method.

The applications of the probability method and the relaxation method to heat conduction problems with phase changes were tested by Haji-Sheikh and Sparrow [65] and Allen and Severn [65], respectively.

3 FREEZING OF WATER WITHOUT MAIN FLOW

3.1 Freezing of Water in Circular Tube

3.1.1. Freezing Phenomena observed by Gilpin.

The freezing of water pipes in cold regions is one of the unavoidable and undesired natural phenomena. It is well known that a pipe with no flow will be blocked by ice formations to the extent that flow can not be restarted much sooner than had been previously predicted on the basis of an annular growth of ice, for instance, by London and Seban [67]. Gilpin [68,69] clarified that this was because dendritic ice forms when there is no main flow through the pipe during the freezing process and showed that there are two stages of ice blockage; one is a blockage of the pipe cross-section by the formation of dendritic ice as a result of supercooling in the water, the other is a subsequent block of the pipe by the inward growth of a solid annulus of ice from the wall.

Gilpin [68] first observed that when quiescent water in a pipe is cooled, ice growth does not nucleate until the water temperature near the pipe wall has been supercooled $4 \sim 6°C$ below freezing. Fig.1(a) shows that the cross-section of the pipe containing water supercooled to $-3°C$ just before the nucleation of ice growth. Fig.1(b)$\sim$(d) indicate the growth of ice dendrites from the nucleation center near the top of the cylinder which will eventually block the cross-section. The total time required to complete this dendritic ice growth phase was about 30 s. The temperature of the water remaining in the pipe returned to $0°C$, caused by the release of fusion latent heat. Lastly, the growth of an annulus ice began to form from the pipe wall.

3.1.2. Effects of Parameters on the Phenomena.

Gilpin [69] also extensively investigated the effect of cooling rate on the formation of dendritic ice within a quiescent water pipe, and found that the slower the cooling rate the more likely the pipe would be blocked. The effect of free convection on the characteristics of dendritic ice growth in a water pipe was also determined by Gilpin [70], who noted that it was an important factor for supercoolings lower than 2 °C.

3.1.3. Freezing Limit.

An interesting study concerning the freezing limit within a water pipe was recently reported by Sugawara et al.[71], who showed that cavity problems would provide further results if the density and the volume were fixed. Their analysis of water freezing inside a closed pipe indicated that as the water freezes, the pressure on the remaining water increases, and lowering the fusion temperature. Eventually, the system could reach an equilibrium state without the liquid completely freezing. Fig.2 shows the maximum thickness of the ice layer, X_{max}, and the maximum pressure, P_{max}, at the freezing limit, where ΔR is the pipe-wall thickness, D is the inner diameter of the pipe, and T_∞ is the environmental temperature. It seems that X_{max} and P_{max} are not affected by ΔR, but the influences of T_∞ and D on the freezing limit are quite strong.

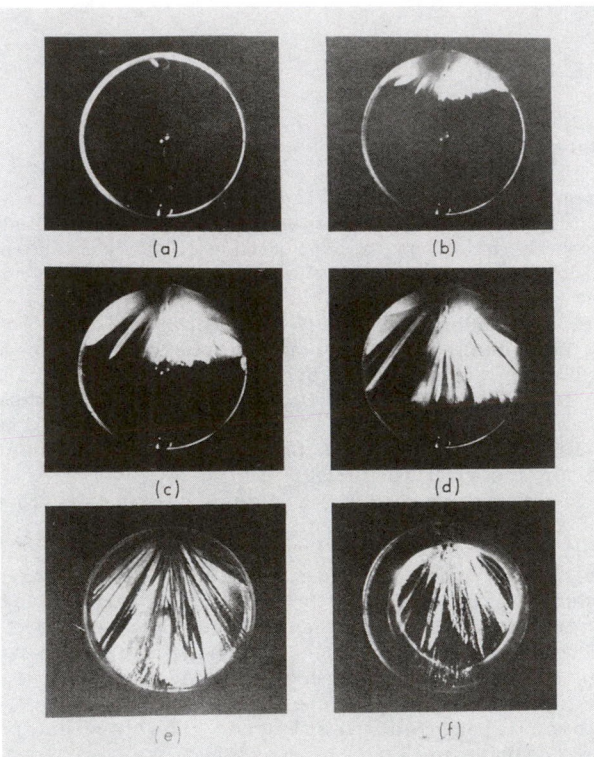

Figure 1 The growth of ice in pipe (a) before nucleation,(b),(c), and (d) during dendritic ice growth, (e) and (f) during annular ice growth [68].

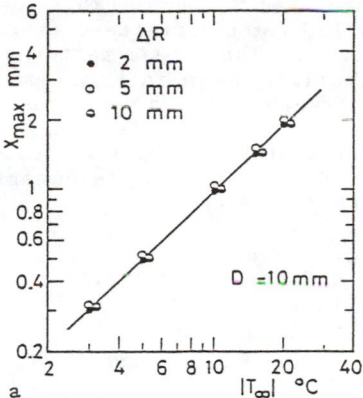 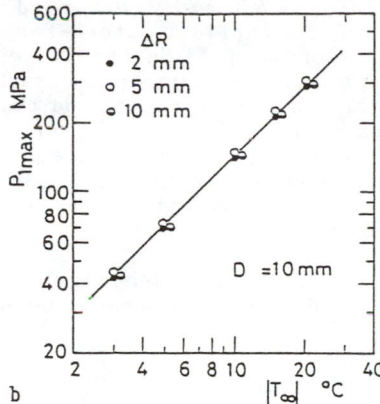

a   b

Figure 2 Maximum thickness of ice layer and maximum pressure increases versus environmental temperature [71].

3.2 Effect of Free Convection on Freezing

Since water has a non-linear relationship between density and temperature, the cooling of water passing through the maximum density point at an initially uniform temperature greater than $4\,°C$ causes a complex free convection, which introduces an additional complication to freezing problems.

3.2.1. Vertical or Horizontal Freezing. The effect of density inversion on free convection along a vertical frozen front adjacent to cold water was studied by Ozaki et al.[72]. They demonstrated both experimentally and analytically that dual flow (up and down) exists in the boundary layer along the vertical ice surface. They noted that the heat-transfer coefficient along a frozen front was affected to a great extent by this peculiar flow motion.

Tankin and Farhadieh [73] used a Mach-Zehnder interferometer to study the role of free convection on the formation of ice when water was cooled from below. They found that the critical Rayleigh number marking the onset of free convection was about 480. Free convection in a horizontal water layer cooled transiently from above to near freezing was numerically investigated for a variety of cooling conditions by Forbes and Cooper [74]. They showed that in all cases for which convection is possible, a $4\,°C$ isothermal line divides the depth of the water into a hydrodynamically unstable region below the isotherm and a stable region of water above the isotherm.

3.2.2. Freezing in Horizontal Tube. Gilpin [75] carried out an experimental study on the cooling of water in a horizontal cylinder, with the wall temperature decreasing at a constant rate through the maximum density point at $4\,°C$. He identified four flow regimes of transient, quasi-steady, inversion, and quasi-steady states which occur during the freezing process. He also studied the phenomena theoretically, using an integral treatment of the boundary layer. Numerical studies on transient free convection of water in a horizontal pipe with a constant cooling

rate through 4 °C were performed by Cheng and Takeuchi [76,77], who employed the explicit Dufort–Fankel method for the finite–difference solution of the full Navier–Stokes equation. The stream patterns and isotherms corresponding to the particular times shown in Fig.3 indicate that the convection motion and thus the free convective heat transfer, behaves in a complex and peculiar manner. Furthermore, Cheng et al.[78] investigated the transient free convection cooling in the same geometry with the convective cooling condition. They presented the numerical solutions both with and without supercooling of water.

3.2.3. Freezing in a Rectangular Cavity. Vasseur and Robillard [79] determined transient free convection flow and heat transfer in a

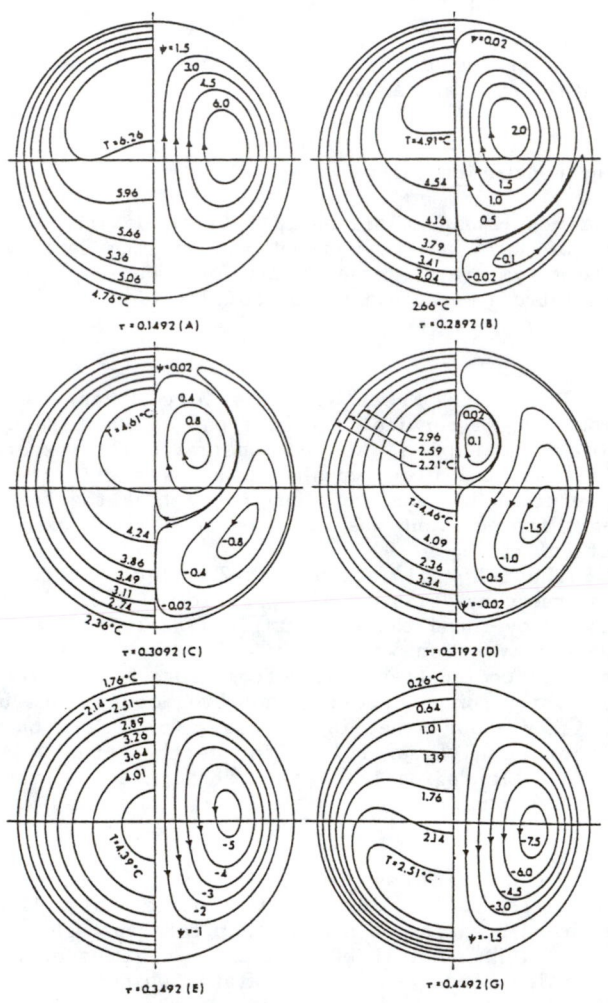

Figure 3 Transient streamline patterns and isotherms for Case 1 [77].

rectangular cavity filled with cold water, where water was assumed to be initially at a uniform temperature T_i ($> 0°C$) and the cavity walls were maintained at a temperature of $0°C$. They numerically solved the system of governing equations for various aspect ratios and initial water temperatures ranging from 4 to $6°C$ and noted that the flow patterns and overall convective heat transfer were markedly influenced by the presence of the density inversion. The transient heat transfer in the same geometry with the supercooling of water was also numerically studied by Robillard and Vasseur [80].

3.3 Freezing of Phase Change Mediums other than Water

In recent years, due to its relevance to phase-change thermal energy storage systems, a number of aspects of the freezing of a phase-change medium other than water have been investigated both experimentally and analytically.

Hsu and Sparrow [81] analytically studied freezing adjacent to a plane wall which was convectively cooled from behind. Two-dimensional freezing along the outside of the tube under convective cooling conditions was also reported by Sparrow and Hsu [82], who used a closed-form analytical solution to start the main numerical solutions. Sparrow et al.[83] conducted an experiment to study the nature of the transition between free-convection-controlled freezing to conduction-controlled-freezing on the outside of a water-cooled vertical tube which was immersed in a phase-change medium (99 percent pure n-eicosane, $T_f = 36.4$ $°C$). Heat-transfer coefficients on the freezing interface between a solid and a liquid along the cooled vertical tube were also measured by Sparrow and Mendes [84] using the same medium. Furthermore, the effect of an inclined sealed cylindrical capsule on the freezing heat-transfer process was experimentally determined by Larson and Sparrow [85]. They found that the inclination of the cylinder played a negligible role in determining the amount of energy released during freezing under any of the studied conditions with initial liquid superheat or cylinder wall supercooling. On the other hand, the inward freezing problems of the spheres with radiation at the sphere surface and with free convection during freezing were numerically investigated by Hill and Kucera [86], and Guenigalt and Poots [87], respectively.

4 FREEZING OF WATER WITH MAIN FLOW

4.1 Freezing with Convection Flow

In freezing problems where a convective flow exists at the phase change interface, two additional complications with the basic non-linearity of the transient phenomenon may take place. The first complication will arise from the fact that the phase change at the interface is equivalent to an effective blowing, since the specific volumes of the solid and liquid phases are different, and may alter the heat-transfer coefficients at the interface between solid and liquid. Merk [88], Yen and Tien [89], Siegel and Savino [90], and Pozvonkov et al.[91] have determined this effect for various free and forced convection geometries. It has been generally found that when the Stefan number St=c $\Delta T/L$ is less than 0.1, where c is the specific heat of liquid, ΔT is the characteristic temperature difference in the liquid, and L is the latent heat of fusion, this effect on the heat-transfer coefficient could be neglected. The second complication is that in convection problems, a mutual interaction takes place among the shape of the interface, the flow field next to it, and the temperature field (namely, the heat transfer from the flow to the interface). The studies of freezing heat-transfer problems with mutual interaction between the flow field and the ice-interface shape have so far been restricted to relatively simple geometry.

4.2 Freezing on Horizontal Flat Plate

Transient freezing of a forced laminar flow over a flat plate was theoretically studied for constant wall temperature by Lapadula and Mueller [92] and Beaubouff and Chapman [49], and for constant heat removal by Miller and Jiji [93], for a heat conducting wall by Epstein [94] and El-Genk and Cronenberg [95]. Savino and Siegel [96,97] carried out analytical and experimental investigations on the transient freezing of a warm liquid flowing over a convectively cooled flat plate.

In these studies, it was generally assumed that the ice layers that formed were thin enough for stream-wise heat conduction within the ice to be neglected and so the effects of the ice layer on the flow over the plate could be neglected. However, the effects produced interesting and complex phenomena. Hirata et al.[98,99] and Gilpin et al.[100] carried out a series of investigation on ice formation along a constant temperature horizontal flat plate in a forced convection flow in laminar, transition, and turbulent regimes. They used a closed loop water tunnel having a test section with dimensions 25.4 cm (width) x 47.5 cm (height) x 213.4 cm (length), with a copper plate 6.35 cm thick, 24.1 cm wide, and 152 cm long installed horizontally.

Fig.4 shows schematically the laminar, transition, and turbulent flow regimes and the shape of the ice surface in each regime. In the laminar regime, Hirata et al.[98] observed that under steady-state conditions the ice layer thickness increased monotonically with an increase in distance from the leading edge of the plate. This is caused by a decrease in the heat-transfer coefficient, which occurred when a laminar boundary layer developed. They also clarified that the steady-state ice profile, namely, the heat-transfer coefficient along the ice-water

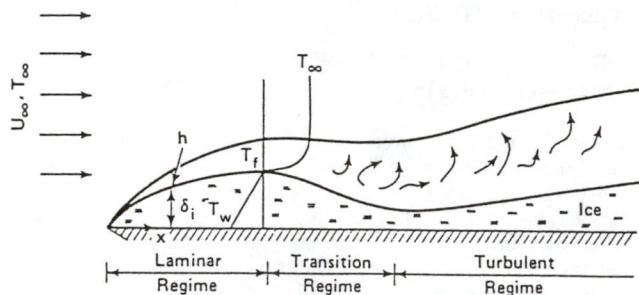

Figure 4 Schematic representation of the ice profile and the flow regimes [99].

interface, could be correlated by a modified Reynolds number of the form Re_x/θ_C^2, where Re_x is the Reynolds number defined as $U_\infty X/\nu$ and θ_C is the cooling temperature ratio defined as ($T_f - T_w)/(T_\infty - T_f)$.

In the transition regime, Hirata et al.[99] found that there were two distinctly different transition modes based on the mutual interactions of the shape of the ice surface, the fluid flow over the surface, and the heat transfer along the surface. One was the "smooth" transition mode where the ice thickness decreased smoothly, and the other was the "step" transition where the ice thickness had a steplike decrease in thickness. Generally, the "smooth" transition took place on thin ice layers, while the "step" transition took place on thick ice layers.

In the turbulent regime, Gilpin et al.[100] observed that for values of θ_C greater than 12, the wavy ice-water interface became unstable due to an interaction between the wavy ice surface and the turbulent flow over it. He noted that the phenomena were quite similar to those observed by Ashton and Kennedy [101] on the bottom of ice sheets in rivers.

4.3 Freezing between Horizontal Parallel Plates

The freezing of liquids in a forced laminar flow between parallel plates was studied analytically by Lee and Zerkle [102] for uniform wall temperature conditions. They used the basic assumption of a parabolic axial velocity profile: that the liquid accelerates due to a constriction of the flow area by the solid phase, and then reduced the problem to Graetz type analysis. Cheng and Wong [103] extended and generalized the analysis of Lee and Zerkle [102] to a uniform convective cooling condition by the eigen-function expansion method using the confluent hypergeometric function. The same geometry with different, uniform external convective cooling at the upper and lower plates was also considered by Cheng and Wong [104]. Shibani and Özisik [105], analyzed the freezing heat-transfer problem under turbulent flow conditions using the matched asymptotic technique. In the turbulent flow prediction, as was done in the laminar cases mentioned above, it was tacitly assumed that a smooth and stable ice-water interface would exist.

Recently, Seki et al.[106-108] carried out a series of experimental investigations on ice-formation phenomena and heat-transfer characteristics for water flow between horizontal parallel plates. The experiments were performed using a test section with a variety of width-height and length-height ratios. When the upper and lower plates were cooled to the same uniform temperature, Seki et al.[106] observed two different types of ice formations. In the range of Reynolds numbers varied from 3.8×10^3 to 3.2×10^4 ; one was the transition ice-formation type, and the other was the smooth ice-formation type, as shown in Fig.5, where Re_H is the Reynolds number defined as $2HU_\infty/\nu_\infty$, and θ_C is the cooling temperature ratio. It was found that the transition ice-formation type took place for $Re_H / \theta_C^{0.741} < 10^4$, while the smooth ice-formation type took place for $Re_H / \theta_C^{0.741} > 10^4$. They also determined the dimensionless ice-transition locations $(X_{tr}/H)_{on}$ and $(X_{tr}/H)_{st}$ at the onset and steady-state conditions as a function of $Re_H^a \theta_C^b$, respectively.

On the other hand, when the upper plate is cooled and maintained at a uniform temperature less than the freezing temperature of water, while the lower plate is kept at a uniform temperature lower than the temperature of water flow, Seki et al.[107] observed that two ice-transition modes occurred based on the different mechanism of the boundary layers

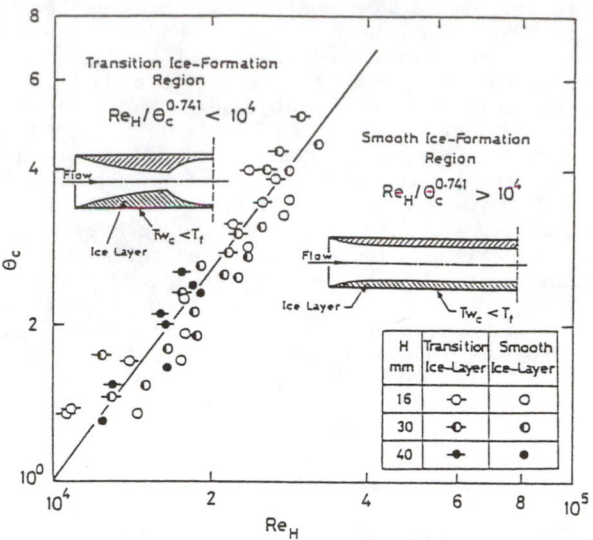

Figure 5 Different region of ice–formation [106].

for Reynolds numbers varying from 3.2×10^2 to 1.5×10^4 and Grashof numbers varying from 3.4×10^3 to 8.9×10^6. They reported that the first ice-transition mode, which may be occurred by a contact between upper and lower thermal boundary layers, occurred for values of $Re_H/(Gr_H \cdot \theta_h)^{0.23}$ less than 1.6×10^{-3} and the second mode, which may be based on an instability of the laminar boundary layer being formed on water-ice interface, occurred for values of $Re_H/(Gr_H \cdot \theta_h)^{0.23}$ greater than 2.3×10^{-3}, where θ_h is a non-dimensional heating temperature defined as $(T_{wh} - T_\infty)/(T_\infty - T_f)$.

The effect of the orifice installed at the leading edge of the horizontal parallel plates on the ice-formation and pressure-loss characteristics were also experimentally studied by Seki et al.[108]. They found that a relative increase in pressure loss resulting from ice formations in the duct was to a great extent decreased in comparison with that for the one without the orifice.

4.4 Freezing in Circular Tube

4.4.1. Laminar Flow. The problem of steady-state freezing and pressure drop in a horizontal circular tube with laminar flow is of great technical importance. This was investigated extensively by Zerkle and Sunderland [109], DesRuisseaux and Zerkle [110], Depew and Zenter [101], Hwang and Sheu [112], and Mulligan and Jones [113] for uniform wall temperatures, and by Lock et al.[114] for external uniform convection cooling. Most of the analysis mentioned above assumed that entrance effects could be neglected, that a parabolic velocity profile was maintained, and that the problem could be reduced to the classical Graetz problem. Özisik and Mulligan [115] extended their analysis to the transient freezing of a liquid flowing in a circular tube by assuming a slug-flow velocity profile and a quasi-steady state heat conduction in

354

the solid layer.

4.4.2. Turbulent Flow. The steady-state freezing of liquids flowing turbulently inside a tube kept at a uniform temperature, lower than the fusion temperature of the liquid, was studied by Genthner [116], Seki et al.[117], Arosa and Howell [118], Shibani and Özisik[119], and Thomason et al.[120]. In these studies, the analysis, most of which were obtained in the framework of Zerkle and Sunderland [109], were compared reasonably with the experimental data of heat transfer and pressure drop. Transient freezing of liquids flowing turbulently inside a circular tube was analytically treated by Cho and Özisik [121]. Furthermore, Epstein et al.[122] investigated the effect of transient solidification along the inner surfaces of a rod bundle upon turbulent axial flow for the case of constant bundle wall temperature and liquid at its fusion temperature, both experimentally and analytically. In the analysis, it was assumed that the rate of reduction of the mean hydraulic diameter of the bundle due to solidification was approximately equal to that of a parallel-plate channel having the same initial hydraulic diameter and temperature.

4.4.3. Freezing Phenomena observed by Gilpin. In the analysis mentioned above, one notes that even for turbulent flow cases, a smooth and stable ice-water interface was assumed to exist. Recently, Gilpin [123-125] extensively observed the ice-formation phenomena in a cold pipe containing flows. Fig.6 presents sketches of the three basic modes of ice formation in water pipes with a main flow. The first growth mode, or "annular" ice growth mode (Fig.6(a)) is the simplest, and was observed for the case of relatively high main flow. This mode is very similar to that analytically predicted by previous investigators. In this mode all the ice growth appears as a thin shell attached to the inside of the

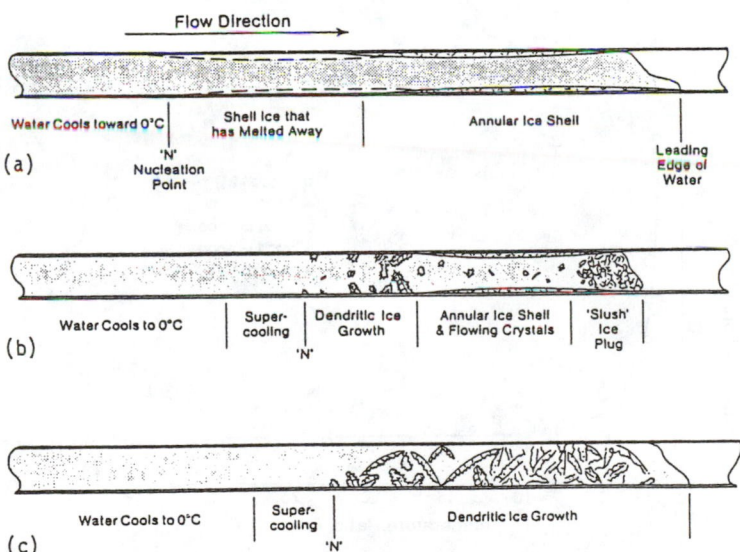

Figure 6 Sketches of the three ice formation mode [124].

pipe. The second growth mode, or the mixed growth mode (Fig.6(b)), was observed for the case of moderate main flow. During the progress of the water through the pipe there is zone of dendritic ice growth and a region in which an annular ice shell with some "frazil" ice particles or broken dendrites exists in the water. Behind the advancing front there is a plug of "slush" ice. The third growth mode (Fig.6(c)) was observed for a very low main flow, and was capable of causing a pipe blockage. In this growth mode, dendritic ice crystals were seen to grow on the pipe wall at one or more sites along the length of the pipe. This dendritic ice growth spread rapidly within the pipe until a large section was filled with a dense matrix of ice crystals interspersed in the water. The growth of these crystals was reported to be similar in all aspects to the growth of dendritic ice in supercooled quiescent water (see Fig.1).

In the first ice-growth mode, Gilpin [125] reported an excellent visual observation of the ice growth in a pipe containing a flow near the transition Reynolds number. He found that the final steady-state ice did not exhibit a uniformly tapered flow passage, as was predicted by most previous theories, but that a flow passage with a cyclic variation occurred in the cross-section along the length of the tube. Fig.7 demonstrates the steady-state ice profiles for various cooling temperature ratios θ_C. For $\theta_C = 11.8$, there were 5 cycles of expansion and contraction of the flow passage (which was called an "ice-band" structure by Gilpin [123]) along the length of the test section. The characteristics of the ice-band structure for Reynolds numbers ranging from 370 to 1.4×10^4 are shown in Fig.8, where S is the distance between separation points on consecutive ice-band and D is the inner tube diameter. In Fig.8, ice-band spacing (wave length) decreases sharply with θ_C for small values of θ_C and then approaches the constant value of about 6 for $\theta_C > 10$.

Periodic oscillations of overall pressure drop, which were observed using copper tubes 116.0 cm in length and 1.45 cm in diameter and which seemed to be really related to the phenomena mentioned above, were reported by Thomason and Mulligan [126].

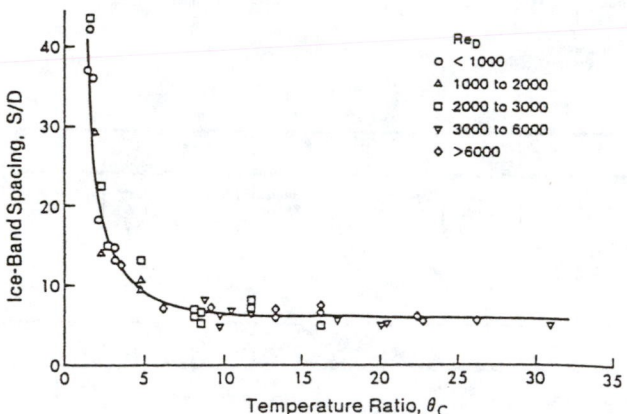

Figure 7 The steady-state ice band spacing in a pipe [125].

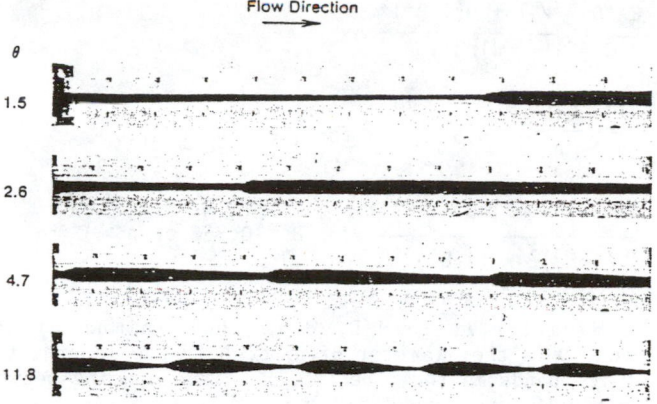

Figure 8 Steady-state ice profiles for successively lower temperatures (that is larger values of θ). Re_D=3025 [125].

4.4.4. Freeze–Off Condition.

A fundamental understanding of the mechanism and the condition for the onset of freeze-off in a pipe are of great importance not only for designing utility systems for northern climates, but also are useful for fast nuclear reactor safety analysis. Epstein et al.[127] analytically and experimentally studied the solidification of a liquid at its fusion temperature as it penetrates an initially empty tube cooled on the outside. Their analytical solutions were later numerically verified by Epstein and Hauser [128]. Sampson and Gibson [129] proposed a mathematical model for the criteria that predicts the conditions under which freeze-off will take place in a pipe containing a laminar flow of liquid. The same problem under turbulent flow conditions was also considered by Sampson and Gibson [130]. Recently, the effect of heat conduction and capacity within the wall of a channel or a tube containing a flow of liquid on transient solidification of the liquid was analytically investigated by Schneider [131], who predicted that the solidification layer eventually disappeared locally if the wall temperature increased to values above the fusion temperature of the liquid, based on the heat transferred from the solidified layer to the wall.

More recently, Hirata and Isihara [132] analytically and experimentally studied the conditions for the onset of freeze-off in a pipe containing a flow of water using copper (16.6 and 19.9 cm in inner diameter) and steel (36 mm in inner diameter) tubes 697 mm in length. They clarified that the pipe freeze-off first took place at the contraction region of the ice-bands and the conditions were given as a relationship between the cooling temperature ratio θ_c and the modified Reynolds number Re_p based on total pressure. In Fig.9, the data are plotted in a logarithmic form using θ_c as the ordinate and Re_p as the abscissa. The solid and open points denote a freeze-off and a steady-state ice-bands, respectively. It appears that the effect of L/D on freeze-off conditions may be small since the factor of pipe length is implicitly contained in θ_c, as will be shown later. In the figure, the predicted results for freeze-off conditions were obtained from the following equations by eliminating n and B

$$Re_p = -\frac{2\lambda^*\theta_c}{0.0045}\frac{\sqrt{1 + n(1-B^2)^2}}{B\ln B} \tag{1}$$

$$\theta_c = \left[\frac{1}{75}\left(\frac{L/D}{n} - 4.5\right)\right]^{-2/3} \quad \text{for } 1.6 \leq \theta_c \leq 31.0 \tag{2}$$

$$n = -\frac{1 + \ln B}{(1-B^2)^2 + (1-B^4)\ln B} \quad \text{for } B > 0.368 \text{ or } n > 0 \tag{3}$$

where B is the diameter ratio d/D, d is the minimum diameter at contraction region, D is the diameter of pipe, L is the length of pipe, n is the number of ice-bands in the pipe L/S, S is the spacing between ice-bands, and λ^* is the thermal conductivity ratio λ_i/λ_w.

4.5 Freezing around Cylinder

The problems of steady-state ice formations around a cylinder cooled below the fusion temperature ($T_w < 0\,°C$) and immersed horizontally in flowing water at a uniform temperature ($T_\infty > 0\,°C$) were studied experimentally by Carlson [133] and Okada et al.[134] for Reynolds numbers lower than 2×10^3 and by Cheng et al.[135] for higher Reynolds numbers up to 8×10^4. For cooling temperature ratios θ_c ranging from 6.3 to 75.8, Cheng et al.[135] determined the local heat-transfer coefficient at the ice-water interface from the measured profile using a series solution of the Laplace equation within the ice. Fig.10 shows the estimated local heat-transfer coefficients along with the measured

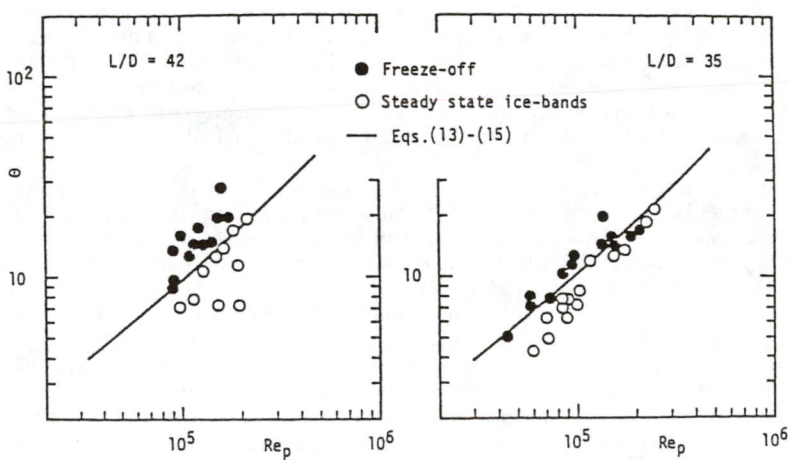

Figure 9 A comparison between predicted and measured freeze-off conditions in a pipe [132].

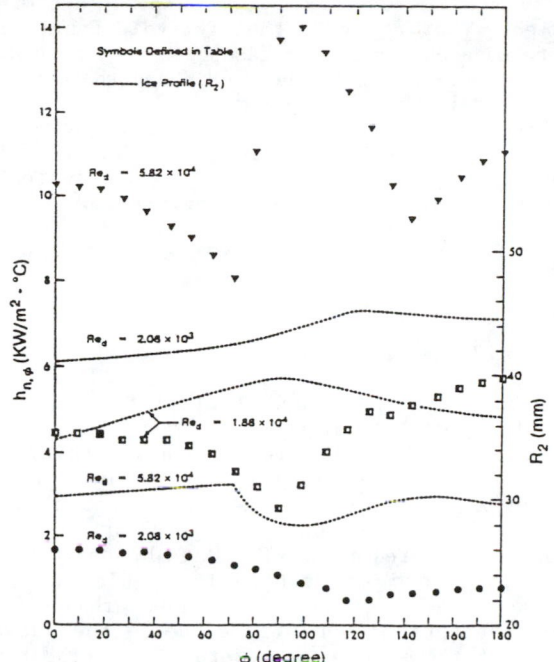

Figure 10 Variation around the cylinder of the thickness of the ice layer and the local heat transfer coefficient that was calculated from it [135].

typical ice profiles for low, moderate, and high Reynolds numbers, respectively. At a lower Reynolds number (Re_d=2.08 x 10^3), a laminar boundary layer appears to develop from the front stagnation point and to separate at $\phi \approx 120$ deg. The laminar separation seems to be clearly evident on the ice surface because it results in an apparently discontinuous change in the slope of the ice layer. On the other hand, at a higher Reynolds number (Re_d=5.82 x 10^4) a laminar boundary layer exists from ϕ = 0 to 70 deg. At $\phi \simeq 70$ deg a laminar separation occurs due to a separation bubble from $\phi \simeq 70$ to 100 deg. At $\phi \simeq 100$ deg, the flow seems to reattach as a turbulent boundary layer, developing along the ice until $\phi \simeq 140$ deg, where it separates, as shown by the ice profile for Re_d=5.82 x 10^4.

4.6 Freezing in a Two–Phase Mixture

The prediction of freezing rates in a two–phase mixture is technically of interest in safety investigations for the fast breeder reactor, in the design of air bubblers as ice control devices for cold–water ports and harbors, and in preconcentration and dehydration of sludges. Petrie et al.[136] experimentally studied this problem by directly measuring the growth of a vertical ice layer in a water–nitrogen gas mixture, where nitrogen gas bubbles were formed at the bottom of a column (square cross section of 50.8 mm by 50.8 mm) where a cooled copper plate of 76.2 mm in length was installed. They reported that for void fractions up to 90 percent, the presence of a discontinuous gas phase in a saturated flowing liquid did not affect the freezing of the liquid and that the ice surface remained smooth. The void in the two-phase mixture was not trapped in the ice. This contradicts previous results, where the gas evolved due to difference in solubility between the liquid and solid. Seki et al. [137] conducted a fundamental investigation on the preconcentration of sludge using a water-air-kaolin mixture as the freezing medium.

5 MELTING OF ICE

5.1 Melting of Horizontal Ice Layer

5.1.1 Melting Forms.
It is well known that at the beginning of the melting process, estimating the thickness of the horizontal melt layer of ice may be not so difficult if the layer is quite thin. However, a somewhat complicated situation will arise as the melt layer becomes thick. In a horizontal melt layer of ice heated from above or below, four typical situations will occur due to a peculiar characteristic of water having a density inversion at 4°C. When heated from above, the fluid layer consists of both a potentially stable and potentially unstable layer if the surface temperature is higher than 4°C; while the entire fluid layer is potentially unstable due to the buoyancy force existing in the layer if the surface temperature is in the range of 0 to 4°C, as shown in Figs.11(b) and (a), respectively. On the other hand, when heated from below, there will be an unstable liquid layer only when the lower boundary temperature is greater than 4 °C, as shown in Fig.11(d).

5.1.2 Melting with Free Liquid Surfaces.
When a horizontal melt layer of lake or river ice is heated from above, for instance by solar energy, it corresponds to Figs.11(a) and 11(b). In the case of Fig.11(b), the free surface effects of both the hydrodynamic boundary condition, including the surface tension and the thermal boundary condition, are only felt indirectly by the potentially unstable layer. If the thickness of the stable layer decreases relative to that of the unstable layer, it is obvious that these effects will become more significant. This then corresponds to an increased effect of surface boundary conditions on the onset of free convection and on heat transfer in the unstable layer. Thus the problems of the onset of free convection and the resulting heat-transfer characteristic in such a system is quite complicated.

Sugawara et al.[138] carried out an experimental investigation concerning the system mentioned above, and reported that the critical value of Rayleigh number Ra for the case of Fig.11(b) was about 500 for T_2 ranging from 9 to 65°C, where Ra was defined as below :

$$Ra = \frac{g|\beta|(T_m - T_1)h^3}{\nu\kappa} \qquad T_2 > 4°C \qquad (4)$$

$$Ra = \frac{g|\beta|(T_2 - T_1)H^3}{\nu\kappa} \qquad T_2 \leq 4°C \qquad (5)$$

Wu and Cheng [139] and Seki et al.[140] analytically determined the critical Rayleigh number marking the onset of free convection in a horizontal melt water layer with a free surface. A graph comparing the experimental and theoretical critical Rayleigh number Ra_c is indicated in Fig.12, where Ma is the Marangoni number representing a measure of a surface-tension force at the free surface and is defined as :

$$Ma = \frac{\sigma_0 \Phi (T_2 - T_1) H}{\rho \nu \kappa} \tag{6}$$

where σ_0 is a surface-tension force of water at $0°C$ and Φ is a constant. For $T_2 > 8°C$, the surface-tension at the free surface has little effect on the potentially unstable layer because of the considerable thickness of the upper stable layer. Therefore, in the figure the values of Ra_c are nearly uniform in this range (note the fact that the theoretical values without surface-tension force (Ma=0) are in good agreement to those with Ma≠0). However, as is clearly shown in the figure, the critical Rayleigh number for $T_2 < 8°C$ increases and takes the value of $Ra_c = 3000$ for $T_2 = 4°C$. It is generally known that the surface tension may motivate the onset of free convection within an unstable layer, contrary to the current results. The reason for this fact will be considered in the following way. Fig.13 shows a flow pattern at the onset of free convection in a melt layer and the effect of surface-

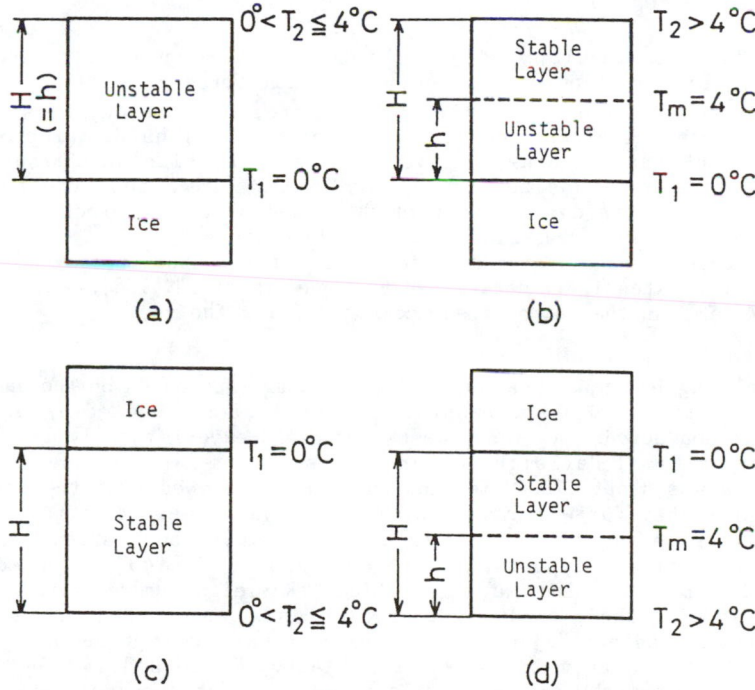

Figure 11 Four situations of horizontal melt water layer [152].

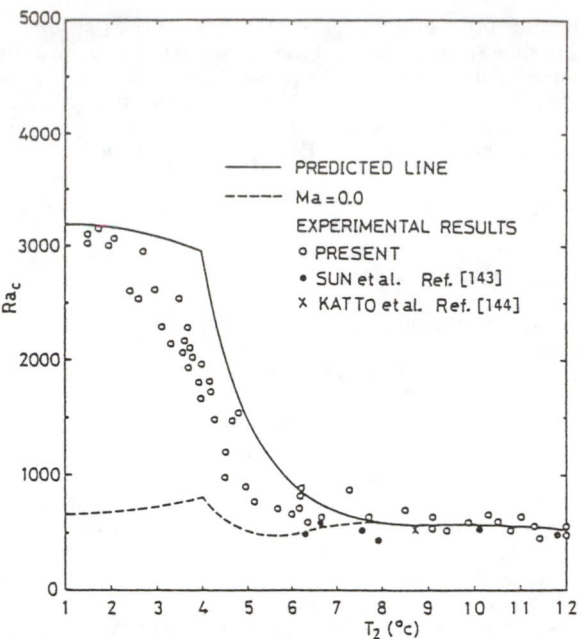

Figure 12 Comparison of experimental and analytical results [140].

tension force on the liquid flow. For the normal liquid without density inversion,the direction of the surface-tension force is similar to that of the liquid flow based on the onset of free convection, thus it may promote the onset of free convection. On the other hand, for the melt water layer with the density inversion,the direction of the surface-tension force is opposite that of the liquid flow, thus the surface tension may be considered to retard the onset of free convection, which corresponds to an increase in Ra_c with a decease in T_2.

The heat-transfer characteristics of the melt water layer with a free surface were extensively determined by Sugawara et al.[138], who used an infrared lamp as the energy resource for melting the ice.

5.1.3 Melting by Upper Heating Wall. A number of studies on the onset of free convection and heat transfer in a melt of ice heated by an upper wall were conducted. Boger and Westwater [141] experimentally determined that the critical Rayleigh number corresponding to the onset of free convection was about 1700. Yen and Galea [142] showed that the critical Rayleigh number for water was not a single value, as it is for common fluids having a monotonic relationship of density, but varied with the temperature of the upper wall surface. Sun et al.[143] reported that their experimental values of the critical Rayleigh number were in good agreement with their analytical predictions. As pointed out earlier, Tankin and Farhadieh [73] found that the critical Rayleigh number for the case of freezing from below was about 480. Katto and Iwanaga [144] demonstrated that the critical Rayleigh number that they and Seki et al. [145] obtained, agreed well with that given by a modification of the analytical results obtained by Sun et al.[143]. Seki et al.[145] found both experimentally and numerically the onset of free convection and the free convection heat transfer were markedly affected by the temperature of upper wall for $T_2 < 8\,°C$, unlike the results obtained for common fluids without a density inversion. Merker et al.[146,147] determined the critical Rayleigh number by defining a modified Rayleigh number which used the height of the complete fluid layer rather than the height of the unstable layer.

Fig.14 shows a graph comparing the experimental and analytical values of Ra_c. In the figure, a full line denotes the analytical results predicted under the physical system, while a dotted line shows the ones corresponding to that with an upper free surface (without surface tension, i.e. Ma=0). The appearance of refraction in the predicted Ra_c at $4\,°C$ results from the mathematically different definition of Ra_c between $T_2 \le 4\,°C$ and $T_2 > 4\,°C$. For $T_2 > 8\,°C$, the upper surface condition may only indirectly effect the potentially unstable layer because there exists a stable layer between upper wall and the unstable layer.
Thus, it is reasonable that the Ra_c values are nearly constant for $T_2 > 8\,°C$. On the other hand, the Ra_c increases with decreasing T_2 for $T_2 < 8\,°C$. This may be based on the fact that as the temperature of the upper wall decreases, the thickness of the stable layer decreases, which corresponds to the increased effect of the upper boundary condition on the onset of free convection within an unstable layer. Note that the value of Ra_c at $T_2 = 4\,°C$ is approximately 1700.

Yen [148] experimented with the melting of ice by an upper wall heated ranging from 4.60 to $25.10\,°C$ and found that the heat flux at the ice-water interface was about 700 W/m^2 regardless of the upper wall

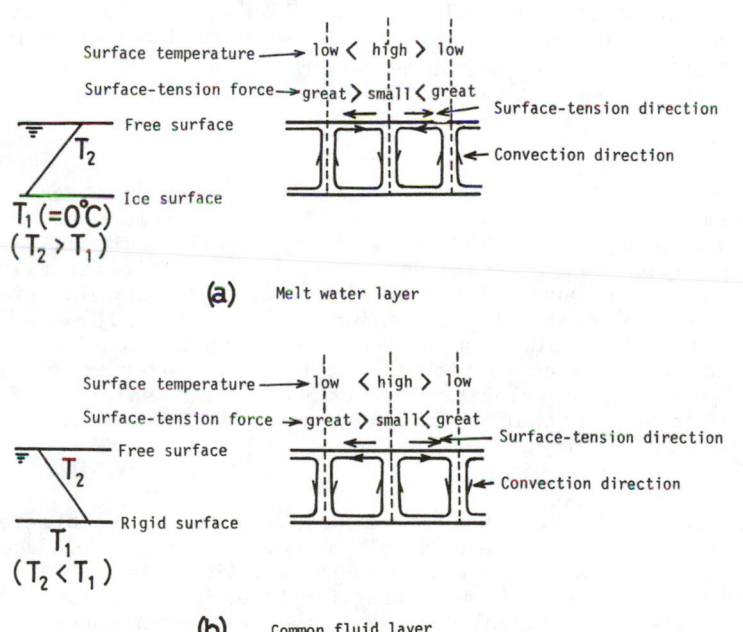

(a)   Melt water layer

(b)   Common fluid layer

Figure 13 Effect of surface-tension force [152].

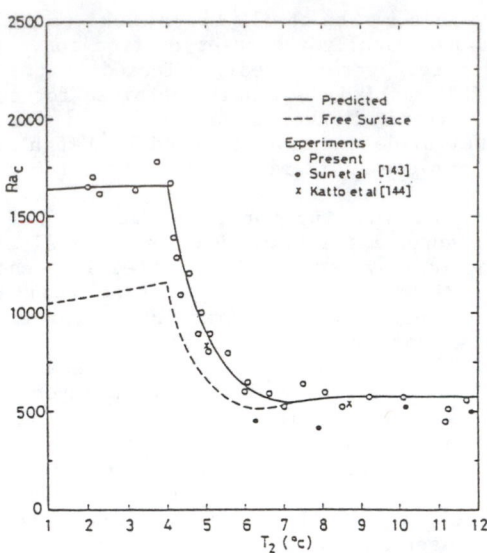

Figure 14 Critical Rayleigh number, comparison of experimental and analytical results [145].

temperature. Tien et al.[149] analytically estimated the heat-transfer rate using Turcotte's convection model. Seki et al.[145] numerically determined the heat-transfer rate using a two-dimensional cell-structure model. Fig.15 shows a comparison between the experimental and analytical results, where the Nusselt number is defined as follows:

$$Nu = \alpha \cdot h/\lambda \tag{7}$$

For $T_2 \gtrsim 9\,^\circ C$, the Nu value at a certain Ra exhibits the same value regardless of the T_2 value because the free convection heat transfer induced by buoyancy is independent of the upper wall condition. However, the Nu value decreases with a decrease in T_2 , and eventually show characteristics similar to common fluids without a density inversion for $T_2 \lesssim 4\,^\circ C$ when the stable layer disappears. Adrian [150] measured the vertical velocity and simultaneous temperature fluctuation in turbulent, statistically steady convection in water flowing over an ice surface using a Laser-Doppler velocimeter with a frequency-shifted reference beam. He reported that there were over 80 percent mean-square fluctuations in velocity and temperature.

5.1.4 Melting by Heating Wall from Below. Boger [141] observed an oscillating water-ice interface in melting experiments and concluded that the critical Reynolds number was approximately 1700. Yen [151] observed a regular cell structure in a melting ice layer heated from below and pointed out that the critical Reynolds number was not a single constant, but depended upon the lower heated-wall temperature. Sun et al.[143] also presented a stability-diagram valid in temperature range of C to 35 $^\circ C$ and reported that the experimental values for the onset of free convection for T_2 ranged from 7.7 to 25.2 $^\circ C$,in good agreement with their analytical results.

A comparison between the experimental results by Yen [151] and Sun et al.[143] and the analytical predictions by Fukusako et al.[152] is shown in Fig.16, where Ra" is a modified Rayleigh number and A is the ratio of the unstable layer thickness to the whole liquid layer and they are defined below.

$$Ra^{''} = g\beta(T_2-T_m)h^3/(\nu\kappa) \qquad (8)$$

$$A = (T_2-T_m)/(T_2-T_1) \qquad (9)$$

It appears that for $A \lesssim 0.6$ the Ra_c value may approximately approach the uniform value of 500. This suggests that the effect of the ice surface on the onset of free convection decreases with a decrease in T_2. On the other hand, as A increases (corresponding to an increase in T_2) the flow induced by buoyancy within an unstable layer tends to extend into the stable layer, and eventually the flow becomes influenced by the shear stress at the ice surface. Thus, the critical Rayleigh number increases for $A \gtrsim 0.6$. When A approaches 1.0 the Ra_c seems to approach the value of 1700, the same as for common fluid without density inversion.

A number of studies on heat transfer in a melted water layer heated from below were reported by Yen and Tien [148,149,153–155]. Fig.17 shows the experimental results by Tien et al.[149] along with the ones by Schmidt et al.[156] for common fluids. It is clearly shown in the figure that when the Ra" is less than about 10^4, the Nu values for $T_1 = 0\,°C$ are greater than those for $T_1 = 4\,°C$ and $8\,°C$, thus physically indicating that the fluid flow in a melted water layer having a stable layer between an unstable layer and an ice surface could be easily performed. However, when the Ra" increases (corresponding to an increase in T_2) the heat-transfer rate approaches those for common fluids, since the flow circulation passes through the stable layer and eventually reaches the ice surface.

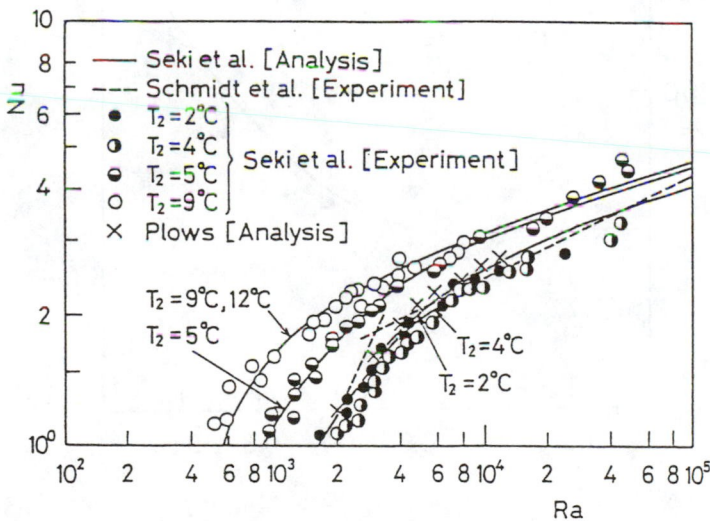

Figure 15 Heat transfer in melted water layer heated from upper wall [145].

5.1.5 Melting by a Radiant Heat Source. The melting problem of a horizontal ice layer sticking to a substrate was studied experimentally and analytically by Gilpin et al.[157], Seki et al.[158,159], and Cho and Ozisik [160] using short wave radiant energy. Gilpin et al.[157] examined the internal absorption and scattering in radiative heating of cloudy and clear ices. Seki et al.[158,159] observed the phenomenon of "back" melting, caused by radiant energy penetrating through the ice layer. They concluded that the behavior of radiation transfer in a cloudy ice layer depended greatly on the density of the containing air bubbles, which may cause scattering radiation.

Fig.18 shows a typical pattern of radiative melting in a horizontal clear ice layer sticking to insulation with a radiatively black surface. It is clearly seen in the figure that there exists both an upper melted layer and a lower melted layer based on "back" melting. Recently, Chan et al.[161] proposed a melting model to account for two-phase zones induced by internal thermal radiation between the liquid layer and the pure solid layers. They did not, however, take the "back" melting phenomenon into consideration.

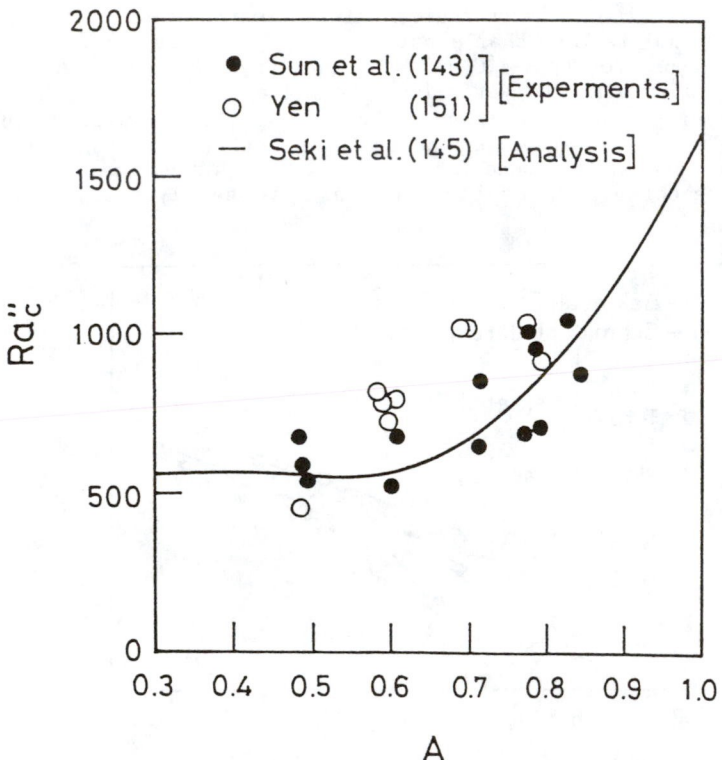

Figure 16 Criterion of onset of free convection in a melt water layer heated from below [152].

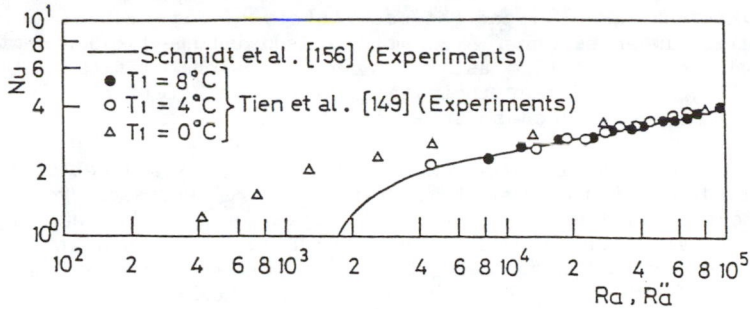

Figure 17 Heat transfer in a melt water layer heated from below [152].

5.2 Melting of a Vertical Ice Layer

5.2.1 Melting in Cold Water. An experimental study of melting heat transfer along a vertical ice cylinder immersed in cold water was carried out by Tkachev [162], who found a minimum heat-transfer coefficient at $T_\infty = 5.5\,^\circ C$. Vanier and Tien [163] investigated analytically and experimentally the effect of both density inversion and melting of free convective heat transfer along a vertical surface and claimed that the effect of melting on the heat-transfer rate was small. They also observed in detail the directional tendencies of flow across the boundary layer region formed adjacent to the vertical surface, in ambient water at temperatures around maximum density, and noted that a dual flow existed near the wall for $T_w = 0\,^\circ C$. Bendell and Gebhart [164] determined the heat-transfer coefficients from the melting rate of vertical ice sheets in ambient water at temperatures from 2.2 to 25.2 $^\circ C$. They reported that a minimum Nusselt number occurred at $T_\infty = 5.6\,^\circ C$ and the experimental results were in good agreement with those obtained from the boundary-layer computations by Gebhart and Mollendorf [165]. Measurements of velocity distributions near a vertical ice surface melting into distilled water at temperatures ranging from 2 to 7 $^\circ C$ were carried out by Wilson and Vyas [165], who observed the steady-state motion upward for $T_\infty < 4.7$

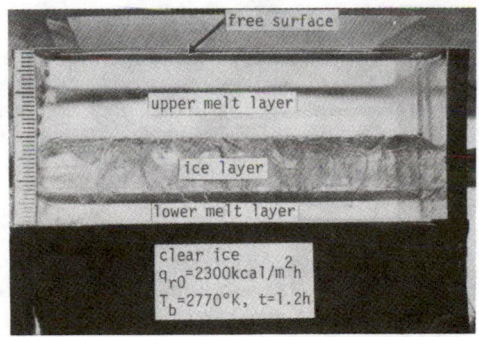

Figure 18 Visual observation of radiative melting [158].

°C, downward for $T_\infty > 7°C$, and the oscillatory bi-directional flow for intermediate temperatures. Recently, Wilson and Lee [167] numerically considered the same problem using the two-dimensional finite difference method. They reported that their analysis did not yield stable results for $4.5°C < T_\infty < 5.7°C$ because this flow regime might be transitory in nature. A visualization study of the flow adjacent to a vertical ice surface melting in cold water was performed by Carey and Gebhart [168], who observed that as the ambient water temperature increased from 3.9 to 8.4 °C, the regimes of upward, locally bi-directional, and downward flows took place. They suggested that the bi-directional flow might result from the reversal of part or all of the upward wake above the top of the ice surface. They also claimed that the experimentally obtained velocity profile and heat transfer for $T_\infty < 4.05°C$ and $T_\infty > 5.9°C$ were in good agreement with analytical results by Carey et al.[169].

5.2.2 Heat Transfer from a Heated Wall in Cold Water. Heat-Transfer problems for a heated wall immersed in cold water at its maximum density were studied by a number of investigators. Ede [170] obtained heat-transfer measurements along an electrically heated vertical wall immersed in cold water. An experimental and analytical study on the free convection heat transfer from a vertical isothermal flat plate to ambient cold water at maximum density was carried out by Schechter and Isbin [171], who reported that bi-directional flow occurred in the boundary layer. Goren [172] analyzed free convection along a heated vertical surface immersed in ambient water at 4°C using an analog computer. Carey et al.[169] studied numerically the effect of the density-inversion presence on the laminar flows driven by thermal transport to or from a vertical isothermal wall in cold water. They concluded that an incipient local flow reversal occurred both outside of the flow region for R (=$(T_m - T_\infty)/(T_w - T_\infty)$)=0.152 and at the surface for R=0.326. Recently, Pop and Raptis [173] analytically examined the transient free convection along a doubly infinite vertical porous plate in water at 4°C, which was subject to a sudden temperature change, by suitably transforming the distance and velocity variables into a set of self-similar equation. More recently, a similar transient free convection flow adjacent to a suddenly heated vertical wall in cold water at its maximum density was numerically studied by Joshi and Gebhart [174].

5.2.3 Melting with Vapor Condensation. When a vertical ice surface is exposed to hot air containing some amount of vapor, melting of the ice will occur, caused by the heat released due to vapor condensation on the ice surface and the heat transferred from the ambient hot air. Thus, the heat-transfer situation becomes somewhat complicated because of simultaneous vapor condensation and melting of the condensing surface. Simultaneous vapor condensation-ice melting on a vertical surface was first analyzed by Tien and Yen [175], who assumed that the condensing vapor and melting solid were of same material. They used an integral technique to obtain closed form solutions for the local and average heat-transfer coefficients for a high Prandtl number liquid. The effect of non-condensable gases on film condensation-melting heat transfer along a vertical ice slab was also considered by Yen et al.[176]. Epstein and Cho [177] studied analytically steady-state vapor condensation onto a vertical melting surface, accounting for the effects of both liquid film inertia and shear at the condensing vapor-liquid film interface. They indicated that the parameters governing the condensing-melting process were Pr, $N_1 = c_\ell(T_r - T_f)/L_v$, and $N_2 = c_\ell(T_r - T_f)/[L_f + c_s(T_f - T_w)]$, and when $N_1 \ll N_2$ the vapor condensation rate was negligibly lower than the melting rate. Laminar film condensation of a vapor flowing over a horizontal melting surface was analytically studied by Cho and Epstein [178], who used an approximate treatment of the shear stress at the vapor-liquid interface and obtained the similarity solutions.

Recently, Taghavi-Tafreshi and Dhir [179] conducted an analytical and experimental study on laminar film condensation on a vertical melting surface when melt and condensate were two immiscible substances. The change in shape of an initially vertical surface undergoing vapor condensation-driven melting was also determined both analytically and experimentally by Taghavi-Tafreshi and Dhir [180], who reported that the steady-state shape was the one which yielded a constant melting rate on the entire surface and that the total melting rate for the steady-state shape was about 35 percent more than for the vertical surface.

5.2.4 Melting by A Radiant Heat Source.
Radiative melting of a vertical ice layer adhering to a substrate was studied both experimentally and analytically by Seki et al.[181], who used a halogen lamp possessing a large fraction of short wave beams or a nichrome heater having a large fraction of long wave beams as the radiant heat source, respectively. They found that heating by short wave radiation produced a peculiar melting pattern on the rough melting surface, caused by the internal melting at the grain boundary of the ice surface, as shown in Fig.19. But for long wave radiation, the melting surface remained smooth. They also reported that the melting rate of the ice layer could be reasonably estimated by an analytical model based on the band model of the extinction coefficient or the absorption coefficient(see Fig.20).

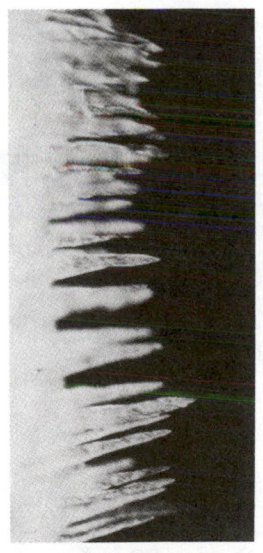

Figure 19 Side view of strongly rough melting-surface for cloudy ice [181].

5.3 Melting of Spherical or Cylindrical Ice

Merk [182] used an integral method to analyse the effect of maximum water density on heat transfer between a melting sphere of ice and the adjacent water. He calculated the local heat transfer around the spherical and predicted that for about $T_\infty = 5\,°C$, the Nusselt number was at a minimum and the direction of the flows changed from upward at lower temperatures to downward at higher temperatures. Dumore et al.[183] did an experiment with melting ice spheres and found that the flow was upward for $T_\infty < 4.8$ °C and downward for $T_\infty > 4.8\,°C$ and that a convective inversion took place at $T_\infty = 4.8\,°C$. Measurements of free convective heat transfer for an ice sphere for T_∞ ranging from 0 to 10 °C were carried out by Schenk and Schenkels [184], who found that a dual flow (upward flow near the wall and downward flow at some distance away from the wall) existed for T_∞ ranging from 4 to 6 °C. Vanier and Tien [185] also conducted an experimental study on the melting characteristics of spherical ice in water at temperatures ranging from 0 to 20 °C and found that a convective inversion occurred at 5.35 °C. Saitoh [186] studied both numerically and experimentally the problem of free convection heat transfer of a horizontal ice cylinder immersed in water at its maximum density and reported that a minimum Nusselt number occurred above $T_\infty = 6\,°C$.

5.4 Effect of Free Convection on Melting in Cavity

5.4.1 Free Convection in Rectangular Cavity. Free convective heat-

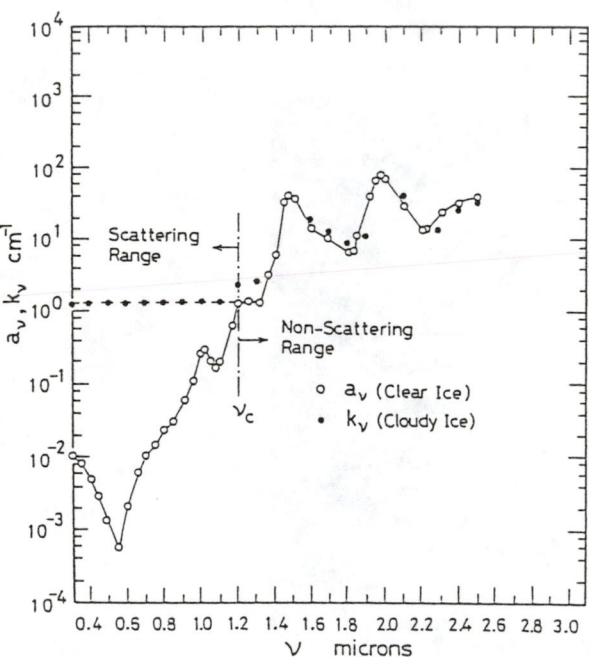

Figure 20 Extinction and absorption coefficient, k_ν, a_ν [181].

transfer problems with water at its maximum density in a confined cavity is of great importance since it is closely related to the melting of an ice layer within various confined vessels. Watson [187] did a numerical analysis for the free convective heat transfer with the density inversion of water in a confined rectangular cavity for the special case of aspect ratio H/W = 1 (H: height of the cavity, W: width of the cavity) and suggested that the effect of physical-property variations, such as viscosity or thermal conductivity, on the flow field and convection heat transfer could not be neglected. His numerical computations, however, were restricted to a rectangular vessel having a small W due to the convergence problem of their solutions. The effect of maximum density on convective heat transfer of water in a rectangular vessel, in which the temperature of one wall was maintained at $0\,^{\circ}C$ and that of the opposing wall varied from 1 to 12 $^{\circ}C$, was determined both experimentally and analytically by Seki et al.[188]. In their analysis, all the thermophysical properties were assumed to depend upon temperature. They demonstrated that the various flow patterns induced by the density inversion were influenced to a great extent by the convection heat transfer, and the mean Nusselt number was at a minimum at $T_h = 8\,^{\circ}C$. In Fig.21, typical flow patterns under various T_h are shown together with the predicted patterns of stream lines. In these experiments, it was reported that the flow pattern was stable and two-dimensional flow was attained.

Recently, Inaba and Fukuda [189] studied experimentally the effect of inclination on the heat transfer with density inversion for the same geometry mentioned above. They observed that the flow patterns in the cavity were changed not only by the effect of the density inversion but also by the inclination angle of the cavity and suggested that two counter eddies might disturb the heat transfer within the cavity.

5.4.2 Free Convection in Annular Cavity. Seki et al.[190] carried out an experimental study of free convection heat transfer with the density inversion of water between two horizontal concentric cylinders with various diameter ratios, keeping the inner cylinder surface at the fusion temperature of water and maintaining the outer cylinder surface at various temperatures above the fusion temperature. They found that the distributions of the local heat-transfer coefficients along the inner and outer cylinder surfaces were changed to a great extent, on the basis of

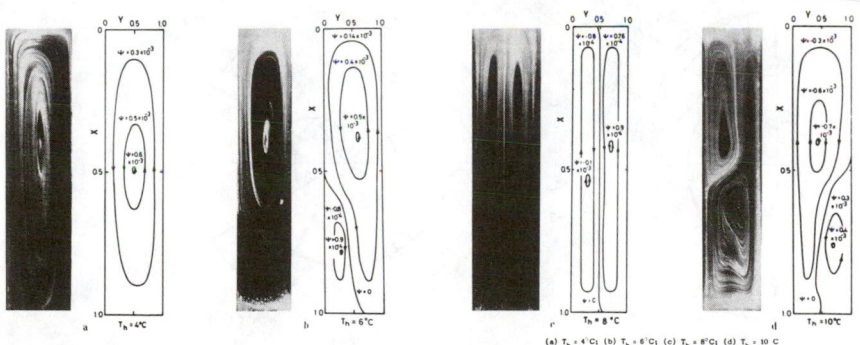

Figure 21 A visual photograph and predicted stream-lines of the flow pattern for H/W=5, W=20 mm [188].

the complicated patterns of standing eddies due to the density inversion of water within the annulus. Seki et al.[191] did also analysis for the same geometry, in which all the thermophysical properties were assumed to be a function of temperature, respectively. Fig.22 shows a typical pattern of the calculated streamline and a corresponding local Nusselt number on the inner and outer cylinders for $T_h = 6\,°C$, along with the experimental data by [190]. The discrepancies between the analytical heat transfer results and the experimental data, particularly over the surface of the inner cylinder, may be due to the fact that the local heat flux for the outer cylinder was evaluated from the electric input to each main heater (consisting of twelve independently controllable heaters with individual guard heaters), while the local heat flux for the inner cylinder was obtained from measured temperature gradients adjacent to the surface.

5.5 Melting of Ice by Forced Convection

Heat transfer from a warm, laminar liquid flow to a melting flat surface was determined analytically by Yen and Tien [192], Pozvonkov et al.[193], and Epstein and Cho [194]. Yen and Tien [192] analyzed the governing equation obtained as an extension of the classical Leveque solution to the melting process, while Pozvokov et al.[193] treated the problem using a heat balance integral method. Epstein and Cho [194] showed that exact similarity solutions for melting heat transfer in a steady laminar flow over a flat plate were given by the solutions which satisfy the same boundary value problem already solved in the frame work of diffusional mass transfer or transpiration cooling.

Yen and Zehnder [195] conducted an experimental study on the melting of an ice block by a water jet. They used hot water in the temperature range of 17.5 to 56.0$\,°C$ with two sizes of nozzles (15.9 and 23.8 mm in diameter) and found that the melting rate was approximately constant during the short duration of the experiments. Recently, Szekely et al.[196] measured the melting rate of an ice rod immersed in a turbulent, recirculating flow induced by the agitation of a gas stream injected at

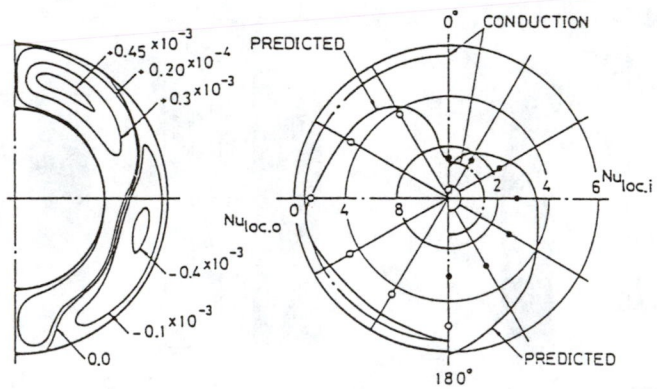

Figure 22 Streamline pattern and local Nusselt number, d_o=121.5 mm, d_i=69.6 mm, and T_o=6$\,°C$ [191].

the bottom. An ice rod with 2.54 cm in diameter and 8 cm in length was used in the experiments. They reported that both the linear velocities and the turbulence levels in the vicinity of the melting surface were likely to play an important role and proposed a correlational equation for the heat-transfer coefficient determined from the measured melting rate.

5.6 Melting of Ice in Saline Water

5.6.1 Melting Problem in Sea Water. In recent years, the feasibility of towing Antarctic icebergs and melting them to provide a supply of fresh water has been investigated in various parts of the world such as Saudi Arabia, Australia, and California. The study of an ice surface melting in flowing or quiescent sea water is of particular interest, since it relates closely to the melting of ice packs and icebergs in various towing processes.

5.6.2 Melting by Forced Convection. Griffin[197] used the heat balance integral method to analyze the forced convection heat-transfer process near a melting surface of pure ice in saline water under laminar flow conditions . He estimated the relative thickness of the momentum, and mass diffusion layers for a variety of flow and thermal parameters, and used the results to determine the sound speed profiles within the temperature and salinity boundary layers near the melting glacial ice for free-stream water temperatures of $5\,^{\circ}C$ and $10\,^{\circ}C$.

5.6.3 Melting by Free Convection. Huppert and Turner [198] studied experimentally the effect of ambient stratification on the flow and transport adjacent to a vertical ice surface melting in sea water. Marschall [199] analytically investigated free convection melting of a vertical ice surface immersed in saline water. He obtained the similarity solutions of the boundary-layer equations governing momentum, heat, and mass transfer by using a co-ordinate system fixed to the ice-saline water interface and then deriving an expression for the blowing velocity at the ice surface due to the melting. Johnson [200] determined experimentally the transport characteristics of a vertical flat ice slab melting in saline water of $S_{\infty} = 35\%$ for ambient temperatures ranging from -1.08 to $24.4\,^{\circ}C$. He used a Schlieren system with a differential thermocouple to determine flow direction and reported that the flow was upwards and laminar at low ambient temperatures, while at higher ambient temperatures, transition to turbulence took place. Josberger and Martin [201] conducted an excellent and extensive observation of the flow along the melting vertical ice surface in saline water. Illuminating suspended particles in the saline water and injecting dye to visualize the flow, they observed the flow characteristics near the ice surface in water at near oceanic salinity for ambient temperatures ranging from -1.15 to 26 $^{\circ}C$. They found that for T_{∞} less than about $18\,^{\circ}C$, the flow was laminar near the surface and bi-directional near the bottom, while near the top of vertical ice the flow was fully upward and turbulent. They also reported that as T_{∞} was increased from about 18 and 23 $^{\circ}C$ flow inversion took place, namely, at about $23\,^{\circ}C$ the flow near the top of the ice was laminar and bi-directional, while near the lower portion of the surface the flow was turbulent and fully downward. They also conducted some experiments at lower salinities (S_∞ = 14.2 and 8.0 %) for lower ambient temperatures (T_∞ = 0.05 and 1.80°C) and reported that fully laminar upward flow was observed under these conditions.

Numerical calculations for laminar free convection along a vertical ice surface melting in saline water were performed by Carey and Gebhart [202], who noted that solutions were possible only to some limited ambient salinity and temperature conditions. Carey and Gebhart [203] also carried out vertical ice melting experiments at a salinity level of 10 percent for ambient water temperatures ranging from 1 to 15°C and showed the nature of free convection flows using a time exposure photographic technique. Recently, the same problem was also studied more extensively by Sammakia and Gebhart [204] and Johnson and Mollendorf [205].

5.7 Melting of Phase Change Mediums other than Ice Block

5.7.1 Melting along Horizontal or Vertical Surfaces.
Recently, the feasibility of using the latent heat produced during a phase change in thermal,solar or waste energy storage systems has motivated an extensive investigation. The majority of the literature reports the results of experimental and analytical studies for a variety of geometric arrangements.

Experimental and analytical studies of the heat transfer phenomena during the phase change process at a vertical surface and above or below a heated horizontal flat plate were carried out by Hale and Viskanta [206,207]. They used various materials as a phase change medium and concluded that free convection in the liquid had to be accounted for in the prediction of a phase-change motion. Sparrow et al.[208] analyzed multi-dimensional melting from a vertical cylinder embedded in a solid, taking into account free convection in the melted region. They also pointed out that free convection had to be considered since it was of first-order importance. Observations of melting phenomena along a vertical heated cylinder embedded in a phase change material were performed by Ramsey and Sparrow [209]. They used naphthalene as the phase change medium.

5.7.2 Melting Around Horizontal Cylinder
The heat transport mechanism around a heated horizontal cylinder embedded in a phase change material was studied by a number of investigators. Sparrow et al.[210] did an experiment on the role of free convection in the melting of a solid surrounding a cylinder using a eutectic mixture of sodium nitrate and sodium hydroxide ($T_f \doteq 517$ K). Melting from an electrically heated horizontal cylinder embedded in n-octadecane (T_f = 301.15 K) was investigated experimentally by Bathelt et al.[211], who photographically determined the shape of the solid-liquid interface and assessed the local heat-transfer coefficient using a shadowgraph technique. They concluded that free convection should be considered in the analysis and design of systems involving a phase change. Also, a photographic study of melting characteristics around a horizontal cylinder within the eutectic mixture was done by Abdel-Wahed et al.[212]. Bathelt et al.[213] and Bathelt and Viskanta [214] also extended their experiments to melting from an array of three staggered, electrically heated cylinders and to melting from a cylinder with uniform surface temperature, respectively.

Yao and Chen [215] analytically determined the effect of free convection on the melting of a solid around a hot horizontal cylinder by developing a series solution in which the free convection effect was treated as a perturbation quantity. A similar analysis which took into account the subcooling effect on the phase-change process was also conducted by Yao and Cherney [216]. Their results, however, seem to be unable to satisfactorily predict the actual process. In order to overcome difficulties resulting from the complex structure of the timewise changing physical domain of the melt region, a numerical mapping technique (numerically generated coordinate systems) was devised by Rieger et al.[217] for an analysis of melting processes around a horizontal heating cylinder.

5.7.3 Melting inside Horizontal Tubes.

The melting process inside a heated horizontal tube was studied numerically by Pannu et al.[218] and Saitoh and Hirose [219]. Both applied the two-dimensional Landau transformation method to immobilize the liquid-solid interface. Recently, Rieger et al.[220] extended their analysis [217] to melting within a horizontal tube. They also conducted an experiment using n-octadecane as a phase change medium and measured the solid-liquid interface positions as well as the local heat-transfer coefficients as a function of time. They reported that the experimental data are in good agreement with their numerical results.

6 MELTING OF SNOW

6.1 Melting Problem in Snow Layer

In cold regions, snow presents a number of problems, most of which are related to transportation. There arises, for instance, the problem of snow clearance on highways, railways, and airport runways. Melting problems of snow layers are of great importance in connection with the occurrence mechanisms of avalanches, road heating, and predictions of both water resources due to the melting of snow on mountains and the water-flow rate of rivers. Very little attention, however, has been focused on the melting of snow layers from the viewpoint of heat transfer. Yoshida et al.[221] proposed a correlation equation to estimate the melting rate of snow layers based on their experiments. Kojima [222] studied the effect of evaporation at the snow surface on the melting of natural snow layers and reported that snow melting was accelerated by artificially protecting the evaporation from the snow surface without changing the heat supplied by both radiation and atmospheric heat transfer.

6.2 Melting of Snow Layers Heated from Below

The melting phenomenon of a snow layer heated from below is similar to that for snow melting during "road-heating". Aoki et al.[223] extensively studied the melting process and proposed an analytical model based on their observations. Fig.23 presents a schematic diagram of the melting process when the packed snow layer is heated from below. As heat is supplied (a), melting occurs. The resulting melted water rapidly permeates into the porous snow layer. The body becomes composed of two layers; the water percolation layer and the dry porous layer (b). As the

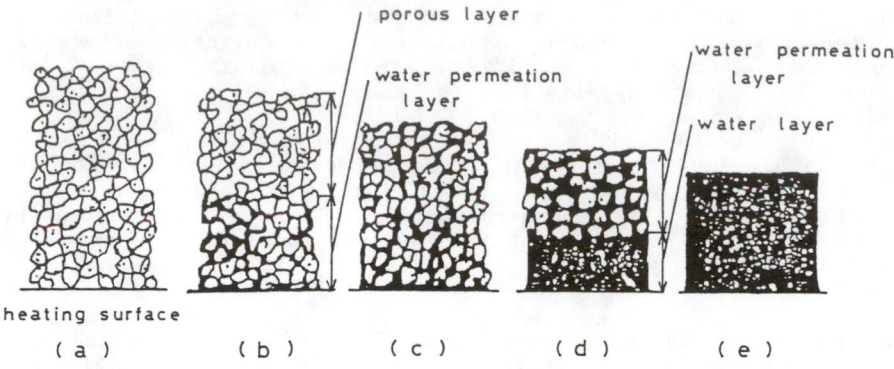

Figure 23 Sketches of snow melt heated from below [223].

melting progresses, the water content within the water-percolation layer increases and the region of the layer becomes greater. The whole body quickly becomes a water-percolation layer (c). After this, the water content within the layer increases with additional melting. The body again becomes composed of two layers; the water layer and the water-percolation layer (d). Following this, the water-percolation layer disappears and the melting process is finished, leaving a water layer (e).

It should be noted that the snow layer is, generally, a porous layer composed of ice and air, thus the melting front seems to progress with both the void-ratio change based on the ice-particle decay and the seepage of the melt water into the remaining porous area. It should also be noted that the seepage water will additionally melt the snow layer at the seepage location, while some amount of the water there may freeze. In these processes, the most important phenomenon is the seepage of the melt water. It is not yet clear whether the motivating force is the capillary action or the osmotic action, because these actions seem to be markedly different and complicated on the basis of the snow structure (for instance, ice-particle size, void ratio, and moisture content).

6.3 Melting of Snow Layer by Radiant Heat Source

The melting of the snow layer by radiative heating, which may correspond to snow melt by solar energy, demonstrates a complicated behavior pattern for heat and mass transfer. Sugawara et al.[224] studied both experimentally and analytically the melting of a snow layer heated by a radiative heat source possessing a comparatively short wave length when the snow-layer temperature is initially cooled below the fusion temperature of ice. They used an analytical model in which the percolation rate of the melt water was assumed to be fundamentally governed by capillary suction pressure.

Fig.24 shows a typical feature of the measured temperature distribution within a melting ice layer. Melt water based on both the surface and internal meltings, which are caused by absorbing the radiative energy, percolates into snow layer, creating two snow layers, one wet and the other dry. As seen in the figure, the temperature in the wet snow zone appears to be uniformly maintained at $0\,°C$. On the other hand,

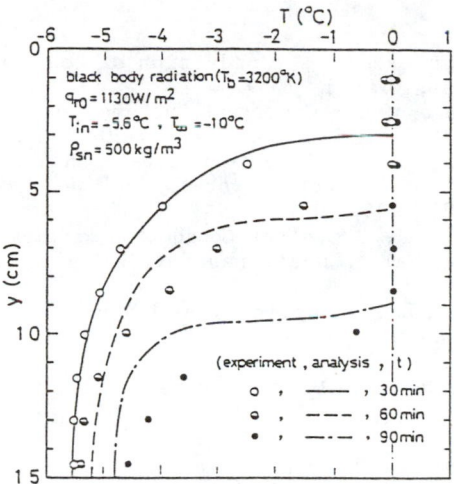

Figure 24 Temperature distribution of snow layer heated from short wave radiation [224].

the temperature in the dry snow zone increases monotonically with an increase in melting time. Thus, it is considered that the melt water percolates through the wet zone and seeps into the dry zone, while some amount of the melt water may happen to freeze at the seepage front, which generally corresponds to the interface between the wet and the dry zones.

7 CONCLUDING REMARKS

This review has considered the heat-transfer problems of ice and snow layers. The article was prepared with the intention of introducing the results of previous experiments performed particularly on the heat transfer during water-freezing or ice-melting phenomena. Consideration was also give to heat-transfer problems with respect to phase changes of mediums other than ice or water. A variety of problems still remain to be investigated, especially on heat transfer in the snow layer with a change of phase. Further development of research in this field is to be expected.

ACKNOWLEDGEMENT

The authors are deeply indebted to Professor Wilhelm Schneider of Institut für Strömungslehre and Wärmeübertragung, Technische Universität Wien for his helpful advice and suggestions throughout the present review work.

REFERENCES

1. Carslaw, H.S., and Jaeger, J.C., Conduction of Heat in Solids, 2nd Ed. Clarendon Press, Oxford, 1959.
2. Yang, W.J., Phase Change of One-Component Systems in a Container, Chem. Eng. Progr. Symp. Ser., Vol.61-57, pp.26-30, 1965.
3. Tien, R.H., and Geiger, G.E., A Heat-Transfer Analysis of the Solidification of a Binary Eutectic System, ASME Journal of Heat Transfer, Vol.89, pp.230-234, 1967.
4. Cho, S.H., and Sunderland, J.E., Heat-Conduction Problems Melting or Freezing, ASME Journal of Heat Transfer, Vol.91, pp.421-426, 1969.
5. Komori, T., and Hirai, E., An Application of Stefan's Problem to the Freezing of a Cylindrical Food-Stuff, J. Chemical Engineering of Japan, Vol.3, No.1, pp.39-44. 1970.
6. Churchill, S.W., and Gupta, J.P., Approximations for Conduction with Freezing or Melting, Int. J. Heat Mass Transfer, Vol.20, pp.1251-1253, 1977.
7. Cho, S.H., and Sunderland, J.E., Approximate Temperature Distribution for Phase Change of a Semi-Infinite Body, ASME Journal of Heat Transfer,Vol.103, pp.401-403, 1981.
8. Budhia,H., and Kreith, F., Heat Transfer with Melting or Freezing in a Wedge, Int. J. Heat Mass Transfer, Vol.16, pp.195-211, 1973.
9. Langford, D., Pseudo Similarity Solutions of the One-dimensional Diffusion Equation with Application to the Phase Change Problem, Quarterly of Applied Mathematics, Vol.25, pp.45-52, 1967.
10. Bluman, G.W., Applications of the General Similarity Solutions of the Heat Equation to Boundary Value Problems, Quarterly of Applied Mathematics, Vol.31, pp.403-415, 1974.
11. Chuang, Y.K., and Szekely, J., On the Use of Green's Function for Solving Melting or Solidification Problems, Int. J. Heat Mass Transfer, Vol.14,pp.1285-1294, 1971.
12. Chuang,Y.K., and Szekely, J., The Use of Green's Functions for Solving Melting or Solidification Problems in the Cylindrical Coordinate System,Int. J. Heat Mass Transfer, Vol.15, pp.1171-1174, 1972.
13. Hassanein, A.M., and Kulcinski, G.L., Simulation of Rapid Heating in Fusion Reactor First Walls Using the Green's Function Approach, ASME Journal of Heat Transfer, Vol.106, pp.486-490, 1984.
14. Goodman,T.R., The Heat Balance Integral and its Application to Problems Involving Change of Phase, ASME Journal of Heat Transfer, Vol.80, pp.335-341, 1958.
15. Libby, P.A., and Chen, S., The Growth of a Deposited Layer on a Cold Surface, Int. J. Heat Mass Transfer, Vol.8, pp.395-402, 1965.
16. Tien, R.H., and Geiger, G.H., The Unidimensional Solidification of a Binary Eutectic System with a Time Dependent Surface Temperature, ASME Journal of Heat Transfer, Vol.90, pp.27-31, 1968.
17. Gutfinger, C., and Chen, W.H., Heat Transfer with a Moving Boundary-Application to Fluidized-Bed Coating, Int. J. Heat Mass Transfer, Vol.12, pp.1097-1108, 1969.
18. Imber, M., and Huang, P.N.S., Phase Change in a Semi-Infinite Solid with Temperature Dependent Thermal Properties, Int. J. Heat Mass Transfer,Vol.16, pp.1951-1954, 1973.
19. Bell, G.E., A Refinement of the Heat Balance Integral Method Applied to a Melting Problem, Int. J. Heat Mass Transfer, Vol.21, pp.1357-1362, 1978.

20. El-Genk, M.S., and Cronenberg, A.W., Some Improvements to the Solution of Stefan-Like Problems, Int. J. Heat Mass Transfer, Vol.22, pp.167-170,1979.

21. Yuen, W.W., Application of the Heat Balance Integral to Melting Problems with Initial Subcooling, Int. J. Heat Mass Transfer, Vol.23, pp.1157-1160, 1980.

22. Lapadula, C., and Müller, W., Heat Conduction with Solidification and a Convective Boundary Condition at the Freezing Front, Int. J. Heat Mass Transfer, Vol.9, pp.701-704, 1966.

23. Westphal, K.O., Series Solution of Freezing Problem with the Fixed Surface Radiating into a Medium of Arbitrary Varying Temperature, Int. J. Heat Mass Transfer, Vol.10, pp.195-205, 1967.

24. Evans, G.W., Isaacson, H.E., and Macdonald, K.L., Stefan-Like Problems,Quarterly of Applied Mathematics, Vol.8, pp.312-319, 1950.

25. Rubinsky, B., and Shitzer, A., Analytic Solutions to the Heat Equation Involving a Moving Boundary with Applications to the Change of Phase Problem (the Inverse Stefan Problem), ASME Journal of Heat Transfer,Vol.100, pp.300-304, 1978.

26. Lock, G.S.H., Gunderson, J.R., Quon, D., and Donnelly, J.K., A Study of One-Dimensional Ice Formation with Special Reference to Periodic Growth and Decay, Int. J. Heat Mass Transfer, Vol.12, pp.1343-1352, 1969.

27. Lock, G.S.H., and Nyren, R.H., Analysis of Fully-Developed Ice Formation in a Convectively-Cooled Circular Tube, Int. J. Heat Mass Transfer, Vol.14, pp.825-824, 1971.

28. Huang, C.L., and Shih, Y.P., Perturbation Solution of Planar Diffusion-Controlled Moving-Boundary Problems, Int. J. Heat Mass Transfer, Vol.18,pp.689-695, 1975.

29. Prusa, J., and Yao, L.S., Melting around a Horizontal Heated Cylinder:Part 1-Perturbation and Numerical Solutions for Constant Heat Flux Boundary Condition, ASME Journal of Heat Transfer, Vol.106, pp.376-384, 1984.

30. Murray, W.D., and Landis, F., Numerical and Machine Solutions of Transient Heat-Conduction Problems Involving Melting or Freezing, Trans. Am. Soc. Mech. Engrs, Series C, J. Heat Transfer, Vol.81, pp.106-112,1959.

31. Teller, A.S., and Churchill, S.W., Freezing outside a Sphere, Chem. Eng.Progr. Symp. Ser., Vol.61, pp.185-189, 1965.

32. Lazaridis, A., A Numerical Solution of the Multidimensional Solidification (or Melting) Problem, Int. J. Heat Mass Transfer, Vol.13,pp.1459-1477 ,1970.

33. Seider, W.D., and Churchill, S.W., The Effect of Insulation on Freezing Front Motion, Chem. Eng. Progr. Symp. Ser., Vol.61, pp.179-184, 1965.

34. Heitz, W.I., and Westwater, J.W., Extension of the Numerical Method for Melting and Freezing Problems, Int. J. Heat Mass Transfer, Vol.13,pp.1371-1375, 1970.

35. Shoji, M., Numerical Method of Heat Conduction Problem with Phase Change, Bulletin of JSME, Vol.44, pp.1624-1632, 1978.

36. Rao, P.M., and Sastri, V.M.K., Efficient Numerical Method for Two-Dimensional Phase Change Problems, Int. J. Heat Mass Transfer, Vol.27, pp.2077-2084, 1984.

37. Douglas, J., and Gallie, T.M., On the Numerical Integration of a Parabolic Differential Equation Subject to a Moving Boundary Condition, Duke Math. Jl 22, pp.557-570, 1955.

38. Gupta, R.S., and Kumar, D., Variable Time Step Methods for

One-Dimensional Stefan Problem with Mixed Boundary Condition, Int. J. Heat Mass Transfer, Vol.24, pp.251-259, 1981.

39. Katayama, K., and Hattori, K., A Study of Heat Conduction with Phase Change, Trans. JSME, Vol.40, pp.1404-1411, 1974.

40. Shamsundar, N., and Sparrow, E.M., Analysis of Multidimensional Conduction Phase Change via the Enthalpy Model, ASME Journal of Heat Transfer ,Vol.97, pp.333-340, 1975.

41. Bonacina, C., Comini, G., and Primicerio, M., Numerical Solution of Phase Change Problems, Int. J. Heat Mass Transfer, Vol.16, pp.1825-1832, 1973.

42. Shamusunder, N., and Sparrow, E.M., Effect of Density Change on Multidimensional Condition Phase Change, ASME J.Heat Transfer Vol.98, pp.550-557, 1976.

43. Voller, V.R., and Cross, M., Accurate Solutions of Moving Boundary Problems Using the Enthalpy Method, Int. J. Heat Mass Transfer, Vol.24, pp.545-556,1981

44. Goodrich, L.E., Efficient Numerical Technique for One-Dimensional Thermal Problems with Phase Change, Int. J. Heat Mass Transfer, Vol.21, pp.615-621, 1978.

45. Dix, R.C., and Cizek, J., The Isotherm Migration Method for Transient Heat Conduction Analysis, Proc. 4th Int. Heat Transfer Conf., Paris, Vol.1, ASME, New York, 1971.

46. Crank, J., and Crowley, A.B., On an Implicit Scheme for the Isotherm Migration Method along Orthogonal Flow Lines in Two Dimensions, Int. J. Heat Mass Transfer, Vol.22, pp.1331-1337, 1979.

47. Gupta,R.S., and Kumar, A., An Efficient Approach to Isotherm Migration Method in Two Dimensions, Int. J. Heat Mass Transfer, Vol.27, pp.1939-1942, 1984.

48. Landau, H.G., Heat Conduction in a Melting Solid, Quarterly of Applied Mathematics, Vol.8, pp.81-94, 1950.

49. Beaubouff, R.T., and Chapman, A.J., Freezing of Fluids in Forced Flow, Int. J. Heat Mass Transfer, Vol.10, pp.1581-1587, 1967.

50. Duda, J.L., Malone, M.F., and Notter, R.H., Analysis of Two-Dimensional Diffusion Controlled Moving Boundary Problems, Int. J. Heat Mass Transfer, Vol.18, pp.901-910, 1975.

51. Saitoh, T., Numerical Methods for Multidimensional Freezing Problems in Arbitrary Domain, ASME J. Heat Transfer, Vol.100, pp.294-299, 1978.

52. Sparrow, E.M., Ramadhyani, S., and Patankar, S.V., Effect of Subcooling on Cylindrical Melting, ASME Journal of Heat Transfer, Vol.100, pp.395-402, 1978.

53. Hsu, C.F., Sparrow, E.M., and Patankar, S.V., Numerical Solution of Moving Boundary Problems By Boundary Immobilization and a Control-Volume-Based, Finite Differences Sheme, Int. J. Heat Mass Transfer, Vol.24, pp.1335-1343, 1981.

54. Cheung, S.B., Chawla, T.C., and Pedersen, D.R., The Effect of Heat Generation and Wall Interaction on Freezing and Melting in a Finite Slab, Int. J. Heat Mass Transfer, Vol.27, pp.29-37, 1984.

55. Ho, C.J., and Viskanta, R., Heat Transfer during Melting from an Isothermal Vertical Wall, ASME Journal of Heat Transfer, Vol.106, pp.12-19, 1984

56. Gadgil, A., and Gobin, D., Analysis of Two-Dimensional Melting in Rectangular Enclosures in Presence of Convection, ASME Journal of Heat Transfer, Vol.106, pp.20-26, 1984.

57. Ho, C.J., and Viskanta,R., Heat Transfer during Inward Melting in a Horizontal Tube,Int. J. Heat Mass Transfer, Vol.27, pp.705-716,

1984.

58. Gupta, R.S., and Kumar, A., Treatment of Multi-Dimensional Moving Boundary Problems by Coordinate Transformation, Int. J. Heat Mass Transfer, Vol.28, pp.1355-1366, 1985.

59. Comini, G., and Giudice, S.Del., Thermal Aspects of Cryosurgery, ASME Journal of Heat Transfer, Vol.98, pp.543-549, 1976.

60. Bonnerot, R., and Jamet, P., Numerical Computation of the Free Boundary for the Two-Dimensional Stefan Problem by Space-Time Finite Elements, J. Computational Physics, Vol.25, pp.163-181, 1977.

61. Siegel, R., Conformal Mapping for Steady Two-Dimensional Solidification on a Cold Surface in Flowing Liquid, NASA TN D-4771(1968)

62. Sproston, J.L., Two-Dimensional Solidification in Pipes of Rectangular Section, Int. J. Heat mass Transfer, Vol.24, pp.1493-1501, 1981.

63. Shamsundar, N., and Srinivasan, R., A New Similarity Method for Analysis of Multidimensional Solidification, ASME Journal of Heat Transfer, Vol.101, pp.585-591, 1979.

64. Shamsundar, N., Similarity Rule for Solidification Heat Transfer with Change in Volume, ASME Journal of Heat Transfer, Vol.103, pp.173-175, 1981.

65. Heji-Sheikh, A., and Sparrow, E.M., The Solution of Heat Conduction Problems by Probability Methods, ASME Journal of Heat Transfer, Vol.89, pp.121-131, 1967.

66. Allen, D.N.deG., and Stevern, R.T., The Application of Relaxation Methods to the Solution of Non-Elliptic Partial Differential Equations, Quart. J. Mech. and App. Math., Vol.15, pp.53-62, 1962.

67. London, A.L., and Seban, R.A., Rate of Ice Formation, Trans. Am. Soc. Mech. Engrs., Vol.80, pp.335, 1958.

68. Gilpin, R.R.,The Effects of Dendritic Ice Formation in Water Pipes, Int. J.Heat Mass Transfer, Vol.20, pp.693-699, 1977.

69. Gilpin, R.R., The Effects of Cooling Rate on the Formation of Dendritic Ice in a Pipe with no Main Flow, ASME Journal of Heat Transfer, Vol.99, pp.419-424, 1977.

70. Gilpin, R.R., The Influence of Natural Convection on Dendritic Ice Growth, J. Crystal Growth, Vol.36, pp.101-108, 1976.

71. Sugawara, M., Seki, N., and Kimoto, K., Freezing Limit of Water in a Closed Circular Tube, Warme-und Stoffubertragung, Vol17, pp.187-192, 1983.

72. Ozaki, O., Iwadate, T., Seki, N., Saitoh, T., Free Convection Heat Transfer on a Vertical Frozen Front, Bulletin of Faculty of Engineering, Hokkaido University, No.59, pp.33-42, 1976.

73. Tankin, R.S., and Farhadieh, R., Effects of Thermal Convection Currents on Formation of Ice, Int. J. Heat Mass Transfer, Vol.14, pp.953-961, 1971.

74. Forbes, R.E., and Cooper, J.W., Natural Convection in a Horizontal Layer of Water Cooled from above to Near Freezing, ASME Journal of Heat Transfer, Vol.97, pp.47-53, 1975.

75. Gilpin, R.R., Cooling of a Horizontal Cylinder of Water through its Maximum Density Point at $4\,^{\circ}C$, Int. J. Heat Mass Transfer, Vol.18, pp.1307-1315, 1975.

76. Takeuchi, M., and Cheng, K.C., Transient Natural Convection in Horizontal Cylinders with Constant Cooling Rate, Warme-und Stoffubertragung, Vol.9, pp.215-225, 1976.

77. Cheng, K.C., and Takeuchi, M., Transient Natural Convection of Water in a Horizontal Pipe with Constant Cooling Rate through $4\,^{\circ}C$,

ASME Journal of Heat Transfer, Vol.98, pp.581-587, 1976.

78. Cheng, K.C., Takeuchi, M., and Gilpin, R.R., Transient Natural Convection in Horizontal Water Pipes with Maximum Density Effect and Supercooling, Numerical Heat Transfer, Vol.1, pp.101-105, 1978.

79. Vasseur, P., and Robillard, L., Transient Natural Convection Heat Transfer in a Mass of Water Cooled through $4\,^{\circ}$C, Int. J. Heat Mass Transfer, Vol.23, pp.1195-1205, 1980.

80. Robillard, L., and Vasseur, P., Transient Natural Convection Heat Transfer of Water with Maximum Density Effect and Supercooling, ASME J. Heat Transfer, Vol.103, pp.528-534, 1981.

81. Hsu, C.F., and Sparrow, E.M., A Closed-Form Analytical Solution for Freezing Adjacent to a Plane Wall Cooled by Forced Convection, ASME J. Heat Transfer, Vol.103, pp.596-598, 1981.

82. Sparrow, E.M., and Hsu, C.F., Analysis of Two-Dimensional Freezing on the Outside of a Coolant-Carrying Tube, Int. J. Heat Mass Transfer, Vol.24, pp.1345-1357, 1981.

83. Sparrow, E.M., Ramsey, J.W., and Harris, J.S.,The Transition from Natural-Convection-Controlled Freezing to Conduction-Controlled Freezing, ASME Journal Heat Transfer, Vol.103, pp.7-12, 1981.

84. Sparrow, E.M., and Mendes, P.S., Natural Convection Heat Transfer Coefficients Measured in Experiments on Freezing, Int. J. Heat Mass Transfer, Vol.25, pp.293-297, 1982.

85. Larson, E.D., and Sparrow, E.M., Effect of Inclination on Freezing in a Sealed Cylindrical Capsule, ASME Journal of Heat Transfer, Vol.106, pp.394-401, 1984.

86. Hill, J.M., and Kucera, A., Freezing a Saturated Liquid Inside a Sphere, Int. J. Heat Mass Transfer, Vol.26, pp.1631-1637, 1983.

87. Guenigalt, R., and Poots, G., Effects of Natural Convection on the Inward Solidification of Spheres and Cylinders, Int. J. Heat Mass Transfer, Vol.28, pp.1229-1231, 1985.

88. Merk, H.J., The Influence of Melting and Anomalous Expansion on the Thermal Convection in Laminar Boundary Layers, Appl. Sci. Res., Vol.4,pp.435-452, 1954.

89. Yen, Y.C., and Tien, C., Laminar Heat Transfer over a Melting Plate, the Modified Leveque Problem, J. Geophys. Res., Vol.68, pp.3673-3678, 1963.

90. Siegel, R., and Savino, J.M., An Analysis of the Transient Solidification of a Flowing Warm Liquid on a Convectively Cooled Wall, Proc. 3rd Int. Heat Transfer Conf., New York, Vol.4, pp.141-151, 1966.

91. Pozvonkov, F.M., Shurgalskii, E.F., and Akselrod, L.S., Heat Transfer at a Melting Flat Surface Under Conditions of Forced Convection and Laminar Boundary Layer, Int. J. Heat Mass Transfer, Vol.13, pp.957-962, 1970.

92. Lapadula, C., and Mueller, W.K., Heat Conduction with Solidification and a Convective Boundary Condition at the Freezing Front, Int. J. Heat Mass Transfer, Vol.9, pp.702-704, 1966.

93. Miller, M.L., and Jiji, L.M., Appl. Sci. Res., Vol.22,pp.141, 1970.

94. Epstein, M., The Growth and Decay of a Frozen Layer in Forced Flow, Int. J. Heat Mass Transfer, Vol.19, pp.1281-1288, 1976.

95. El-Genk, M.S., and Cronenberg, A.W., On the Thermal Stability of a Frozen Crust in Forced Flow on an Insulated Finite Wall, Int. J. Heat Mass Transfer, Vol.22, pp.1719-1723, 1979.

96. Savino, J.M., and Siegel, R., An Analytical Solution for Solidification of a Moving Warm Liquid onto an Isothermal Cold

Wall, Int. J. Heat Mass Transfer, Vol.12, pp.803–809, 1969.

97. Savino, J.M., Zumdieck, J.F., and Siegel, R., Experimental Study of Freezing and Melting of Flowing Warm Water at a Stagnation Point on a Cold Plate, Heat Transfer 1970, Elsevier, Amsterdam, Vol.1, Cu 2.10, 1970.

98. Hirata, T., Gilpin, R.R., Cheng, K.C., and Gates, E.M, The Steady State Ice Layer Profile on a Constant Temperature Plate in a Forced Convection Flow: I. The Laminar Regime, Int. J. Heat Mass Transfer, Vol.22, pp1425–1433, 1979.

99. Hirata, T., Gilpin, R.R., and Cheng, K.C., The Steady State Ice Layer Profile on a Constant Temperature Plate in a Forced Convection Flow: II. The Transition and Turbulent Regimes, Int. J. Heat Mass Transfer, Vol.22, pp.1435–1443, 1979.

100. Gilpin, R.R., Hirata, T., and Cheng, K.C., Wave Formation and Heat Transfer at an Ice–Water Interface in the Presence of a Turbulent Flow, J. Fluid Mech., Vol.99, pp.619–640, 1980.

101. Ashton, G.D., and Kennedy, J.F., Ripples on Underside of River Ice Covers, Proc. A.S.C.E., Vol.98(HY9), pp.1603–1624, 1972.

102. Lee, D.G., and Zerkle, R.D., The Effect of Liquid Solidification in a Parallel Plate Channel upon Laminar–Flow Heat Transfer and Pressure Drop, ASME Journal of Heat Transfer, Vol.91, pp.583–585, 1969.

103. Cheng, K.C., and Wong, S.L., Liquid Solidification in a Convectively–Cooled Parallel–Plate Channel, Canadian J. Chem. Engng., Vol.55, pp.149–155, 1977.

104. Cheng, K.C., and Wong, S.L., Asymmetric Solidification of Flowing Liquid in a Convectively Cooled Parallel–Plate Channel, Appl. Sci. Res., Vol.33, pp.309–335, 1977.

105. Shibani, A.A., and Ozisik, M.M., A Solution of Freezing of Liquids of Low Prandtl Number in Turbulent Flow Between Parallel Plate, ASME Journal of Heat Transfer, Vol.99, pp.20–24, 1977.

106. Seki, N., Fukusako, S., and Younan, G.W., Ice–Formation Phenomena for Water Flow Between Two Cooled Parallel Plates, ASME Journal of Heat Transfer, Vol.106, pp.498–505, 1984.

107. Seki, N., Fukusako, S., and Younan, G.W., A Transition Phenomenon of Ice Formation in Water Flow Between Two Horizontal Parallel Plates, Wärme–und Stoffübertragung, Vol.18, pp.117–128, 1984.

108. Seki, N., Fukusako, S., Tanaka, J., Ito, K., and Hirata, T., Transient Freezing of Water Flow Between Horizontal Parallel Plates, Trans. JSME, Vol.49, pp.2172–2179, 1983.

109. Zerkle, R.D., and Sunderland, J.E., The Effect of Liquid Solidification in a Tube Upon the Laminar–Flow Heat Transfer and Pressure Drop, ASME Journal of Heat Transfer, Vol.90, pp.183–190, 1968.

110. DesRuisscaux, N., and Zerkle, R.D., Freezing of Hydraulic Systems, Can. J. of Chem. Eng., Vol.47, pp.233–237, 1969.

111. Depew, C.A., and Zenter, R.C., Laminar Flow Heat Transfer and Pressure Drop with Freezing at the Wall, Int. J. Heat Mass Transfer, Vol.12, pp.1710–1714, 1969.

112. Hwang, G.J., and Sheu, J.P., Liquid Solidification in Combined Hydrodynamic and Thermal Entrance Region of a Circular Tube, Can. J. Chem. Eng., Vol.54, pp.66, 1976.

113. Mulligan, J.C., and Jones, D.D., Experiments on Heat Transfer and Pressure Drop in Horizontal Tube with Internal Solidification, Int. J. Heat Mass Transfer, Vol.19, pp.213–219, 1976.

114. Lock, G.S.H., Freeborn, R.D.J., and Nyren, R.H., Fourth Int. Heat Transfer Conf., Paris–Versailles, Vol.1, Cu.2.9, 1970.

115. Özisik, M.N., and Mulligan, J.C., Transient Freezing of Liquids in Forced Inside Circular Tubes, ASME Journal of Heat Transfer, Vol.91, pp.385, 1969.

116. Genthner, K., Die Erstarrung Turbulent Strömender Flüssigkeiten in Rohre, Kältechnik-Klimatisierung., Vol.22, pp.414–420, 1970.

117. Seki, N., Fukusako, S., and Tokura, I., Turbulent Freezing Heat Transfer Inside a Circular Tube, Proceedings of 8th Japan National Heat Transfer Symposium, II-2.6, pp.201–204, 1971.

118. Arora, A.P.S., and Howell, J.R., An Investigation of the Freezing of Super-Cooled Liquid in Forced Turbulent Flow Inside Circular Tube, Int. J. Heat Mass Transfer, Vol.16, pp.2077–2085, 1973.

119. Shibani, A.A., and Özisik, M.N., Freezing of Liquids in Turbulent Flow Inside Tubes, The Can. J. of Chem. Eng., Vol.55, pp.672–677, 1977.

120. Thomason, S.B., Mulligan, J.C., and Everhart, J., The Effect of Internal Solidification on Turbulent Flow Heat Transfer and Pressure Drop in Horizontal Tube, ASME Journal of Heat Transfer, Vol.100, pp.387–394, 1978.

121. Cho, C., and Özisik, M.N., Transient Freezing of Liquids in Turbulent Flow Inside Tubes, ASME Journal of Heat Transfer, Vol.101, pp.465–468, 1979.

122. Epstein, M., Stachyra, L.J., and Lambert, G.A., Transient Solidification in Flow into a Rod Bundle, ASME Journal of Heat Transfer, Vol.102, pp.330–334, 1980.

123. Gilpin, R.R., The Morphology of Ice Structure in a Pipe at or near Transition Reynolds Number, ASME-AIChE Heat Transfer Symposium Series 189, Vol.75, pp.89–94, Sandiego, Calif., 1979.

124. Gilpin, R.R., Methods of Ice Formation and Flow Blockage that Occur While Filling a Cold Pipe, Cold Regions Science and Technology, Vol.5, pp.163–171, 1981.

125. Gilpin, R.R., Ice Formation in Pipe Containing Flow in the Transition and Turbulent Regimes, ASME Journal of Heat Transfer, Vol.103, pp.363–368, 1981.

126. Thomason, S.B., and Mulligan, J.C., Experimental Observations of Flow Instability during Turbulent Flow Freezing in a Horizontal Tube, ASME Journal of Heat Transfer, Vol.102, pp.782–784, 1980.

127. Epstein, M., Yim, A., and Cheung, F.B., Freezing-Controlled Penetration of a Saturated Liquid into a Cold Tube, ASME Journal of Heat Transfer, Vol.99, pp.233–238, 1977.

128. Epstein, M., and Hauser, G.M., Freezing of an Advancing Tube Flow, ASME Journal of Heat Transfer, Vol.99, pp.687–689, 1977.

129. Sampson, P., and Gibson, R.D., A Mathematical Model of Nozzle Blockage by Freezing, Int. J. Heat Mass Transfer, Vol.24, pp.231–241, 1981.

130. Sampson, P., and Gibson, R.D., A Mathematical Model of Nozzle Blockage by Freezing-II. Turbulent Flow, Int. J. Heat Mass Transfer, Vol.25, pp.119–126, 1982.

131. Schneider, W., Transient Solidification of a Flowing Liquid at a Heat Conducting Wall, Int. J. Heat Mass Transfer, Vol.28, pp.331–337, 1985.

132. Hirata, T., and Ishihara, M., Freeze-Off Conditions of a Pipe Containing a Flow of Water, Int. J. Heat Mass Transfer, Vol.28, pp.331–337, 1985.

133. Carlson, F.M., An Investigation of the Solidification of a Flowing Liquid on a Circular Cylinder in Crossflow, Ph.D. Thesis, University of Connecticut, 1975.

134. Okada, M., Katayama, K., Tarasaki, K., Akimoto, M., and Mabune, M,

Freezing Around a Cooled Pipe in Crossflow, Bulletin of the JSME, Vol.21, No.160, pp.1514-1520, 1978.

135. Cheng, K.C., Inaba, H., and Gilpin, R.R., An Experimental Investigation of Ice Formation Around an Isothermally Cooled Cylinder in Crossflow, ASME Journal of Heat Transfer, Vol.103, pp.733-738, 1981.

136. Petrie, D.J., Linhan, J.H., Epstein, E., Lambert, G.A., and Stachyra, L.J., Solidification in Two-Phase Flow, ASME Journal of Heat Transfer, Vol.102, pp.784-786, 1980.

137. Seki, N., Fukusako, S., and Hayaki, K., Freezing in Water-Air-Kaoline Mixture, Proceedings of 15th Japan National Heat Transfer Symposium, pp. 409-411, 1978.

138. Sugawara, M., Fukusako, S., and Seki, N., Experimental Studies of the Melting of a Horizontal Ice Layer, Bulletin of the JSME, Vol.18, pp.714-721, 1975.

139. Wu, R.S., and Cheng, K.C., Maximum Density Effects on Thermal Instability by Combined Buoyancy and Surface Tension, Int. J. Heat Mass Transfer, Vol.19, pp.559-565, 1976.

140. Seki, N., Fukusako, S., and Sugawara, M., A Criterion of Onset of Free Convection in a Horizontal Melted Water Layer with Free Surface, ASME J. Heat Transfer, Vol.99, pp.92-98, 1977.

141. Boger, D.V., and Westwater, J.W., Effect of Buoyancy on the Melting and Freezing Process, ASME Journal of Heat Transfer, Vol.89, pp81-89, 1967.

142. Yen, Y.C., and Galea, F., Onset of Convection in a Water Layer Formed Continuously by Melting Ice, The Physics of Fluids, Vol.12, pp.509-516, 1969.

143. Sun, Z.S., Tien, C.,and Yen, Y.C., Thermal Instability of a Horizontal Layer of Liquid with Maximum Density, AIChE Journal, Vol.15, pp.910-915, 1969.

144. Katto, Y., and Iwanaga, Y., Onset of Natural Convection in Water Layer with a Upper Stable Layer, Preprints of 12th Japan Heat Transfer Symposium, a311, pp.178-180, 1975.

145. Seki, N., Fukusako, S., and Sugawara, M., Free Convective Heat Transfer and Criterion of Onset of Free Convection in a Horizontal Melt Layer of Ice Heated by Upper Rigid Surface, Wärme-und Stoffübertragung, Vol.10, pp.269-279, 1977.

146. Merker, G.P., Waas, P., and Grigull, U., Einsetzen der Konvektion in einer von unten Gekuhlten Wasserschicht bei Temperaturen unter 4 °C, Wärme -und Stoffübertragung, Vol.9, pp.99-110, 1976.

147. Merker, G.P.,Wass, P., and Grigull, U., Onset of Convection in a Horizontal Water Layer with Maximum Density Effects, Int. J. Heat Mass Transfer, Vol.22, pp.505-515, 1979.

148. Yen, Y.C., On the Effect of Density Inversion on Natural Convection in a Melted Water Layer, Chem. Eng. Symp. Ser., Vol.65, pp.245-253, 1968.

149. Tien, C., Yen, Y.C., and Dotson, J.W., Free Convective Heat Transfer in a Horizontal Layer of Liquid - the Effect of Density Inversion, AIChE, Symp. Ser., Vol.68, pp.101-111,1970.

150. Adrian, R.J., Turbulent Convection in Water Over Ice, J. Fluid Mech., Vol.69, pp.753-781, 1975.

151. Yen, Y.C., Onset of Convection in a Layer of water Formed by Melting Ice from Below, Physics of Fluids, Vol.11, pp.1263-1270, 1968.

152. Fukusako, S., Sugawara, S., and Seki, N., Onset of Free Convection and Heat Transfer in Melt Water Layer, J. JSME, Vol.80, pp.445-450, 1977.

153. Tien, C., and Yen, Y.C., Approximate Solution of a Melting Problem with Natural Convection, Chem. Eng. Symp. Ser., Vol.62, pp.166–172, 1966.

154. Yen ,Y.C., Tien, C., and Sander, G., An Experimental Study of a Melting Problem with Natural Convection, Proc. 3rd. Int. Heat Transfer Conf., Vol.4, pp.159–166, 1966.

155. Yen, Y.C., Further Studies on a Melting Problem with Natural Convection, AIChE. J., Vol.13, pp.824–825, 1967.

156. Schmidt, E., and Silveston, P.L., Natural Convection in Horizontal Liquid Layers, Chem. Eng. Progr. Symp. Ser., Vol.66, pp.163–169, 1959.

157. Gilpin, R.R., Robertson, R.B., and Singh, B., Radiative Heating in Ice, ASME J. Heat Transfer, Vol.99, pp.227–232, 1977.

158. Seki, N., Sugawara, M., and Fukusako, S., Radiative Melting of Horizontal Clear Ice Layer, Wärme–und Stoffübertragung, Vol.11, pp.207–216, 1978.

159. Seki, N., Sugawara, M., and Fukusako, S., Back–Melting of Horizontal Cloudy Ice Layer with Radiative Heating, J. ASME Heat Transfer, Vol.101, pp.90–95, 1979.

160. Cho, C., and Özisik, MN.,Effects of Radiation on Melting of Semi-Infinite Medium, Proc. 6th Int. Heat Transfer Conf., Vol.3, pp.373–378, 1978.

161. Chan, S.H., Cho, D.H., and Kocamustafaogullari, G., Melting and Solidification with Internal Radiative Transfer – A Generalized Phase Change Model, Int. J.Heat Mass Transfer, Vol.26, pp.621–633, 1983.

162. Tkachev, A.G., Heat Exchange in Melting and Freezing of Ice, Problems of Heat Transfer during a Change of State, A Collection of Articles, AEC–TR–3405, Translated from a publication of the State Power Press, Moscow–Leningrad, pp.169–178, 1953.

163. Vanier, C.R., and Tien, C., Effect of Maximum Density and Melting on Natural Convection Heat Transfer from a Vertical Plate, Chem. Eng. Prog. Sym. Ser., Vol.64, pp.240–254, 1968.

164. Bendell, M.S., and Gebhart, B., Heat Transfer and Ice–Melting in Ambient Water near its Density Extremum, Int. J. Heat Mass Transfer, Vol.19, pp.1081–1087, 1976.

165. Gebhart, B., and Mollendorf, J.C., Buoyancy–Induced Flows in Water under Conditions in which Density Extremum may Arise, J. Fluid Mech., Vol.87, pp.673–708, 1978.

166. Wilson, N.W., and Vyas, B.D., Velocity Profiles near a Vertical Ice Surface Melting into Fresh Water, ASME Journal of Heat Transfer, Vol.101, No.2, pp.313–317, 1979.

167. Wilson, N.W., and Lee, J.J., Melting of a Vertical Ice Wall by Free Convection into Fresh Water, ASME Journal of Heat Transfer, Vol.103, pp.13–17, 1981.

168. Carey, V.P., Gebhart, B., Visualizatin of the Flow Adjacent to a Vertical Ice Surface Melting in Cold Pure Water, J. Fluid Mech., Vol.107, pp.37–55, 1981.

169. Carey, V.P., Gebhart, B., and Mollendorf, J.C., Buoyancy Force Reversals in Vertical Natural Convection Flows in Cold Water, Journal of Fluid Mechanics, Vol.97, pp.279–297, 1980.

170. Ede, A.J., The Influence of Anomalous Expansion on Natural Convection in Water, Appl. Sci. Res., Vol.5, pp.548–560, 1955.

171. Schechter, R.S., and Isbin, H.S., Natural –Convection Heat Transfer in Regions of Maximum Fluid Density, AIChE Journal, Vol.14, pp.81–89, 1958.

172. Goren, S.L., On Free Convection in Water at $4\,^{\circ}C$, Chem. Engng Sci.,

Vol.21, pp.515-518, 1966.

173. Pop, I., and Raptis, A., A Note on Transient Free Convection of Water at 4 °C over a Doubly Infinite Vertical Porous Plate, ASME J. Heat Transfer, Vol.104, pp.800-802,, 1982.

174. Joshi, Y., and Gebhart, B., Vertical Transient Natural Convection Flows in Cold Water, Int. J. Heat Mass Transfer, Vol.27, pp.1573-1582, 1984.

175. Tien, C., and Yen, Y.C., Condensation-Melting Heat Transfer, Chem. Engr. Prog. Symp. Ser., Vol.67, No.113, pp.1-9, 1971.

176. Yen, Y.C., Zhender, A., Zavoluk, S., and Tien, C., Condensation-Melting Heat Transfer in the Pressure of Air, Chem. Eng. Progr. Symposium Ser., Vol.69, No.131, pp.23-29, 1973.

177. Epstein, M., and Cho, D.H., Laminar Film Condensation on a Vertical Melting Surface, ASME Journal of Heat Transfer, Vol.98, pp.108-113, 1976.

178. Cho, D.H., and Epstein, M., Laminar Film Condensation of Flowing Vapor on a Horizontal Melting Surface, Int. J. Heat Mass Transfer, Vol.20, pp.23-30, 1977.

179. Taghavi-Tafreshi, K., and Dhir, V.K., Analytical and Experimental Investigation of Simultaneous Melting-Condensation on a Vertical Wall, ASME Journal of Heat Transfer, Vol.104, pp.24-33, 1982.

180. Taghavi-Tafreshi, k., and Dhir, V.K., Shape Change of an Initially Vertical Wall Undergoing Condensation-Driven Melting, ASME Journal of Heat Transfer, Vol.105, pp.235-240, 1982.

181. Seki, N., Sugawara, M., and Fukusako, S., Radiative Melting of Ice Layer Adhering to a Vertical Surface, Warme-und Stoffubertragung, Vol.12, pp.137-144, 1979.

182. Merk, H.J., The Influence of Melting and Anomalous Expansion on the Thermal Convection in Laminar Boundary Layers, Appl. Sci. Res., Vol.4, pp.435-452, 1953.

183. Dumore, J.M., Merk, H.J., and Prints, J.A., Heat Transfer from Water to Ice by Thermal Conduction, Nature, Vol.172, pp.460-461, 1953.

184. Schenk, J., and Schenkels, F.A.M., Thermal Free Convection from Ice Sphere in Water, Appl. Sci. Research, Vol.19, pp.465-476, 1968.

185. Vanier, C.R., and Tien, C., Free Convection Melting of Ice Spheres, American Institute of Chemical Engineers Journal, Vol.16, pp.76-82, 1970.

186. Saitoh, T., Natural Convection Heat Transfer from a Horizontal Ice Cylinder, Appl. Sci. Res., Vol.32, pp.429-451, 1976.

187. Watson, A., The Effect of the Inversion Temperature on the Convection of Water in an Enclosed Rectangular Cavity, Quarterly Journal of Mechanics and Applied Mathematics, Vol.15, pp.423-446, 1972.

188. Seki, N., Fukusako, S., and Inaba, H., Free Convective Heat Transfer with Density Inversion in a Confined Rectangular Vessel, Wärme-und Stoffübertragung, Vol.11, pp.145-156, 1978.

189. Inaba, H., and Fukuda, T., An Experimental Study of Natural Convection in an Inclined Rectangular Cavity Filled with Water at its Density Extremum, ASME Journal of Heat Transfer, Vol.106, pp.109-115, 1984.

190. Seki, N., Fukusako, S., and Nakaoka, M., Experimental Study on Natural Convection Heat Transfer with Density Inversion of Water Between Two Horizontal Concentric Cylinder, ASME Journal of Heat Transfer, Vol.97, pp.556-561, 1975.

191. Seki, N., Fukusako, S., and Nakaoka, M., An Analysis of Free Convective Heat Transfer with Density Inversion of Water Between

Two Horizontal Concentric Cylinders, ASME Journal of Heat Transfer, Vol.98, pp.670-672, 1976.

192. Yen, Y.C., and Tien, C., Laminar Heat Transfer Over a Melting Plate, the Modified Leveque Problem, J. Geophy. Res., Vol.68, No.12, pp.3673-3678, 1963.

193. Pozvonkov, F.M., Shurgalskill, E.F., and Axselrod, L.S., Heat Transfer at a Melting Flat Surface Under Conditions of Forced Convection and Laminar Boundary Layer, Int. J. Heat Mass Transfer, Vol.13, pp.957-962, 1970.

194. Epstein, E., and Cho, D.H., Melting Heat Transfer in Steady Laminar Flow Over a Flat Plate, ASME Journal of Heat Transfer, Vol.98, pp.531-533, 1976.

195. Yen, Y.C., and Zehnder, A., Melting Heat Transfer with Water Jet, Int. J. Heat Mass Transfer, Vol.16, pp.219-223, 1973.

196. Szekely, J., Grevet, H.H., and Kaddah, N.E., Melting Rates in Turbulent Recirculating Flow Systems, Int. J. Heat Mass Transfer, Vol.27, pp.1116-1121, 1984.

197. Griffin, O.M., Heat, Mass, and Momentum Transfer During the Melting of Glacial Ice in Sea Water, ASME Journal of Heat Transfer, Vol.95, pp.317-323, 1973.

198. Huppert, H.E., and Turner, J.S.. On Melting Icebergs, Nature, Vol.271, pp.46-48, 1978.

199. Marschall, E., Free Convection Melting of Glacial Ice in Saline Water, Lett. J. Heat Mass Transfer, Vol.4, pp.381-384, 1977.

200. Johnson, R.S., Transport from a Melting Vertical Ice in Saline Water, M.Sc. thesis, State University of New York at Buffalo, Amherst, New York, 1978.

201. Josberger, E.G., and Martin, S., A Laboratory and Theoretical Study of the Boundary Layer Adjacent to a Vertical Melting Ice Wall in Salt Water, J. Fluid Mech. Vol.111, pp.439-473, 1981.

202 Carey, V.P., and Gebhart, B., Transport near a Vertical Ice Surface Melting in Saline Water: Some Numerical Caluculations, J. Fluid Mech. Vol.117, PP.379-402, 1982.

203. Carey, V.P., and Gebhart, B., Transport near a Vertical Ice Surface Melting in Saline Water: Experiments at Low Salinities, J. Fluid Mech., Vol.117, pp.403-423, 1982.

204. Sammakia, B., and Gebhart, B., Transport near a Vertical Ice Surface Melting in Water of Various Salinity Levels, Int. J. Heat Mass Transfer, Vol.26, pp.1439-1452, 1983.

205. Johnson, R.S., and Mollendorf, J.C., Transportation from a Vertical Ice Surface Melting in Saline Water, Int. J. Heat Transfer, Vol.27, pp.1928-1932, 1984.

206. Hale, N.W., Jr., and Viskanta, R., Photographic Observation of the Solid-Liquid Interface Motion During Melting of a Solid Heated from an Isothermal Vertical Wall, Letters in Heat Transfer, Vol.5, pp.329-337, 1978.

207. Hale, N.W., Jr., and Viskanta, R., Solid-Liquid Phase Change Heat Transfer, and Interface Motion in Materials Cooled or Heated From Above and Below, Int. J. Heat Mass Transfer, Vol.23, pp.283-292, 1980.

208. Sparrow, E.W., Patankar, S.V., and Ramadhyani, S., Analysis of Melting in the Presence of Natural Convection in the Melting Region, ASME Journal of Heat Transfer, Vol.99, pp.520-526, 1977.

209. Ramsey, J.W., and Sparrow, E.M., Melting and Natural Convection due to a Vertical Embedded Heater, ASME J. Heat Transfer, Vol.100, pp.368-370, 1978.

210. Sparrow, E.M., Schmidt, R.R., and Ramsey, J.W., Experiments on the

Role of Natural Convection in the Melting of Solids, ASME Journal of Heat Transfer, Vol. 100, pp.11–16, 1978.

211. Bathelt, A.G., Viskanta, R., and Leidenfrost, W., An Experimental Investigation of Natural Convection in the Melted Region Around a Heated Horizontal Cylinder, J. Fluid Mech., Vol.90, pp.227–239, 1979.

212. Abdel-Wahed, R.M., Ramsey, J.W., and Sparrow, E.M., Photographic Study of Melting about an Embedded Horizontal Heating Cylinder, Int. J. Heat Mass Transfer, Vol.22.pp.453–458, 1979.

213. Bathelt, A.G., Viskanta, R., and Leidenfrost, W., Latent Heat-Of-Fusion Energy Storage: Experimental on Heat Transfer from Cylinders During Melting, ASME J. Heat Transfer, Vol.101, pp.453–458, 1979.

214. Bathelt, A.G., and Viskanta, R., Heat Transfer at the Solid-Liquid Interface During Melting From a Horizontal Cylinder, Int. J. Heat Mass Transfer, Vol.23, pp.1493–1503, 1980.

215. Yao, L.S., and Chen, F.F., Effects of Natural Convection in the Melted Region Around a Heated Horizontal Cylinder, ASME Journal of Heat Transfer, Vol.102, pp.667–672, 1980.

216. Tao, L.S., and Cherney, W., Transient Phase-Change Around a Horizontal Cylinder, Int. J. Heat Mass Transfer, Vol.24, pp.1971–1981, 1981.

217. Rieger, H., Projahn, U., and Beer, H., Analysis of Heat Transport Mechanisms During Melting Around a Horizontal Circular Cylinder, Int. J. Heat Mass Transfer, Vol.25, pp.137–147, 1982.

218. Pannu, J., Joglekar, G., and Rice, P.A., Natural Convection to Cylinders of Phase Change Material Used for Thermal Storage, AIChE Symposium Series, PP.47–55, 1980.

219. Saitoh, T., and Hirose, K., High-Rayleigh Number Solutions to Problems of Latent Heat Thermal Energy Storage in a Horizontal Cylinder Capsule, ASME Journal of Heat Transfer, Vol. 104, pp.545–553, 1982.

220. Rieger, H., Projahn, U., Bareiss, M., and Beer, H., Heat Transfer During Melting Inside a Horizontal Tube, ASME Journal of Heat Transfer, Vol.105, pp.226–234, 1983.

221. Yoshida, Z., Kojima, K., and Aoki, S., Experimental study on Melting of Snow Layer, Low Temperature Science, Vol.53, pp.101–108, 1947.

222. Kojima, K., A Fluid Experiment on the Influence of Evaporation of Snow upon Snow Melt, Low Temperature Science, Ser.A, Vol.25, pp.119–126, 1967.

223. Aoki, K., Hattori, M., Chiba, S., and Hayashi, Y., A Study of the Melting Process in Ice-Air Composite Materials, ASME Paper, 81-WA/HT-26,1981.

224. Sugawara, M., Seki, N., Fukusako, S., and Ota, T., Snow Melting with Radiative Heating, Wärme-und Stoffübertragung, Vol.17, PP.31–38, 1982.

Chapter 12

Snow Melting by Radiative Heating

M. SUGAWARA
Department of Mechanical Engineering, Akita University,
Akita 010, Japan

N. SEKI
Department of Mechanical Engineering, Hokkaido University,
Sapporo 060, Japan

K. KIMOTO
Akita City Office, Akita 010, Japan

ABSTRACT

This paper discusses experiments dealing with the melting of snow by
radiative energy, using blackbody radiation with a source temperature of
3200 K, and short wave radiative energy. The transfer of radiation in the
snow is significantly affected by both the porosity of snow and water
saturation. The internal melting in snow is the unique characteristic of
radiative heating, and is caused by the absorption of comparatively
short wave radiation. In this study, an analysis is used
to predict the variation of snow density, the moving rate of the dry-wet
interface of snow due to the percolation of melt water, and the transient
temperature distribution in the dry snow zone located under the wet snow
zone.

1. INTRODUCTION

The melting of snow by radiative heating produces some of the significant
phenomena of combined heat and mass transfer. For example, snow is a
semi-transparent medium for comparatively short wave radiation. The
temperature in the snow layer initially below the melting point, rises
gradually by absorbing the radiative energy. After the temperature of the
snow has reached its melting point, surface or internal melting occurs
through the absorption of radiative energy. Consequently, melt water
seeps into the dry snow, whose temperature is below $0°C$, and is
maintained uniformly at its melting point. Accordingly, some amount of
melt water freezes at this seepage front, which corresponds to the wet-
dry interface.

The above mentioned phenomena cause a variation in the porosity of the
snow layer due to the growth or decay of snow particles. Note that the
percolation behaviour of water seeping through the snow layer is affected
by this variation in snow porosity. Moreover, radiative heat transfer in
the wet zone is significantly affected by the water saturation of the
snow. It can be seen that the melting of snow by radiative heating
presents very complicated phenomena which are difficult to analyze. The
most important problem in the melting of snow is the mechanism of melt
water percolation. So far, Golbeck et al. [1] investigated water
percolation through homogeneous snow. In their paper, the gravity-flow

theory of water percolation is generalized to include any power-law relationship between permeability to the water phase and effective water saturation. Experimental observations of water percolation through homogeneous snow have also been reported. Wakahama et al. [2] reported their experimental investigation concerning the percolation of melt water into natural snow cover. Yoshida [3] presented a theoretical treatment for the percolation of melt water into dry snow cover. Kuroiwa [4] measured the flow down speed of melt water in a snow layer. Most of the previous investigations for snow melting were mainly concerned with the melt water percolation mechanism in snow.

In this paper, the one-dimensional melting of packed snow by radiative heating was investigated experimentally and analytically. The radiation sources used in this experiment were the blackbody radiation produced by halogen lamps having a filament temperature of 3200 K, and a short wave radiation similar to solar radiation. Also discussed in this paper are the effects of a wavelength band of radiation sources, radiative heat flux, environmental temperature, snow properties, the effect of water saturation on the melting rate at the surface of the snow layer, the moving rate of the seepage front or so-called wet-dry interface, and the temperature distribution in the dry snow zone.

2. EXPERIMENTAL APPARATUS AND PROCEDURE

A snow melting experiment was conducted in a low temperature room controlled at a constant temperature in order to establish the prescribed environmental temperature as shown in Figure 1. The inside dimensions of the room were $0.7 \times 0.7 \times 0.6 \, m^3$, and a fan was used to minimize local temperature deviations. A vessel made of a 10 mm thick lucite plates and having a surface area of 12 cm x 12 cm, was located at the center of the low temperature room. The temperature of the snow layer was measured using Cu-Co thermocouples (50 microns meter diameter) enclosed in

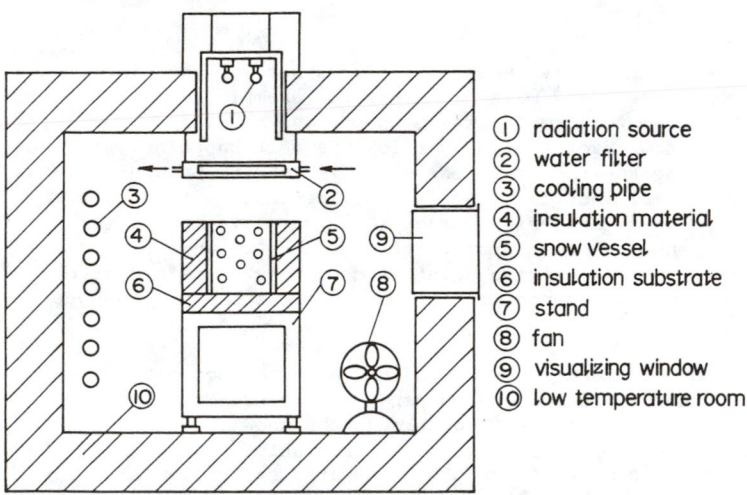

① radiation source
② water filter
③ cooling pipe
④ insulation material
⑤ snow vessel
⑥ insulation substrate
⑦ stand
⑧ fan
⑨ visualizing window
⑩ low temperature room

FIGURE 1. Schematic view of experimental apparatus

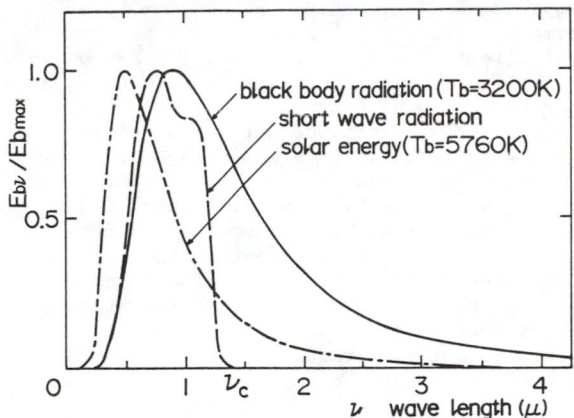

FIGURE 2. Spectral band type of radiation

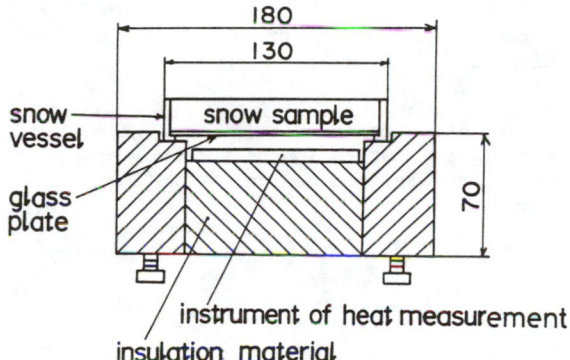

FIGURE 3. Measuring instrument for extinction coefficient of snow

polished stainless tubes to avoid the direct effect of the radiation. The snow layer had 12 thermocouples placed at 15 mm intervals down from the top surface. The bottom and sides of the vessel were surrounded by a 50 mm thick insulation material in order to establish one-dimensional snow melting behaviour.

Two kinds of radiation sources were used in this snow melting experiment. One was the blackbody radiation produced by halogen lamps with tungsten filaments of 3200 K colour temperature. Since the radiative energy obtained by the tungsten lamps was similar to the blackbody radiation as reported by Gilpin et al. [5], the present spectral band type of radiation shows the maximum energy value at a wavelength of 0.9 microns meter, as indicated in Figure 2. The second radiation source was short wave radiation, obtained by cutting a long wave fraction of the blackbody radiation with a 10 mm thick water filter instrument of polished parallel pyrex glass plates, as shown in Fig.1. Short wave radiation was used since it has a similar effect to solar radiation on snow melting patterns, as indicated in Fig.2. All of the snow melting experiments were started after the initial temperature distribution in the snow layer reached a uniform vertical temperature.

Measurement of Extinction Coefficient The extinction coefficient of snow must be obtained in order to analyze snow melting behaviours with internal or surface melting by radiative heating. In this investigation the measurement of the extinction coefficient was attempted under restrictive conditions of short wave radiation, as shown in Fig.2. In general, it is difficult to measure the extinction coefficient of snow because of its strong scattering characteristic. In this study, the following experimental technique was adopted to overcome this difficulty. The instrument for measuring the extinction coefficient is indicated schematically in Figure 3. The vessel used in this experiment had side walls made of 5 mm thick lucite plates, and had an area of 130 mm x 130 mm. The bottom plate of the vessel was made of a well polished 3 mm thick glass plate. The extinction coefficient was evaluated using the following equation, negating the reflection effect at the top surface of the snow sample.

$$k_e = \frac{1}{h_2 - h_1} \ln \frac{q_{r2}}{q_{r1}} \tag{1}$$

where q_{r1} and q_{r2} indicate the radiative heat fluxes at snow sample thicknesses of h_1, h_2, respectively.

q_{r1}, q_{r2} were measured using an instrument utilized in previous work [6]. It is generally noted that the extinction coefficient of snow is mainly affected by both snow porosity and water saturation. A wet snow sample having the prescribed water saturation was produced in the following manner: water at $0°C$ was injected using a syringe into the snow sample, also maintained at $0°C$.

3. ANALYSIS

Figure 4 shows a one-dimensional analytical model concerned with snow melting through radiative heating. q_{r0} is the amount of radiative heat flux onto the top surface of the snow layer. The initial temperature of the snow layer was uniformly maintained at T_{in} below $0°C$. h_s indicates the distance from the initial surface position to the transient surface position due to surface melting. Melt water produced by surface and internal melting percolates into the snow layer, creating two snow layers, one wet and the other dry, as denoted in Fig.4. It may be postulated that the temperature distribution in the wet snow zone was uniformly maintained at $0°C$. Therefore the main problem in the wet snow zone was the water percolation behaviour. The interface of wet and dry snow zones, which is called the seepage front in this paper, moves gradually downward with the freezing of the melt water.

3.1 Basic Equation

3.1.1 Wet snow zone ($h_s < y < h_d$)   In this investigation, the melt water percolation in the snow layer was assumed to be caused by gravity and capillary action. Applying Krisher's capillary model to the law of Poiseuille, the percolation rate of the melt water W is represented in the following equation, which was devised by Seki et al. [7].

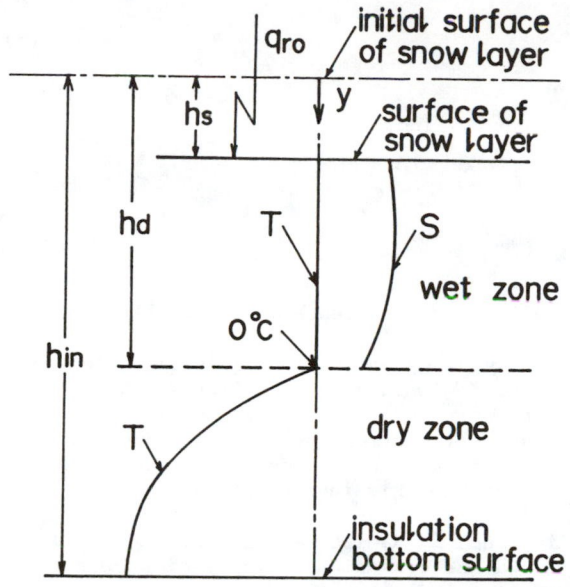

FIGURE 4. Schematic diagram of the analytical model

$$W = -\overline{\gamma}\ \frac{\tau d}{\mu}\ \frac{\rho_w\varepsilon^2}{2(1-\varepsilon)\phi}\ \{-\frac{g\rho_w d}{\tau}\ \frac{\varepsilon}{(1-\varepsilon)\phi}\ -\frac{dP}{dy}\} \qquad (2)$$

where P is the non-dimensional capillary suction pressure, and $\overline{\gamma}$ is the effective way factor. These are defined in the following:

$$P = \frac{P_t - P_c}{\tau \cdot}\ \frac{\varepsilon}{1-\varepsilon}\ \frac{d}{\phi} \qquad (3)$$

$$\overline{\gamma} = \gamma \int_{S_{min}}^{S}\frac{1}{P^2}\ dS \qquad (4)$$

In addition, P is represented as a function of water saturation by the experimental results of Ohtani [8].

$$P = 0.77 + \frac{0.3}{S+0.28} - \exp\{30(S-1)\} \qquad (5)$$

Substituting eq.(5) into eq.(2), the percolation rate of the melt water can be expressed as a function of water saturation S, which is convenient for analysis. Accordingly, a mass balance equation in the wet snow zone can be introduced as follows:

$$\varepsilon\rho_w \frac{\partial S}{\partial t} = -\frac{\partial W}{\partial y} - \frac{1}{L}\frac{\partial q_r\{y\}}{\partial y} \tag{6}$$

where the variation of the porosity ε, induced by the internal melting, is represented as follows:

$$\varepsilon = -\frac{1}{L\rho_i}\int_{t_{in}}^{t} \frac{\partial q_r\{y\}}{\partial y}\, dt + \varepsilon_{in} \tag{7}$$

where t_{in} denotes critical time for the occurrence of internal melting.

The second term on the right hand side of eq.(6) indicates the amount of internal melting, and $q_r\{y\}$ is the downward radiation flux which transmits into the snow layer at position y. Generally, it is difficult to estimate $q_r\{y\}$ exactly because of its strong spectral dependence. In this study, the following simple treatment was produced to consider the effect or radiation sources utilized in this investigation.

(1) Heating by short wave radiation: Radiative heat flux $q_r\{y\}$ in snow is represented   by   a one-dimensional simple relationship, similar to a form reported by Weller [9].

$$q_r\{y\} = (1-Z)\Phi_1 q_0 \exp\{-k_e y\} \tag{8}$$

where $\Phi_1 = q_{r0}/q_0$ and Z is the ratio of upward to downward components of radiation. In addition, q_0 in this investigation may be represented approximately by the following equation (see Fig. 3).

$$q_0 = \int_0^\infty E_{b\nu}\exp\{-2a_{g\nu}h_g - a_{w\nu}h_w\}d\nu \tag{9}$$

where h_g and $a_{g\nu}$, h_w and $a_{w\nu}$ indicate the thickness and the absorption coefficient of the glass and water film, respectively.

(2) Heating by blackbody radiation: For the heating condition characterized by a wide wavelength range $0 \sim \infty$, $q_r\{y\}$ in eq.(6) corresponds to the shortwave energy in the wavelength range $0 \sim \nu_c$.

$$q_r\{y\} = (1-Z)\Phi_2 \int_0^{\nu_c} E_{b\nu} \exp\{-k_e y\}d\nu \tag{10}$$

where $\Phi_2 = q_{r0}/\sigma T_b^4$, ($T_b$:temperature of blackbody)

However, it is difficult to obtain information about the critical opaque wavelength ν_c for snow. To overcome this difficulty, ν_c is postulated as producing 1.4 microns meter, by taking into account both the short wave radiation band as shown in Fig.2, and the characteristic of the extinction coefficient for cloudy ice investigated by Seki et al. [10].

3.1.2 Dry snow zone ($h_d < y < h_{in}$)   In this region, surface and internal melting do not need to be considered, so it is sufficient to solve the following energy equation:

$$\frac{\partial T}{\partial t} = \kappa \frac{\partial^2 T}{\partial y^2} - \frac{\kappa}{\lambda} \frac{\partial q_r\{y\}}{\partial y} \tag{11}$$

The second term on the right hand side of eq.(11) represents the amount of radiation that is absorbed. The treatment of $q_r\{y\}$ is the same as in eq.(6).

3.2 Boundary Conditions

3.2.1 At the top surface ($y=0$ or $y=h_s$)   The boundary condition at the top surface of the snow layer changes depending upon the type of radiation source used, as demonstrated in the previous section.

(1) Heating by short wave radiation: Neglecting the sublimation at the top surface, the energy balance can be derived from the following equation:

$$y = 0; \quad \lambda \frac{\partial T}{\partial y} = \alpha(T_s - T_\infty) \tag{12}$$

(2) Heating by blackbody radiation: Since the effect of radiative energy in the wavelength range of $0 \sim \nu_c$ is included in the basic energy equation of (6) and (11), as indicated in the previous section, this supposition has been adopted in this study. Long wave radiation q_2 in the range of $\nu_c \sim \infty$ is added into the boundary condition equation, as,

$$y = 0; \quad \lambda \frac{\partial T}{\partial y} = \alpha(T_s - T_\infty) - q_2 \tag{13}$$

where

$$q_2 = (1-\beta)\Phi_2 \int_{\nu_c}^{\infty} E_{b\nu} d\nu \tag{14}$$

By defining q_{me} as the net amount of heat necessary to begin melting the snow, the percolation rate W_s of the melt water at $y=h_s$ can be represented as

$$W_s = \frac{(1-\varepsilon)\rho_i + \varepsilon S \rho_w}{(1-\varepsilon)\rho_i} \frac{q_{me}}{L} \tag{15}$$

Assuming that W_s in eq.(15) flows down into the snow layer according to the law represented in eq.(2), the boundary condition at the surface is given below as,

$$y=h_s; \quad \frac{\partial S}{\partial y} = -\{1/(\frac{\partial P}{\partial S})\}\{W_s \frac{2\mu(1-\epsilon)\phi}{\overline{\gamma}\tau d\rho_w\epsilon^2} + \frac{g\rho_w d}{\tau}\frac{\epsilon}{1-\epsilon}\} \tag{16}$$

Another equation is needed for the representation of the downward movement of the top surface position induced by surface melting, and is given as:

$$(1-\epsilon)L\rho_i \frac{dh_s}{dt} = q_{me} \tag{17}$$

3.2.2 At the seepage front $(y=h_d)$ Some of the melt water percolating into the seepage front freezes, while residual amount remains in a liquid state. Therefore, the boundary condition at the seepage front becomes (see Figure 5):

$$y=h_d; \quad \overline{\epsilon}\overline{S}\rho_w \frac{dh_d}{dt} = W_d - (-\frac{\lambda}{L}\frac{\partial T}{\partial y}\Big|_{h_d}) \tag{18}$$

where $\overline{\epsilon}$ and $\overline{S}$ indicate the snow porosity and water saturation, respectively, in a small strip dh_d, as shown in Fig.5. This paper investigates monotonically increasing seepage fronts. Therefore, a restricting condition, given below, is needed:

$$W_d > -\frac{\lambda}{L}\frac{\partial T}{\partial y}\Big|_{h_d} \tag{19}$$

In this analysis it is assumed that the moving rate of the seepage front dh_d/dt can be expressed as:

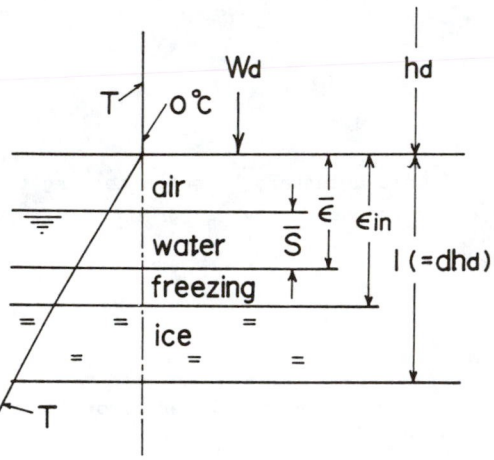

FIGURE 5. Small strip at the seepage front

$$\frac{dh_d}{dt} = \frac{W_d}{\varepsilon_d s_d \rho_w} \tag{20}$$

On the other hand, the temperature at the seepage front can be assumed to be the melting point.

$$y = h_d; \qquad T|_{h_d} = 0 \tag{21}$$

3.2.3 At the bottoms surface ($y=h_{in}$)   The boundary condition at the bottom surface of the dry snow zone is introduced in the following equation and takes into account the insulation substrate used in this experiment.

$$y = h_{in}; \qquad \frac{\partial T}{\partial y}\Big|_{h_{in}} = 0 \tag{22}$$

After h_d arrives at the bottom surface, the entire snow layer becomes a wet zone at $0°C$. Therefore,   eq.(6), which indicates the melt water percolation, can be applied. Accordingly, the boundary condition at the bottom surface can be found using the following equation, taking into account the fact that there is no seepage of water into the substrate.

$$y = h_{in}; \qquad W|_{h_{in}} = 0 \tag{23}$$

3.3   A Consideration of the Boundary Condition for Numerical Analysis

The equation for predicting the rate of snow melting by radiative heating is very complicated to solve. As demonstrated in the previous section, it is generally difficult to obtain a closed form analytical solution. Therefore, the present numerical predictions are carried out by using an implicit finite difference method. From the standpoint of numerical calculation, it is necessary to consider the effect of radiative energy absorbed near the top surface when using short wave radiation $q_r\{y\}$. Accordingly, the energy absorbed in half a mesh of the first difference node is as follows:

$$q_1 = \left(\frac{3}{4}\frac{\partial q_r\{0\}}{\partial y} + \frac{1}{4}\frac{\partial q_r\{\Delta y\}}{\partial y}\right)\frac{\Delta y}{2} \tag{24}$$

The above relation is dealt with in previous work [6]. Therefore, q_1 could be included in the boundary conditions at the top surface by subtracting it from the right hand side of eq.(12) and (13). In addition, q_{me} in eqs (15) and (17)   can be found using the following relation when using blackbody radiative heating.

$$q_{me} = q_1 + q_2 - \alpha(T_s - T_\infty) \tag{25}$$

However, one notes that q_2 (see eq.(14)) in eq.(25) has a value of zero under the heating conditions of short wave radiation.

3.4 Characteristics of Some Properties of Snow Examined in This Investigation

The thermal conductivity λ and the reflectivity β for the density ρ_{in} =500 kg/m^3 used in this investigation were estimated as 0.42W/(mK) and 0.7, respectively, by Yoshida [11]. The mean diameter of snow particle d was measured as being 0.2 mm using micrograph. Generally the shape coefficient of a snow particle has a wide value range of $6<\phi<14$ with the corner angle. In this paper, ϕ=10 was adopted as the mean value. The ratio of upward to downward radiation components, Z, was evaluated as 0.8 by Weller [9].

4 RESULTS AND DISCUSSION

4.1 Behaviour of the Extinction Coefficient of Snow

Figure 6 indicates the measurements of the extinction coefficient k_e versus snow porosity ε. Generally, it is difficult to assess the extinction coefficient of snow because of certain very complicated characteristics of snow, such as the time-dependent variation of the mean diameter of a snow particle. However in this study, the variation of the extinction coefficient due to porosity was considered. Valuable data produced by Oura [12], were employed in this figure for comparison with present data.

The results demonstrate that k_e decreases monotonically with increasing ε, and the scattering in the present data is comparatively small. The data measured by Oura [12] indicates a smaller value than the present data. However, it is clear that the deviation between both data is within one-order. The above mentioned tendency of the extinction coefficient is

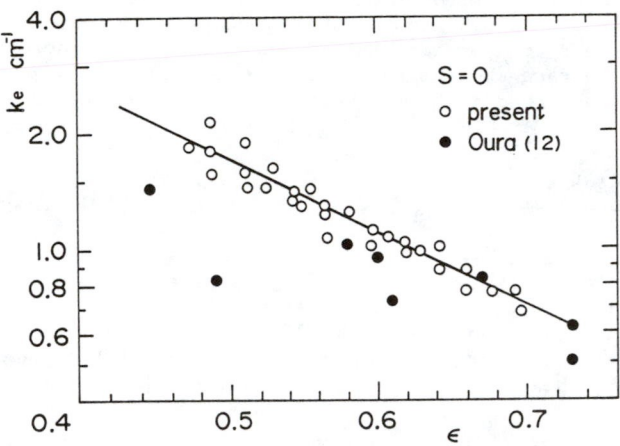

FIGURE 6. Extinction coefficient behaviour with porosity of snow

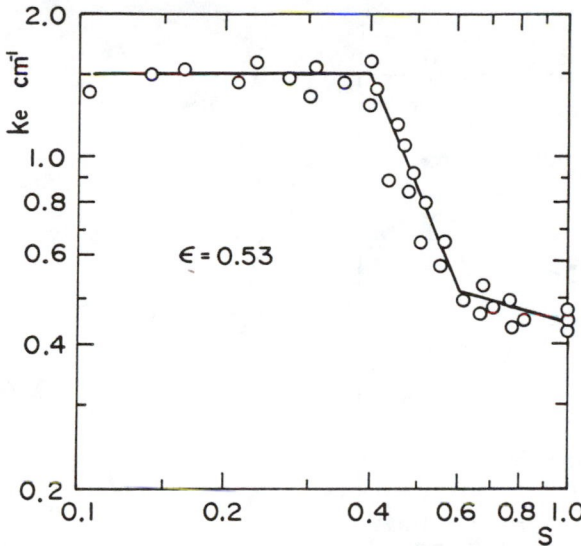

FIGURE 7. Extinction coefficient behaviour with water saturation of snow

applicable only to the specific condition of dry snow, that is, the snow sample does not have a water fraction. Figure 7 denotes the measured results of the extinction coefficient k_e of wet snow characterized by water saturation S in the case of $\varepsilon = 0.53$. k_e in the range of $0.1 < S < 0.4$ produces an almost constant value of 1.5, however, the k_e decreases rapidly in the range of $0.4 < S < 0.6$. Therefore, water in the snow promotes the transmission of radiation. As it is very difficult to measure the k_e over a wide range of porosities ε, the following relationship may facilitate the present analysis of the snow melting problem.

$$k_e = 10^{(-1.9\varepsilon+1.2)} \qquad\qquad \dots\dots \; 0 < S < 0.4$$

$$k_e = 0.092 \times 10^{(-1.9\varepsilon+1.2)} S^{-2.6} \qquad \dots\dots \; 0.4 < S < 0.6 \qquad\qquad (26)$$

$$k_e = 0.302 \times 10^{(-1.9\varepsilon+1.2)} S^{-0.28} \qquad \dots\dots \; 0.6 < S < 1.0$$

4.2 Snow Melting Characteristics

In this section, typical snow melting characteristics are examined by comparing the experimental results with analytical ones. Figures 8 and 9 show the transient temperature distribution under two kinds of heating conditions, shortwave radiation and blackbody radiation, respectively. The other conditions such as radiative heat flux q_{r0} are all the same value.

It was found that the temperature in the dry snow zone increases monotonically over an increasing melting period. The temperature distribution for short wave radiation had a gentle slope. On the other hand, the temperature distributions for blackbody radiation had a steeper

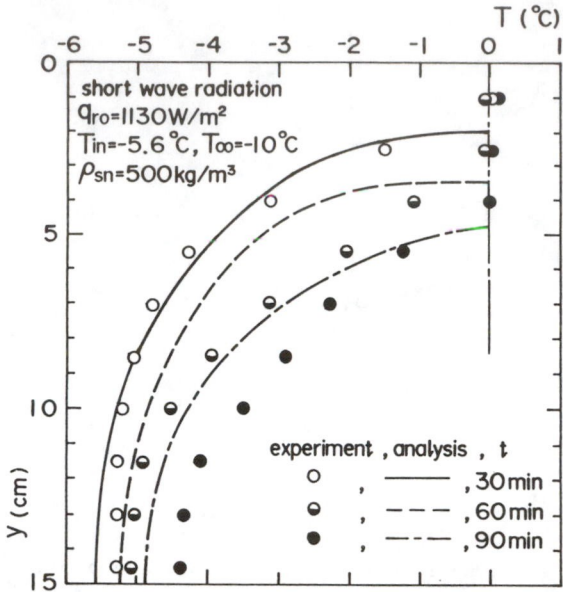

FIGURE 8. Transient temperature distribution by heating short wave radiation

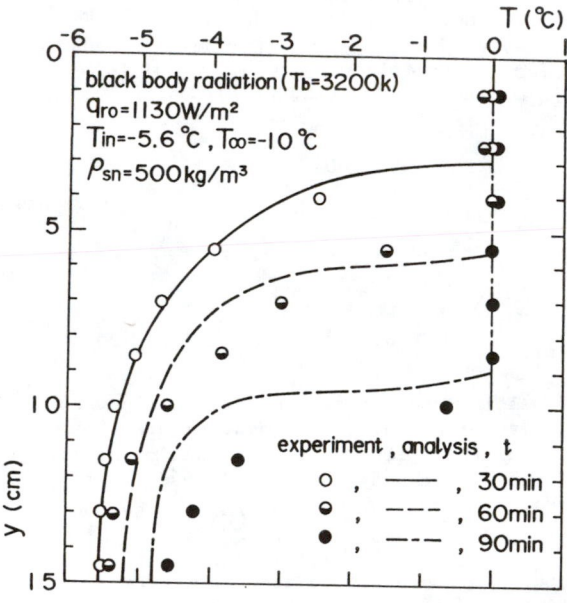

FIGURE 9. Transient temperature distribution by heating blackbody radiation

slope near the dry-wet interface compared to short wave radiation. This behaviour is mainly attributed to the opaque band (long wave radiation) included in the blackbody radiation which enhances the melting rate of snow. In other words, the transient temperature in the dry zone is significantly effected by the melt water percolation rate, which is the moving rate to the seepage front. The agreement between the experimental and analytical results was not always good, however, the present analysis is useful for predicting quantitatively the melting behaviour.

Figure 10 presents the variations in a seepage front h_d over time. The moving rate of the seepage front under blackbody radiation is faster than under short wave radiation. This behaviour is mainly attributed to the fact that the energy in the case of blackbody radiation includes the long wave radiation which enhances the rate of melting as discussed in Figs. 8 and 9. The h_d-t curves in Fig.10 may be of major interest for studies of the process of snow melting due to melt water percolation in the snow

layer. The moving rate is comparatively small at the beginning of the melting process. However, the rate increases gradually with increasing time. This interesting phenomenon mainly results from water saturation S and snow porosity ε, which generally increase with internal or surface melting during the melting process of snow.

Figure 11 shows an analytical prediction of snow density variation during the melting process under blackbody radiation heating conditions. Snow density ρ'_{sn} in this paper is defined as $\rho_{sn} = (1-\varepsilon)\rho_i + S\rho_w$. The result for t=30 min reveals that the density distribution has a higher value

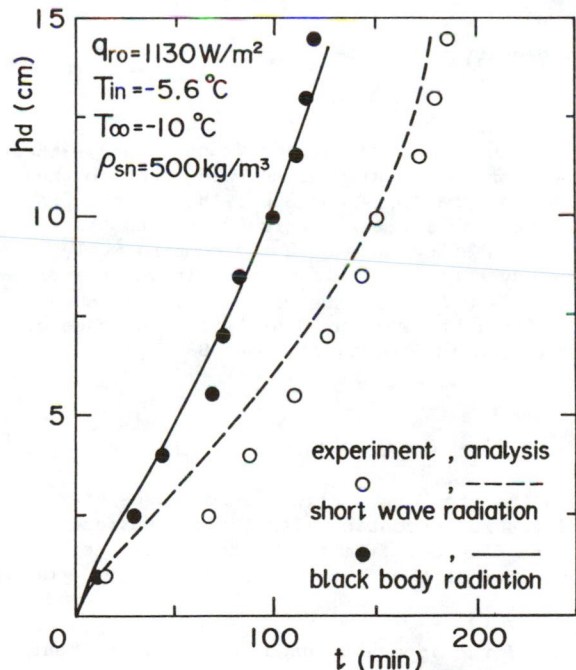

FIGURE 10. Variation of seepage front with time

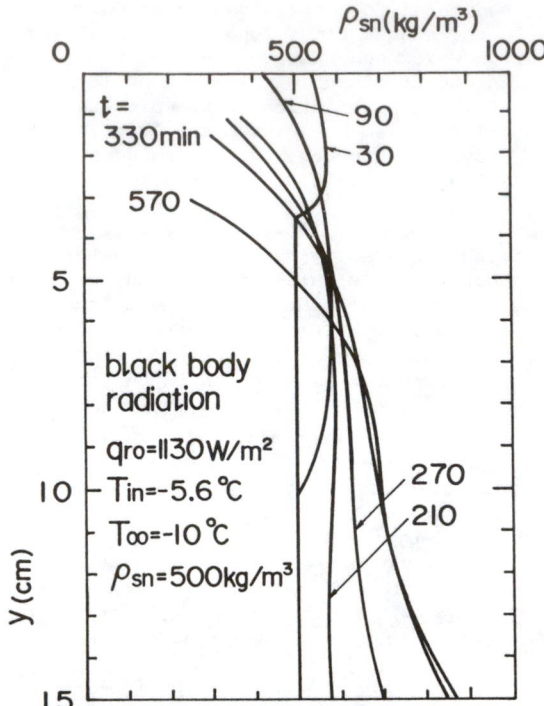

FIGURE 11. Variation of snow density with time

than the initial density of 500 kg/m^3 because of the inclusion of the melt water in the snow. The thickness of the seepage front h_d under these conditions is about 30 mm. However the density of snow adjacent to the top surface at t=90 min is smaller than ρ_{sn}=500 kg/m^3. This interesting behaviour is mainly attributed to the increased porosity due to the internal melting of snow particles after absorbing the comparatively short wave radiation in blackbody radiation. After the melt water arrives at the bottom surface, the density ρ_{sn} increases gradually, due to the retention of water at this surface. The maximum density for the case of the melting condition, as indicated in Fig.10, is about 850 kg/m^3, close to the density of ice.

4.3   Flow out of Melt Water from the Bottom Surface of the Snow Layer

The snow melting behaviour in the previous section is for cases where the bottom surface of the snow layer is in contact with a rigid surface. Therefore, melt water could not flow out from the bottom surface of the snow layer. However, the behaviour of the flow out of melt water needs to be investigated.

Figure 12 shows an adaptation of the apparatus used in Fig.1. The heat source used was infrared radiation from an electrical heater of nichrome wire. A steel net was used as the bottom surface of the snow layer to allow the flow out of the melting water.

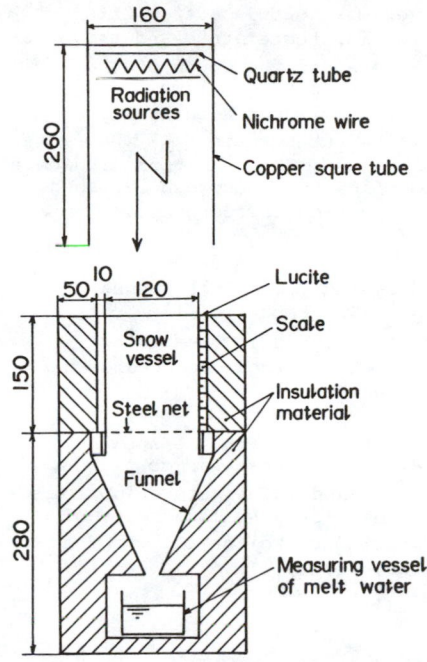

FIGURE 12. Experimental apparatus for flow out amount measuring

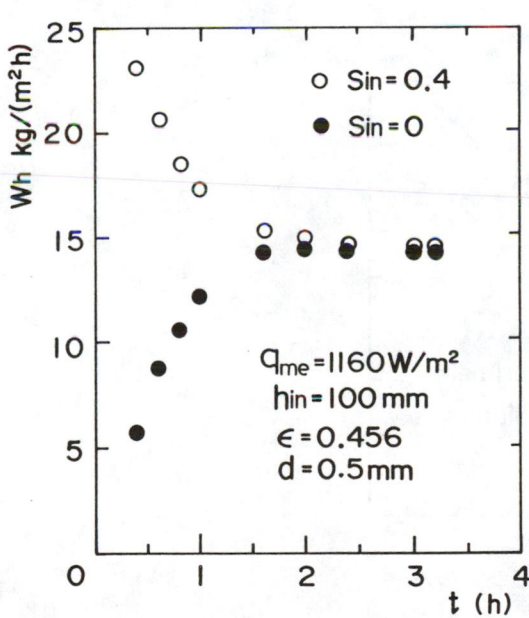

FIGURE 13. Quasi-steady flow out amount from the bottom surface of snow layer

Figure 13 shows the relation predicted between the amount of melted water flowing out W_h, and the time t. From the experimental results, the water saturation at the bottom of the snow layer was postulated for this analysis as having a value of 0.5. The initial temperature and the amount of water saturation S_{in} had a uniform value along the depth of the snow layer. In the initial melting region, the flow out amount was significantly effected by the initial amount of water saturation. However, the deviation of the amount of flow out between S_{in}=0 and S_{in}=0.4 became gradually smaller as time passed. This behaviour indicates the existence of a quasi-steady amount of flow out. Accordingly, the amount of flow out used hereafter in this section corresponds to the quasi-steady amount of flow out.

Figure 14 shows the amount of flow out compared to the net amount of the heat flux utilized for surface melting q_{me}. The amount of flow out increases with increasing q_{me}. The present analysis is useful for predicting the amount of flow out of melt water, since the predicted and the experimental results were in close agreement.

Figure 15 shows the distribution of water saturation in the snow layer and its effect on snow density. Water saturation increases downwards, toward the bottom surface of the snow layer, and water saturation for a large density of snow has a larger value than for a small density of snow. This typical behaviour is mainly attributed to the effect of the water retention capability by the capillary force.

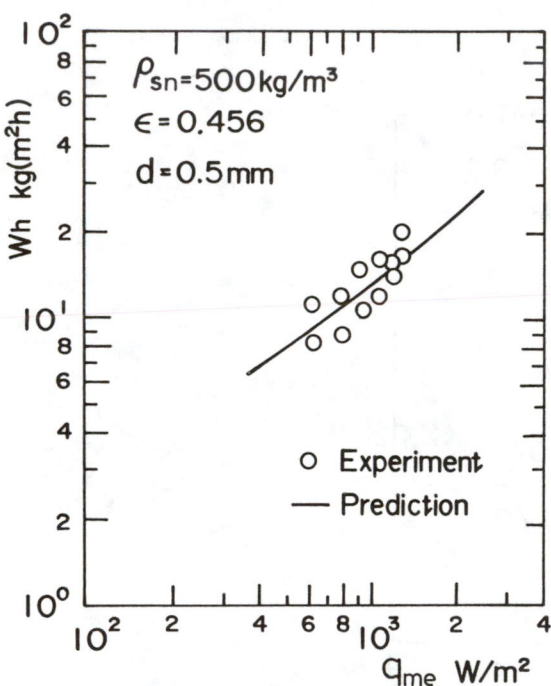

FIGURE 14. Comparison of experimental and prediction

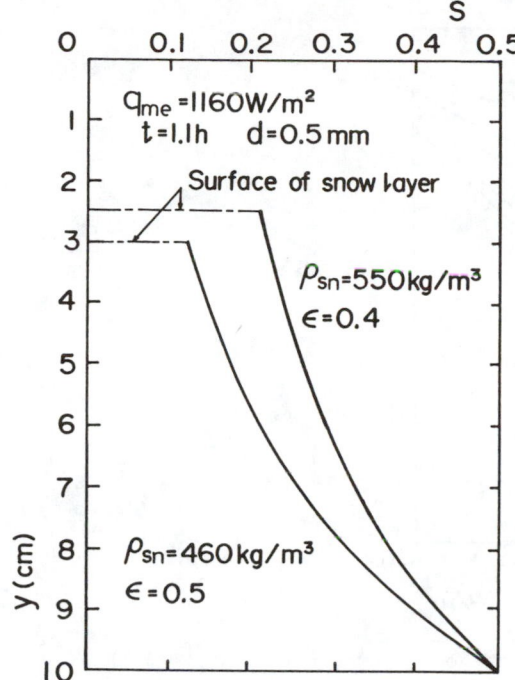

FIGURE 15. Distribution of water saturation in snow layer

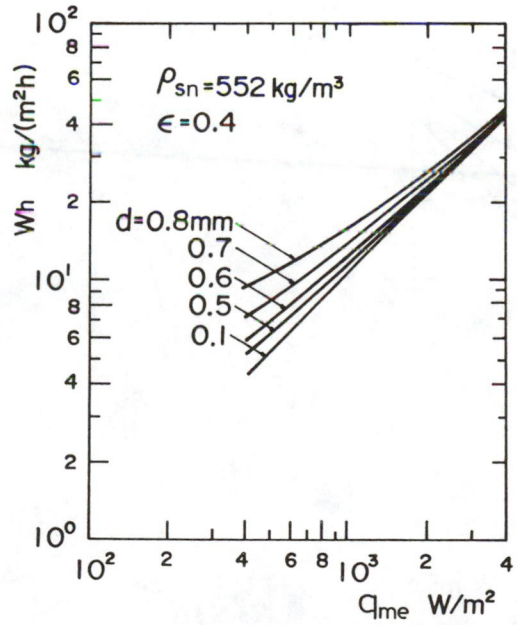

FIGURE 16. Flow out amount against heat flux for melting (ϵ=0.4)

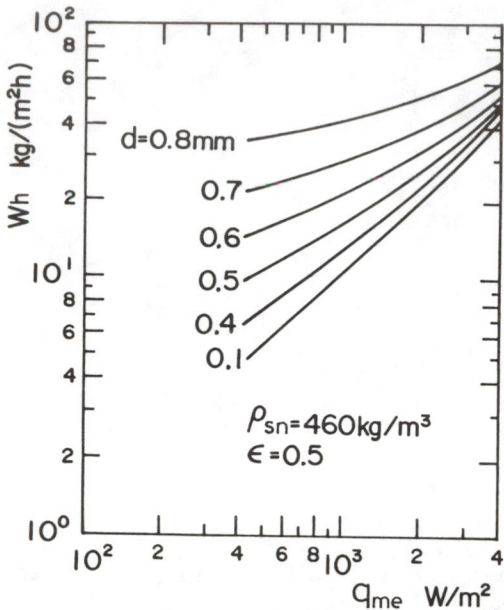

FIGURE 17. Flow out amount against heat flux for melting ($\varepsilon=0.5$)

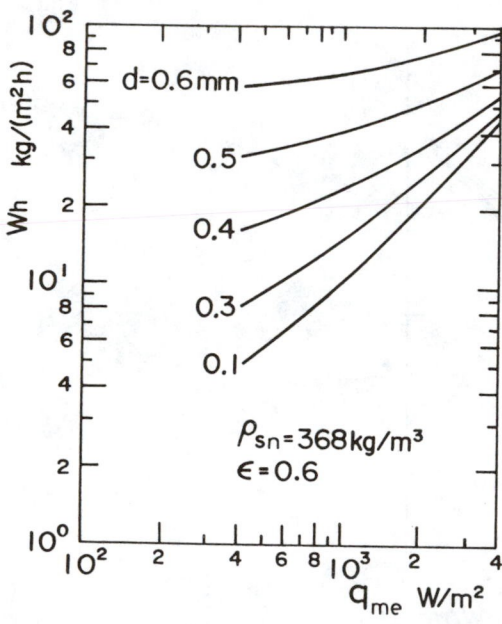

FIGURE 18. Flow out amount against heat flux for melting ($\varepsilon=0.6$)

Figure 16 through Figure 18 compares the flow out amount W_h against the heat flux for melting q_{me} for parametric values of the mean diameter of a snow particle d. The flow out amount for a high density of snow is not greatly affected by the diameter of snow particles (Fig.16). On the other hand, the flow out amount for a low density is strongly affected by the diameter of snow particles (Figs. 17, 18). Moreover the flow out amount increased with an increase in the diameter, and was not dependent on the density of snow. From the above behaviour, the percolation of melt water in the snow layer was seen to be significantly affected by the diameter of the snow particles, the density of the snow, and the heat flux for melting.

5. CONCLUSION

These experimental and analytical investigations were conducted in order to clarify snow melting behaviours caused by radiative heating. The following conclusions can be drawn from the results of this study.
1. The energy transfer to short wave radiation in snow is affected mainly by snow porosity and water saturation.
2. It is necessary to consider the wave band of incident radiation when evaluating the amount of melted snow in a snow layer.
3. The transient temperature distribution in the dry snow zone located under the wet snow zone is affected by the moving rate of the seepage front, which predominantly results from the amount of melt water.
4. The moving rate of the seepage front can be predicted using the present analysis, based on the law of gravity and capillary suction pressure.
5. The present analysis makes it possible to predict the increase and decrease of snow density during the melting process.
6. The amount of melt water flowing out from the bottom surface is strongly affected by the diameter of snow particles, the density of the snow, and the heat flux for melting.

ACKNOWLEDGMENT

It is a pleasure to acknowledge the work of undergraduate students K. Tateno and the late T. Sugawara in conducting the experiment presented herein and in preparing the figures.

NOMENCLATURE

d	mean value of snow particle diameter
$E_{b\nu}$	monochromatic emissive power of blackbody radiation
g	acceleration of gravity
h	length from initial top surface of the snow layer
k_e	extinction coefficient of snow
L	latent heat of melting
P	non-dimensional capillary suction pressure
pc	capillary suction pressure
P_t	total pressure
$q_r\{y\}$	short wave radiation transferred through the snow layer
q_{r0}	radiative heat flux impinging on the top surface of the snow layer
q_0	short wave radiation obtained using water filter, eq.(9)

q_1	short wave radiation absorbed at the top surface from the standpoint of numerical analysis, eq.(24)
q_2	long wave radiation absorbed in the top surface
q_{me}	net amount of heat flux utilized for surface melting, eq.(25)
S	water saturation
S_{min}	critical water saturation of capillary force action
t	time
T	temperature
W	percolation rate of melt water
y	coordinate

Greek symbols

α	heat transfer coefficient
β	reflectance
γ	way factor
ε	porosity of snow
κ	thermal diffusivity of snow
λ	thermal conductivity of snow
μ	coefficient of viscosity
ν	wavelength
ν_c	critical wavelength
ρ	density
σ	Stefan-Boltzmann constant
τ	surface tension
ϕ	shape coefficient

Subscript

d	seepage front
i	ice
in	initial
s	surface of snow layer
sn	snow
w	water
∞	environment

REFERENCES

1. Golbeck, S. G. and Gail Devidson, Water percolation through homogeneous snow, The role of snow and ice in hydrology, Proceedings of the Banff Symposia, vol.1, Geneva Switzerland, pp.242-257, Sept. 1972.

2. Wakahama, G., Infiltration of melt water into snow cover I, Low Temperature Science (Japan), Ser. A, vol.21, pp.45-71, 1963.

3. Yoshida, J., Infiltration of thaw water into a dry snow cover, Low Temperature Science (Japan). Ser. A, vol.31, pp.117-133, 1973.

4. Kuroiwa, D., Liquid permeability of snow, Low Temperature Science (Japan), Ser. A, vol.26, pp.29-52, 1968.

5. Gilpin, R. R., Heat transfer in a horizontal water layer with radiative heating, Transactions of the CSME, vol.1, pp.213-218, 1972.

6. Seki, N., Sugawara, M. and Fukusako, S., Radiative melting of a horizontal clear ice layer, Wärme-und Stoffübertragung, vol.11, pp.207-216, 1978.

7. Seki, N., Fukusako, S. and Tanaka, M., Drying of porous media included water under the heating condition of high heat flux, Transactions of JSME, vol.43, pp.1086-1095, 1977.

8. Ohtani, S., Drying and heat transfer, Research of Mechanical Engineering (Japan), vol.21, pp.129-139, 1969.

9. Weller, G. E., The heat budget and heat transfer processes in antarctic plateau ice and sea ice, Anare Science Reports, Ser.A(IV) Glaciology, Publication No.102, 1969.

10. Seki, N., Sugawara, M. and Fukusako, S., Back melting of a horizontal cloudy ice layer with radiative heating, ASME Journal of Heat Transfer, vol.101, pp.90-95, 1979.

11. Yoshida, J., Internal melting of snow due to the penetrating sunlight, Low Temperature Science (Japan), Ser. A, vol.19, pp.97-110, 1960.

12. Oura, H., Reflection and transmission of light by snow cover II, Low Temperature Science (Japan), vol.6, pp.35-40, 1951.

CONTENTS

1. Introduction

2. Experimental Apparatus and Procedure

3. Analysis
   3.1 Basic Equation
   3.1.1 Wet snow zone ($h_s < y < h_d$)
   3.1.2 Dry snow zone ($h_d < y < h_{in}$)
   3.2 Boundary Conditions
   3.2.1 At the top surface ($y = 0$ or $y = h_s$)
   3.2.2 At the seepage front ($y = h_d$)
   3.3 A Consideration of the Boundary Condition for Numerical Analysis
   3.4 Characteristics of Some Properties of Snow Examined in this Investigation

4. Results and Discussion
   4.1 Behaviour of the Extinction Coefficient of Snow
   4.2 Snow Melting Characteristics
   4.3 Flow Out of Melt Water from the Bottom Surface of the Snow Layer

5. Conclusion

411

FROST FORMATION

Chapter 13

Heat and Mass Transfer in the Frosting Process on Cold Surfaces

HAKARU SAITO and IKUO TOKURA
Department of Mechanical Engineering, Muroran Institute of Technology, Muroran 050, Japan

ABSTRACT

A review work concerning the studies on heat and mass transfer in frost formation on cold surfaces and the investigations in the related fields has been performed. The literature was reviewed, classifying into nine categories; frost formation in forced and natural convection, in cryogenic temperatures, heat and mass transfer coefficients on the surfaces of frost layers, thermal properties of frost and methods for the prediction, and recent studies on frosting phenomena. Also, paying attention to the incipient process of frost formation, the mechanisms of growth of frost layers were clarified and dimensionless parameters associated with the phenomena were derived.

CONTENTS

NOMENCLATURE

C_p : specific heat of humid air
d : diameter of circular cylinder
H_f : thickness of frost layer

$\bar{H}_f$: thickness of frost layer, averaged around a cylinder
h : heat transfer coefficient
h_D : mass transfer coefficient
$\bar{h}$: mean value of heat transfer coefficient
$\bar{h}_D$: mean value of mass transfer coefficient
L : latent heat of sublimation of ice
$\dot{m}$: mass flow rate of water vapor to cold surface
m_f : mass of frost layer
P : porosity
p_∞ : partial pressure of water vapor in free stream
p_s : partial pressure of water vapor saturated at surface temperature of frost layer
$p_{\infty s}$: partial pressure of water vapor saturated at temperature of free stream
u_∞ : air velocity of free stream
Pr : Prandtl number of humid air
Sc : Schmidt number
T_m : mean temperature of frost layer
T_w : temperature of cooling surface
T_s : surface temperature of frost layer
T_∞ : temperature of free stream
T_o : melting temperature of ice
x : characteristic length
ρ_a : density of humid air
$\rho_{1\infty}$: density of water vapor in surrounding air
ρ_{1f} : density of water vapor on surface of frost layer
ρ_1 : density of water vapor on cooling wall
ρ_f : density of frost layer
ρ_{ice} : density of ice
λ_a : thermal conductivity of air
λ_c : thermal conductivity of composite material
λ_{cmax} : maximum value of thermal conductivity of composite material
λ_{cmin} : minimum value of thermal conductivity of composite material
λ_f : thermal conductivity of frost layer
λ_g : thermal conductivity of gas
λ_s : thermal conductivity of solid phase
λ_{ice} : thermal conductivity of ice
τ : time
ϕ : relative humidity

1. INTRODUCTION

Frost formation is a phenomenon in which water vapor in the air condenses on a cold surface and forms a porous layer of ice-crystals, and many examples can be seen both in nature and in industry. In northern areas, cold morning air causes frost damage in agricultural crops. In northern countries, it has been reported that humidity underground moves toward the ground surface, freezing close to the surface and causing deformation in buildings, which is known as frost- heaving.

In industry, frost layers growing on low temperature heat transfer devices increases resistance to air–flow and then lowers the thermal performance of heat exchangers. Therefore, defrosting is necessary to maintain the prescribed performance of the heat exchangers. Since successful methods to prevent frosting on cold surfaces have not been proposed up to now, it is necessary to predict frost growth and its thermal properties to establish effective measures to control defrosting timing.

In some engineering applications, frosting is used to separate a component from a multicomponent mixture. However, useful examples of frosting for industrial purposes are quite few.

Frosting is a simultaneous transfer of heat and mass. The frost layer increases in thickness and density since water vapor condenses not only on the surface of the layer, but also inside the layer because of its porous construction. Furthermore, frosting is a transient phenomenon, varying its characteristics depending upon the ambient temperature, water vapor concentration and the surface temperature of the cooling plates. Therefore, frosting problems have to be treated differently from heat transfer problems associated with moving boundaries of some constant temperatures, such as the freezing of water on a surface cooled below zero degrees. In this case, the moving rate of the boundary is directly proportional to the amount of heat required for a phase change. However, a frost layer is composed of numerous ice crystals which originate from nuclei on the wall and grow inward, interfering with adjoining ones. Therefore, the problem cannot be simplified as in the case of the growth of snow crystals, in which the ice crystal develops from a single nucleus.

Studies on frosting were performed by Piening [74] in 1933 for a horizontal cylinder in natural convection and by Hiltz [31] in 1940 for the case of forced convection in a doubled tube heat exchanger. These are mainly concerned with qualitative characteristics of frost layers. Observations were made of development and external appearance of frost layers. A more precise study was conducted by Kamei et al. [45] in 1950, and they found that an analogy exists between heat and mass transfer. This study is probably one of the earliest reports which treat the frosting problem quantitatively. So, the frosting problems is relatively new as a subject of engineering study.

Most early investigations of frost formation were carried out experimentally and concerned with such properties of the frost layer as density or thermal conductivity. To solve the problem analytically, the factors governing the growth rate and physical properties of frost layers had to be known and predictable theoretically. This is the main problem which investigators encountered and are making efforts to solve even now.

Since many different types of heat exchangers are used in engineering, frost formation is studied for various geometries of surfaces such as flat plate, in-tube flow, concentric tube heat exchangers, cross flow tubes or tube-banks, fin-coil heat exchangers across convective flow, vertical or horizontal plates or tubes or tube-banks under natural convection conditions.

In most previous work on frost formation, the thick frost layers were investigated and the heat and mass transfer coefficients were defined in the same manner as in the case of flat solid surfaces. The results deviated greatly between investigators. Therefore, it can be said that much uncertainty still exists even in the fundamental data such as coefficients of heat and mass transfer in the frosting process.

The prediction of frost thickness is very important in engineering for estimating heat transfer rates of low-temperature heat exchangers, for designing air flow spaces in air conditioning apparatus, and for determining the rational termination of defrosting in heat pump systems. There are very few studies on the growth of frost layers, but these may be classified into two types of approaches. One is to predict frost thickness by mathematical modeling of frost growth. The major focus is, in this case, on examining the mechanisms of heat and mass transfer. In the second approach, empirical formula associated with some of the physical quantities involved are used, and are applicable to very limited frosting conditions such as frosting in highly humid ambient atmospheres.

In this paper, studies on frost formation are reviewed and summarized from the viewpoint of heat transfer engineering. This study presents some aspects for new approaches to investigating the frosting phenomena.

2. CLASSIFICATION OF STUDY

Much attention has been concentrated on the frosting process since it causes a decrease in heat transfer rates and an increase in the power required for driving air flow through heat exchangers operating at temperatures below the freezing point of water vapor.

Investigations on the frosting problem can be classified into three categories [72], depending upon the main target of the investigation.

Most of reports are concerned with heat and mass transfer during frost deposition. These are classified further by shape of heat transfer surfaces, into cases rather simple surfaces such as plates or tubes, and cases with more complicated configurations such as tube-banks or heat exchangers in actual use. In the second category, properties of frost layers such as density, thermal conductivity and radiative emissivity are the main focus of the investigation. In the third category, attention is focused on the prediction of growing rate of frost layers.

The physical properties are very complicated since water vapor in air diffuses and condenses, some on the surface of the frost layer and the remainder penetrates inside the frost layer. In the early stages of frost growth, heat transfer can be regarded as a heat transfer on a rough surface. As the frost layer grows, the heat transfer must be regarded as a coupled transfer of heat and mass both on the surface and inside the frost layer. Since the change occurs gradually and continuously with time, the two stages mentioned here have to be treated not as individual problems but as problems related to each other.

3. FROSTING UNDER FORCED CONVECTION

3.1 Frost Formation on Flat Plates

Many reports have been published on heat and mass transfer in frost formation on plates which are kept at a relatively high temperatures (higher than $-30\ °C$) [32, 36, 38, 43, 71, 110, 113]. Heat and mass transfer coefficients on flat plates in forced convection were measured by Yamakawa et al. [113] and Hayashi et al. [36, 37]. Fig. 1 shows the schematic view of their apparatuses, which mainly consisted of ducts and cooling plates, cooled by brines, on which frost layers developed. They pointed out that the coefficients of heat and mass do not vary with time except immediately after the beginning of the frost formation, where those coefficients were greatly influenced by the degree of roughness of the surface.

Hayashi and his colleagues in their report [38] classified the frosting process into three stages:(1) crystal growth, (2) frost layer growth, and (3) frost layer full growth. This classification was based on the experimental observation of frost layers in terms of the orientation and structure of ice crystals in the layers and in the manner of their development over time. Fig. 2 shows the classification of the structure and its map, expressed on a concentration-temperature

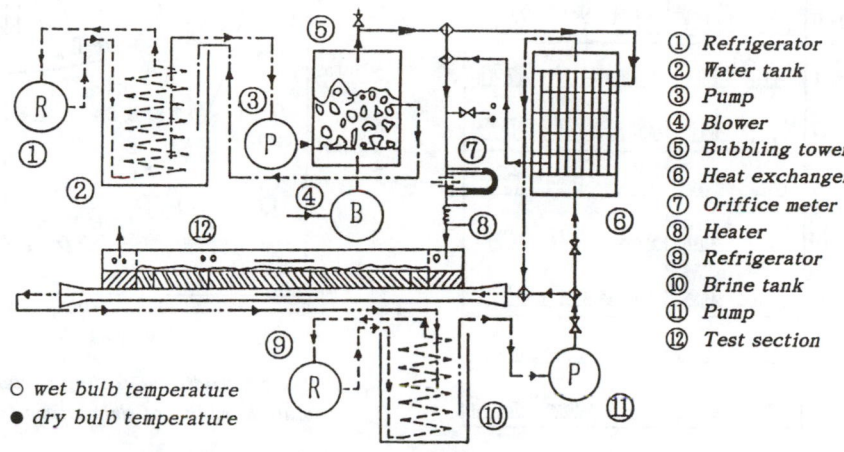

①	Refrigerator
②	Water tank
③	Pump
④	Blower
⑤	Bubbling tower
⑥	Heat exchanger
⑦	Oriffice meter
⑧	Heater
⑨	Refrigerator
⑩	Brine tank
⑪	Pump
⑫	Test section

○ wet bulb temperature
● dry bulb temperature

(b) Yamakawa's Testing Apparatus [113]

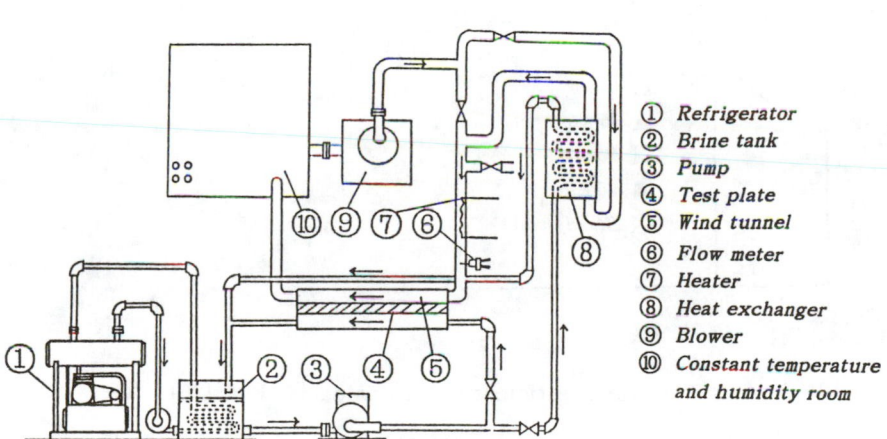

①	Refrigerator
②	Brine tank
③	Pump
④	Test plate
⑤	Wind tunnel
⑥	Flow meter
⑦	Heater
⑧	Heat exchanger
⑨	Blower
⑩	Constant temperature and humidity room

(b) Hayashi's Testing Apparatus [36, 37]

FIGURE 1. Yamakawa and Hayashi's Testing Apparatus

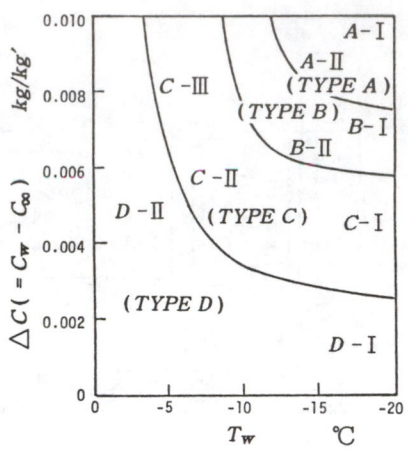

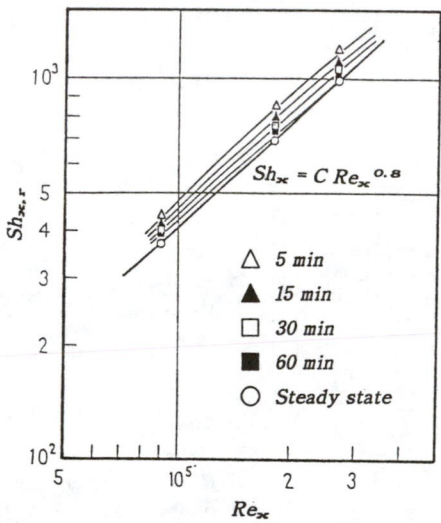

FIGURE 2. Hayashi's Classification of Frost Formation Type and Its Map[38]

C_W: absolute humidity at cooling surface
C_∞: absolute humidity of main stream

FIGURE 3. Mass Transfer Coefficient in Dimensionless Form[36]

diagram. They explained the periodical change in density, mass transfer coefficient, in relation to the above classification. This study can be evaluated because it proposes a generalized approach to the frosting problem, rather than an individual treatment of the problem depending upon the frosting conditions or the geometry of heat transfer surfaces. Fig. 3 shows the mass transfer on plates under forced convection.

3.2 Frost Formation on Cylinders

Heat and mass transfer on an tube set across convection flows was studied by Katsuta et al. [50] and Chung et al. [19]. These reports showed that the heat transfer rate decreases to a quasi-steady value in 1 to 2 hours, as shown in Fig. 4. This finding was also reported by Shah [89] and White [107, 108] for the frost formation in forced convection conditions, and by Cremers and Mehra [25] in free convection conditions. The quasi-steady state was attributed to the increasing thermal conductivity of the frost nullifying the effect of the increasing thickness of frost layers. Chung et al. also found that a frost layer with uniform thickness formed, usually soon after starting the frosting, along the entire surface of the cylinder except near the sides (Fig. 5). Furthermore, Schneider [90, 91, 92] reported that air velocities have little effect on the thickness of frost layers. This fact had been pointed out previously by Kamei et al. [45] and Yonko and Sepsy [110].

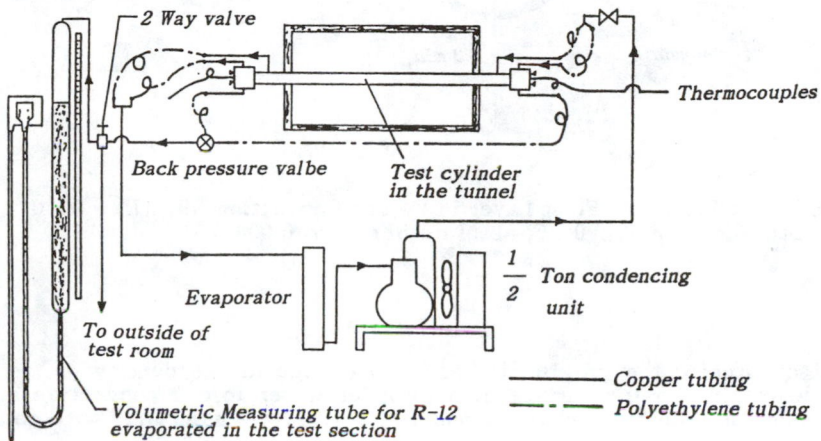

(a) Experimental Apparatus

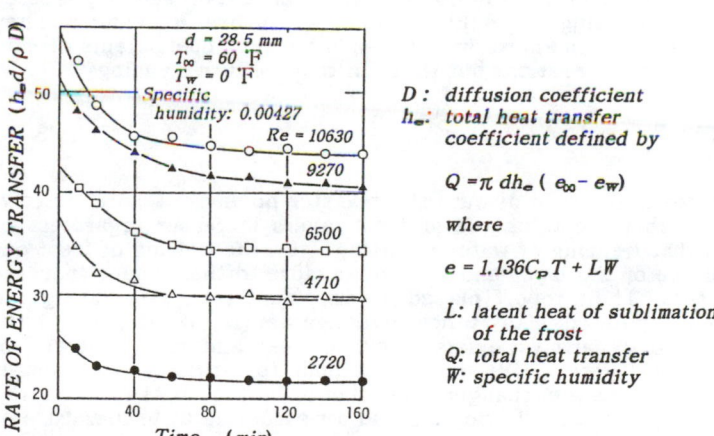

D : diffusion coefficient

h_e : total heat transfer coefficient defined by

$$Q = \pi\, dh_e\, (\,e_\infty - e_w)$$

where

$$e = 1.136 C_p T + LW$$

L: latent heat of sublimation of the frost
Q: total heat transfer
W: specific humidity

(b) Experimental Result:Heat Transfer Coefficient

FIGURE 4. Heat Transfer Coefficient on Frost Layers Growing on A Circular Cylinder in Cross Flow [19]

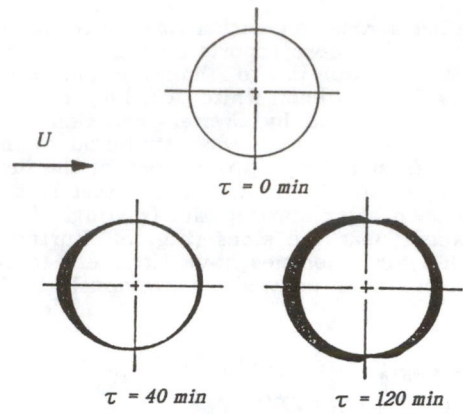

FIGURE 5. Distribution of Frost Layer in Forced Convection [19] (Re = 9270, d = 28.5 mm, T_∞=60 ° F, T_w=0 ° F, absolute humidity: 0.00427)

Aoki et al. studied the growth rate and the change in the density of frost layers developing on the surface of a cylinder under forced convective air flow[7] and also proposed a physical model of frost layers associated with condensation of water vapor on the surface [8].

3.3 Frost Formation in Tube Flow

The study by Kamei et al. [45], an early investigation, experimentally treated the forced convection frosting in a doubled tube type of heat exchanger. They found that the Lewis' relation approximately holds between coefficients of heat and mass transfer in frost formation, but that Chilton-Colburn's analogy

$$\frac{h}{h_D} = \rho \, C_p (Sc/Pr)^{2/3} \tag{3.1}$$

has to be considered in order to discuss the transfer phenomena more strictly. They also reported that the thickness of frost layers increases regardless of flow velocity and that freezing of water resulting from the melting of frost occurs near the surface of the layer and that areas close to the surface increase in density over time. Their report played an important role, stimulating the investigations which followed. Many other investigators [17, 49, 113, etc.] have verified that an analogy always exists between heat and mass transfer in frosting processes. Beatty et al. [9] studied methods to estimate heat transfer rates in double tube heat exchangers, as shown in Fig. 6, based on the difference in enthalpy between flowing air and air saturated at temperature of cold surfaces.

3.4 Frost Formation in Heat Exchangers

The flow of air in fin-coil heat exchangers, in narrow paths and in tubes was investigated under frosting conditions to find the increase in pressure drop to air flow through the exchangers and the change in heat transfer characteristics [3 ,4 , 10, 23, 33, 35, 39, 41, 42, 56, 57, 61, 63, 75, 112].

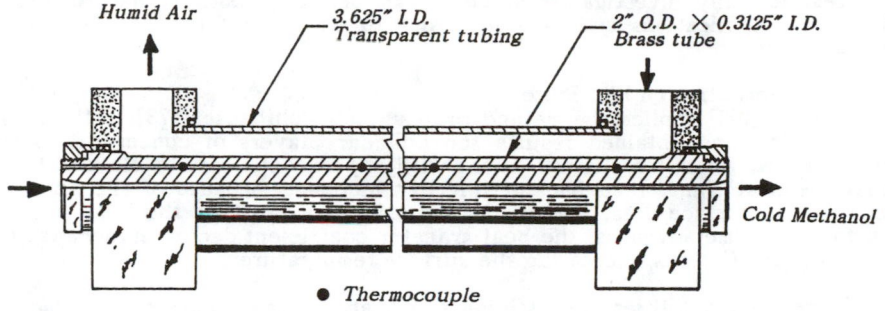

(a) Schematic View of Double–Tube Heat Exchanger

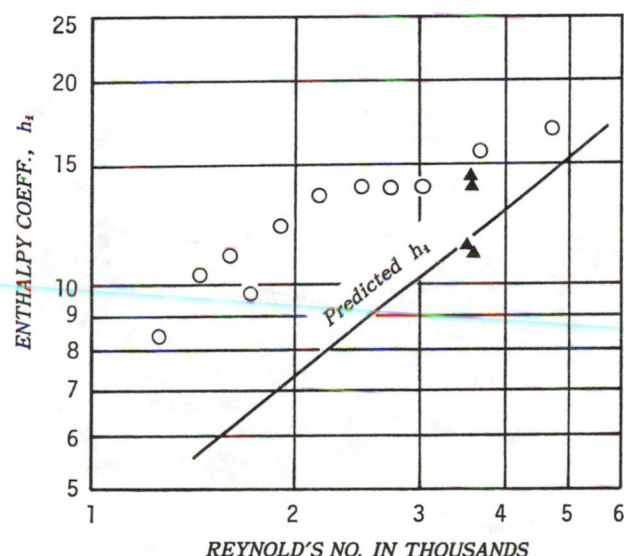

(b) Heat Transfer Correlation Based on Enthalpy Difference in Double–Tube Heat Exchanger

FIGURE 6. Heat Transfer Coefficient in Double–Tube Heat Exchanger [9]

4. FROSTING UNDER NATURAL CONVECTION

Heat and mass transfer in frost formation by natural convection has been of interest to many investigators since it can be found both in nature and in industrial applications and also is easier to use than forced convection in experiments.

4.1 Frost Formation on Flat Plate

Whitehurst [106] applied the method proposed by Pohlhausen [73] to the frosting problem and obtained results for boundary layers of concentration and temperature, which were in agreement with experimental results. He also found that the heat transfer coefficient initially increases because of the apparent increase in surface area, due to the deposition of frost on the heat transfer surface. As time advances, the heat transfer coefficient decreases because the frost surface flattens, increasing the surface temperature .

The time dependabilities of coefficients of heat and mass transfer were also examined by Goodman et al. [29] by using frosting cylinders with relatively large radii. They found that both coefficients increased slightly with time (Figs. 7 and 8), due to the increase in the surface area associating with heat and mass transfer. Furthermore, after analyzing temperature fields obtained by Mach–Zehnder interferometers, they concluded that the surface roughness of the frost layers had no effect on the temperature boundary layers. However, this conclusion is not always appropriate because the information obtained from interferograms is the average behavior through the light path and any local behavior near the surface cannot be obtained from this. An interesting fact observed by Goodman is the periodic fluctuations in surface temperature within a range of $\pm 1.5°C$.

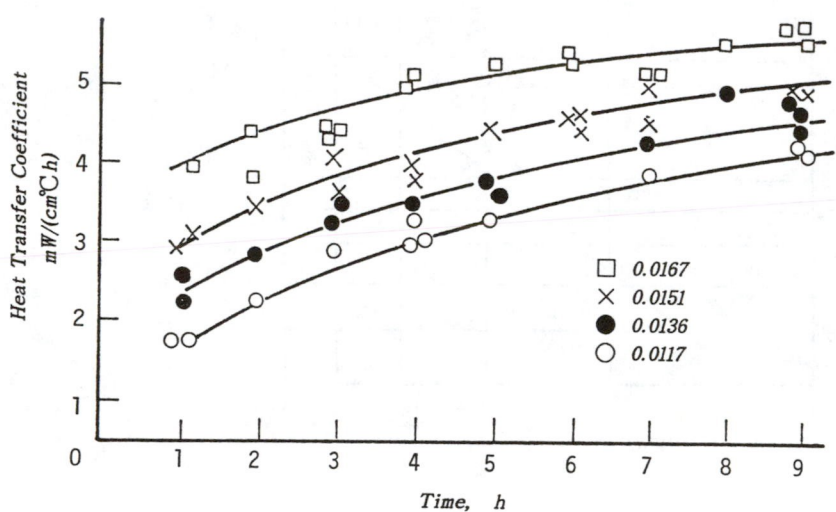

FIGURE 7. Variation of Heat Transfer Coefficient with Time [29]

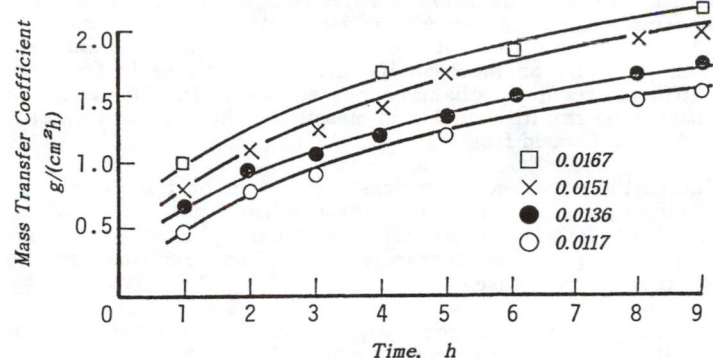

FIGURE 8. Variation of Mass Transfer Coefficient with Time [29]

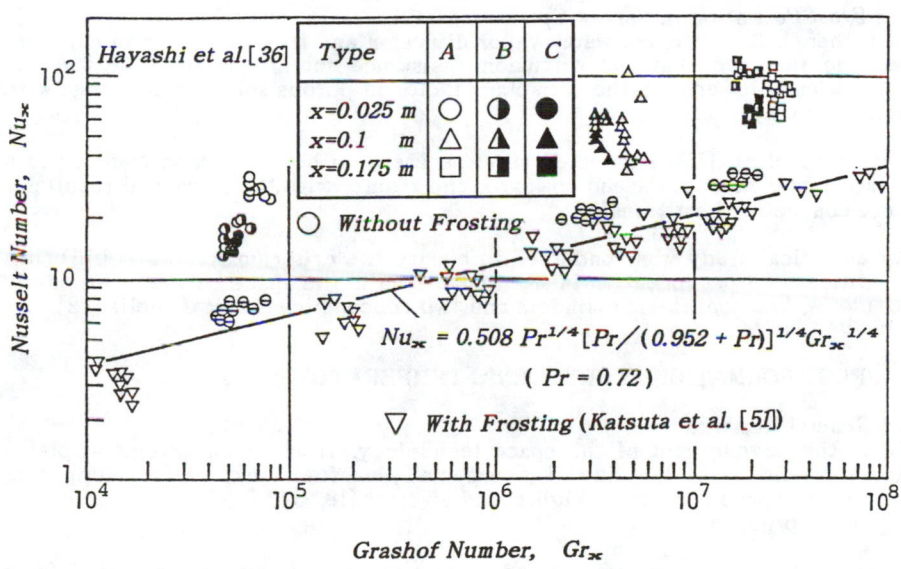

$$Nu_x = 0.508\ Pr^{1/4}[Pr/(0.952 + Pr)]^{1/4}Gr_x^{1/4}$$

$$(Pr = 0.72)$$

FIGURE 9. Heat Transfer Coefficients on a Flat Plate under Natural Convection

It was also reported that the heat transfer coefficients for vertical plates [51, 68] has the same value with or without frost formation [82]. These results have been plotted in Fig. 9. Contrary to these results, Hayashi et al. [36] conducted a similar experiment and found that the heat transfer coefficients remarkably increases in comparison to pure convective heat transfer (also being plotted in Fig. 9). Hayashi concluded that the increase in heat transfer coefficients was caused by the change in temperature neat the frost surface due to the delivery of latent heat, rather than by an increase in surface roughness in frosting. These completely different results probably occurred due to the difference in experimental conditions and the difficulties in measuring the surface temperatures of frost [111] and heat flux in frosting.

Tajima et al. conducted the experiments on frost formations on flat plates [98, 99, 100, 101, 102] facing upward, or downward, or vertically positioned. Okino and Tajima studied frost formations around a vertical cylinder [69]. Both researchers obtained the heat and mass transfer rates, the densities and the thickness of a frost layer for each case.

Considering the actual applications, defrosting processes were investigated for frost on a vertical plate [54], and the reduction rate of frost layer thickness by melting was compared with the analytical results, based on a model consisting of layers of porous ice and ice-water mixtures [55].

4.2 Frost Formations on Horizontal Cylinders
Schropp [84] measured the average values of heat and mass transfer rates and Katsuta et al. [48] obtained the local values. Both were almost in agreement with pure convective heat transfer values reported by Levy [58] and Hermann [30]. Stoeker [87] and Katsuta [52, 53] discussed the interaction of frosting with neighboring tubes in horizontally arranged tube banks.

4.3 Frost Formation in Narrow Spaces
Auracher [1, 2, 5] studied water vapor diffusion and frost formation in capillaries. He reported that the diffusion resistance factor for frost growing in capillaries is lower than the resistance factor in porous solids or packings with equal porosity.

Yamakawa et al. [115] conducted an experiment on heat and mass transfer in a closed rectangular space and compared the results with the numerical results of pure convective heat transfer.

An analytical study was conducted to clarify the criterion for the equilibrium conditions of frost formation in fluid contained in the space between two parallel plates, two concentric cylinders and two concentric spherical shells [62].

5. FROST FORMATION IN CRYOGENIC TEMPERATURES

5.1 General Remarks
Since the development of the space technology, frosting on cryogenic plates under conditions of low humidity, especially on fuel tanks or fuel supplying systems of space rockets or high speed aircraft [18, 34, 78, 85] has become an important problem to solve.

When frosting occurs around cylinders in a cross flow, sufficiently aged frost layers consist of two different types of frost: a coarse frost layer and a compacted frost layer similar to pure ice. This multilayered construction is created by a repeating cycle, where water melting at the surface soaks into the layer inside and freezes, increasing the thermal conductivity of the frost layer, and resulting in a drop in surface temperature which allows an additional growth of the frost layer [79].

Loper [60] calculated the transient heat transfer on a liquid oxygen container, and Smith et al. [88] derived a formula to predict mass transfer on a surface cooled to $-190°C$ by liquid nitrogen.

$$\dot{m} = [\, 0.1197\, u_{\infty} + 25.87\, (p_{\infty} - p_{s})^{1/2}][\, u^{1/2}(p_{\infty} - p_{s})]\quad \mathrm{lb_m/hft^2} \tag{5.1}$$

Barron and Han [12] conducted frosting experiments on nitrogen cooled vertical plates and compared those results with calculated results obtained by the integral method. Their results showed   good agreement for heat transfer rates, while measured mass transfer coefficients were one tenth of the theoretical expectation. They concluded that this discrepancy was caused by very fine crystals created in the boundary layer acting as obstacles to water vapor diffusion to the frost surface. Okubo and his colleagues [70] tested the hypothesis in their experiments and found the same reduction in mass transfer rate in their experiment using vertical plates with temperature below $-80°C$. This implies that   some singularities may exist in the cryogenic frosting process. There are, therefore, many problems to solve for these cryogenic frosting phenomena, including experimental verification for the above.

5.2 Transfer Coefficients under Fog Formation Conditions

In cryogenic frosting, fog formation along the cooling surfaces may play a role in total transfer phenomena. In such cases, mass transfer coefficients cannot simply be defined based on the difference in concentrations between the surface and the main stream. For fog formation along a low temperature surface, the so-called homogeneous nucleation theory with critical super saturation model has been proposed by Epstein and Rosner [28, 80]. However, for heat and mass transfer accompanying fog formation, few analytical reports have been published [81].

6. EFFECT OF FROST SURFACE ROUGHNESS ON TRANSFER COEFFICIENTS

Chen et al. [21] clarified the theory that heat transfer rates increase immediately after the beginning of frosting, due to the increase in the roughness of the surface. They derived an experimental formula for predicting heat transfer coefficient as a function of the surface-roughness of frost layers, based on a theory by Nikuradse [66, 67] and von Karman [44].

Hayashi et al. [36] analyzed the growing process of frost layers by using their own model and discussed the effect of the surface roughness on mass transfer in frost formation. They found that the mass transfer coefficients on the actual frost surface differ from those obtained by regarding the surface to be flat for two reasons: 30–40% occurs through concentration difference and 60–70% occurs from the increase in flow turbulence induced by the surface roughness. The frost model in this case consists of ice columns of the same height uniformly distributed on the cold surface. Using this model, the calculated frost layer grows almost linearly with time, which is not the case in actual experiments. Therefore, the conclusion they obtained for the mass transfer coefficient seems to have a very limited meaning, though it does provide a physical interpretation concerning the effect of surface roughness on mass transfer in frosting.

As mentioned above, the effect of the surface roughness of frost layers on the heat and mass transfer is still unknown. The most important step to obtaining a solution is to find a rational numerization of frost surface roughness and to include it in the calculation.

7. PROPERTIES OF FROST LAYERS

7.1 Theoretical Prediction of Thermal Conductivity

Thermal properties such as the thermal conductivity of frost have been investigated for many cases since it relates directly to heat transfer through the frost layers. Frost layers can be considered as consisting of a mixed material of ice crystals and the humid air surrounding them. Accordingly, many types of formulas have been proposed for the prediction of thermal conductivity, depending upon how the mixed construction is modeled.

For example, Brailsford [11] proposed formulas for estimating the possible minimum and maximum values of the mixture of a solid and fluid.

$$\lambda_{cmax} = (1-P) \lambda_s + P\lambda_g \tag{7.1}$$

$$\frac{1}{\lambda_{cmin}} = \frac{1-P}{\lambda_s} + \frac{P}{\lambda_g} \tag{7.2}$$

Equations (7.1) and (7.2) correspond to the cases where two components are arranged in laminated form, parallel and normal to the direction of heat flow, respectively.

Thermal conductivities of actual mixture material fall between two extreme values calculated from Eqs. (7.1) and (7.2), depending upon the manner of scattering of the solid phase and the distribution of sizes or shapes in the solid material. For materials containing isolated voids, Maxwell [64], Cheng [22] and Kunii [47] proposed formulas based on their different mathematical models. However, these are not always applicable to the prediction of frost properties, since frost layers are considered to have significantly different structures from those of their mathematical models.

Woodside [105] introduced a prediction equation using a model of thermal conductivity for a medium consisting of cubic lattice of solid spherical particles uniformly distributed in a gas.

$$\frac{\lambda_g}{\lambda_c} = 1 - (\frac{6s}{\pi})^{1/3} \{1 - (\frac{a^2-1}{a}) \ln (\frac{a+1}{a-1}) \} \tag{7.3}$$

$$s = 1 - P, \qquad 0 < s < 0.5236$$

$$a = 1 + \{\frac{4}{\pi (\lambda_s/\lambda_g - 1) (6s/\pi)^{2/3}}\}^{1/2}$$

This equation yields a good approximation for thermal conductivity of snow and frost by using the effective thermal conductivity of air which takes latent heat transfer due to the movement of water vapor into account.

The values calculated from the formulas presented above vary widely, in most cases by the construction of mixed materials. The formulas were derived without considering the thermal contact of the neighboring ice-crystals and the tortuosity of the thermal flow in the solid phase of the frost layer.

Some studies have used models of frost construction to analyze the mechanisms of change in thermal conductivity as the frost layer increases in thickness with time. Yamakawa et al. [114] calculated this change by solving energy and mass balancing equations based on their frost-column model. The essential part of this model is, as shown in Fig. 10, to simplify the actual frost layer into circular columns of frost uniformly distributed on a cold surface. There are some problems in the calculation with this model, since the density and temperature which have to be obtained by integrating the equations are not clear in relation to the frosting conditions. Also, a correcting factor they used

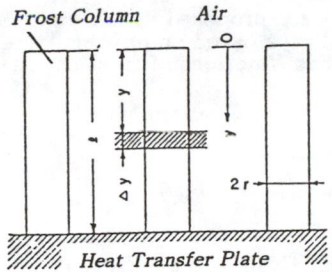

FIGURE 10. Yamakawa's Frost Model [114]

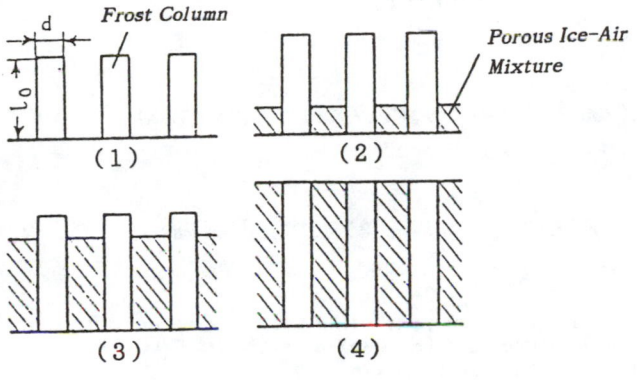

FIGURE 11. Hayashi's Frost Model [36]

to correct the heat conduction is not clear in the physical sense. To this, Hayashi et al. [36] considered a model in which frost columns of the same height were distributed parallel to the heat flow in a porous ice-air mixture (Fig. 11). Using the integration calculation, they concluded that thermal conductivity can be estimated by using Woodside's Eq. (7.3) and was in good agreement with measured data. This model provides uniform density distribution in the direction of heat flow. Therefore, the thermal conductivity is determined by a ratio of areas on a cold surface covered by columns and porous ice.

7.2 Empirical Formulas of the Thermal Conductivity of Frost Layers
In practical engineering, it is necessary to predict thermal properties of frost

layers. Therefore, experimental formulas were proposed which yield more precise predictions, although most were not always based on physically reasonable explanations. They were mainly expressed as functions of frost density and are summarized as follows.

Abels [110]:

$$\lambda_f = 2.88 \times 10^{-6} \rho_f^2 \qquad W/(mK) \tag{7.4}$$

$$139 < \rho_f < 340 \qquad kg/m^3$$

Jansson [40]:

$$\lambda_f = 0.028 + 7.94 \times 10^{-4} \rho_f + 2.58 \times 10^{-12} \rho_f^4 \quad W/(mK) \tag{7.5}$$

$$\rho_f: kg/m^3$$

Devaux [110]:

$$\lambda_f = 0.0293 + 2.93 \times 10^{-6} \rho_f^2 \qquad W/(mK) \tag{7.6}$$

$$99 < \rho_f < 597 \qquad kg/m^3$$

Kondrat'eva [46]:

$$\lambda_f = 3.56 \times 10^{-6} \rho_f^2 \qquad W/(mK) \tag{7.7}$$

$$350 < \rho_f < 500 \qquad kg/m^3$$

Van Dusen [110]:

$$\lambda_f = 0.0209 + 4.03 \times 10^{-4} \rho_f + 2.37 \times 10^{-9} \rho_f^3 \quad W/(mK) \tag{7.8}$$

$$243 < T_m < 273 \ K$$

Yonko and Sepsy [110]:

$$\lambda_f = 0.0242 + 7.22 \times 10^{-4} \rho_f + 1.18 \times 10^{-6} \rho_f^2 \quad W/(mK) \tag{7.9}$$

$$\rho_f < 577 \quad kg/m^3$$

The relations listed above are valid for predicting thermal conductivities in the range when the mean temperature of a frost layer is above $-30°C$, but are not always reliable for the other temperature range.

Sakatsume and Seki [93] proposed the following empirical formula to predict the thermal conductivity:

$$\frac{\lambda_c}{\lambda_a} = \frac{1-P}{(1-P^{1/3}) + \dfrac{P^{1/3}}{(1-P^{2/3}) + kP^{2/3}}}$$

$$+ \frac{kP}{\{1-(1-P)^{1/3}\} + \dfrac{(1-P)^{1/3}}{\{1-(1-P)^{1/3}\} + (1/k)(1-P)^{2/3}}} \tag{7.10}$$

$$k = \lambda_g / \lambda_s$$

The formula (7.10) is a good method for predicting the thermal conductivity of snow and fine crushed ice over the wide range of porosities, and is therefore, applicable to the problem of frost layers.

The predictions obtained from the above formulas are plotted in Fig. 12 as a function of frost density. As seen in the figure, the predictions differ considerably with each other. This difference can be explained by the fact that each of the fundamental models is based on the different mathematical considerations and the data used for deducing the formulas were obtained from experiments conducted under different conditions. Furthermore, even though the parameters associated with frosting process such as the ambient and surface temperatures and humidities are all kept constant, subtle differences in surrounding conditions, such as the surface temperature of the walls in the experimental rooms, possibly influenced frosting characteristics.

7.3 Thermal Conductivity of Frost Layer in Cryogenic Frosting

Thermal conductivity is estimated uniquely as a function of density. However, subsequent studies on the cryogenic frosting process [12, 14, 76] showed that the thermal conductivity of frost changes with the average temperature of a frost layer as well as with its density.

Brian et al. [15], Dietenberger [26, 27] and Biguria [13, 16] proposed the prediction expressions of thermal conductivity in terms of both density and average temperature throughout the layer. The reason that the thermal conductivity changes with the layer's average temperature, even though its density remains constant, is attributed to the construction of the frost layer. Thermal conductivity of ice and air change with temperature, and fog formation in the boundary layer and its deposition on the frost surfaces affects heat transfer, especially in the cryogenic frosting.

Lisovsky and Pavlov derived a formula for predicting the thermal conductivity of frost layers [59], based on the models proposed by Biguria [16] and Dietenberger [26].

Marinyuk [65] devised the following relation from the measurement of thermal conductivity of frost layers on a vertical cylinder in natural convection.

$$\lambda_f = 1.3(T_S - T_W)^{-1}[0.156 \ \{\exp(0.0137\,T_S) - \exp(0.0137\,T_W)\}$$

$$+5.59 \times 10^{-5} \, \rho_f \{\exp(0.0214\,T_S) - \exp(0.0214\,T_W)\}] \quad W/(mK) \quad (7.11)$$

$$60 < \rho_f < 300 \ kg/m^3, \quad 93 < T_S < 233 \ K, \quad T_W: K$$

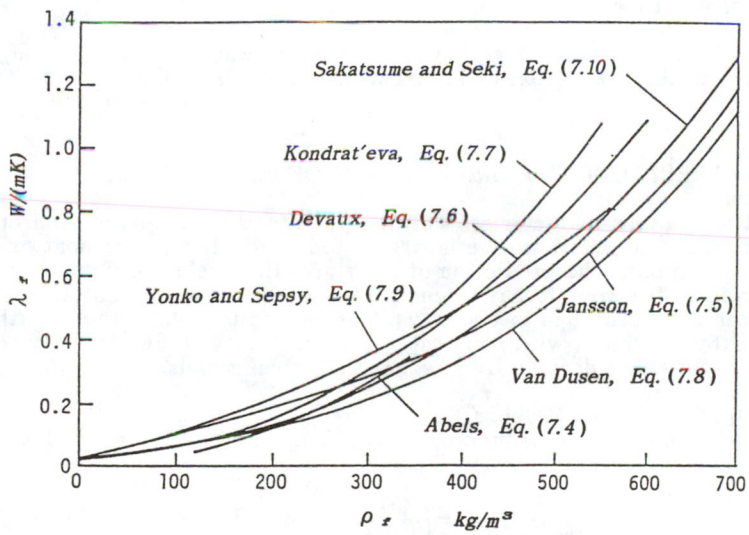

FIGURE 12. Thermal Conductivities Predicted by Equations (7.4) to (7.5)

7.4 Density Distribution in Frost Layers

In the investigations reviewed in the preceding sections, the density of frost layers were considered to be uniform in the lateral direction. However, recent studies treated frost density as a variable with respect to time and position in frost layers. Trammel et al. [97] reported that the density of a frost layer decreases towards the surface. They measured the density variation by detecting radio activity from a radio source installed on the surface of a cold plate.

Also, Cremers et al. [24] calculated the variation of frost density deposited on a circular cylinder by measuring the absorption of gamma rays irradiated on the surface of the frost layer. It was then found that the density can be distributed in the layer in two different manners: it uniquely decreases toward the surface of the frost layer, or it has one or two peak values in the layer. Contrary to this, Brian et al. [15] directly measured the density of the frost layer on a surface cooled by liquid nitrogen, and noted that there were no remarkable variations in density. However, according to Hayashi's study [36], the density tends to decrease near the cold wall and to increase toward the surface.

As reviewed above, there have been many different results concerning the density distribution in frost layers. Frost density is estimated from mass of a frost layer divided by its thickness. Therefore, these differences may come from the difficulties in measuring the thickness of frost layers which is usually very thin. In any case, much more precise experiments should be performed and more exact physical models should be considered.

7.5 Radiative and Other Properties of Frost Layers

For thermal properties other than density and thermal conductivity, Schmidt [83] conducted experiments on radiative emissivity and obtained a value of 0.985 ± 0.003 for a frost layer at $-9.6°C$ and 0.1 to 0.2 mm thick. Cunningham[20] measured the radiative absorption of frost layers and determined the value to be 0.95 ± 0.04 for frost layers greater than 0.6 mm thick. They also reported that the emissivity of substratum has no influence upon the emissivity of frost layers thicker than 0.1 mm.

The adhesion ability of an ice layer to metal surface was investigated [86], which is related to the prevention of ice/frost formation on various devices of aircraft.

8. STUDIES ON PREDICTING THE GROWTH RATE OF FROST LAYERS

It is important to examine characteristics of heat transfer and the pressure drop of air flow in order to solve problems associated with the use of heat exchangers in cold regions. The prediction of frost growth is relatively new area of study: therefore, few reports have been published. The prediction of frost layer growth can be approached in two ways. The first is to predict the growth of a frost layer by relating it with physical quantities involved, and the second involves solving equations derived from assumed physical models.

8.1 Empirical Approach to the Problem

For frost growth on a cylinder in cross-flow, Schneider [92] proposed the following expression:

$$H_f = 0.465 \ \{\frac{\lambda_{ice}}{L\rho_{ice}} \tau \, (T_S - T_w)^{1/2} \tau^{-0.03} (T_S - T_w)^{-0.01} \ [\frac{p_\infty - p_S}{p_{\infty S} - p_S}]$$

$$\times \ [1 + 0.052 \frac{T_\infty - T_o}{T_o - T_w}] \ \} \qquad H_f: \text{m}, \ \tau: \text{hr} \qquad (8.1)$$

This equation can predict the growth rate of the frost layer within uncertainties of $\pm 10\%$ in the range $1 < \tau < 8$hr. Schneider performed his experiment under conditions of relatively high ambient humidities, so the surface temperature of the frost layer was considered to be close to $0°C$. Therefore, the frosting conditions where the equation is applicable are rather limited. Generally, the surface temperature of frost layers becomes low, especially when both humidity and wall temperatures are low. So Eq. (8.1) is valid only for frosting under ambient humidity close to its saturation.

Cremers and Mehra [25] measured the frost layer developing on the surface of a cylinder in natural convection conditions, and correlated the data by using least square method to obtain:

$$H_t = 0.20 \{ \tau (T_s - T_w) \}^{0.40} \qquad H_t: \text{mm}, \quad \tau: \text{min.}, \quad T_s, \quad T_w: \text{K} \qquad (8.2)$$

This equation includes the temperature difference between the wall and the surface of the frost layer. The surface temperature has to be determined from heat balance at the surface of the frost and, therefore, it cannot be obtained if the the thermal properties of the frost layer have not been confirmed. Accordingly, it is impossible to predict the actual surface temperature of the frost layer without determining its thickness. Therefore, it can be said that Eq. (8.2) is applicable only when the ambient humidity is relatively high and the surface temperature is close to $0°C$, as in the case of the Schneider's formula, Eq. (8.1).

8.2 Analytical Calculation of Frost Layer Growth

The Brian and Jones method [15] estimated the thickness of frost layers by integrating the equations derived from models in which the density is assumed to be constant throughout the layer but changes with time. Their physical models are based on the assumption that water vapor coming into the layer through its surface accumulates uniformly in the layer and its thermal conductivity can be predicted by Brian's empirical formula. The models show that frost layer grows more linearly in general than actual. However, their models have an advantage in that the construction factor is not in the calculation.

Yamakawa [114] approximated the actual frost construction by circular frost columns and introduced basic heat and mass transfer equations based on it. The numerical integration yielded a good approximation for predicting the growth by substituting the measured surface temperature, which is not predictable as mentioned before.

Hayashi et al. [36] and Aoki [6] proposed models which replaced the air space between the frost columns of Yamakawa's model with porous ice. Aoki proposed a model which consisted of ice columns and porous ice arranged parallel to the heat flow. In order to calculate the thickness of a frost layer by using these models, the procedure is basically as follows: a set of heat and mass balancing equations are derived with respect to the transfer of heat and mass in the frost layer, being modeled as cylindrical columns. Assuming the temperature distribution in the layer be rectilinear in the vertical direction, a mass transfer equation around the column must be solved to obtain its concentration distribution. The next step is to determine the ratio of water vapor transferred to the top and the base of the column by using the concentration distribution obtained previously. Thus, the growth rate can be estimated. The results showed that frost layers grow almost linearly with time, so the density distribution through the layer have to be modified to obtain more accurate results.

Rostami [77] conducted a numerical calculation to predict the growth rate and the density of frost layers, taking into consideration the variation of heat transfer coefficients due to the frost formation. In this investigation, the spatial variation of frost density was assumed to be uniform and the frost layer was modeled as a porous material growing on a cold surface.

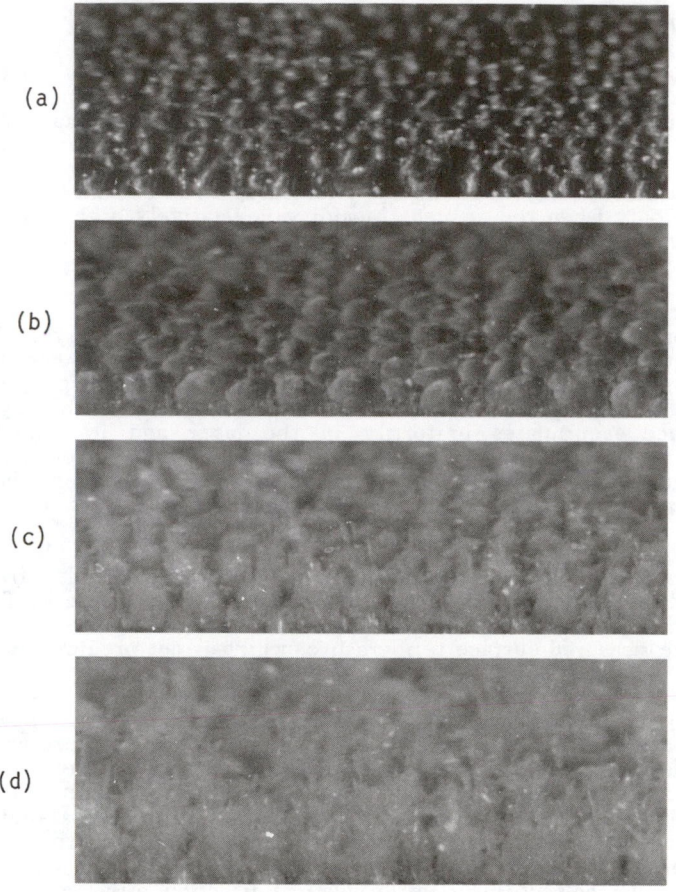

FIGURE 13. Microscopic Photograph of Frost At The Beginning of Its Formation (free stream temp.=21 °C, wall temp.= -7.5 °C, relative humidity=73%, stream velocity=1.6 m/s, ×31) [103]

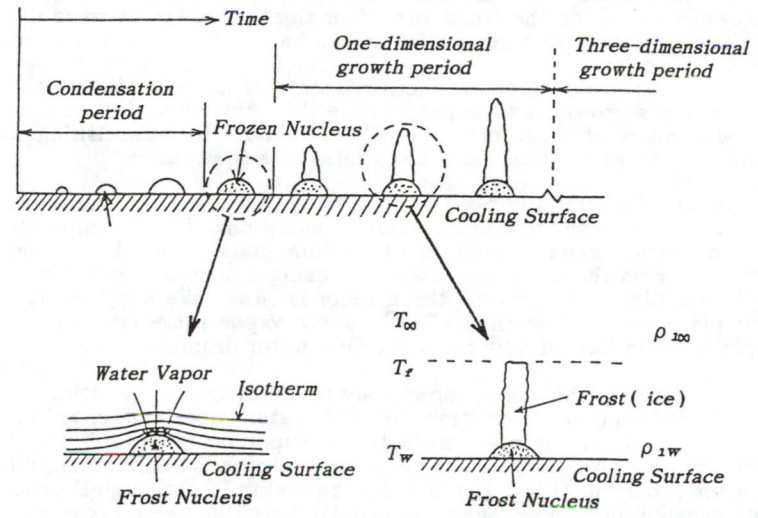

FIGURE 14. Illustration of Frost Deposition in Early Stage of Formation [103]

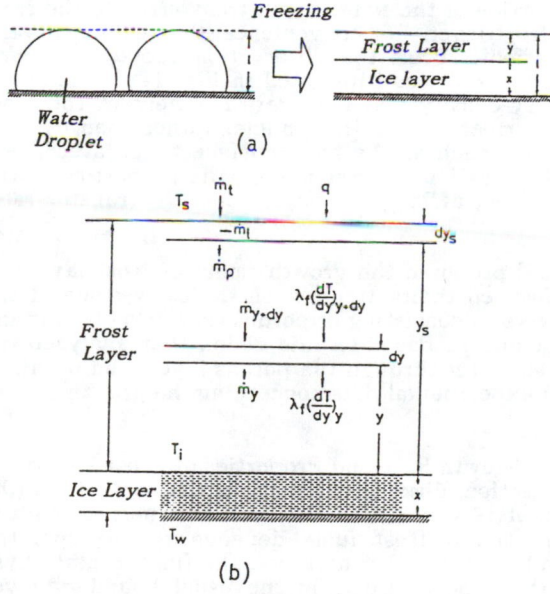

FIGURE 15. Ice Sublayer Model [96]

9. RECENT STUDIES ON FROST FORMATION

9.1 Generation of Frost Layers from Bare Surfaces

In most of the previous work on the frost formation, thick frost layers were investigated. The mechanism of frost generation from a bare surface has not been clarified. Since the porosity of a frost layer is usually very large, it is reasonable to expect that some evidence of the incipient state of the frost layer still remains even after it has grown to some perceptible thickness. For this reason, clarifying the mechanisms of frost creation will contribute to establishing a method for predicting the properties and growth rates of a frost layer.

Recently, Saito et al. [95, 103] and Seki et al. [94, 96] studied the incipient process of frost formation under forced convection conditions. They used small wind tunnels with bottom walls consisting of cooling plates. The deposition process of water vapor on the plate was observed using a microscope and was recorded on a 35-mm film every two or three seconds. When the surface temperature of the plate was higher than $-30\,°C$, water vapor immediately condensed on the plate in the form of supercooled liquid water droplets.

Fig. 13 shows an example of the microscopic observation of frost deposition on a cold wall [103]. Immediately after starting the test, water vapor condensed on the wall in the form of microscopic droplets of supercooled liquid water (Fig.13(a)). The droplets grew larger and began to coalesce. Then, the liquid droplets froze after a certain time (Fig. 13(b)), becoming "frost nuclei" from which micro-ice crystals developed, and grew until a visible frost layer was formed (Fig. 13(c,d)). In this period of frost deposition, water vapor sublimated mainly on the top of the nuclei and the crystals developing on them grew mainly in the direction normal to the cooling wall. Therefore, this period of frost formation is called "one-dimensional growth period". In the succeeding period, the frost layer grew not only in the normal direction but also in the direction parallel to the cooling wall. In this period, the frost grew more slowly than in the former period since some portion of the water vapor transferred to the frost surface sublimated inside the frost layer. For convenience, they called this the "three- dimensional growth period". From the observation above, Saito and Tokura introduced the frost layer model as illustrated in Fig. 14. In the one-dimensional growth period, frost density and the growth pattern of the frost layer are determined by the number of the frost nuclei rather than by conditions at the frost growth front, such as the frost-surface temperature, etc. Many factors contribute to the formation of frost nuclei; the properties of the surface on which frost deposit (i.e., affinity of water), the heat transfer rate, the mass transfer rate, etc.

To this, Seki and Fukusako [96] proposed the growth model of frost layers as shown in Fig. 15. They also focused their attention on the early stage of the formation and modeled frost layers as consisting of porous materials of frost and very thin sublayer of the frost nuclei. Based on this model, they analyzed the heat and mass transfer to the sublayer through the porous layer, and obtained results in good agreement with experimental data concerning the growth rate of frost layers.

9.2 Dimensionless Correlation of Growth Rate and Properties of Frost Layers

On the basis on the preceding section, dimensional analysis was carried out [95] to find the dimensionless parameters which correlate with the thermal conductivity of a frost layer. The formation of frost nuclei depends strongly upon the heat and mass transfer rates to the cold wall. Therefore, the fundamental physical quantities are basically the same as those in the usual boundary layer problems, i.e., h, $T_\infty - T_w$, h_D, $\rho_{1\infty} - \rho_{1w}$, L, x, τ, λ_a, and λ_f. The result of dimensional analysis is,

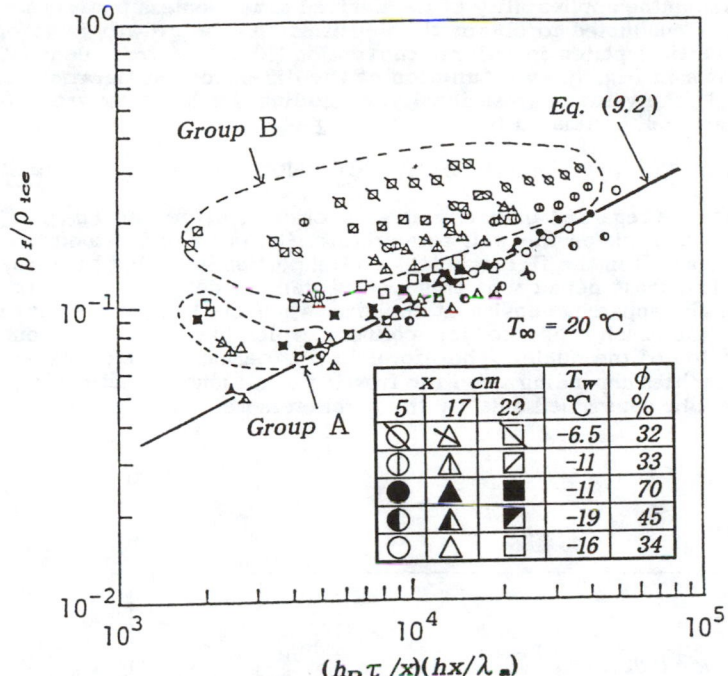

FIGURE 16. Frost Density As A Function of $(h_D \tau /x)(hx/ \lambda_a)$ [103]

$$\frac{\lambda_f}{\lambda_a} = F \left\{ \frac{h_D(\rho_{1\infty} - \rho_{1w})L}{h(T_\infty - T_w)} , \frac{h_D \tau}{x} , \frac{hx}{\lambda_a} \right\} \tag{9.1}$$

where the first term in F{ } represents the ratio of latent heat of the phase change to sensible heat transferred by convection. The second term is the ratio of the quantity of water vapor transferred to the surface of the frost layer to that contained in the ambient air. The third is the Nusselt number.

The dimensional analysis was based on the fact that the density of a frost layer is determined mainly by the number density of frost nuclei in the early period of frost deposition and that the formation of frost nuclei depends upon the heat and mass transfer rates at the beginning of the frost deposition. The values of the dimensionless parameters in F{ } of Eq. (9.1) are determined by the initial frosting conditions, i.e., T_∞, T_w, ϕ and τ together with the measured values of h and h_D. Similar parameters were used to correlate the density and the thickness of a frost layer by White and Cremers [108, 109]. The surface temperatures were involved in the parameters since those were derived from heat and mass balance through a two-layered model.

In order to determine the applicability of the derived dimensionless parameters, an experiment was conducted to obtain the densities and the growth rates of frost layers on vertical plates in natural convection [103]. The frost density, ρ_f / ρ_{ice}, is shown in Fig. 16 as a function of the dimensionless parameter, $(h_D \tau / x)(hx/ \lambda_a)$. The value of frost density, excluding the data from group A and B in Fig. 16 are well correlated by the following equation:

$$\rho_f / \rho_{ice} = 0.001 Z_1^{1/2}, \quad Z_1 = (h_D \tau / x) (hx/ \lambda_a) > 5 \times 10^3 \qquad (9.2)$$

Therefore, ρ_f can be regarded primarily as a function of parameter, $(h_D \tau / x)(hx/ \lambda_a)$, but also depends on the wall temperature. It can be understood that the data groups A and B in the figure belong to the portion indicated by a line (a) in Fig. 17. The frost density in this period can be determined by the relation between the apparent density at the time when the frost nuclei have just formed and the density of the ice columns which have grown up one dimensionally on top of the nuclei. Therefore, in this period, the frost layers have different densities depending upon the frosting conditions, as indicated in Fig. 16, and cannot be determined only by the parameters mentioned above.

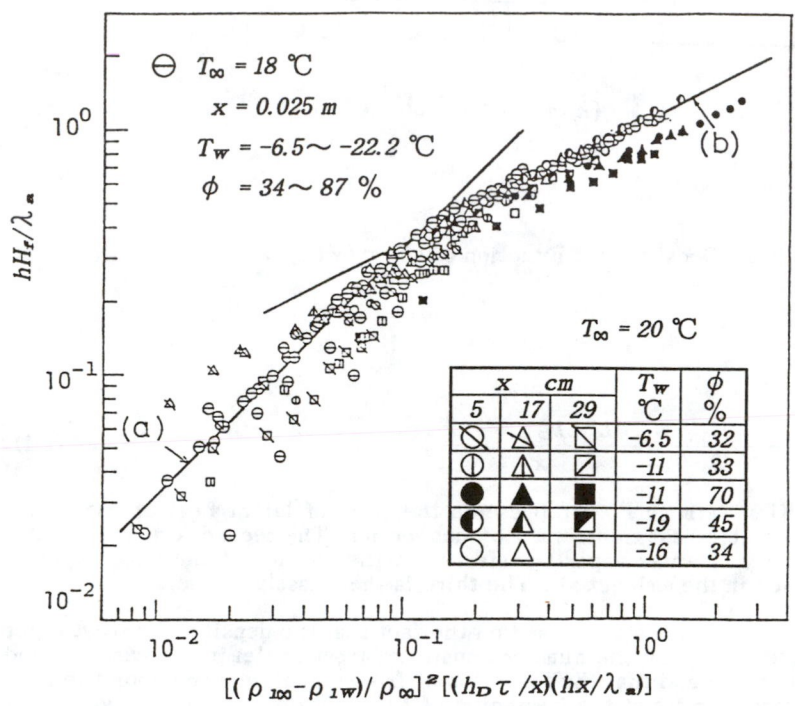

FIGURE 17. Dimensionless Thickness of Frost As A Function of $[(\rho_{100} - \rho_{1w})/ \rho_\infty]^2 [(h_D \tau / x)(hx/ \lambda_a)]$ [103]

438

The mass of a frost layer can be approximated using the following equation:

$$m_f = \rho_f H_f = h_D(\rho_{100} - \rho_{1f})\tau \tag{9.3}$$

Usually, it is very difficult to predict the values of ρ_{1f} since the surface temperature of a frost layer must be determined by a heat balance at the air-frost interface. Therefore, by using ρ_{1w} instead of ρ_{1f} for convenience, the thickness of a frost layer can be written in the form

$$\frac{hH_f}{\lambda_a} \propto \frac{\rho_{100} - \rho_{1w}}{\rho_\infty} \left\{ \frac{h_D \tau}{x} \cdot \frac{hx}{\lambda_a} \right\}^{1/2} \tag{9.4}$$

Fig. 17 shows the dimensionless thickness hH_f/λ_a as a function of the dimensionless parameter $[(\rho_{100} - \rho_{1w})/\rho_\infty]^2[(h_D \tau/x)(hx/\lambda_a)]$. The experimental values are approximated by the solid lines (a) and (b) in the figure. This figure shows that the process of the frost growth can be divided into two categories, with a division at 10^{-1} of the abscissa. One portion of the process is the period in which the frost thickness increases linearly with time and the other is the period that the thickness increases in proportion to the square root of time. Those regions are represented by the lines (a) and (b), respectively. Referring to the observation of the microscopic frost deposition, the region indicated by the line (a) corresponds to the region where ice-crystals grow one-dimensionally on the frost nuclei in a direction normal to the plate. In the succeeding period of time indicated by the line (b), the frost layers grow more slowly since some portion of the water vapor transferred to the frost surface sublimates inside the frost layer. Consequently, the crystals grow not only in a normal direction but also in a direction parallel to the plate. The thickness of frost layer in these periods is correlated by the following equations:

$$hH_f/\lambda_a = 3.23 Z_2, \qquad Z_2 < 0.11 \tag{9.5}$$

$$= 1.08 Z_2^{1/2}, \qquad Z_2 > 0.11 \tag{9.6}$$

$$Z_2 = [(\rho_{100} - \rho_{1w})/\rho_\infty]^2 [(h_D \tau/x)(hx/\lambda_a)]$$

Saito and Tokura also conducted the experiments to examine the thermal and physical properties of frost layers which grow on horizontal cylinders in a vertical array in free convection [104]. Five horizontal cylinders made of copper tubes 28.5mm in diameter, 1400mm long and 1mm thick were set in a vertical line between baffle plates and were located in a large air conditioned room. The cylinders were cooled by brine flowing inside the tubes so that frost layers develop on them. Using a photographic technique, the thickness of frost layers was measured for every $15°$ of the angular position, from the top stagnation point of each cylinder. The frost deposition rate was measured by scraping off small amounts of frost from the cylinders and weighed using a precision balance.

The thickness of frost layers on the cylinders in an array is not always uniform, but are thinnest near the rear-stagnation points. When the spacing between the cylinders is extremely close, the layers adjacent cylinders come into contact with each other. However, experimental measurements of frost thickness have shown that there are some portions on each cylinder where the shape of the frost surface is circular, concentric to the cylinder as shown in Fig. 18. It should be noted that the thickness of frost layers maintaining this circular shape is almost same for any cylinder, although the mean transfer coefficients

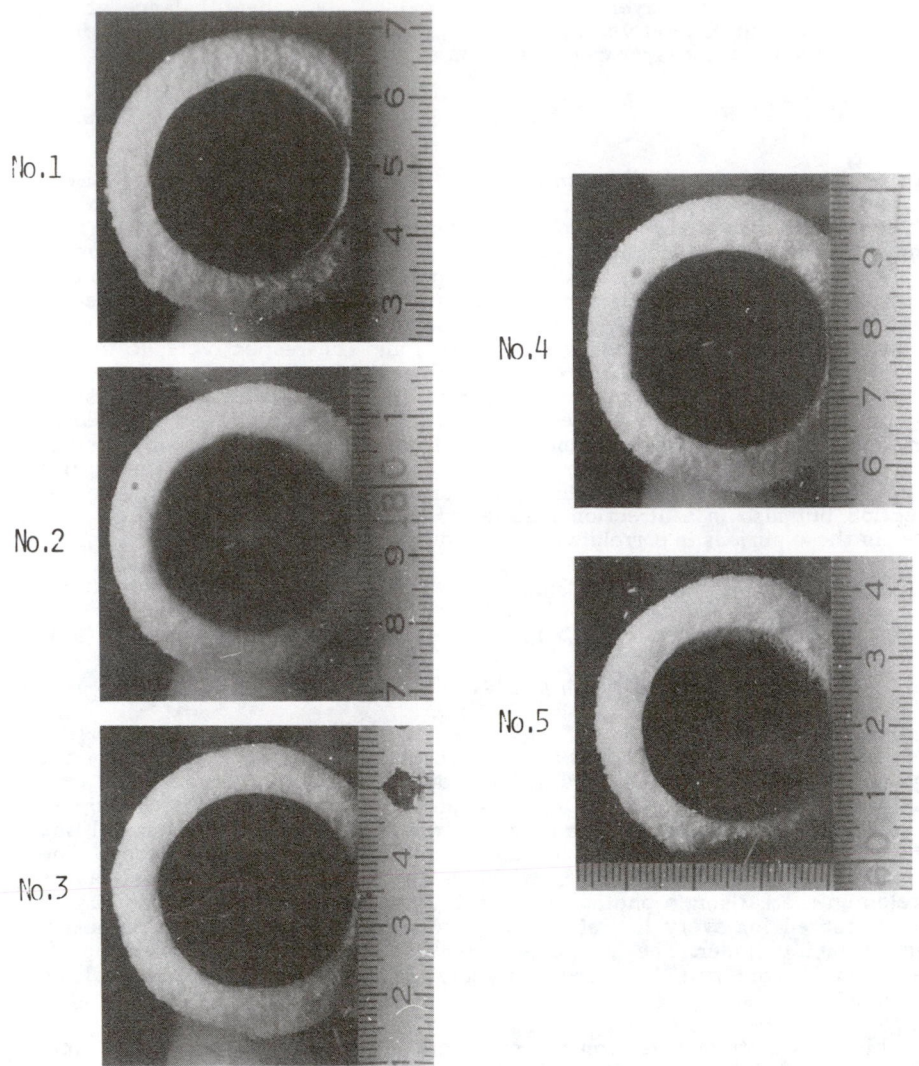

FIGURE 18.   Frost Layers on Circular Cylinder In An Array
(d=28.5 mm, T_∞=20 ℃, T_w= -24 ℃, relative humidity=40 %,
after 9 hours) [104]

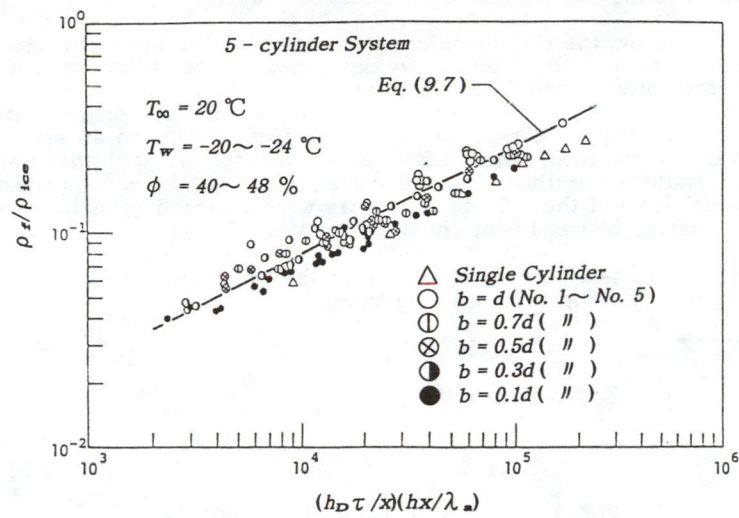

FIGURE 19. Apparent Mean Frost Density As A Function of $(\bar{h}_D \tau / d)(\bar{h}d/ \lambda_a)$ [104]

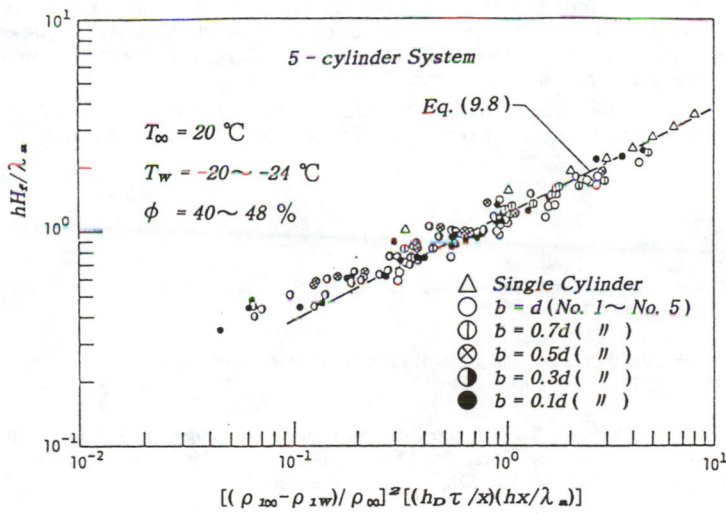

FIGURE 20. Dimensionless Frost Layer Thickness As A Function of $[(\rho_{100} - \rho_{1w})/\rho_\infty]^2[(\bar{h}_D \tau / d)(\bar{h}d/ \lambda_a)]$ [104]

are different for each cylinder in an array. Therefore, this frost layer thickness can be defined as the characteristic thickness of a layer.

Fig. 19 and Fig. 20 show the dimensionless correlation of the apparent mean density and the thickness of the frost layers developed on the cylinders in an array. The apparent mean density was calculated by using the characteristic thickness and mass of the frost layer deposited around the entire periphery of each cylinder. In the figures, b represents the cylinder spacing in an array. The measured value of the mean mass transfer coefficient for each cylinder was used in order to evaluate the dimensionless variables in the figure. The mean heat transfer coefficients of the cylinders in array were obtained by using the Chilton–Colburn analogy between heat and mass transfer.

The density and thickness of the frost layers on the cylinders can be correlated from the figures by using the following equations:

$$\rho_f/\rho_{ice} = 8 \times 10^{-4} Z_3, \qquad Z_3 > 3 \times 10^3 \qquad (9.7)$$

$$Z_3 = (\bar{h}_D \tau / d)(\bar{h} d / \lambda_a)$$

$$\bar{h} \bar{H}_f / \lambda_a = 1.2 Z_4^{1/2}, \qquad Z_4 > 0.1 \qquad (9.8)$$

$$Z_4 = [(\rho_{1\infty} - \rho_{1w})/\rho_\infty]^2 [(\bar{h}_D \tau / d)(\bar{h} d / \lambda_a)]$$

10. CLOSURE

This article was prepared as a review of previous studies on heat and mass transfer during the frosting process and also as a search for more advanced approaches to solving the frosting problem. The problem of melting or frost layers or cyclic melting and frosting in actual heat exchangers such as fin-tubes and fin-coil heat exchangers was not dealt with in detail. However, solutions to these kinds of problems will be very important since they are related closely to the industrial applications.

The authors hope that this review article will be beneficial in considering new subjects in frosting problems and in developing new approaches at more advanced levels.

REFERENCES

1. Auracher, H., "Water Vapor Diffusion and Frost in Porous Materials," ASTM Spec. Tech. Publ., No. 544, 1974, pp. 49-67.

2. Auracher, H., "Wasserdampfdiffusion und Reifbildung in porösen Stoffen, " VDI-Forschungsheft No. 566, 1974, pp. 1-44.

3. Adachi, M., Inoue, S., and Aizawa, T., "On the Refrigeration Cycle Property of Heat Pump Air Conditioner Operating with Frost Formation, Part 1 Effect of the Air Conditions," Refrigeration(in Japanese),   Vol. 50, No. 576, 1975, pp. 812-820.

4. Adachi, M., Inoue, S., and Inoda, K., "On the Refrigeration Cycle Property of Heat-Pump Air Conditioner Operating with Frost Formation,   Part 2 The Optimum Defrost Conditions of Refrigeration Cycle Operating by the Reverse Cycle Defrosting Method," Refrigeration(in Japanese),   Vol. 52, No. 598, 1977, pp. 715-727.

5. Auracher, H., "Heat Transfer in Frost and Snow," Int. Heat Transfer Conf. Vol. 6th, No. 3, 1978, pp. 25-30.

6. Aoki, K., Katayama, K, Hayashi, Y., and Adachi, S., "Study of Frost Formation (Theory of Growth of Frost Layer),"(in Japanese), Trans. JSME, Vol. 45, No. 394, 1979, pp. 869-876.

7. Aoki, H., Yamakawa, N., and Ohtani, S., "Forced Convection Heat Transfer around a Vertical Cylinder under Frosting Conditions," Kagaku Kogaku Ronbunshu (in Japanese), Vol. 6, No. 1, 1980, pp. 8-14.

8. Aoki, K, Katayama, K., and Hayashi, Y., "Study on Frost Formation (Full Growth Period of Frost Layer with Water Penetration and Solidification)," (in Japanese), Trans. JSME, Vol. 48, No.429, 1982, p   952-961.

9. Beatty, K. O., Finch, E. B., and Schoenborn, E. M., "Heat Transfer From Humid Air to Metal Under Frosting Conditions," Refrigerating Engineering, 1951, pp.1203-1207.

10. Bryan, W. L., "Heat and Mass Transfer in dehumidifying surface coils," ASHRAE Journal, 1961, pp. 51-54, 91.

11. Brailsford, A. D. ,and Major, K. G., "Thermal Conductivity of Aggregates of Several Including Porous Materials," Brit. J. Applied Physics, Vol. 15, 1964, pp. 313.

12. Barron, R. F. ,and Han, L. S., "Heat and Mass Transfer to Cryosurface in Free Convection," Trans. ASME, Journal of Heat Transfer, Vol. 87, No. 4, 1965, pp. 499-506.

13. Biguria, G., "The Moving Boundary Problem with Frost Deposition to Flat Plate at Subfreezing Temperature and Forced Convection Conditions. The Measurement and Correlation of Water Frost Properties," Ph. D. Thesis, Leihigh University, 1968.

14. Brian, P. L. T., Reid, R. C., and Brazinsky, I., "Cryogenic Frost Properties," Cryogenic Technology, Vol. 5, 1969, pp. 205-212.

15. Brian, P. L. T., Reid, R. C., and Shah, Y. T., "Frost Deposition on Cold Surfaces," Industrial and Engineering Chemistry Fundamentals, Vol. 9, 1970, pp. 375-380.

16. Biguria, G., and Wenzel, L. A., "Measurement and Correlation of Water Frost Thermal Conductivity and Density," Industrial and Engineering Chemistry Fundamentals, Vol.9, 1970, pp. 129-138.

17. Coles, W. D., and Ruggeri, R. S., "Experimental Investigation of Sublimation of Ice at Subsonic and Supersonic Speed and Its Relation to Heat Transfer," NACA Techn. Note, 3104, 1954.

18. Coles, W. D., "Experimental Determination of Thermal Conductivity of Low-Density Ice," NACA Techn. Note, 3143, 1954.

19. Chung, P. M., and Algren, A. B., "Frost Formation and Heat Transfer on a Cylinder Surface in Humid Air Crossflow, Part 1 and 2," ASHRAE Transactions, Vol. 65, 1959, pp. 213-244.

20. Cunningham, T. M., and Young, R. L., "The Absorption of Water Cryodeposit at 77 deg K for 350 deg K Radiation," Arnold Engineering Development Center, TDR-63-155, 1963.

21. Chen, M. M., and Rosenow, W., "Heat, Mass and Momentum Transfer Inside Frost Tubes. Experiment and Theory," Trans. ASME, Journal of Heat Transfer, Vol. 86, 1964, pp. 334-340.

22. Cheng, S. C., and Vachon, R. I., "The Prediction of the Thermal Conductivity of Two and Three Phase Solid Heterogeneous Mixtures," International Journal of Heat and Mass Transfer, Vol. 12, 1969, pp. 249.

23. Chuang, M. C., "The Frost Formation on Parallel Plate at very Low Temperature in a Humid Stream," Pap. ASME, No. 76-WA/HT-60, 1976, pp. 1-5.

24. Cremers, C. J., Hahn, O. J., and Skorupski, J. H., "Frost Density Measurement of Vertical Cylinders by Gamma-Ray Attenuation," Advance in Cryogenic Engineering, Vol. 23, 1978, pp. 371-375.

25. Cremers, C. J.,and Mehra, V. K., "Frost Formation on Vertical Cylinders in Free Convection," Trans. ASME, Journal of Heat Transfer, Vol. 104, 1982, pp. 3-7.

26. Dietenberger, M. A., Kumar, A., and Luers, J., "Frost Formation on an Airfoil: Mathematical Model 1," NASA Contractors Report 3129, 1979.

27. Dietenberger, M. A., "Generalized Correlation of the Water Frost Thermal Conductivity," International Journal of Heat and Mass Transfer, Vol. 21, 1978, pp. 607-619.

28. Epstein, M., and Rosner, D. E., "Enhancement of Diffusion-limited Vaporization Rates by Condensation within the Thermal Boundary Layer: 2. Comparison of Homogeneous Nucleation Theory with the Critical Supersaturation Model," International Journal of Heat and Mass Transfer, Vol. 13, 1970, pp. 1393-1413.

29. Goodman, J., and Kennedy, L. A., "Free Convection Frost Formation of Cool Surfaces," Proceedings of 1972 Heat Transfer and Fluid Mechanics Institute, Stanford University, Palo Alto, 1972, pp. 338-352.

30. Hermann, R., "Wärmeübertragung bei freie Strömung am waagerechten Zylinder in zwei-atomigen Gasen," VDI-Forschungsheft, Nr. 379, 1936.

31. Hilz, R., "Vershiedene Arten des Ausfrierens einer Komponente aus binaren, strömenden Gasgemishen," Zeitschr. f. d. ges. Kälteindustrie, Vol. 47, 1940, pp. 34-37, 74-78 and 88-92.

32. Hofmann, E., "Wärmedurchgangsversuche an einem Plattenluftkühler unter besonderer Berücksichtigung der Reifschicht," Die Kälte, No. 2, 1948, pp. 25-31.

33. Huffmann, G. D., and Sepsy, C. F.,"Heat Transfer and Pressure Loss in Extended Surface Heat Exchangers Operating under Frosting Conditions, Part 2: Data Analysis and Correlation," Trans.-Nr. 2045 zur 74, ASHRAE Jahrestagung, Minneapolis, USA, 1967.

34. Holten, D. C., "A Study of Heat and Mass Transfer to Uninsulated Liquid Oxygen Containers," Advances in Cryogenic Engineering, Vol. 6, 1961, pp. 499-508.

35. Hausmann, H., "Wärmeübertragung und Druckverlus in bereifenden Spalten, " Klima + Kälte - Ingenieur, Vol. 4, No. 7/8, 1976, pp. 271-274.

36. Hayashi, Y., Aoki, K., and Yuhara, H., "Study on Frost Formation in Forced Convection," Trans. JSME, Vol. 42, No. 355, 1976, pp. 885-899. (Heat Transfer-Japanese Research, Vol. 6, No. 3, 1977, PP. 79-94.)

37. Hayashi, Y., and Aoki, K., "Study on Frost Formation (Classification of Growth by Construction of Frost Layer)," (in Japanese), Trans. JSME, Vol. 43, No. 368, 1977,pp. 1384-1391.

38. Hayashi, Y., Aoki, K., Adachi, S., and Hori, K., "Study on Frost Properties Correlating With Frost Formation Types," Trans. ASME, Journal of Heat Transfer, Vol. 99, 1977, pp. 239-245.

39. Ivanova, V. S., and Gatchlov, T. S., "Some Aspects on Frost Formation on Extended Surface Air Coolers," Int. Congr. Refrig., No. 2, 1978, pp. 625-632.

40. Jansson, M., "Über die Wärmeleitungsfähigkeit des Schnee," Ofversigt auf Kongl. Vetenskaps-Akademien Forhandlingar, Vol. 58, 1901, pp. 207-222.

41. Joffe, D., "Kühlhaus-Kühlkörper aus Rippenrohren," Proizvodstroi Technika, Vol. 32, 1955, pp. 23-31.

42. Javnel, B. K., "Die Wärmeübertragung druch eine Reifschicht," Cholodilnaja Technika, 1969, No. 5, pp. 34-37.

43. Jones, B. W., and Parker, J. D., "Frost Formation with Varying Environmental Parameters," Trans. ASME, Journal of Heat Transfer, Vol. 97, 1975, pp. 255-259.

44. Karman, T. von., " Mechanische Ähnlichkeit und Turbulenz," Proc. 3rd. Intern. Congr. Appl. Mech., Stockholm, Prt. 1, 1931, pp. 85.

45. Kamei, S., Mizushina, T., Kifune, S. and Koto, T., "Research on Frost Formation in a Low Temperature Dehumidifier," (in Japanese), Chemi. Eng. Japan, Vol. 14, 1950, pp. 53-60.

46. Kondrat'eva, A. S., "Thermal Conductivity of the Snow Cover and Physical Processes Caused by the Temperature Gradient," Snow, Ice and Permafrost Research Establishment, Translation, No. 22, 1954.

47. Kunii, T., "Radiation Heat Transfer in Porous Media," (in Japanese), Journal of JSME, Vol. 65, No. 525, 1962, pp. 1447.

48. Katsuta, K., and Ishihara, I., "Heat Transfer by Natural Convection with Frosting Simultaneously on a single Horizontal Cylinder," Refrigeration (in Japanese), Vol. 48, No. 552, 1973, pp. 923.

49. Kennedy, L. A., and Goodmann, J., "Free Convection Heat Transfer under Conditions of Frost Deposition," International Journal of Heat and Mass Transfer, Vol. 17, 1974, pp. 477-484.

50. Katsuta, K., and Ishihara, I., "Heat Transfer by Forced Convection with Simultaneous Frosting on a single Cylinder," Refrigeration (in Japanese), Vol. 50, No. 570, 1975, pp. 262-270.

51. Katsuta, K., and Ishihara, I., "Heat Transfer by Natural Convection with Simultaneous Frosting on Vertical Flat Plates," Refrigeration(in Japanese), Vol. 52, No. 593, 1977, pp.283-291.

52. Katsuta, K., and Ishihara, I., "Heat Transfer by Natural Convection with Simultaneous Frosting on Horizontal and Parallel Cylinders," Refrigeration (in Japanese), Vol. 52, No. 601, 1977, pp. 977-985.

53. Katsuta, K.,and Ishihara, I., "Heat Transfer by Natural Convection with Simultaneous Frosting on Horizontal Cylinders in a Vertical Array," Refrigeration (in Japanese), Vol. 54, No. 625, pp. 899-273.

54. Katsuta, K., Ishihara, I., and Mukai, T., "Experimental Studies on Defrosting by the Ambient Air under Natural Convection," Refrigeration (in Japanese), Vol. 58, No. 665, 1983, pp. 229-238.

55. Katsuta, K., Ishihara, I., and Mukai, T., "An Analytical Study on Defrosting by the Ambient Air under Natural Convection," Refrigeration (in Japanese), Vol. 58, No. 669, 1983, pp. 645-650.

56. Katsuta, K., Ishihara, I.,and Ikeno, T., "A Study on Pressure Drops in Air Flowing through Inside Tubes with Frosting," Refrigeration (in Japanese), Vol. 59, No. 677, pp. 263-273.

57. Kondepudi, S. N., O'Neal, D. L., "Performance of Triangular Spine Fins Under Frosting Conditions", Heat Recovery Systems & CHP Vol.8, No.1 1988, pp.1-7

58. Levy, S., "Integral Method in Natural-Convection Flow," Trans. ASME, Journal of Applied Mechanics, Vol. 22, 1955, pp. 515-522.

59. Lisovsky, V. M., Pavlov, B. M., "Solid Phase Condensation Heat Transfer on a Cold Surface," Heat Transfer Soviet Research, Vol.19, No.1, January-February 1987, p. 18-22

60. Loper, J. L., "Frost Formation upon a Thin Aluminum Tank Containing Liquid Oxygen," ASHRAE Transactions, Vol. 66, 1960, pp. 104-113.

61. Lotz, H., "Wärme- und Stoffaustauschvorgänge in bereifenden Lamellenrippen - Luftkühlern im Zusammenhang mit deren Betriebsverhalten," Kältetechnik-Klimatisierung, Vol. 23, No. 7, 1971, pp. 208-217.

62. Lin, S., "A Criterion for the Equilibrium Condition of Frost Formation with Heat Supply," Pap. Am. Inst. Aeronaut, Astronaut, No. 74-747, 1974, pp. i, 1-5.

63. Ledermann, H.,"Einfluss der Bereifung des Luftkühlers auf den Betrieb von Wärme-Pumpen mit Aussenluft als Wärmequelle," Temp. Tech., Vol. 16, No. 2, 1978, pp. 86-88.

64. Maxwell, J. C., "A Treatise on Electricity and Magnetism," Vol. 1, p. 435, 3rd. edn., Oxford University Press, London, 1904.

65. Marinyuk, B. T., "Heat and Mass Transfer under Frosting Conditions," International Journal of Refrigeration, Vol. 3, No. 6, 1980, pp. 366-368.

66. Nikuradse, J., "Gesetzmässkeiten der Turbulente Reibungsvorgängen in Flüssigkeiten," VDI-Forschungsheft, No. 356, 1932.

67. Nikuradse, J., "Strömungsgesetze in rauhen Rohren," VDI-Forschungsheft, 1933.

68. Nakamura, H., "Free Convection Heat Transfer from Humid Air to a Vertical Plate under Frosting Conditions," Bulletin of the Japanese Society of Mechanical Engineers, Vol. 17, 1974, pp. 75-82.

69. Okino, Y., and Tajima, O., "Heat and Mass Transfer by Free Convection under Frosting Conditions," Refrigeration (in Japanese), Vol. 48, No. 546, 1973, pp. 338-349.

70. Okubo, H., and Tajima, O., "The Frosting Phenomena to the Vertical Plate in Natural Convection Flow (Effects of surface temperature)," Refrigeration (in Japanese), Vol. 58, No.663, 1983, pp. 3-11.

71. O'Neal, D. L., Tree, D. R., "Experimental Measurement of Frost Thickness on A Flat Plate", IIR XVth International Congress of Refrigeration (Paris 1983), Proceedings Tome II, p.533-539.

72. O'Neal, D. L., Tree, D. R.,"A Review of Frost Formation in Simple Geometries," Trans. ASHRAE, p. 267-281

73. Pohlhausen, E., "Der Wärmeaustausch zwischen fester Körpern und Flüssigkeiten mit kleiner Reibung und kleiner Wärmeleitung," ZAMM, Vol. 1, 1921, pp. 115.

74. Piening, W., "Der Wärmeübergang an Rohren bei freier Strömung unter Berucksichtigung der Bildung von Schwitzwasser und Reif," Gesundheitsingenieur, Vol. 56, 1933, pp. 493-497.

75. Prins, L., "Wärme- und Stoffübertragung in einem querangeströmten, bereifender Luftkühler," Kältetechnik, Vol. 8, No. 5, 1956, PP. 160-164, 182-187.

76. Pitman, D. and Zucherman, B., "Effective Thermal Conductivity of Snow at -88, -27, and -5 ℃," Journal of Applied Physics, Vol. 38, 1967, pp. 2698-2699.

77. Rostami, A. A., "Prediction of Frost Growth Rate," ASME Heat Transfer Division, vol. 66, 1986, p. 65-71.

78. Ruccia, F. E.,and Mohr, C. M., "Atmospheric Heat Transfer to a Vertical Tank Filled with Liquid Oxygen," Advances in Cryogenic Engineering, Vol. 4, 1960, pp. 307-318.

79. Richard, R. J., Edmonds, K., and Jacobs, R. B., "Heat Transfer between a Cryo-Surface and a Controlled Atmosphere," Inter. Institutional Refrigeration Annex. 1962-1 Supplementary Bulletin, 1962, pp. 89-110.

80. Rosner, D. E., "Enhancement of Diffusion-limited Vaporization rates by Condensation within the Thermal Boundary Layer: 1. The Critical Supersaturation Approximation," International Journal of Heat and Mass Transfer, Vol. 10, 1967, pp. 1267-1279.

81. Rosner, D. E., and Epstein, M., "Fog Formation Conditions near Cool Surfaces," Journal of Colloid Interface Sci., Vol. 28, 1968, pp. 60-65.

82. Schmidt, E., and Beckmann, W., "Das Temperatur und Geschwindigkeitsfeld von einer Wärme Abgebenden senkrechten Platte bei naturilicher Konvektion," Forsch. Ing.-Wes., Vol. 1, 1930, pp. 391.

83. Schmidt, E., "Die Wärmestrahlung von Wasser und Eis, von bereiften und benetzten Oberflächen," Forshung, Vol. 5, No. 1, 1934, pp. 1-5.

84. Schropp, K., "Untersuchungen über die Tau- und Reifbildung an Kühlrohren in ruhende Luft und ihr Einfluss auf die Kälteübertragung," Zeitschrift für die gesamte Kälte -Industrie, Vol. 42, 1935, No. 5, pp. 81-85, No. 7, pp. 126-131, No. 8, pp. 151-154.

85. Seclic, D. P., "Heat and Mass Transfer to Cryogenically Cooled Surface under Frosting Conditions.--A Survey of Research Efforts and Analysis-Frosting of Air Coolers Part II," IIR XVth International Congress of Refrigeration (Paris 1983) Proceedings Tome II, p.541-550.

86. Sibbitt, W. L., Fontain, W. E. and Dotson, J. P., "Ice Formation on Metal Surfaces," Refrigerating Engineering, December 1954, pp. 49-94.

87. Stoeker, W. F., "Frost Formation on Refrigeration Coils," ASHRAE Transactions, Vol. 66, 1960, pp. 91-103.

88. Smith, R. V., Edmonds, D. K., Brentari, E. G. F. and Richard, R. J., "Analysis of the Frost Phenomena on a Cryo-Surface," Advances in Cryogenic Engineering, Vol. 9, 1964, pp. 88-97.

89. Shah, Y. T., "Theory of Frost Formation," Sc. D. Thesis, M.I.T., Cambridge, 1968.

90. Schneider, H. W., "Transferred Mass and Density of Frost Formed on a Cylindrical Tube in Cross Flow," Inter. Institutional Refrigeration Annex. 1972-1 Supplementary Bulletin, 1972, pp. 149-155.

91. Schneider, H. W., "Einfluss der Reifbildung auf den Wärmeübergang eines quer angeströmten Rohres," Forsch. Ingenieurwes., Vol. 42, No. 5, 1976, pp. 145-148.

92. Schneider, H. W., "Equation of the Growth Rate of Frost Forming on Cooled Surfaces," International Journal of Heat and Mass Transfer, Vol. 21, 1978, pp. 1019-1024.

93. Sakatsume, S., and Seki, N., "Heat Transfer Characteristics of Ice and Snow in Low Temperature Region", Trans. JSME, Vol. 44, No. 382, 1987, pp. 2059-2069.

94. Seki, N., Fukusako, S, Matsuo, K., and Uemura, S., "Incipient Phenomena of Frost Formation," Trans. JSME, Vol. 50, No. 451, pp. 825-831.

95. Saito, H., Tokura, I., Kishinami, K. and Uemura, S, "A Study on Frost Formation (On Dimensionless Parameters Correlating Density and Thickness of Frost Layer)," Trans. JSME, Vol. 50, No. 452, 1984, pp.1190-1196. (Heat Transfer Japanese Research, Vol. 13, No. 4, 1984, pp. 76-88.)

96. Seki, N., Fukusako, S, Matsuo, K., and Uemura, S., "An Analysis of Incipient Frost Formation," Wärme- und Stoffübertragung, Vol. 19, 1985, pp. 9-18.

97. Trammel, G. J., Little, D. C., and Killgore, E. M.,"A Study on Frost Formed on a Flat Plate Held at Sub-Freezing Temperature," ASHRAE Journal, July 1968, pp. 42-47.

98. Tajima, O., Yamada, H., Kobayashi, U., and Mizutani, L., "Frost Formation of Air Coolers, Part 1: Natural Convection for a Flat Plate Facing Upwards," Refrigeration (in Japanese), Vol. 46, 1971, pp. 333-341. (Heat Transfer Japanese Research, Vol.1, No. 2, 1972, pp. 39-48.)

99. Tajima, O., Naito, E., Tsutsumi, Y., and Yoshida, H., "Frost Formation of Air Coolers, Part 2: Natural Convection for a Cooled Plate Facing Downwards," Refrigeration (in Japanese), Vol. 47, 1972, pp. 350-358. (Heat Transfer Japanese Research, Vol. 2, No. 2, 1973, pp. 55-76.)

100. Tajima, O., Naito, E., Nakashima, K., and Yamamoto, H., "Frost Formation of Air Coolers, Part 3: Natural Convection for the Cooled Vertical Plate," Refrigeration (in Japanese), Vol. 48, 1973, pp. 395-402. (Heat Transfer Japanese Research, Vol. 3, No. 4, 1974, pp. 55-66.)

101. Tajima, O., Naito, E., Goto, T., Segawa, S., and Nishimura, K., "Frost Formation of Air Coolers, Part 4: Natural Convection for two Cooled Vertical Plates," Refrigeration (in Japanese), Vol. 49, 1974, pp. 95-103. (Heat Transfer Japanese Research, Vol. 4, No. 3, 1975, pp. 21-36.)

102. Tajima, O., Nishio, H., and Morino, S., "Frost Formation of Air Coolers, Part 5: Natural Convection for a Cooled Vertical Plate Opposed to an Insulated Vertical Plate," Refrigeration (in Japanese), Vol. 50, No. 574, 1975, pp. 589-596.

103. Tokura, I., Saito, H., and Kishinami, K., "Study on Properties and Growth Rates of Frost Layers on Cold Surfaces," Trans. ASME, Journal of Heat Transfer, Vol. 105, No. 4, 1983, pp. 895–901.

104. Tokura, I, Saito, H., and Kishinami, K., "An Experimental Study on Heat and Mass Transfer of Frost Layers on Vertically-arranged Horizontal Cylinders in Natural Convection," Trans. JSME, Vol. 50, No. 449, 1984, pp. 173–178.

105. Woodside, W., "Calculation of the Thermal Conductivity of Porous Media," Canadian Journal of Physics, Vol. 36, 1958, pp. 815–823.

106. Whitehurst, C. A., "Heat and Mass Transfer by Free Convection from Humid Air to Metal Plate under Frosting Conditions," ASHRAE Journal, No. 5, 1962, pp. 58–69.

107. White, J. E., "Heat and Mass Transfer in Thick Frost Layers," Ph. D. Dissertation, University of Kentucky, 1972.

108. White, J. E., and Cremers, C. J., "Prediction of Growth Parameters of Frost Deposits in Forced Convection," Pap. Am. Inst. Aeronaut Astronaut, No. 74-746, 1974, pp. i,1–9.

109. White, J. E., and Cremers, C. J., "Prediction of Growth Parameter of Frost Deposits in Forced Convection," Trans. ASME, Journal of Heat Transfer, Vol. 103, 1981, pp. 3–6.

110. Yonko, J. D., and Sepsy, C. F., "An Investigation of the Thermal Conductivity of Frost while Forming on a Flat Horizontal Plate," ASHRAE Transactions, Vol. 73, 1967, pp. I1.1–1.11.

111. Yamakawa, N., Takahashi, N., and Ohtani, S., "On the Measurement of the Surface Temperature under the Frosting Conditions," Kagaku Kogaku (in Japanese), Vol. 33, No. 7, 1969, pp. 699–701.

112. Yanagida, T., "Performance of Frosted Coils," Refrigeration (in Japanese), Vol. 44, No. 506, 1969, pp. 1182–1189.

113. Yamakawa, N., Takahashi, N., and Ohtani, S., "Heat and Mass Transfer by Forced Convection under the Frosting Conditions," Kagaku Kogaku (in Japanese), Vol. 35, No. 3, 1971, pp. 328–334.

114. Yamakawa, N., and Ohtani, S., "Heat and Mass Transfer in the Frost Layer," Kagaku Kogaku (in Japanese), Vol. 36, No. 2, 1972, pp.197–203. (Heat Transfer Japanese Research, Vol. 1, No. 3, 1972, pp. 75–82.)

115. Yamakawa, N., Kawamura, F.,and Ohtani, S., "Heat Transfer by Natural Convection under the Frosting Condition in an Enclosed Cavity: Bottom Plate is Heated and one side plate is cooled," Kagaku Kogaku (in Japanese), Vol. 37, No. 4, 1973, pp. 373–379.

Chapter 14

A Study of Frost Formation

KAZUO AOKI and MASARU HATTORI
Department of Mechanical Engineering, Nagaoka University of Technology,
Nagaoka 940-21, Japan

YUJIRO HAYASHI
Department of Mechanical Engineering, Kanazawa University,
Kanazawa 920, Japan

ABSTRACT

Frost formation has become an important problem relating to air-source heat pump systems. The frost formed on a heat exchanger causes a lower performance because of the increase in the thermal resistance and in the pressure drop through the heat exchanger. In this article, we present some general remarks on the fundamental aspects of frost formation and the performance of heat exchangers under frosting conditions.

CONTENTS

NOMENCLATURES

A	area, m^2
A_{min}	minimum free flow area, m^2
A_T	transfer area on air side, m^2
a	thermal diffusivity, m^2/s
C	concentration of water vapor, kg/kg'
ΔC	concentration difference of water vapor between a bulk and a cryosurface, kg/kg'
c_p	specific heat, $J/(kg\ K)$
D	diffusion coefficient, m^2/s
d_p	tube diameter, m
d_e	modified diameter defined by Eq.(46), m
G	mass flow rate of air, kg/h
H_F	fin pitch, m
H_1	tube row pitch, m
H_2	tube line pitch, m
h	heat transfer coefficient, $W/(m^2 K)$
h_D	mass transfer coefficient, $kg/(m^2 h \Delta C)$
kg/kg'	absolute humidity
l_f	frost thickness, m or mm
L_H	latent heat of sublimation, kJ/kg
P	pressure, Pa
ΔP	pressure drop through heat exchanger, Pa
p	constant number defined by Eq.(28)
Q	heat flux, W/m
Q_0	supplied heat flux, W/m^2
q	constant number defined by Eq.(28)
R	gas constant, $J/(kg\ K)$
r_1	radius of frost column, m
r_2	radius of frost unit shown in Fig.9, m
r_p	tube radius, m
Re	Reynolds number
Re_δ	roughness Reynolds number
S	water content ratio
S_F	fin thickness, m
Sh	Sherwood number
T	absolute temperature, K
t	temperature, $°C$
t_m	fusion temperature, $°C$
V	air velocity, m/s
w	rate of frost deposition, $kg/(m^2\ h)$
w_1	rate of frost deposition at the top of frost columns, $kg/(m^2\ h)$
w_2	rate of frost deposition at the imaginary surface, $kg/(m^2\ h)$
$\overline{w}_1$	rate of frost deposition relating to frost thickness, $kg/(m^2\ h)$
$\overline{w}_2$	rate of frost deposition relating to frost density, $kg/(m^2\ h)$
w_{diff}	rate of mass transfer due to diffusion in frost layer, $kg/(m^2 h)$

Greek Symboles

α_f	proportion of frost columns occupying each unit area of frost layer
α_{ice}	proportion of ice columns occupying each unit area of frost layer
δ	surface roughness of frost layer, m
ε	area efficiency
ζ	saturation ratio of the imaginary surface to the frost column

η	efficiency of heat transfer
$\bar{\eta}$	efficiency of mass transfer
θ	time, h
λ	thermal conductivity, W/(m K)
λ_{eff}	effective thermal conductivity, W/(m K)
ν	Kinematic viscosity, m^2/s
ξ_F	fin efficiency
ρ	density, kg/m^3

Subscripts

a	air
comp	ice-air composite materials
F	fin
f	frost
g	gas phase
ice	ice
p	tube
s	cryosurface
sat	saturation
T	total
∞	main stream
o	initial condition

1 INTRODUCTION

Frost formation occurs when humid air is exposed to a cold surface at temperatures below 0 °C. Frost formation is a problem often encountered in the refrigeration industry. Recently, it has become an important problem not only with cryogenic equipments but also with air conditioners connected to heat pump systems. When heat pump systems are used to supply thermal energy from air for the purpose of heating in winter, the surface temperature of the evaporator often falls below 0 °C, because it gains heat from the cold air outdoors and frost formation occurs.

The frost formed on a heat exchanger surface acts as a thermal insulator, reducing the ability of the surface to transfer heat. Also, the accumulation of frost often becomes thick enough to restrict air flow. In extreme cases, the frost will completely block the flow of air. Accordingly, frost formation causes a lower performance in heat exchangers.

Frost formation should be treated as a problem of a moving boundary with heat and mass transfer between the air stream and a deposit surface. It cannot, however, be readily treated as a growth of uniform materials, such as ice, because of the complicated nature of frost layers. Accordingly, frost formation has been studied so far from the following various points of view :

(1) Frost growth and frost structure.
(2) Frost properties.
(3) Heat and mass transfer with a moving boundary or with a rough surface.
(4) Theory or empirical relation for the growth of a frost layer.
(5) Frosting on air coolers.
(6) Defrosting.

Table 1. Classification of the study with regard to
surface geometry and flow state

	forced convection	natural convection
flat plate	[2] [3] [6] [15] [16] [31-33] [59]	[10] [21-23] [25] [69-76]
cylinders	[29] [30] [60-64]	[77-82]
in tube	[24] [65-67]	
finned tube	[36-46] [50] [51] [68]	

In addition, as shown in Table 1, many studies have been done concerning the various geometries and states of air flow.

In this article, we will make some general remarks on frost formation based on our studies, paying attention to the foregoing items. With regards to the review of references on frost formation, the paper of Sekulić[1] is available.

2 GROWTH OF A FROST LAYER

The growth of a frost layer on a cold surface can be considered as a consecutive growth of the nucleation, the crystal growth, the transition and the interaction of each frost crystal, and the internal diffusion in the process of sublimation and desublimation. The structure of a frost layer changes for various frosting conditions. To begin studying frost formation, the growth and the structure of a frost layer should be made clear.

2.1 Frost Formation Process

The mechanisms of frost formation are complicated, due to the effects of many environmental parameters, that is, cryosurface temperature, water vapor concentration in an air stream, air velocity and so on. Hayashi et al.[2,3] have observed the growth of frost layers under various frosting conditions and have shown that the frost formation process can be divided into three periods. Figure 1 shows a schematic diagram of the observed frost formation process as described below.

Crystal growth period.   When frosting begins, a thin frost layer covers the cryosurface.   Frost crystals which are relatively distant from each other, generate on the thin frost layer, and grow in a vertical direction at about the same rate. The frost formation in this period is characterized as a crystal growth in a linear direction. Saito et al.[4,5] and Seki et al.[6,7] have shown that the thin frost layer formed during the early stage of frosting is generated by the freezing of supercooled liquid water droplets.

Frost layer growth period.   The rough frost formed in the crystal growth period changes its shape due to the generation of branches around the top of a crystal or due to the interaction of each crystal, and then grows gradually into a meshed and more uniform frost layer . The frost density

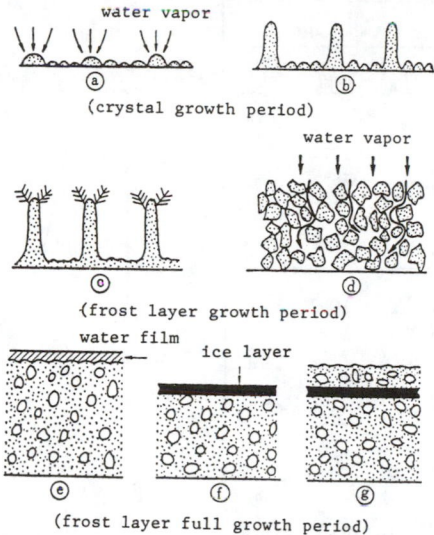

(crystal growth period)

water vapor

(frost layer growth period)

water film   ice layer

(frost layer full growth period)

Fig.1. Frost formation process

increases with the growth of a frost layer because of three-dimensional growth and the internal diffusion of water vapor in the frost layer.

Frost layer full growth period.   Aoki et al.[8] have investigated this period in detail. After the surface temperature of the frost layer reaches $0\,^{\circ}C$, water vapor moving from the bulk to the frost surface cannot be crystallized as a frost any longer, but forms water droplets on the frost surface. Before long, the water permeates into the frost layer and freezes due to the subcooled frost layer, and an ice layer is formed. This freezing causes a sudden increase in the frost density and thus a decrease in its thermal resistance, and the frost deposition occurs again. Then a cyclic process of the forming of water droplets, permeating, freezing and deposition often continues periodically.

The time required for each period varies with the frosting conditions. A higher rate of frost deposition results in a shorter length of time for each period. Note that this division proposes a generalized approach to frost formation instead of an individual treatment for the problem depending on the frosting conditions and the surface geometries, and is effective for comparing the complex nature of the frost layer whose growth rate depends on the frosting conditions.

2.2 Frost Formation Types

As mentioned above, the process of frost formation is divided into the three characteristic periods from a macroscopic viewpoint. The detailed mechanism of frost formation, however, is different from a microscopic viewpoint in the various frosting conditions. Hayashi and Aoki[3,9] have

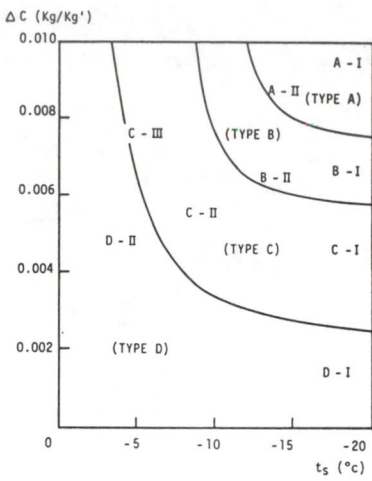

Fig.2. Classification of frost formation types

found that the concentration difference between the bulk and the cryosurface, ΔC, and the cryosurface temperature, t_s, are the most significant parameters for frost formation from a crystal growth viewpoint. They have classified frost formation into four types on a ΔC-t_s plane as shown in Fig. 2, considering the shapes of the crystal, and its transition and interaction. We shall not mention it here in detail. The main characteristic is that a frost layer has a denser structure along the direction from type A to type D on this figure, corresponding to the change of frost crystals from the rod-type crystals to the plane-type crystals.

Air velocity is independent from the frost formation types themselves, and depends on the number of frost columns formed on the cryosurface. A higher air velocity results in a denser structure, due to the increase in the number of frost columns. In conclusion, a frost layer has a denser structure for the conditions of the frost formation type D on a ΔC-t_s plane and a higher velocity. The effect of air temperature on frost formation is insignificant to consider.

3 FROST PROPERTIES

3.1 Frost Thickness

Frost properties may be closely connected to the structure of a frost layer, especially concerning frost formation processes and frost formation types. Figure 3 shows typical changes of frost thickness with time. Frost thickness rapidly increases with time during the early stage of frost formation due to the linear growth of frost crystals, and then slowly

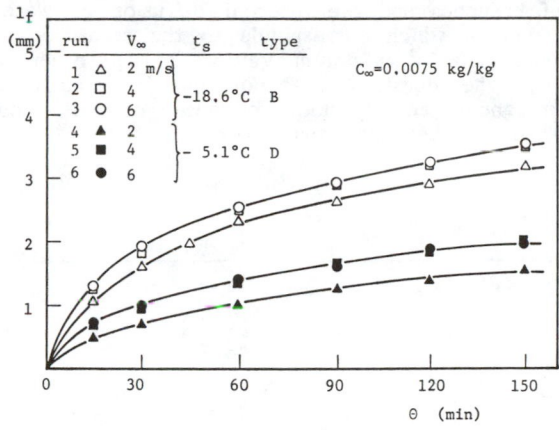

Fig.3. Variation of frost thickness with time

increases with time due to three-dimensional growth. As a results, the curves of frost thickness have a square-root of time dependence, as noted by a number of investigators. The higher rate of frost deposition does not always result in a higher frost thickness. In comparison between run No.1 and No.6, for example, a higher rate of frost deposition results in a thinner frost thickness, and in comparison between run No.2 and No.3 frost thickness appears to be nearly the same in spite of the large difference in the rate of frost deposition. These results can be explained by the differences in the structure of frost layers described above.

3.2 Frost Density

Trammel et al.[10] have investigated the distribution of frost density in a frost layer by using the absorption of a radio source and have reported that frost density decreases towards the surface of a frost layer. Also, Cremers et al.[11] have shown that the frost density is distributed in the layer in two different ways: it uniquely decreases towards the surface of a frost layer, or it has one or two peak values in the layer. Here, we will discuss only the average frost density over a frost layer.

   Figure 4 shows the variation of frost density with time. The density increases with time due to the three-dimensional growth and the internal diffusion of water vapor in a frost layer. As is expected, frost density becomes larger for the condition corresponding to frost formation type D and a higher velocity. In order to clarify the relation between frost formation and frost properties, Figure 5 shows the variation of frost densities with time during the early stage at various cryosurface temperatures. In spite of the small deference in the rate of frost deposition, the frost density depends strongly on the frosting conditions. The density corresponding to types A and B changes greatly with time. For the case of run No.3, after the generation of a thin frost layer(point a), the frost density decreases in the crystal growth period due to linear dimensional growth, but increases in the frost layer growth period due to the generation of branches and the internal diffusion of water vapor in a frost layer. At point b, which corresponds to the transition point of these periods, the density takes a minimum value. On the other hand, for the case of type D, the density does not change greatly and increases monotonously throughout two periods because of the nearly constant accumulation of the homogeneous frost layer.

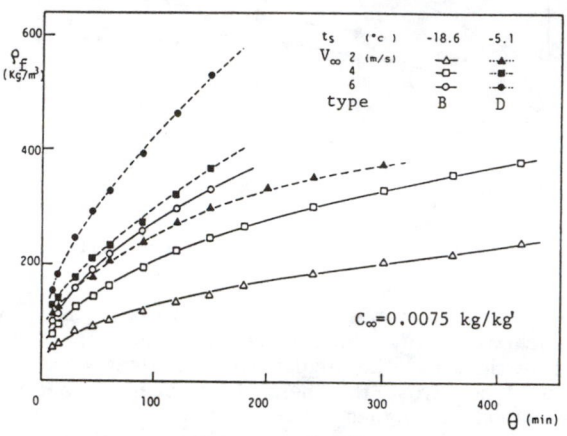

Fig.4. Variation of frost density with time

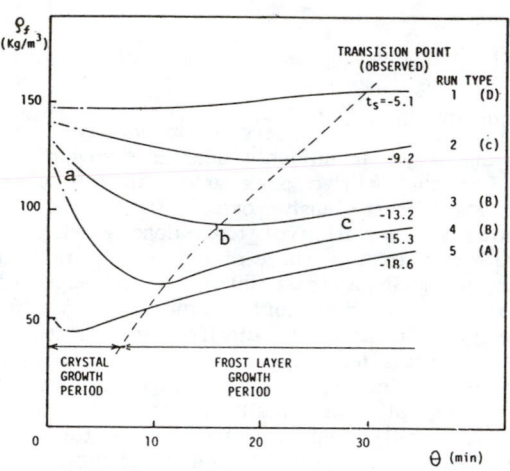

Fig.5. Variation of frost density with time during the early stage of frosting

458

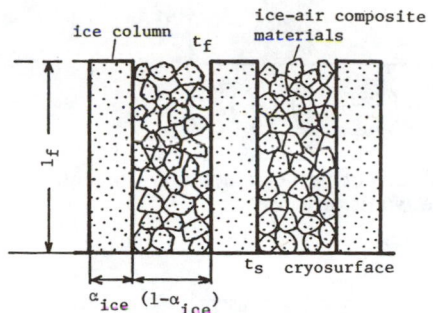

Fig.6. Structural model of frost layer

3.3 Frost Thermal Conductivity

Thermal conductivity is one of the most important properties. It is considered to be directly related to frost density and the relationship between both has been represented as empirical formulas by many investigators[12,13]. The thermal conductivity of a frost layer, however, cannot be expressed only in terms of frost density, since a frost layer does not have a uniform structure, but rather is formed of complicated structures depending on various frosting conditions. The internal diffusion of water vapor in a frost layer plays an important part for heat transport through a frost layer. Reid et al.[14] and Yamakawa and Ohtani[15] have predicted the thermal conductivity considering internal diffusion in a frost layer based on the uniform frost structure. Biguria and Wenzel[16] and Auracher[17,18] have discussed it considering the non-uniform structure of a frost layer.

Also, Hayashi et al.[3] have predicted frost thermal conductivity on the basis of the frost structural model shown in Fig. 6. This model is a parallel model which is composed of two parts, ice columns and ice-air composite material. The proportion of ice columns, with which frost columns are replaced, changes with the frosting conditions; the number of ice columns depends on air velocity and their size is related to ΔC and ts concerning the frost formation types. By this structural model, effective frost thermal conductivity is represented as follows, using the thermal conductivity of ice, λ_{ice}, and ice-air composite material, λ_{comp}, and the proportion of ice columns occupying each unit area of a frost layer, α_{ice},

$$\lambda_{eff} = \alpha_{ice} \cdot \lambda_{ice} + (1 - \alpha_{ice})\lambda_{comp} \tag{1}$$

Predicting the thermal conductivity of the ice-air composite material, Woodside's model[19], which consists of a cubic lattice of uniform solid spherical particles in a gas, is adopted when taking into account the contribution of water vapor diffusion to air conductivity. Woodside's equation is

$$\frac{\lambda_g}{\lambda_{comp}} = 1 - \left(\frac{6s}{\pi}\right)^{1/3} \left[1 - \left(\frac{a^2 - 1}{a}\right) \ln \left(\frac{a + 1}{a - 1}\right)\right] \tag{2}$$

459

where

$$a = \left\{ 1 + \frac{4}{\pi \left(\frac{\lambda_{ice}}{\lambda_g} - 1\right)\left(\frac{6s}{\pi}\right)^{2/3}} \right\}^{1/2} \quad , \qquad s = \frac{\rho - \rho_{air}}{\rho_{ice} - \rho_{air}} \tag{3}$$

and is valid for $0 < s < 0.5236$. Krisher[20] has devised the following expression for the effective conductivity of air inside the pore of a material whose pore walls are wet,

$$\lambda_g = \lambda_{air} + \frac{D}{RT}\left(\frac{P}{P - P_{sat}}\right)\frac{dP_{sat}}{dT} \cdot L_H \tag{4}$$

where

$$D = 0.086 \frac{10000}{P}\left(\frac{T}{273}\right)^{2/3} \tag{5}$$

The comparison between the predictions and the experimental results is shown in Fig. 7. It is found that the effective thermal conductivity has various values for the same frost densities and cannot be represented only in respect to frost density. This model of the difference in frost structures expresses the practical frost layer and gives a good prediction of the effective thermal conductivity. On the other hand, as shown in the same figure, Yamakawa's results[15] assumed a uniform frost structure, and Biguria's calculation[16], obtained on the basis of Woodside's model and the random mixture model, represent only the average relation to frost density. The discrepancy between the predictions and the data during the early stage of frost formation is assumed to be caused by neglecting the air turbulence contribution to the effective thermal conductivity in this model.

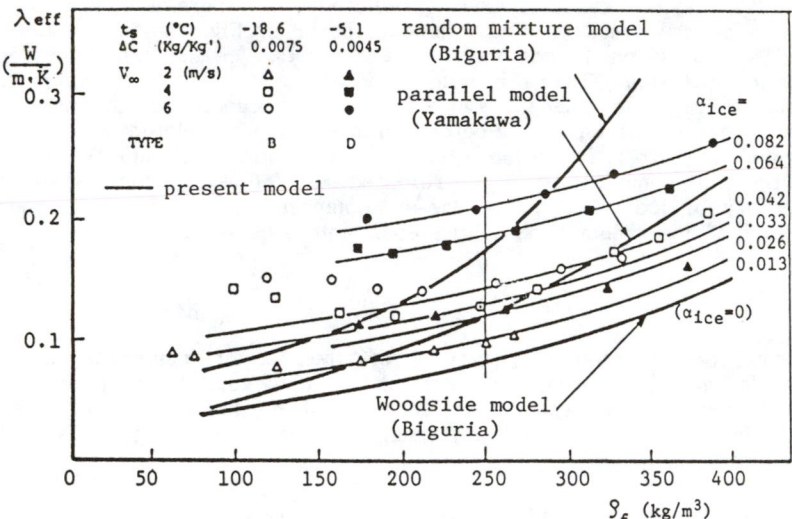

Fig.7. Relationship between frost thermal conductivity and frost density

4 HEAT AND MASS TRANSFER WITH FROSTING

4.1 General Remarks

A number of investigators have studied frost formation as a heat and mass transfer problem between the bulk and the frost surface. Whitehurst[21], Barron and Han[22], and Okino and Tajima[23] have predicted heat and mass transfer coefficients based on the boundary layer approximation with the assumption of the flat frost surface. On the other hand, Chen and Rohsenow[24] have investigated the roughness of the frost surface and pointed out the increase in heat transfer and pressure drop due to its roughness. Kennedy and Goodman[25] have also shown that the increase in the heat transfer coefficient results from the increase in the actual surface area due to frost roughness.

Hayashi et al.[26] have considered the effect of frost surface roughness on mass transfer rate and divided it into the following two factors:

(1) driving force of mass transfer due to the surface roughness;
(2) mass transfer coefficient due to turbulence generated by the surface roughness.

4.2 Effects of frost surface roughness on mass transfer

<u>Effect of factor(1).</u> In order to clarify the effects of the foregoing two factors on mass transfer rate, Hayashi et al.[26] have predicted the effect of factor(1) on the basis of a frost surface model shown in Fig.8. The model is composed of frost columns and valleys among the columns. The rates of mass transfer, which differ between the frost columns and valleys, are represented along with the driving forces of water vapor concentration between the bulk and the frost surface as follows:

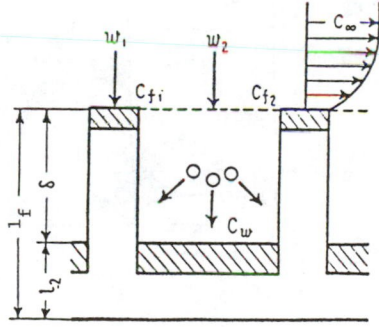

Fig.8. Mechanism of frost formation due to
frost surface model

$$w_1 = h_D(C_\infty - C_{f1}) \tag{6}$$

$$w_2 = h_D(C_\infty - C_{f2m}) = h_D(C_\infty - \zeta C_{f1}) \tag{7}$$

where C_{f1} is the concentration of water vapor at the surface of the frost column tops, and C_{f2m} is the mean value at the imaginary surface among the frost columns. ζ is the ratio of C_{f2m} to C_{f1} and represents the saturation ratio of the imaginary surface to the frost column surface. The total mass transfer rate is

$$w = \alpha_f w_1 + (1 - \alpha_f) w_2 \tag{8}$$

where α_f is the proportion of the frost columns occupying each unit area of a frost layer.

In order to obtain the unknown parameter, ζ, the diffusion rate of the water vapor passing through the imaginary frost surface must be analyzed. Figure 9 shows the unit model of the frost layer. Water vapor molecules pass through the imaginary surface, diffuse into the space of the unit and deposit on the surface. Therefore, this problem can be treated as a three-dimensional diffusion problem in a finite hollow cylinder. The governing equation and the boundary conditions are written as follows:

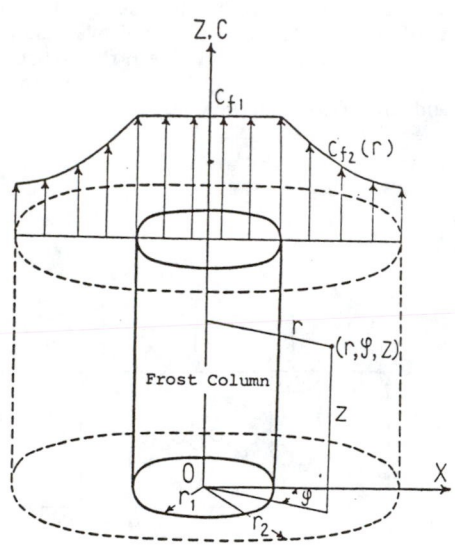

Fig.9. Unit model of frost layer

governing equation:

$$\frac{1}{r}\frac{\partial}{\partial r}(r\frac{\partial C}{\partial r}) + \frac{\partial^2 C}{\partial z^2} = 0 \tag{9}$$

boundary conditions:

$$r = r_1 , \quad C = f(z) \tag{10}$$

$$r = r_2 , \quad \frac{\partial C}{\partial r} = 0 \tag{11}$$

$$z = 0 , \quad C = C_w \tag{12}$$

$$z = 1_f , \quad C = g(r,\zeta) \tag{13}$$

where

$$g(r,\zeta) = C_{f1}\left\{6(1-\zeta)\frac{(r_1+r_2)(r+r_1-2r_2)}{(r_2-r_1)^2(3r_1+5r_2)} + 1\right\} \tag{14}$$

The concentration of water vapor at the surface of the structural unit is assumed to be at saturation. The function $f(z)$ is determined from the rectilinear temperature distribution in the vertical direction. At the outer boundary of the unit is symmetric with the neighboring units, and water vapor does not transfer in a radial direction. At the imaginary surface, the distribution of the water vapor concentration is assumed to be expressed as a parabolic function with respect to radius and is represented by Eq.(14) in terms of C_{f1} and ζ .

By solving the governing equation with the boundary conditions, the distribution of the water vapor concentration in a finite hollow cylinder is obtained, and then the transfer rate of the water vapor molecules which pass through the imaginary surface of the frost layer is calculated by integrating the diffusion rates over the imaginary surface. We shall omit further analysis here.

Figure 10 shows the comparison for ζ between the calculated results and the data. ζ increases with a decrease in the frost surface roughness, and converges to unity when the surface becomes flat. This means that the rougher the frost surface becomes, the more frost deposits at the bottom of the frost valley. In other words, frost with a rough surface changes to frost with a flat surface. The difference between the calculated results and the data seems to be generated by factor(2) neglected in this model.

Effect of factor(2). It is very difficult to directly estimate the effect of factor(2) on mass transfer rates, that is, the increase in mass transfer coefficients due to turbulence generated by the frost roughness. However, it may be possible to deduce it from the comparison between the data which includes the effects of factor(1),(2) and the predictions for the effect of factor(1). Figure 11 shows a comparison of mass transfer coefficients between the data and the deduced results including only the effect of factor(2) by excluding the effect of factor(1) on the foregoing

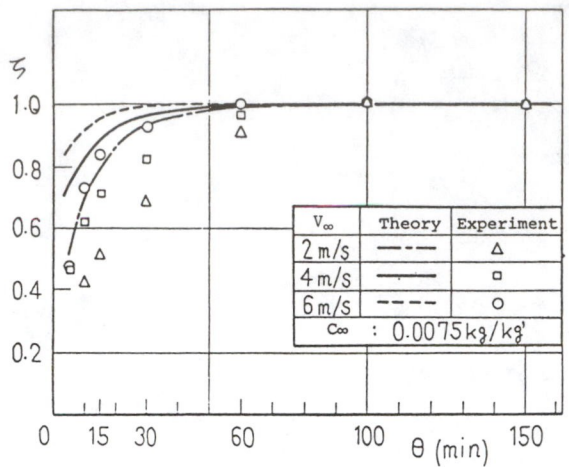

Fig.10. Variation of saturation ratio of imaginary
surface to frost column surface with time

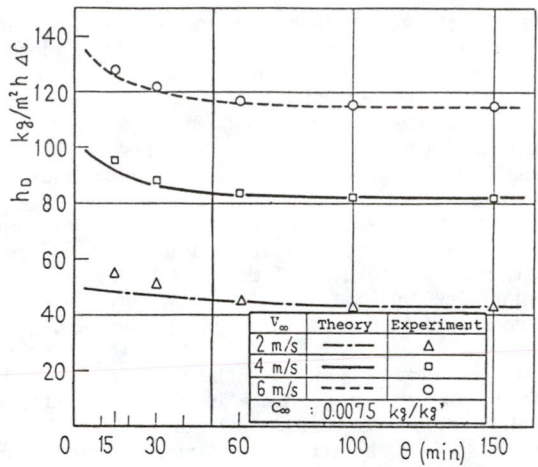

Fig.11. Variation of mass transfer coefficient
with time

model. From these results, it can be considered that the initial increase in
mass transfer coefficients for the calculated results is produced by the
turbulence generated by the frost surface roughness. After the frost
surface becomes flat, the Sherwood number, which is the non-dimensional
mass transfer number, is found to converge with the constant relation
with respect to the Reynolds number, which is defined by using the
distance measured from the edge of the plate as a characteristic length.

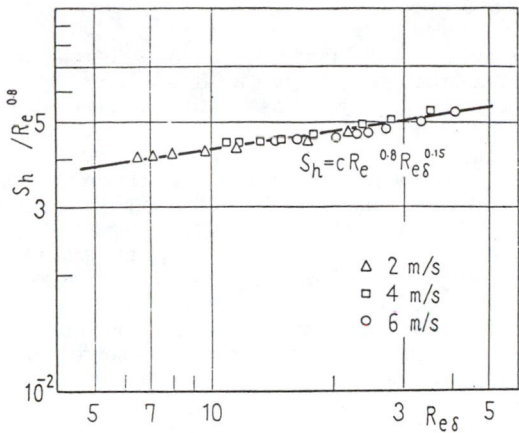

Fig.12. Relationship between $Sh/Re^{0.8}$ and Re_δ

Then, as shown in Fig.12, the effect of this factor(2) on mass transfer coefficients is represented by the following relation as a function of the roughness Reynolds number, Re_δ , which is defined by using the surface roughness as a characteristic length,

$$Sh=c\cdot Re^{0.8}Re_\delta^{0.15} \qquad (15)$$

5 FROST GROWTH THEORY

5.1 General Remarks

It is very important to estimate the growth of a frost layer under the various frosting conditions. The approach can be classified into two categories. One is to represent the frost growth using the empirical relation as a function of frosting conditions and time, as studied by Schneider[27,28], White and Cremers[29] and Tokura et al.[4,5]. The other is to predict frost growth using a diffusion theory which considers the mechanism of internal diffusion in the process of simultaneous sublimation and desublimation as well as the diffusion of water vapor towards deeper layers from the surface. Chung and Algren[30], Brian et al.[31], Jones and Paker[32] have calculated the growth of a frost layer using the model based on the assumption that water vapor coming into the layer through its surface deposits uniformly in the layer. Brian et al.[33] have also calculated the growth using the non-uniform model, which considers the distribution in the density in the frost layer. Yamakawa and Ohtani[15] have calculated the growth of a frost layer on the basis of the model composed only of ice columns which replace a frost layer. Further, Aoki et al.[34] have progressed the growth theory of a frost layer by considering not only internal diffusion in a frost layer, but also frost structures dependent on frosting conditions. This theory includes the calculation of frost thermal conductivity under the same model, and the outline is described below.

5.2 Analysis of Frost Growth

<u>Model.</u> The purpose of this model, which is based on the diffusion theory, is to predict the growth of a frost layer by considering the effects of the crystal growth as initial frost structures in frost formation. Figures 13 and 14 show the frost structural model and the mechanism of heat and mass transfer in a frost layer. In this structural model, ice columns, with which frost columns are replaced, play a basic part in the frost structure, and the ice-air composite materials play an additional part in the frost growth due to water vapor diffusion.

The frost layer at the end of the crystal growth period, which corresponds phenomenologically to one of common representative time, is modeled as an initial condition. The proportion of the ice columns occupying each unit area of a frost layer and the initial frost densities depend on the frosting conditions, and are represented by the following empirical equations, respectively[35],

$$\alpha_{ice}=0.0067(h_D)^{0.3}(\Delta C)^{-0.5}(t_m-t_s+8.0)^{-0.56}-0.0395 \tag{16}$$

$$\rho_{f0}=5.9(h_D)^{0.3}(\Delta C)^{-0.5}(t_m-t_s+8.0)^{-0.56} \tag{17}$$

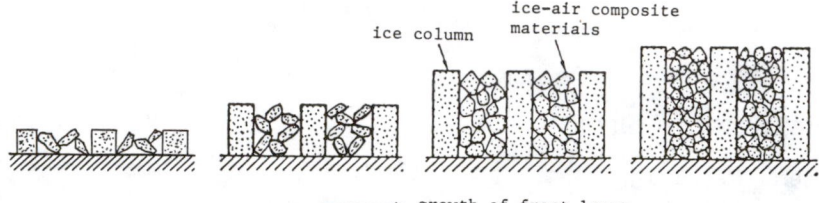

Fig.13. Structural model of frost layer

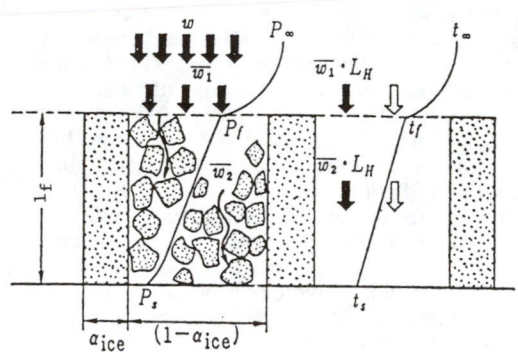

Fig.14. Mechanism of frost formation relating to frost structural model

In these equations, the initial structure of the frost layer is represented as functions of the temperature difference between the fusion and the cryosurface temperature, the concentration difference between the bulk and the cryosurface, and the mass transfer coefficient. Referring to the frost growth stages mentioned before, the former two affect the size of the ice columns depending on the frost formation types and the latter affects the number of ice columns.

This model is applicable to the growth of a frost layer during the frost layer growth period, with the exception of the growth in the frost layer full growth period, which includes the formation of a film of water on the frost surface, and its permeation and freezing in a frost layer.

Mass balance. The rate of frost deposition, w , transported to the frost surface can be divided into two parts. A part of the rate, $\overline{w}_1$, freezes out at the frost surface, which serves to increase the frost thickness. The other part, $\overline{w}_2$, diffuses into the frost layer through the surface, which serves to increase the frost density. It is assumed that water vapor deposits uniformly in a frost layer and the frost density is usually uniform over the layer. The increase in frost thickness and in frost density are represented as follows:

$$\overline{w}_1 = \rho_f \frac{dl_f}{d\theta} = \alpha_{ice}\rho_{ice} + (1 - \alpha_{ice})\rho_{comp} \frac{dl_f}{d\theta} \tag{18}$$

$$\overline{w}_2 = l_f \frac{d\rho_f}{d\theta} = (1 - \alpha_{ice}) \overline{w}_{diff} \tag{19}$$

where, w_{diff} denotes the mass transfer rate into the ice-air composite materials and is represented by

$$\overline{w}_{diff} = (1 - \frac{\rho_{comp}}{\rho_{ice}})D \frac{1}{RT_f} \frac{P}{P - P_{sat,f}} (\frac{dP_{sat}}{dx})_{x=1_f} \tag{20}$$

Energy balance. The energy transport in a frost layer is comprised of heat conduction and energy transport due to the diffusion of water vapor. In this model, the vapor diffusion occurs only in part of the ice-air composite materials. The energy balance equation in a frost layer is represented as

$$\frac{d}{dx} (\lambda_{eff} \frac{dt}{dx}) = 0 \tag{21}$$

Where λ_{eff} denotes an effective thermal conductivity as shown by Eq.(1). The boundary conditions for the energy equation under the condition of constant cryosurface temperature are given as follows:

$$x = 0 , \quad t = t_s \tag{22}$$

$$x = 0 , \quad \lambda_{eff} \frac{dt}{dx} = h(t_\infty - t_f) + h_D(C_\infty - C_f) L_H \tag{23}$$

467

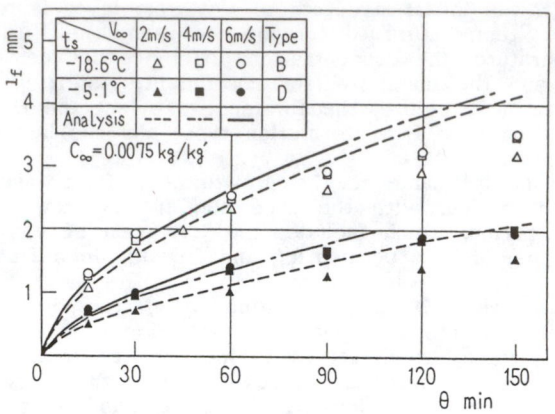

Fig.15. Comparison for frost thickness between
experimental and predicted results

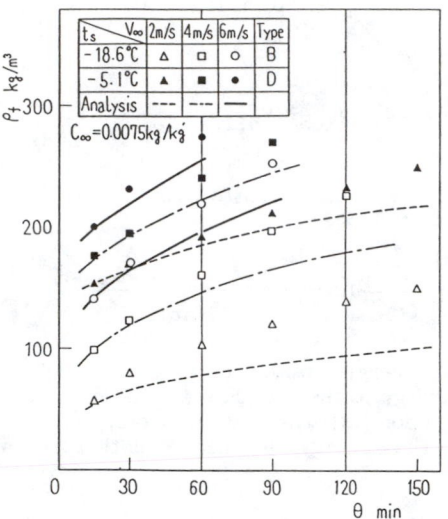

Fig.16. Comparison for frost density between
experimental and predicted results

<u>Results and discussion.</u> The frost growth is calculated in time steps by simultaneously solving the mass and energy balance equations using effective thermal conductivity. Figures 15 and 16 show the comparison between the predictions and the data for frost thickness and frost density obtained by constant cryosurface temperatures on a flat plate. The calculations are made until the frost surface temperature attains 0 °C. These predictions correspond to the data which are inexplicable when the

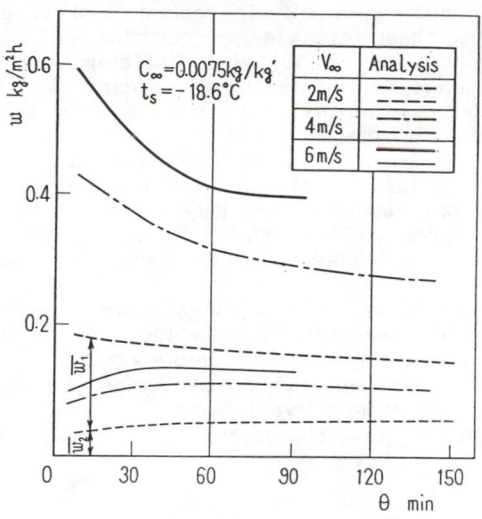

Fig.17. Variation of w, $\overline{w}_1$ and $\overline{w}_2$ with time

predictions which take only internal diffusion into account are made. Namely, a higher rate of frost deposition does not always result in a thicker frost thickness, and the frost thickness and the frost density vary even under the nearly same rate of frost deposition. Figure 17 shows the time variation of the frost deposition rate, w, and the diffusion rate into the frost layer through the surface, $\overline{w}_2$. The latter increases with time due to the rise of the frost surface temperature during the early stage of frosting, and then decreases due to the reduction of the diffusion space corresponding to the increase in the frost density. Also, a faster air velocity results in a larger internal diffusion rate due to the increase in the temperature gradient in the frost layer.

In this model, it is assumed that the frost surface is usually flat. A theory which would include the effect of the frost roughness on the growth of a frost layer is left as a future problem to be solved.

6 PERFORMACE OF EXTENDED SURFACE HEAT EXCHANGERS UNDER FROSTING CONDITIONS

6.1 General Remarks

There are some characteristics of heat exchangers under frosting conditions which differ from those of ordinary heat exchangers. That is:

(1) Thermal resistance not only increases with the growth of a frost layer but also depends on the position of the heat exchanger.
(2) Heat and mass transfer coefficient changes due to the change in geometry of heat exchanger with frosting.
(3) Pressure drop increases with the growth of the frost layer.
(4) It is necessary to take the characteristics into account in the cycles of frosting and defrosting.

The most common types of air coolers used in practice are the extended surface type heat exchanger, consisting of tube bundles with continuous fins. Gates et al.[36] and Huffman and Sepsy[37] have investigated 9 air coolers with different fin spacings and with a different number of rows. The experimental data have been arranged in terms of dimensionless numbers characterizing heat transfer, mass transfer, pressure drop and time. Hosoda and Uzuhashi[38] have given relatively simple semi-empirical relations for the overall heat transfer coefficient and the air-side pressure drop. Further, Javel[39,40], Yanagida[41], Lotz[42,43], Sanders[44], and Ivanova and Gatchilov[45] have given the calculation methods on total heat and mass transfer considering the change of fin efficiency due to frosting.

In spite of these works, until recently, there have been no practical predictions which could adequately express the dynamic characteristics in heat exchangers with frosting, taking into account the time-dependent properties of a frost layer and the heat transfer coefficient in order to obtain precise and usable quantitative indexes. The following is a review of our attempts to predict the characteristics of heat exchangers under frosting conditions.

6.2 Efficiency of Heat and Mass Transfer

In a finned tube type heat exchanger the effect of frost formation on heat transfer is comprised of two parts, that is, the increase in thermal resistance and the change in the geometry of the heat exchanger due to the growth of a frost layer. In addition, these effects have to be taken into consideration on each part of tube and fin. Figure 18 shows a model which predicts the efficiency of heat and mass transfer[46]. It is assumed that the frost layer grows uniformly on both the tube and fin surface, except that some areas on the tube and fin have a double layer of frost. These areas are regarded as adiabatic.

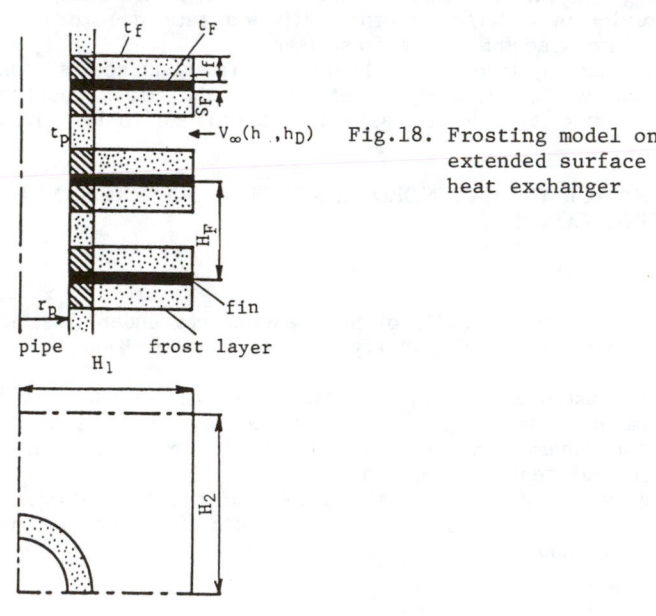

Fig.18. Frosting model on extended surface heat exchanger

<u>Tube.</u> The efficiency of the heat transfer under frosting conditions on a tube, η_{pf} , is defined as ratio of the resulting heat transfer rate to the heat transfer rate of a bare surface. The following equation considers the change of the transfer area due to frosting,

$$\eta_{pf} \cdot h_a(t_a - t_p)A_p = h_a(t_a - t_f)\varepsilon_p A_p \tag{24}$$

Accordingly, η_{pf} becomes

$$\eta_{pf} = \frac{t_a - t_f}{t_a - t_p} \varepsilon_p \tag{25}$$

where the area efficiency, ε_p , is defined as the ratio of the frost surface area to the bare tube surface area and is given by

$$\varepsilon_p = (\frac{H_F - S_F - 2l_f}{H_F - S_F})(\frac{r_p + l_f}{r_p}) \tag{26}$$

Based on the energy balance of the frost layer, the following equation is obtained,

$$h_a(t_a - t_f) + h_D L_H(C_a - C_f) = \lambda_f(t_f - t_p)/l_f \tag{27}$$

The concentration of water vapor on the frost at saturated conditions is assumed to be linear with respect to temperature as follow:

$$C_f = p \cdot t_f + q \tag{28}$$

The slope, n, of a straight line, which is formed between the surrounding state(ta,Ca) and the tube surface state(tp,Cp) is introduced as follows:

$$\frac{C_a - C_p}{t_a - t_p} = n \tag{29}$$

If Eq.(25) is rewritten using Eqs.(27),(28) and (29), it becomes,

$$\eta_{pf} = \frac{\lambda_f + h_D L_H l_f(p - n)}{\lambda_f + (h_a + h_D L_H p)l_f} \varepsilon_p \tag{30}$$

<u>Fin.</u> When it is assumed that heat transfer occurs two-dimensionally on the fin and one-dimensionally in the frost layer, the energy balance equations are given as follows, respectively,

$$\frac{d^2 t_F}{dy^2} + \frac{d^2 t_F}{dz^2} = \frac{2}{\lambda_F S_F} h_a(t_f - t_a) + h_D L_H(C_f - C_a) \tag{31}$$

$$h_a(t_a - t_f) + h_D L_H(C_a - C_f) = \lambda_f(t_f - t_F)/l_f \tag{32}$$

Referring to the study of Iuchi et al.[47] dealing with a finned tube heat exchanger, it is assumed that the concentration of water vapor on the fin surface at saturation can also be represented by the same straight line which is formed between the surrounding state and the tube surface because of the small temperature difference on the fin. Then, the relation between the temperature and concentration on the fin can be expressed as,

$$\frac{C_a - C_F}{t_a - t_F} = n \tag{33}$$

Substituting Eqs.(32) and (33) into Eq.(31) yields the following equation,

$$\frac{d^2 t_F}{dy^2} + \frac{d^2 t_F}{dz^2} = - \frac{2\lambda_f(h_a + h_D L_H n)/l_f}{\lambda_F S_F(\lambda_f/l_f + h_a + h_D L_H p)}(t_a - t_F) \tag{34}$$

Using Schmidt's approximate relation for a continuous fin[48], the fin efficiency, ξ_F , considering the growth of a frost layer is represented as follows:

$$\xi_F = \frac{\tanh(m_f K_f)}{m_f K_f} \tag{35}$$

where

$$m_f = \sqrt{\frac{2\lambda_f(h_a + h_D L_H n)/l_f}{\lambda_F S_F(\lambda_f/l_f + h_a + h_D L_H p)}} \tag{36}$$

$$K_f = \left\{ \sqrt{\frac{H_1 H_2}{\pi}} - (r_p + l_f) \right\} \left\{ 1 + 0.35 \ln(\frac{\sqrt{H_1 H_2/\pi}}{r_p + l_f}) \right\} \tag{37}$$

Using the efficiency of heat transfer with frosting on the fin, η_{Ff} , yields the following equation,

$$\eta_{Ff} h_a(t_a - t_p) A_F = h_a(t_a - t_f)\varepsilon_F A_F \tag{38}$$

Accordingly, η_{Ff} is expressed as,

$$\eta_{Ff} = \frac{t_a - t_f}{t_a - t_p} \varepsilon_F \tag{39}$$

where the area efficiency on fin, ε_F , is given by

$$\varepsilon_F = \frac{H_1 H_2 - \pi(r_p + l_f)^2}{H_1 H_2 - \pi r_p^2} \tag{40}$$

Table 2. Specification of the test heat exchanger

type: 2-row (or 4-row.), 23-line finned tube banks			
pipe arrangement	:	in line	
pipe outer diameter (d_p)	:	11	(mm)
pipe row pitch (H_1)	:	25	(mm)
pipe line pitch (H_2)	:	25	(mm)
fin thickness (S_F)	:	0.2	(mm)
fin pitch (H_F)	:	4	(mm)

If Eq.(39) is rewritten using Eqs.(28),(32),(33) and (35), in terms of the fin efficiency it becomes

$$\eta_{Ff} = \xi_F \frac{\lambda_f + h_D L_H l_f (p - n)}{\lambda_f + (h_a + h_D L_H p) l_f} \varepsilon_F \tag{41}$$

Overall. The overall efficiency for heat transfer with frosting is represented by

$$\eta_{Tf} = \frac{\eta_{pf} A_p + \eta_{Ff} A_F}{A_p + A_F} \tag{42}$$

Similarly, the efficiency for mass transfer is given by

$$\overline{\eta}_{Tf} = \frac{\overline{\eta}_{pf} A_p + \overline{\eta}_{Ff} A_F}{A_p + A_F} \tag{43}$$

The apparent heat and mass transfer coefficients, h_a^* and h_D^*, which are defined as a basis of the tube surface is represented by the following equations using the real heat and mass transfer coefficients and the overall efficiencies for heat and mass transfer.

$$h_a^* = \overline{\eta}_{Tf} \cdot h_a \tag{44}$$

$$h_D^* = \overline{\eta}_{Tf} \cdot h_D \tag{45}$$

Results. The specifications of the heat exchanger used are given in Table 2. We will discuss first the case in which the mass flow rate of air is held constant, in spite of an increasing pressure drop through the heat exchanger due to frosting. Figure 19 shows the change in the apparent heat transfer coefficients with time. In this figure, the dashed lines show

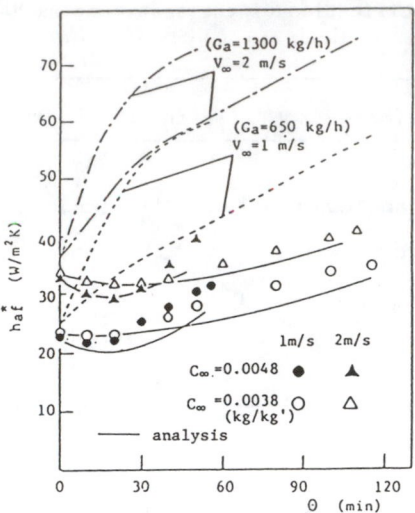

Fig.19. Comparison for apparent heat transfer coefficient
between experimental and predicted results
(constant mass flow rate of air)

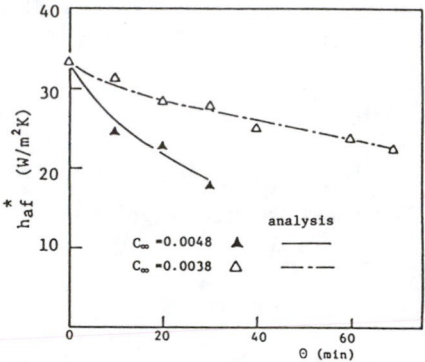

Fig.20. Comparison for apparent heat transfer coefficient
between experimental and predicted results
(constant rotation number of blower)

the real heat transfer coefficient, calculated by considering the increase in
the flow rate due to frosting. In early stage of frosting the apparent heat
transfer coefficient decreases with time, and then begins to increase with
time. It can be considered to be the combined result of the increase in
the real heat transfer coefficient due to the increasing air velocity and
the increase in the thermal resistance of the frost layer. As a result, in
the early stage of frosting the latter plays a larger part, while the former
becomes more significant as the frost layer grows. The solid lines show
the predictions described above.

Usually, when devising a system coupled to a heat exchanger under frosting conditions, it is difficult to hold the flow rate constant against the increase in pressure drop. Then, the mass flow rate of air will depend on the characteristics of the blower used. Next, we consider the case of maintaining the rotation number of the blower at a constant. The apparent heat transfer coefficients for this case, shown in Fig.20, decreases continuously with time. This is caused by the increase in the thermal resistance and the decrease in the real heat transfer coefficient as a result of the decrease in the mass flow rate of air.

6.3 Change in Pressure Drop due to Frosting

Usually, an increase in pressure drop through heat exchangers under frosting conditions occurs due to the decrease in air flow area and the roughness of the frost surface. It is, however, not easy to separate two effects occurring simultaneously, since the roughness of frost surface varying with its location on heat exchangers can not be correctly measured. As has been pointed out by Sanders[44], the effect of the frost roughness on pressure loss has not been established yet. Here, we will focus only on the effect of the decrease in flow area due to frosting on pressure drop through heat exchangers.

According to Kays and London[49], the modified diameter, d_e , and the Reynolds number, R_e , for the extended surface heat exchanger shown in Table 2, are defined as follows:

$$d_e = \frac{4A_{min} \cdot H_1}{A_T} \tag{46}$$

$$R_e = \frac{V_{max} \cdot d_e}{\nu} \tag{47}$$

where, A_{min} , A_T and V_{max} denote the minimum free flow area, the total transfer area on the air side and the maximum air velocity, and are represented as follows, respectively,

$$A_{min} = (H_F - S_F)(H_2 - d_p) \tag{48}$$

$$A_T = 2(H_1 H_2 - \pi d_p^2/4) + \pi d_p(H_F - S_F) \tag{49}$$

$$V_{max} = \frac{H_F H_2}{(H_F - S_F)(H_2 - d_p)} \cdot V_\infty \tag{50}$$

If the contraction and the expansion effects at inlet and outlet are included in the friction factor, the friction factor is defined by

$$f = \frac{\Delta P}{\frac{\rho V_{max}^2}{2} \frac{4H_1}{d_e}} \tag{51}$$

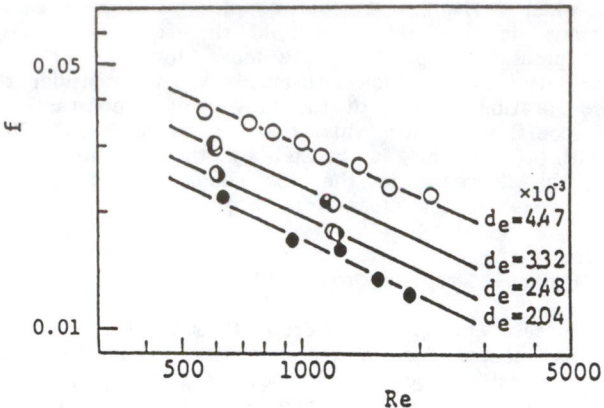

Fig. 21. Relationship between friction factor and Reynolds
number as a parameter of modified diameter

Figure 21 shows the relationship between the friction factor and the
Reynolds number as a parameter of modified diameter, which is obtained
from the experiments using the uniformly frosted heat exchanger[50]. The
modified diameter, the Reynolds number and the friction factor are
corrected by considering the frost thickness. The difference in the modified
diameter means the difference in the frost thickness. Note that the
friction factor is not only related to the Reynolds number, but depends on
the modified diameter, and so the friction factor is represented by

$$f = 58.7 \cdot Re^{-0.44} d_e^{0.83} \tag{52}$$

Since the type of heat exchanger used here is very limited, we
cannot conclude that pressure drop through various types of heat
exchangers under frosting conditions, is usually predicted by the method
described above.

6.4 Performance of Heat Exchangers Under Frosting Conditions

The performance of extended surface heat exchangers under frosting
conditions may be predicted by considering the changes of the
efficiency of heat and mass transfer and the pressure drop due to frosting
into the analysis of frost growth, based on the characteristics of used heat
exchangers and blowers[51].
We will compare the predictions with the data for the heat
exchanger composed of four rows. Figure 22 shows the variation of frost
thickness for each row with time. The concentration of water vapor
decreases along the direction of air flow due to mass transport at the
upstream part, so that the frost layer is thick on the 1st row and
becomes thinner along the direction of air flow. Figure 23 shows the

476

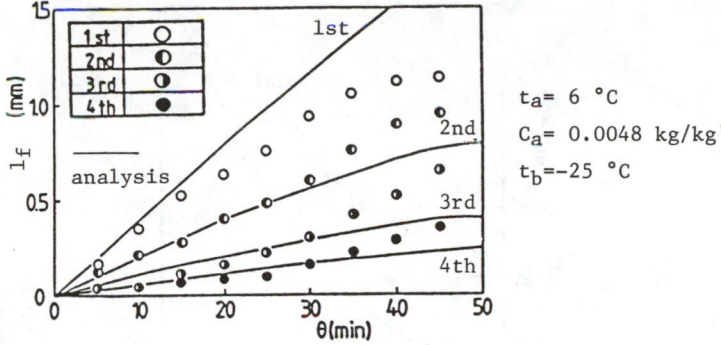

Fig.22. Comparison for frost thickness for each row
between experimental and predicted results

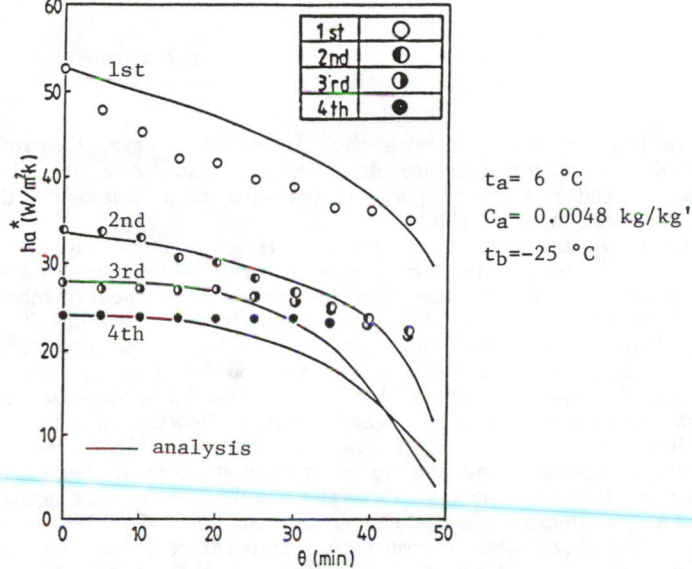

Fig.23. Comparison for apparent heat transfer coefficient
for each row between experimental and predicted
results

variation of apparent heat transfer coefficient for each row with time.
The coefficient decreases with time because of the decrease in the mass
flow rate of air and the increase in the frost thermal resistance due to
frosting. Also, it rapidly decreases on 1st row where frost layer grows
fast. Figure 24 shows the variation of pressure drop with time. Pressure
loss rapidly increases with time due to the reduction of free flow area by
frosting. The predictions calculated as a sum total of pressure drop for

477

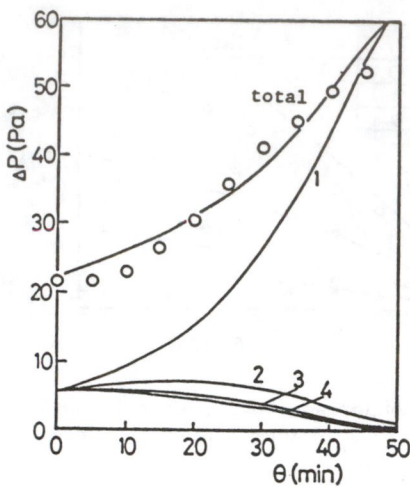

Fig.24. Comparison for pressure drop through heat exchanger
between experimental and predicted results

each row are in agreement with the experimental data. For reference, the
calculated results for pressure drop through each row are also shown in
this figure, and the major part of pressure drop arises on the 1st row
where the frost layer is thick.

Usually, the growth of a frost layer depends on the cryosurface
temperature and the water vapor concentration deference between the bulk
and the cryosurface, so that the performance of heat exchangers under
frosting conditions differs between parallel and counter flows. Where,
parallel flow means that both air and coolant flow the same direction
from the 1st row to the 4th row, and counter flow means that coolant
flow from the 4th row to the 1st row. Figure 25 shows the comparison in
frost thickness between parallel and counter flows. Referring to the frost
formation types shown in Fig.2, in parallel flow the cryosurface
temperature increases and the concentration difference decreases along the
direction of the air flow, so that the frost thickness at upstream of air
flow becomes thicker than that at downstream. On the other hand, in
counter flow both the cryosurface temperature and the concentration
difference decrease along the direction of the air flow, so that the
distribution in frost thickness becomes nearly uniform. Figure 26 shows the
comparison in pressure drop between parallel and counter flows. The
pressure drop in counter flow slowly increases with time as compared with
that in parallel flow because of the uniform distribution in the frost
thickness. Also, the time required to attain to the state of blocked air
flow becomes shorter in counter flow. To make the performance of heat
exchanger clear, the comparison in heat transfer rate between parallel and
counter flows are shown in Figure 27. It is found that heat transfer rate
in counter flow ia always larger than that in parallel flow, and keep a
larger value because of a lower pressure drop.

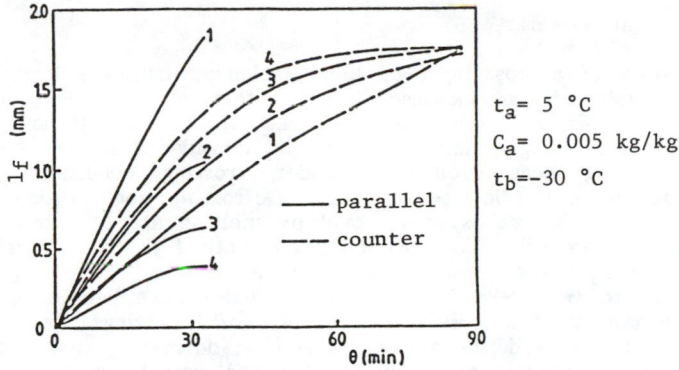

Fig.25. Comparison for frost thickness for each row
between parallel and counter flow types

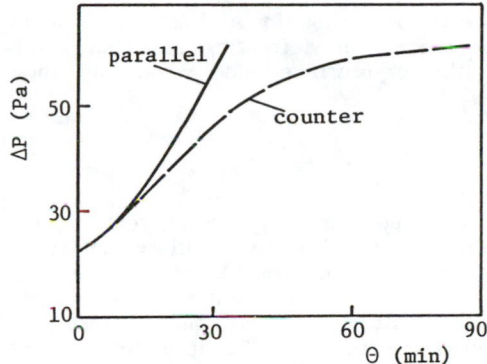

Fig.26. Comparison for pressure drop through heat exchanger
between parallel and counter flow types

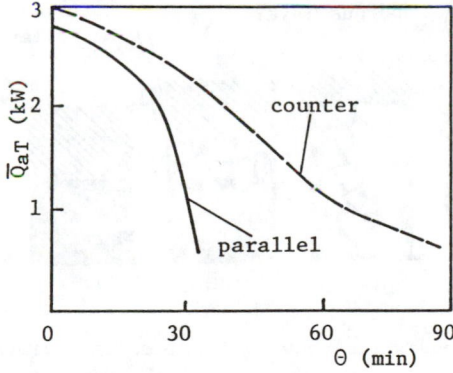

Fig.27. Comparison for heat transfer rate between
parallel and counter flow types

7. MELTING OF A FROST LAYER

7.1 General Remarks

The growth of a frost layer on heat exchanger causes a lower performance as a result of an increase in the thermal resistance and air flow resistance. No successful method free from frosting has so far been proposed. Accordingly, defrosting may be considered as an important aspect for operating heat exchangers under frosting conditions in order to maintain the prescribed performance. Defrosting can be done in different ways, that is, by means of removal by melting and by sublimation ,and by mechanical removal. The most common method is frost melting, supplying heat by using hot gas refrigerant system.

Up to the present time, a few studies have been considered on the melting process of a frost layer, in spite of it having important effects on thermal efficiency during defrosting and on defrosting time. Sanders[44] has classified the process of defrosting for the various defrosting systems and has evaluated the process based on a simple model. Katsuta et al.[52,53] have investigated the melting from the frost surface due to warm air. Adachi et al.[54] have investigated the process of defrosting relating to hot gas defrosting systems. Zakrzewski[55] has discussed the optimum defrost cycle. Further, Aoki et al.[56,57] have studied the permeation and refreezing of melted water occurring in a frost layer during defrosting, and have calculated the process of defrosting in fundamental cases where heat is supplied from the cryosurface side. The outline is described below.

7.2 Melting Process

The melting of porous materials such as frost layers is characterized by the permeation of melted water due to capillary action or gravity. In addition, when the temperature in a frost layer is lower than the fusion temperature, refreezing of the permeating water occurs in the frost layer.

Figure 28 shows the melting process when a frost layer is exposed to the surrounding temperature below 0 °C. It is characterized by water permeation and refreezing due to the heat loss from the frost surface to

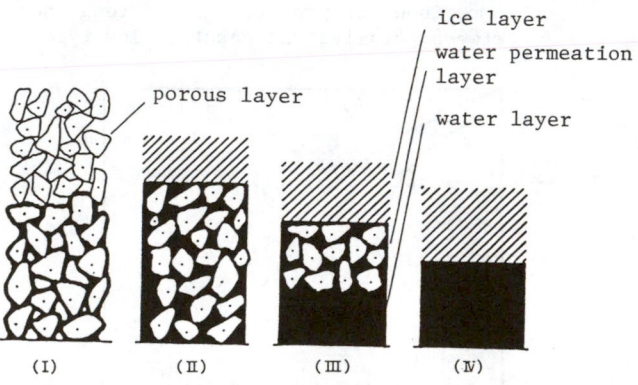

Fig.28. Schematic diagram showing frost melting process

the surroundings. The melting process is divided into four periods. In period (I) the body is composed of a porous layer and a water permeation layer formed by the permeation of the melted water. As the water permeation layer grows, a part of the permeating water refreezes on the water permeation front because the temperature of the porous layer is lower than the fusion temperature, and the porous structure on the water permeation front changes with time. In period (II) after the water permeation front reaches the frost surface, refreezing occurs at the frost surface and an ice layer starts to form due to successive permeation and refreezing. Accordingly, the body is composed of the ice layer and the water permeation layer. In period (III) a water layer begins to form from the bottom of the body because the melting rate at the melting surface becomes greater than the water permeating rate away from the surface. The body is composed of the ice layer, the water permeation layer and the water layer. In period (IV) after the water permeation layer disappears, the body is composed of the ice layer and the water layer. Finally, the ice layer also disappears and the melting process is completed, leaving the water layer only.

7.3 Results and Discussion

The melting process was calculated by using the model which is composed of the water layer, the water permeation layer, the porous layer and the ice layer. A temperature gradient exists in the water layer, the porous layer and the ice layer. In the water permeation layer, a water content gradient exists.

The temperature distribution in the frost body changes with time in a complicated manner due to the water permeation and its refreezing in the frost body. Figure 29 shows an example of the variation of the heat loss with time calculated from the frost surface temperature during melting. The heat loss is characteristic of the melting process. Namely, in period (I) the water permeation causes the temperature to rise in the body so that the heat loss increases with time due to the increase in the temperature's driving force between the surface and the surroundings. This loss attains a maximum value at the end of this period. In periods (II) and (III), the growth of the body causes a recession of the water permeation front, so that the heat loss gradually decreases as a result of the decrease in the temperature's driving force. In period (IV), the melting of the ice layer causes an increase in the driving force, so that heat loss gradually increases again. The results calculated for the classical Stefan problem which neglects water permeation are also included in this figure as a dash-dot line and they show that the heat loss increases monotonically with time due to melting. From a comparison between both it is shown that water permeation into porous materials results in an increase in heat loss for a constant heat transfer coefficient.

Figure 30 shows a comparison in melting time between the predictions considering water permeation and the classical Stefan results on cases where heat loss exists and in the absence of heat loss. In the case where heat loss is neglected, the predictions considering water permeation yield a shorter melting time in comparison with the classical results, because water permeation in porous materials delay the formation of a water layer which is supposed to introduce a thermal resistance. On the contrary, in the case where heat loss exists, the predictions yield a longer melting time in comparison with the classical Stefan solutions. This is because the increase in a heat loss due to water permeation and its refreezing plays a more important part in comparison with the decrease in the thermal

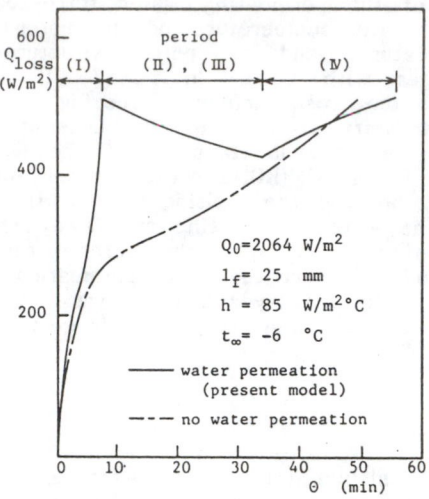

Fig.29. Variation of heat loss from frost surface during melting

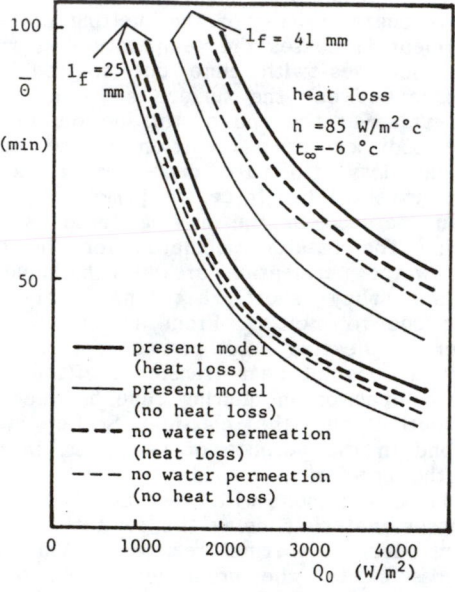

Fig.30. Comparison for melting time between cases with and without heat loss

resistance of a water layer due to water permeation.

We introduced here only the fundamental case for defrosting. Actually, it may be necessary to consider also the direction of the heating surface, the drainage of the melting water[58] and the formation of an air gap which can occur between the melting surface and the heating surface due to the change in density. These problems with the defrosting process are left as future studies.

8 CONCLUDING REMARKS

This article has been prepared with the intention of offering an introduction to frost formation on a cold surface. We have mainly focused on the various characteristics of frost formation based on our studies, and in order to avoid excessive confusion, we have not touched on the difference in frost formations generated under more complex geometries and various flow states.

As this article shows, our knowledge of the subject has expanded during the last two decades. In particular, an understanding of the basic frost formation has been accomplished. However, there are some gaps between the basic study and the applied study for the characteristics of various types of heat exchangers under frosting conditions. Accordingly, in order to obtain an insight into the most suitable heat exchanger shape and the most suitable defrosting cycle, some additional research efforts may be needed.

REFERENCE

1. Sekulić,D. P., XVIth Int. Cong. of Refrigeration, Paris, B1-231,1983.

2. Hayashi, Y., Yuhara, H., and Aoki, K., Trans. JSME, vol.42, no.355, pp.885-892, 1976. (in Japanese)

3. Hayashi, Y., Aoki, K., Adachi, S., and Hori, K., Trans. ASME, J. Heat Transfer, vol.99, no.2, pp.239-245, 1977.

4. Tokura,I., Saito, H., and Kishinami, S., Trans. ASME, J. Heat Transfer, vol.105, no.4, pp.895-901.

5. Saito, H., Tokura, I., Kishinami, S., and Uemura, S., Trans. JSME, vol.50, no.452, pp.1190-1196, 1984. (Heat Transfer-Japanese Research, vol.13, no.4, pp.76-88, 1985.)

6. Seki, N., Fukusako, S., Matsuo, K., and Uemura, S., Trans. JSME, vol.50, no.451, pp.825-831, 1984. (in Japanese)

7. Seki, N., Fukusako, S., Matsuo, K., and Uemura, S., Wärme-und Stoffübertragung, vol.19, pp.9-18, 1985.

8. Aoki, K., Katayama, K., and Hayashi, Y., Trans. JSME, vol.48, no.429, pp.952-961, 1982. (Bull. of JSME, vol.26, no.211, pp.87-93, 1983.)

9. Hayashi, Y., and Aoki, K., Trans. JSME, vol.43, no.368, pp.1384-1391, 1977. (in Japanese)

10. Trammel, G. T., Little, D. C., and Killgore, E. M., ASHRAE Journal, vol.6, pp.42-47, 1968.

11. Cremers, C. J., Han, O. J., and Skorupski, J. H., Advaces in Cryogenic Engineering, vol.23, pp.371-375, 1978.

12. Coles, W. D., NASA TN 3143, 1954.

13. Yonko, J. D., and Sepsy, C. F., ASHRAE Trans., vol.73, pp.1-11, 1967.

14. Reid, R. C., Brian, P. L.T., and Wever, M. E., AIChE Journal, vol.12, no.6, pp.1190-1195, 1966.

15. Yamakawa, N., and Ohtani, S., Kagaku Kogaku, vol.36, no.2, pp.197-203, 1972. (Heat Transfer-Japanese Research, vol.1, no.3, pp.75-82, 1972.)

16. Biguria, G., and Wenzel, L. A., Ind. Eng. Chem., Fundamentals,vol.9, pp.129-138, 1970.

17. Auracher, H., VDI-Forschungscheft, no.566, pp.1-44, 1974.

18. Auracher, H., 6th Int. Cong. of Heat Transfer, Toronto, vol.3, EN-5, pp.25-30, 1978.

19. Woodside, W., Can. J. Phys., vol.36, pp.815-823, 1958.

20. Krisher, O., and Rohnalter, H., VDI-Forschungscheft, vol.11, no.402, pp.1, 1940.

21. Whitehurst, C. A., ASHRAE Journal, no.5, pp.58-69, 1962.

22. Barron, R. F., and Han, L. S., Trans. ASME, J. Heat Transfer, vol.87, no.4, pp.499-506, 1965.

23. Okino, Y., and Tajima, O., Reito, vol.48, no.546, pp.338-349, 1973. (Heat Transfer-Japanese Research, vol.3, pp.45-61, 1974.)

24. Chen, M. M., and Rohsenow, W., Trans. ASME, J. Heat Transfer, vol.86, pp.334-340, 1964.

25. Kennedy, L. A., and Goodmann, J., Int. J. Heat Mass Transfer, vol.17, pp.477-484, 1974.

26. Hayashi, Y., Aoki, K., and Yuhara, H., Trans. JSME, vol.42, no.355, pp.893-901, 1976. (Heat Transfer-Japanese Research, vol.6, no.3, pp.79-94, 1977.)

27. Schneider, H. W., XIIIth Int. Cong. of Refrigeration, Washington, paper 2.47, vol.2, pp.235-241, 1972.

28. Schneider, H. W., Int. J. Heat Mass Transfer, vol.21, pp.1019-1024, 1978.

29. White, J. E., and Cremers, C. J., Trans. ASME, J. Heat Transfer, vol.103, pp.3-6, 1981.

30. Chung P. M., and Algren, A. B., Heating, Piping and Air Conditioning, vol.30, no.10, pp.115-122, 1958.

31. Brian, P. L. T., Reid, R. C., and Brazinsky, I., Cryogenic Technology, vol.5, pp.205-212, 1969.

32. Jones, B. W., and Parker, J. D., Trans. ASME, J. Heat Transfer, vol.97, no.2, pp.255-259, 1975.

33. Brian, P. L. T., Reid, R. C., and Shah, Y. T., Ind. Eng. Chem., Found., vol.9, pp.375-380, 1970.

34. Aoki, K., Katayama, K., Hayashi, Y., and Adachi, S., Trans. JSME, vol.45, no.394, pp.869-876, 1979. (in Japanese)

35. Aoki, K., Hattori, M., and Mizuno, S., Trans. JSME, vol.54, no.503, pp.1754-1759, 1988. (in Japanese)

36. Gates, R. R., Sepsy, F. C., and Huffman, G. D., 74th ASHRAE Conf., Mineapolis, Trans. no.2044, 1967.

37. Huffman G. D., and Sepsy, F. C., 74th ASHRAE Conf., Mineapolis, Trans. no.2045, 1967.

38. Hosoda, Y., and Uzuhashi, H., Hitachi Hyoron, vol.49, no.6, pp.647-651, 1967. (in Japanese)

39. Javnel, B. K., Holodilynaja Tehnika, no.9, pp.1-18, 1969.

40. Javnel, B. K., Holodilynaja Tehnika, no.9, pp.25-29, 1970.

41. Yanagida, T., Reito, vol.44, no.506, pp.1182-1189, 1969. (in Japanese)

42. Lotz, H., Kältetechnik-Klimatisierung, vol.23, pp.208-217, 1971.

43. Lotz, H., Kältetechnik-Klimatisierung, vol.24, pp.275-285, 1972.

44. Sanders, C. Th., Ph. D. Diss., TH Delft, 1974.

45. Ivanova, V. S., and Gatchilov, T. S., XIVth Int. Congress of Refrigeration, Moskow, paper B1.96, 1975.

46. Aoki, K., Hattori, M., and Itoh, T., Trans. JSME, vol.51, no.469, pp.3048-3054, 1985. (Bull. JSME, vol.29, no.251, pp.1499-1505, 1986.)

47. Iuchi, S., Oshima, T., Kagawa, S., and Yamamoto, N., Kagaku Kogaku, vol.32, no.2, pp.159-164, 1968. (in Japanese)

48. Schmidt, T. E., Abhandl. Deutsch. Kaltetechn. Vereins, nr.4, 1950.

49. Kays, W. M., and London, A. L., Compact Heat Exchangers, McGraw-Hill Book Co., New York, 1964.

50. Aoki, K., Hattori, M., and Akita, K., Trans. JSME, to be published.

51. Aoki, K., Hattori, M., and Hiramatsu, T., Trans JSME, to be published.

52. Katsuta, K., Ishihara, I., and Mukai, T., Reito, vol.58, no.665, pp.229-238, 1983. (in Japanese)

53. Katsuta, K., Ishihara, I., and Mukai, T., Reito, vol.58, no.669, pp.645-650, 1983. (in Japanese)

54. Adachi, M., Inoue, S., and Inoda, K., Reito, vol.52, no.598, pp.715-727, 1977. (in Japanese)

55. Zakrzewski, B., Int. J. of Refrigeration, vol.7, no.1, pp.41-45, 1984.

56. Aoki, K., Hattori, M., Chiba, S., and Hayashi, Y., ASME Paper, 81-WA/HT, 1981.

57. Aoki, K., Hattori, M., and Chiba, S., Trans. JSME, vol.51, no.471, pp.3567-3573, 1985. (Bull. JSME, vol.29, no.253, pp.2138-2144, 1986.)

58. Aoki, K., Hattori, M., and Ujiie, T., Trans. JSME, vol.53, no.495, pp.3352-3357, 1987. (JSME Int. J., vol.31, no.2, pp.269-275, 1988.)

59. Yamakawa, N., Takahashi, N., and Ohtani, S., Kagaku Kogaku, vol.35, no.3, pp.328-334, 1971. (in Japanese)

60. Chung, P.M., and Algren, A. B., Heating, Piping and Air Conditioning, vol.30, no.9, pp.171-178, 1958.

61. Katsuta, K., and Ishihara, I., Reito, vol.50, no.570, pp.262-270, 1975. (in Japanese)

62. Sekulić, D. P., Dvozdenac, D. D., and Bačlić, B. S., XVth Int. Cong. of Refrigeration, Venezia, B1-15, 1979.

63. Aoki, K., Hattori, M., and Hayashi, Y., XVth Int . Congress of Refrigeration, Venezia, B1-55, 1979.

64. Aoki, K., Katayama, K., Watanabe, S., and Hayashi, Y., Trans. JSME, vol.47, no.423, pp.2171-2180, 1981. (in Japanese)

65. Kamei, S., Mizushina, Y., Kifune, S., and Koto, T., Kagaku Kogaku, vol.14, pp.53-60, 1950. (in Japanese)

66. Beatty, K. O., Finch, E. B., and Schoenborn, E. M., Refrigerating Engineering, no.12, pp.1203-1207, 1951.

67. Aoki, K., Hattori, M., and Yanagihara, Y., Trans. JSME, vol.51, no.462, pp.699-704, 1985. (in Japanese)

68. Stoeker, W. F., Refrigerating Engineering, no.2, pp.42-46, 1957.

69. Tajima, O., Yamada, H., Kobayashi, Y., and Mizutani, S., Reito, vol.46, no.522, pp.333-341, 1971. (Heat Transfer-Japanese Research, vol.1, no.2, pp.39-48, 1972.)

70. Tajima, O., Naito, E., Tsutsumi, Y., and Yoshida, H., Reito, vol.47, no.534, pp.350-358, 1972. (Heat Transfer-Japanese Research, vol.2, no.2, pp.55-76, 1973.)

71. Tajima, O., Naito, E., Nakasima, K., and Yamamoto, H., Reito, vol.48, no.547, pp.395-402, 1973. (Heat Transfer-Japanese Research, vol.3, no.4, pp.55-66, 1974.)

72. Tajima, O., Naito, E., Goto, T., Segawa, S., and Nishimura, K., Reito, vol.49, no.556, pp.95-103, 1974. (Heat Transfer-Japanese Research, vol.4, no.3, pp.21-36, 1975.)

73. Katsuta, K., and Ishihara, I., Reito, vol.52, no.593, pp.283-291, 1977. (in Japanese)

74. Nakamura, H., Trans. JSME, vol.39, no.324, pp.2522-2529, 1973. (in Japanese)

75. Hayashi, Y., Adachi, S., and Yamaguchi, K., Reito, vol.52, no.598, pp.707-714, 1977. (in Japanese)

76. Hayashi, Y., and Aoki, K., Meeting of Comm. B1, B2 and E1 of IIR, Belgrade, pp.37-42, 1977.

77. Stoeker, W. F., ASHRAE Trans., vol.66, pp.91-103, 1960.

78. Katsuta, K., and Ishihara, I., Reito, vol.48, no.552, pp.923-928, 1973. (in Japanese)

79. Katsuta, K., and Ishihara, I., Reito, vol.52, no.601, pp.977-985, 1977. (in Japanese).

80. Katsuta, K., and Ishihara, I., Reito, vol.54, no.625, pp.899-905, 1979. (in Japanese)

81. Cremers, C. J., and Mehra, V. K., Trans. ASME, J. Heat Transfer, vol.104, no.1, pp.3-7, 1982.

82. Tokura, I., Saito, H., and Kishinami, K., Trans. JSME, vol.50, no.449, pp.173-178, 1984. (in Japanese)

SOLIDIFICATION AND MELTING OF SOLUTIONS

Chapter 15

Solidification and Melting of Solutions

YUJIRO HAYASHI
Department of Mechanical Engineering, Kanazawa University,
Kanazawa 920, Japan

ABSTRACT

In the solidification of solutions, the redistribution of solute occurs and this induces very complicated behaviors that are distinctly different from the case for a pure material. The purpose of this article is to provide an updated review concerning the basic concepts and the most advanced theories of solidification and melting heat transfer of solutions. Explanation of the solidification processes is especially concentrated on the appearance of the solid-liquid phase region, the interface morphology in relating with relaxation of the temperature and solute concentration fields. Some topics are also discussed as example of practical applications.

CONTENTS

NOMENCLATURE

C	solute concentration
c	specific heat capacity
D	diffusion coefficient
f	solid fraction
G	temperature gradient
G_m	mass velocity of working fluid
h	heat transfer coefficient
K_o	equilibrium distribution coefficient
L_H	latent heat-of-fusion

l	characteristic length
m	slope of liquidus line
q	heat flux
q'	heat generation
R	growth rate of molten zone
r	space coordinate in radial direction
T	temperature
T'	liquidus temperature
t	time
x	space coordinate in x-direction
z	space coordinate in z-direction
Greek symbols	
η	interface position between solid and liquid regions
η_1	interface position between solid and solid-liquid regions
η_2	interface position between solid-liquid and liquid regions
κ	thermal diffusivity
λ	thermal conductivity
ρ	density
ϕ	error function
Subscripts	
1	solid phase region
2	solid-liquid phase region
3	liquid phase region
A, B	denotes cell group
e	eutectic point
i	initial state

1 INTRODUCTION

Transient heat transfer problems of solidification and melting of solutions are relevant to practical industrial processes such as the freezing of foodstuffs, the preservation of biological specimens, the desalination of sea water, crystal growth from melts, solidification of casting, thermal energy storage and many other systems. In the solidification of solutions, the redistribution of a solute results from the difference in solubility in the solid and liquid phases. This induces some characteristic phenomena that are distinctly different from the case for a pure material.

Firstly, the phase change occurs at the equilibrium temperature corresponding to the solute concentration at the solid-liquid interface, rather than isothermally. Thus, the solidification process should be treated as a coupled heat and mass transfer problem, taking into account the thermodynamic relation at the solid-liquid interface. Secondly, when a constitutional supercooling occurs in the liquid adjacent to the interface, it induces interface instabilities to form a solid-liquid phase made of numerous dendritic crystals with the concentrated solution trapped among them. The physical representation can be constructed by separating the liquid and the solid with a two-phase region. Solidification and melting are companion processes that occur consecutively in heat transfer applications. Finally, the history of the freezing process must be considered in the treatment of the melting problem, since the solid produced by the solidification of the solution is not homogeneous in its composition.

The primary purpose of this article is to present the basic concept for predicting heat transfer during the solidification and melting of solutions. Topics are chosen from what is presently accepted as the most advanced theories on the solidification and melting of solutions.

2 SOLIDIFICATION OF SOLUTIONS

2.1 General Remarks

The equilibrium relationship between a solid and a liquid is conveniently represented on a binary equilibrium diagram by means of liquidus and solidus lines, as shown in Fig. 1. When the composition of the liquid, C_L, changes into that of the solid, C_s, by solidification, the solute is rejected or incorporated at the solid-liquid interface by the difference between C_L and C_s. The type shown in Fig. 1-(a), where the slopes of the liquidus and solidus lines are negative, is much more common than that shown in Fig. 1-(b), so the discussion will be restricted to the former case. The ratio, $C_L/C_s(= K_o)$, termed by the equilibrium distribution coefficient, takes a value between zero and one. The value of zero corresponds to the complete rejection of the solute, while a value of one corresponds to the complete incorporation of the solute into the solid phase. The changes of the concentration and temperature distributions during the solidification process are shown schematically in Fig. 2. During solidification some of the solute is rejected and diffuses into the liquid to form a solute-rich layer in front of the advancing interface. The concentration distribution in the liquid is not always controlled by the diffusion, but is sometimes controlled by the convection resulting from density gradients. In the solid, however, diffusion is much slower than in the liquid, and the distribution may be considered to be fixed in the space. In Fig. 2-(b), the actual and liquidus temperatures are represented by the solid and dotted lines, respectively. The phenomenon of the concentration polarization in the liquid gives rise to the depression of the freezing temperature, as shown by the chained line, and sometimes causes interface instabilities due to the constitutional supercooling in the region just ahead of the interface. Since the constitutional supercooling is thermodynamically unstable, this collapses easily by the generation of dendritic crystals at the interface, and finally relaxes to form the thermodynamically stable solid-liquid phase, as shown in Fig. 2-(iii). Chalmer[1] first recognized this phenomenon and clarified the condition of interface instability[2]. This was verified experimentally by Walton et al. [3], and by Tiller and Rutter[4] for the Pb-Sn alloy system.

Thus, the solidification of solutions that proceeds in a stable state can be classified largely into two types: (1) The solid phase with a planner surface grows into the liquid according to the temperature depression at the interface. Because of the coupling phenomena of heat-and mass-transfer, it is necessary to solve the problem to determine the spatial and time dependence of the temperature and concentration fields, taking into account their equilibrium relationship at the interface. (2) The solid-liquid phase

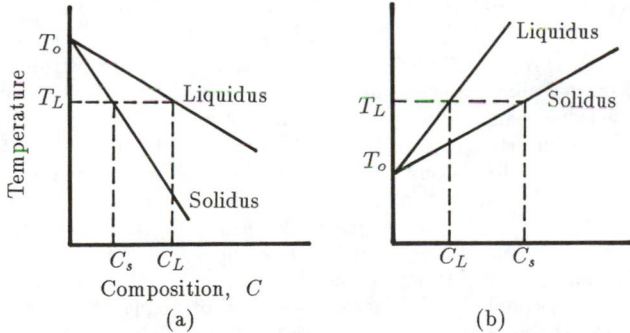

Figure 1 Relationship between equilibrium compositions of solid and liquid.

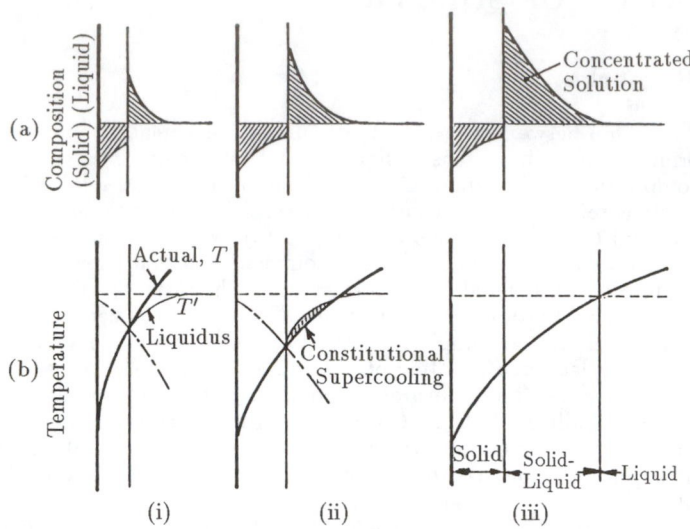

Figure 2 Changes of concentration and temperature distribution during solidification.

capturing the concentrated solution among numerous crystals grow into the liquid forming a rough surface. Since the thermodynamical equilibrium is maintained in this region, the problem in this case can be treated simply as a pure conduction with a phase change over a range of temperatures. The temperature at the freezing front is the liquidus temperature corresponding to the initial solute concentration, and the solid fraction will be determined by the temperature distribution. Mathematical treatments for these freezing problems will be described in the following section.

In practical industrial processes, these two types of solidification occur consecutively under certain circumstances. This reveals the necessity of discussing the transient process between two stable processes. In this sense, a supercooling which sometimes occurs at the beginning of freezing, is not less significant than the constitutional supercooling. To this end, the freezing of saline aqueous solutions will be described in the next section.

2.2 Unstable Solidification [5]

In order to study the relaxation process of unstable solidification, precise experimental works may be essential. Figure 3 shows a schematic diagram of a sample freezing cell. The cell is made of thin acrylic resin plates, and a copper plate serves as a cooling plate by using thermo module inserted into one end of the cell. Thermocouples for temperature measurements are installed along the center line of the bottom plate. Although a transparent aqueous solution such as saline water is appropriate for visual observations, the change in the freezing process proceeds too rapidly to obtain data. This is due to the complete rejection of the solute to freezing. A video-microscope is very useful for observing not only the fast motion of a freezing front, but also the micro-change of the interface morphology. In this section, the principle of the relaxation mechanism of unstable freezing will be discussed on the basis of the experimental studies performed for pure water and for sodium chloride solutions of 2, 5, and 10 % by weight composition.

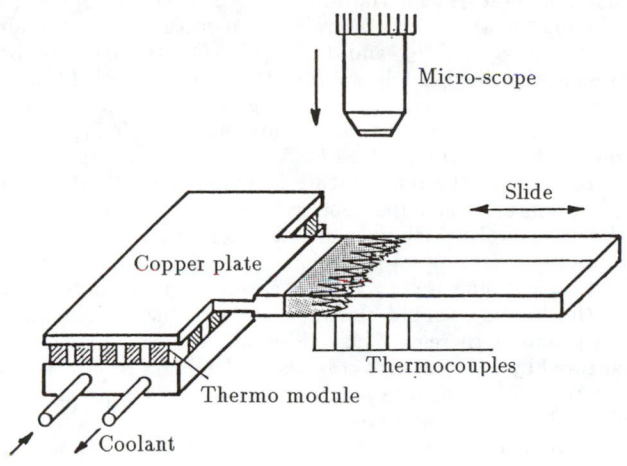

Figure 3 Freezing test cell.

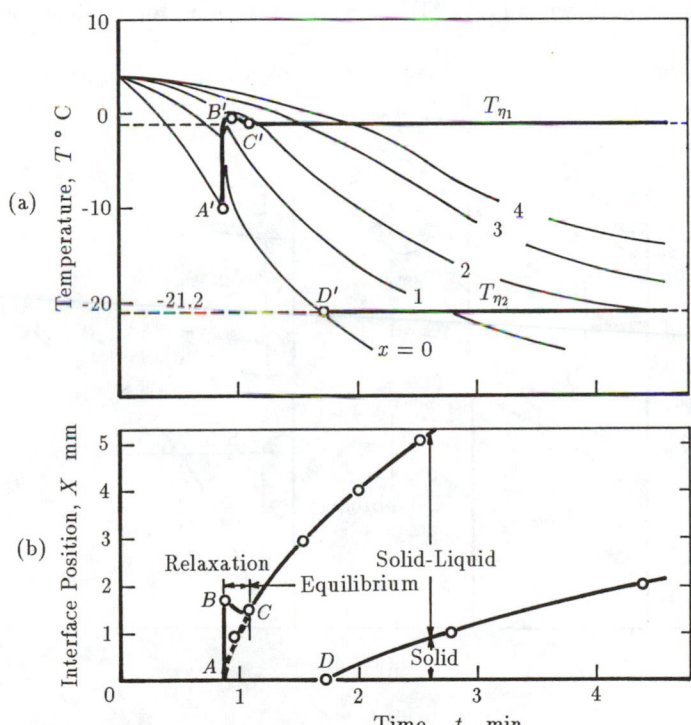

Figure 4 Changes of temperature and advancing interface positions.

<u>Unstable solidification with supercooling.</u> In Fig.4, sample results of the change of interface positions and their temperatures for the solution of 2 %wt are represented. Points A', B', C' and D' in Fig. 4-(a) are the interface temperatures corresponding to the interface positions of points A, B, C and D in (b). On the basis of both results, the relaxation process is explained as follows. The test solution cooled from the cryosurface decreases its temperature, and supercooled zone appears in the vicinity of the surface. Freezing starts at the supercooled temperature, about $-10°C$(point A'), which is lower than the liquidus temperature, $-1.15°C$. According to the rapid growth of ice crystal from point A to point B, the temperature increases abruptly due to the liberation of latent heat of fusion, and the supercooling is supposed to relax to some extent. Next, the tips of the ice crystals which have grown into position at the liquidus temperature, melt again from point B to point C. This is also due to the recalescence temperature above $0°C$. After exceeding point C, the freezing front advances at the liquidus temperature, and the freezing should continue under a thermodynamically stable state. Between points C and D, there exists a solid-liquid phase region where the concentrated solution is captured by numerous ice crystals, and a liquid region. Beyond point C the temperature at the cooling surface reaches the eutectic temperature. At this point solid, solid-liquid and liquid regions exist.

In Fig.5, the timewise evolution of the freezing-front morphology are shown schematically along with the actual and liquidus temperature distributions. From this figure, the mechanism of the freezing process can be understood more clearly. First, the supercooling state, represented by the hatched zone between the thin and solid lines, collapses due to the generation of ice crystals on the cooling surface in the opposite direction of heat flow. The ice crystals reject the solute and latent heat of fusion, and the supercooled field relaxes to some extent since the actual and liquidus temperatures are close each other, as shown in Fig. 5-(ii). Next, these crystals grow gradually in

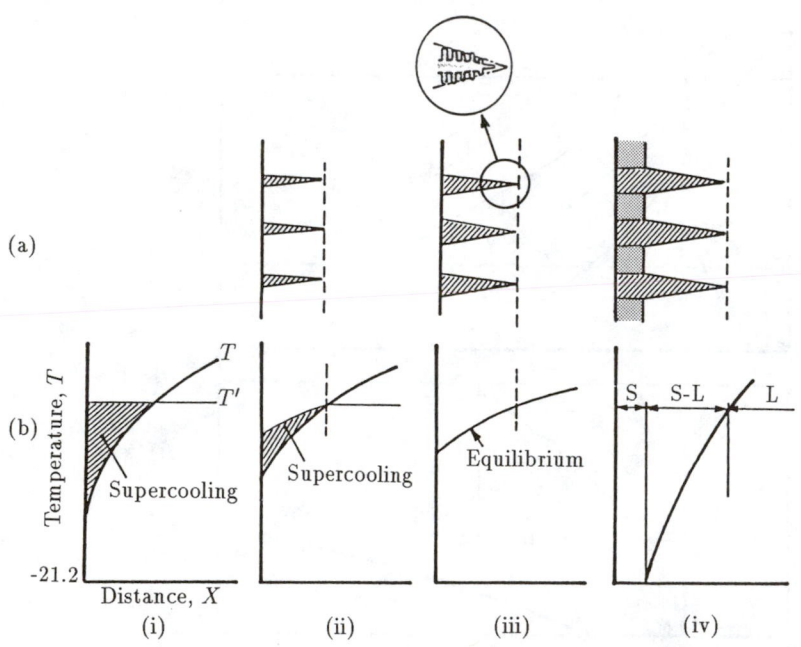

Figure 5 Schematic of freezing-front morphology and temperature distributions during relaxation process of thermal and constitutional supercooling.

width from the root to top by the generation of secunder dendrites, and capture the concentrated solution among them to form a equilibrium solid-liquid phase region(Fig. 5-(iii)). The dotted line between points A and C in Fig. 4 represents this relaxation process. The freezing after point C is the same as described above.

In Fig. 6, the freezing process for pure water is shown in order to demonstrate characteristics of the freezing of solutions. First, ice crystallization occurs in almost the same manner as for the case of solutions. However, the supercooled state remains much as shown in Fig. 6-(ii), since ice crystallization affects only the actual temperature. Next, a dense solid phase develops from the cryosurface, and finally the solid layer covers the ice crystals, as shown in Fig. 6-(iv). After this point, the freezing process becomes a stable process having a planar freezing front maintained at 0°C. Harrison and Tiller[6] observed the interface morphology and crystal texture produced during the freezing of pure water. Saito[7] studied the freezing of water by supercooling as a conduction problem and assumed that the solid-liquid phase region was always in a state of equilibrium.

Unstable solidification with constitutional supercooling. The relaxation process of the freezing of a constitutionally supercooling solution will be discussed. To this end, a small piece of ice was put into the liquid just ahead of the cooling surface, and the occurrence of supercooling at the beginning of the experiment was restrained artificially. Using Fig. 7, which is schematically depicted on the basis of the experimental results, the freezing process may be explained as follows.

Freezing first occurs around the nucleus of the ice piece, and extends over the cold surface to form a thin ice layer segregating the solute. Namely, the layer having a planar surface advances one-dimensionally, keeping the liquidus temperature at the interface, Fig. 7-(i). After a short period, the constitutional supercooling occurs as shown by the hatched region in Fig. 7-(ii). This makes the planar interface unstable and induces the dendritic crystal growth with needle-shaped structures perpendicular to the interface. This is due to the fact that the growth rate will be increased in any localized region that advances ahead of the general interface. Mullins and Sekerka[8] introduced the variation of sinusoidal perturbations into the interface configuration, and determined the critical condition for interface instability. These crystals grow gradually both in height and in width, capturing the concentrated solution between them to relax the thermodynamically unstable state, Figs. 7-(ii) and (iii). After this period, freezing advances in a stable state, however, the interface between the solid and the solid-liquid regions remains stationary at the same location. Finally, this interface begins to move quasi-steadily in all three regions, Fig. 7-(iv), as the freezing continues.

2.3 Stable Solidification

As previously mentioned, the solidification of solutions proceeding in a thermodynamically stable state is classified into two types; the advance of the solid phase region with a planar surface, and that of the solid-liquid phase region with a rough surface. Though the former type is similar to freezing of a pure material on the point that solidification proceeds in the state of the solid and liquid regions, the mathematical treatment is more complicated due to the necessity of including the coupling heat- and mass-transfer. Studies of this problem have been conducted to discover the fundamental mechanism of the instability of a planar freezing front, rather than the process of stable freezing itself. On the other hand, the latter problem has been investigated as a basic study to determine the proper handling of industrial processes, for which the accurate prediction of heat transfer during freezing is very useful. From these standpoints, literature concerning the mathematical treatment of the solidification of solutions will be reviewed in this section.

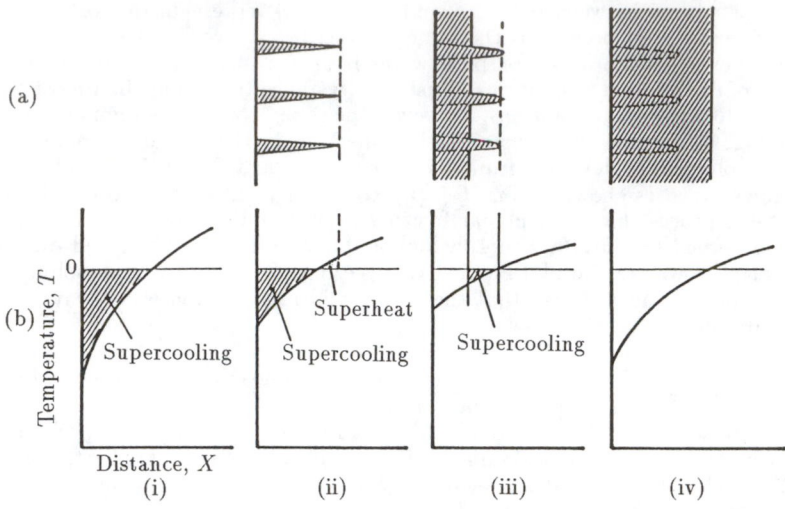

Figure 6 Schematic of freezing-front morphology and temperature distributions during relaxation process of supercooling.

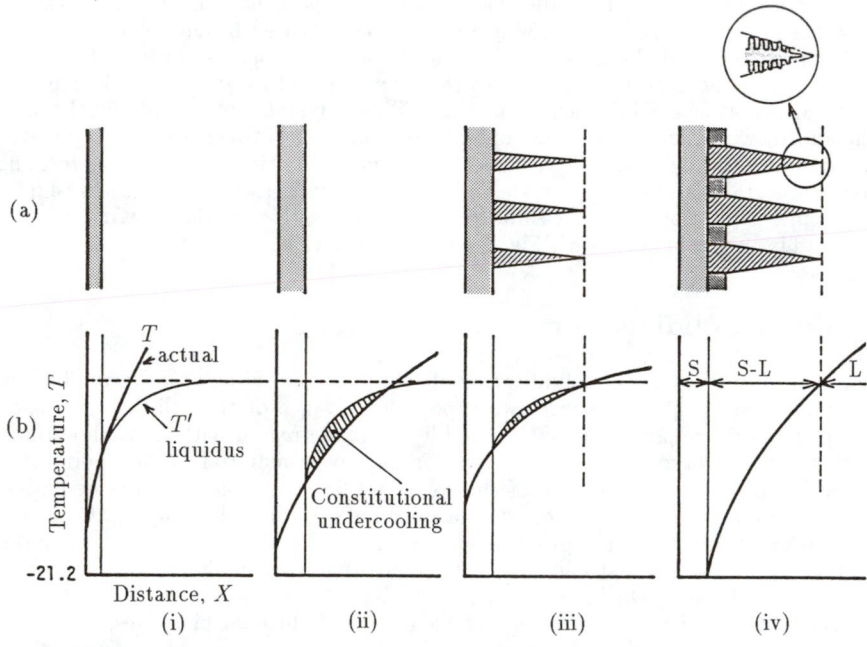

Figure 7 Schematic of freezing-front morphology and temperature distributions during relaxation process of constitutional supercooling.

Solidification with a planar interface and its instability. Constitutional supercooling induces instabilities on the planar interface to form more complicated yet more stable dendritic structures. However, constitutional supercooling is a necessary condition for morphology instability. Mullins and Sekerka[8] calculated the variation of sinusoidal perturbations introduced into the shape of the interface, and obtained a sufficient condition, taking into account the temperature and concentration distributions in the vicinity of the interface. The condition for stability for an aqueous binary solution is expressed by

$$\frac{1}{2}(g_1 + g_3) - mG_c > 0 \tag{1}$$

$$g_n = \frac{2\lambda_n}{\lambda_1 + \lambda_3} \frac{dT_n}{dx} \qquad (n = 1, 3)$$

where $n = 1, 3$ refers to the solid and liquid regions. Equation (1) represents the "conductivity-weighted" temperature gradient, and mG_c is the gradient of liquidus temperature at the interface.

O'Callagham et al. [9] discussed the instability of the planar interface during the freezing of binary aqueous solutions by using the Mullins-Sekerka stability criterion. Their results indicate that the planar interface rapidly becomes unstable except at very low freezing rates or solute concentrations. The basic equations and interface conditions, essential for determining the one-dimensional temperature and concentration fields, are represented in Table 1. The freezing geometry for a slab with specific conditions is shown schematically in Fig. 8-(a). Adding appropriate boundary and initial conditions, solutions can generally be obtained for various systems. Terwilliger and Dizio[10] analyzed solute rejection from a planar solid-liquid interface, and obtained the criterion for the onset of constitutional supercooling. Levin[11][12] presented a generalized analysis of the unidirectional freezing assuming that the solid-liquid interface remains planar in spite of the appearance of the constitutional supercooling.

Table 1.

Basic Equations

$$\frac{\partial T_1}{\partial t} = \kappa_1 \frac{\partial^2 T_1}{\partial x^2} \tag{2}$$

$$\frac{\partial T_3}{\partial t} = \kappa_3 \frac{\partial^2 T_3}{\partial x^2} - (1 - \frac{\rho_1}{\rho_3}) \frac{d\eta}{dt} \frac{\partial T_3}{\partial x} \tag{3}$$

$$\frac{\partial C_1}{\partial t} = D_1 \frac{\partial^2 C_1}{\partial x^2} \quad (D_1 << D_3) \tag{4}$$

$$\frac{\partial C_3}{\partial t} = D_3 \frac{\partial^2 C_3}{\partial x^2} - (1 - \frac{\rho_1}{\rho_3}) \frac{d\eta}{dt} \frac{\partial C_3}{\partial x} \tag{5}$$

Interface Conditions

$$T_1(\eta, t) = T_3(\eta, t) = T'(C_f) \tag{6}$$

$$\lambda_1 (\frac{\partial T_1}{\partial x})_\eta - \lambda_3 (\frac{\partial T_3}{\partial x})_\eta = \rho L_H \frac{d\eta}{dt} \tag{7}$$

$$C_3(\eta, t) = C_f \tag{8}$$

$$D_1 (\frac{\partial C_1}{\partial x})_\eta - D_3 (\frac{\partial C_3}{\partial x})_\eta = (1 - K_o) C_{3\eta} \frac{d\eta}{dt} \tag{9}$$

Furthermore, O'Callagham et al. [13][14] analyzed energy and mass transport near the dendrite-tip and -basal plane regions, and predicted the dendrites shapes, lengths and spacing, together with the temperature and concentration fields. Kurz and Fisher [15] studied dendrite growth using a simplified solution for the wavelength of instability, and obtained a relationship between growth conditions and primary arm spacing. Based on their calculations, Roosz et al. [16] developed a method of solute distribution in the liquid between dendrite arms in order to produce alloys with uniform properties by controlling the process.

Table 2.

Basic Equations

$$\frac{\partial T_n}{\partial t} = \kappa_n \frac{\partial^2 T_n}{\partial x^2} \qquad\qquad (10), (11), (12)$$

where $\kappa_2 = \lambda_2/\{(c\rho)_2 + \rho_3 L_H k\}$

Interface Conditions

$$T_1(\eta_1, t) = T_2(\eta_1, t) = T_e' \qquad\qquad (13)$$

$$T_2(\eta_2, t) = T_3(\eta_2, t) = T_i' \qquad\qquad (14)$$

$$\lambda_1 \left(\frac{\partial T_1}{\partial x}\right)_{\eta_1} - \lambda_2 \left(\frac{\partial T_2}{\partial x}\right)_{\eta_1} = \rho_3 L_H (1 - f_e) \frac{d\eta_1}{dt} \qquad\qquad (15)$$

$$\lambda_2 \left(\frac{\partial T_2}{\partial x}\right)_{\eta_2} - \lambda_3 \left(\frac{\partial T_3}{\partial x}\right)_{\eta_2} = 0 \qquad\qquad (16)$$

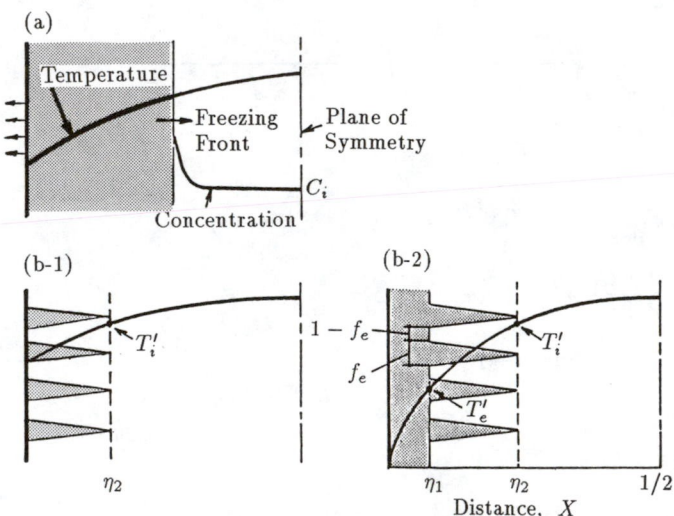

Figure 8 Physical and coordinate systems of stable solidification of solutions.

Solidification with a rough interface. Many studies have been conducted concerning transient heat transfer problems with a phase change over a range of temperatures, since this type of solidification is not only dominant during the whole solidification process, but also comparatively simple for mathematical treatment. In many situations, the position of the advancing interface is not known a priori, but is determined by the heat transfer mechanism.

The basic equations and interface conditions relevant to this type of solidification problem are represented in Table 2. Physical and coordinate systems are shown in Fig. 8-(b). The term on the right side of Eq. (15) represents the latent heat effect due to the freezing of the remaining liquid at the eutectic temperature. The second term of κ_2 on the left side of Eq. (11) is also heat generation, which is dependent upon the solid fraction distributed in the two-phase region. For the solid fraction distribution, Tien and Geiger[17] postulated a linear relationship with distance, and this was used for obtaining an exact solution by Cho and Sunderland[21].

$$f = f_e(1 - \bar{x}) \tag{17}$$

where $\bar{x}$ is a dimensionless space variable in the freezing zone, and is defined by $(x - \eta_1)/(\eta_2 - \eta_1)$. Tien and Geiger[18] also applied the following solid-fraction temperature relationship in their analysis of the normal non-equilibrium mode of freezing by Pfann[19].

$$f = 1 - (\frac{mC_i}{T_o - T})^{\frac{1}{1-K_o}} \tag{18}$$

Hayashi et al. [20] used the following relationship based on a liquidus line, $C = g(T)$, and presented the amount of internal heat generation, q', as a function of the temperature only.

$$f = 1 - \frac{C_i}{C} = 1 - \frac{g(T_i')}{g(T)} \tag{19}$$

$$q' = -\rho_3 L_H \frac{\partial}{\partial T}(1 - f) = -\rho_3 L_H \frac{\partial}{\partial T}\{\frac{g(T_i')}{g(T)}\}\frac{\partial T}{\partial t} \tag{20}$$

Using this relationship, the energy equation for the solid-liquid region is obtained.

$$\frac{\partial T_2}{\partial t} = \kappa_2 \frac{\partial^2 T_2}{\partial x_2} \tag{21}$$

$$\kappa_2 = \lambda_2/\{(c\rho)_2 + \rho_3 L_H k\} \tag{22}$$

where, $k = \partial/\partial T\{g(T_i')/g(T)\}$.

Adding appropriate boundary and initial conditions to the equations in Table 2, solutions for the temperature distribution and rate of planar change may be obtained for various systems.

Cho and Sunderland[21] presented an exact solution for a semi-infinite body. However, no exact closed form solutions have been found for a finite body, except in special cases[22]. Thus, the approximate method is useful for describing the process analytically. Muehlbauer et al. [23] used the heat-balance integral method developed by Goodman[24] to obtain approximate solutions for the transient one-dimensional solidification of a finite slab of a binary alloy. Those approximate methods and solutions were reviewed by Muehlbauer and Sunderland[25]. The application of approximate solutions to the solidification of casting was reviewed by Jones[26]. Stephan et al.[27] performed calculations and discussed the solidification of iron and salt solutions. Grange et al. [28], Shingu et al. [29] and Takeshita [30] developed approximated analytical methods, and confirmed the existence of a thin solute-rich layer just ahead of the freezing front. Szekely and Jassal[31] presented a mathematical model for the 3-region system, taking into account a fluid motion for both the liquid and two phase regions due to the buoyancy effect.

3 COMPANION PROCESSES OF FREEZING AND MELTING

Freezing and melting are companion processes that occur consecutively in numerous heat transfer applications[32]. For example, thermal energy storage using latent heat of fusion and the freezing of foodstuffs are accompanied by the cyclic or reciprocal processes of freezing and melting. In the case of pure materials, both the phase change processes and the mathematical treatments of it are basically the same. However, the solid produced by the freezing of a solution is not homogeneous in its composition, and a heat transfer analysis of its melting process should be performed. Cho and Sunderland[14] and Stephan et al. [22] dealt with melting problems for slab. Chiesa and Guthrie[33] analyzed heat transfer rates associated with Benard convections during the melting of metals and alloy systems. In order to explicitly understand these companion processes, the freezing and melting processes are analyzed for a one-dimensional slab consisting of water-sodium chloride solution, and these results are compared with the experimental results[34].

Schematics of the freezing and melting processes are shown in Fig. 9. In the first, a two-phase region appears and advances from both sides of the slab(a). When the wall surface temperature descends to the eutectic temperature, the growth of the solid starts and freezing advances in three regions(b). This state continues until the advancing solid-liquid phase regions join each other at the center of the slab(c). The liquid region vanishes at this point, and finally the slab is completely occupied by the solid. Schematics of (d), (e) and (f) correspond to the melting process. Since the frozen material is assumed to consist of pure ice layer and eutectic layers, the process is defined by the melting temperature of each layer: 0°C, and the eutectic temperature. First, the eutectic part begins to melt, and there exists the solid and solid(pure ice)-

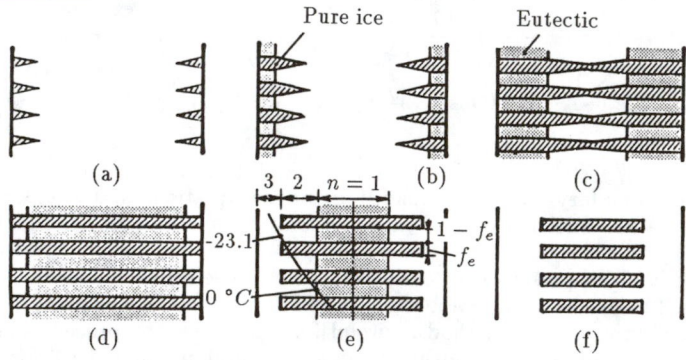

Figure 9 Schematic of companion processes of freezing and melting.

liquid regions(d). After the temperature of the heating plate surface rises to 0°C, the melting of the ice layers starts and the melting process advances in three regions(e). Finally the melting process goes to and through the process advanced in the liquid and liquid-solid regions(f). In the melting process, however, the progression of interface contours were not kept flat. Specifically, the contour of the interface between the liquid and two-phase regions displayed the tapered shape. This is due to the presence of buoyancy-induced convection in the liquid region.

An analytical treatment of the freezing process, whose basic equations are represented in Table 2, is omitted here. An analysis of the melting process will be described for a solid composed of two kinds of parallel parts in the direction of the heat flow. As shown in the physical model, Fig. 9-(e), the fractions of the pure ice and the eutectic ice layers are determined from the initial and eutectic solute concentrations.

The basic equations are:

$$\frac{\partial T_n}{\partial t} = \kappa_n \frac{\partial^2 T_n}{\partial x^2} \qquad (n = 1, 2, 3) \qquad\qquad (23),(24),(25)$$

where, $\kappa_n = \lambda_n/(c\rho)_n$.
The initial and boundary conditions are:

$$\lambda_n(\frac{\partial T_n}{\partial x})_{x=0} = h(T_n - T_\infty)_{x=0} \qquad\qquad (26)$$

$$T_3(\eta_2, t) = T_2(\eta_2, t) = 0 \qquad\qquad (27)$$
$$T_2(\eta_1, t) = T_1(\eta_1, t) = T_e' \qquad\qquad (28)$$
$$T_1(x, 0) = T_i \qquad\qquad (29)$$

$$\frac{\partial T}{\partial x}(l/2, t) = 0 \qquad\qquad (30)$$

$$\lambda_3(\frac{\partial T_3}{\partial x})_{\eta_2} - \lambda_2(\frac{\partial T_2}{\partial x})_{\eta_2} = \rho_3 L_H f_e \frac{d\eta_2}{dt} \qquad\qquad (31)$$

$$\lambda_2(\frac{\partial T_2}{\partial x})_{\eta_1} - \lambda_1(\frac{\partial T_1}{\partial x})_{\eta_1} = \rho_3 L_H (1 - f_e) \frac{d\eta_1}{dt} \qquad\qquad (32)$$

The equations presented here apply for the case of the melt advancing in three regions. When the eutectic ice layer melts away and the solid region vanishes in the slab, equations (23), (28) and (30) do not exist and the condition, $\partial T_2/\partial x = 0$, deduced from equation (32) replaces equation (30).

Comparisons of the experimentally measured and analytically predicted results for the freezing and melting processes will be shown and discussed. Exact solutions for temperature distributions and interface positions cannot be obtained due to the inherent non-linearities. Analytical results presented here were obtained numerically using a finite-difference method. Figure 10-(a) shows the change of freezing interface positions, η_1 and η_2, with time for initial sodium chloride solutions of 2, 5 and 10 %wt. Experimental results were obtained by visual observation, as was stated previously. In this case, the solid and solid-liquid regions didn't appear simultaneously. A little before or after the interface η_2 reached the mid-point of the slab(x=15 mm), the solid region began to appear and grew rapidly. The measured and predicted results agree quite closely, including the effect of difference of initial solute concentration. Typical temperature distributions for solute concentrations of 10 %wt are shown as a function of position with time in Fig. 11. The temperature at η_1 is associated with the eutectic temperature, -21.2°C, and the temperature at η_2 is equivalent to the liquidus temperature, -7°C. Next, the predicted and measured results of the interface positions and temperature distributions for the melting processes will be discussed. The measured

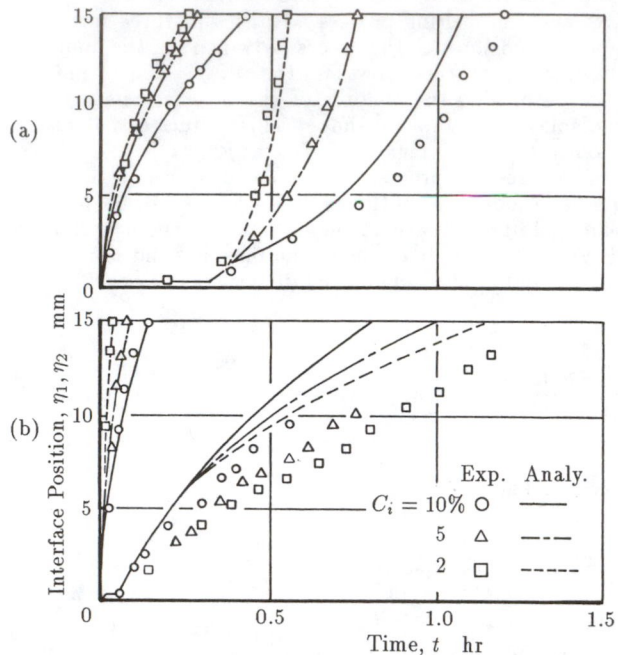

Figure 10 Comparison of predicted and measured interface positions during (a) freezing process, and (b) melting process.

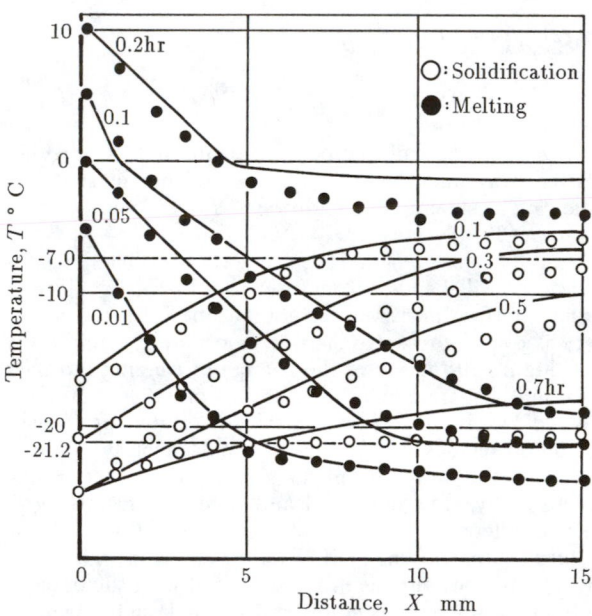

Figure 11 Comparison of predicted and measured temperature distribution.

values are at a mid-point height in the system. Although both results agree fairly well on their tendencies, the predicted results show considerable discrepancy in comparison to the experimental data. The temperature distributions measured experimentally are lower than the predicted results at later times, and this tendency appears especially in the discrepancies of interface position η_2. Though the sources of the discrepancies cannot be explained explicitly, one possible cause could be due to the occurrence of natural convection in the liquid region and its effect in the two phase region, as noted by previous investigators[31]. This suggests that one must take into account the presence of buoyancy-induced convection when a large temperature difference exists in liquid region during two-dimensional melting.

4 PRACTICAL APPLICATIONS

There are many industrial systems in which the solidification of solutions occurs and/or is applied artificially. In this chapter, the following topics are discussed as examples of practical applications: the freezing of foodstuffs involving aqueous solutions, the discharge of heat from thermal energy storage devices due to the solidification of solutions used as the phase change material, and the directional solidification of solutions with eutectic composition.

4.1 Freezing of Foodstuffs

-Freezing of Cellular Materials Involving Aqueous Solutions-
Generally, water included in food exists in the state of aqueous solutions and colloidal solutions. Freezing of colloidal solutions is important in connection with thawing reversibility and/or the life and death of living cells, but its freezing temperature range is lower than that of usual food. If a foodstuff is assumed to be a cellular material, the subject of this section can be replaced by the following subtitle: Freezing of cellular materials involving aqueous solutions. During the freezing of aqueous solutions, the solute is rejected at the solid-liquid interface and diffuses into the liquid area of cells to increase the concentration. This induces many characteristic phenomena in the freezing of food, such as the depression of the freezing point, ice crystallization due to a constitutional supercooling, an appearance of a freezing zone, etc. Reviews of heat transfer with freezing and defrosting of food are available[36].

Physical Model. [37,38]   Figure 12 shows schematically the freezing process of an aqueous solution contained in cells. As the model of cellular material presented here, two kinds of one-dimensional cells in which two different solutions are contained, are arranged alternately. A denotes the inside of cells and B denotes the outside of cells in the model of cell groups. For simplicity, the freezing process of one of the cell groups, A, is considered first. When the temperature of the liquid at the cell wall descends to the liquidus temperature T'_{Ai} corresponding to the initial solute concentration C_{Ai} by cooling, the cell begins to freeze and the freezing front grows into the cell. As the freezing advances, the solute concentration in the liquid of the cell increases with time by the rejected solute. Assuming impermeable cell walls, the concentration gradient of the solute is zero at the end of each cell, and the concentration profiles are shown by solid lines in Fig. 12-(b). Thus the freezing of cells continues with a depression of the freezing temperature until the solute-rich region reaches the eutectic point, as shown by the chained line. These matters suggest that freezing starts from one point to another in the adjacent cells before freezing is completed in the foremost cell. Namely, the solid-liquid region where the cells are partially frozen appears over a range of temperature extending from the liquidus temperature T'_{Ai} to the eutectic temperature T'_{Ae}.

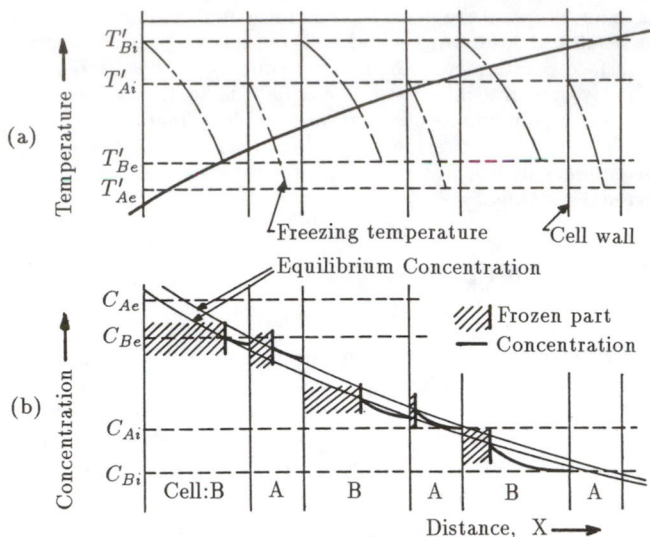

Figure 12 Schematic of temperature and concentration distribution in freezing of cellular materials involving aqueous solutions.

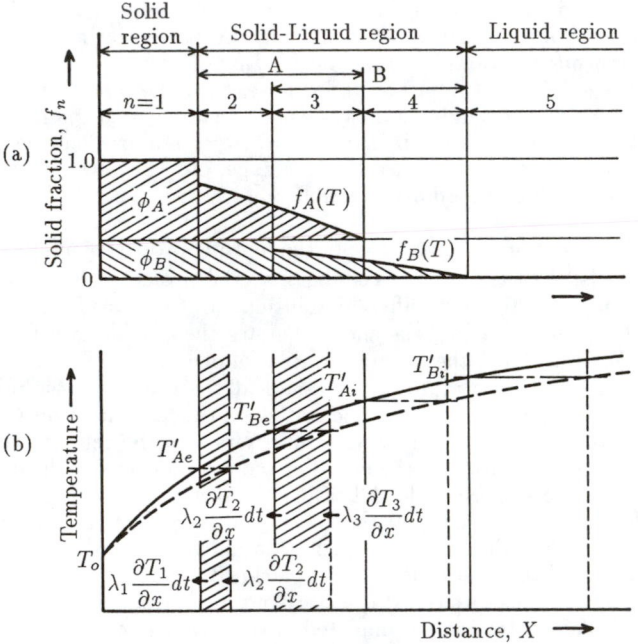

Figure 13 Physical and coordinate systems.

The process of freezing mentioned above is the same in case of the other cell group, B.

The size of a cell is small in most foods, and the distribution of the solid fraction in the solid-liquid region is considered to be continuous. Furthermore, it is assumed that an equilibrium state is established locally between the solute concentration and the temperature. As the equilibrium relation in cell A and in cell B is given by $C_A = g_A(T)$, $C_B = g_B(T)$ respectively, the solid fraction, f_A, f_B is represented by the function of temperature as the following equations:

$$f_A = 1 - \frac{C_{Ai}}{C_A} = 1 - \frac{g_A(T'_{Ai})}{g_A(T)} \tag{33}$$

$$f_B = 1 - \frac{C_{Bi}}{C_B} = 1 - \frac{g_B(T'_{Bi})}{g_B(T)} \tag{34}$$

The distribution of the solid fractions in sample materials is varied in each region distinguished by the fixed temperatures, $T'_{Ai}, T'_{Bi}, T'_{Ae}$, and T'_{Be}, as shown in Fig. 13. The amount of internal heat generation corresponding to the released latent heat, q_n, and the energy balance equations are obtained by using the solid fraction f_n as follows:

$$q_n = -\rho_3 L_H \frac{\partial}{\partial t}\{1 - f_n(T)\} = -\rho_3 L_H \frac{\partial}{\partial T}\{1 - f_n(T)\}\frac{\partial T}{\partial t} \tag{35}$$

$$\{(c\rho)_n + \rho_3 L_H \frac{\partial}{\partial T}(1 - f_n)\}\frac{\partial T_n}{\partial t} = \lambda_n \frac{\partial^2 T_n}{\partial x^2} \tag{36}$$

where suffix n denotes region n.

The values of $\partial(1 - f_n)/\partial T$ vary in each region of the freezing zone ($n=2,3,4$), and are obtained from the relationship between the internal solute concentration and the temperature of the field, as shown later. For simplicity, taking into account the solid fraction in each region, the average values of $\lambda_n, (c\rho)_n$, and $\partial(1 - f_n)/\partial T$ are used for the analytical calculation.

Analysis. An exact solution is presented in this section for a semi-infinite body which is divided into the five regions: the solid or frozen region, the three solid-liquid regions where freezing is taking place, and the liquid region. Fig. 13-(b) shows the schematic physical system and the coordinate.

To simplify the mathematical manipulation for this problem, the following assumptions are made:

1. The temperature and the solute concentration are initially constant, moreover, the temperature of the cooling surface is fixed during the freezing process.

2. Heat flows along the x-direction only; i. e. , a one-dimensional conduction problem is assumed.

3. Convection in cells is not considered.

4. All physical properties of each region are assumed to remain constant, but may be different for different regions.

Upon these assumptions, the corresponding energy equations for the five regions are:

$$\frac{\partial T_n}{\partial t} = \kappa_n \frac{\partial^2 T_2}{\partial x^2} \quad (n = 1, 2, 3, 4, 5) \tag{37-41}$$

where,

$$\kappa_n = \lambda_n / \{(c\rho)_n + \rho L_H k_n\} \ , \quad k_n = \frac{\partial}{\partial T}\{1 - f_n(T)\}$$

$$f_1 = 1.0 \ , \quad f_2 = f_A(T)\phi_A + \phi_B$$
$$f_3 = f_A\phi_A + f_B(T)\phi_B$$
$$f_4 = f_B(T)\phi_B \ , \quad f_5 = 0.0$$

The initial and boundary conditions for this problem are

$$T_1(0,t) = T_o \tag{42}$$
$$T_1(\eta_1,t) = T_2(\eta_1,t) = T'_{Ae} \tag{43}$$
$$T_2(\eta_2,t) = T_2(\eta_2,t) = T'_{Be} \tag{44}$$
$$T_3(\eta_3,t) = T_4(\eta_3,t) = T'_{Ai} \tag{45}$$
$$T_4(\eta_4,t) = T_5(\eta_4,t) = T'_{Bi} \tag{46}$$
$$T_5(x,t) = T_i \tag{47}$$

$$\rho_3 L_H \{f_{n,\eta_n} - f_{n+1,\eta_n}\}\frac{d\eta_n}{dt} = \lambda_n(\frac{\partial T_n}{\partial x})_{\eta_n} - \lambda_{n+1}(\frac{\partial T_{n+1}}{\partial x})_{\eta_n} \quad (n = 1,2,3,4) \tag{48}-(51)$$

General solutions may be obtained by applying the boundary conditions at the interfaces. Suppose that $\eta_n(n = 1,2,3,4)$ are proportional to the square root of time t, that is, $\eta_n = \alpha_n\sqrt{t}$, the solutions for this problem are

$$T_1 = \frac{(T_o - T'_{Ai})\phi(x/2\sqrt{\kappa_1 t}) - T_o\phi(\alpha_1/2\sqrt{\kappa_1}) + T'_{Ae}}{1 - \phi(\alpha_1/2\sqrt{\kappa_1})} \tag{52}$$

$$T_n = \frac{\begin{pmatrix} T'_{Be} \\ T'_{Ai} \\ T'_{Bi} \end{pmatrix}\phi(\alpha_{n-1}/2\sqrt{\kappa_n}) - \begin{pmatrix} T'_{Ae} \\ T'_{Be} \\ T'_{Ai} \end{pmatrix}\phi(\alpha_n/2\sqrt{\kappa_n}) + \begin{pmatrix} (T'_{Ae} - T'_{Be}) \\ (T'_{Be} - T'_{Ai}) \\ (T'_{Ai} - T'_{Bi}) \end{pmatrix}\phi(x/2\sqrt{\kappa_n t})}{\phi(\alpha_{n-1}/2\sqrt{\kappa_n}) - \phi(\alpha_n/2\sqrt{\kappa_n})}$$

$$(n = 2,3,4) \tag{53}-(55)$$

$$T_5 = T_i + \frac{(T'_{Bi} - T_i)\phi(x/2\sqrt{\kappa_5 t})}{\phi(\alpha_4/2\sqrt{\kappa_5})} \tag{56} \qquad \phi(x) = \frac{2}{\sqrt{\pi}}\int_x^\infty e^{-\xi^2}d\xi$$

Furthermore, $\alpha_n(n = 1,2,3,4)$ are determined from the following simultaneous equations

$$\frac{-\lambda_1(T_o - T'_{Ae})\exp\{-(\alpha_1/2\sqrt{\kappa_1})^2\}}{\sqrt{\kappa_1}\{1 - \phi(\alpha_1/2\sqrt{\kappa_1})\}} + \frac{\lambda_2(T'_{Ae} - T'_{Be})\exp\{-(\alpha_1/2\sqrt{\kappa_2})^2\}}{\sqrt{\kappa_2}\{\phi(\alpha_1/2\sqrt{\kappa_2}) - \phi(\alpha_2/2\sqrt{\kappa_2})\}} = \frac{1}{2}\sqrt{\pi}\alpha_1\rho_3 L_H(1 - f_{2\eta_2})$$

$$\frac{-\lambda_2(T'_{Ae} - T'_{Be})\exp\{-(\alpha_2/2\sqrt{\kappa_2})^2\}}{\sqrt{\kappa_2}\{\phi(\alpha_1/2\sqrt{\kappa_2}) - \phi(\alpha_2/2\sqrt{\kappa_2})\}} + \frac{\lambda_3(T'_{Be} - T'_{Ae})\exp\{-(\alpha_2/2\sqrt{\kappa_3})^2\}}{\sqrt{\kappa_3}\{\phi(\alpha_2/2\sqrt{\kappa_3}) - \phi(\alpha_3/2\sqrt{\kappa_3})\}} = \frac{1}{2}\sqrt{\pi}\alpha_2\rho_3 L_H(f_{2\eta_2} - f_{3\eta_2})$$

$$\frac{-\lambda_3(T'_{Be} - T'_{Ai})\exp\{-(\alpha_3/2\sqrt{\kappa_3})^2\}}{\sqrt{\kappa_3}\{\phi(\alpha_2/2\sqrt{\kappa_3}) - \phi(\alpha_3/2\sqrt{\kappa_3})\}} + \frac{\lambda_4(T'_{Ai} - T'_{Bi})\exp\{-(\alpha_3/2\sqrt{\kappa_4})^2\}}{\sqrt{\kappa_4}\{\phi(\alpha_3/2\sqrt{\kappa_4}) - \phi(\alpha_4/2\sqrt{\kappa_4})\}} = 0$$

$$\frac{-\lambda_4(T'_{Ai} - T'_{Bi})\exp\{-(\alpha_4/2\sqrt{\kappa_4})^2\}}{\sqrt{\kappa_4}\{\phi(\alpha_3/2\sqrt{\kappa_4}) - \phi(\alpha_4/2\sqrt{\kappa_4})\}} + \frac{\lambda_5(T'_{Bi} - T_i)\exp\{-(\alpha_4/2\sqrt{\kappa_5})^2\}}{\sqrt{\kappa_5}\phi(\alpha_4/2\sqrt{\kappa_5})} = 0$$

$$(57)-(60)$$

Results. For experimental convenience, the width of each cell was the same and salt solutions of two kinds of composition were used. Then the eutectic temperature and the interface position associated with this temperature of cell group A coincided with that of B: $T'_{Ae} = T'_{Be}$ and $\eta_1 = \eta_2$. Comparison of the experimentally measured and analytically predicted results for the interface motion and the temperature distribution are shown in Figs. 14 and 15 respectively. The freezing fronts, η_3 and η_4, were obtained by measuring the position of the initial freeze in the cell of each group. Practically, these positions vary discontinuously with time, however, it is admissible that η_3, η_4 change continuously in terms of the cell width. The temperature at η_1 is associated with the eutectic temperature, while the temperature at η_3, η_4 is equivalent to the liquidus temperatures corresponding to the initial concentrations of solute in each cell group. In both figures, the experimental and analytical results are in good agreement, except during the early stage.

According to experimental observation, the solid region produced by a complete segregation of ice and salt looks cloudy and can be easily distinguished from the solid-liquid region. Figure 16 shows a photograph of the local discontinuous freezing in the solid-liquid phase region. In this region, the stepwise freezing in each cell does not appear up to expectations, but freezing fronts could be discerned within several cells. From this observation, it is clear that the solid fraction in cells declines in the opposite direction of the heat flow. Other neighboring cells, including those with no visible freezing front, may be considered to be incompletely frozen.

Figure 17 shows a schematic representation of a concentration profile of a solute in a cell. In the case of (a), the concentration of the solute is higher than the equilibrium concentration shown by the dotted line at the time t_1, and the liquid adjacent to the interface is stable thermodynamically, so freezing advances with a planar surface. At the time t_2, however, the equilibrium concentration becomes higher and the interface becomes unstable due to constitutional supercooling. This suggests that the cell wall cannot suppress the diffusion of the solute efficiently enough to prevent local constitutional supercooling of the solution, and this results in formation of a rough surface (Fig.18-b). On the other hand, in the case of (b) where the freezing rate is slow, constitutional supercooling can be suppressed, as there is enough time for the solute to diffuse.

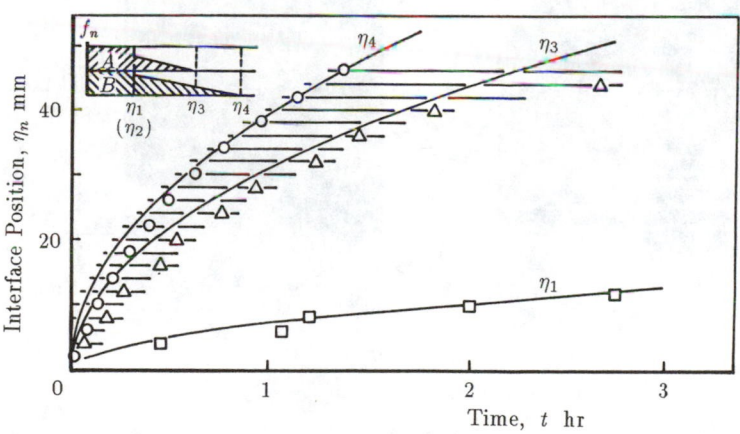

Figure 14 Comparison of predicted and measured interface positions.

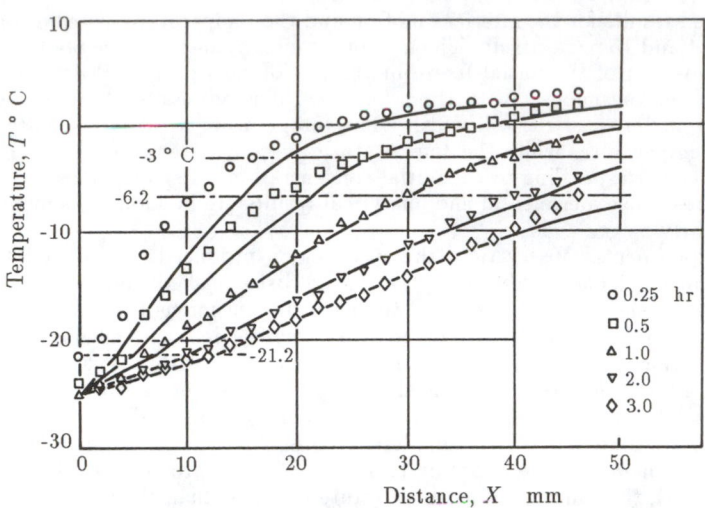

Figure 15 Comparison of predicted and measured temperature distribution.

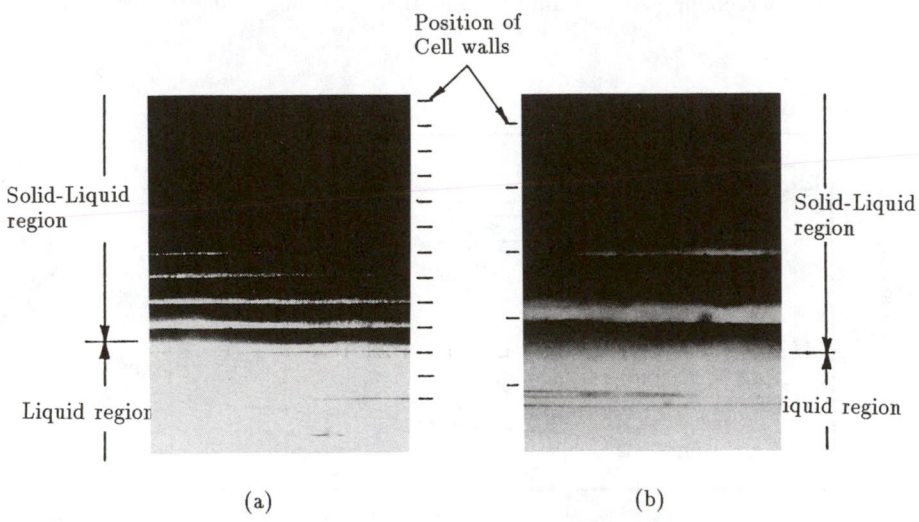

Figure 16 Photograph of local discontinuous freezing in solid-liquid phase region for cell width, (a) 2 mm and (b) 6 mm.

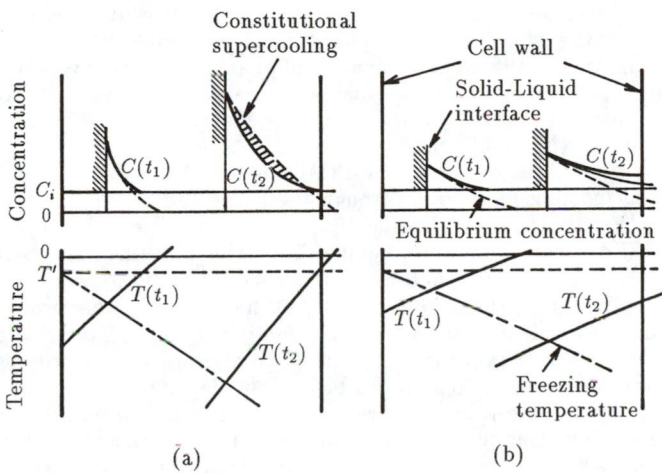

Figure 17 Schematic representation of solute and temperature distribution in a cell.

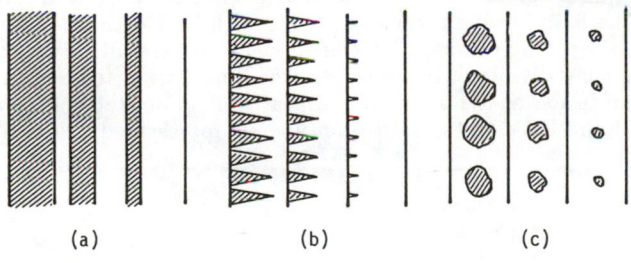

Figure 18 Schematic of freezing state appeared in cellular material.

4.2 Thermal Eenergy Storage using Solution as Phase Change Material

Efficient energy storage is an important contemporary problem. Phase-change energy storage is the most efficient, having many significant advantages: the ability to store and recover a large amount of thermal energy per unit mass of a storing medium, and having the thermal storage process occurring at a nearly constant temperature[39]. However, this type of thermal energy storage has an inherent disadvantage, that is, it takes a long time to discharge because of the deposition of the phase change materials(PCM) on the inner surface of the storing device. Practical application of phase-change energy storage concepts requires further advancement in the thermal performance of the system. From this standpoint, many efforts have been made for the improvement of heat transfer due to the insulation properties of the PCM's used.

In the shell-and-tube type heat exchanger, the PCM fills the shell-side of the heat exchanger while a heating/cooling fluid flows through the tubes and serves to convey the stored energy from the storage unit. During the heat recovery period, the fluid exchanges heat with PCM and changes its bulk temperature which, in turn, affects the phase change process. This induces the decrease in thickness of the solid PCM forming on the tubes in the flow direction. Shamsunder et al. [40] have studied this problem by introducing a two-dimensional analysis of heat conduction with a phase change. The outcome of these calculations was presented in the form of graphs of effectiveness versus the number of transfer units. As mentioned before, the deposition of the PCM on the tube surface restrains the rate of heat recovery because of an increase in thermal resistance. In order to improve this inherent disadvantage, installation of metal fins on the shell-side is fairly effective. Sasaguchi et al. [41] studied the heat transfer characteristics of a finned-tube-type latent heat storage unit. On the other hand, the use of a PCM such as a solution has been proposed to improve the high performance of thermal energy storage. In comparison with a pure substance, the solution has the advantage of flexibility of operating temperature, and of maintaining an effective heat transfer surface[42].

In Fig. 19 the physical and coordinate systems are shown schematically, concentrating on the element of the shell-and-tube heat exchanger. If an aqueous solution like a salt solution is used for the PCM, the solid-liquid phase appears over a range of temperatures in the shell-side tube. Then heat conduction in the radial direction is important. In practice, heat conduction in the axial direction is negligible, since the tube length is very large in comparison the tube diameter and tube spacing. The difference in the freezing progress in the axial direction is mainly a result of changes in the temperature of the fluid flowing through the tube with heat exchange. The shell-and-tube heat exchanger is sub-divided into small elements of width z in the axial direction. Assuming constant physical properties, one dimensional temperature variation, and the heat of fusion as the only internal heat generation, the energy balance equation for the fluid and PCM in the shell-side tube are represented, respectively, as follows:

$$q_z + q_w = q_{z+\Delta z} \tag{61}$$

$$\frac{d}{dt}\{(c\rho)_1 \int_{r_1}^{\eta_1} (T_i - T_1)rdr + (c\rho)_2 \int_{\eta_1}^{\eta_2} (T_i - T_2)rdr + (c\rho)_3 \int_{\eta_2}^{r_2} (T_i - T_3)rdr$$

$$+\rho_1 L_H \{\eta_1 \frac{d\eta_1}{dt} + \frac{d}{dt} \int_{\eta_1}^{\eta_2} frdr\} = q_w \tag{62}$$

Using the appropriate initial and boundary conditions, solutions for temperature distribution and the motion of the interface were obtained.

A comparison of experimentally measured and analytically predicted results for the freezing of a salt solution is shown. Figure 20 shows the typical temperature distributions at the three representative positions along the tube as a function of position in the radial direction with time. The temperature at η_1 is associated with the eutectic temperature, $-21.2°C$, and the temperature at η_2 is equivalent to the liquidus temperature, $-11.2°C$. The changes of the freezing interface position with time are shown for initial solute concentrations, 5 and 10 %wt, in Fig.21. From these two figures, it is clear that the freezing beginning in the solid-liquid phase shifts to the eutectic solid, according with the temperature decrease due to the extraction of latent heat of fusion. Furthermore, the tendency of the freezing to progress, developing in the axial and radial directions, depends on the solute concentration as shown in Figs. 21(a) and (b).

In Fig. 22, the result of the heat recovery rate are shown as a function of time. From this result, it appears that the change of heat recovery is classified into three characteristic periods with the duration of time. The second period is characterized by the extraction of latent heat of fusion which obtains a nearly uniform heat recovery.

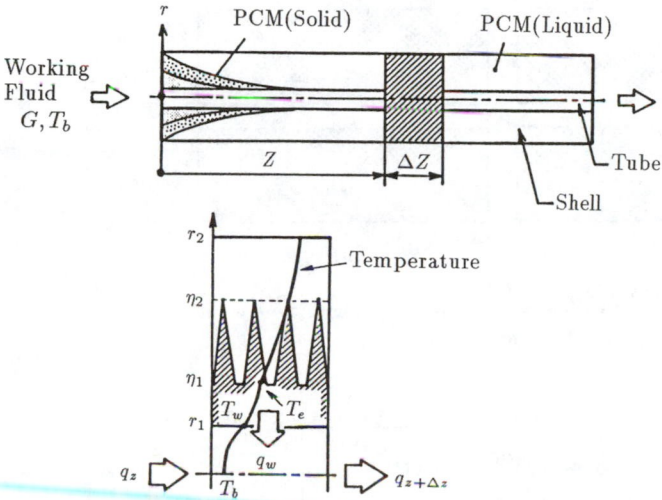

Figure 19 Physical and coordinate systems.

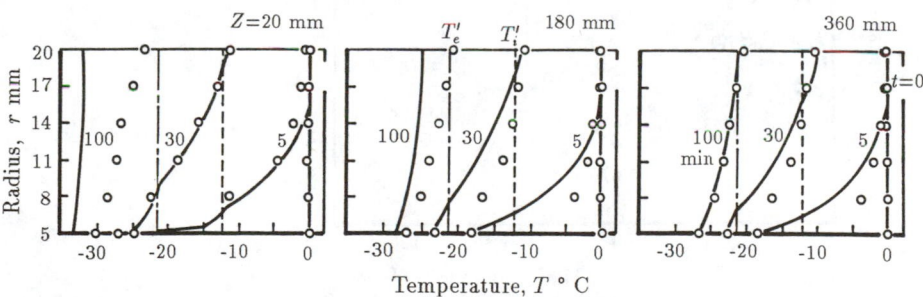

Figure 20 Comparison of predicted and measured temperature distribution.

On the other hand, the first and third periods are characterized by the sensible heat of the liquid and solid respectively. Comparing the results for C_i=5 and 15 %wt, it is clear that the difference of the progression of the freezing is strongly effected by the tendency of heat recovery. In conclusion, the larger the temperature range of the phase change, the more uniform the freezing in the shell side of the tube in the axial direction, and the heat exchange surface is effectively maintained during the discharging period.

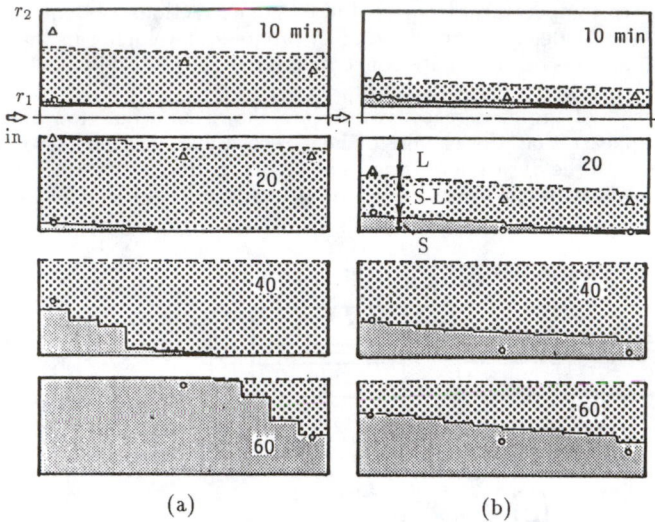

Figure 21 Changes of freezing interface positions with time for initial solute concentration, (a) 5 % wt and (b) 10 % wt.

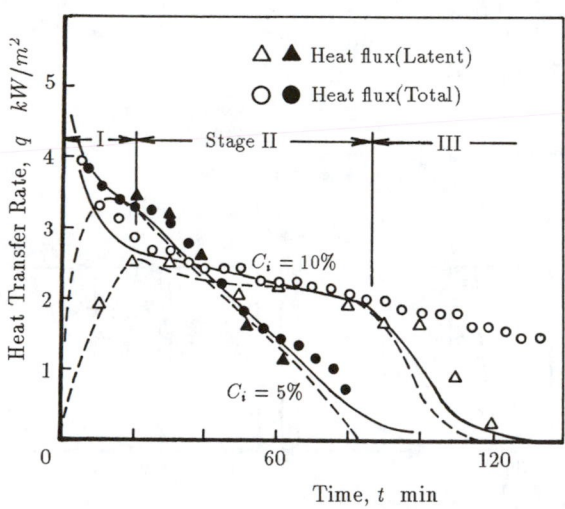

Figure 22 Comparison of predicted and measured heat recovery rate during discharging process.

4.3 Directional Solidification

Eutectic composite metal alloys have received considerable attention as "new materials". When solutions of eutectic composition are solidified under near equilibrium conditions, phases which include each component grow at some interlameller spacing into the melt to form a solid material with an aligned structure. Since this composite microstructure results in anisotropic properties, eutectic metal alloys are used as structural material and electronic devices. The industrial process to produce such material with a uniform and aligned structure is termed by the "directional solidification", and is useful for restricted operating conditions; the steady-state solidification and the plane-front solidification. In practice, most eutectics which contain some solute are really off-eutectic compositions; the solidification induces the concentration polarization and sometimes gives rise to interface instabilities due to constitutional supercooling. Ashbrook[43] reviewed literature relevant to directional solidification. Here the condition of the plane-front growth, effects of some solidification parameters on microstructure, and other important matters in connection with the directional solidification are described on the basis of representative papers[44],[45]. First, the criterion for plane-front growth according to Rutter and Chalmers[1], is briefly explained. In order to progress the steady-state solidification, a molten zone produced by the local heating of a cylindric rod moves at a constant speed. In Fig. 23, the temperature and concentration distributions in a molten zone are shown schematically. If the solid-liquid interface is taken at the origin, the mass balance equation for the steady-state is

$$D\frac{d^2C}{dx^2} + R\frac{dC}{dx} = 0 \tag{63}$$

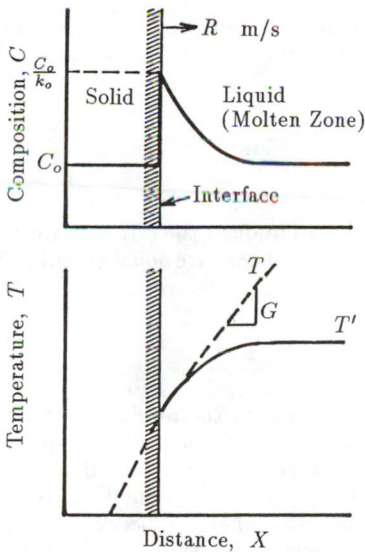

Figure 23 Schematic of concentration and temperature distributions in a molten zone.

where D is the diffusion coefficient of the solute in the melt, R is the growth rate of the molten zone. In the steady-state condition, the composition of the solid must be maintained at the initial concentration of melt, C_i, and the concentration of the melt at the solid-liquid interface is C_i/K_o. Then the concentration C at the distance X from the interface is

$$C = C_i\{1 + \frac{1 - K_o}{K_o}\exp(-\frac{R}{D}x)\} \tag{64}$$

Assuming the linear relationship between temperature and concentration in the constitutional diagram, the liquidus line is represented by

$$T' = T_o - mC \tag{65}$$

where T_o is the equilibrium temperature for the pure solution, and m is the slope of the liquidus. Using Eq. (3), the distribution of the equilibrium temperature in the molten zone is

$$T' = T_o - mC_i\{1 + \frac{1 - K_o}{K_o}\exp(-\frac{R}{D}x)\} \tag{66}$$

On the other hand, the temperature distribution can be expressed by using the temperature gradient G as follows:

$$T = T_o - m\frac{C_i}{K_o} + Gx \tag{67}$$

If the actual temperature T is lower than the equilibrium temperature T', thermodynamical instability termed by constitutional supercooling appears in the region from the interface up to X_s,

$$1 - \exp(\frac{R}{D}X_s) = \frac{G}{mC_i(1 - K_o)/K_o}X_s \tag{68}$$

Mollard and Fleming[44] developed the equation and obtained the criterion for plane-front growth as the critical condition for the existence of constitutional supercooling.

$$\frac{G}{R} = -\frac{m}{D}(C_e - C_i) \tag{69}$$

where C_e is the eutectic composition. Figure 24 represents the results for structures of directionally solidified lead-tin ingots as a function of G/R and C_i. In the figure, the solid line shows an experimental boundary between the dendrite and plane front structure, and the dashed line boundary predicted by the criterion due to constituting supercooling, Eq.(69). It was found that the smaller the difference between the initial and eutectic compositions, the more the fine composite structure may be produced at very small G/R ratios. However, as the amount of the difference increases, an increasing amount of G/R is required to achieve aligned composite growth[43].

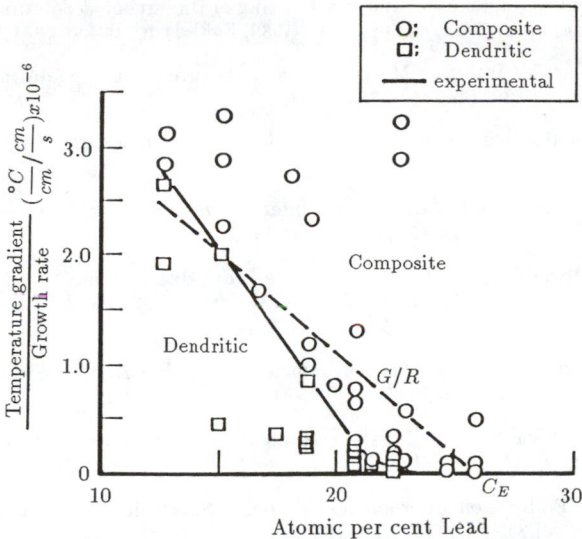

Figure 24 Results for structures of directionally solidified lead-tin ingots as a function of G/R and C_i[43].

5 CONCLUDING REMARKS

Recently, increasing attention has been focused on the solidification and melting heat transfer of solutions concerning with the advancing thermal technologies. In material processing, for instance, novel materials having micro structure and/or metastable structure may be produced by highly controlling the cooling rate and temperature distribution.

The basic concepts and the most advanced theories reviewed in this article may be expected to contribute for more development of practical industrial processes.

ACKNOWLEDGEMENT

The author is deeply indebted to Associate Professor Akira Takimoto of Department of Mechanical Systems Engineering, Kanazawa University for his many helpful advices through the present work.

REFERENCES

1. B.Chamlers, *Principle of Solidification*, John Wiley, New York, 1964.

2. M.C.Flemings, *Solidification Processing*, McGraw-Hill, New York, 1974.

3. D.Walton, W.A.Tiller, J.W.Rutter and W.C.Winegard, Instability of a Smooth Solid-Liquid Interface during Freezing, *Trans. Am. Inst. Min. Metall. Engrs.*, vol.1, pp.1023-1028, 1955.

4. W.A.Tiller, and J.W.Rutter, The Effect of Growth Conditions upon the Solidification of a Binary Alloy, *Can. J. Phys.*, vol.34, pp.96-107, 1956.

5. Y.Hayashi, K.Kunimine, and T.Nagamoto, Study on Freezing of Undercooled Solution, *Int. Sympo. on Cold Regions Heat Transfer*, pp.85-90, 1989, Hokkaido University.

6. J.D.Harrison, and W.A.Tiller, Ice Interface Morphology and Texture Developed during Freezing, *J. Applied Physics*, vol.34, pp.3349-3355, 1963.

7. A.Saito, et al., Heat Conduction with Supercooled Solidification, *Bull. JSME*, vol.25, pp.591-598, 1982.

8. W.W.Mullins, and R.F.Sekerka, Stability of a Planar Interface during Solidification of a Dilute Binary Alloy, *J. Appl.Phys.*, vol.35, pp.126-131, 1963.

9. M.G.O'Callagham, et al., Instability of the Planar Freeze Front during Solidification of an Aqueous Binary Solution, *ASME Journal of Heat Transfer*, vol.102, pp.673-677, 1980.

10. J.P.Terwilliger, and S.F.Dizio, Salt Rejection Phenomena in the Freezing of Saline Solutions, *Chem. Engng. Sci.*, vol.25, pp.1331-1349, 1970.

11. R.L.Levin, Generalized Analytical Solution for the Freezing of a Super-cooled Aqueous Solution in a Finite Domain, *Int.J. Heat Mass Transfer*, vol.23, pp.951-959, 1980.

12. R.L.Levin, The Freezing of Finite Domain Aqueous Solutions: Solute Redistribution, *Int. J. Heat Mass Transfer*, vol.24, pp.1443-1455, 1981.

13. M.G.O'Callagham, et al., An Analysis of the Heat and Solute Transport during Solidification of an Aqueous Binary Solution-I. Basal Plane Region, *Int. J. Heat Mass Transfer*, vol.23, pp.553-561, 1982.

14. M.G.O'Callagham, et al., An Analysis of the Heat and Solute Transport during Solidification of an Aqueous Binary Solution-II. Dendrite Tip Region, *Int. J. Heat Mass Transfer*, vol.25, pp.563-573, 1982.

15. W.Kurz, and D.J.Fisher, Dendrite Growth at the Limit of Stability: Tip Radius and Spacing, *Acta Metall.*, vol.29, pp.11-20, 1981.

16. A.Roosz, et al., Solute Redistribution during Solidification and Homogenization of Binary Solid Solution, *Acta Metall.*, vol.32, pp.1745-1754, 1984.

17. R.H.Tien, and G.E.Geiger, A Heat-Transfer Analysis of the Solidification of a Binary Eutectic System, *ASME Journal of Heat Transfer*, vol.89, pp.230-234, 1967.

18. R.H.Tien, and G.E.Geiger, The Unidimensional Solidification of a Binary Eutectic System with a Time-Dependent Surface Temperature, *ASME Journal of Heat Transfer*, vol.90, pp.27-31, 1968.

19. W.G.Pfann, Principle of Zone-Melting, *Trans. AIME*, vol.135, pp.85, 1952.

20. Y.Hayashi and K.Yamaguchi, Mathematical Treatment of Freezing of Food, *Refrigeration*, vol.54, pp.575-583, 1979.

21. S.H.Cho, and J.E.Sunderland, Heat-Conduction Problem with Melting or Freezing, *ASME Journal of Heat Transfer*, vol.91, pp.421-426, 1969.

22. L.I.Rubinstein, The Stefan Problem, *Trans. Math. Monogr.*, vol.27, pp.52-59, 1971.

23. J.C.Muehlbauer, Transient Heat Transfer Analysis of Alloy Solidification, *ASME Journal of Heat Transfer*, vol.102, pp.324-331, 1973.

24. T.R.Goodman, The Heat Balance Integral and its Application to Problems involving Change of Phase, *ASME Journal of Heat Transfer*, vol.80, pp.335, 1958.

25. J.C.Muehlbauer, and J.E.Sunderland, Heat Conduction with Freezing or Melting, *Applied Mechanics Reviews*, vol.18, pp.951, 1965.

26. H.Jones, A Comparison of Approximate Analytical Solutions of Freezing from a Plane Chill, *J. Institute of Metals*, vol.97, pp.38, 1969.

27. K.Stephan, et al., Heat Conduction in Solidification coupled with Phase Transformation in the Solid, *Fifth International Heat Transfer Conf.*, vol.1, pp.235-239, 1974.

28. B.W.Grange, R.Viskanta, and W.H.Stevenson, Diffusion of Heat and Solute during Freezing of Salt Solutions, *Int. J. Heat Mass Transfer*, vol.19, pp.373-384, 1975.

29. H.Shingu, et al., An Analysis of the Solidification of a Binary Eutectic System considering the Temperature and Solute Distribution, *J. of Japan. Institute of Metals*, vol.42, pp.172-179, 1978.

30. K.Takeshita, An Analysis of the Solidification of a Binary Eutectic System in Consideration of Both Heat and Solute Diffusion, *J. of Japan. Institute. of Metals*, vol.47, pp.647-653, 1983.

31. J.Szekely and A.S.Jassal, An Experimental and Analytical Study of the Solidification of a Binary Dendritic System, *Metallurgical Transactions*, vol.9-B, pp.389-398, 1978.

32. E.M.Sparrow, et al., Freezing Controlled by Natural Convection, *ASME Journal of Heat Transfer*, vol.101, pp.578-584, 1979.

33. F.M.Chiesa and R.I.L.Guthrie, Natural Convective Heat Transfer Rates during the Solidification and Melting of Metals and Alloy Systems, *ASME Journal of Heat Transfer*, vol.96, pp.377-384, 1974.

34. Y.Hayashi, et al. , Heat Conduction during Freezing and Melting of a Binary Aqueous Solution, *ASME Paper*, No.81-WA/HT-38.

35. Ch.Koerber, M.W.Scheiwe and K.Wollhoever, Solute Polarization during Planar Freezing of Aqueous Solutions, *Int. J. Heat Mass Transfer*, vol.26, pp.1241-1253, 1983.

36. K.Hayakawa, Estimation of Heat Transfer during Freezing of Defrosting of Food, *Int. Congress of Refrigeration*, Germany, 1977.

37. Y.Hayashi and T.Komori, Investigation of Freezing of Salt Solution in Cells, *ASME Journal of Heat Transfer*, vol.101, pp.459-465, 1979.

38. Y.Hayashi, S.Kato and K.Hattori, Fundamental Study on Freezing of Food, *Trans. JSME*, vol.47-B, pp.361-368, 1981.

39. R.Viskanta, *Solar Heat Storage*, vol.1, CRC Press, 1983.

40. N.Shamsunder and R.Srinivasan, Effectiveness-NTU Charts for Heat Recovery from Latent Heat Storage Units, *ASME Journal of Solar Energy Engng.*, vol.102, pp.263-271, 1980.

41. K.Sasaguchi, H.Imura and H.Furusho, Heat Transfer Characteristic of a Latent Heat Storage Unit with Finned Tube, *Trans. JSME*, vol.52, pp.159-166, 1986.

42. Y.Hayashi, K.Kunimine and K.Yamaguchi, Study on Latent Heat Thermal Energy Storage Using Aqueous Solution as PCM, *Trans. JSME*, vol. 54-B, pp. 452-458, 1988.

43. R.L.Ashbrook, Directional Solidified Ceramic Eutectics, *J. American Ceramic Society*, vol.60, pp.428-435, 1977.

44. F.R.Mollard and M.C.Flemings, Growth of Composites from the Metal: I, *Trans. AIME*, vol.239, pp.1526-1533, 1967.

45. F.R.Mollard and M.C.Flemings, Growth of Composites from the Metal:II, *Trans. AIME*, vol.239, pp.1534-1543, 1967.

FRAZIL ICE

Chapter 16

Frazil Ice

STEVEN F. DALY
U.S. Army Cold Regions Research and Engineering Laboratory,
Hanover, New Hampshire 03755-1290, USA

ABSTRACT

A physically based quantitative model of frazil ice in natural water
bodies, which describes the dynamic evolution of the frazil crystal size
distribution function, is developed. The crystal number continuity
equation and the heat balance for a differential volume serve as the
basis for the model. The crystal growth rate and secondary nucleation
rate are the major parameters that appear in these equations. Expressions for both are derived. The crystal growth rate is controlled by
the heat transfer rate from the crystal to the supercooled water, which
is shown to be a function of the crystal size, the fluid turbulence and
the fluid properties. Secondary nucleation is assumed to be caused by
collisions, which cause fragments of the crystals to shear off. These
fragments become new crystals. The rate of secondary nucleation is
assumed to be limited only by the collision rate, and can be described
by the product of the collision energy rate per unit volume and a
material property that describes the number of crystals created per unit
of collision energy. The collision energy depends on the mechanism
causing the collision; expressions are derived for collisions caused by
the turbulent shear, differences in the bouyancy of the crystals and
collisions of the crystals with solid boundaries. Three basic environmental parameters, and one material property, control the evolution of
the crystal size distribution function. These environmental parameters
are the heat loss rate, the turbulent energy dissipation rate of the
fluid and the crystal seeding rate. The material property is the number
of crystals produced per unit of collision energy. The environmental
parameters can be measured or controlled in experiments, and the resulting measured size distributions of crystals would allow the values of
the parameters to be estimated. The model could be extended to simulate
the influence of impurities (for example salt). In addition, differential rise velocities and the diffusion of crystals could be included in
the basic equations to extend the applicability of the model.

TABLE OF CONTENTS

INTRODUCTION

BASIC EQUATIONS

Crystal Number Continuity Equation
Heat Balance
Parameters in the Basic Equations

ICE CRYSTAL GROWTH RATES

Morphology
Intrinsic Kinetic Growth Rate
Heat Transfer from Ice Crystals Suspended in Turbulent Water

NUCLEATION

Initial Nucleation
Secondary Nucleation
Application of the Model

SUMMARY

NOMENCLATURE

B	Birth function
C	Heat capacity of fluid
C_T	Fluid impurity concentration
C_i	Heat capacity of ice
D	Death function
E_{rb}	Collision energy created by crystal-boundary collisions
$E(r_1,r_2)$	Collision energy created by collisions of crystals of size r_1 and r_2
$\dot{E}_t$	Rate of energy transfer by collision
F_1	Number of particles generated per unit of collision energy
F_2	Fraction of particles surviving to become crystals
g	Gravity
g'	Reduced gravity
G	Crystal growth rate
h	Heat transfer coefficient
k	Heat conductivity of fluid
K_v	Crystal shape factor
L	Mean latent heat of fusion of ice

m*	Nondimensional crystal size
$m(r)$	Mass of crystal of size r
n	Size distribution function
N	Total number of crystals per unit volume
$\dot{N}_I$	Rate of introduction of new crystals
$\dot{N}_T$	Total secondary nucleation
Nu	Nusselt number
Nu_T	Turbulent Nusselt number
Nu_o	Nusselt number for a particle in a stationary fluid
Pe	Péclet number
Pr	Prandtl number
q_{rb}	Frequency of collisions between crystals and boundary
$q(r_1,r_2)$	Frequency of collisions between crystals of size r_1 and r_2
Q*	Net heat transfer
r	Major linear dimension of frazil crystals
r_f	Face dimension of disk
R	Region of phase space
R_c	Collision radius
Re	Reynolds number
R_h	Hydraulic radius
S_N	$(F_1)(F_2)$
t	Time
T_f	Temperature of fluid
u'	Turbulent fluctuation of velocity
U_{rb}	Relative velocity of crystal and boundary
$\nu(r_1,r_2)$	Relative velocity of crystals of size r_1 and r_2
$\vec{V}_e(R,t)$	External phase space convective velocity
$\vec{V}_f$	Fluid velocity
$\vec{V}_i(R,t)$	Internal phase space convective velocity
$\vec{V}(R,t)$	Phase space velocity
α	Thermal diffusivity
α_T	Turbulent intensity
ψ	Collision efficiency
ϵ	Turbulent dissipation rate
η	Dissipation length scale
θ	Bulk supercooling
ν	Fluid kinematic viscosity
ρ_f	Density of water
$\hat{\rho}_f$	Mass concentration of water in mixture
$\hat{\rho}_i$	Mass concentration of ice in mixture
ρ_i	Density of ice
ℓ_m	Length scale of maximum eddy size

INTRODUCTION

Frazil ice is formed in turbulent supercooled water. It is found in lakes and oceans and is the predominant type of ice formed in rivers and streams. Characterized by a disk shape during its initial formation, frazil evolves through several stages that can vary widely in length scales, time scales and material properties. The evolution of frazil ice in rivers and streams follows a progression whose broad outlines are well known, but only now has a quantitative understanding begun to emerge.

Three rather general stages of frazil evolution can be identified. The first is the dynamic, nonequilibrium stage, characterized by supercooled water, turbulent flow, rapid growth of disk-shaped crystals and the creation of new crystals by secondary nucleation. The length scales of the ice associated with this stage (Fig. 1) range from several microns to perhaps a few centimeters. This stage may last for a relatively short time and occurs during very cold periods when there is maximum heat loss from the open water surface. Typical sequences and locations are described by Michel (1971). Associated with this stage is the freezeup of water intakes by active frazil, that is, frazil actively growing in the supercooled water.

The second stage is the evolution and transport stage. This stage follows the first in time, and results from the rapid formation and production of the first stage. The second stage is characterized by water more or less at the equilibrium temperature and frazil in the form of flocs, anchor ice and floes. The length scales of the ice associated with this stage (Fig. 1) range from several millimeters to many meters.

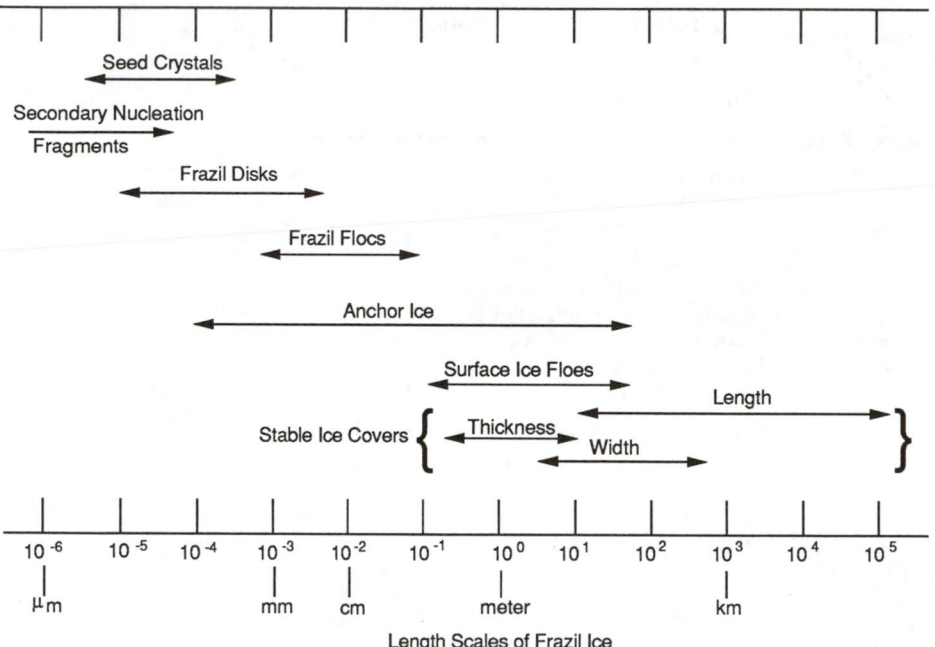

FIGURE 1. Length scales of frazil ice in rivers and streams.

The frazil is now in motion, moving under the influence of the river or stream, generally at the surface. This ice may travel long distances and move for many days. After cold nights, it is very typical to see slush, formed of frazil flocs, moving along at the water surface of northern rivers and streams. This slush may eventually form large moving floes through a variety of processes described by Osterkamp and Gosink (1982).

The third stage is characterized by stationary, floating ice covers that may be quite large and last for the entire winter season. These ice covers are formed by a variety of mechanisms, depending on the form of the frazil ice when it arrives at the stationary ice cover and the hydraulic conditions at its leading edge. These floating covers may raise stages and cause flooding, cause excessive head loss that can disrupt power production and interfere with navigation. However, an intact floating ice cover will always prevent the excessive heat loss associated with the first stage of frazil growth, and, therefore, will suppress the formation of frazil. Frazil is only produced on rivers before a stable ice cover forms, after it breaks up, or at locations where ice covers cannot form because of the hydraulic conditions. Therefore, a successful strategy for preventing frazil formation is to foster the formation of an intact ice cover as quickly as possible by use of ice control structures in reaches where frazil is produced.

In this chapter we will discuss the development of a physically based quantitative model for frazil ice in natural water bodies that describes the dynamic evolution of the frazil crystal size distribution function during the first stage of frazil evolution. In the present work two equations serve as the basis for the model: the crystal number continuity equation and the heat balance for a differential volume. This simplifies the presentation but it is not meant to suggest any limitation; conservation of impurities (such as salt), for example, could be added. The crystal growth rate and secondary nucleation rate are the major parameters that appear in these equations. Expressions for both are derived. We will see that there are three basic environmental parameters: the heat loss rate and the crystal seeding rate, which appear directly in the basic equations, and the turbulent energy dissipation rate, which influences both the crystal growth rate and the rate of secondary nucleation. One material property, the number of crystals produced per unit of collision energy, directly controls the rate of secondary nucleation. Its value is, at present, unknown. The model described in this chapter is meant as a first step towards solution of practical problems caused by frazil. The deficiencies in the current state of knowledge are "severely hindering development of rational design methods for avoidance, or alleviation of, frazil ice problems" (ASCE Task Committee 1974).

BASIC EQUATIONS

Crystal Number Continuity Equation

The crystal distribution will be described in a space termed the crystal phase space or more generally the particle phase space. Particle phase space is defined by the least number of independent coordinates that provides a complete and useful description of the properties of the crystal distribution. It is convenient, if somewhat arbitrary, to divide particle phase space into two subregions defined by external coordinates and internal coordinates. The external coordinates describe the spatial distribution of the crystals. Internal coordinates refer to properties attached to each individual crystal, which quantitatively measure its state, and are independent of its position.

To begin, a crystal distribution function $n(R,t)$ will be considered. This function is defined over a region R of the particle phase space consisting of the three spatial dimensions (the external coordinates) plus any number of internal property coordinates. In all further cases the internal coordinates will be restricted to one, which will corre-spond to a major linear dimension, r, of the ice crystals. The function $n(R,t)$ is defined as the population density of crystals in the region R. At a time t the number of crystals in an incremental region of the particle phase space dR is given by

$$dN = ndR \tag{1}$$

and the total number of crystals in a region R at time t is

$$N(R) = \int_R ndR \ . \tag{2}$$

Individual crystals can continuously change their position in the particle phase space. If these changes are regular, that is, if they are the result of gradual and continuous movement, the convective crystal velocity along a respective particle coordinate can be defined. Let $\vec{V}_e$ be the external convective velocity and $\vec{V}_i$ be the internal con-vective velocity.

It may also be necessary to deal with the sudden appearance (birth) or disappearance (death) of crystals at a point in the particle phase space. The net appearance in an incremental region dR at a time would be

$$(B-D)dR \tag{3}$$

where $B(R,t)$ and $D(R,t)$ represent birth and death functions at a point in the phase space.

The population balance of crystals in some fixed region R, which moves convectively with the particle phase space velocity $\vec{V}$, can be defined as

$$\frac{d}{dt} \int_R ndR = \int (B-D)dR \ . \tag{4}$$

Expanding the first term using Leibnitz's rule, noting that the region R was arbitrary, and that only a single internal coordinate r is con-sidered, where $G(r,t)$ is the convective velocity along r or simply the growth rate of the ice crystal, we then see that

$$\frac{\partial n}{\partial t} + \frac{\partial}{\partial r} (Gn) + D - B + \nabla(\vec{V}_e n) = 0 \ . \tag{5}$$

This is the number continuity equation in general form. Further exten-tions can be made by considering the diffusion of crystals, and the rise velocity of the crystals.

Heat Balance

The general expression for the heat balance of the frazil-ice/water system will be developed strictly for frazil crystals suspended in fresh water.

Consider a differential volume in which $\hat{\rho}_f$ is the mass concentration of water (grams of water per cubic centimeter of mixture), $\hat{\rho}_i$ is the mass concentration of ice and the temperature of the water is T_f. It is assumed that, to good approximation, $\hat{\rho}_f$ can be set equal to ρ_f, the density of water, and

$$\hat{\rho}_f + \hat{\rho}_i \simeq \hat{\rho}_f \simeq \rho_f .$$

A second assumption is then

$$C_i \hat{\rho}_i / C \rho_f \ll 1 .$$

Additionally, heat conduction can be neglected and heat capacities and the latent heat can be considered constant because of the small variation in temperature. Therefore, based on the above assumptions, the heat balance for a mixture of frazil and water can be written

$$\frac{\partial T_f}{\partial t} + \nabla(\vec{V}_f T_f) = \frac{L}{C\rho_f} \left[\frac{\partial}{\partial t} \hat{\rho}_i + \nabla(\vec{V}_e \hat{\rho}_i) \right] + \frac{Q*}{C\rho_f} . \qquad (6)$$

where

C = heat capacity of liquid water
$\vec{V}_f$ = convective velocity of the fluid
$Q*$ = net heat transfer from the mixture
L = latent heat of fusion of the ice at the equilibrium temperature of the mixture.

Parameters in the Basic Equations

The two basic equations are the crystal number continuity (eq. 5), and the heat balance (eq. 6). The various parameters that appear in these two equations will now be discussed.

1. G - The growth rate of the major linear dimension of the crystals is a function of the heat transfer and the intrinsic kinematics of the ice crystal. G is effectively determined by the heat transfer rate. Thus, in general

$$G = \frac{h(r,\epsilon)}{\rho_i L} \theta$$

where h is the heat transfer coefficient and is a function of the crystal size r and the level of turbulence ϵ; θ is the supercooling of the mixture.

2. D - The death function can be set to zero for all sizes of crystals.

This is equivalent to assuming that there is no large-scale breakage of the crystals.

3. B - The birth function is determined by the rate of the sudden appearance of new crystals. New crystals can appear as a result of spontaneous nucleation, secondary nucleation and the introduction of crystals; however, spontaneous nucleation is not possible under frazil-forming conditions. Therefore, B will be determined by the rate at which new crystals are introduced and the rate of secondary nucleation.

Let $\dot{N}_T$ be the rate of secondary nucleation. $\dot{N}_T$ is a function of the crystal size distribution n, the turbulence dissipation rate ϵ, the supercooling of the mixture θ and perhaps other parameters. Let $\dot{N}_I$ be the rate at which new crystals are introduced. We assume that new crystals are created and introduced at a size small enough that the radius of new crystals can be approximated as zero. Thus

$$B = [\dot{N}_T(\theta, n, \epsilon) + \dot{N}_I] \, \delta(r-0)$$

where $\delta(r-0)$ is the dirac delta function $[\delta(r=0) = 1, \delta(r \neq 0) = 0]$.

4. $\hat{\rho}_i$ - The mass of ice per unit volume of mixture can conveniently be determined using the moment equation

$$\hat{\rho}_i = \rho_i K_v \int_0^\infty r^3 \, n(r) dr \ .$$

5. Q* - This is determined by the environment of the water body of interest and, in particular, the meteorologic and hydraulic conditions.

6. $\vec{V}_e$, $\vec{V}_f$ - The convective velocity of the ice crystals and the fluid will generally be very similar. The action of buoyancy and inertial forces on the ice crystals may cause the ice crystal velocity to differ from the fluid velocity if these forces become large compared to the fluid drag force.

Substituting the above expressions into eqs. 5 and 6 gives us

$$\frac{\partial n}{\partial t} + \frac{\theta}{\rho_i L} \frac{\partial}{\partial r} (hn) + \nabla(\vec{V}_e n) = (\dot{N}_T + \dot{N}_I) \delta(r-0) \tag{7a}$$

$$\frac{\partial \theta}{\partial t} + \nabla(\vec{V}_f \theta) = \frac{L \rho_i K_v}{C \rho_f} \left[\frac{\partial}{\partial t} \int_0^\infty r^3 \, n \, dr + \nabla \left(\vec{V}_e \int_0^\infty r^3 \, n \, dr \right) \right] + \frac{Q*}{C \rho_f} \tag{7b}$$

where

$$n = n(x,y,z,r,t)$$
$$\theta = \theta(x,y,z,t)$$
$$h = h(r,\epsilon)$$
$$\vec{V}_e = \vec{V}_e(x,y,z,r,t)$$
$$\dot{N}_T = \dot{N}_T(\theta, \epsilon, n)$$
$$\vec{V}_f = \vec{V}_f(x,y,z,t)$$
$$Q* = Q*(x,y,z,t) \ .$$

Writing the equations in this form emphasizes the dynamic way in which they interact. To determine θ and n uniquely, both equations must be solved simultaneously, and the boundary conditions and initial conditions of θ and n must be known. Difficulties arise because θ and n are dimensionally incompatible.

ICE CRYSTAL GROWTH RATES

Morphology

A description of the morphology of ice is not simple. The various shapes of ice crystals appear to result from a complex interaction of the imposed heat transfer conditions and the intrinsic crystallography of ice. We know from observations that the dominant shape of ice crystals that grow at the supercooling levels found in turbulent water bodies is a flat disk. Virtually all field observations of frazil ice note that the crystals are disk shaped. It has been reported that ice crystals in the shape of six-pointed stars, hexagonal plates or spheres, and small pieces of dendritic ice all evolve into the disk shape in natural water bodies. Researchers have studied the morphology of large numbers of frazil ice crystals because of interest in desalination by freezing. The observations of Margolis (1969) indicate that the thickness of the frazil disk is 0.68 r $\pm$ 16.7%, where r is the major radius of the disk. Smith and Sarofim (1979) say that the maximum radius is approximately 0.8 mm for disk crystals produced in turbulent crystallizers. The aspect ratios of frazil crystals grown in a laboratory flume (Daly and Colbeck 1986) are shown in Figure 2.

Disk-shaped crystals have been studied in the laboratory by Kumai and Itagaki (1953), Arakawa (1954) and Williamson and Chalmers (1966). The disk shape taken by ice at low supercooling levels is apparently the result of the anisotropic growth kinetics of ice. The hexagonal ice molecule has two growth axes and they are identifiable by their optical properties. The hexagonal axis is the c-axis, and the three axes normal

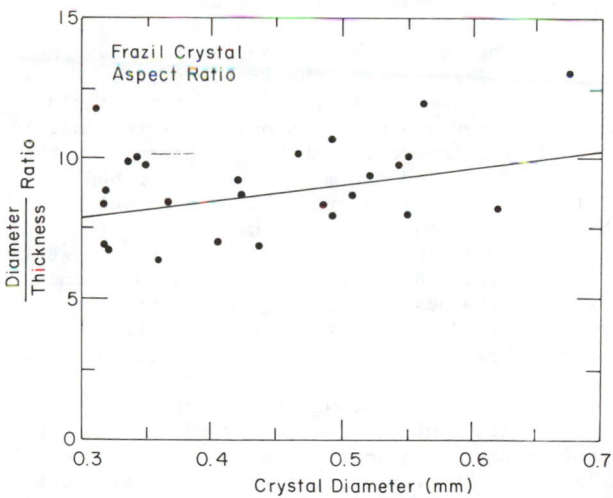

FIGURE 2. Aspect ratio of frazil crystals grown in a laboratroy flume.

to the c-axis are the a-axes. The a-axes are all equivalent because the crystal is symmetrical about the c-axis. The plane that contains the a-axes is the basal plane, and the growth rate in it is tens of times faster than that parallel to the c-axis.

Intrinsic Kinetic Growth Rate

The mechanisms that determine the rate at which an ice crystal can grow are transport of water molecules to the crystal surface (for ice grown in the pure water this is not a consideration), their incorporation into the crystal surface and the transport of latent heat away from the surface. The incorporation of molecules is controlled by the crystallization or interface kinetics of ice and the heat transfer reflects the particular physical situation of the system under consideration.

The interface kinetics of ice have been studied both theoretically and experimentally. As noted in the previous section, ice has two principal growth directions: a-axis growth and c-axis growth. The interface kinetics of each growth direction appear to be different. Growth in the c-axis probably proceeds by surface nucleation for perfect crystals and by a dislocation mechanism for damaged crystals. The interface kinetics of a-axis growth have not been completely defined; the mechanism is probably that of continuous growth. However, it appears that the kinetics are very fast and that for practical purposes the growth rate of the a-axis is totally controlled by the rate at which heat is transported away from the interface. Growth along the c-axis is much slower than a-axis growth for all sizes of crystals. This implies that c-axis growth is controlled by the intrinsic kinetics. The c-axis growth rate does not appear in the number continuity equation. The latent heat released by growth along the c-axis may contribute somewhat to the overall heat balance; however, it may be that the latent heat released by c-axis growth is effectively negligible. Therefore, only the growth along the a-axis will be considered.

Heat Transfer from Ice Crystals Suspended in Turbulent Water

In this section expressions for the rate of heat transfer from suspended ice crystals will be formulated. To determine the transfer rate, it is necessary to describe the ambient velocity distributions of the fluid about the crystal. Frazil is created and develops only in water that is turbulent. Rivers and channels are inherently turbulent because of the instability of their bulk currents. Wind can make large water bodies become turbulent. Frazil is also created in crystallizers in which the water is made turbulent by impellers, turbines or other means. To describe the velocity distribution of the water surrounding the crystals requires knowledge of the properties and characteristics of turbulence. Turbulence can be visualized as numerous interacting eddies of all possible scales. The very largest eddies originate directly from the instabilities of the mean bulk flow. The scale and orientation of these largest eddies are imposed by the geometry of the flow situation. Energy is extracted from the large eddies through the inertial interaction of these eddies with smaller eddies. The energy cascade is not affected by the fluid viscosity until the smallest scales are reached, where this energy is dissipated by the viscosity. The dissipation rate must equal the rate at which energy is supplied to the small-scale eddies. The dissipation rate ϵ and the fluid viscosity ν form a length scale η such that

$$\eta \sim (\nu^3/\epsilon)^{1/4}. \tag{8}$$

η is the dissipation length scale or the Kolmogorov scale.

The range of scales larger than the dissipation scale but smaller than the scale of the large energy-containing eddies is the inertial sub-range. It is impossible to predict a priori the maximum scale at which the inertial subrange will begin. Within the inertial subrange the fluid viscosity has no effect and all energy dissipation results from the inertial interactions between eddies of different sizes. Therefore, the only scaling parameter available is ϵ, which can be interpreted as the rate (per unit mass of fluid) at which energy cascades through the spectrum of eddies sizes within the inertial subrange.

The range of scale that is smaller than the dissipation scale is the dissipation subrange. The fluid viscosity plays an important role in the subrange and acts quickly to dampen and dissipate the fluid motion. This small-scale motion automatically adjusts itself to the value of the viscosity and the rate of energy transfer.

If the crystal size is small relative to the Kolmogorov length scale, it is in the dissipative regime. In the dissipative regime the fluid eddies are strongly dampened and dissipated by the fluid viscosity. In effect, the crystal is smaller than the smallest scales of the turbulent eddies. It does not experience the turbulence as interacting eddies but rather as a fluid motion that varies linearly with position. The Reynolds number of particle motion will be (assuming the particles experience only this shear)

$$Re = (r^2 \epsilon^{1/2})/\nu^{3/2} \tag{9}$$

where r is the major disk radius and the Péclet number will be

$$Pe = (r^2 \epsilon^{1/2})/(\alpha\nu^{1/2}) \tag{10}$$

where α is the thermal diffusivity of the fluid. As $r < \eta$ in the dissipative region, the Reynolds number defined by eq.9 is less than 1, and

$$r < (1/Pr^{1/2})\eta; \quad Pe < 1$$

$$r > (1/Pr^{1/2})\eta; \quad Pe > 1 \quad .$$

where Pr is the Prandtl number.

The heat transfer rate at a small Péclet number can be written (Batchelor 1980)

$$\frac{Nu-Nu_0}{Nu_0} = 0.17 \ Nu_0 \left[\frac{r^2}{\alpha} \frac{\epsilon^{1/2}}{\nu^{1/2}}\right]^{1/2} . \tag{12}$$

The large Péclet number rate can be written

$$Nu = Nu_0 + 0.55 \left[\frac{r^2 \epsilon^{1/2}}{\alpha\nu^{1/2}}\right]^{1/3} \tag{13}$$

where

$$Nu = \frac{hr}{k}$$

and h is the heat transfer coefficient, k is the thermal conductivity of the fluid and Nu_0 is the Nusselt number for pure diffusion. These

expressions are for a particle immersed in a shear flow produced by the turbulence of the fluid. It is assumed that there are no inertial forces or buoyancy forces acting on the particle that would cause additional movement of it relative to the fluid.

If the crystal size is large relative to the Kolmogorov length scale, it is in the inertial regime. The ambient velocity can be characterized in many different ways, each corresponding to a different eddy size. It seems reasonable to assume, following Wadia (1974), that the predominant shear that the particle will experience will be produced by eddies closest to the particle that are of the same size as the particle. Eddies that are significantly larger than the particle will entrain both the particle and the fluid around it. Very small eddies relative to the particle size may enhance the overall transport by some mechanism of renewal of the boundary layer surrounding the particles, but it is eddies of a size comparable to the size of the particle that will cause the most significant gradients near the crystal surface. The Reynolds number of the particle motion will be

$$Re = \frac{r^{4/3} \epsilon^{1/3}}{\nu} \tag{14}$$

and the Péclet number

$$Pe = \frac{r^{4/3} \epsilon^{1/3}}{\alpha} \ . \tag{15}$$

As $r > \eta$ in the inertial regime, the Reynolds number defined by eq. 14 is larger than 1, and the Peclet number is greater than the Prandtl number.

In the inertial subrange, a linear ambient velocity distribution does not exist. The Péclet number is large; therefore, gradients of temperature will exist only in small boundary layers near the particles. So, to determine the heat transfer from the particles, the Frössling equation will be used

$$Nu = Nu_0 + 0.70 \left(\frac{r^{4/3} \epsilon^{1/3}}{\nu} \right)^{1/2} Pr^{1/3} \ . \tag{16}$$

The small-scale motion, smaller than the size of the crystal, may enhance the heat transport from the crystal by penetrating the boundary layer around the crystal. It is difficult to quantify this process but this enhancement has been successfully accounted for (although empirically) by correlation with the turbulent intensity, α_T, of the fluid. α_T is defined as

$$\alpha_T = \sqrt{u'^2}/\bar{v}_f \tag{17}$$

where $\sqrt{u'^2}$ is the rms value of the velocity deviation from the mean velocity $\bar{v}_f$. The experiments of Lavender and Pei (1967) demonstrated that the Frössling equation could be written as

$$Nu = Nu_0 + 0.44 \, \alpha_T^a \, Re^{(1/2+\alpha)} Pr^{1/3} \qquad (18)$$

where a is an experimentally derived coefficient. They found that for

$\alpha_T \, Re < 1000$, $a \approx 0.035$

and

$\alpha_T \, Re > 1000$, $a \approx 0.25$

with the break occurring at the point at which the boundary layer becomes turbulent. Therefore, having $\alpha_T \, Re < 1000$ gives

$$Nu = Nu_0 + 0.70 \, \alpha_T^{0.035} \, Re^{0.535} \, Pr^{1/3} \qquad (19a)$$

and having $\alpha_T \, Re > 1000$ gives

$$Nu = Nu_0 + 0.70 \, \alpha_T^{0.025} \, Re^{0.75} \, Pr^{1/3} \quad . \qquad (19b)$$

For $\alpha_T \, Re > 1000$, Nu is essentially independent of the crystal size. This range is where industrial crystallizers generally operate and the independence of the transfer rate and particle size is often seen.

At this point the heat transfer equations developed earlier will be applied to ice crystals suspended in turbulent water. As the density of ice crystals is different from that of water $[(\rho_i - \rho_f/\rho_f) \approx 8\%]$, the crystals are subject to gravitational and inertial forces that give them a translational motion relative to the fluid. Daly (1984) showed that the magnitude of the translational motion due to inertial forces was not significant. The motion due to gravity may be significant for large crystals at low levels of turbulence.

The Nusselt number for heat transfer from suspended crystals is shown in Figure 3. The Nusselt number is calculated using the properties of water at 0°C, and a turbulent intensity of 0.2 is assumed. The heat transfer from spheres and disks is shown. The relevant size of the particle has been nondimensionalized by the dissipative scale η. Also shown are the heat transfer rates that would occur from a particle at its terminal velocity under the influence of gravity.

A method of determining the Nusselt number that provides an intuitively easier means of seeing the relative value of the actual heat transfer coefficient is as follows. Let Nu_T be the turbulent Nusselt number defined as

$$Nu_T = \frac{h\eta}{k} = \frac{h}{k} \left(\frac{\nu^3}{\epsilon}\right)^{1/4} \quad .$$

Let $m^* = r/\eta$. The heat transfer relationships are then: for $m^* < 1/(Pr)^{1/2}$

$$Nu_T = (1/m^*) + 0.17 \, Pr^{1/2} \qquad (20a)$$

for $1/(Pr)^{1/2} < m^* < \approx 10$

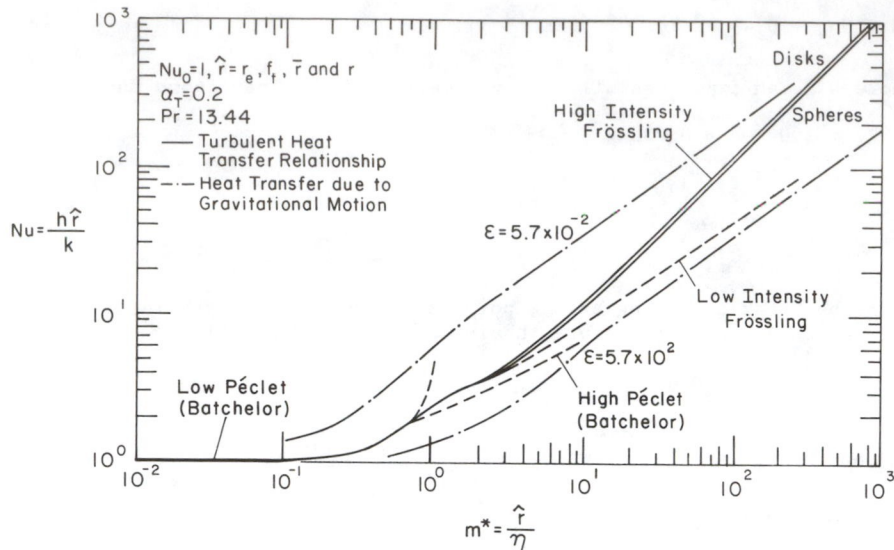

FIGURE 3. Nondimensional heat transfer correlation.

$$Nu_T = [(1/m^*) + 0.55 \ (Pr/m)^{1/3}] \tag{20b}$$

for $m^* > 1$, with a low intensity $\alpha_T m^{*4/3} < 1000$

$$Nu_T = [(1/m^*) + 0.70\alpha_T^{0.035} \ (Pr/m)^{1/3}]\beta \tag{20c}$$

for $m^* > 1$, with a high intensity $\alpha_T m^{*4/3} > 1000$

$$Nu_T = (1/m^*) + 0.70 \ \alpha_T^{0.25} \ Pr^{1/3} \tag{20d}$$

where $\beta = 1.0$ for a sphere and 1.1 for a disk. These Nusselt number relationships are shown in Figure 4.

NUCLEATION

Initial Nucleation

Researchers have thought that frazil ice may form by three types of nucleation: homogeneous nucleation, heterogeneous nucleation and secondary nucleation. Secondary nucleation results irrespective of its mechanism, only because of the presence of ice crystals in the super-cooled liquid. The importance of secondary nucleation to frazil ice formation has long been recognized (Altberg 1936). It is begging the question, however, to use secondary nucleation as the entire explanation for the existence of frazil ice. Undoubtedly, secondary nucleation plays the major role in increasing the total numbers of frazil crystals; it will be discussed later. The object here is to discuss the source of the original frazil crystals. The original crystals added to industrial crystallizers to begin secondary nucleation are called seed crystals.

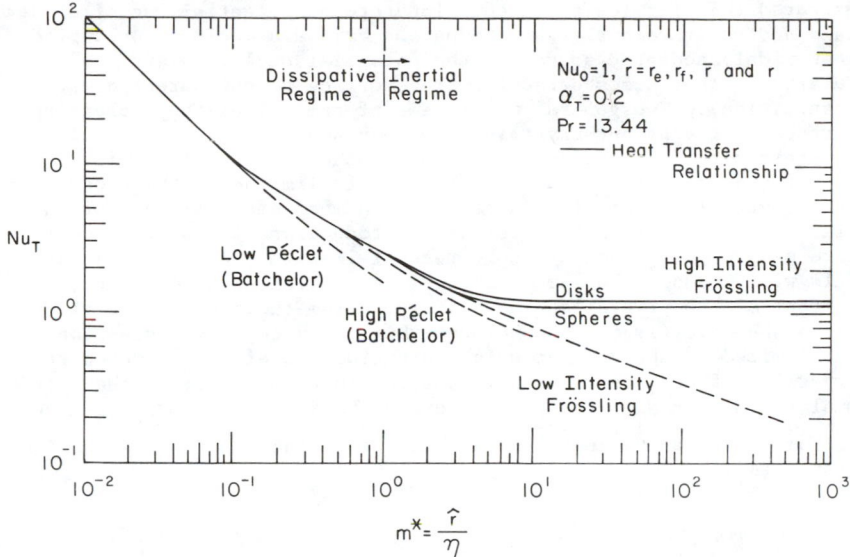

FIGURE 4. Nondimensional heat transfer correlation based on a turbulent Nusselt number.

Are seed crystals necessary to start the formation of frazil ice in natural water bodies? If so, where and how do they originate?

All the available data indicate that spontaneous nucleation of ice is not possible in natural water bodies that are producing frazil; therefore, seed crystals are necessary. The seed crystals may come from outside the water body or from ice already in it. There are many situations where frazil ice has been observed in waters in which ice had not existed previously or where the ice was far from the zone producing frazil. Osterkamp (1977) proposed a mass exchange across the air/water interface as the mechanism providing the seed crystals in these cases. The seed crystals are ice crystals that have a variety of origins. Water that originated in the water body and was thrown into the air by bubble bursting, splashing, wind spray, evaporation, etc., can freeze and return in the form of ice crystals. It is interesting to note that often minimum air temperatures of -9 to -8°C are reported as necessary for the production of frazil. These temperatures correspond to the minimum temperatures at which spontaneous heterogeneous nucleation could be expected in water particles suspended in air. Ice particles that originated at some distances from the water body, such as snow, frost, ice particles from trees, shrubs, etc., could be effective seed crystals. Very cold soil particles and cold organic materials at temperatures lower than the supercooling level necessary to cause spontaneous nucleation can also be introduced across the air/water interface and may serve to nucleate ice, although their effectiveness is not known.

Secondary Nucleation

The processes that govern the rates of secondary nucleation are poorly understood. However, a partial modeling of the kinetics of secondary nucleation is possible based on the work of Evans et al. (1974a,b), who demonstrated that for ice the production rate of potential nuclei of new crystals and their removal from the parent crystals could be uncoupled. The most widely accepted source of the potential nuclei is surface irregularities that are sheared from the surface of the parent crystals (microattrition). Two general mechanisms of removal of the nuclei from the surface of the parent crystals have been suggested: collisions of the crystals with hard surfaces (including other crystals) and fluid shear. If the rate of secondary nucleation is limited by the production rate of potential nuclei, increases in the number of collisions of an individual crystal will not increase the production of new crystals. If the rate of secondary nucleation is removal-limited, however, the parent crystals will produce the same number of new nuclei each collision, independent of the crystal's time history. From their experimental work, Evans et al. (1974a,b) concluded that the secondary nucleation of ice was limited by the rate at which potential nuclei were removed from the crystal surface. Therefore, it was possible to determine the overall nucleation rate $\dot{N}_T$, with two or more mechanisms of removal, as the linear sum of the actual nucleation rate attributable to each mechanism of removal ($\dot{N}_i$),

$$\dot{N}_T = \dot{N}_1 + \dot{N}_2 + \ldots \dot{N}_i \quad . \tag{21}$$

The nucleation rate of each mechanism of collision can be expressed as the product of three functions (Botsaris 1976)

$$\dot{N}_T = (\dot{E}_t)(F_1)(F_2) \tag{22}$$

where

$\dot{E}_t$ = rate of energy transfer to crystals by collision

F_1 = number of particles generated per unit of collision energy

F_2 = fraction of particles surviving to become nuclei.

At this time the values of F_1 and F_2 must be determined empirically. Therefore, to simplify matters let F_1 and F_2 be combined and eq. 22 be rewritten as

$$\dot{N}_T = \dot{E}_t S_N \tag{23}$$

where $S_N = (F_1)(F_2)$.

We expect that S_N is a function of all the parameters that govern the surface morphology and the crystal growth, including supercooling θ, impurity concentrations C_T, turbulence level ϵ, etc. The total nucleation rate can be expressed as

$$\dot{N}_T = S_N(\theta, \epsilon, C_T, \text{ etc.})(\dot{E}_{t1} + \dot{E}_{t2} + \dot{E}_{t3} \ldots) \quad . \tag{24}$$

The next parts of this section will focus on determining the rate of energy transfer for each mechanism of collision. Two general classes of crystal collisions can be identified: collisions between crystals in suspensions (crystal-crystal collisions) and collisions between crystals and external boundaries (crystal-boundary collisions).

We will assume that the nuclei produced by collisions are effectively at zero size. Therefore, the internal coordinate of the parent crystal remains unchanged during a collision. We also assume that large-scale breakage of crystals, which has not been observed in natural water bodies or agitated crystallizers, does not happen during a collision.

We will assume all crystal-crystal collisions to be between two crystals only. Three types of collisions between crystals can be identified:

1. Collisions from the crystals moving with the fluid. These collisions are caused by spatial variations in the fluid motions.

2. Collisions from the crystals moving relative to the fluid. These are collisions caused by the crystals' inertia. It has been shown (Daly 1984) that the velocity of the crystal caused by inertia forces was always much less than the velocity determined by the shear rate. Therefore, collisions caused by inertia will not be considered further.

3. Collisions caused by buoyancy.

Note that only type 1 will cause collisions between crystals of similar size.

The rate of energy transfer to the crystals by collision can be determined following the method of Evans et al. (1974b). $\dot{E}_t$ is the product of the collision energy $E(r_1, r_2)$ and the frequency of collision $q(r_1, r_2)$ between crystals of size r_1 and r_2, integrated over the crystal size distribution. Thus

$$\dot{E}_t = \int_0^\infty \int_0^\infty q(r_1, r_2) E(r_1, r_2) n(r_1) n(r_2) dr_1 \, dr_2 \ . \tag{25}$$

For simplicity, the crystal size distribution function, n, will be written as a function of the crystal size only. We determine the collision frequency per unit volume between crystals of size r_1 to $r_1 + dr_1$ and size r_2 to $r_2 + dr_2$ as follows. Let $R_c = (r_1 + r_2)$ be the "collision radius," and $\psi(r_1, r_2)$ be the collision efficiency of crystals of size r_1 and r_2. The collision efficiency is defined as the portion of crystals that would have collided if the fluid flow field hadn't been distorted by the crystals. Assume that the coordinate system is centered on one particle and is moving with the steady velocity of the particle. Let $\nu(r_1, r_2)$ be the relative motion between the crystals in this coordinate system. Thus

$$q(r_1, r_2) = \pi (r_1 + r_2)^2 \, \nu(r_1, r_2) \psi(r_1, r_2) \tag{26}$$

and by this definition, $\nu(r_1, r_1) \geq 0$.

The rate of energy transfer can now be expressed assuming totally inelastic collisions as

$$\dot{E}_t \int_0^\infty \int_0^\infty \frac{\pi}{2} (r_1+r_2)^2 \frac{m(r_1)m(r_2)}{[m(r_1)+m(r_2)]} \psi(r_1,r_2)\nu^3(r_1,r_2)n(r_1)n(r_2)dr_1dr_2 \quad (27)$$

and it now remains for us to determine the relative velocity and colli- sion efficiency applicable to each mechanism of collision.

First we will determine the energy resulting from collisions from the crystals moving with the fluid. For this analysis we will assume that the crystals exactly follow the fluid motion, that is, all inertial buoyancy and other effects are ignored. It will be necessary to analyze the dissipative and inertial subranges separately.

If $R_c < \eta$, then we can estimate the rate of energy transfer, assuming that both crystals are in the dissipative subrange. The mean square relative velocity between two points separated by a distance R_c, where $R_c < \eta$, is

$$v(r_1,r_2) = 0.13 \ R_c(\epsilon/\nu)^{1/2} \quad . \quad\quad (28)$$

If $R_c < \eta$, then the relative velocity cannot be estimated as neatly as above. It may be that r_1 or r_2 or both are greater than η. We will assume that if $R_c > \eta$, then the crystals will move with a relative velocity appropriate to the inertial subrange, regardless of each individual crystal's size. The mean square velocity difference between points separated by a distance R_c, where $R_c > \eta$, is

$$\nu(r_1,r_2) = 2.7 \ (\epsilon R_c)^{1/3} \quad . \quad\quad (29)$$

The collision efficiency $\psi(r_1,r_2)$ is difficult to estimate. For a dis- cussion of the collision efficiency of suspended particles see Mercier (1984). For particles in a shear flow, the efficiency is a function of the radius ratio of the particles, the strain rate of the fluid, the dynamic viscosity of the fluid and the strength of the Van der Waal's force.

Substituting the above expressions into the rate of energy transfer, we see that for $R_c < \eta$

$$\dot{E}_t = (0.003)(\epsilon/\nu)^{3/2} \int_0^\infty \int_0^\infty (r_1+r_2)^5 \frac{m(r_1)m(r_2)}{[m(r_1)+m(r_2)]} n(r_1)n(r_2)dr_1dr_2 \quad (30)$$

and for $R_c > \eta$

$$\dot{E}_t = (30.92)(\epsilon) \int_0^\infty \int_0^\infty (r_1+r_2)^3 \frac{m(r_1)m(r_2)}{[m(r_1)+m(r_2)]} n(r_1)n(r_2)dr_1dr_2 \quad . \quad (31)$$

When the density of the crystals is not identical to that of the fluid, the crystals will move relative to the fluid under the influence of gravity. This motion will be in addition to all other motions of the crystals. Again, for a discussion of the collision efficiency of crystals' collisions attributable to differential rising, see Mercier (1984). The efficiency is a function of the particle radius ratio and the radius of the smaller particles. Using the intermediate law for calculating the drag coefficient, we find that

$$\dot{E}_t = (0.16 \ g'^{0.715}/\nu^{0.428})^3 (\pi/2) \int_0^\infty \int_0^\infty (r_1+r_2)^2 \ \frac{m(r_1)m(r_2)}{[m(r_1)+m(r_2)]}$$

$$(|r_1^{1.14} - r_2^{1.14}|)^3 \ n(r_1)n(r_2)dr_1dr_2 \quad . \tag{32}$$

The second general class of collisions are those between the crystals and the external boundaries. To determine the frequency of crystal-boundary collisions requires knowledge of the size and shape of the water body of interest. As this chapter is intended to be as general as possible, this knowledge cannot be assumed; however, this section is presented in the interest of completeness.

The experiments of Evans et al. (1974b) demonstrated that collisions between ice crystals and the walls, baffles and impeller of the crystal-lizer were a significant cause of secondary nucleation. They assumed that any crystal moving with the bulk flow could potentially collide with the impeller, and any crystal closer than an eddy size away from the wall could collide with the wall. They were unable to determine if either collision mechanism dominated, but they did find that coating the metal surfaces of the impeller and crystallizer with a soft material substantially reduced the nucleation rate.

In natural water bodies, collisions with external boundaries are probably not a significant cause of secondary nucleation, except in some circumstances. First of all, the ratio of surface area to volume of most natural water bodies is much smaller than that of a crystallizer. Also, the boundaries of natural water bodies tend to be very rough compared to crystallizers. Crystals colliding with these rough boundaries tend to stick and remain at the boundary, which results in the buildup of anchor ice. Therefore, only in locations such as shallow, rocky rapids could crystal-boundary collisions be important.

The rate of energy transfer during the collision of a crystal with a boundary may be estimated in the following manner. This analysis does not account for the effects of a boundary layer, inertia, etc. Let ℓ_m be the maximum eddy size that can bring crystals into contact with the boundary. Assume that the number density of crystals is uniform throughout the region. The probability that a crystal of size r exists in the region within a distance close enough to collide with a boundary is

$$\text{prob.} \approx \frac{\ell_m}{R_h} \int_0^\infty n(r)dr \tag{33}$$

where R_h is the hydraulic radius, or the ratio of volume to surface area. The relative velocity of the crystal and the boundary can be estimated as

$$U_{rb} \approx (\epsilon \, \ell_m)^{1/3} \qquad\qquad (34)$$

and the energy at collision

$$E_{rb} = 1/2 \; m(r)\nu^2_{rb} = 1/2 \; m(r)(\epsilon \, \ell_m)^{2/3} \; . \qquad\qquad (35)$$

Now the rate of collisions will be proportional to

$$q_{rb} \approx \frac{(\epsilon \, \ell_m)^{1/3}}{\ell_m} \; \frac{\ell_m}{R_h} \int_0^\infty n(r)dr \qquad\qquad (36)$$

and thus

$$\dot{E}_t \approx \frac{\epsilon}{2} \frac{\ell_m}{R_h} \int_0^\infty m(r)n(r)dr \; . \qquad\qquad (37)$$

Application of Model

Mercier (1984) has developed a general comprehensive model of the reactive transport of suspended particles in a turbulent fluid. In a particular application of this model, he developed a dynamic model of frazil ice growth based on the formulation presented in this chapter. To his general reactive transport model, Mercier added separate sub-models for the effects of seeding, secondary nucleation, crystal growth and flocculation. To model the flocculation of frazil crystals, a general coagulation equation was employed with turbulent shear and differential rising considered as collision mechanisms. The collision efficiency was determined based on theoretical considerations. A zero dimensional model was developed initially. This model assumes that the frazil formation occurs in a well-mixed region without gradients of any parameter or variable. The initial seeding rate, and the number of crystals produced per collision, were optimized based on the laboratory results of Michel (1963). Without modifying these parameters, Mercier was able to reproduce the data of both Michel and Carstens (1966) very well.

Mercier also modeled the cases of vertical mixing with unequal transport of heat and mass in the vertical direction, and vertical distributions of the turbulent kinetic energy, the energy dissipation rate and vertical eddy diffusivity. He found that compared to the zero dimen-sional case, vertical mixing delays the peak in maximum supercooling and causes it to be larger in magnitude. Mercier determined the crystal size distributions and profiles of mass concentrations at various times and depths. His simulations are by far the most complete to date.

SUMMARY

In this chapter a physically based quantitative model of frazil ice was described. Two equations serve as the basis of the model: the crystal number continuity equation and the heat balance for a differential volume. One focus of the chapter has been to describe in detail two of the major parameters that appear in this model: the crystal growth rate and the secondary nucleation rate. Both involve complex interaction of the crystals with the fluid turbulence. Three basic environmental parameters and one material property control the evolution of the crystal size distribution function. These environmental parameters are the heat loss rate, the turbulent energy dissipation rate of the fluid, and the crystal seeding rate; the material property is the number of crystal produced per unit of collision energy. A tremendous amount of literature describes the first two of these parameters, and virtually none the third; a modest literature is available on secondary nucleation, with very little describing ice.

The importance of a quantitative model such as described here is that it allows predictions to be made that can be tested in laboratory experiments. By controlling or measuring the environmental parameters in experiments, knowledge of the resulting size distributions would allow estimates of the growth rate and secondary nucleation rate to be tested. Unfortunately, the measurement of the crystal size distribution is quite difficult, even in the laboratory. However, such tests would advance our fundamental understanding of frazil ice evolution, and the efficient solution of practical problems could not be far behind.

REFERENCES

Altberg, W.J. (1936) Twenty years of work in the domain of underwater ice formation (1915-1935). International Union of Geodesy and Geophysics, International Association of Scientific Hydrology, Bulletin No. 23, pp. 373-407.

Arakawa, K. (1954) Studies on the freezing of water. II. Formation of disc crystals. Journal of the Faculty of Science, Hokkaido University, Series II, IV(5):310-339.

ASCE Task Committee on Hydromechanics of Ice of the Committee on Hydromechanics (1974) River ice problems: A state-of-the-art survey and assessment of research needs. Journal of the Hydraulics Division, ASCE, HY1: 1-15.

Batchelor, G.K. (1980) Mass transfer from small particles suspended in turbulent fluid. Journal of Fluid Mechanics, 98(3):609-623.

Botsaris, G.D. (1976) Secondary nucleation - A review. In Industrial Crystallization (J.W. Mullin, Ed.). New York: Plenum Press.

Carstens, T. (1966) Experiments with supercooling and ice formation in flowing water. Geofysiske Publikasjoner, 26(9):3-18.

Daly, S.F. (1984) Frazil ice dynamics. USA Cold Regions Research and Engineering Laboratory, CRREL Monograph 84-1.

Daly, S.F. and S. Colbeck (1986) Frazil ice measurements in CRREL's flume facility. International Association for Hydraulic Research Ice Symposium, 1986, Iowa City, Iowa.

Evans, T.W., G. Margolis and A.F. Sarofim (1974a) Mechanisms of secondary nucleation in agitated crystallizers. American Institute of Chemical Engineers Journal, 20(5):950-958.

Evans, T.W., A.F. Sarofim and G. Margolis (1974b) Models of secondary nucleation attributable to crystal-crystallizer and crystal-crystal collisions. _American Institute of Chemical Engineers Journal_, 20(5):950-958.

Kumai, M. and K. Itagaki (1953) Cinematographic study of ice crystal formation in water. _Journal of the Faculty of Science, Hokkaido University_, Series II, IV(4):234-246.

Lavender, W.J. and D.C.T. Pei (1967) The effect of fluid turbulence on the rate of heat transfer from spheres. _International Journal of Heat and Mass Transfer_, 10:529-539.

Margolis, G. (1969) The nucleation and growth rates of ice in a well-stirred crystallizer. Ph.D. Dissertation. Department of Chemical Engineering. Cambridge: Massachusetts Institute of Technology.

Mercier, R. (1984) The reactive transport of suspended particles: Mechanisms and modeling. Ph.D. Dissertation, Joint Committee on Oceanographic Engineering. Cambridge: Massachusetts Institute of Technology.

Michel, B. (1963) Theory of formation and deposit of frazil ice. _Eastern snow Conference, Proceedings of the 1963 Annual Meeting,_ Quebec City.

Michel, B. (1971) Winter regime of rivers and lakes. USA Cold Regions Research and Engineering Laboratory, Cold Regions Science and Engineering Monograph III-B1a.

Osterkamp, T.E. (1977) Frazil-ice nucleation by mass-exchange processes at the air/water interface. _Journal of Glaciology_, 19(81):619-625.

Osterkamp, T.E. and J.P. Gosink (1982) A photographic study of frazil and anchor ice formation, frazil ice evolution and ice cover development in interior Alaska stream. Fairbanks: Geophysical Institute, University of Alaska.

Smith, K.A. and A.F. Sarofim (1979) Fundamental studies of desalination by freezing. Final Report to Office of Water Research and Technology, U.S. Department of Interior, Washington, D.C.

Wadia, P.H. (1974) Mass transfer from spheres and discs in turbulent agitated vessels. Ph.D. Dissertation. Department of Chemical Engineering. Cambridge: Massachusetts Institute of Technology.

Williamson, R.B. and B. Chalmers (1966) Morphology of ice solidified in undercooled water. In _Crystal Growth_ (H.S. Peiser, Ed.). New York: Pergamon Press.

GROUND FREEZING
AND FROST HEAVE

Chapter 17

Frost Heaving in Artificial Ground Freezing

TAKAHIRO OHRAI and HIDEO YAMAMOTO
Seiken Co., Ltd., 2-11-16, Kawarayamachi,
Chuo-ku, Osaka, Japan

ABSTRACT

Artificial ground freezing methods have been applied to geotechnical
construction projects for stabilizing earth materials and controlling
water seepage into the ground. However, this can result in frost heaving
and causes the same engineering problems as encountered with the natural
freezing of soil.

In natural freezing, the ground freezes from the surface downward. When
artificial ground freezing is applied at a deep location, however,
freezing is limited locally. The soil condition differs between them as
follows:
Natural freezing — unsaturated and without overburden pressure,
Artificial freezing -- saturated and under overburden pressure.
The authors investigated the practical application of artificial ground
freezing and examined the frost behaviour of a saturated soil under
overburden pressure. This paper presents the results obtained from
experiments concerning frost heaving and discusses frost heaving at the
freezing site.

CONTENTS

1. INTRODUCTION

Cold regions, where permafrost prevails and the ground thaws only to a
shallow depth in summer, occupy about 20 per cent of the world's land
area. The depth that is subjected to a freeze-thaw cycle is called the
active layer. Frost heaving in cold regions make development difficult
and cause damage to construction including roads, railways, airports,
and petroleum pipelines. Many scientists and technologists conducting
research on permafrost have looked into its fundamentals as well as ways
to prevent damage and take countermeasures against frost heaving.

In Japan, artificial ground freezing has been applied to geotechnical
construction projects in order to stabilize earth materials and control
ground water seepage. It is frequently applied to urban civil engineering
in which buildings are crowded and underground structures abound, i.e.,
gas and sewage pipes, groups of buildings, etc.; so allowable
displacement against them must not exceed 1 cm. Artificial ground
freezing utilizes the strength and practical impermeability of frozen
ground, but can result in frost heaving in the same way as in natural
freezing.

When the ground freezes from the surface downward, the amount of frost
heave appears mostly on the surface. When artificial ground freezing is
applied at a deep location, however, freezing is limited locally,
resulting in a compound behaviour; that is, the increase in pressure in
the unfrozen soil around the frozen location is partly absorbed in the
unfrozen soil and partly appears as a heave of the ground surface.
Accordingly, much information should be available for predicting or
estimating the displacement in an area of unfrozen soil apart from this
location, the increase in soil~pressure, and the heave amount at the
surface. The principle of frost heaving remains the same whether it is
caused by natural or artificial freezing of the ground. Therefore, when
the ground is frozen artificially, analyzed data obtained from the
results of research into the natural freezing of the ground can be
applied to engineering. However, there are problems with some engineering
applications which require more investigation.

The following are the three major differences between natural and
artificial ground freezing:
i) Saturation of soil
    natural freezing : unsaturated.
    artificial freezing: saturated because the soil to be frozen is below
                    the ground water table.
ii) Overburden pressure
    natural freezing : < 1 kgf/cm^2 because the active layer is
                    above a depth of several meters.
    artificial freezing : from zero to several tens of kgf/cm^2.
iii) Penetration rate
    natural freezing : < 2.5 mm/h
    artificial freezing : < 50 mm/h.

The authors are engaged in the practical application of artificial ground freezing, and have been investigating the behaviour of a soil saturated with water and under pressure. This paper presents the results of their investigation. Section Two outlines artificial freezing and the freezing process in the ground; Section Three shows the results of laboratory experiments concerning frost heaving; and Section Four discusses frost heaving at the freezing site.

Most of the research described in this paper was conducted under the guidance of the late Dr. Takashi, whose assistance is gratefully acknowledged.

2. DESCRIPTION OF ARTIFICIAL GROUND FREEZING

2.1 Refrigeration Plant

The installation for artificial ground freezing consists of two parts: freeze pipes and a refrigeration system. The freeze pipe, which is a metal pipe usually 10 cm in diameter, is installed at appropriate spacings in the ground to be frozen, and is cooled to a temperature lower than the freezing point of the water in the soil. Soil surrounding the freeze pipe is cooled and then frozen gradually as the temperature of the soil water falls below the freezing temperature, thus forming a thick frozen soil column. As time goes by, the frozen soil columns merge with the adjacent ones, forming a solid frozen soil wall. Since the frozen soil wall is impermeable and mechanically strong, it can be used as a temporary logging for conducting civil engineering work. Upon completion of the work, the cooling of the freeze pipes is stopped, allowing the frozen soil wall to thaw gradually. When necessary, the frozen soil wall may be thawed rapidly by circulating hot water through the freeze pipes.

Two methods are available to cool the freeze pipes: the brine method and liquid nitrogen (LN_2) method, as shown in Figure 1.

The brine method employs a solution of brine, usually Calcium Chloride

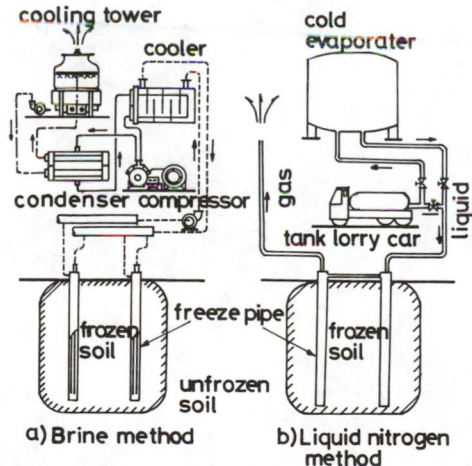

FIGURE 1. Freezing installations.

(CaCl$_2$). The solution is cooled from -20 °C to -30 °C by a refrigeration
unit, and then circulated through the freeze pipes by a pumping system,
which removes heat from the surrounding soil. The brine is returned by
collector pipes to the refrigeration unit, cooled and then sent to the
freeze pipe again. The refrigeration unit consists of a compressor, a
condenser and a cooler; the evaporating liquid usually used in Japan is
Freon Gas (R22). The volume of frozen soil per a refrigeration unit with
75 kw capacity, ranges from 400 m^3 to 1000 m^3. This system is widely
applied in large-scale freezing work which requires from 200 m^3 to 35000
m^3 in volume of frozen soil (Ohrai et al., 1985).

The liquid nitrogen method, shown in Figure 1b, uses liquid nitrogen,
which is supplied to the freeze pipes through a distributor pipe from an
LN$_2$ tank or tank lorry loaded with LN$_2$. The soil surrounding the freeze
pipes is frozen mainly by the latent heat (38.5 kcal/l) of LN$_2$ and
partly by the sensitive heat of gas (boiling point=-196 °C). Nitrogen gas
from the freeze pipes is exhausted directly into the atmosphere. Mainly
for economic reasons, this method is applied to small scale, short-term
work, in which the volume of soil subjected to freezing is less than 200
m^3.

2.2 Thermal Analysis of Artificial Freezing of Ground

<u>Freezing of soil surrounding freeze pipes.</u> The freezing of soil
surrounding the freeze pipes advances along a cylindrical front until the
frozen soil columns merge with adjacent ones, and then freezing advances
along a plane. When the soil freezes, frost heaving or water migration or
both should occur, whether the soil is frost-susceptible or not. Thermal
analysis involving both water and heat presents a complex problem for the
freezing of soil surrounding the freeze pipes. For many practical
applications of artificial ground freezing, we use a model of heat
conduction which neglects water migration, since there is little water
migration relative to overburden pressure.

The equations of heat conduction, when expressed using the cylindrical
coordinates as shown in Figure 2, are:

$$\frac{\partial \Theta_1}{\partial t} = \kappa_1 \left(\frac{\partial^2 \Theta_1}{\partial r^2} + \frac{1}{r} \frac{\partial \Theta_1}{\partial r} \right) \quad \text{for } a \leq r \leq R \tag{1}$$

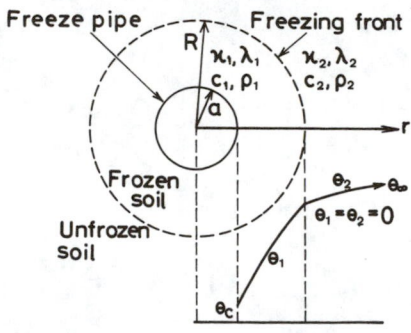

FIGURE 2. Calculation model.

$$\frac{\partial \Theta_2}{\partial t} = \kappa_2 \left(\frac{\partial^2 \Theta_2}{\partial r^2} + \frac{1}{r} \frac{\partial \Theta_2}{\partial r} \right) \quad \text{for } R \leq r \tag{2}$$

Initial condition is:

$$\Theta_2 = \Theta_\infty \quad \text{at } t = 0 \tag{3}$$

Boundary conditions are:

$$\Theta_1 = \Theta_c \quad \text{at } r = a \tag{4}$$

$$\Theta_2 = \Theta_\infty \quad \text{at } r = \infty \tag{5}$$

Boundary condition at the freezing front is:

$$\left(\lambda_1 \frac{\partial \Theta_1}{\partial r} - \lambda_2 \frac{\partial \Theta_2}{\partial r} \right)_{r=R} = L \rho_1 \frac{dR}{dt} \tag{6}$$

Where Θ: temperature
   r: radius coordinate,
   R: radius of frozen soil,
   t: time,
   κ: coefficient of thermal diffusivity,
   λ: thermal conductivity,
   L: latent heat,
   ρ: density of soil,
   suffixes 1, 2 and ∞: frozen soil, unfrozen soil and infinitive,
         respectively.

Takashi (1961) gave an exact solution for these equations. The radius of the frozen soil, R, is given by the following equation:

$$\lambda_1 \Theta_c \left[\frac{\frac{\partial F(\kappa_1, a; r, t)}{\partial r}}{F(\kappa_1, a; r, t)} \right]_{r=R} + \lambda_2 \Theta_\infty \left[\frac{\frac{\partial F(\kappa_2, a; r, t)}{\partial r}}{1 - F(\kappa_2, a; r, t)} \right]_{r=R} = L \rho_1 \left\{ \frac{\frac{\partial F(\kappa_1, a; r, t)}{\partial t}}{\frac{\partial F(\kappa_1, a; r, t)}{\partial r}} \right\}_{r=R} \tag{7}$$

where

$$F(\kappa, a; r, t) = -\frac{2}{\pi} \int_0^\infty e^{-(x/a)^2 t} \frac{J_0(xr/a) Y_0(x) - J_0(x) Y_0(xr/a)}{J_0^2(x) + Y_0^2(x)} \frac{dx}{x} \tag{8}$$

where J_0 and Y_0 are Bessel functions of the order zero and the integral order zero, respectively.

An example of this calculation is shown in Figure 3. According to this, a freeze pipe 10.2 cm in diameter will form a frozen soil column about 50 cm in radius and will be formed after one month when $\Theta_\infty = 14\,°C$ and $\Theta_c = -20\,°C$.

When a frozen wall is formed by the merging of adjacent columns, it advances forward as a freezing front. A practical example was given by Miyoshi et. al. (1975) where a structure was built below the bottom of a river at times by freezing the soil into the shape of a board. The freezing of a semi-infinite material by a freezing front in the shape of

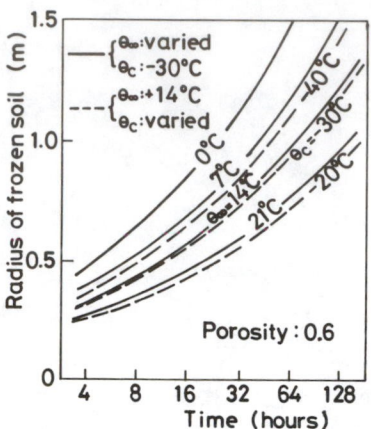

FIGURE 3. Change in radius of frozen column with time.

a board was examined by Neuman. He investigated the freezing of the still water surface of a lake in winter, and arrived at a solution now known as Neuman's solution (Carslow and Jaeger, 1959). In this case, the frost penetration length, X, is proportional to $\sqrt{t}$:

$$X = \alpha \sqrt{t} \tag{9}$$

where α is the constant determined from thermal properties of frozen and unfrozen soil, and temperature. For this equation, the soil type and water content must be taken into account. For example, when $\theta_c = -30\,°C$ and $\theta_\infty = 14\,°C$:

$$X = 0.0455 \sqrt{t} \quad [m] \tag{10}$$

That is, the soil freezes to the thickness of 1.2 m after about one month.

Freezing efficiency in soil freezing. We define the "freezing efficiency, η_f" as follows:

$$\eta_f = Q_L / Q_T \tag{11}$$

where Q_L: total quantity of latent heat (the quantity of latent heat used to freeze water contained in soil subtracted from the quantity of heat removed from the freezing front during a period from the beginning of cooling to the time of measurement)

Q_T: total quantity of heat consumed by the freezing of the soil. It was found that η_f is independent of the cooling interval when the freezing surface is a flat plane (Takashi, 1965). Figure 4 shows the relation between η_f and the temperature of the cooling surface. From this figure, we can find the maximum value of η_f, which increases with the decreasing temperature of the ground, θ_∞. Since the ground temperature at a shallow depth is from 14 to 18 °C in Japan, η_f reaches its maximum at the temperature of the cooling surface ranging from -20 to -30 °C. Therefore, the brine method is thermally more advantageous than the liquid Nitrogen method since the latter's freezing temperature is -196 °C.

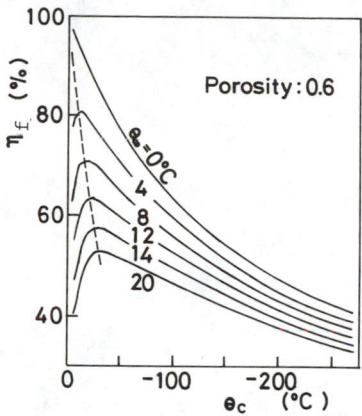

FIGURE 4. Freezing efficiency η_f as a function of temperature of cooling surface Θ_c.

Influence of seepage flow on artificial freezing of soil. In artificial soil freezing, the presence of a seepage stream disturbs the development and the merging process of frozen soil columns around the freeze pipes, due to heat gain from the seepage stream. This phenomenon has frequently occurred, especially where there are a large number of sites where the ground has frozen artificially.

Takashi (1969) has demonstrated a criterion for determining whether the frozen soil columns can merge with one another. As shown in Figure 5, a finite number of freeze pipes were arranged in a row, and the seepage flow was perpendicular to the plane formed by the pipes. As the soil froze, the flow velocity of the seepage increased as a result of an increase in the hydraulic gradient caused by the phenomenon of dam-up. The following equation presents the critical value, $U_{\infty \, crit}$, of the velocity of natural seepage in merging frozen soil columns, taking into account the dam-up phenomenon:

$$U_{\infty \, crit} = \frac{6 \pi M^2 \lambda_1}{(n/100)l_f \, \gamma_w c_w} \frac{\Theta_f - \Theta_c}{\Theta_\infty - \Theta_f} F(\frac{2a}{P_i}, b_{crit}) \tag{12}$$

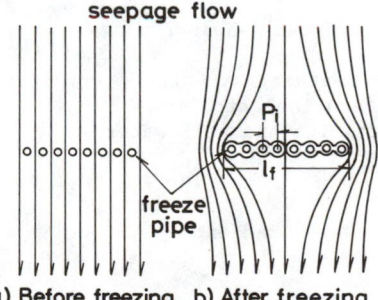

a) Before freezing   b) After freezing

FIGURE 5. Model of seepage flow during freezing.

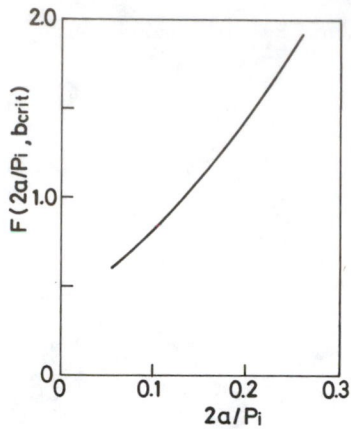

FIGURE 6. Relation between $F(2a/P_i, b_{crit})$ and $2a/P_i$

where $U_{\infty\ crit}$: critical flow velocity of seepage,
      M: number of rows of freeze pipes,
      λ_1: thermal conductivity of frozen soil,
      n: porosity,
      l_f: width of a frozen soil wall perpendicular to the direction
        of seepage flow,
      γ_w: unit weight of soil water,
      c_w: specific heat of soil water,
      Θ_f: freezing temperature of soil water,
      Θ_c: temperature of brine,
      Θ_∞ : temperature of soil water
      $F(2a/P_i, b_{crit})$: the function of radius, a, and spacing, P_i, of
                  the freeze pipes.
Figure 6 shows the relation between F and $2a/P_i$. Function F increases
linearly with an increase in $2a/P_i$. In Japan, $U_{\infty\ crit} = 2$ m/day for the
brine method and 10 m/day for the liquid nitrogen method.

3. FROST HEAVING IN SATURATED SOIL

3.1 Factors Affecting Frost Heaving

The following are the factors which affect frost heaving.
(A) Internal factors
    (1) physical properties of soil --- texture, structure, specific
        surface area, permeability, saturation, etc.
    (2) chemical properties of soil particles and soil water
    (3) history of stress and freeze-thaw
(B) External factors
    (1) overburden pressure
    (2) penetration rate and/or temperature gradient
    (3) pore water pressure

While scientists and technologists engaged in cold region research have
investigated natural freezing (group A), the authors have experimentally
investigated frost heaving with artificial freezing (group B), since this
will determine the effects of external factors on frost heaving. An

understanding of these effects is necessary before artificial ground freezing can be applied to civil engineering. This Section shows the results of their investigation.

The following are the three main definitions for describing the results of the quantities obtained from the experiments:

(1) Frost heave ratio, ξ:

$$\xi = \Delta V/V = \Delta h/H \tag{13}$$

where ΔV:frost expanded volume during freezing,
  V:volume of soil before freezing,
  Δh:length of frost heave,
  H:length of soil before freezing.

(2) Water intake or discharge ratio, ξ_w:

$$\xi_w = \Delta V_w/V = \Delta h_w/H \tag{14}$$

where ΔV_w = total volume of water intake or discharge during freezing,
  $\Delta h_w = \Delta V_w/A$ (A is the cross-sectional area of specimen).

(3) Frost penetration rate, U:
Provided that unfrozen soil of ΔH freezes during unit time, Δt:

$$U = \Delta H/ \Delta t \tag{15}$$

On the other hand, the thickening rate of the frozen soil, U_f, is:

$$U_f = (\Delta H + \Delta h)/ \Delta t = U + v_h \tag{16}$$

where v_h is the heave rate.
Although U and U_f have been frequently used without distinction, in this paper U defined by equation 15 is always used as the penetration rate or freezing rate.

3.2 Experimental Procedure

Apparatus. The experimental apparatus for determining the factors of group (B) which control frost heaving, must allow the overburden pressure, the penetration rate and the pore water pressure to take given values, and must also satisfy conditions which offer an open system to water migration.

Figure 7 shows the experimental apparatus which satisfies these conditions. The authors have conducted a series of experiments on frost heaving in a saturated soil using this apparatus. A soil specimen was contained in a transparent acrylite cylinder of $4 \sim 10$ cm inside diameter. The lower end was fixed on a cooling plate (negative temperature side) and the upper end was in contact with a porous plate mounted on a piston (positive temperature side), movable in the cylinder. Thermoelectric cooling devices (Peltier batteries) mounted on the cooling plate and the piston, allowed the temperature of each end plate to be readily controlled. Water intake to or discharge from the specimen was measured by the change in the water level in the water-supply container connected to the piston through an access pipe. Total pressure was applied using a spring. Pore water pressure was applied using air pressure in a pressure tank. A recorder or computer was used to record the temperature at each end plate, the amount of frost heave and the amount of water intake or discharge.

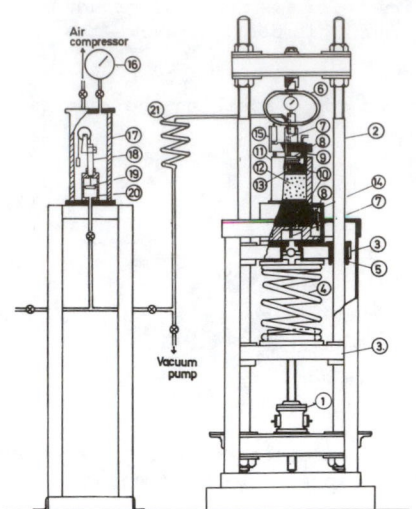

1 Jack
2 Support
3 Guide arm
4 Spring
5 Slide bearing
6 Proving ring
7 Thermister conduit
8 Thermo electric module
9 Piston
10 O-ring
11 Cylinder
12 Specimen
13 Porous plate
14 Cooling plate
15 Differential transformer
16 Bourdon's gauge
17 Pressure container
18 Differential transformer
19 Float
20 Water container
21 Cu-pipe

Fig. 7 Schematic of test apparatus.

Control and confirmation of penetration rate. Frost heaving depends on the penetration rate, as will be described later. When the temperatures of the cooling plate and piston are constant but at different temperatures, the frost penetration rate dX/dt is obtained from equation 9:

$$dX/dt = \alpha/2\sqrt{t} \tag{17}$$

In this case dX/dt changes with time. In an experiment studying the relation between frost heave ratio and penetration rate, it is desirable to make the penetration rate constant, since making the temperature at each end plate constant is not suitable for this experiment.

It was possible to make the penetration rate constant, upon analysis of the temperature conditions for this purpose, by changing θ_1 and θ_2 with the lapse of time according to the following equation, where θ_1 is the temperature at one end of the sample in contact with the cooling plate and θ_2 is the temperature at the other end of it (Takashi and Masuda, 1975):

$$\theta_1 = \frac{\kappa_1}{\lambda_1} \left(\frac{\lambda_2}{\kappa_2} \frac{U + v}{U + v_h} \theta_\infty + \gamma_w L_w \frac{n_f U + v_w}{U + v_h} \right)[1 - \exp\{\frac{(U + v_h)^2}{\kappa_1} t\}] \tag{18}$$

$$\theta_2 = \theta_\infty [1 - \exp\{- \frac{U + v}{\kappa_2} (H - Ut)\}] \tag{19}$$

where U:penetration rate,
    v_h:heave rate,
    v_w:water intake or discharge rate,
    λ:thermal conductivity,
    κ:thermal diffusivity,
    θ_∞ :initial temperature of soil,
    H:specimen height,

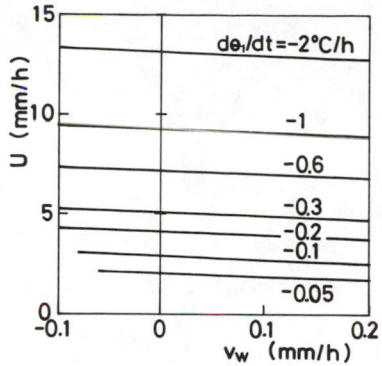

FIGURE 8. Penetration rate U as a function of water intake or discharge rate v_w.

$$v : (c_w \gamma_w / c_s \gamma_s) v_w, \text{ where } c = \text{specific heat and } \gamma = \text{unit weight}$$

suffixes 1 and 2: frozen and unfrozen soil, respectively
suffixes w and s: pore water and saturated soil, respectively

In the experiment, the value of U and H were made small and the temperature of the soil before freezing was adjusted to become equal to the freezing temperature of soil water; then $\theta_2 = 0$. As a result, equations 18 and 19 become approximately as follows:

$$\theta_1 = -\frac{\gamma_w L_w}{\lambda_1}(n_f U + v_w)(U + v_h)t \qquad (20)$$

$$\theta_2 = 0 \qquad (21)$$

Figure 8 shows the relation between v_w and U. Since the effect of U on v_w is relatively small and it is difficult to predict v_w before the experiment, we can control the temperature practically according to the following equation, ignoring v_w and v_h:

$$\theta_1 = -\frac{\gamma_w L_w n_f}{\lambda_1} U^2 t \qquad (22)$$

<u>Specimen's height in unidirectional freezing test of soil.</u> Takashi et al. (1976) analyzed the effect on ζ of resistance in soil water migration through the unfrozen part within a freezing specimen, and found that the frost heave ratio, ζ, is a function of the specimen's height, H

On the other hand, it has recently been considered that ice segregation occurs within a certain limited zone in frozen soil (Dirksen and Miller, 1966; Hoekstra, 1966; Miller, 1972; Radd and Oertle, 1973; Loch and Kay, 1978; Takashi et al., 1979; Fukuda et al., 1980; Penner and Goodlich; 1980). The author's view on the position of the zone where ice segregation occurs is shown schematically in Figure 9, where θ_f is the freezing point of bulk water, θ_{init} is the temperature at which ice segregation is initiated and θ_{crit} is the critical temperature at which ice segregation is completed. The plane $\theta = \theta_f$ is defined as a freezing front. The zone $\theta_{init} < \theta < \theta_f$ is an "in situ water freezing zone" or "frozen

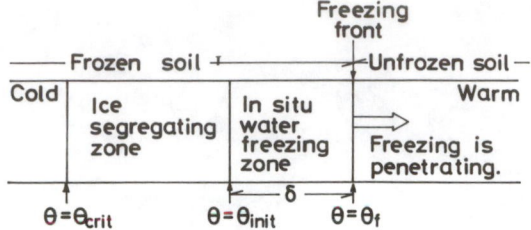

FIGURE 9. Schematic representation of unidirectional freezing of soil.

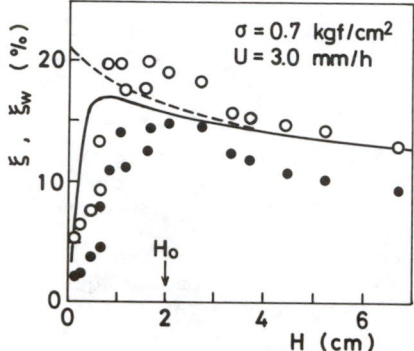

FIGURE 10. Dependence of frost heave ratio ζ and water intake or discharge ratio ζ_w and specimen height H (Manaitabashi clay).

fringe" defined by Miller (1972). According to this view point, the specimen's height has some influence on the frost heave ratio.

From the experiment using various heights of a specimen, Takashi et al. (1982) determined the relation between ζ, ζ_w and H, as shown in Figure 10. Here ζ reaches its maximum at a certain value H_O of the specimen's height H . When $H > H_O$, ζ decreases gradually with increasing H. Meanwhile, ζ decreases rapidly with decreasing H when $H < H_O$. When $H > H_O$, it is assumed that the effect on ζ of resistance in soil water migration through the unfrozen soil appears as H becomes smaller than H_O; the reason for the decrease in ζ may be explained qualitatively from the existence of a frozen fringe.

From the experiment concerning various values of σ and U, it was found that the value of H_O increased with increasing σ and with decreasing U. It is suggested from the experiment that the optimum H in laboratory frost heave tests is near H_O. At this point both the resistance in soil water movement through the unfrozen soil and the existence of an in-situ water freezing zone within the frozen soil have the minimum compound effect on ζ.

3.3 Effect of overburden pressure on frost heave ratio.

It is not unusual that the frost heave ratio, ζ, exceeds 100 % in

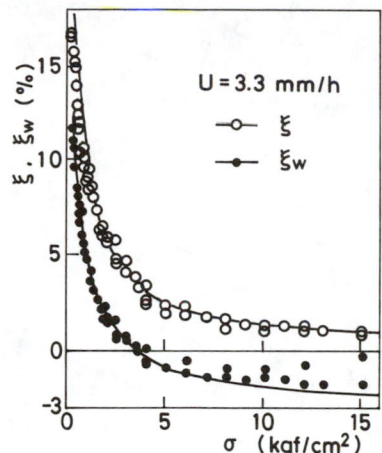

FIGURE 11. Dependence of frost heave ratio ξ and water intake or discharge ratio ξ_w on stress σ (Negishi silt).

natural freezing when the ground freezes from its surface downward. It is unusual, however, that ξ exceeds 10 % in artificial freezing applied to a fairly deep location underground. A reason that ξ differs between natural and artificial freezing is supposed to be caused by a difference in confining stress, σ.

The relation between frost heave ratio, ξ, water intake or discharge ratio, ξ_w, and confining stress, σ, for a Neghishi silt are shown in Figure 11 (Takashi et al., 1978). Values of ξ and ξ_w decrease with increasing σ and are in inverse proportion to σ, as shown by the following relation:

$$\xi = \xi_0 + \frac{C}{\sigma} \tag{23}$$

where ξ_0 and C are the constants; ξ_0 may be considered to represent the amount of expansion caused by the gradual freezing of unfrozen water in the frozen soil; C depends on the penetration rate, as will be mentioned later. Accordingly, it is of interest to note that with increasing σ the nature of water migration changes from water intake to water discharge.

As seen from Figure 11, when σ is smaller than 1 kgf/cm^2, both ξ and ξ_w have somewhat smaller values than those obtained from equation 23. It has become almost obvious, from the analysis and experiment, that such phenomena are due to an increase in the effective stress caused by a drop in pore water pressure in unfrozen soil as a result of intense water intake occurring at the freezing front (Takashi et al, 1976).

3.4 Effect of Penetration Rate on Frost Heave Ratio

Taber (1929, 1930) suggested qualitatively that with a decrease in frost penetration rate, U, the frost heave ratio increases. Since his work, many researchers have pursued this subject, arriving at varying

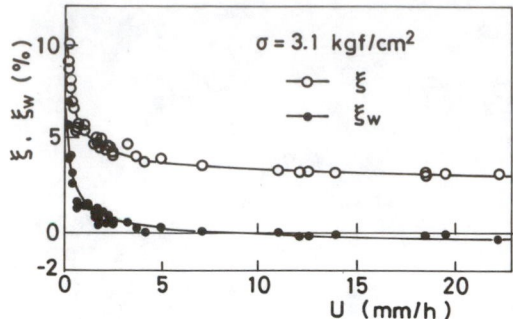

FIGURE 12. Dependence of frost heave ratio ξ and water intake or discharge ratio ξ_w on penetration rate U (Negishi silt).

conclusions, and in extreme cases, contradictory results.

The authors conducted an experiment to systematically investigate the effect of U on ξ, keeping U constant during the freezing of a specimen (Takashi et al., 1979). Figure 12 represents the relation between ξ, ξ_w and U for $\sigma=3.1$ kgf/cm^2 using a Negishi silt. Values of ξ and ξ_w decrease with increasing U. Especially, ξ_w changes from the water intake type to the water discharge type with increasing U. From this experiment, the following empirical equation was obtained:

$$\xi = A + \frac{B}{\sqrt{U}} \tag{24}$$

where A and B are the constants. The solid line in Figure 12 represents equation 24, which coincides with experimental results for a wide range of U.

When tests were conducted by changing U, it was found that ξ is inversely proportional to σ, but there was a difference in the value of C. It is predicted that C depends upon U, that is:

$$\sigma(\xi - \xi_0) = C = f(U) \tag{25}$$

All data of Figures 11 and 12 expressed by plotting $1/\sqrt{U}$ against $\sigma(\xi-\xi_0)$, as shown in Figure 13, lead to a proportional relation between them. Therefore, we obtain:

$$\sigma(\xi - \xi_0) = C_1 - C_2/\sqrt{U} \tag{26}$$

where C_1 and C_2 are the constants dependent upon a soil type, but not on σ and U. Arranging the equation by putting $C_1=\sigma_0$ and $C_2= \sigma_0\sqrt{U_0}$, equation 26 is rewritten as:

$$\xi = \xi_0 + \frac{\sigma_0}{\sigma} (1 + \sqrt{\frac{U_0}{U}}) \tag{27}$$

where ξ_0, σ_0 and U_0 are the constants. The values of the soils used in the authors' experiments are given in Table 1.

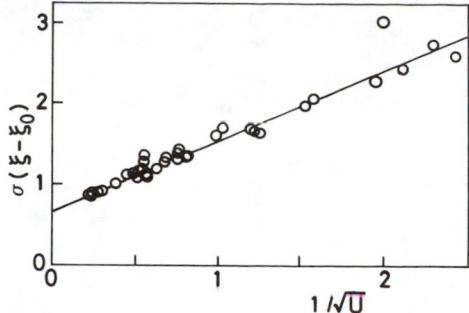

FIGURE 13. Relation between $\sigma(\xi - \xi_0)$ and $1/\sqrt{U}$.

TABLE 1. Properties of specimen.

		Nanao silt	Negishi silt	Manaitabashi clay
soil fraction				
sand	(%)	28	27	0
silt	(%)	65	50	33.5
clay	(%)	7	23	66.5
frost heaving				
ξ_0	(−)	0.0055	0.003	0.01
σ_0	(kgf/cm^2)	0.0255	0.0667	0.0179
U_0	(mm/h)	2.941	1.71	137.02
n_f	(−)	0.372	0.348	0.167
consolidation				
k	(cm/s)	4×10^{-8}	3.3×10^{-8}	7×10^{-8}
m_v	(cm^2/kgf)	3.3×10^{-3}	4×10^{-3}	13×10^{-3}
c_v	(cm^2/s)	12.1×10^{-3}	8.25×10^{-3}	5.38×10^{-3}

3.5 Data Deduced from the Experiment for Stress and Penetration Rate

Relation between frost heave ratio and water intake or discharge ratio.
As mentioned above, soil has a tendency to heave during the freezing
process, not only when it takes in water, but also when it discharges
water, corresponding to the level of U and σ. For obtaining a certain
relation between ξ and ξ_w, all points of Figures 11 and 12 were plotted
against ξ and ξ_w, as shown in Figure 14. In this figure, all the points
lie on a straight line having a gradient of $1+\Gamma$ (Γ is the volume
expansion ratio of about 0.09 for water when its phase changes to ice).
When $\xi_w=0$, the value of ξ results in the freeze expansion of the part
of the soil water that became frozen and is given by the following, based
on the results of the analysis:

$$\xi = \xi_0 + n_f \Gamma \tag{28}$$

where ξ_0 is the freeze expansion ratio of unfrozen water at a certain

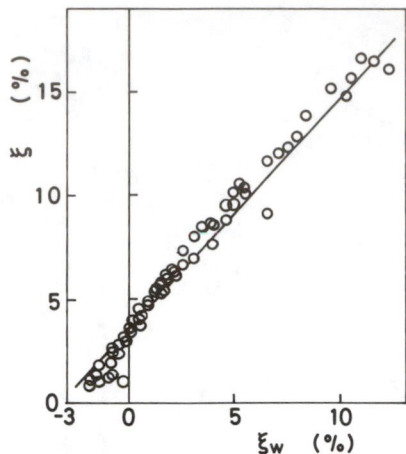

FIGURE 14. Relation between frost heave ratio ξ and water intake or discharge ratio ξ_w.

negative temperature, and n_f is the volumetric free water content which freezes at the freeze temperature. From these relations, ξ is equal to

$$\xi = \xi_0 + n_f \Gamma + (1 + \Gamma) \xi_w \tag{29}$$

Equation 29 shows the case of a soil saturated with water that ξ is the sum of ξ_w, the freeze expansion ratio of ξ_w, and the freeze expansion ratio of the water filling pores between soil particles which become frozen at the lowest temperature during the experiment. The principle of conservation of volume holds true for the case of saturated soil freezing. The maximum water discharge ratio, $\xi_{w\ max}$, is obtained as follows when $\xi=0$:

$$\xi_{w\ max} = -\frac{\xi_0 + n_f \Gamma}{1 + \Gamma} < n\frac{\Gamma}{1 + \Gamma} \tag{30}$$

where n is the porosity of the soil. When the soil consisting of sand or gravel freezes, ξ_w becomes around $\xi_{w\ max}$.

We obtained the relation between ξ, σ and U from equations 27 and 29 as follows:

$$\xi_w = \frac{1}{1 + \Gamma}\frac{\sigma_0}{\sigma}(1 + \sqrt{\frac{U_0}{U}}) - n_f\frac{\Gamma}{1 + \Gamma} \tag{31}$$

Calculated results are shown in Figures 11 and 12. These results coincide well with the experimental results.

Effect of σ and U on heave rate and water intake or discharge rate. When dh denotes the heave amount during a lapse of time dt, then the heave rate is given by dh/dt. Meanwhile, dX denotes the distance that the freezing front advanced during dt, then the frost penetration rate is given by dX/dt. Accordingly, the following equation is derived from the causal fact that dh arises from the penetration of dX in time dt, together with the condition, maintained at constant during the experiment:

$$\frac{dh}{dt} = \frac{dh}{dX}\frac{dX}{dt} = \frac{dh}{dX}U = \zeta U \qquad (32)$$

Substituting equation 27 into ζ of equation 32, we obtain the heave rate under σ and U:

$$\frac{dh}{dt} = (\zeta_0 + \frac{\sigma_0}{\sigma})U + \frac{\sigma_0}{\sigma}\sqrt{U_0 U} \qquad (33)$$

In the same way, we obtain the water intake or discharge rate, dw/dt, under σ and U:

$$\frac{dw}{dt} = \frac{U}{1+\Gamma}\frac{\sigma_0}{\sigma}(1 + \sqrt{\frac{U_0}{U}}) - n_f\frac{\Gamma}{1+\Gamma}U \qquad (34)$$

Figures 15 and 16 show dh/dt and dw/dt, respectively. The heave rate, dh/dt, increases monotonically with increasing U when σ is constant. The water intake or discharge rate, dw/dt, increases monotonically with increasing U when σ is smaller than the critical value of σ, σ_c. When $\sigma > \sigma_c$, however, dw/dt has a peak. In order to examine the behaviour of dw/dt, we differentiate dw/dt as to U and equate this to zero:

$$\frac{d}{dU}(\frac{dw}{dt}) = \frac{1}{1+\Gamma}\frac{\sigma_0}{\sigma}(1 + \frac{1}{2}\sqrt{\frac{U_0}{U}}) - n_f\frac{\Gamma}{1+\Gamma} = 0 \qquad (35)$$

Solving this equation with regard to U we obtain:

$$\sqrt{U_1} = \frac{1}{2}\frac{U_0 \sigma_0}{n_f\Gamma\sigma - \sigma_0} \qquad (36)$$

Here U_1 is the penetration rate which represents the maximum value of dw/dt, dw/dt_{max}, as shown in Figure 16 when σ is constant. Further, σ_c is obtained from equation 36 as:

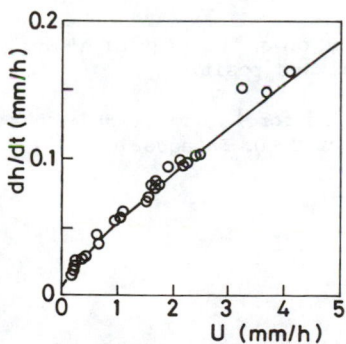

Figure 15. Heave rate dh/dt vs. penetration rate U.

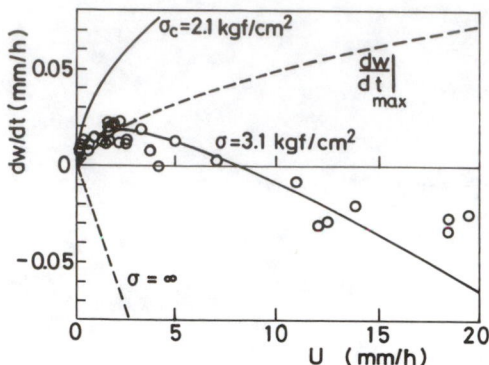

FIGURE 16. Water intake or discharge rate dw/dt penetration rate U.

$$\sigma_c = \frac{\sigma_0}{n_f \Gamma} \tag{37}$$

The following relation between the temperature gradient, α, of the frozen soil and U, the penetration rate, is kept constant by maintaining a constant falling rate of temperature at the cooling plate:

$$\alpha \propto U \tag{38}$$

Rearranging equation 34, we have:

$$\frac{dw}{dt} = \frac{1}{1 + \Gamma} (\frac{\sigma_0}{\sigma} - n_f \Gamma)U + \frac{\sigma_0\sqrt{U_0}}{\sigma(1 + \Gamma)}\sqrt{U}$$

$$= AU + B\sqrt{U} \tag{34'}$$

Since dw/dt becomes the sum of the terms proportional to U and $\sqrt{U}$, we have the following relation between dw/dt and α :

$$\frac{dw}{dt} = A'\alpha + B'\sqrt{\alpha} \tag{39}$$

where A' and B' are the constants which depend on σ. The sign of A' may become negative when σ increases, but B' is always positive.

It is problematical whether or not the equations for ζ_w and dw/dt mentioned above can be applied to the case that $U \to 0$. In equation 31

$$\zeta_w \to \infty \quad \text{when } U \to 0$$

On the other hand, from equation 34

$$dw/dt \to 0 \text{ when } U \to 0$$

These relations are incompatible. When $U \to 0$, frost heaving continues, eventually growing into an ice lens; that is, dw/dt≠0 and $\zeta_w \to \infty$, since the freezing front does not penetrate into the unfrozen soil. Therefore, it should be noted that the empirical formulas obtained in this paper must be applied for cases where the freezing front always propagates into the unfrozen soil, and may not be applied to the critical case when $U \rightarrow 0$.

The   case where ice lenses form in the final stage as $U \rightarrow 0$ will be described later since this is related to the long-term freezing property of soil.

3.6 Effect of Pore Water Pressure on Frost Heave Ratio

In a saturated soil, the "effective stress" is defined as being equal to the total pressure, P, minus the pore water pressure, P_w:

$$\sigma = P - P_w \qquad\qquad (40)$$

The mechanical behaviour of the soil is correlated with the effective stress instead of   either total pressure or pore water pressure.

In the author's experiment, the pore water pressure is approximately equal to the atmospheric pressure. Therefore, the effective stress is always equal to the total pressure, that is:

$$\sigma = P$$

On the other hand, the pore water pressure of a soil to be frozen by artificial ground freezing is usually larger than the atmospheric pressure. This presents a problem as to whether or not the obtained results described in this paper can be applied to the case of $P_w \neq 0$.

The effect of the effective stress on the frost heave was examined by changing the combination of the total pressure and pore water pressure while maintaining the effective stress at constant, as shown in Figure 17. Experimental results, shown in Figure 18, suggest that the frost heave ratio, ζ, changed little or not at all when σ was constant, in spite of an increase in total pressure and pore water pressure. In particular, when σ=2 kgf/cm^2 little change occurred in ζ when P_w was less than the atmospheric pressure. It is clear then that ζ was dependent on σ but not on P or P_w when the penetration rate was kept constant.

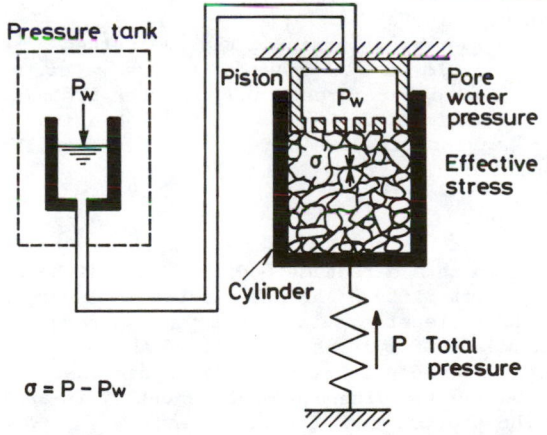

FIGURE 17. Schematic explanation of the pressure equilibrium in soil.

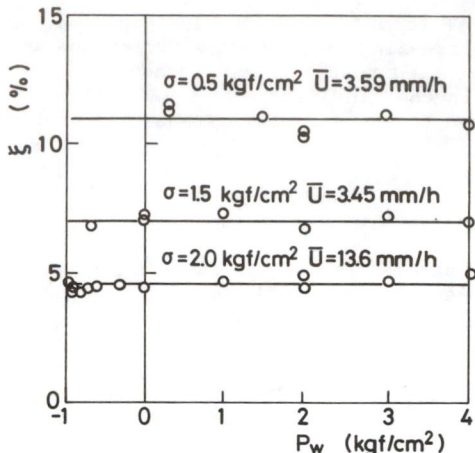

FIGURE 18. Dependence of frost heave ratio ξ on pore water pressure P_w(Negishi silt).

Therefore, equations given in this paper can be expressed by using the effective stress.

3.7 Maximum Heaving Pressure

Since frost heaving pressure causes damage to structures in cold regions, and in the vicinity of a site in a warm area to which artificial ground freezing has been applied, the understanding of its mechanism has been treated as an important subject of experimental and theoretical study. Experimental methods of studying the frost heaving phenomenon are divided into the following two groups (Sutherland and Gaskin, 1973):
1) Measurement of heaving pressure
   A soil specimen is partially frozen in system open with respect to pore water so that the pore water pressure is atmospheric, not allowing frost heaving in order that the maximum heaving pressure developed can be observed.
2) Measurement of a drop in water pressure
   A soil specimen is partially frozen by maintaining the overburden pressure at constant in a system closed with respect to pore water. By measuring a drop in pressure of pore water resulting from the freezing of the specimen, we can obtain water intake pressure (frost heaving pressure). Then, the maximum heaving pressure can be determined as the maximum effective stress as follows:

$$\sigma_{max} = P - P_w$$

Measurement 1) can be considered to be a direct method, however, it has the disadvantage of causing the possibility of an undesirable influence on the result of the measurement. While stress increases, the perfect confinement of the specimen becomes difficult as a result of the consolidation of the specimen and the occurrence of a strain necessary for the detection of pressure. Meanwhile, although measurement 2) is an indirect method, it minimizes the shortcoming of measurement 1).

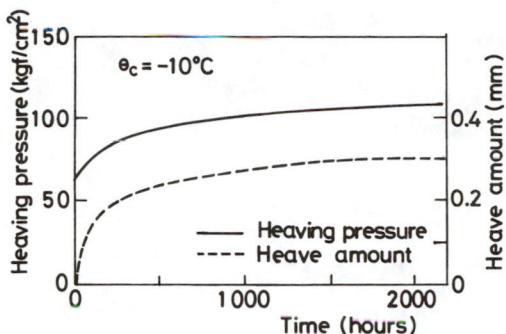

FIGURE 19. Heaving pressure and frost heave amount with time for Manaitabashi clay.

The authors conducted an experiment according to measurement 2) using four soils differing in frost-susceptibility. Typical curves of heaving pressure developed over a lapse of time, indicating that an equilibrium state was attained when P_w fell very slowly, as shown in Figure 19. It seems that the small quantity of water required to bring about a change in P_w migrated into the frozen part from the measuring system. Consequently, it became possible for the frozen part to attract water, even when its stress exceeded 100 kgf/cm^2; and frost heaving occurred, corresponding to the amount of water taken in. In fact, the authors were able to observe that an ice lens grew in contact with the cooling plate while the specimen was partially frozen. Since a freezing isotherm had to come to rest at the midway point of the specimen, it appears that the unfrozen water was allowed to migrate toward the ice front passing through the frozen soil. From this it was assumed that the frozen water continues to remain and builds up a network of veins in the frozen part, which combines the ice front with pore water in the unfrozen part, under a temperature gradient imposed on the partially frozen soil. When a change in σ ceased, the growth of the ice lens due to water intake also ceased, i.e. frost heaving was restrained. We define σ at this stage as the "upper limit of heaving pressure, σ_u", under the temperature condition. Figure 20 shows σ_u vs. the temperature of the cooling plate, θ_c. Until a specific value of θ_c is reached depending on the soil type, σ_u increases linearly with decreasing θ_c. An empirical formula was then obtained as follows:

$$\sigma_u = -11.4 \ \theta_c \quad [\text{kgf/cm}^2] \tag{41}$$

which coincides with the result obtained by Radd and Oertle (1973). The relation between σ_u and θ_c expressed by equation 41 coincides with the generalized Clausius-Clapeyron equation from the analysis using thermodynamics as follows:

$$P = -\frac{L_w}{V_i} \frac{\Delta T}{T} = -11.4 \ \Delta T \quad [\text{kgf/cm}^2] \tag{42}$$

where L_w is the latent heat of water, V_i is the specific volume of ice, T is the Kelvin temperature and ΔT is the freezing point depression. Although σ_u depends linearly on θ_c, as shown in equation 41, it is difficult to believe that this relation holds unlimitedly with decreasing

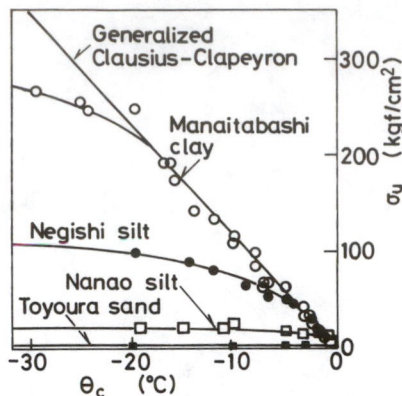

FIGURE 20. Dependence of upper limit of heaving pressure σ_u on temperature of cooling plate θ_c.

θ_c. There is a phenomenon that if $\theta_c < -25\,°C$, then the σ_u obtained deviates from a straight line toward the lower side, as in case for a Manaitabashi clay, as shown in Figure 20 (Takashi et al. 1981a; 1981b). This tendency becomes more pronounced for a Negishi silt. If $\theta_c < -4\,°C$, then σ_u deviates from the straight line and it appears that σ_u converges to its maximum value with decreasing θ_c. We will call it "maximum heaving pressure, σ_u max" that is:

$$\lim_{\theta_c \to \infty} \sigma_u = \sigma_u \text{ max}$$

It is assumed that σ_u max is one of the constants which determines the frost susceptibility of soil.

3.8 Long-term Frost Heaving in Partially Frozen Soil

In large-scale artificial ground freezing work and in the freezing of ground surrounding an underground LNG storage tank, it has been observed that the ground continues to freeze for a very long time and the temperature profile in the ground is maintained at almost constant. Takashi et al. (1981a) have shown experimentally that frost heaving continues for a fairly long time, even though the temperature profile attains an equilibrium state. It is important then to estimate how long the frost heaving will continue and how much it will amount to eventually.

The behaviour of ice lenses was observed in an experiment on long-term frost heaving for which a constant thermal regime was maintained in an open system, whereby the total heave, and the location and thickness of the ice lenses were measured for 6044 hours (Ohrai and Yamamoto, 1985). The temperatures at each end were $-8\,°C$ and $4\,°C$, respectively. No overburden pressure was applied. Figure 21 shows the elapsed change in frost heave amount, h, and water intake amount, w. Although h and w increased during the partial freezing, their rates decreased with time. It was observed that several ice lenses grew simultaneously in the frozen part of the specimen. The position and thickness of each ice lens and the height of the specimen vs. time are shown in Figure 22. A serial number

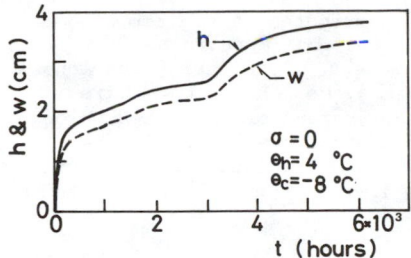

FIGURE 21. Total frost heave h and total water intake w with time for Manaitabashi clay.

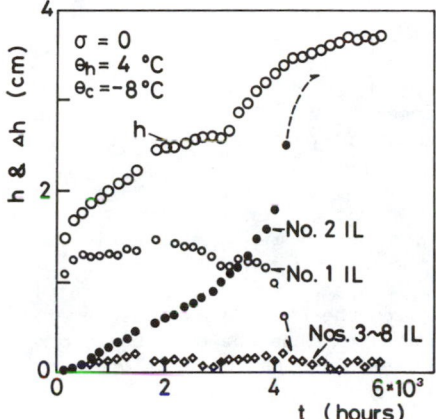

FIGURE 22. Total frost heave h; and increment Δh of thickness of ice lens with time t for the first (No. 1 IL), second (No. 2 IL) and remaining ice lenses (Nos. 3 ~ 8).

was assigned to each ice lens beginning with the one nearest to the freezing front. Ice lenses Nos. 3 ~ 8 grew little because the cumulative h was 0.1 ~ 0.2 mm. However, ice lenses Nos. 1 and 2 displayed an interesting behaviour as follows: ice lens No. 1 attained a thickness of 13 mm after 200 hours after which no significant increase was noted. Meanwhile, after 20 hours ice lens No. 2 grew at a constant heave rate. After about 3000 hours passed, the thickness of ice lens No. 1 began to decrease as ice lens No. 2 grew; and disappeared after about 4000 hours.

When ice lens No. 1 grows to form a layer, water migrating through unfrozen soil should segregate on the higher temperature side of the ice lens, since water cannot migrate through an ice layer (Horiguchi and Miller, 1980). However, the thickness of ice lens No. 1 did not increase corresponding to the amount of water taken in, but decreased conversely. Following explanation for this result was considered: When ice lens No. 1 grows and becomes a layer, water through the unfrozen soil segregates on its higher temperature side. However, melting takes place at the bottom lower temperature side of ice lens No. 1, so that the quantity of melted water becomes equal to that of the frozen water which is necessary for ice lens No. 2 to grow. The melted water migrates through the frozen soil between ice lenses Nos. 1 and 2, and then segregates at the top higher temperature side of ice lens No. 1. As a result, it appears as if water through the unfrozen soil passed the ice lenses. Considering ice lenses Nos. 1 and 2, the frozen soil between them may be likened to a wire in "wire regelation". The phenomenon where the frozen soil migrates through the ice lens may be regarded as "regelation" in frozen soil. The estimate based on the rate of regelation showed that the hydraulic conductivity of the frozen soil ranged from 2×10^{-11} to 1.5×10^{-12} cm/s at temperatures ranging from -1.5 to $-2.2\,°C$.

From this experiment, frost heaving accompanied by water intake occurred, and the frost heave amount did not converge to a constant value during the experiment, even though the temperature profile of the specimen attained an equilibrium state.

4. PREDICTION OF FROST HEAVING AT FREEZING SITE

The influence of frost heaving on the freezing of soil were examined in relation to artificial soil freezing and the preparation of a site for a shield machine from a shaft, as shown in Figure 23.

When the ground to be frozen consists of silty soil and/or clay soil, the soil may heave as water is taken in. The resistance against water migration is relatively small if the dimensions of the frozen soil are of the same magnitude as the lengths of specimens used in laboratory tests. However, at a construction site where the stratum exceeds several meters, the frost heave ratio is always smaller than the one obtained from the laboratory tests since the resistance against water migration is large and the supply of water is reduced.

Moreover, when the ground is soft and weak, a drop in pore water pressure as a result of the intake of water causes an increase in effective stress, leading to the discharge of water from the surrounding ground and the resultant consolidation of it. On the other hand, when the ground is

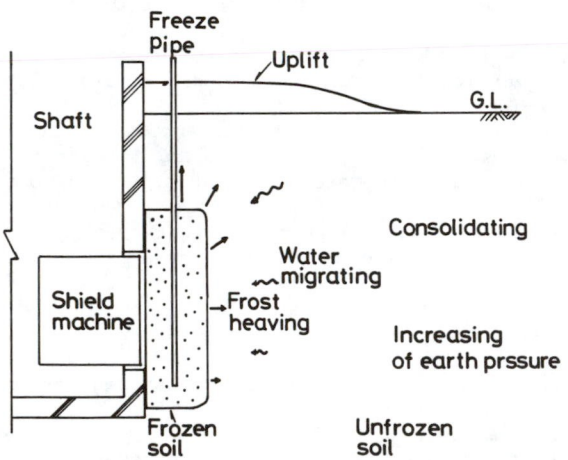

FIGURE 23. Schematic of phenomena resulting from frost heaving in ground.

densely compact as in the case of diluvial soil, soil pressure is increased by frost heaving.

As a consequence of these compound effects, the heave of the ground surface takes place with the unfrozen soil as a medium.

Discussed in this section are the "influence of permeability of unfrozen soil on frost heave", the "consolidation of unfrozen soil during freezing", the "change in earth pressure due to freezing" and the "uplift of the ground surface". A countermeasure against frost heaving during artificial freezing is briefly mentioned.

4.1 Influence of Permeability of Unfrozen Soil on Frost Heave

Expressing all data of Figure 11 by plotting $1/\sigma$, we obtain Figure 24.

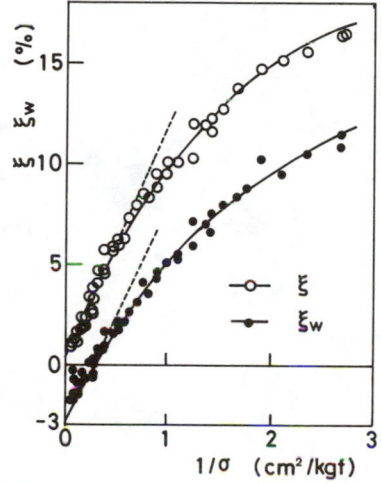

FIGURE 24. Relation between ξ, ξ_w and $1/\sigma$.

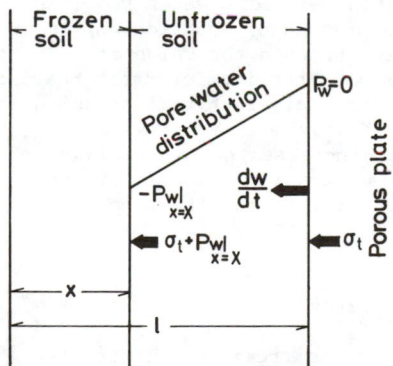

FIGURE 25. Freezing model when steady state is attained concerning water intake in unfrozen soil.

The experimental data is in good agreement with equation 27 for a wide range of large stresses, but does not provide good agreement for the range of small stresses ($\sigma < 1$ kgf/cm^2). The authors attribute this behaviour to the increment of effective stress arising from the resistance of the water movement drawn to the freezing front from the unfrozen region.

When water is taken in during the freezing of saturated soil, the pore water pressure decreases and then the effective stress increases, as shown in Figure 25. Considering that the unfrozen soil zone cannot consolidate, the pore water pressure gradient becomes linear. The following equation was obtained on the basis of equations 27 and 31 by Takashi et al. (1976) as a result of an analysis of the effect on ζ of resistance in soil-water movement through the unfrozen part within a freezing soil specimen:

$$\zeta = \zeta_0 + n_f \Gamma + \frac{(1 + \Gamma)}{2U}\left\{\sqrt{(K\sigma - B)^2 + 4AK} - K\sigma - B\right.$$

$$+ \frac{K(2A - B\sigma)}{B}\log\frac{B\sqrt{(K\sigma - B)^2 + 4AK} + B^2 + K(2A - B\sigma)}{2AK}$$

$$\left. - K\sigma\log\frac{\sqrt{(K\sigma - B)^2 + 4AK} + (2A - B\sigma)/\sigma + K\sigma}{2K\sigma}\right\} \qquad (43)$$

where

$$A = \frac{U}{1 + \Gamma}\sigma_0\left(1 + \sqrt{\frac{U_0}{U}}\right) \qquad (44)$$

$$B = n_f\frac{\Gamma}{1 + \Gamma}U \qquad (45)$$

$$K = k/\gamma_w l \qquad (46)$$

The calculated result is shown in Figure 24 by solid lines. Theoretically the difference between the experimental results and equations 27 and 31 were due to the effect on ζ of the resistance in soil water movement. The value of K in equation 43, which is expressed by equation 46, is the ratio of permeability to the specimens length. When the value of K is the same, the frost heave ratio is also the same. That is, for frost heaving, increasing the specimen's height 10 times is equivalent to increasing its permeability by one-tenth. Therefore, frost heaving due to the water intake can be reduced if it is possible to increase the permeability of the soil near the freezing front by some method, when the soil to be frozen is near the aquifer which supplies water to the freezing front passing through the unfrozen soil.

4.2 Consolidation of Unfrozen Soil During Freezing

Practical cases abound where it is necessary to freeze a silt or clay that is soft, weak and low in permeability. Shown in Figure 26 is a water content profile near the freezing front at a site subjected to artificial freezing. The water content decreases toward the freezing front in the

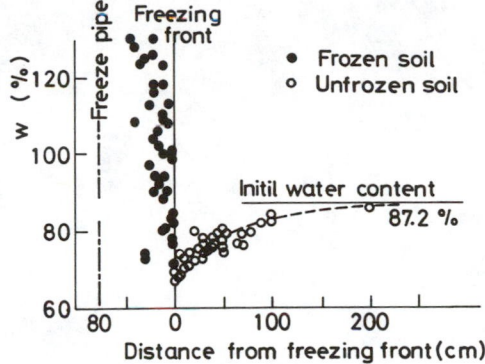

FIGURE 26. Distribution of water content, w, near the freezing front in an artificial frozen silty ground.

unfrozen soil because the consolidation of the soil occurs due to an increase in effective stress. However, the water content increases in the frozen soil before freezing when water is taken in. That is, from a macroscopic point of view, using the freezing front as a boundary, the soil expands on the side of the frozen soil and shrinks on the side of the unfrozen soil; then, if viewed from the side of the frozen soil, the result of these compound effects occurs on the side of the unfrozen soil.

Takashi et al. (1977) developed this equation theoretically. The pore water pressure was made to follow Terzaghi's consolidation equation within the unfrozen soil; and equation 34 was used to express the water intake rate at the freezing front, with σ as the effective stress at the freezing front. The following differential equation was obtained:

$$\frac{\partial P_w(x,t)}{\partial t} = c_v \frac{\partial^2 P_w(x,t)}{\partial x^2} \tag{47}$$

where P_w is the pore water pressure and c_v is the coefficient of consolidation:

$$c_v = \frac{k}{\gamma_w m_v} \tag{48}$$

where k and m_v are the permeability and coefficient of the volume change of unfrozen soil, respectively. It is difficult to obtain an analytical solution from this equation, since the question is concerned with a boundary condition of nonstationary movement. We will demonstrate a solution for the case in which the phenomenon becomes stationary after a lapse of sufficient time from the beginning of freezing. The pore water pressure in the stationary state in the unfrozen soil is given by:

$$P_w = \frac{(1+\Gamma)m_v\sigma + n_f\Gamma - \sqrt{\{(1+\Gamma)m_v\sigma - n_f\Gamma\}^2 + 4(1+\Gamma)m_v\sigma_0(1+\sqrt{U_0/U})}}{2(1+\Gamma)m_v} e^{-U\zeta/c_v} \tag{49}$$

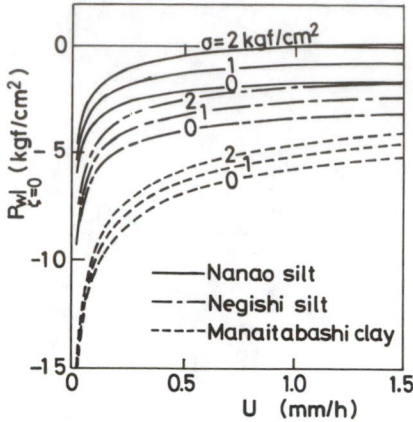

FIGURE 27. Pore water depletion $P_w \big|_{\zeta=0}$ at freezing front as a function of penetration rate U.

where ζ is the distance from the freezing front. Shown in Figure 27 is the relation between pore water pressure at the freezing front, $P_w \big|_{\zeta=0}$, and the frost penetration rate under the effective stress $\sigma=0$, 1 and 2 kgf/cm^2. It is known that a large drop generates in pore water pressure at the freezing front. Therefore, it is predicted that the drop in pore water pressure in the unfrozen soil leads to intense consolidation, increasing the effective stress when the soil is soft. The region where the consolidation occurs, δ_s, is:

$$\delta_s = c_v/U \tag{50}$$

Figure 28 shows the relation between δ_s and U. In the freezing site δ_s is unexpectedly large. Moreover, $\zeta \big|_{\zeta > \delta_s}$ which is observed at a point distant from δ_s is:

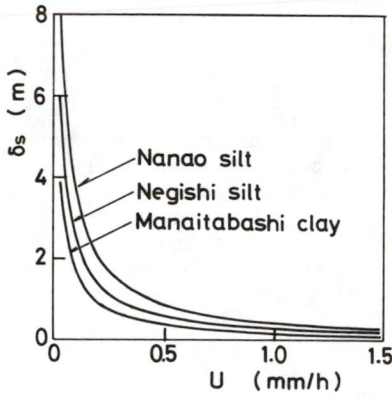

FIGURE 28. Consolidation region δ_s as a function of penetration rate U.

$$\xi\,|_{\,\zeta>\delta_s} = \zeta_0 + n_f \Gamma \qquad\qquad (51)$$

since at the freezing front the frost heave amount in the frozen soil
cancels the amount of shrinkage resulting from water discharge and
consolidation in the unfrozen soil. This equation shows that at a point
distant from δ_s, the frost heave amount is approximately of the same
magnitude as the amount of the expansion of water when frozen. This means
that when the ground consists of a silt or clay that is low in permeability, besides being soft and weak, artificial freezing must be applied in
a tightly closed system, differently from an open system used in laboratory experiments. It can be stated that the frost heave amount observed
in the artificial freezing method is always smaller than that obtained
from laboratory experiments as a result of the effects of consolidation
and/or resistance against the water intake of the unfrozen soil.

4.3 Change in Earth Pressure due to Freezing

While soil is artificially frozen, the earth pressure increases in the
unfrozen soil due to frost heaving when the soil is frost susceptible. A
simple case in which the soil surrounding a vertical freeze pipe freezes
was considered, as shown in Figure 29. When frost heaving occurred, the
unfrozen soil surrounding the frozen column deformed by δ_1 at the
outside, and the freeze pipe deformed by δ_0 at the inside. The
deformation, δ_0, which is caused by shrinkages due to decreasing
temperature and increasing pressure, was less than δ_1 and may be
calculated easily. Within the unfrozen region, the earth pressure may
increase against the deformation, δ_1. This antagonistic earth pressure
becomes maximum at the freezing front and decreases with the increasing
distance from the interface. Unfrozen soil behaves elastically when
increments of earth pressure and deformation are relatively small, but
when they increase a plastic yield or rupture occurs in the unfrozen soil
zone. Takashi (1972) has shown this change in earth pressure theoretically, using equation 23 for frost heaving under overburden pressure. In
the soft soil, the increment of earth pressure is small, but the deformation is large. On the other hand, in the hard soil the former is fairly
large but the latter is small as shown in Figure 30. In designing artificial ground freezing, especially in the case of hard ground, the increase
of earth pressure during freezing must be considered.

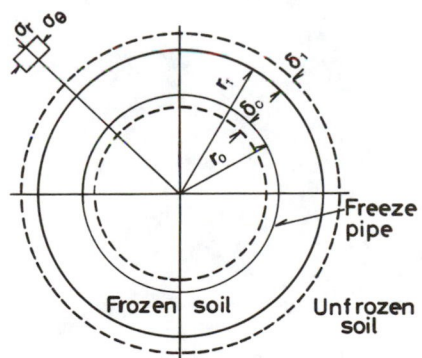

FIGURE 29. Freezing model.

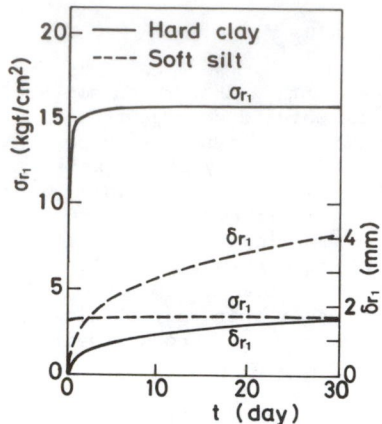

Figure 30. Stress change σ_{r1} and displacement δ_{r1} at freezing front due to frost heaving.

4.4 Uplift of Ground Surface

All frost heaving is not absorbed by horizontal displacement and/or consolidation of unfrozen soil, but appears partly as an uplift of the ground surface. The uplift presents a practical problem when artificial ground freezing is used in urban areas, where buildings are crowded and underground structures abound.

We were able to estimate the uplift distribution on the ground surface using calculations based on the assumption that the distribution may be approximated in the shape of the Gauss distribution curve. In the case of the model shown in Figure 31, it is given by the following equation (Tobe et al., 1979):

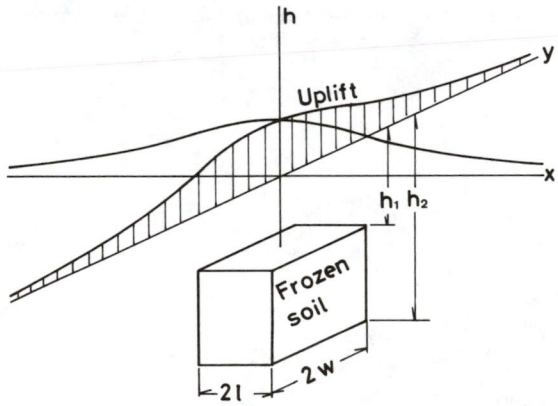

FIGURE 31. Model for the uplift of the ground surface

$$G(h,x,y)=\frac{\xi'}{4}\int_{h_1}^{h_2}[\{erf(\frac{1+x}{ah})+erf(\frac{1-x}{ah})\}\{erf(\frac{w+y}{ah})+erf(\frac{w-y}{ah})\}]\,dh \qquad (52)$$

where $G(h,x,y)$: uplift amount,
h_2-h_1, 1 and w: lengths of the sides of a rectangular frozen soil,
h: depth of the frozen zone,
a: a constant concerned with the influence angle equal
  approximately to $a=\tan(45 + \Phi/2)$,
  Φ: friction angle,
  ξ': modified frost heave ratio.
Equation 52 allows us to estimate in advance the amount of uplift at a
freezing site with a maximum error of ± 30 %.

4.5 Countermeasures Against Frost Heaving in Artificial Ground Freezing

In natural freezing, the supply of select granular soils and/or
insulating materials are used generally as a countermeasure against frost
heaving. However, it is difficult to use them for artificial ground
freezing, since the area to be frozen is deep under the ground surface.

Countermeasures for artificial ground freezing, which the authors have
used in practice, are briefly described as follows:

Increasing the viscosity of pore water   The resistance of the unfrozen
soil against water migration toward the freezing front may decrease the
frost heave amount (see 4.1). That is, it is advantageous to try to
decrease the permeability of the soil by some means. Takashi et al.
(1980) considered that one can obtain some effect by increasing the
viscosity of the pore water as a mechanical means of decreasing the
permeability of the soil. Figure 32 shows the relation between the frost
heave ratio, ξ, and the viscosity of pore water, η. Carboxy Methyl
Cellulose (CMC) was used to increase η. As ξ decreases remarkably with
increasing η, it is clear that the frost heave amount in a frost
susceptible soil can be reduced by increasing the viscosity of the water
attracted toward the freezing front during soil freezing.

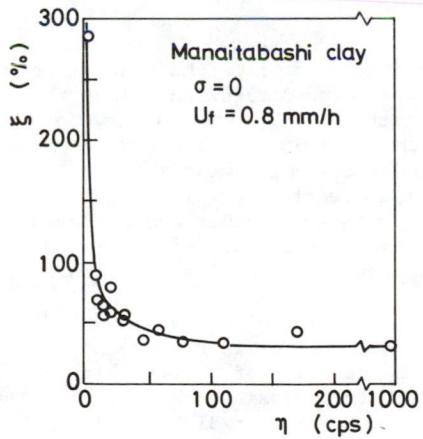

FIGURE 32.   Dependence of frost heave ratio ξ on viscosity of pore water η.

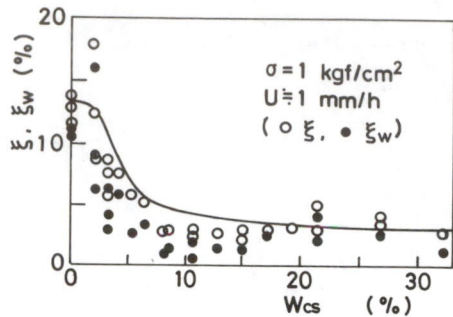

FIGURE 33. Dependence of frost heave ratio ξ and cement content W_{cs}.

Restraint of frost heaving by cement addition Thanks to the development
of techniques of civil engineering techniques, it becomes easy to locally
mix soil with cement underground. If the frost heave amount of the soil
decreases from the addition of cement, this method is effective for
artificial ground freezing. Accordingly, laboratory tests were conducted
to investigate how much portland cement restrained the frost heave amount
of soil, using a Fujinomori clay collected in Kyoto (Ohrai et al., 1984).
As shown in Figure 33, the frost heave ratio, ξ, decreased 1/8 - 1/4
times that of cement free specimens when W_{cs}, which denotes that the
weight ratio in the percentage of added cement to oven dried soil is 10%.
    The decrease in ξ is small when $W_{cs} > 10$ %. Theoretical consideration
shows that a decrease in the permeability of an unfrozen soil as a result
of the addition of cement leads to little decrease in ξ. It is
suggested that a decrease in ξ comes about from a decrease in the
driving force to absorb water, due to the presence of cement in the soil.

It has been examined experimentally that thaw-settlement decreases and
the mechanical strength of frozen soil increases when the proposed method
is used.

5. CONCLUSIONS AND SUMMARY

This paper described the experimentally obtained effects of factors on
frost heaving, and then discussed the prediction of frost heaving at the
site of construction subjected to artificial ground freezing. Since these
factors do not include all the factors which affect frost heaving, and
analyses were made on assumptions as to the method of prediction, we must
state that the results obtained give only approximate solutions.
Nevertheless, in practical cases somewhat successful results have been
obtained. This may be due to the fact that the experiments and analyses
conducted by the authors and others on frost heaving have not deviated
much from the fundamentals, since the results from practical applications
of artificial ground freezing were taken into account.

The use of freezing as a ground stabilizing technique increases the range
of applications to civil engineering, with the shape of soil frozen
varied for each purpose. In these cases, it is impossible to rely on
analytical approaches for a precise and quantitative prediction of frost
heaving and its resultant phenomena. Fortunately, marked progress in
computer technology is making such analytical approaches possible, since

we can now receive abundant input data, and calculate it with some speed. However, it is the authors' opinion that more established data on frost heaving and unfrozen soil must be collected before full advantage can be taken of computer technology. In order to predict frost heaving more precisely at the site of actual construction, we must continue to pursue experimental and theoretical research.

REFERENCES

Carslow, H.S. and J.C. Jager, 1959: Conduction of heat in solids, 2nd edition, Oxford at the Clarendon Press.

Dirksen, C. and R.D. Miller, 1966: Closed-system freezing of unsaturated soil. Soil Sci. Soc. Am. Proc., 30, 163-173.

Fukuda, M., A. Orhaum and N. Luthin, 1980: Experimental studies of coupled heat and moisture transfer in soils during freezing, Cold Regions Sci. and Tech., Vol. 3, 223-232.

Hoekstra, P., 1966: Moisture movement in soils under temperature gradients with the cold side temperature below freezing. Water Resources Research Board Special Report, 103, 78-90.

Horiguchi, K. and R.D. Miller, 1980: Experimental studies with frozen soil in "Ice sandwich" permeameter. Cold Reg. Sci. Technol. Vol. 3, 177-183.

Loch, J.P.G. and B.D. Kay, 1978: Water redistribution in partially frozen, saturated silt under several temperature gradients and overburden loads. Soil Sci. Soc. Am. J., 42, 400-406.

Miller, R.D., 1972: Freezing and heaving of saturated and unsaturated soils. Highway Res. Record, 393,1-11.

Miyoshi, M., T. Tsukamoto and S. Kiriyama, 1975: Large-scale freezing work for subway construction in Japan. Engineering Geology, 13, 397-415.

Ohrai, T., H. Yamamoto, J. Okamoto and H. Izuta, 1984:Restraint of frost heaving and thaw settlement of soil by cement addition. SEPPYO, J. Japanese Soc. Snow and Ice, Vol. 46, No. 4, 189-197.

Ohrai, T. and H. Yamamoto, 1985: Growth and migration of ice lenses in partially frozen soil. Proceedings of the 4th International Symposium on Ground Freezing, 79-84.

Ohrai, T., Y. Ishikawa and Y. Kushida, 1985: Actual results of ground freezing in Japan. Proceedings of the 4th International Symposium on Ground Freezing, Vol. 2, 289-294.

Penner, E., 1960: The importance of freezing rate in frost action in soils. Proceedings Am. Soc. Test. Mater. 60, 1151-1165.

Penner, E. and L.E. Goodrich, 1980: Location of segregated ice in frost susceptible soil. Proceedings of the 2nd Int. Symp. Ground Freezing, 626-639.

Radd, F.J. and D.H. Oertle, 1973: Experimental pressure studies of frost heave mechanism and the growth fusion behavior of ice. Permafrost, 2nd Int. Conf., North Am. Contribu., Washington D.C., Nat. Acad. Sci., 377-384.

Sutherland, H.B. and P.N. Gaskin, 1973: Pore water and heaving pressure developed in partially frozen soils. Permafrost, 2nd Int. Conf., North Am. Contribu., Washington D.C., Nat. Acad. Sci., 409-419.

Taber, S.,1929: Frost heaving. J. Geol. 37, 428-461.

Taber, S. 1930: The mechanics of frost heaving. J. Geol., 38, 303-317.

Takashi, T. and S. Wada, 1961: The soil freezing method in engineering construction (I). 36, No. 408, 1-15.

Takashi, T., 1965: On the freezing efficiency in soil freezing method. (in Japanese) Refrigeration, Japanese Association of Refrigeration, Vol. 40, No. 456, 1-7.

Takashi, T., 1969: Influence of seepage stream on the joining of frozen soil zones in artificial soil freezing. Special Rep. 103, Highway Research Board, Washington, D.C., 273-286.

Takashi, T., 1972: On the stress and displacement in unfrozen soil zone around artificial frozen soil. Proceedings of JSCE, No. 200, 49-62.

Takashi, T. and M. Masuda, 1975: On an exact solution of heat transfer equation in freezing soil with constant speed, accompanying uniform form of suction water to the freezing front.

Takashi, T., M. Masuda and H. Yamamoto, 1976: Influence of permeability of unfrozen soil on frost heave. (in Japanese) SEPPYO, Journal of the Japanese Society of Snow and Ice, Vol. 38, No. 1, 1-10.

Takashi, T., T. Ohrai and H. Yamamoto, 1977: Pore water pressure and consolidation in unfrozen soil near the freezing front, SEPPYO, J. Japanese Soc. Snow and Ice, Vol. 39, No. 2, 53-64.

Takashi, T., H. Yamamoto, T. Ohrai and M. Masuda, 1978: Effect of penetration rate of freezing and confining stress on the frost heave ratio of soil, Permafrost, 3rd Int. Conf., Vol. 1, 737-742.

Takashi, T., T. Ohrai and H. Yamamoto, 1980: Decrease of frost heave amount by increasing the viscosity of pore water, Proceedings of JSCE, No. 298, 77-85.

Takashi, T., T. Ohrai, H. Yamamoto and J. Okamoto, 1981a: Upper limit of heaving pressure derived by pore water pressure measurements of partially frozen soil, Eng. Geol., 18, 245-257.

Takashi, T., T. Ohrai, H. Yamamoto and J. Okamoto, 1981b: An experimental study on maximum heaving pressure of soil, SEPPYO, J. Japanese Soc. Snow and Ice, Vol. 43, No. 4, 207-215.

Takashi, T., T. Ohrai, H. Yamamoto and J. Okamoto, 1982: Effect of specimen height on frost heave ratio in unidirectional freezing test of soil. Proceedings of the 3rd Int. Symp. on Ground Freezing, 247-254.

Tobe, N. and O. Akimoto, 1979: Calculating method of frost heaving. 34th Symp. of JSCE, III, 243-244.

580

Chapter 18

The Physics of Freezing Soils and an Engineering Frost Heave Approach

J.-M. KONRAD
Department of Civil Engineering, Université Laval,
Ste Foy, Québec, Canada G1K 7P4

ABSTRACT

A review of the fundamental aspects of heat and mass transfer applied to one-dimensional freezing in porous media is presented in this chapter. Equilibrium conditions for the water-ice system is discussed in terms of classical reversible thermodynamics for isothermal closed system freezing. Freezing under an imposed temperature gradient is then presented for open systems, i.e. free access to water. Changes in soil freezing characteristics during transient heat flow are also considered.

The second part of the chapter deals with the application of the fundamentals to engineering problems, and introduces the concept of the segregation potential, SP. SP is the ratio of water migration rate and temperature gradient in the frozen zone near the frost front. It depends upon many factors such as the suction in the water films, the rate of cooling, the applied overburden pressure, pore water salinity, and many others. However, for a given soil at a given porosity, the functional relationship between SP and these factors is unique.

CONTENTS

1. INTRODUCTION

When a frost-susceptible soil is subjected to freezing, heave occurs as a result of the development of ice lenses formed from water supplied from the unfrozen soil or possibly an external source. This phenomenon has been known for a long time and has been the subject of numerous studies. Notwithstanding the very considerable research devoted in the past to the frost heave process, there has been no general agreement on a predictive theory. The delivery of natural gas both from the Arctic and elsewhere is improved by operating the pipeline at freezing temperatures. Artificial freezing is also finding increasing use in civil and mining engineering works. Therefore there is a substantial incentive for developing a theory of frost heave capable of making predictions of heave and moisture transfer that are reliable enough for engineering needs.

In the first part, the paper discusses the equilibrium conditions for the water-ice system in terms of classical reversible thermodynamics for conditions of isothermal closed system freezing. Then, the freezing of saturated soils subjected to a temperature gradient which is changing with time is analysed. Effects of thermal imbalance leading to various rates of cooling, overburden pressure, and suction in the unfrozen water films will also be considered.

The second part deals with the application of the physics of freezing soils to engineering problems, and introduces the concept of the segregation potential during transient freezing.

2. PART 1. THE PHYSICS OF FREEZING

2.1. Isothermal Freezing in Porous Media - (Closed System)

2.1.1   Unfrozen Water in Frozen Soils

When a fine-grained soil is frozen not all the water within the soil pores freezes at $0^{O}C$ (Bouyocos, 1916; Lovell, 1957). In some clay soils, up to 50% of the moisture may exist as a liquid at temperatures of $-2^{O}C$ (Nersesova and Tsytovich, 1963). The water remains unfrozen mainly as a result of capillary and surface adsorption effects. The unfrozen water content, W_u, is the amount of liquid water in a soil at a given sub-zero temperature. It is usually expressed as a percentage of the dry weight of the soil.

In general, the liquid phase may be considered to consist of an infinite number of subphases because of the direct influence of the mineral matrix on the liquid water.

As shown in fig. 1, the unfrozen water can be separated mainly into two different types. The unfrozen water that is relatively far from the soil particle surfaces can be considered as capillary or bulk water. It is not affected significantly by the surface activity of the soil particles. The pressure in the bulk water would correspond to the value measured by a tensiometer provided that such a device would not alter the system.

When the unfrozen water is close to the particles, it can be considered as being strongly influenced by the solid surface of the soil particles. Gilpin (1980) proposed a model that includes the effect of a solid surface in the formulation of the free energy of that type of water. It was demonstrated that pressure and temperature affect the thickness of that liquid film layer.

Measurements of unfrozen water contents in soil have been made by dilatometry, adiabatic calorimetry, X-ray diffraction, heat capacity, nuclear magnetic resonance, differential thermal analysis, and several indirect techniques (Anderson and Morgenstern, 1973). Although each involves its own set of assumptions and approximations, the results obtained on the same soils are remarkably consistent. All these experimental procedures have two common factors: 1) the samples are frozen in a closed system, i.e. no water supply during freezing, and 2) measurements are taken at thermal equilibrium, which in turn means that no further phase change can occur at a given temperature.

2.1.2 Phase Equilibrium

For each of the phases separately one may write the Gibbs-Duhem equation (Kay and Groenevelt, 1974).

$$d\mu_w = - S_w dT + V_w dP_w \qquad (1) \text{ bulk water}$$

$$d\mu_e = - S_e dT + V_e dP_e \qquad (2) \text{ Adsorbed water}$$

$$d\mu_i = - S_i dT + V_i dP_i \qquad (3) \text{ ice}$$

At equilibrium, the chemical potential, μ, of all phases of the water are equal which results in the following well known relationships:

$$V_w dP_w - V_i dP_i = (S_w - S_i) dT = H_f \ dT/T \qquad (4)$$

$$V_w dP_w - V_e dP_e = (S_w - S_e) dT = H_w \ dT/T \qquad (5)$$

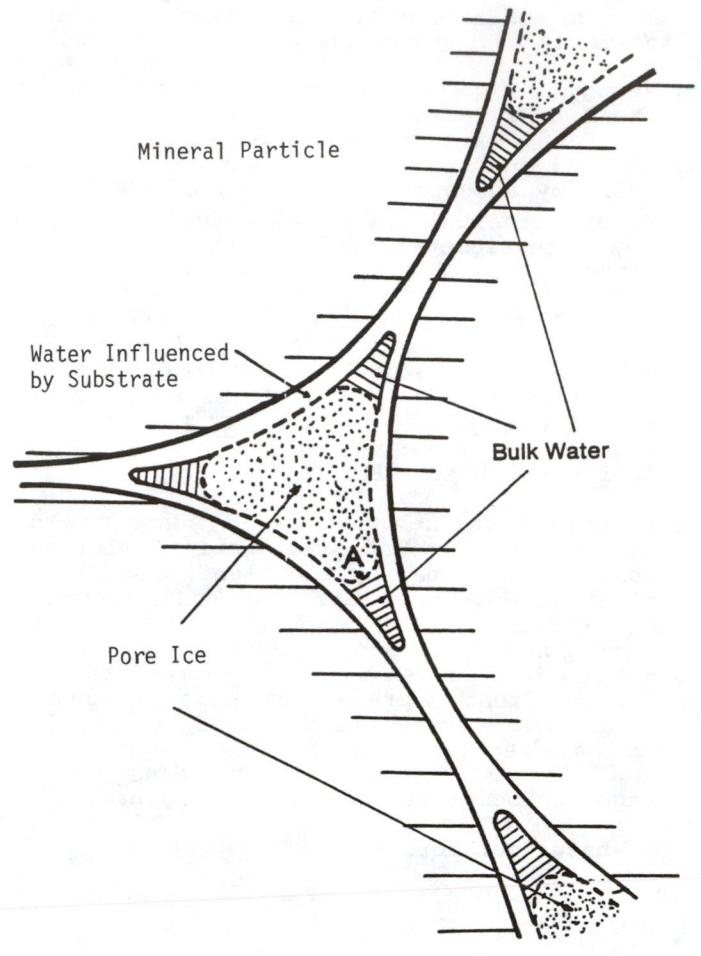

Figure 1:   The water phase in frozen soil.

$$V_f dP_f - V_i dP_i = (S_f - S_i) dT = (H_f - H_w) \, dT/T \qquad (6)$$

where H_f is the specific latent heat of fusion of free water H_w is the specific heat of wetting.

Although the bulk water is per definition in equilibrium with the water near the solid surfaces, μ_w (bulk) = μ_e (adsorbed) the actual pressure in the bulk water P_w is not necessarily equal to the actual pressure P of the liquid phases under the influence of the mineral surfaces. In contrast, the pressures in the bulk ice and the ice near the solid substrates are considered to be indistinguishable and will be referred to P_i.

The unfrozen water is thus in equilibrium with the pore ice at given temperature and pressure conditions. Under constant pressure, the chemical potential of the ice decreases with decreasing temperature inducing a concomittant decrease of the chemical potential of the unfrozen water. This is achieved by reducing the water content in the system as a consequence of an equal increase in ice content at constant overall moisture content. Under constant temperature, the application or removal of pressure changes the chemical potential of the ice, and maintaining equilibrium requires an equal change in the chemical potential of the unfrozen water. Since the process is isothermal, $dT = 0$, eq. 4 indicates that the change in chemical potential of the ice is $V_i \Delta P$ for a pressure change ΔP. However, according to eq. 1, the chemical potential change in the bulk water is $V_w \Delta P_w$ and thermodynamic equilibrium yields:

$$\Delta P_w = (V_i/V_w) \, \Delta P \qquad (7)$$

Eq. 7 suggests that ΔP_w should theoretically be about 9% greater than ΔP which is unlikely. Another way to maintain phase equilibrium is to allow freezing or melting which in turn affects the chemical potential of the unfrozen water. Indeed, if we assume that both phases experience the same change in pressure ΔP, the additional change in the chemical potential of the bulk water is:

$$\Delta (\Delta \mu_w) = \Delta P (V_i - V_w) \sim 0.09 \, \Delta P \qquad (8)$$

This analysis demonstrates that a change in external pressure on a frozen soil is also associated with a change in the unfrozen moisture content. Increasing pressures result in increasing unfrozen water contents. Experimental evidence for this relationship was given by Williams (1976) (see fig. 2). However, only slight variations in W_u were observed in response to substantial pressure changes, supporting eq. 8 which states that the change in chemical potential of the bulk water, hence W_u, corresponds only to 0.09 ΔP.

In summary, a unique relationship between unfrozen water content and changes in its chemical potential can be obtained for each soil. Under constant ice pressure, $\Delta P_i = 0$ there is a known relation between the changes in chemical potential

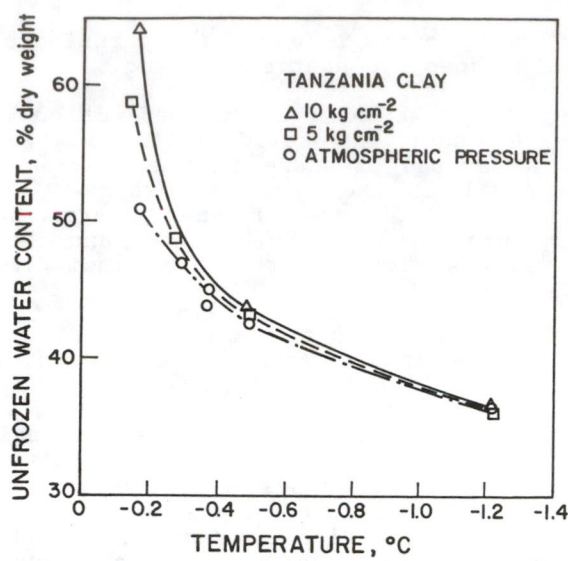

Figure 2: Changes in water content of frozen Tawzania clay as a function of temperature and pressure. After Williams, 1976.

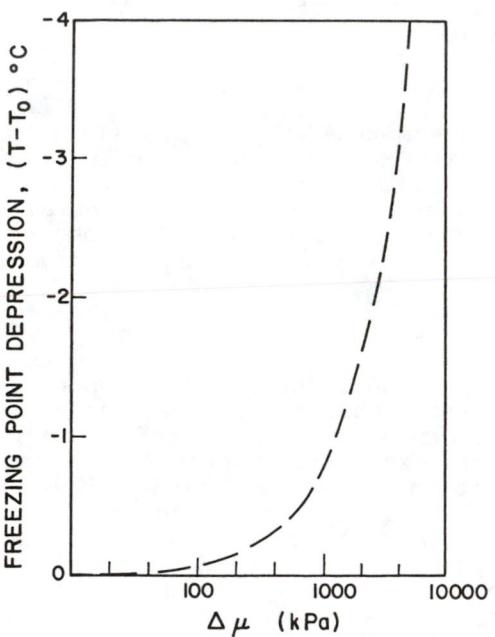

Figure 3: Relationship between change in chemical potential and freezing point depression.

and in the temperature T given in fig. 3:

$$\Delta\mu = (H_f \ (T_O - T))/V_w \ T_O \qquad (9)$$

Under isothermal condition, the relation between $\Delta\mu$ and ΔP is given by eq. 8. Thus, all the curves shown in fig. 2 can be reduced to a unique curve if W_u is expressed as a function of the changes in chemical potential induced by a change in both temperature and pressure as shown in fig. 4.

2.1.3   Effect of Curved Interfaces

One of the characteristics of porous media is that the pore ice will have to penetrate into a pore neck and adopts therefore a curved interface as shown by A in fig. 1. The average radius of curvature, r, of the ice at A is given by (Everett, 1961).

$$P_i - P_w \ = \ 2 \ \sigma_{iw}/r \qquad (10)$$

where σ_{iw} is the surface energy of the ice-water interface. It should be noted that eq. (10) can also be written as:

$$P_i - P_w \ = \ \sigma_{iw}f \ (W_u) \qquad (11)$$

where $1/r$ has been substituted by a function f relating W_u to the radius of the ice-water interface.

Phase equilibrium exists when:

$$V_i (P_w + 2 \ \sigma_{iw}/r) - V_w P_w \ = \ - H_f \ \Delta T/T_O \qquad (12)$$

which is a generalized form of the Laplace equation. Eq. (10) and (11) define the conditions of temperature and pressure for which ice will propagate through all pore openings of average radius greater or equal to r.

Unfortunately, only T can be determined with fair accuracy; but to-date no satisfactory measurements of P_i and P_w were made. Eq. (12) provides only the value of the difference between the ice and liquid pressures, thus an infinity of combinations of P_i and P_w may be obtained at any given W_u.

As a solid, ice is capable of maintaining a complex state of stress but it must be kept in mind that for temperature near 0^oC, the solid ice is also extremely close to its homologous temperature. This, in turn, signifies that ice will tend to creep under stress and relieve any stress concentrations. Therefore, when a weightless porous solid is subjected to isothermal freezing and is free to expand under atmospheric conditions, it is thought that the pressure in the ice at equilibrium is also atmospheric. Again, because frost heave is driven by (P_i-P_w), the precise knowledge of the value of P_i is not too critical.

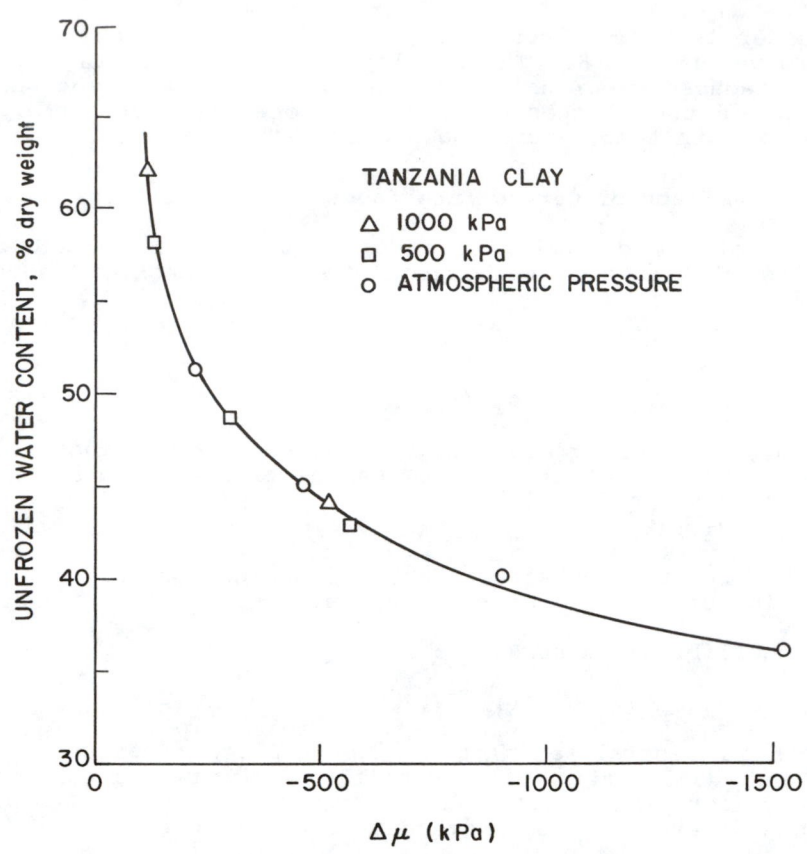

Figure 4:   Relationship between unfrozen water content and changes in chemical potential.

2.2 Freezing of Saturated Porous Media Subjected to Temperature Gradients.

This section describes the physical processes associated with freezing of soils subjected to a temperature gradient. This results in non-equilibrium effects which influence the relationships established for isothermal, closed system freezing. Firstly, we establish the basis for heat and mass transfer during transient freezing for no applied surcharge. Then we discuss the effects of suction, rate of cooling and applied load on the phase composition of the frozen fringe.

2.2.1 Heat and Mass Transfer in Freezing Soils

When a temperature below freezing is applied to the surface of a soil sample, unsteady heat flow is initiated. The $0^{\circ}C$ isotherm progresses into the soil as a function of the imbalance of the heat supplied to the heat removed. Thus, if the surface temperature is constant, the freezing is driven by a temperature gradient across the sample which decreases continuously with increasing frost depth. The temperature gradient can also remain fairly constant or even increase with time if the surface temperature varies with time.

The fact that not all the water in a fine-grained soil freezes at a unique temperature, but rather over a range dependent on soil type has a significant impact on the liquid – solid phase change in porous media. Firstly, internal heat is released over a range of negative temperatures as pore water freezes. Furthermore, the amount of heat generation and the extent of the zone of phase change is soil type dependent. Once nucleated, the ice crystals propagate into and through pore space according to eqs. 10 and 12. As the $0^{\circ}C$ isotherm advances through the soil, the position of the freezing front will be at the temperature at which ice will propagate through the maximum pore size openings. This warmest temperature at which ice can grow in the soil pores, termed the in-situ freezing temperature T_i, is mainly a function of soil type, applied pressure and solute concentration. In many soils T_i is very close to $0^{\circ}C$. Secondly, because the unfrozen water channels form an interconnected network, water from the unfrozen soil, or possibly an external source, migrates through the frozen soil and freezes to create an actively growing ice lens that does not coincide with the T_i isotherm (front at which ice begins to form), but is on the cold side of it (Miller, 1972). The zone between the frost front (T_i) and the segregation front (T_s) has been referred to as the frozen fringe.

2.2.1.1 Heat Transfer

Fourier's general equation expresses the conditions that govern the flow of heat in the transient state. Convective transport is neglected here. For one-dimensional heat flow see fig. 5, the equation reduces to:

$$\delta/\delta z\,(\lambda\ \delta T/\delta z)\ +\ Q\ =\ C\ \delta T/\delta t \qquad (13)$$

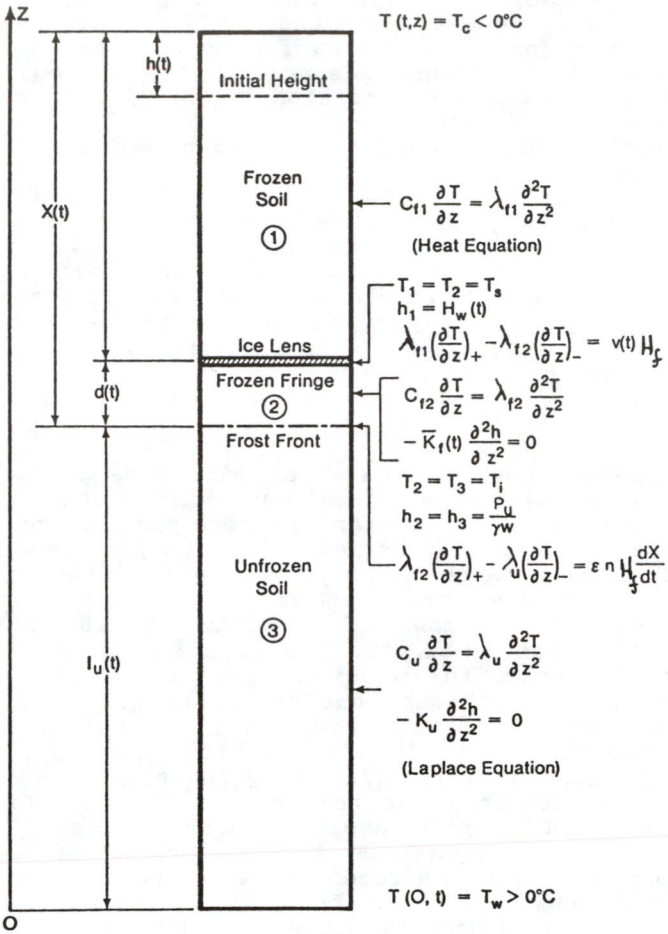

Figure 5: Equations for one dimensional freezing in soils.

when C is the volumetric heat capacity and λ is the thermal conductivity, both temperature dependent. Q is an internal heat generation term per unit volume and per unit time.

The internal heat is liberated as the pore water freezes, Q_1, and when the migratory water freezes at the segregation freezing front, i.e. T_s isotherm, Q_2. Since the pore water freezes over a range of temperatures, one may simplify the equations by lumping the process at the freezing front, T_i isotherm, where most of the pore water is frozen. This assumption is realistic for sandy and silty soils not so much for clays where the unfrozen water content is still substantial at $-1^{\circ}C$. The internal heat in freezing soils is then given as:

$$Q_1 = \varepsilon.n.H_f.(dX/dt) \qquad (14)$$

and

$$Q_2 = v.H_f \qquad (15)$$

where ε is a dimensionless factor taking into account the proportion of unfrozen water remaining in the sample and lumped to T_i

    n   is the porosity of the soil
 dX/dt   is the rate of frost front advance
    v   is the migratory water flux

2.2.1.2   Mass Transfer

Temperature conditions associated with the presence of an ice lens are illustrated in fig. 6. Once an ice lens is formed it acts like a cutoff with regard to water migration (Hoekstra, 1969) and the whole freezing system can be separated into a passive one, defined as the frozen soil between the surface where freezing is initiated and the cold side of the warmest ice lens and an active system composed of the frozen fringe and the unfrozen soil. In the passive system moisture transfer is much reduced due to very low frozen soil permeabilities and the contribution to the final heave has been found negligible under both laboratory conditions (Mageau and Morgenstern, 1979) and under field conditions, at least over a period of several years (Slusarchuk et al., 1978).

Assuming that there is no accumulation within the frozen fringe, water flow within both the frozen fringe and the unfrozen soil are governed independently by the Laplace equation, though it can be extended to embrace soil compressibility. Water pressure and discharge are continuous across the frozen-unfrozen interface, T_i isotherm, but at the base of the warmest ice lens the Clausius-Clapeyron equation (see eqs. 4 and 9) holds as shown by Vignes and Dijkema (1974) and Biermans et al. (1978). Thus, in the frozen fringe,

$$- K_f (T) \delta^2 \Phi / \delta z^2 = 0 \qquad (16)$$

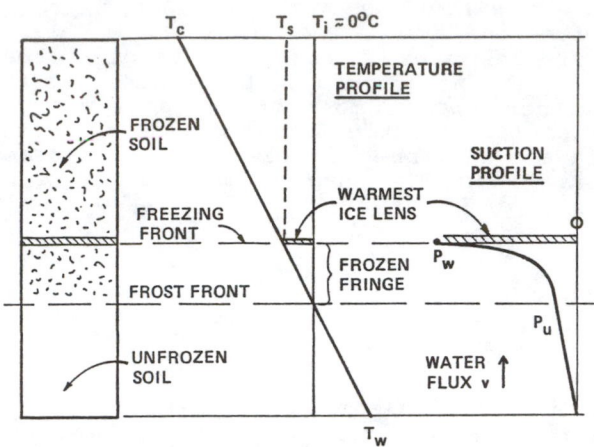

Figure 6:   Conditions associated with the frozen fringe.

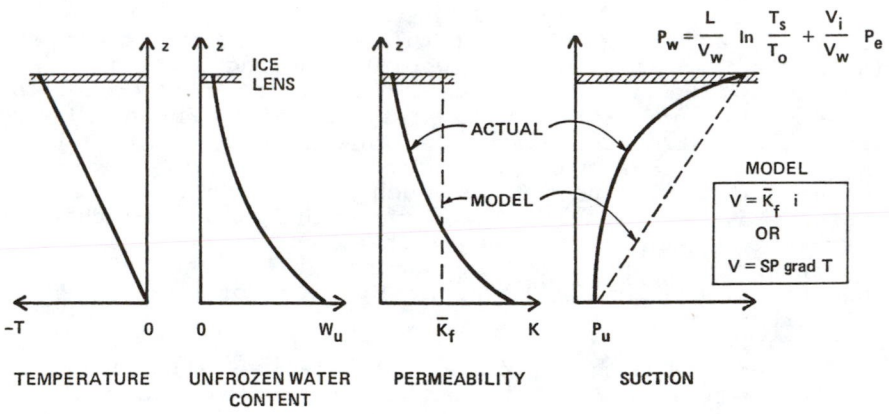

Figure 7:   Characteristics of the frozen fringe.

The average flux in the frozen fringe is given by:

$$v = - K_f (T) \; \delta\Phi/\delta z \qquad (17)$$

where $K_f(T)$ is the hydraulic conductivity of the frozen fringe as a function of temperature. Φ is the driving potential expressed in metres of water.

The actual characteristics of the frozen fringe are summarized in fig. 7.

2.2.2 Phase Composition of the Frozen Fringe During Transient Freezing

2.2.2.1 Effect of Suction

This section is devoted to the analysis of the phase composition in the frozen fringe during transient freezing since the hydraulic conductivity is related to the unfrozen water content and thus any change in phase composition will affect the frost heave process. From a thermodynamic point of view, W_u is solely related to change in chemical potential of the water as shown in fig. 4. However, state variables such as T and P affect the chemical potential which in turn influences W_u and thus the hydraulic conductivity of the frozen fringe.

Open system freezing is fundamentally different from a closed system freezing since water flowing through the frozen fringe interacts with the surrounding medium. In order to maintain continuity of flow within the frozen fringe subjected to a given temperature gradient requires melting or freezing some of the pore ice at each location of the frozen fringe to ensure compatibility between permeability distribution (i.e. unfrozen water content) and suction profile. A negative pore pressure change at the same temperature T, i.e. an increase in suction causes a decrease in W_u; hence in local hydraulic conductivity. Identically, a positive pore water pressure change at a given temperature, i.e. a decrease in suction, results in an increase in W_u. It should be noted that the pore ice pressure is free to change as the pore water pressure changes, however as stated earlier, the magnitude of the change is not easily predictable nor measureable. This is best illustrated by applying Kelvin's equation to point A in fig. 1.

$$P_i - P_w = 2\sigma_{iw}/r_A \qquad (18)$$

If the suction increases, it is plausible to assume that the pressure in the pore ice increases also and eq. (18) indicates that $P_i - P_w$ becomes larger and r_A must decrease, which in turn signifies that the unfrozen water content decreases as established earlier. However, we do not know how much unfrozen water will freeze to maintain thermodynamic equilibrium at the temperature corresponding to the location of point A in the frozen fringe.

This statement is further supported by Forland and Ratkje (1980) who derived the following expression for the water flux in the frozen fringe:

$$v = - L_{22} (L \Delta T/T + \Delta\mu_{WT}) \qquad (19)$$

When L_{22} is a coefficient that can be viewed as an expression of the hydraulic conductivity of the frozen fringe, $\Delta\mu_{WT}$ is the difference in chemical potential of the water due to changes other than T, thus composition and pressure.

2.2.2.2   The Mechanism of Ice Lens Formation

Martin (1959) presented a theory to explain rhythmic ice banding. It is based on unsteady heat flow conditions combined with a nucleation temperature appreciably lower than the temperature required for freezing the most easily frozen soil water. Furthermore, he stipulates that the permeability of the unfrozen soil is such that a steady-state heat flow condition can be maintained temporarily, thus allowing the frost front to stabilize for a while and the ice lens to thicken. Further frost front penetration is initiated as the water flux to the ice lens decreases in response to to decreasing permeability of the unfrozen soil. Ice nucleates again when the actual temperature is lowered to the nucleation temperature.

Spontaneous filling of saturated pores by ice has very significant implications for soils. As the zero degree isotherm front advances through the material, the position of the in-situ freezing front will be at the temperature at which ice will propagate through the maximum pore openings. Konrad and Morgenstern (1980) proposed a new mechanistic theory of ice lens formation in fine-grained soils based upon continuous freezing front penetration during unsteady heat flow and on the physical properties within the frozen fringe that vary in response to temperature changes.

Figure 8 represents schematically the process of initiation and growth of a single ice lens of the formation of the subsequent ice lens. Let us consider the time t at which one or more ice lenses have formed in the specimen. At this time, another ice lens is initiated at a given location defined by its particular segregation-freezing temperature T_{sf}. Because unsteady heat flow is still occurring, the frost front advances at some rate dX/dt. This, in turn, produces changes in the temperature profile across the sample and more specifically across the current frozen fringe. During a time interval Δt the base of the ice lens cools below T_{sf} and the extent of the frozen fringe increases by ΔX. In other words, the frozen fringe existing at the onset of formation of the ice lens increases in length with time and the temperature across it becomes colder. Consequently, the overall permeability of the fringe decreases, the flow path increases, and the suction potential developed at the base of the ice lens becomes greater. For temperatures close to $0^{\circ}C$, the suction at the base of the ice lens varies

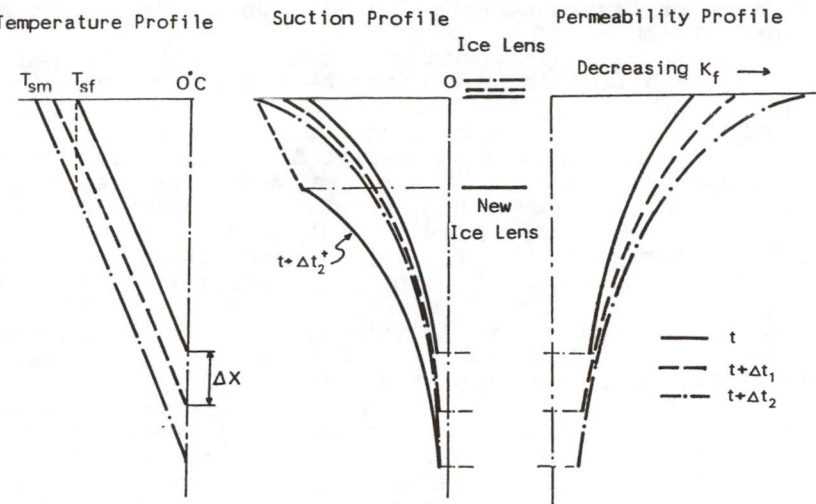

Figure 8:   Changes of the physical characteristics of the current frozen fringe during transient freezing.

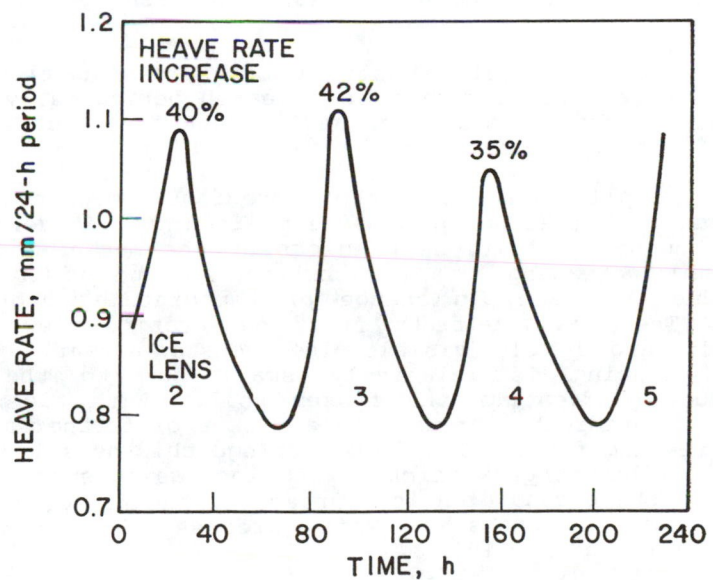

Figure 9:   Increase in heave rate during growth history of ice lenses. After Penner (1986).

linearly with temperature below $0^{\circ}C$. In contrast, the unfrozen water content and hence the frozen soil permeability decay more or less exponentially with decreasing temperature (Johansen 1977).

As long as the relation between suction and frozen fringe permeability is such that the water flow entering the fringe is able to traverse it, water will be drawn to the base of the ice lens and thus contribute to its growth. Therefore, during the time interval Δt the ice lens grows to a thickness equal to $1.09v\ \Delta t$ where v is the average water intake flux. Because the hydraulic conductivity of the frozen fringe decreases more rapidly than the suction with decreasing temperature, it follows that the average water flux decreases continually with time. Further penetration of the frost front results in further chilling of the current frozen fringe. After a while, the temperature at the base of the warmest ice lens reaches a value T_{sm} at which the permeability of the upper part of the fringe is so small that water flow is essentially stopped in the zone of extremely low permeability. Water now accumulates somewhere below the base of the former ice lens. The new level of accumulation is governed by the local permeability of the current frozen fringe, which can be associated with a segregation-freezing temperature of ice lens formation, T_{sf}. Because T_{sf} is warmer than T_{sm}, the average permeability of the new current frozen fringe is larger than the one associated with the former ice lens. The rate of ice lens growth is therefore higher at the initiation of a new ice lens, but steadily decreases during its growth. Experimental data reported by Penner (1986) support the above described mechanism as shown on fig. 9.

In a test with fixed thermal boundary conditions the process of ice lens formation is repeated periodically until steady-state conditions are reached and the final ice lens is formed at a temperature T_{so}.

The application of the previous mechanism to unidirectional freezing of a fine-grained soil under fixed thermal boundary conditions predicts at the outset a frozen zone with increasing ice enrichment, but no visible ice lenses due to the rapid change of temperature across the sample. Then, as the front frost penetration slows down, very thin and barely visible ice lenses appear. Their vertical spacing is relatively small due to the high temperature gradient in the frozen soil. With time, this temperature gradient decreases as a result of further advance of the freezing front. The frozen fringe thickness increases and the ice lenses grow thicker with increasing spacing. As the steady-state condition is approached, the fringe tends to its maximum thickness. This process is represented schematically in fig. 10.

2.2.2.3 Effect of Cooling Rate

It is worth recalling that many attempts have been made by different researchers (Beskow, 1935; Penner, 1968, 1972; Kaplar, 1968, 1970) to establish a relationship between

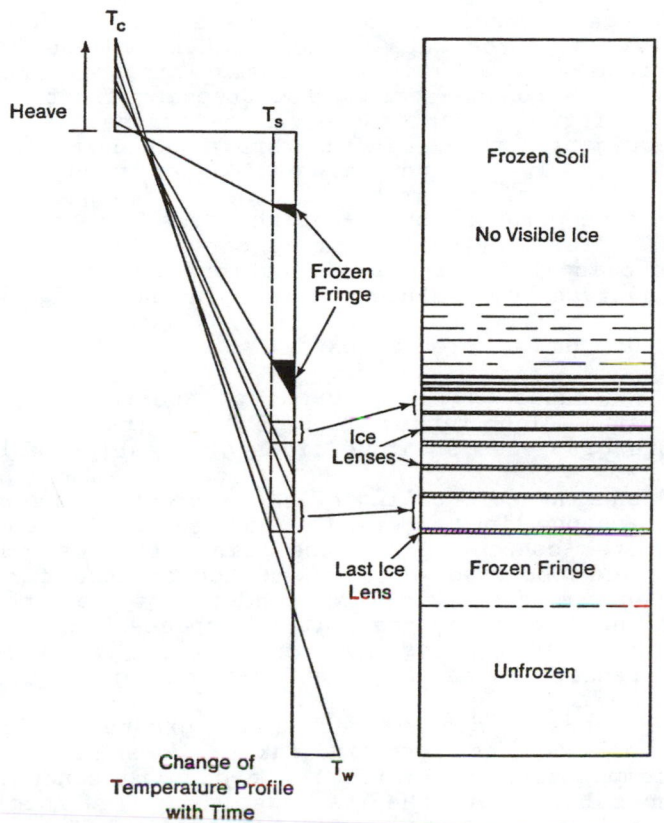

Figure 10: Schematic rhythmic ice lens formation with fixed temperature boundary conditions.

variables such as rate of heat extraction or rate of frost penetration and rate of heave. Unfortunately, no unification of data in all cases were achieved; although for some soils under specific freezing conditions the correlations appeared quite convincing.

The above described mechanism of ice lens formation stresses the importance of temperature changes within the frozen fringe, especially at the base of the warmest ice lens. Obviously, the rate of temperature change within the frozen fringe will also influence significantly the rate of water migration to the segregation-freezing front. The rate of cooling of the frozen fringe can be defined as the change in the segregation freezing temperature per unit time. Since the value of T_s is not always known precisely during transient freezing, the rate of cooling is approximated by the rate of temperature change at the freezing front which is close to $0^\circ C$ for many soils. It is convenient to express the rate of cooling in soils as a function of frost penetration rate and average temperature gradient in the fringe as:

$$dT_f/dt = grad\ T.(dX/dt) \tag{20}$$

Eq. (20) indicates that high rates of cooling can either be achieved with high rates of frost penetration, or large temperature gradients or a judicious combination of both.

When the rate of cooling of the current frozen fringe is high, the change in temperature can be so large that the unfrozen water content within the fringe reaches quickly its minimum value obtained at a given temperature for a given soil. Because the hydraulic conductivity of the frozen fringe is then extremely small it is expected that there is a limiting rate of cooling for which there is very little moisture transfer to the segregation-freezing front.

Small rates of cooling permit the growth of the current ice lens as long as its temperature does not exceed the maximum temperature for which the hydraulic conductivity of the frozen soil beneath the ice lens is too low forcing a new ice lens to initiate in the warmer zone of the frozen fringe. For these conditions, the average water flux to the growing ice lens decreases steadily during the cooling process until a new ice lens is initiated. As discussed earlier the rate of water transfer to a new ice lens is considerably higher at least at the beginning of the process, but with continuous cooling the water flux decreases steadily. It appears therefore that the rate of ice lenses initiation must also affect the overall frost heave rate and that a maximum rate of water migration may occur for intermediate rates of cooling.

2.2.2.4 Effect of Applied Load

Beskow (1935) was likely the first to document that externally applied pressures cause a decrease in the rates of heave of silty soils subjected to freezing. It has already been established that the application or removal of pressure on a frozen soil changes the chemical potential of the ice which in turn causes a change in the unfrozen water content. The application of load will also affect the segregation-freezing temperature since, for a given temperature, the unfrozen water content increases with increasing pressure in the ice. Plate I illustrates elegantly the validity of the above statement by showing that different ice lenses were created under the same temperature gradient simply by changing the applied pressure. The segregation-freezing temperature becomes colder with increasing applied pressure. The same phenomenon was also observed by Penner and Goodrich (1980) using the X-Ray technique.

The fact that the segregation-freezing temperature decreases with increasing pressure results in colder average temperatures in the fringe. Therefore, the average unfrozen water content in the fringe decreases as pressure is applied, despite the moderating effect of increasing pressures in the ice. The combination of reduced overall permeability of the frozen fringe and reduced average hydraulic gradient due to larger fringes and smaller suction potentials (see eq. 4) is then responsible for substantially reduced heave rates as load is applied.

Freezing tests with applied pressure conducted in the laboratory under fixed temperature boundary conditions, i.e. relatively small samples, often show that water may be first expelled from the freezing front and later attracted to it. These observations further indicate the importance of the rate of cooling of the current frozen fringe in the freezing process of soils. At the base of the ice lens the integration of eq. (4) gives:

$$P_w = (H_f \, T_s/V_w \, T_o^2) + V_i \, P_e/V_w \qquad (21)$$

where P_e is the pressure acting on the warmest ice lens.

The no-flow condition can be obtained from eq. 21 by setting $P_w = 0$ neglecting elevation head which yields:

$$T_s^* = T_o \, \exp(P_e \, V_i/L) \qquad (22)$$

Therefore if the actual segregation-freezing temperature which is dependent upon the rate of cooling of the fringe is warmer than T_s^* for a given applied pressure, the pore pressure in the water film adjacent to pore ice which is at a pressure equal to P_e is positive and expulsion of water occurs. Data obtained by Konrad and Morgenstern (1982) clearly establish a unique relationship between the rate of cooling and the applied pressure which water will be drawn into the sample during freezing (see fig. 11). The results also indicate that the higher the applied pressure, the lower the cooling rate to induce water migration towards the segregation-freezing front.

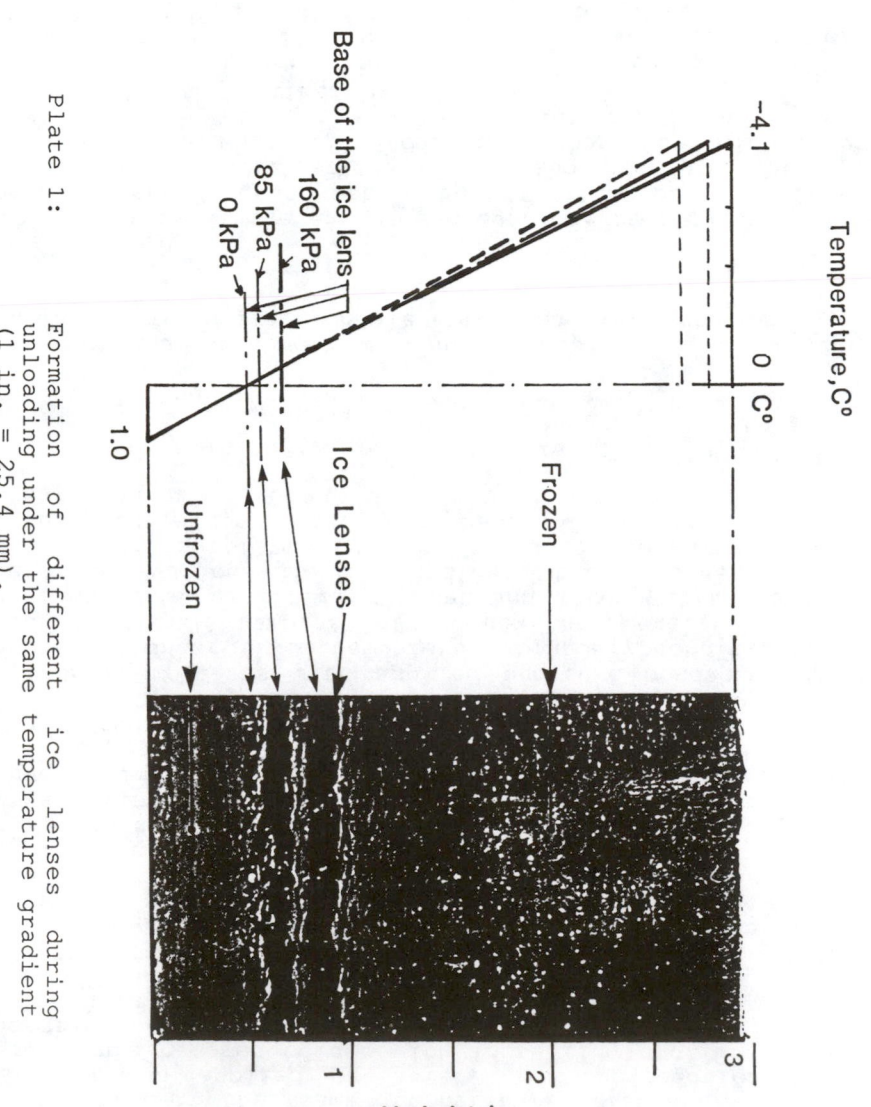

Temperature,C°

−4.1   0   C°

Base of the ice lens

160 kPa

85 kPa

0 kPa

1.0

Unfrozen

Ice Lenses

Frozen

3   2   1

Height,in.

Plate 1:   Formation of different ice lenses during
unloading under the same temperature gradient
(1 in. = 25.4 mm).

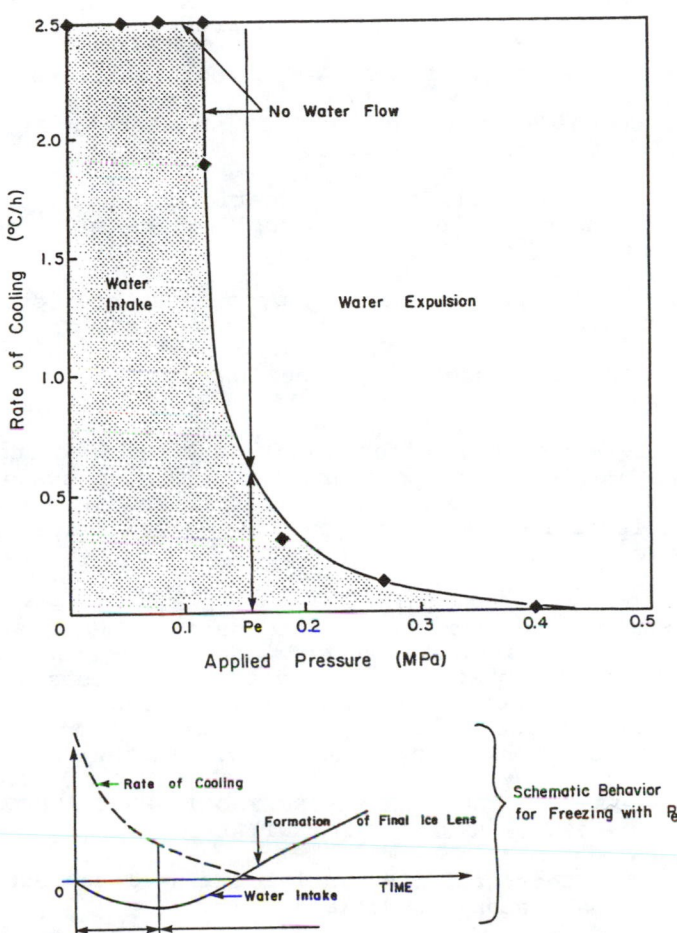

Figure 11:   Critical rate of cooling during freezing with applied load.

2.3 SUMMARY

A model of the freezing process in porous materials must therefore account for the following observations:

1. Not all the water freezes at $0^{o}C$ for a given soil, the unfrozen water content will depend upon temperature, suction in the water films and pressure in the pore ice.

2. The location of the segregation-freezing front (corresponding to the base of a growing ice lens) does not coincide with the frost front where pore ice first begins to form in the pores.

3. For a given soil, the segregation-freezing temperature is dependent at least upon applied pressure and rate of cooling of the frozen fringe.

Unfortunately, the freezing processes occur over a small temperature range of 0 to $-0.1^{o}C$ in sandy soils; 0 to $-0.5^{o}C$ in silts and 0 to $-2^{o}C$ in clays which makes precise measurements of segregation-freezing temperature a rather difficult task.

Furthermore, the effects of rate of cooling on the segregation-freezing temperature has not been systematically studied to-date mainly because of experimental difficulties. It has neither been accounted for by any existing predictive frost heave models.

Rather than determining these basic parameters of a freezing soil, Konrad and Morgenstern (1980, 1981, 1982) proposed a new parameter termed the segregation potential (SP) that can be obtained from any freezing tests without the need of local measurements of high accuracy and most importantly without requiring the determination of pore ice pressure or suction at any location within the frozen fringe.

3.1 PART 2. AN ENGINEERING FROST HEAVE APPROACH BASED ON THE SEGREGATION POTENTIAL

3.1.1 The Segregation Potential During Transient Freezing – Laboratory Conditions

From a phenomenological point of view, the mechanics of frost heave can be regarded as a problem of impeded drainage to an ice-water interface at the segregation front. Substantial suctions are generated at this interface, but the reduced permeability of the frozen fringe impedes the flow of water to the segregation front, particularly for fully saturated soils. Thus, it appears that the frost heave characteristics of freezing soils are related to the characteristics of the frozen fringe that, in turn, depend on the unfrozen water content. It is well known that the relationship between unfrozen water content and temperature for soils of different textures depends mainly on (1) specific surface of the soil, (2) gradation, grain size, amount of fines, type of minerals, (3) quantity and type of exchangeable cations, (4) pressure in ice and water phase, (5) solute concentration, and (6) density of the unfrozen soil.

Furthermore, in open system freezing the unfrozen water content depends also on the suction in the liquid phase as well as on the pressure in the pore ice.

If one considers that water migration occurs through a layered medium composed of saturated unfrozen soil and the frozen fringe, the freezing soil would be best characterized by suction at the segregation-freezing front and permeability of the frozen fringe (fig. 7). Both parameters are temperature and pressure dependent. Although these parameters provide a better insight into the physical processes involved during freezing, from a practical point of view they are difficult to measure.

As discussed by Konrad and Morgenstern (1980, 1981, 1982a, b), the characteristics of a freezing soil can also be represented by the segregation potential, SP, defined as the ratio of the rate of water migration, v, and the overall temperature gradient in the frozen fringe, grad T_f.

Thus, at any time t during transient freezing:

$$SP(t) = v(t)/gradT_f(t) \tag{23}$$

The use of SP is advantageous because it can be determined directly from a freezing test in which water intake rate and temperature profile with time are measured. Since the physical size of the frozen fringe is relatively small, its temperature gradient is approximated by the measured overall temperature gradient in the frozen soil near the 0°C isotherm.

It should be stressed that grad T_f in eq. 23 is introduced to define the thickness of the frozen fringe. The temperature distribution in a soil freezing unidirectionally, is a function of the thermal properties of the media and of the internal heat generation at the frost front and the ice lens, as a result of phase change. In laboratory freezing tests, quasi-linear temperature profiles in the frozen soil are obtained during the transient phase, and it is reasonable to approximate the overall temperature gradient in the frozen fringe by that measured over a distance of about 1 cm (Konrad, 1987). It should be noticed that during transient freezing grad T_f is always different from the temperature gradient in the unfrozen soil (grad T_u) depending on the amount of in-situ freezing and, of course, on the difference between the thermal properties of the two zones. However, as the frost front approaches its steady state position (i.e., no further in situ freezing in the pores), grad T_f will be much closer to grad T_u and might in some cases be equal to grad T_u if the difference between the thermal properties is small.

Both theoretical (see part I) and experimental results reported by Konrad and Morgenstern (Op. Cit.) indicate that SP is dependent on the average suction in the frozen fringe, the pressure applied on the ice lens and the rate of cooling of the frozen fringe.

It is difficult to estimate the average suction in the fringe because the distribution of pressure is not linear, as a result of decreasing hydraulic conductivity with decreasing temperatures. It has been argued that the suction at the frozen - unfrozen interface (P_u) gives an indication of the average suction in the fringe. The lower P_u is, the lower is the average suction. P_u is relatively easy to determine if Darcy's equation is applied to the unfrozen soil, once its permeability (K_u) is known and the velocity of moisture migration measured in a freezing test.

$$P_{atm} - P_u = v \, l_u/K_u \tag{24}$$

where l_u is the length of flow in the unfrozen soil.

The rate of cooling of the frozen fringe has been determined from eq. (20).

Fig. 12 represents a three-dimensional view of the engineering freezing characteristics of saturated Devon silt which is unique for a given initial porosity freezing under no applied load. As discussed in Part I, evidence of a limiting rate of cooling of the fringe is provided by the experimental results. For Devon silt, the limiting rate of cooling is approximately 2.5°C/h at which water flow to the freezing front does not occur.

Kaplar (1970) and Penner (1972) suspected that a limiting rate of frost penetration, at which only in-situ water freezing should occur, exists for any soil. Kaplar found that for gravelly sand this rate was about 1.6 cm/h. If one assumes that the temperature gradient was about 1°C/cm, the limiting rate of cooling becomes 1.6°C/h which is of the same order of magnitude as the critical value for Devon silt. Indeed, the limiting rate of cooling is soil type dependent.

Moreover, fig. 12 supports the view that there is a maximum value for SP obtained for a rate of cooling of about 0.2°C/h (4.8°C/day).

The characteristic surface shown in fig. 12 reveals also that the suction at the frozen - unfrozen interface cannot exceed a limiting value of about -80 kPa. This is thought to be the critical suction at which cavitation occurs in unfrozen Devon silt.

Fig. 13 shows that the effect of applied load on the segregation potential for a rate of cooling close to zero, i.e. when the frost front becomes stationary, can be accounted for in a simple manner.

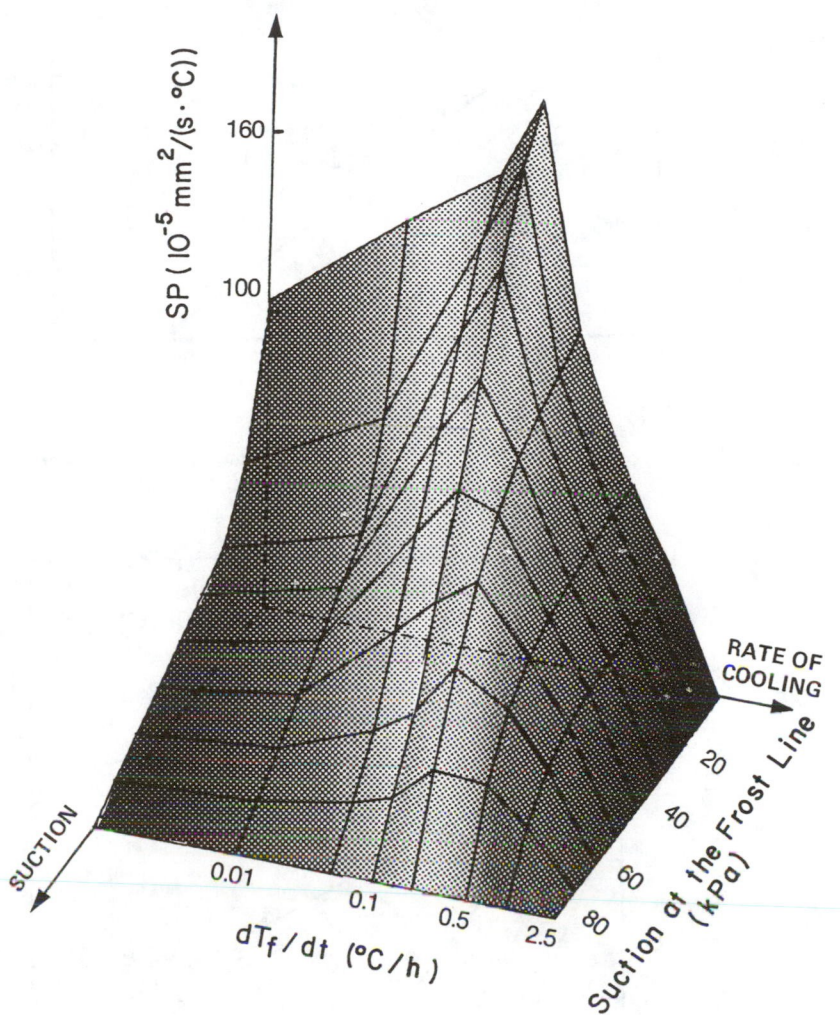

Figure 12:   Freezing characteristics of Devon silt for no applied load in terms of the segregation potential.

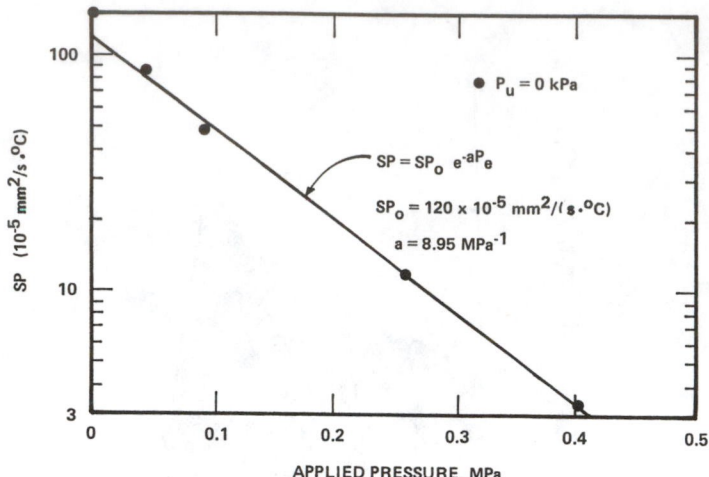

Figure 13: Effect of applied load on the segregation potential for Devon silt.

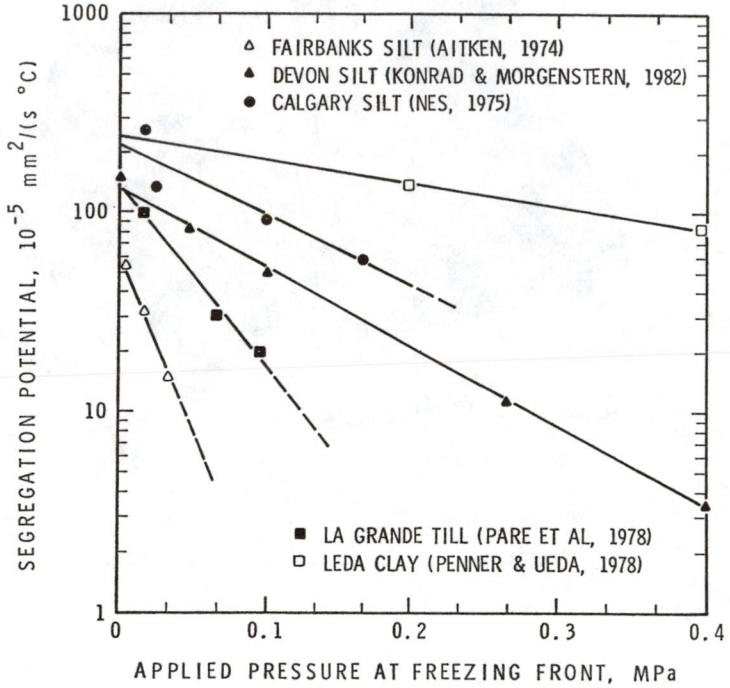

Figure 14: The segregation potential - pressure relationship for various soils.

$$SP\ (T_f\sim o)\ =\ SP_o\ exp\ (-aP_e) \tag{25}$$

where SP_o corresponds to the segregation potential obtained with no applied load and a is constant for a given soil. Again, experimental data show that the frost heave characteristics are strongly influenced by applied load. For Devon silt, the application of 250 kPa reduces the segregation potential by more than one order of magnitude.

The results of a literature survey on well documented cases of frost heave data are shown in fig. 14. The segregation potential for various applied pressures can be expressed by using eq. (25) for many soils ranging from sandy silts to clays, thereby giving strong support for the general validity of this relationship to characterize any freezing soil.

Fig. 14 also reveals the importance of including the overburden pressure in any frost susceptibility classification and any predictive frost heave model. In general, the segregation potential of soils with low unfrozen water contents (silty sands) will be very sensitive to small changes in overburden pressures; this is not the case for soils with higher unfrozen water contents such as clays.

Konrad and Morgenstern (1982) proposed an engineering frost heave model during transient freezing which considers the relationship between SP, T_f, P_u and P_e as input to the general formulation of heat and mass transfer outlined in fig. 6. The model performed well in all simulations including non-conventional freezing tests.

3.2 Prediction of Frost Heave for Field Conditions

Laboratory studies, have shown that the rate of segregation frost heave h_s can be expressed as:

$$\overset{\bullet}{h}_s\ =\ 1.09\ v\ =\ SP(P_u,\ P_e,\ T_f)\ grad\ T_f \tag{26}$$

Thus to successfully predict frost heave in the field during transient freezing using eq. (26), SP must be determined from laboratory tests using representative values of P_u, P_e and T_f. Evaluation of these parameters appears, at first sight, to require a substantial number of tests and this might seem to be a serious limitation on the practical use of the frost heave model that has been developed. These problems can be overcome by seeking an upper-bound value to frost heave and by considering the following simplications inherent to field freezing conditions.

In the field, frost penetration rates are usually small. It is common to observe in the field frost penetration rates of about 3 cm/day or less. Furthermore, the actual average temperature gradients in the frozen zone near the frost front are much smaller than in laboratory conditions owing to the greater thickness of frozen material.

It is current to have gradients of about $0.1^\circ C/cm$ or less. Eq. 20 then implies that the rate of cooling of the frozen fringe is likely to be less than $0.3^\circ C/day$ ($0.013^\circ C/h$) which is an extremely small rate of cooling.

Freezing experiments on Devon silt under fixed thermal boundary conditions have shown that the final ice lens was initiated when the rate of cooling was approximately $0.01^\circ C/h$. The small rate of cooling in the field, however, does not imply that a single ice lens is growing in the field. Because the frost front is not stationary, several ice lenses may form.

The segregation potential decreases with increasing suctions at the frost front. Therefore, an upper-bound of frost heave can be computed for the case where P_u is very small. In a laboratory test this would correspond to a warm plate temperature close to $0^\circ C$ to ensure small values of P_u since the length of unfrozen soil at the formation of the final ice lens would be very small. This assumption is not too conservative, because in many field freezing conditions P_u is usually small owing to high mass permeability of the unfrozen soil and small water migration rates. The upper-bound solution also assumes that water is abundantly available at the freezing front. If the soil is unsaturated, the rate of frost heave is considerably reduced.

In the field, the influence of overburden cannot be neglected. The effect of overburden becomes even more important if additional loading from berms or other types of surcharges are introduced. From the previous laboratory studies and considering that an upper-bound to frost heave is obtained by neglecting transient cooling of the frozen fringe and assuming zero suction at the frost front, the field frost heave characteristics of a given soil simply reduce to eq. (25).

It appears, therefore, that only a limited number of well-controlled freezing tests may be required to adequately characterize the segregation potential of homogeneous soil in the field over a wide range of overburden pressures. In practice, three freezing tests using constant temperature boundary conditions and different applied surcharges covering the expected range in the field suffice to define the field frost heave characteristics.

The predictive power of the field frost heave model in which the freezing soil is characterized by eq. (25), has been shown through the successful detailed analysis of the performance, over several years, of the chilled pipeline sections at the test facility at Calgary, Alberta. The results of the simulation for the deep burial section using the characteristics of Calgary silt given in fig. 14, is given in fig. 15 and compared with actual field measurements over about 2,000 days of freezing.

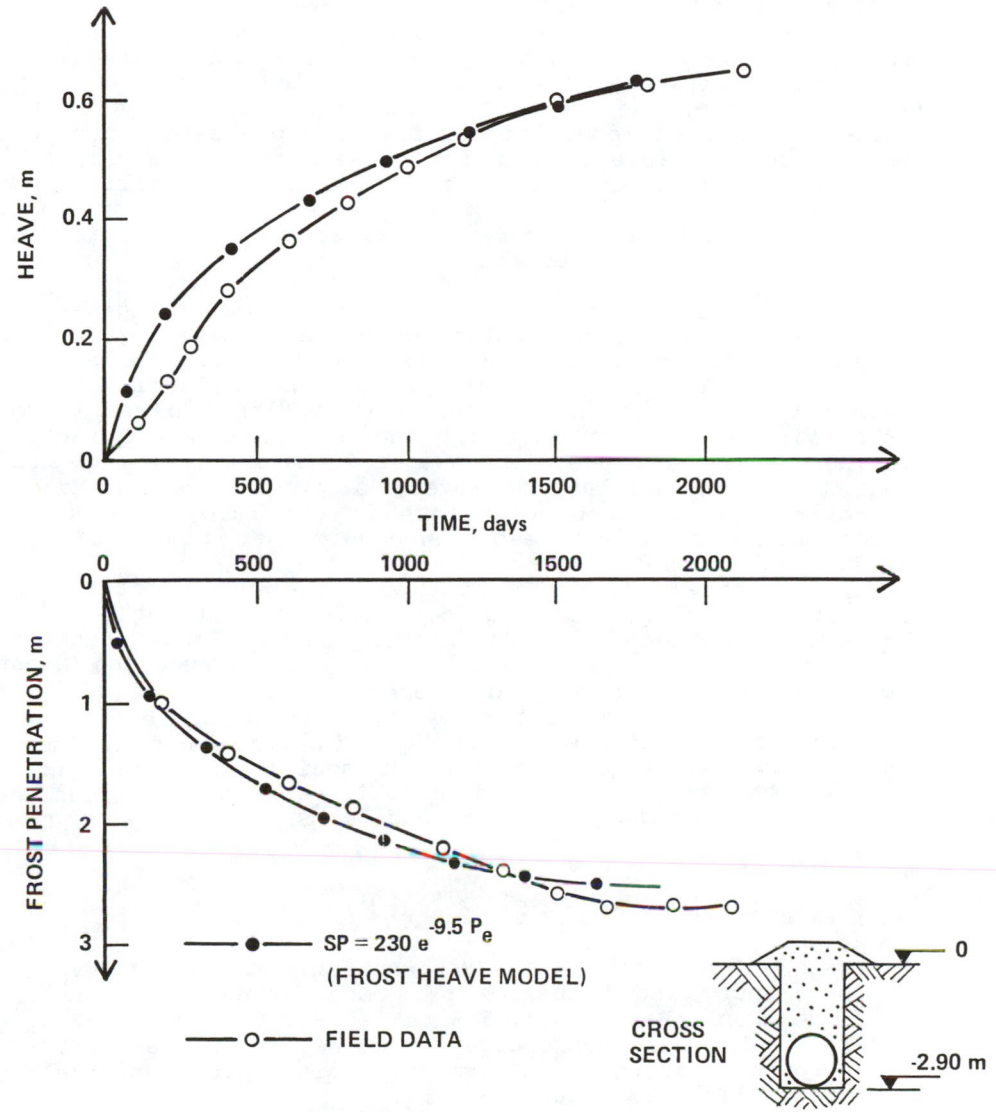

Figure 15:   Predictions and observations of frost heave for the deep burial pipeline test section at Calgary, Canada.

4   CONCLUSIONS

In this article the physics of the freezing of water in porous materials is reviewed in order to gain insight in the major processes controlling segregational frost heaving.

The application of reversible thermodynamics to open system freezing in a given soil indicates that segregational frost heave rate is at least dependent upon the segregation-freezing temperature, the pressure applied to the ice lens and the length, average suction and rate of cooling of the frozen fringe. A model for the freezing processes in porous media must therefore account for these facts. However, from a practical point of view, these parameters are difficult to measure and it is not surprising that to-date no model accounts for changes in segregation-freezing temperature with rate of cooling or in suction.

Based on the physics of the frozen fringe an engineering parameter termed the segregation potential was proposed as an alternative to detailed measurements of the more fundamental parameters. Because it is defined as the ratio of measured water migration rate and temperature gradient in the frozen fringe it embraces in fact all the above mentioned factors influencing frost heave. The use of SP as input in a general heat transfer formulation proved to yield excellent first approximation of frost heave both in the laboratory and in the field for transient heat flow conditions.

Baker et al. (1987) point out in the summary of "Ground Freezing '85" that it is desirable to strengthen the cooperation between scientist and engineers. This statement is fully endorsed by the writer as the engineering frost heave approach was only developed as a result of well-conducted scientific research by many reputed investigators. The scientists, however, must keep in mind that engineers must design and require models set against a background of theory, however with sound simplifications that can be readily used for many freezing conditions.

REFERENCES

Anderson, D.M. and N.R. Morgenstern, 1973. Physics, Chemistry and Mechanics of Frozen Ground. Proc. 2nd Int. Conf. Permafrost, Yakutsk, U.S.S.R., pp. 257-288.

Baker, T.H.W., H.L. Jesseberger, B.D. Kay and N. Maeno, 1987. Ground freezing '85-A Summary. Cold Regions Science and Technology, 131, 301-306.

Beskow, G., 1935. Soil freezing and frost heaving with special application to roads and railroads, (translated by J. Osterberg). Northwestern University Tech. Inst., 1947.

Biermans, M., K. Dijkema and D.A. de Vries, 1976. Water movement in porous media towards an ice front. Nature, Vol. 264, pp. 166-167.

Biermans, M., K. Dijkema and D.A. de Vries, 1978. Water movement in porous media towards an ice front. J. Hydrology, Vol. 37, pp. 137-148.

Bouyoucos, G.J., 1916. The freezing point method as a new means of measuring the concentration of the soil solution directly in the soil. Mich. Agric. College Exp. Station, Tech. Bull. 24, pp. 1-44.

Everett, D.H., 1961. The thermodynamics of frost damage to porous solids. Trans. Faraday Soc., Vol. 57, pp. 1541-1551.

Forland, T. and Ratkje, S.K., 1980. On the theory of frost heave. Frost i Jord, 21, pp. 45-49.

Gilpin, R.R., 1980. A model for the prediction of ice lensing and frost heave in soils. J. Water Resources Research.Vol. 16, No 5, pp 918-930.
A key reference

Gilpin, R.R., 1979. A model of the "liquid-like" layer between ice and a substrate with application to wire regelation and particle migration. J. of Colloid and Inter.

Hoekstra, P., 1969. Water movement and freezing pressures. Soil Sci. Soc. Amer. Proc., Vol. 33, pp. 512-518.

Hoekstra, P. and E. Chamberlain, 1964. Electro-osmosis in frozen soils. Nature, Vol. 203, pp. 1406-1407.

Johansen, O., 1977. Frost penetration and ice accumulation in soils. International Symposium on Frost Action in Soils, University of Lulea (Sweden), Feb. 16-18, Proceedings, Vol. 1, pp. 102-111.

Kaplar, C.W., 1968. New experiments to simplify frost susceptibility testing of soils. Highw. Res. Rec., 215, pp. 48-59.

Kaplar, C.W., 1970. Phenomenon and mechanism of frost heaving. Highway Research Record No. 304, pp. 1-13.

Kay, B.B. and P.H. Groenevelt, 1974. On the interaction of water and heat transport in frozen and unfrozen soils: I. Basic theory; the vapor phase. Soil Sci. Soc. Am. Proc., 38, pp. 395-400.

Konrad, J.-M., 1987. Procedure for determining the segregation potential of freezing soils. Geotechnical Testing Journal, Vol. 10, No. 2, pp. 51-58.

Konrad, J.-M., 1987. The influence of heat extraction rate in freezing soils. Cold Regions Science and Technology, June issue.

Konrad, J.-M. and N.R. Morgenstern, 1984. Frost heave prediction of chilled pipelines buried in unfrozen soils. Canadian Geotechnical Journal, Vol. 21, No. 1, February, pp. 100-115.

Konrad, J.-M. and N.R. Morgenstern, 1982. Effects of applied pressure on freezing soils. Canadian Geotechnical Journal, Vol. 19, No. 4, 1982, pp. 494-505.

Konrad, J.-M. and N.R. Morgenstern, 1982. Prediction of frost heave in the laboratory during transient freezing. Canadian Geotechical Journal, 19, pp. 250-259.

Konrad, J.-M. and N.R. Morgenstern, 1981. The segregation potential of a freezing soil. Canadian Geotechnical Journal, 18, pp. 482-491.

Konrad, J.-M. and N.R. Morgenstern, 1980. A mechanistic theory of ice lens formation in fine grained soils. Canadian Geotechnical Journal, 17, pp. 473-486.

Lovell, C., 1957. Temperature effects on phase composition and strength of partially frozen soil. Hwy. Res. Board Bull 168, pp. 74-95.

Mageau, D. and N.R. Morgenstern, 1979. Observations on moisture migration in frozen soils. Canadian Geotechnical Journal, Vol. 17, No. 1, Feb. 1980, pp. 54-60.

Martin, R.T., 1959. Rhythmic Ice Banding in Soil. Highway Research Board Bulletin 218, NAS-NRC, Washington, pp. 11-23.

Miller, R.D., 1972. Freezing and heaving of saturated and unsaturated soils. Highway Res. Rec. 393, pp. 1-11.

Nersesova, Z., and N.A. Tsytovich, 1963. Unfrozen water in frozen soils. Proc. 1st Int. Conf. Permafrost, Purdue Univ., Lafayette, Indiana, pp. 230-234.

Penner, E., 1986. Aspects of ice lens growth in soils. Cold Regions Science and Technology, Vol. 13, pp. 91-100.

Penner, E., 1972. Influence of freezing rate on frost heaving. Highway Res. Rec. 393, pp. 56-64.

Penner, E., 1968. Particle size as a basis for predicting frost action in soils. Soils and Foundations, Vol. 8, No. 4, pp. 21-28.

Penner, E. and L.E. Goodrich, 1980. Location of segregated ice in frost susceptible soil. Proc. 2nd Int. Symp. on Ground Freezing Trondheim, Norway, pp. 626-639.

Slusarchuk, W., J. Clark, J.F. Nixon, N.R. Morgenstern, P. Gaskin, 1978. Field test results of a chilled pipeline buried in unfrozen ground. Proc. 3rd Int. Conf. Permafrost, Edmonton, Alberta, pp. 878-890.

Vignes, M. and K. Dijkema, 1974. A model for the freezing of water in a dispersed medium. J. Colloid Interface Sci., Vol. 49, pp. 165-172.

Williams, P.J., 1968. Properties and behaviour of freezing soils. Norwegian Geotech. Inst. Publ., No. 72, Oslo.

Williams, P.J., 1976. Volume change in frozen soils. Laurits Bjerrum Memorial Volume, Norwegian Geotech. Inst., pp. 233-246.

ATMOSPHERIC AND MARINE ICINGS

Chapter 19

On the Modeling of Ice Accretion

E. P. LOZOWSKI
Division of Meteorology, Department of Geography,
University of Alberta, Edmonton, Alberta, Canada T6G 2H4

E. M. GATES
Formerly, Department of Mechanical Engineering, University of Alberta,
Edmonton, Alberta, Canada T6G 2G8

ABSTRACT

Ice accretion is a phenomenon that takes on many forms depending upon
where it occurs. The accretion of supercooled cloud droplets leads to
hailstone formation, ice on aircraft which can affect their performance
and control, and ice on transmission lines, which, in extreme
circumstances, can cause the collapse of their supporting towers. At
sea, ice accumulates on ships from comparatively large spray drops
generated by ship–wave collisions. This also affects their performance
and stability. Since the thermodynamics and much of the physics
associated with ice accretion under these different circumstances is
similar, numerical models of these various types of ice accretion have
much in common.

In this review paper, we begin by examining the underlying principles
which are common to all types of ice accretion. We then proceed to
examine the unique characteristics of each in turn. We conclude that
while a single numerical ice accretion model which covers all
possibilities might be feasible in principle, the differing geometries
and environmental conditions associated with each type are sufficiently
diverse that models tuned to a specific purpose and phenomenon are more
practical. We also emphasize the importance of both laboratory and
field measurements for model formulation and verification, and we
encourage scientists who specialize in the various sub–disciplines
(meteorological, aircraft, transmission line and marine) to get to know
each other's work.

CONTENTS

NOMENCLATURE

a	radiation linearization constant
$\vec{B}$	Bassett or history term in droplet equation of motion
c_i	specific heat capacity of ice
c_p	specific heat capacity of air at constant pressure
c_w	specific heat capacity of water or brine
c_{10}	atmospheric boundary layer drag coefficient at 10 m
C_D	droplet drag coefficient
D	droplet diameter
$\vec{D}$	drag force term in droplet equation of motion
e_a	environmental vapor pressure
e_d	saturation vapor pressure at t_d
e_s	saturation vapor pressure over brine
e_{so}	saturation vapor pressure over pure water
h	local heat transfer coefficient
H	wave height
I	ice accretion rate (expressed as a flux density)
k	thermal conductivity of ice substrate
k_a	thermal conductivity of air
ℓ_f	specific latent heat of freezing at t_f
ℓ_{vs}	specific latent heat of vaporization or sublimation at t_d
ℓ_{eff}	effective specific latent heat of formation of spongy ice
m	droplet mass
$\vec{M}$	apparent mass term in droplet equation of motion
n_f	freezing fraction - fraction of impinging water mass which freezes and remains as part of the accretion
n_i	icing fraction - fraction of impinging water mass which remains as part of the accretion whether as ice or as incorporated liquid = I/R
n_ℓ	"sponginess" or liquid fraction of the accretion
$n_{\ell max}$	postulated maximum liquid fraction of the accretion
Nu	Nusselt number
p	atmospheric pressure
$\vec{P}$	pressure gradient term in droplet equation of motion
Pr	Prandtl number
q_c	heat flux density by forced convection
q_e	heat flux density due to evaporation or sublimation
q_i	heat flux density due to cooling of ice to effective surface temperature, t_d
q_k	heat flux density by internal conduction

q_l	heat flux density due to release of latent heat of freezing
q_r	heat flux density due to net radiation
q_v	heat flux density due to aerodynamic heating
q_w	heat flux density due to warming or cooling directly impinging water to freezing point
q_w^*	heat flux density due to warming or cooling the shed water to the shedding temperature
r	local recovery factor
$\vec{r}$	position vector of droplet centre of mass
R	incoming mass flux density of water or brine
R_d	droplet radius
Re	Reynolds number
s_b	surface brine salinity
s_i	accretion salinity
s_w	impinging brine salinity
Sc	Schmidt number
t	time
t_a	air temperature in environment
t_d	surface temperature of the accretion
t_f	equilibrium freezing temperature of impinging water or brine
t_i	local ice thickness
t_s	mean temperature of the locally shed and/or runback water
t_w	mean temperature of the incoming water or brine
$\partial T/\partial n$	local temperature gradient along the direction normal to the accretion surface
$u(z)$	mean wind speed in boundary layer
u_{10}	mean wind speed at 10 m
u_*	friction velocity
v	free stream airspeed
$\vec{V}_a$	free stream air velocity vector
$\vec{V}_d$	droplet velocity vector
w	liquid water content of airstream
$\vec{W}$	weight of droplet
X_t	evaporation factor
z	height above mean ocean surface
z_o	roughness height
ϵ	ratio of molecular weights of water vapor and dry air
κ	von Karman's constant
ρ	ice density
ρ_a	air density

ρ_w brine density

σ Stefan-Boltzmann constant

τ time interval

1   INTRODUCTION

1.1   Background

Ice accretion in the atmosphere takes on many forms. In all of these, ice grows on some substrate as the result of the impingement and freezing of water drops or snow. Examples include the build-up of ice on ships due to spray (Figure 1), on aircraft flying through clouds (Figure 2), on transmission lines immersed in fog, cloud, snow, or freezing rain (Figure 3), and on precipitation particles such as graupel or hail (Figure 4).

Models of ice accretion are used for research, design, regulatory and forecasting purposes. But whatever their use, all models begin with certain quantities, usually the environmental conditions and the details of the icing substrate, and use these to predict certain characteristics (e.g., shape, mass or aerodynamic characteristics) of the accreted ice. Models can take various forms, but we will place the emphasis here on physical models, incorporating mathematical formulations which are amenable to solution on digital computers.

The goal of this paper is to provide a state-of-the-art review of the modelling of ice accretion on land, at sea, and in the air. It is intended to provide a synoptic perspective both for those scientists new to the field, and for those already working in a particular branch of icing modelling who may wish to get an idea of what is going on in related fields of research. It is not intended as a cook-book for creating an icing model, although by consulting the various references, the reader will have access to the resources necessary to do so.

FIGURE 1.   Ice accretion on the bow of the chemical tanker Anna Broere.

FIGURE 2. AS 332 Super Puma following a natural icing encounter (from AGARD, 1986).

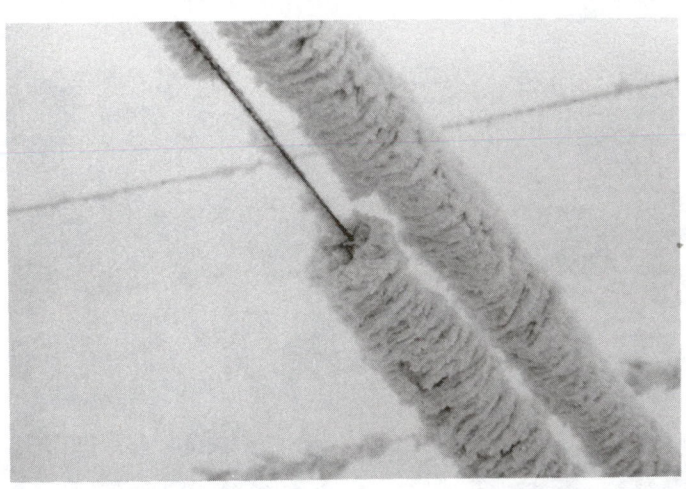

FIGURE 3. Ice accretion on a transmission line, observed during a three-day episode of natural icing near Grenoble, France. (Photo courtesy of P. Admirat.)

FIGURE 4. Cross-section (in reflected light) of a giant hailstone which fell near Cedoux, Saskatchewan, August 27, 1973.

Some of the early efforts to model ice accretion include those of Schumann (1938) and Ludlam (1951) for hail, Hardy (1946), Tribus (1952), and Messinger (1953) for aircraft, Mertins (1968) for ships, and Imai (1953) for transmission lines. Since these early days, ice accretion models have proliferated. Recent reviews in the areas of hail and marine icing include those of Lozowski and Gates (1986), Brunet (1986), Jessup (1984), List (1985), Macklin (1977), and Parrish and Heymsfield (1985). No comparable up-to-date reviews of transmission line icing or aircraft icing in general are available. For a review of transmission line icing the reader would do well to consult the proceedings of the three international workshops on atmospheric icing of structures (Minsk 1983, Ervik 1984, Welsh 1987). Makkonen (1984) is also a useful reference. For aircraft icing modelling the reports of Lozowski and Oleskiw (1983) and of Cansdale and Gent (1983) are a good starting point. Helicopter icing, specifically, has been reviewed recently by the NATO Advisory Group for Aerospace Research and Development (AGARD 1986).

There isn't always a clear distinction amongst models intended for research, design, regulatory and forecasting purposes. Research models tend to be more complex and forecasting models simpler, but the spectrum is continuous rather than discrete. There is a clearer distinction between physical and statistical models, the latter being used mostly for forecasting. Statistical models often contain some physical insight, but they rely primarily on statistical correlations between the input and output parameters. This limits them to specific situations to which they can be tailored and thereby produce accurate forecasts. One must be cautious, however, about applying them beyond the range of conditions for which they were derived. Physical models, relying as they do on fundamental physical laws, are, in principle at least, more general. However, our lack of knowledge of the phenomena, and limitations in computing power, usually introduce assumptions or parameterisations into these models, which can be very restrictive and give rise to errors. Consequently, the predictions of physical models may have to be adjusted or tuned using a statistical procedure if they are to be useful in forecasting.

The chief advantage of physical models is that they represent a way to synthesize our knowledge about the ice accretion process into a convenient package which can be used for analysis, design, and forecasting. With such models, it is possible, for example, to test the sensitivity of the accretion process to variations in the environmental parameters. Anti-icing or de-icing measures can be simulated and evaluated. And, given sufficient environmental information, normal and extreme value icing statistics can be generated. Finally, physical models can also be used to assess some of the consequences of ice accretion. Combined with aerodynamic and structural models, for example, the effects of ice loads may be evaluated. The chief disadvantage of physical models is that we don't usually know all of the physical details that the model requires. This is where statistical models have the edge, but, of course, these cannot be applied in situations where there has been no opportunity to collect the relevant icing statistics.

Some examples of the use of models for design and regulation can be found in the papers of Makkonen (1986) for transmission lines, Werner (1973) for helicopters, and Horjen and Vefsnmo (1985b) for marine structures. Forecasting models have been described by Anonymous (1980) for aircraft icing and by Overland et al. (1985) and MacDonald and Jessup (1985) for marine icing.

Ice accretion modelling is very much an international undertaking. Table 1 gives some idea of just how extensive it is. We have not attempted to be comprehensive in compiling this list, and we apologize to those individuals or groups whom we may have inadvertently omitted. The point of presenting this table is to make groups doing ice accretion modelling in one area, aware of groups doing similar things in another area. Heretofore, these groups have tended to remain isolated, attending their own specialist conferences, and publishing in widely diverse journals. The proprietary nature of some of the research has further compounded the intercommunication difficulties.

In what follows, we will begin our considerations by taking a brief look at the settings in which ice accretion occurs. We will then discuss the common physical principles which underlie all types of ice accretion modelling. After this, we will consider in turn the specific requirements and problems which are associated with the various modelling applications - transmission line icing, marine icing, aircraft icing, and meteorological icing. We will conclude with some recommendations for the improvement of ice accretion models in general.

1.2   The Geometry of Ice Accretion

Supercooled water droplets in an airstream can and do impinge and freeze on objects of just about any size and shape. Transmission lines, ships, aircraft, and hailstones are typical examples. Because the geometry of most of these "substrates" is quite complex, most of the effort in ice accretion modelling has so far been directed towards the modelling of icing on simple shapes - cylinders, plates, airfoils, and oblate spheroids. These have the advantage that information on airflow and heat transfer for such shapes is available in the literature. Although it is possible, in principle, using panel methods and Navier-Stokes solvers to

TABLE 1. Selected groups and individuals around the world, who are, or have been, involved in ice accretion model development.

AUSTRALIA
Prof. W.C. Macklin, Department of Physics; University of Western Australia, Nedlands, Western Australia, (hail).

CANADA
Mr. R. Brown, Mr. R. Jessup, Mr. L. Welsh; Atmospheric Environment Service, 4905 Dufferin Street, Downsview, Ontario, Canada M3H 5T4, (transmission lines, marine).

Prof. E.M. Gates, Prof. E.P. Lozowski, Dr. W.P. Zakrzewski, Dr. K. Szilder, Dr. K. Finstad; Department of Geography and Department of Mechanical Engineering, University of Alberta, Edmonton, Alberta, Canada, T6G 2H4, (hail, aircraft, transmission lines, marine).

Dr. S. Krishnasamy, Dr. D. Havard; The Hydro Commission of Ontario, Research Division, 800 Kipling Avenue, Toronto, Canada, M8Z 5S4, (transmission lines).

Prof. R. List; Department of Physics, University of Toronto, Toronto, Ontario, Canada, M5S 1A7, (hail).

Prof. P. McComber; Department of Applied Sciences, University of Quebec at Chicoutimi, Chicoutimi, Quebec, Canada, (transmission lines).

Dr. M.M. Oleskiw; Resource Technology Department, Alberta Research Council, Box 8330, Station F, Edmonton, Alberta, T6H 5X2, (aircraft).

Mr. J.R. Stallabrass; Low Temperature Laboratory, National Research Council of Canada, Montreal Road, Ottawa, Ontario, Canada, K1A 0R6, (aircraft, marine).

FINLAND
Prof. J. Launiainen; Department of Geophysics, University of Helsinki, Fabianinkatu 24A, SF-00100, Helsinki 10, Finland, (marine).

Dr. L. Makkonen; Laboratory of Structural Engineering, Technical Research Center of Finland, Betonimiehenkuja 3, SF-02150, Espoo, Finland, (transmission lines, marine).

FRANCE
Mr. J.J. Cassaing; ONERA, 29 Avenue de la Division Leclerc, 92320 Châtillon, France, (aircraft).

Prof. J-F. Gayet; Laboratorie Associé de Météorologie Physique, Complexe Scientifique des Cézeaux, Université de Clermont II, B.P. 45, F-63170, Aubiere, France, (transmission lines, aircraft).

Dr. J.C. Grenier and Dr. P. Admirat; EDF/CNRS/LGGE, B.P. 96, 11 Rue Diderot, STE/DER C.D. Grenoble, 38000 Grenoble, France, (transmission lines).

FEDERAL REPUBLIC OF GERMANY
Mr. H-E. Hoffmann, Mr. W. Fuchs, Mr. K-P. Schickel; Deutsche Forschungs- und Versuchsanstalt für Luft- und Raumfahrt, 8031 Wessling, Oberpfaffenhofen, West Germany, (aircraft).

NORWAY
Dr. M. Ervik; The Norwegian Research Institute of Electricity Supply (EFI), N-7034, Trondheim - NTH, Norway, (transmission lines).

Table 1, continued:

Dr. I. Horjen; River and Harbour Laboratory, Norwegian Hydrodynamic Laboratories, Klaebuveien 153, P.O.B. 4118 - Valentinlyst, N-7001 Trondheim, Norway, (marine).

UNITED KINGDOM
Mr. J.T. Cansdale, Mr. R.W. Gent; Royal Aircraft Establishment, Procurement Executive, Ministry of Defence, Farnborough, Hants, England, (aircraft).

Mr. A.E.W. Ford; Engineering Department, The Electricity Council, 30 Millbank Rd., London SW1P 4RD, England, (transmission lines).

Prof. G. Poots, Dr. J.W. Elliott; Department of Applied Mathematics, University of Hull, Cottingham Road, Hull, HU6 7RX, England, (transmission lines).

U.S.A.
Dr. S. Ackley; U.S. Army Cold Regions Research and Engineering Laboratory, 72 Lyme Road, Hanover, New Hampshire, U.S.A., 03755, (aircraft, transmission lines).

Prof. M.B. Bragg; Aeronautical and Astronautical Engineering Research Laboratory, The Ohio State University, 2300 West Case Road, Columbus, Ohio, U.S.A., 43220, (aircraft).

Dr. A.J. Heymsfield; Convective Storms Division, National Center for Atmospheric Research, P.O. Box 3000, Boulder, Colorado, U.S.A., 80303, (hail).

Mr. J.J. Kim; Boeing Military Airplane Company, P.O. Box 7730, Wichita, Kansas, U.S.A., 67277-7730, (aircraft).

Prof. C.D. MacArthur; University of Dayton Research Institute, 300 College Park, Dayton, Ohio, U.S.A., 45469, (aircraft).

Dr. G.J. Mulvey; Meteorology Research Incorporated, 464 West Woodbury Road, Altadena, California, U.S.A., 91001, (transmission lines).

Mr. J.E. Overland, Mr. A.L. Comiskey; Pacific Marine Environmental Laboratory, National Oceanic and Atmospheric Administration, Seattle, Washington, U.S.A., 98115, (marine).

Dr. J. Shaw, Dr. W.V. Olsen; NASA Lewis Research Centre, 2100 Brookpark Rd., Cleveland, Ohio, U.S.A., 44135, (aircraft).

Mr. J.B. Werner; Lockheed California Company, P.O. Box 551, Burbank, California, U.S.A., 91520, (aircraft).

USSR
Dr. E.P. Borisenkov, Dr. V.V. Panov, Prof. L.G. Kachurin; Arkticheskii i Antarkticheskii, Nauchno-Issledovatiel'skii Institut, Leningrad, Bering 38, USSR, (aircraft, marine).

---

calculate the flow around more complex geometries, this has not found much application in icing modelling so far. Instead, modellers have tried to break down the more complex structures into elements which can be approximated by some of the simple shapes mentioned above. However, the question of how to account for the orientations of the structural elements and their interactions, and the resulting effects on the ice accretion has not yet been addressed.

Even for the simple substrate shapes, the geometry of the ice accretion is not usually simple except at the outset. Because of the nature of the accretion process, the shape of the accreting surface is a function of time. Moreover, the shapes which develop are often convoluted, especially for lengthy accretions, with large roughness elements and icicles complicating the geometry. Certain general shapes, however, occur with sufficient frequency that terms such as "horned" and "spearhead" ice are in common usage (Figure 5). Only a few models currently predict the shape of the accretion. In most, the shape is assumed a priori. The reasons for this are not simply convenience and economy of modelling. At present, there is no simple way to numerically model the flow and heat transfer as a function of time for complicated, rough, three-dimensional shapes at the high Reynolds numbers typical of icing. Furthermore, in models where this has been attempted in a limited way, there are usually physical and numerical feedbacks which lead to artificial instabilities in the computed shapes.

1.3 Environmental Conditions Conducive to Icing

Generally speaking, ice accretion will occur if there is a flux of water impinging upon an object and heat transfer from the object sufficient to allow ice to form on it. The source of the water flux varies depending upon the location and type of icing. Supercooled cloud or fog droplets play an important role in aircraft, transmission line, and meteorological icing, with freezing raindrops also being of some significance. Marine

FIGURE 5a. Rime accretion at 1220 m on Cairngorm, Scotland. Rime accretions frequently adopt this "spearhead" shape.

FIGURE 5b. Ice accretion on a transmission line near Arna, Norway. Because the line rotates as the ice builds up, the resulting accretion has an almost circular symmetry.

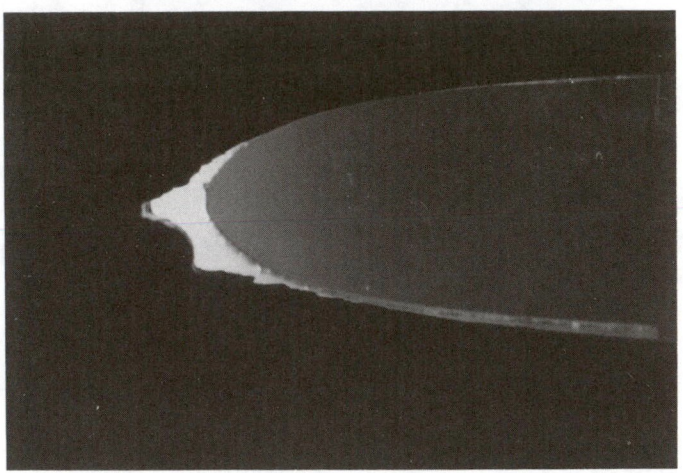

FIGURE 5c. Horned accretion on an NACA 0015 airfoil with 21.3 cm chord, as grown in the National Research Council icing wind tunnel. The growth conditions were -2°C, 110 m s^{-1}, liquid water content 0.28 g m^{-3}, ice crystal content 0.48 g m^{-3}.

icing, on the other hand, at least in its most severe manifestations, is almost exclusively the result of spray generated by wave impacts. Finally, the impinging "water" does not need to be entirely liquid. Melting ice particles (wet snow or graupel) can also be an important source of ice accretion, particularly for transmission lines. Ice accretions formed by direct deposition from the vapor (frost), while important in some circumstances, will not concern us here.

The heat flux requirements normally mean that the air and the substrate are at a temperature below the freezing point, although a temperature somewhat above freezing can give rise to ice accretion under the right circumstances (if the water is supercooled, for example, or if the air undergoes adiabatic cooling due to expansion, as in the carburetor icing). It is not, however, necessary for the impinging water to be supercooled, and it usually isn't in marine icing.

Naturally, there must be some mechanism for transporting the water towards the icing substrate. The impingement velocity of the water on the substrate will usually be a function of the wind velocity, the velocity of the icing substrate, and the dynamics of the water itself. These factors will assume differing degrees of importance depending upon the context. In aircraft icing, for example, the substrate velocity is all-important, for transmission lines, it is chiefly the wind speed, while for marine icing and meteorological icing, all three factors must be considered. The effects of turbulence on this transport are largely unknown.

When viewed on a microscopic scale, the water transport is a discrete phenomenon, the water arriving in individual drops. Few models take this into account. Most instead treat the problem as a continuous one, with variability at longer time and space scales associated with the spray generation process (marine icing) or the cloud properties.

The liquid water impingement is probably best characterized in terms of the liquid water flux in the air. This is the product of the liquid water content (spatial density) and the transport velocity of the water. Because smaller drops have a tendency to flow around objects with the airstream rather than impacting upon them, the drop size distribution in the cloud or spray is also important. Other characteristics of the water flux which will affect the ice accretion are the temperature and the salinity of the water.

In summary then, the main environmental variables controlling the rate of ice accretion are the three temperatures (air, water and substrate), the corresponding three velocities, and the other physical characteristics of the water flux (liquid water content, drop size distribution, and salinity). The important physical variables associated with the substrate are its size, shape, orientation, and thermal conductivity. Finally, the importance of icing duration cannot be overstated. Thus the envelope for the occurrence of icing lies in a thirteen-dimensional space. The envelope for possible icing will span the ranges of each of the variables. Because of correlations among them, however, the envelope for the actual occurrence of natural icing will be rather smaller. Finally, by neglecting the inconsequential and commonplace events, one could in principle determine an even smaller envelope for extreme icing events only.

There have been few attempts to determine such envelopes. One example is the FAR 29 "atmosphere" for the occurrence of aircraft icing. A discussion of icing atmospheres is included in the AGARD (1986) report. The nomograms which have been devised for forecasting marine icing also implicitly define such an envelope (Kachurin et al. 1974, Comiskey et al. 1984). More work needs to be done, however, to specify and understand the meteorology of icing, so that the envelope for natural icing occurrence, and in particular for extreme icing events, is neither overestimated nor underestimated.

2   PRINCIPLES OF ICE ACCRETION MODELLING

In this section, we will discuss some general principles that apply to all ice accretion models. There are essentially five aspects to the problem:

1. determination of the air flow about the icing substrate,
2. determination of the trajectories of the water drops,
3. thermodynamic calculations for the accretion to determine how much of the impinging water freezes,
4. splashing, shedding, surface flow and sponginess computations to determine what happens to the unfrozen water, and
5. growth physics.

We will consider these in turn.

2.1   Flow Calculations

The flow around the icing substrate needs to be determined because it controls both the water trajectory and the thermodynamics and heat transfer of the accretion. Accurate flow calculations are therefore critical. Detailed numerical solutions of the Navier-Stokes equation in the complex geometries and at the high Reynolds numbers typical of many icing situations, are impossible at present and unlikely in the foreseeable future. Simplifications have to be introduced in order to make the problem tractable.

A common simplification is to approximate the shape of the icing substrate. Circular cylinders are a common choice. If the substrate or the accretion is more or less cylindrical, one can apply the model results directly. If not, one must attempt to develop a correlation (usually based on icing experiments) between the cylindrical model predictions and the ice accretion on the actual substrate.

Another widely employed simplification is the use of potential flow to approximate the high Reynolds number flows which are typical of icing situations. When supplemented with boundary layer models, this has proved to be quite successful for airfoils (Gent and Cansdale 1985). However, some questions have been raised about the validity of this approximation when employed for bluff bodies with large wakes (Joe et al. 1976). K. Wong (personal communication) has attempted to simulate the effect of a cylinder wake on the water droplet trajectories. He finds an effect, but it is not large. Makkonen (1985) has devised a boundary layer model for a circular cylinder and used it to estimate the local heat transfer coefficient quite successfully. He uses the measured surface pressure distribution to drive the model, rather than that given by potential flow theory. However, it is unlikely that the boundary layer at high Reynolds number will have much of an effect on impacting drop trajectories, although it could be of importance for bouncing drops.

One complication of ice accretion modelling is that the interface between the accretion and the airstream is itself in motion and continually changing. Fortunately, the interface velocity is so very much smaller than the airstream velocity that it can be ignored. However, it is difficult to ignore the changing shape and dimensions of the accretion. In some models (e.g., Ackley and Templeton 1978 and 1979, Makkonen 1984) a simplification is introduced by assuming the accretion shape to be fixed and calculating its change in size only. In order to take into account the changing shape explicitly, one really needs to employ some methods for calculating potential flow about general accretion shapes. Such methods could also be employed for calculating the flow about more complicated single-element icing substrates. There are a number of such techniques now extant, deriving from the early work of Hess and Smith (1967), and going under the generic name of "panel methods". It is not our intention to describe these in detail here. Some general books which deal with this subject are those of Hewitt et al. (1976), Wirz and Smolderen (1978), Kollmann (1980), Hunt (1980), and Roe (1982). Applications to airfoils have been described by Kraus (1978) and Hunt (1980). Finally, use of these techniques in ice accretion modelling has been described by Oleskiw (1982), Bragg and Gregorek (1981), and Gent and Cansdale (1985). Only Oleskiw and Bragg have attempted to use these methods to follow the development of the icing shape with time, and the effects of these changes on the airflow, in a step-wise fashion. Only Gent and Cansdale have taken into account the effects of the compressibility of the airstream, which are important for icing at Mach numbers close to unity.

Some limited work has been undertaken so far to deal with complex multi-element structures. The work of Norment (1980, 1985) for aircraft fuselages, of Elliott and Poots (1986) for conductor cables, and of Lehtonen et al. (1986) for antenna masts are examples. Lozowski and Gates (1986) have alluded to the possibility of treating offshore drilling rigs in this way and Horjen and Vefsnmo (1985) have described some preliminary work along these lines. By ignoring fluid dynamical interactions among the structural elements, and decomposing the complex structures into arrays of elements with simple shapes, some progress has been made. However, where interaction is important, progress has been slow.

To complicate the picture still further, ice accretion often occurs in situations where the motion of the accretion relative to the airstream may be quite complicated. Oscillating helicopter rotor blades, galloping transmission lines, the rolling and heaving of ships, and the tumbling of hailstones are a few of the possibilities. In each of these cases, the general motion consists of a secondary motion superimposed on an overall translational motion. In most models, the effect of these secondary motions on the ice accretion have simply been ignored. However, there have been one or two attempts to take them into account in a simple way, without trying to solve for the full transient flow field (e.g., Szilder et al. 1987).

Finally, turbulence in the airstream and its effect on the accretion processes have been largely ignored, although Laforte (1984) has shown that relatively long time scale fluctuations in the wind speed (gustiness) can be important because of the non-linearity of the problem.

2.2 Drop Trajectory Calculations

The equation of motion for a non-evaporating, single drop may be written (Soo 1967):

$$m \frac{d^2 \vec{r}}{dt^2} = \vec{D} + \vec{P} + \vec{M} + \vec{B} + \vec{W} \tag{2.2.1}$$

where $\vec{r}$ is the position vector of the centre of mass, m is the total mass, t is the time, $\vec{D}$ is the drag force, $\vec{P}$ is the pressure gradient term, $\vec{M}$ is the apparent mass term, $\vec{B}$ is the Basset or "history" term, and $\vec{W}$ is the weight of the drop. Pearcey and Hill (1956), Landau and Lifshitz (1959), Clift et al. (1978), and King (1985) have discussed various aspects of the details of the right-hand side. For the purpose of icing, the terms $\vec{P}$, $\vec{M}$, and $\vec{B}$ may usually be neglected, since the drop density is much greater than that of the airstream, and since they are not as a rule undergoing large accelerations. Norment (1980) has shown, for example, that the Basset term may be neglected if the acceleration modulus is less than about 0.01. This is defined as:

$$\frac{D \left| \dfrac{d^2 \vec{r}}{dt} \right|}{\left| \dfrac{d\vec{r}}{dt} \right|^2}$$

where D is the drop diameter.

Making these simplifications leads to the equations of motion used by Langmuir and Blodgett (1946):

$$\frac{d\vec{V}_d}{dt} = - \frac{3}{8} \frac{\rho_a C_D}{\rho_d R_d} \left| \vec{V}_d - \vec{V}_a \right| (\vec{V}_d - \vec{V}_a) \tag{2.2.2}$$

where C_D is the drop drag coefficient, R_d is the drop radius, ρ_d is the drop density, $\vec{V}_d$ is the drop velocity vector, $\vec{V}_a$ is the air velocity vector, and ρ_a is the air density.

By further assuming the drop to be a rigid sphere with no internal circulation, one may make use of empirical data for the steady-state drag coefficient of spheres. Clift et al. (1978) provide a table of "recommended drag correlations" for this purpose.

Several authors have solved Equation (2.2.2) by various means (Figure 6), starting with Langmuir and Blodgett (1946), who used a differential analyzer. McComber and Touzot (1981) used a novel approach, treating the equations from an Eulerian rather than a Lagrangian point of view. However the solution is obtained, it is usually time-consuming. Consequently, Finstad et al. (1987) after checking Langmuir and Blodgett's results using a modern digital computer, proposed a series of approximating formulae for calculating the collision efficiency and related parameters, without having to solve the trajectory equation.

Unfortunately, there has been rather little experimental verification of the various trajectory calculations. Makkonen and Stallabrass (private

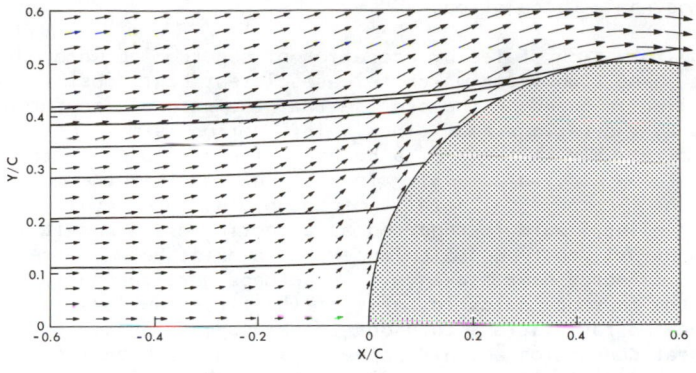

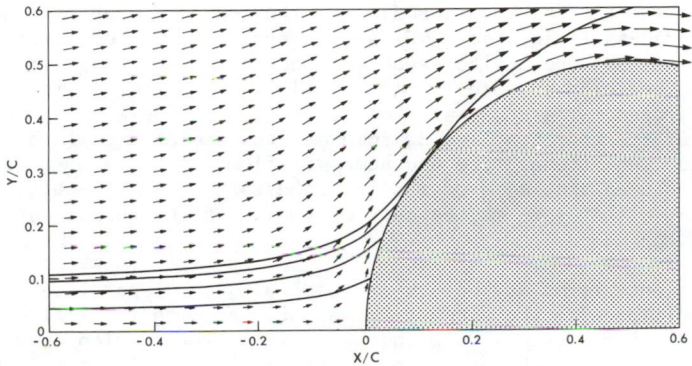

FIGURE 6. Airflow vectors (arrows) and droplet trajectories (solid curve) for a 15 cm diameter cylinder, in a freestream wind of 114.3 m s^{-1}. The droplet diameter is 84.2 µm in the upper figure and 14.0 µm in the lower one. (After Lozowski and Oleskiw, 1983.)

communication) have undertaken some wind tunnel experiments, the results of which are consistent with the trajectory computations of Finstad et al. (1987), under conditions simulating transmission line icing. There have also been some experimental verifications of trajectory computed collision efficiencies for airfoils (Gelder et al. 1956). Although the number of such studies has been rather few, they do generally indicate that the trajectory calculations are capable of yielding good predictions of collision efficiency, at least for relatively smooth icing surfaces. If the surface becomes quite convoluted, as can occur in transmission line icing or in hail, there has been a recent suggestion (Personne and Gayet 1987) that the effect of roughness on the flow and trajectories may need to be taken into account.

2.3 Thermodynamic Calculations

Because the formation of ice by accretion of supercooled water drops involves a phase transition which releases latent heat, most ice accretion modelling is based on the premise that the rate of ice formation is controlled by the various heat sinks. This notion is most often formulated quantitatively in terms of a balance of heat fluxes at the icing interface (Lozowski et al. 1983). In keeping with this thermodynamic approach, the microphysical details of the process are usually ignored, and the impinging liquid is regarded as a continuum. Since it is not obvious, a priori, that such a balance should exist between the inward and outward directed heat fluxes, it is perhaps more appropriate to formulate the thermodynamics in terms of the first law (Szilder et al. 1987). The two formulations turn out to be essentially equivalent when internal heat conduction and high speed transients can be ignored. Consequently, for simplicity, we will adopt this formulation here.

Before attempting to discuss the details of the heat balance, a brief overview of the current understanding of the general nature of the ice accretion process is in order. First, a certain amount of water (which may or may not be supercooled depending upon the situation) impinges locally on the icing surface. Some of this is converted to ice and some remains unfrozen. There may also be splashing, but we will ignore it for now. Of the unfrozen liquid, some may be incorporated into the ice matrix producing "spongy" ice and some may flow elsewhere. It is generally assumed that most of the ice adheres to the surface, although it is possible that some may be carried away with the runback or runoff water.

The heat balance terms may be sub-divided into four general categories: terms associated with the flow, terms associated with the incident water which stays, terms associated with the incident water which leaves, and terms associated with the accretion substrate. These are detailed below. We will discuss them in the context of the *local* heat balance on an icing object. However, with some minor adjustment, a similar set of terms is appropriate for discussing the *overall* heat transfer of the accretion.

Terms associated with the flow
- conduction and convection q_c
- evaporation and sublimation q_e
- dynamic heating q_v

Terms associated with the substrate
- internal conduction q_k
- radiation q_r

Terms associated with the impinging water which stays
- warming or cooling the water to the freezing point q_w
- latent heat of freezing q_l
- cooling the ice to the effective surface temperature q_i

Terms associated with the impinging water which leaves
- warming or cooling to the shedding temperature q_w^*

The heat balance equation may be written:

$$\sum_j q_j = 0 \qquad\qquad (2.3.1)$$

where we consider positive terms to be "sources" (heat transferred to the accretion and negative ones to be "sinks" (heat transferred away from the accretion). The detailed formulation of the various terms is given below:

$$q_c = h(t_a - t_d)$$

$$q_e = -h\left(\frac{Pr}{Sc}\right)^{.63} \frac{\varepsilon\ell_{vs}}{pc_p} (e_a - e_d)$$

$$q_r = -\sigma a(t_a - t_d)$$

$$q_v = \frac{hrv^2}{2c_p}$$

$$q_k = k\frac{\partial T}{\partial n}$$

$$q_w = Rn_i c_w(t_w - t_f)$$

$$q_w^* = R(1 - n_i)(t_w - t_s)$$

$$q_1 = Rn_f\ell_f(t_f)$$

$$q_i = Rc_i(t_f - t_d)$$

where a is the radiation linearization constant $8.1 \times 10^7 K^3$

c_i is the specific heat capacity of ice over the indicated temperature range

c_w is the specific heat capacity of water over the indicated temperature range

c_p is the specific heat capacity of air at constant pressure

e_a is the environmental vapor pressure

e_d is the saturation vapor pressure at t_d

h is the local heat transfer coefficient

k is the thermal conductivity of the substrate

$\ell_f(t_f)$ is the latent heat of freezing at t_f

ℓ_{vs} is the latent heat of sublimation or vaporization at t_d

n_i is the icing fraction (the fraction of the *impinging water mass* which remains as part of the accretion whether as ice or as incorporated liquid)

n_f is the freezing fraction (the fraction of the impinging water mass which freezes and remains as part of the accretion)

p is the static pressure

Pr is the Prandtl number for air

r is the local recovery factor

R   is the incoming mass flux equal to the sum of the directly impinging flux and the incoming surface flux (runback)

Sc   is the Schmidt number for air

t_a is the airstream temperature

t_d is the effective surface temperature of the ice deposit

t_f is the freezing temperature of the impinging water

t_s is the mean temperature of the locally shed or runback water

t_w is the mean temperature of the incoming liquid

$\frac{\partial T}{\partial n}$ is the local temperature gradient along the n-direction normal to the surface

v   is the freestream airspeed

ε   is the molecular weight ratio of water vapor and dry air

σ   is the Stefan-Boltzmann constant.

We have omitted here two terms which are exceedingly poorly understood: the kinetic energy of the impinging water, and the latent heat released by any of the shed water which freezes. They are not likely to play an important role except in certain special circumstances. For more discussion of these particular terms and of the others, the reader should consult Lozowski et al. (1979) or List (1985).

At this stage, it is appropriate to make some general remarks about the solution of Equation 2.3.1. Given the environmental conditions and a way of determining quantities such as the heat transfer coefficient, the internal temperature gradient, and the temperature of the shed water, the equation appears to be a function of three unknowns t_d, n_i, and n_f. The number of unknowns is reduced to two if one realizes that if n_i or $n_f < 1$, then t_d must equal t_f, and conversely, that if $n_i = n_f = 1$, then $t_d < t_f$. Thus Equation (2.3.1) can be seen as an equation either for determining t_d *or* for determining n_i, n_f. However, it cannot be used to determine n_i and n_f without another relation which specifies the "sponginess" of the accretion. The sponginess or liquid fraction of the accretion n_ℓ is related to n_i and n_f through:

$$n_\ell = 1 - \frac{n_f}{n_i} \qquad (2.3.2)$$

Thus if n_ℓ can be specified in some way, Equations (2.3.1) and (2.3.2) form a system of two equations in two unknowns. More will be said about this matter in section 4.1.

The different heat transfer terms vary in importance depending upon the application and the environmental conditions. List et al. (1965) have indicated their relative importance for hail. One may say with some generality that q_c and q_e are usually the most important heat sinks under a wide range of conditions. For this reason, the heat transfer coefficient, h, exercises a very strong control over the constitution and growth of the ice accretion.

The value of h which should be used in Equation (2.3.1) will depend on
the geometry of the ice substrate and on its roughness. Unfortunately,
there are no general theoretical means for determining h, a priori.
Consequently, it must either be determined experimentally, or in particularly amenable cases, determined through the solution of a boundary
layer flow model. Both approaches are fraught with problems. There
have been very few measurements of heat and mass transfer from actual
icing surfaces, or even from objects which closely approximate their
shape. The work of Schuepp and List (1969), Van Fossen et al. (1984),
and Arimilli et al. (1984) are the only examples known to the authors
(Figure 7). Consequently, modellers have had to rely on empirical heat
transfer correlations for very simple geometrical shapes, which more or
less approximate the ice accretions. These include, depending upon the
application, cylinders, spheres, and airfoils. Two typical correlations
which have seen widespread use for the *overall* heat transfer coefficient
are:

$$Nu = 0.6 \; Pr^{.33} \; Re^{.50} \qquad \text{for spheres (Ranz and Marshall 1952)} \qquad (2.3.3)$$

and

$$Nu = 0.26 \; Re^{.60} \qquad \text{for cylinders (Zukauskas 1972,} \qquad (2.3.4)$$
$$\text{Zukauskas and Ziugzda 1985)}$$

where

$$h = \frac{k_a Nu}{D}$$

and Nu is the Nusselt number, k_a is the thermal conductivity of air, Re
is the Reynolds number, and D is the sphere or cylinder diameter.

For some applications the local heat transfer coefficient rather than
its global average value is required. For cylinders, the work of
Achenbach (1977) is relevant. He has also examined the effect of roughness. This and other influences on the heat transfer have also been discussed by Narten et al. (1986).

Boundary layer models for determining the local heat transfer coefficient
in certain cases have been presented by Makkonen (1985) and Cansdale and
Gent (1983). These methods are computationally time consuming even in
two dimensions. Nevertheless, they hold out the possibility of general-
izing the empirical heat transfer data, and allowing it to be applied to
situations different from those of the original measurements. Since the
Reynolds number is not the only similarity parameter for heat transfer
from a rough surface, such models could play a very important role in
the future. For a more detailed review of heat transfer in icing, the
reader should consult Lozowski et al. (1987).

2.4 Surface Conditions

We have already alluded to the fact that the icing surface may well have
a flow of unfrozen or partly frozen "runback" water. Such a situation
is generally referred to as "wet" icing. If all of the impinging water
freezes in situ, the process is usually referred to as "dry" icing. In
either case, the surface can be rough, although the nature of the roughness elements differs in the two cases. In the dry case, the roughness

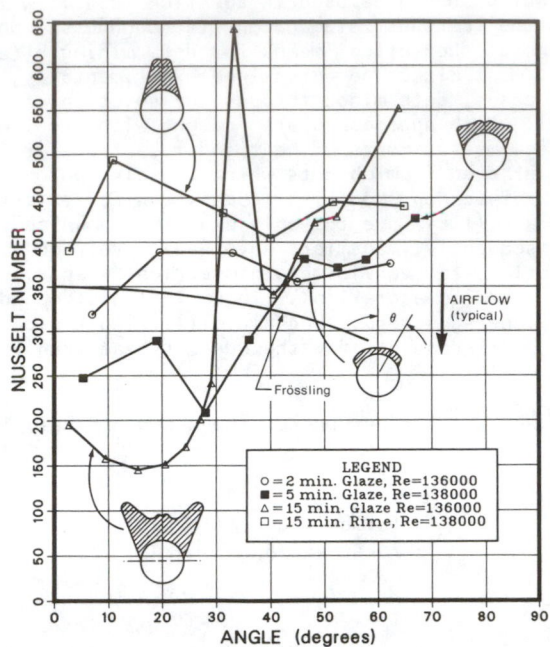

FIGURE 7. Heat transfer distribution around typical ice accretion shapes (after van Fossen et al. 1984).

elements are probably attributable to the stochastic process which give rise to rime feathers (Gates et al. 1987). In the wet case, they may result initially from surface tension effects (Olsen and Walker 1986) which produce small roughness elements that then grow due to a locally favorable heat and mass transfer. Naturally, the ensemble of roughness elements taken together will affect the global heat and mass transfer also. We referred to this possibility earlier in section 2.3.

Simply because all of the impinging water may not freeze locally does not mean that one can ignore the unfrozen portion. It may find its way along the surface to a region where the local collision efficiency is lower or the heat transfer coefficient higher, and consequently freeze there. This water may also give rise to icicle formation. Thus it can be important to keep track of this "runback" water. There are two basic approaches which have been followed in order to do this. The first is based on a simple mass balance as exemplified in the work of Lozowski et al. (1983). The second approach (Launiainen and Lyyra 1986) is based on a numerical solution of the boundary layer equations for the thin surface film. For two-dimensional applications where the surface flow is essentially unidirectional, the two approaches will give equivalent results in the steady state, since the thickness of the liquid layer must adjust, so that the aerodynamic stresses will give rise to a velocity field, which is capable of carrying the requisite mass flux. However, for three-dimensional situations, it is important to know where the water flows on the surface, and this cannot be determined from a mass balance alone. Unfortunately, to do this by solving the thin film equations in this case, would be computationally very expensive, and require a knowledge of the airflow about the detailed surface topography, which is simply unavailable in the present state of the art.

Loss of unfrozen water due to splashing, and the associated heat transfer, may also need to be taken into account under certain circumstances. However, Obreiter (1987) has shown, by modelling the trajectories of splashed drops around cylinders, that most splashed drops will re-impinge upon the ice accretion surface.

To complicate matters still further, it is conceivable that when the accretion is spongy, there may be an internal flow of liquid through the porous ice matrix, driven by a combination of gravity and the aerodynamic pressure gradients. It is largely unknown whether such a flow exists; however, if it does, it could be of some consequence, particularly for marine ice accretions (Zakrzewski, private communication).

2.5 Growth Physics

After determining the drop dynamics and the heat transfer, and calculating the ice accretion rate, I, the next step is to calculate the actual growth of the ice. This will lead to the shape of the ice accretion, or, if the shape is assumed, its dimensions. For some purposes, such as aerodynamics, the detailed shape of the accretion is important, and this must be determined by the model, through the calculation of local growth rates on the accreting surface (Lozowski et al. 1983). In other situations, such as transmission line ice accretion, certain aspects of the structural or environmental mechanics (e.g., wire rotation or changes in wind direction) may tend to give rise to an ice accretion with a simple cylindrical shape. In such cases, if the accretion shape is assumed a priori (e.g., Ackley and Templeton 1978, Makkonen 1984), then it is only necessary to use the model to calculate the overall ice accretion rate.

In either event, the dimensions of the accretion can only be determined with a knowledge of the ice density. Thus, given the density, ρ, the local ice thickness (normal to the surface) is:

$$t_i = \int_0^\tau \frac{I}{\rho}\, dt \doteq \frac{I}{\rho}\, \tau \qquad (2.5.1)$$

where τ is the time interval of the growth.

Most ice density formulations in numerical ice accretion models (e.g., Bain and Gayet 1983) have been based on the work of Macklin (1962) which gives the overall density of ice accretion on rotating cylinders. Farley (1987) reviews this and other formulations. Recent work by Makkonen and Stallabrass (private communication) has improved Macklin's formulation, by correcting for the bias in drop size measurement using oiled slides. Liu (1986) and Finstad (1986) have also extended Macklin's work by devising a formulation for the local density on non-rotating cylinders.

The accuracy of the approximation in Equation (2.5.1) is naturally dependent on how constant I and ρ are with time. The shorter τ is, the better the approximation should be. Consequently, in order to model ice accretion accurately over extended periods, it is necessary to break the entire interval into a number of shorter steps. When employing such a multi-step method, it is best to take into account the feedback effect of the developing ice accretion on the airflow, the droplet trajectories, and the heat transfer. The first two of these feedbacks can be taken into account most readily if the ice accretion shape is assumed a priori, or if panel methods are used to determine the flow (Oleskiw 1982). Because panel methods are computationally intensive, Finstad (1986) has devised some approximate methods for taking this feedback into account. The heat transfer feedback can only be taken into account adequately at present, if the ice accretion shape is assumed. This is an important limitation. It means that for now, it is not possible to calculate the detailed growth of wet ice accretion over an extended period of time. It is possible, however, to calculate wet growth over short periods (Gent and Cansdale 1985). It is also possible to calculate wet growth over extended periods, if either the expected growth symmetry or the level of accuracy required, allow a simple assumption to be made about the ice shape.

In an attempt to shorten the computation time by using long growth steps, τ, various authors have tried to make allowances for the approximation in Equation (2.5.1). These include a "correction" for the growth thickness (Lozowski et al. 1983), and an allowance for the growth direction (de Lorenzis 1979). While there may be some justification for both of these procedures, especially allowing for the growth direction, which is so apparent from bubble lines and rime feathers in actual accretion cross-sections, it remains preferable to use Equation (2.5.1) as it stands, with t_i normal to the surface and τ as short as practicable.

In the overall scheme of an ice accretion model, the importance of getting the growth right cannot be overemphasized. The feedback effect of the growth on the collision efficiency and heat transfer means that small growth errors can amplify quickly with time, thereby leading to very large errors over extended growth periods.

3   ICING ON TRANSMISSION LINES AND TOWERS

3.1   Icing Models

The modelling of icing on transmission lines and towers presents a number of specific problems. First among these is that the environmental conditions under which the icing occurs are usually unknown (especially in the design stage). Consequently, a large part of the overall modelling effort has to go into modelling the detailed meteorology of icing situations. Space limitations do not allow us to delve into such models. Suffice it to say that there has been some progress in this area. Ervik and Fikke (1986) give an idea of the current state of the art.

For given environmental conditions (wind speed, structure geometry, temperature, liquid water content, cloud droplet size distribution, fluxes of freezing rain or snow), the modelling proceeds much as described in Section 2. For transmission lines, the geometry is usually assumed to be that of a single circular cylinder, although Elliott and Poots (1986) have attempted to examine the interactions among bundled cable elements. For towers, Lehtonen et al. (1986) have had success in modelling ice growth by decomposing the tower into individual structural elements, and ignoring the interaction among them, at least until the ice growth closes the inter-elemental gaps. Another problem specific to transmission lines, with great importance for modelling, is the twisting of the line under torsional load of the accretion. Because of this phenomenon, lengthy icing events often lead to ice completely surrounding the cable.

Some modellers have attempted to take the cable rotation into account explicitly (McComber 1984, Egelhofer et al. 1984, Finstad 1986). Others have assumed the accretion to have cylindrical symmetry a priori (Makkonen 1984). Clearly, the level of computational effort required in the former is much greater than in the latter. Moreover, the models with explicit rotation are pretty much limited to an examination of rime accretion cases.

For practical modelling of ice accretion on transmission lines and towers, a number of other effects must also be accounted for. These will only be mentioned briefly. The variation of wind with height in the boundary layer must certainly be accounted for. The non-linear effects of gustiness have been discussed by Laforte et al. (1984). Finally, there is some evidence (Makkonen, private communication and our own unpublished experiments) that only the wind component normal to the cable is significant for icing. This result needs further verification, however.

3.2 Freezing Rain and Wet Snow Accretion Models

Although there are some similarities, there are sufficient differences between in-cloud icing and icing due to freezing rain or wet snow that it isn't really profitable to try to develop a single icing model which can handle all three cases. Moreover, the three phenomena are very unlikely to occur simultaneously, although they could occur in succession.

Some freezing precipitation models are based on simple correlations (Krishnasamy and Brown 1986), while others are more physically based (Stallabrass 1983, Makkonen 1986). Kolomeychuk et al. (1986) have reviewed several models for freezing precipitation accretion on transmission lines. After some initial controversy (Makkonen 1981, Kemp (1980), the physical modelling of wet snow accretion on transmission lines has been successfully tackled recently by Admirat et al. (1985). Their approach is based on a heat balance similar to that for in-cloud icing, which takes into account such factors as snow melting and Joule heating. No account however is taken of the mechanical properties of the snow (cohesion). Progress in modelling wet snow accretion continues to rely on the results of laboratory experiments and field measurements (Wakahama et al. 1977).

3.3 Ice Loads

The principal reason for wanting to model ice accretion on transmission lines, towers, and other structures is to be able to predict ice loads, particularly extreme loads, for design and regulatory purposes. An ice accretion model by itself will yield the static load due to the weight of the ice build-up. However, supplementary models are required to determine the static and dynamic wind loads. Although the dynamic loads can be the most significant in terms of structural failure, we will not attempt to examine such models here. While the phenomenon of "galloping" has been investigated for some time, (Pohlman and Havard 1983), the general aerodynamics of ice-covered structures has become a subject of more intense research in recent years as evidenced by an entire session on the subject at the Third International Workshop on Atmospheric Icing of Structures. Some insightful papers dealing with the modelling of aerodynamics include those of McComber and Bouchard (1986) and Davenport (1986).

One phenomenon which has a very important impact on the question of ice loads is the shedding of ice. Unfortunately, it has so far been resistant to attempts to model it. Some success has been achieved in modelling deliberate shedding induced by Joule heating of transmission lines (Elliott and Poots 1986). The modelling of shedding due to a combination of aerodynamic forces and mechanical fracture is largely terra incognita, although there is some experimental evidence that cable torsional stiffness can play an important role (Govoni and Ackley 1986, Lapeyre and Admirat 1986).

3.4 Future Progress

It is tempting to speculate in what direction research in this field will go in the near future. The severity of the problem and the commitment of various electrical utilities around the world to its solution, suggest that research will be ongoing for several years to come. However, rather than trying to forecast even the near future, we will content ourselves with pointing out a few critical areas related to modelling which are likely to yield profitable results. These are:

1. A comprehensive field measurement program for model verification. While many of the models have been wind tunnel tested, there have been no field measurements of ice accretion in which a sufficient number of environmental parameters has been continuously monitored during an entire icing storm, so that a meaningful comparison with a physical ice accretion model can be made. Since everything else hinges on the validity of existing models, their rigorous testing in the field is perhaps the most pressing requirement at present.

2. Combined meteorology/ice accretion/aerodynamic models. By combining ice accretion models with meteorological models which predict the environmental conditions and aerodynamic models which predict the resulting loads, it will be possible to establish a climatology of ice accretion, including extreme value statistics, and to attempt to forecast severe ice accretion events in advance, so that appropriate measures may be taken to mitigate the deleterious effects.

3. Ice accretion model improvements. A few areas stand out as likely candidates for further research to improve the accuracy of existing ice accretion models. The first is the introduction of time-dependence to allow for (slowly varying) environmental conditions. In this connection, the effects of a transition from very dry to wet growth yielding "soaked rime" (Pflaum 1984, Prodi et al. 1986) could be an important area for the effect of shape and roughness on collision efficiency and heat transfer. A third is the growth of rime feathers (Gates et al. 1987).

4. MARINE ICING

Two of the principal features of marine icing which distinguish it from other types of ice accretion are salinity and intermittency of the impinging spray. Makkonen (1987) has proposed a method for modelling salinity effects which will be discussed below. As for intermittency, there has been very little progress in models towards taking this phenomenon into account, even though the experimental evidence (Gates 1985) indicates that it can have a major effect. Perhaps the only current icing model which is capable of taking intermittency explicitly into account is the time-dependent model of Szilder et al. (1987). However, this model is probably too cumbersome to use operationally. Consequently, a suitable parameterization of intermittency effects will have to be developed, which can be employed in existing steady-state models.

4.1 Salinity Effects

Following Makkonen (1987), and treating the brine flux as continuous in time, one may write the salt balance equation for the deposit as:

$$Rs_w = Is_i + (R - I)s_b \qquad (4.1.1)$$

or

$$s_w = n_i s_i + (1 - n_i)s_b$$

where R is the brine flux, I is the icing rate expressed as a flux, s_w, s_i, s_b are, respectively, the average salinity of the impinging brine, of the accretion, and of the surface brine, and $n_i = I/R$ is the "icing fraction".

The picture of the accretion on which Equation (4.1.1) is based, is a two-component system, consisting of the "ice" surmounted by a layer of unfrozen brine. As the accretion forms, it traps within itself pockets of brine, giving rise to "spongy ice". I is the rate of formation of this spongy ice. The ice lattice itself rejects salt as it forms, thereby increasing the salinity of the unfrozen brine, both in the trapped pockets and on the surface. A simple and reasonable assumption is that the salinity of the brine pockets is the same as that of the surface brine. This assumption readily leads to the result that

$$\frac{s_i}{s_b} = n_\ell \qquad (4.1.2)$$

where n_ℓ is the liquid fraction of the spongy accretion (ratio of the mass of trapped brine to the total mass of the spongy accretion).

Combining Equations (4.1.1) and (4.1.2) leads to the relation:

$$s_b = \frac{s_w}{1 - n_i(1 - n_\ell)} \qquad (4.1.3)$$

The importance of Equation (4.1.3) is that $s_b > s_w$ and that s_b/s_w

becomes larger as $n_\ell \to 0$ and $n_i \to 1$, that is, as the liquid fraction of the accretion is reduced, and the heat transfer becomes capable of increasing the ice fraction towards unity.

If the accretion is assumed to be in equilibrium with the surface brine, then the interface temperature will be the freezing point of the brine. If the brine is well-mixed, there will be no salinity or temperature gradients, and the effective surface temperature, which determines the heat transfer, will be the equilibrium freezing temperature of the brine, t_f. Since t_f is a monotonically decreasing function of s_b (Schwerdtfeger 1963), a tendency to increase s_b will result in a tendency to reduce t_f, and hence reduce the convective and evaporative heat transfer. This in turn will tend to increase n_ℓ and decrease n_i, thereby reducing s_b. Because of the existence of this negative feedback, one expects that the system of equations which consists of the heat balance equation, $t_f(s_b)$, and Equation (4.1.3) will have a stable solution. Two additional equations are also required in the system to allow for the influence of sponginess on the effective specific latent heat of formation of spongy ice, viz:

$$\ell_{eff} = \ell_f (1 - n_\ell) \qquad\qquad (4.1.4)$$

and the influence of salt on the equilibrium vapor pressure:

$$e_s = e_{so}(1 - 0.537 \, s_b) \qquad\qquad (4.1.5)$$

where e_{so} is the saturation vapor pressure for pure water. By substituting Equations (4.1.3), (4.1.4), (4.1.5), and $t_f(s_b)$ into the heat balance equation, we obtain an equation in two unknowns, n_i and n_ℓ. In order to solve this equation, Makkonen (1987) presents some evidence to indicate that n_ℓ is approximately constant at 0.26. An alternative approach is as follows. If we can assume that the spongy matrix traps all the unfrozen brine up to a certain limit, $n_{\ell max}$, then n_i and n_ℓ are related as follows:

$$\text{if } n_\ell = n_{\ell max} \qquad \text{then } n_i < 1 \qquad\qquad (4.1.6)$$

$$\text{if } n_i = 1 \qquad \text{then } n_\ell < n_{\ell max} \qquad\qquad (4.1.7)$$

Condition (4.1.6) may be called the "soaked shedding" regime, while condition (4.1.7) corresponds to the "spongy regime". Lesins and List (1986) have suggested that for ice accretions subject to centrifugal forces (e.g., hailstones), there also exists an intermediate "spongy shedding" regime. In view of the previous remarks, there may be no "dry" regime ($n_i = 1$, $n_\ell = 0$) for marine ice accretion.

Using the approach described above, the effect of salinity on the thermodynamics of the accretion can be readily taken into account in a model. How to model the effect of salinity on the mechanical properties of the ice is not so obvious. The increase of the ice density due to sponginess can be easily accounted for. However, in order to model the effect on ice strength and adhesion, more experimental work will have to be undertaken.

4.2 The Marine Boundary Layer

The rate of icing will be a strong function of height in the marine boundary layer, because of the vertical variation of air temperature, wind, and particularly liquid water content. Under icing conditions (cold air over a warm sea), the boundary layer near the sea surface will be unstable, but, under the strong winds often associated with severe icing, its stability may not be far from neutral. Thus a logarithmic wind profile and isothermal condition is often assumed (Zakrzewski 1986). This may be written in the usual form:

$$u(z) = \frac{u_*}{\kappa} \ln \frac{z}{z_o} \tag{4.2.1}$$

where u_* and z_o may be related to the 10 m drag coefficient c_{10} as follows:

$$u_* = u_{10} c_{10}^{0.5} \tag{4.2.2}$$

$$z_o = 10 \exp(-\kappa c_{10}^{0.5}) \tag{4.2.3}$$

The drag coefficient c_{10} can in turn be related empirically to the 10 m wind speed u_{10} (Smith 1980):

$$10^3 c_{10} = 0.61 + 0.063 u_{10} \tag{4.2.4}$$

The vertical profile of liquid water content (LWC) is more problematical, because there are very few measurements and little, if any, theory to permit generalizations to be made. There is one point on which most workers agree, however, and that is that wind generated spray plays a rather minor role on the whole, compared with spray generated by wave impacts (Figure 8). One of the major difficulties is that the vertical profile of liquid water content in the spray cloud will undoubtedly be controlled by the precise mechanics of the impaction process. Thus, in addition to wind and wave conditions, the architecture, speed, and heading of the vessel will also play a role. Moreover, spray fallout will mean that the LWC profile will be a function of horizontal position as well. This lack of horizontal homogeneity could significantly increase the complexity of the icing process at sea. Nevertheless, observations of severe vessel icing generally show much stronger vertical gradients than horizontal ones. Consequently, the largest effort so far has been to model the vertical variation of LWC with height.

The Norwegian Hydrodynamic Laboratories have measured LWC in spray underneath the platforms of drilling vessels (Carstens, private communication), but unfortunately these data are still proprietary. Consequently, the only significant source of spray data, available in the open literature at present, is that of Borisenkov et al. (1975) for Soviet trawlers. Zakrzewski (1986), after making unit conversions, gives the relation as:

$$w = 2.30 \times 10^{-2} \exp(-z/1.82) \tag{4.2.5}$$

where z is the height in m above the deck of the trawler and w is LWC expressed in kg m^{-3}. Because this result is specific to a particular vessel and set of conditions, Zakrzewski (1986) has attempted to generalize the coefficient in this relation to allow for arbitrary wave heights

FIGURE 8. A supply vessel, heading towards a drilling platform off the coast of Norway, created its own cloud of spray through collision with waves. (Photo courtesy of I. Horjen.)

and ship speeds. The result given is:

$$w = 6.15 \times 10^{-5} Hv^2 \exp(-z/1.82) \qquad (4.2.6)$$

where H is the wave height in m and v the ship speed in m s^{-1} relative to the wave. Much more work needs to be done, however, to verify whether the Hv^2 proportionality is appropriate. It will also be very important to determine how the penetration scale height (1.82 m in Equation (4.2.5) and (4.2.6)) varies with the impact conditions.

Intermittency is another problematical feature in the modelling of marine icing. Most modellers seem to have tackled this problem by assuming continuous icing with a constant liquid water content equal to the time average value, even though Gates (1985) has demonstrated experimentally that this is not valid, at least under the conditions of his investigation. A similar assumption used by others, is that marine icing is always wet and there is always a liquid layer on the surface even between sprays. Thus once again, the growth may be treated as being continuous. Zakrzewski (1986) has argued that this may not be so, and that the drying of the surface due to runoff between sprays needs to be taken into account. More experimental data on this phenomenon will be needed, however, if this process is going to be modelled adequately.

Another feature unique to marine icing is the initially warm temperature of the spray and its subsequent cooling in the air. Injected into the air at the ocean surface temperature, it will begin to cool towards the air temperature. The amount of cooling will depend inter alia on the duration of the trajectory before impact. Stallabrass (1980) has proposed the following relation for the time-dependence of the drop temperature:

$$T_d = T_a + (T_w - T_a)e^{-t/\tau} \qquad (4.2.7)$$

644

where T_d, T_a, T_w are the drop, air, and ocean surface temperatures, respectively, and τ is a time constant given by

$$\tau = \frac{\rho_w c_w D^2}{6 k_a Nu X_t}$$

where ρ_w is brine density, c_w is specific heat capacity of the brine, D is drop diameter, k_a is thermal conductivity of air, Nu is Nusselt number for the drop, X_t is an evaporation factor (see Stallabrass (1980) for details).

Because of the generally high liquid water contents associated with marine icing, the sensible heat of the impinging brine can also play an important role in the heat balance. It is unfortunate therefore, that this term is so dependent upon the details of the drop history. In principle, Equation (4.2.7) can be used to try to take this effect into account in a model. However, it is questionable whether the basic premise of this equation is valid, since the spray drops are not generated as individual drops at the ocean surface. Their history is a complicated one, and a simple cooling equation such as (4.2.7) may not adequately describe it.

4.3 Modelling the Icing of Cylinders

There are several existing physical models of marine icing on cylinders, and more are under development. It is impossible here to discuss them in detail. Perhaps the earliest is that of Kachurin et al. (1974). Although the original Russian paper has been translated into English, Jessup's (1984) exposition is easier to read since he clarifies the model assumptions and details. What makes the model unique is that the physics of the surface liquid film are taken into account explicitly. Consequently, there are three heat balance equations to satisfy, two for the phase boundaries, and one for the substrate. It isn't clear that this complication is really necessary, unless one wishes to take into account the surface flux of brine. Such an attempt has been made recently in a model by Horjen and Vefsnmo (1985). Launiainen and Lyyra (1986) have also recently modelled the surface brine film explicitly, in an effort to investigate its effect on the heat transfer.

In view of the present state of ocean and wind tunnel experiments in this field, such complex models as these may not be justified, because it is very difficult to verify the results of incorporating liquid film effects. Consequently, the simpler models of Stallabrass (1980) and Makkonen (1987) should suffice for now. A careful and detailed model intercomparison would be beneficial at this stage, to help determine the direction for future research.

4.4 Modelling the Icing of Drilling Vessels

The only operational models of marine ice accretion on offshore structures that we are aware of are the ICEMOD model of the Norwegian Hydrodynamic Laboratories (Loset and Vefsnmo 1986) and the Canadian AES model (Roebber and Mitten 1987). These are menu-oriented computer programs for analyzing both ice accretion and thermal control methods.

Both of these models are based on the notion of structural decomposition, that is, breaking down the entire structure into a number of simple elements (usually cylinders or flat plates), for which the icing rate may be calculated using one of the existing models. While certainly justified at the present stage, such models do not take into account interaction effects among the structural elements. Although from the model point of view, each structural element is exposed to identical environmental conditions, in practice, the various elements and the structure as a whole will alter the undisturbed wind, wave, and spray conditions. While certainly important, it is unlikely that these inter-action effects will ever be taken into account through detailed modelling of the hydrodynamic of the atmospheric and oceanic interactions with the structure. Some form of parameterization will have to be used, and that will require experimental measurements. However, since a comprehensive field measurement program under all possible icing conditions could be very expensive, the greatest payoff will likely come from concentrating on the extreme events.

Future developments will probably tie these icing models into atmospheric and oceanographic forecasting models, so that significant icing events for offshore structures can be anticipated and suitable action taken. In this context, a third type of model which relates the icing to its effects on vessel operation and stability would be useful. Such detailed models do not exist at present.

4.5 Modelling the Icing of Ships

Only Zakrzewski (1986) and Zakrzewski and Lozowski (1987) have so far attempted a full simulation of ice accretion on ships. In this case, the structural decomposition method used for drilling vessels doesn't work, because most of the surfaces susceptible to icing are contiguous and generally not of simple geometry. Thus it is necessary to begin with a model for ship spraying (Zakrzewski 1986), (Figure 9). This model

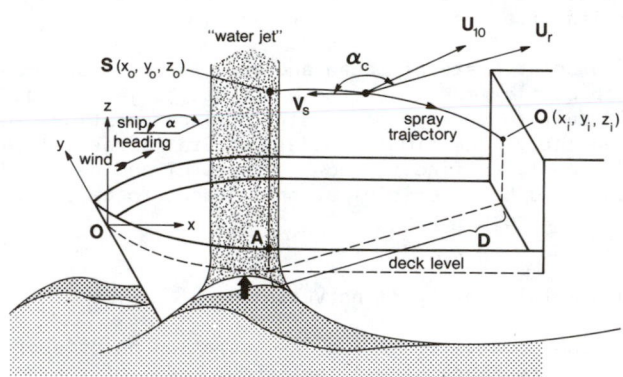

FIGURE 9. The model of Zakrzewski and Lozowski (1987) considers the details of spray trajectories over a computer modelled, three-dimensional ship.

neglects the effect of the ship on the wind field, and parameterizes the spray cloud generation process based on field measurements. In this way, it is possible to calculate the spray flux to each part of the vessel, taking into account the state of the atmosphere and ocean, and the archi-tecture and navigational practice of the vessel.

The critical difficulty in this model, and indeed in any model of vessel icing, is the lack of measurements of heat transfer from ships. Since in wet growth the icing rate is entirely controlled by the heat transfer coefficient, such measurements should receive a high priority. For the time being, however, approximations based on flat plates or cylinders are used.

Because of the difficulties and uncertainties of such a detailed model of ship icing, most other ship icing models have been of the model output statistics (MOS) variety. That is, a relatively simple physical icing model is used to predict marine ice accretion on a standard cylinder, under given environmental conditions. Statistical methods are then used to relate these model predictions under a range of conditions to observed vessel icing under the same conditions (Kachurin et al. 1974, Overland et al. 1986). Such an approach would be ideal if a single vessel type could be used to obtain the observational statistics, since marine ice accretion is vessel dependent and sensitive to navigation practices. However, in order to achieve a big enough statistical data base, ice accretion observations from various ships using various navigational procedures are usually assembled. Consequently, a lot of noise is intro-duced into the MOS procedure, and this reduces the accuracy of the fore-cast. Nevertheless, at present, this is the only means for forecasting vessel icing operationally. Future developments, however, will likely lead to the use of full simulation models to make vessel-specific icing forecasts. They will also allow the precise determination of the effect of icing on ship stability. Even more importantly, such models will allow optimum routing, handling, and navigation decisions to be made in order to minimize ice accretion effects.

5   AIRCRAFT ICING

The modelling of icing on aircraft has perhaps a longer history than any other aspect of ice accretion modelling. Moreover, aircraft ice accret-ion models have been able to draw upon the advances in aerodynamics generally, so that modelling in this field is relatively advanced. Nevertheless, aircraft ice accretion modelling presents a number of unique features which make it both interesting and challenging. Chief among these is perhaps aerodynamic heating, by which we mean the combined effects of viscous heating, and adiabatic compression (or expansion) in a compressible flow. Naturally, the geometry of aircraft icing is unique, involving various types of structures ranging from quasi-two-dimensional cylinders and airfoils to the fully three-dimensional fuse-lage. Finally, icing effects for aircraft are more complicated than for other icing situations. Here the mass of the ice is usually less important than its effect on lift, stability, and the operation of con-trol surfaces. Moreover, in some circumstances, ice which is removed and ingested into an engine, for example, may be more of a hazard than ice which stays in situ.

5.1 Icing on Airfoils

The modelling of icing on airfoils follows the general principles outlined in Section 2. Droplet trajectories may be calculated readily using panel methods (Kraus 1978) to determine the potential flow (Norment 1980 and 1985). Although the velocities are high, the droplets are usually quite small, so that the small aircraft components usually have the highest collision efficiency. Thus the parts most susceptible to icing are engine inlets, control surfaces, and the airfoils of small aircraft (especially helicopter rotor blades). In principle, panel methods should allow one to take into account both two- and three-dimensionality, multi-component airfoils (Kennedy and Marsden 1976), and swept wings. However, they are not suitable for handling turbulent or separated flows, such as may occur when the accretion becomes large.

As in most other applications of icing modelling, heat transfer is perhaps the major stumbling block. While boundary layer models exist which allow the estimation of the heat transfer coefficient on an airfoil in compressible flow, there are at present no general methods which will permit the determination of the heat transfer from arbitrary ice accretion shapes, short of attempting to solve the full Navier-Stokes equation (Potapczuk and Gerhart 1985). Some experimental data are available on heat transfer from simulated ice accretion shapes (Arimilli et al. 1984, van Fossen et al. 1984), but so far there has been insufficient data obtained to make general parameterizations for modelling purposes.

On the other hand, the modelling of aircraft icing is, in some ways, simpler than modelling other types of ice accretion. If thermal anti-icing or de-icing is used, the accretions will normally be small, the simulation times short, and the geometry of the substrate and flow will not be substantially changed by the ice. Consequently, time-dependent feedback effects can be ignored under these conditions.

Airfoil icing models could in principle be used for forecasting purposes, but most of them are at present too cumbersome for operational use. Instead, they are used chiefly for design, sensitivity testing, and certification purposes. They are particularly valuable as an adjunct to the certification procedure (Masters 1985), because of the difficulty of finding the extreme conditions of the icing envelope for full-scale, in situ flight testing. Airfoil icing models can also aid in the interpretation of icing instrumentation measurements, in terms of the actual icing conditions on the aircraft component.

There are now several models of airfoil icing being used, including a number in the private sector which are not well known. The ones in the public domain include those of Bragg and Gregorek (1981), Lozowski and Oleskiw (1981), Finstad (1986), and Gent and Cansdale (1985). The first three of these are rime ice accretion models, which do not take heat transfer into account. However, they are multi-step models which allow for feedbacks between the growing accretion and the airflow/droplet trajectories. Thus, they can be used to simulate quite large rime accretions. The Gent and Cansdale model is the only one of the group that calculates the heat transfer coefficient around the airfoil, thereby allowing a consideration of wet icing on airfoils. However, we are unaware of any models which explicitly take into account the feedback between the growth of the accretion and the heat transfer. Consequently, Gent and Cansdale's model is limited to calculating the initial icing rate and extrapolating in time for a single step. This limitation notwithstanding, the ability to take into account wet icing is a very important feature of the Gent and Cansdale model, since wet icing is likely to impose the most severe aerodynamic penalties (Sand et al. 1984), especially when drizzle size drops are involved, which can roughen the unprotected undersurface of the airfoil. Figure 10 shows a comparison between model and experiment for this model.

Although it shares many features with other types of airfil ice accretion, the modelling of icing on helicopter rotor blades involves several unique characteristics, mainly associated with the oscillatory nature of the flow. Airfoil icing models used for this purpose must really be time-dependent so that allowance can be made for the oscillation of both airspeed and angle of attack in forward flight. Although not an airfoil model, the model presented by Szilder et al. (1988) is able to handle such temporal oscillations.

A helicopter rotor blade icing model can be computationally quite expensive, since it will need to be applied at several points along the span of the blade. Centrifugal and related effects have not been incorporated into any models so far, even though there are experimental indications that they could be important (Ackley et al. 1979, Saunders and Zhang 1987). Naturally, the computation of ice accretion on propellers will also share some of these difficulties.

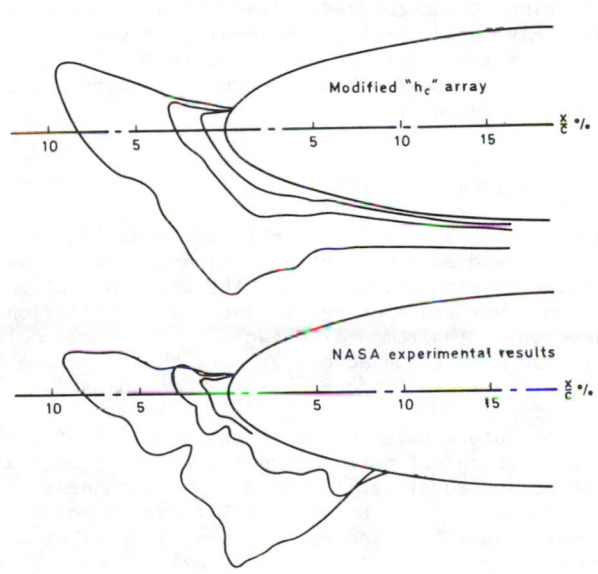

FIGURE 10. A comparison of accretion shapes at 2, 5, and 15 minutes for a NACA 0012 airfoil of 0.53 m chord. The conditions are Mach number 0.175, -9.5°C, LWC 2.1 g m^{-3}, angle of attack 8°. The upper figure is the model prediction, the lower the NASA experimental result.

5.2 Modelling the Effects of Aircraft Icing

One of the principal reasons for modelling icing on aircraft is to predict its effect on aircraft performance. Some progress has been made in the development of coupled icing/performance models, at least for air- foils (Bragg and Gregorek 1982, Potapczuk and Gerhart 1985). However, it may be some time before the performance of the entire aircraft can be predicted based on icing modelling alone. Actual icing flights will still be necessary for this purpose for some time to come (Cooper et al. 1984).

Although it is possible in principle to model the effects of icing on the drag, lift, stall, and possibly also the control characteristics of an aircraft, one of the major icing concerns, namely ice shedding, may never be successfully modelled because its behaviour is too random. Nevertheless, this can be a significant hazard if engine ingestion or collision with the airframe occurs.

5.3 Engine Inlet Icing

Next to airfoil icing, the modelling of engine inlet icing has received the greatest attention (Rosen and Potash 1981). There has been very little work done on the modelling of icing on other parts of the air- frame, particularly such sensitive areas as windows and instrumentation, perhaps because such small, specific portions of the aircraft are relatively easy to keep ice-free.

Engine inlet icing introduces several additional complications. First, the flow is likely to be three-dimensional, and second, the heat transfer characteristics of the inlet may not be well-known. The first of these problems seems to have been largely overcome (Albers and Stockman 1975, Kim 1985), but the second remains.

5.4 Future Directions

If significant progress is to be made in the modelling of aircraft icing, future work will need to concentrate on time-dependence and the feed- back between the growing accretion and the airflow. Of particular import- ance will be the heat transfer regime and its modification as the ice accretion develops. The temporal changes in the local collision effic- iencies will also have to be accounted for, but there are already indications that this can be done (Lozowski and Oleskiw 1981).

Another area for future development is the coupling of airframe icing models with meteorological forecasting models to produce aircraft- specific forecasts. Model verification in wind tunnels will continue to be important, but it will be necessary first to resolve a couple of recent controversies (Olsen and Walker 1986, Itagaki et al. 1986). The first has to do with the significance (or existence) of surface water flow for aerodynamic icing, the second with the importance of environ- mental relative humidity and possible differences between natural clouds and icing wind tunnels. Finally, model verification and application will depend ultimately on an improved worldwide aircraft icing data base (Jeck 1986).

All forms of icing are "meteorological" in the sense that the weather
is the controlling element. However, we use the term "meteorological
icing" here to mean the accretional growth of several types of frozen
precipitation in clouds. In many ways, the modelling of the growth of
frozen hydrometeors in clouds (rimed snow, graupel, hail) is the most
challenging type of ice accretion modelling. The geometry of the
phenomenon is fully three-dimensional, and one cannot take much advantage
of symmetry, except to employ the frequently-invoked approximation of
sphericity. There are some interesting feedbacks which influence the
modelling process as well. For example, the airspeed relative to the
growing particle (or equivalently its fallspeed) is not an independent
variable. Rather it depends on the size, density, shape, and roughness
of the particle and hence on the growth process itself. Thus, growth
even under specified environmental conditions (air temperature, liquid
water content, and drop size spectrum), is complicated enough. Even
greater complexity arises, however, when one realizes that another order
of feedback controls the very environmental conditions under which the
particle grows. The fallspeed of the growing particle determines
whether it ascends or descends and hence whether the air temperature it
encounters rises or falls with time. The release of the latent heat of
freezing, sweepout of liquid water, and shedding of water by the growing
particle introduce a collective feedback process, since one is dealing
not with individual particles, but with ensembles. At this level of
feedback, the collective accretion process influences the dynamics of
the cloud, which in turn controls the growth environment. When turbulent
fluctuations are superimposed on all of this, it is a wonder that any
advances at all have been made in this field. The fact is, however,
that much progress has been made, and that, although there have been
some ad hoc assumptions and empirical approximations employed, the most
advanced numerical cloud models are capable of handling the development
of ice particles, starting from single crystals and proceeding to the
formation of giant, irregular hailstones.

6.1 Detailed Frozen Hydrometeor Growth Models

Parrish and Heymsfield (1985) have outlined a model framework in which
the entire life history of an ice hydrometeor can be followed. Diffusional growth is included as well as accretion. Some of the details are
supplied by Rasmussen and Heymsfield (1987). It is necessary to para-
meterize the collision efficiency, heat and mass transfer, terminal
velocity and density calculations, and these authors have searched the
literature to determine the most appropriate formulations. This growth
model is intended for use with kinematic airflow models, either as prescribed mathematically or as derived from Doppler radar measurements of
storms. Similar, though simpler, growth and trajectory models for single
frozen hydrometeors have also been described by Nelson (1983), Xu (1983),
and Foote (1984).

These groups are not the only ones, however, to work on the detailed
growth and trajectories of frozen hydrometeors. Farley (1985) has
described the "hail category model" developed by Orville and his group
at the South Dakota Institute of Mines and Technology. While the
individual particle microphysics are not as detailed as those of Parrish
and Heymsfield, the modelling is performed in the context of a dynamical
cloud model, so that the collective feedbacks alluded to above are taken
into account.

Computer limitations often require simpler parameterizations of the growth/accretion process than those described above. One can also argue that measurement limitations preclude the verification of detailed microphysical parameterizations, and consequently, that there is some justification for assuming microphysical simplicity. For these reasons, some investigators have used simpler growth/accretion models. Examples are included in the work of Lin et al. (1983), Proctor (1985), and Knight and Knupp (1986).

All of this work owes a great deal to some of the original research on hail growth modelling by Ludlam (1958), Macklin (1963), List et al. (1965), and English (1973). More recent expositions are to be found in Macklin (1977), Ludlam (1980), and List (1985).

Workers in other areas of ice accretion modelling will find it profitable to study the recent work of List (1985) in particular, since it concentrates more on simulating the physics of the accretion process itself, rather than on the trajectory of the growing hailstone in the cloud. Unfortunately, the work of List and his students does not suggest that we will quickly solve all of the problems of hailstone ice accretion modelling. The complications introduced by the aerodynamics of the falling hailstone, coupled with the heat transfer associated with shed water, will provide a fertile field of research for some time to come.

7   CONCLUSIONS AND RECOMMENDATIONS

In this all-too-brief overview of ice accretion modelling on land, at sea, and in the air, we have touched on so many aspects of the problem that it is really impossible to devise a brief summary for presentation here. There are, however, one conclusion and three recommendations which we wish to emphasize.

The conclusion: Ice accretion modelling is sufficiently complex that it would not be profitable to attempt to produce a single grand unified model, which could simulate all aspects of the phenomenon.

The recommendations:

1.   Because these models are dependent on experimental results for parameterization and verification, much more attention should be paid to carefully designed laboratory experiments for the purpose of obtaining the necessary data.

2.   Models are of little use unless they can be verified. Consequently, there is a need for comprehensive field measurement programs, which include measurements of all the environmental parameters necessary for running the accretion models.

3.   Last, but by no means least, it is important for scientists in the various areas of ice accretion research to talk to each other, or, at the very least, to get to know each other's work. This is not simply to avoid unnecessary duplication, but rather to provide for the cross-fertilization of ideas and methods, which will accrue to the benefit of all of the individual research areas, as well as to the field as a whole. It is the hope of the authors that this review will contribute in some measure to the achievement of this goal.

8 ACKNOWLEDGEMENTS

This work was funded through NSERC strategic and operating grants. One of us (EPL) was the recipient of a McCalla Professorship from the University of Alberta. We wish especially to thank Ms. L. Smith and Ms. S. Datoo who typed the manuscript, and Mr. G. Lester and his cartographic staff who prepared the figures.

9 REFERENCES

Achenbach, E. The effect of surface roughness on the heat transfer from a circular cylinder to the cross flow of air, *International Journal of Heat and Mass Transfer*, 20, pp. 359-369, 1977.

Ackley, S.F. and M.K. Templeton, Numerical simulation of atmospheric ice accretion. In "Snow removal and ice control research". Special Report 185, Transportation Research Board, National Academy of Sciences, pp. 44-52, 1978.

Ackley, S.F. and M.K. Templeton, Computer modelling of atmospheric ice accretion, U.S. Army Cold Regions Research and Engineering Laboratory Report CRREL 79-4, 36 pp, 1979.

Ackley, S.F., G.E. Lemieux, K. Itagaki, and J. O'Keefe, Laboratory experiments on icing of rotating blades. In "Snow removal and ice control research". Special Report 185, Transportation Research Board, National Academy of Sciences, pp. 85-92, 1979.

Admirat, P., J.C. Grenier, and M. Maccagnan, Theory and modelling of the formation of wet snow on cylinders, Technical Report EDF/DER/ERMEL/TA/HM725288 available from Electricité de France, 1 avenue de Général de Gaulle, 92141 Clamarat Cedex, France, 1985.

AGARD, Rotorcraft Icing - Progress and Potential. AGARD Advisory Report No. 223, 138 pp.

Albers, J.A. and H.O. Stockman, Calculation procedures for potential and viscous flow solutions for engine inlets, *Journal of Engineering for Power*, pp. 1-10, January 1975.

Anonymous, Forecasters' guide on aircraft icing, United States Air Weather Service Report AWS/TR-80/001, 58 pp., 1980.

Arimilli, R.V., E.G. Keshock, and M.E. Smith, Measurement of local convective heat transfer coefficients on ice accretion shapes, AIAA paper 84-0018, 12 pp., 1984.

Bain, M. and J.-F. Gayet, Contribution to the modelling of the ice accretion process: ice density variation with the impacted surface angle. Proceedings First International Workshop on Atmospheric Icing of Structures, U.S. Army Cold Regions Research and Engineering Laboratory, Special Report 83-17, pp. 13-20, 1983.

Borisenkov, Ye.P., G.A. Zablockiy, A.P. Mikshtas, A.I. Migulin, and V.V. Panov, On the approximation of the spray cloud dimensions. In: Arkticheskii i Antarkticheskii Nauchno-Issledovatielskii Institut, Trudy 317, Leningrad: Gidrometeoizdat, pp. 121-126, 1975 (in Russian).

Bragg, M.B. and G.M. Gregorek, Aerodynamic characteristics of airfoils with ice accretions, AIAA paper, AIAA 82-0282, 14 pp., 1982.

Brunet, L., Conception et discussion d'un modele de formation du givre sur des obstacles varies. These de 3^e cycle. Published by: ONERA, B.P. 72, 92322 Chatillon, France, 75 pp., 1986.

Cansdale, J.T. and R.W. Gent, Ice accretion on aerofoils in two-dimensional compressible flow - a theoretical model, Royal Aircraft Establishment Technical Report 82128, 64 pp., 1983.

Clift, R., J.R. Grace, and M.E. Weber, *Bubbles Drops and Particles*, Academic press, 380 pp., 1978.

Comiskey, A.L., L.D. Leslie, and J.L. Wise, Superstructure icing and forecasting in Alaskan waters. Unpublished draft report submitted by the Arctic Environmental Information and Data Centre to the Pacific Marine Environmental Laboratory (NOAA) Seattle, Washington, U.S.A., 39 pp., 1984.

Cooper, W.A., W.R. Sand, M.K. Politovich, and D.L. Veal, Effects of icing on performance of a research aircraft, *J. Aircraft*, 21, pp. 708-715, 1984.

Davenport, A., Interaction of ice and wind loading on guyed towers. Proceedings Third International Workshop on Atmospheric Icing of Structures, Vancouver, May 1986 (in press).

de Lorenzis, B.Q., Time-dependent behaviour of ice accretion on a non-rotating cylinder, M.Sc. Thesis, University of Alberta, 92 pp., 1979.

Egelhofer, K.Z., S.F. Ackley, and D.R. Lynch, Computer modelling of atmospheric ice accretion and aerodynamic loading of transmission lines. Proceedings Second International Workshop on Atmospheric Icing of Structures, Trondheim, June 1984 (in press).

Elliott, J.W. and G. Poots, Accretion and shedding of ice on cables incorporating free streamline theory and the Joule effect, Proceedings Third International Workshop on Atmospheric Icing of Structures, Vancouver, May 1986 (in press).

English, M., Alberta hailstorms. Part II: growth of large hail in the storm, *Meteorological Monographs*, Vol. 14, No. 36, American Meteorological Society, pp. 37-98, 1973.

Ervik, M., ed., Proceedings Second International Workshop on Atmospheric Icing of Structures, 1984 (in press).

Ervik, M. and S.M. Fikke, Extended use of the icing model to estimate combined ice and wind loads, Proceedings Third International Workshop on Atmospheric Icing of Structures, Vancouver, May 1986 (in press).

Farley, R.D., Extended summary of the IAS two-dimensional hail category model, In: Notes for the International Cloud Modelling Workshop/Conference, World Meteorological Organization Report WMO/TD No. 57, 14 pp., 1985.

Farley, R.D., Numerical modelling of hailstorms and hailstone growth. Part II: the role of low-density riming growth in hail production, *Journal of the Atmospheric Sciences*, 26, pp. 234-254, 1987.

Finstad, K., Numerical and experimental studies of rime ice accretion on cylinders and airfoils, Ph.D. Thesis, University of Alberta, 229 pp., 1986.

Finstad, K., E.P. Lozowski, and E.M. Gates, A computational investigation of water droplet trajectories, *Journal of Atmospheric and Oceanic Technology* (in press).

Foote, G.B., A study of hail growth utilizing observed storm conditions. *Journal of Climate and Applied Meteorology*, 23, pp. 84-101, 1984.

Gates, E.M., Simulated marine icing in an icing wind tunnel, Dept. of Mechanical Engineering, Univ. of Alberta, Report No. 49, August 1985.

Gates, E.M., A. Liu, and E.P. Lozowski, A stochastic model of atmospheric rime icing, *Journal of Glaciology* (in press), 1987.

Gelder, T.F., W.H. Smyers, and U. von Glahn, Experimental droplet impingement on several two-dimensional airfoils with thickness ratios of 6 to 16 percent, NACA Technical Note NACA TN-3839, 1956.

Gent, R.W. and J.T. Cansdale, The development of mathematical modelling techniques for helicopter rotor icing, AIAA 23rd Aerospace Sciences Meeting, Paper AIAA 85-0336, 10 pp., 1985.

Govoni, J.W. and S.F. Ackley, Conductor twisting resistance effects on ice build-up and ice shedding, Proceedings Third International Workshop on Atmospheric Icing of Structures, Vancouver, May 1986 (in press).

Hardy, J.K., Protection of aircraft against ice, Royal Aircraft Establishment Report SME 3380, 1946.

Hess, J.L. and A.M.O. Smith, Calculation of potential flow about arbitrary bodies, *Progress in Aeronautical Sciences* 8, pp. 1-138, 1967.

Hewitt, B.L. et al., eds., *Computational Methods and Problems in Aeronautical Fluid Dynamics*, Academic Press, 525 pp., 1976.

Horjen, I. and S. Vefsnmo, A numerical sea spray icing model including the effects of a moving water film, Proceedings International Workshop on Offshore Winds and Icing, Halifax, T.A. Agnew and V.R. Swail, eds., pp. 152-164, 1985a.

Horjen, I. and S. Vefsnmo, Computer modelling of sea spray icing on marine structures, Proceedings Symposium on Automation for Safety in Offshore Operations, Trondheim, 1985b (in press).

Hunt, B., ed., *Numerical methods in applied fluid dynamics*, Academic Press, 651 pp., 1980.

Hunt, B., The panel method for subsonic aerodynamic flows: a survey of mathematical formulations and numerical models and an outline of the new British Aerospace scheme, In: Computational Fluid Dynamics, W. Kollmann, ed., 1980.

Imai, I., Studies on ice accretion, *Research on Snow and Ice*, 1, pp. 35-44, 1953 (in Japanese).

Itagaki, K., G.E. Limieux, and H.W. Bosworth, Natural rotor icing on Mount Washington, New Hampshire, U.S. Army Cold Regions Research and Engineering Laboratory Report CRREL 86-10, 62 pp., 1986.

Jeck, R.K., Airborne cloud-physics projects from 1974 through 1984, *Bulletin American Meteorological Society*, 67, pp. 1473-1477, 1986.

Jessup, R.G., Forecast techniques for ice accretion on different types of marine structures including ships, platforms and coastal facilities, Draft Report presented to WMO Commission for Marine Meteorology, October 1984, 90 pp.

Joe, P.I. and 12 others, Loss of accreted waters from growing hailstones, *Proceedings International Conference on Cloud Physics*, Boulder Colorado, pp. 264-269, 1976.

Kachurin, L.G., L.I. Gashin, and I.A. Smirnov, Icing rate of small displacement fishing boats under various hydrometeorological conditions, *Meteorology and Hydrology*, 3, pp. 58-71, 1974.

Kemp, A.K., The formation of ice on electrical conductors during heavy falls of wet snow, *Meteorological Magazine*, 109, pp. 69-74, 1980.

655

Kennedy, J.L and D.J. Marsden, Potential flow velocity distribution on multi-component airfoils sections, *Canadian Aeronautics and Space Journal*, 22, pp. 243-256, 1976.

Kim, J.J., Computational particle trajectory analysis on a 3-dimensional engine inlet, AIAA paper, AIAA 85-0411, 9 pp., 1985.

King, W.D., Air flow and particle trajectories around aircraft fuselages, Part III: extensions to particles of arbitrary shape, *Journal of Atmospheric and Oceanic Technology*, 2, pp. 539-547, 1985.

Knight, C.A. and K.R. Knupp, Precipitation growth trajectories in a CCOPE storm, *Journal of the Atmospheric Sciences*, 43, pp. 1057-1073, 1986.

Kollman, W., ed., *Computational fluid dynamics*, Hemisphere Publishing Corp., 1980.

Kolomeychuk, R.J., G.C. Castonguay, and L.E. Welsh, Ice accretion data for model evaluation, *Proceedings Third International Workshop on Atmospheric Icing of Structures*, Vancouver, May 1986 (in press).

Kraus, W., Panel methods in aerodynamics, In *Numerical Methods in Fluid Dynamics*, H.J. Wirz and J.J. Smolderens, eds., McGraw-Hill, pp. 237-297, 1978.

Krishnasamy, S. and R. Brown, Extreme value analysis of glaze ice accretion in Southern Ontario, *Proceedings Third International Workshop on Atmospheric Icing of Structures*, Vancouver, May 1986 (in press).

Laforte, J.-L., L.C. Phan, and N.D. Du, Preliminary investigation of effects of wind speed fluctuations on ice accretion grown on fixed and rotating aluminum conductor, *Proceedings Second International Workshop on Atmospheric Icing of Structures*, Trondheim, June 1984 (in press).

Landau, L. and E.M. Lifshitz, *Fluid Mechanics*, Pergamon Press, 536 pp., 1959.

Langmuir, I. and K.M. Blodgett, A mathematical investigation of water droplet trajectories, In *Collected Works of I. Langmuir*, Pergamon Press, 10, pp. 348-393, 1946.

Lapeyre, J.L. and P. Admirat, Mise en évidence de l'effet préventif d'une forte rigidité en torsion des conducteurs sur l'accrétion de "neige collante", Electricité de France Note Technique HM/725516, available from Electricité de France, 1 avenue Général de Gaulle, 92141 Clamart Cedex, France, 1986.

Launiainen, J. and M. Lyyra, Icing on a non-rotating cylinder under conditions of high liquid water content in the air; II: Heat transfer and rate of ice growth, *Journal of Glaciology*, 32, pp. 12-19, 1986.

Lesins, G.B. and R. List, Sponginess and drop shedding of gyrating hailstones in a pressure-controlled icing wind tunnel, *Journal of the Atmospheric Sciences*, 43, pp. 2813-2825, 1986.

Lin, Y.-L., R.D. Farley, and H.D. Orville, Bulk parameterization of the snow field in a cloud model, *Journal of Climate and Applied Meteorology*, 22, pp. 1065-1092, 1983.

List, R., P.H. Schuepp, and R.G. Methot, Heat exchange ratios of hailstones in a model cloud and their simulation in a laboratory, *Journal of the Atmospheric Science*, 22, pp. 710-718, 1965.

List, R. and J.-G. Dussault, Quasi steady state icing and melting conditions and heat and mass transfer of spherical and spheroidal hailstones, *Journal of Atmospheric Sciences*, 24, pp. 522-529, 1967.

List, R., Properties and growth of hailstones, In *Thunderstorm Morphology and Dynamics*, E. Kessler, ed., University of Oklahoma Press, pp. 259-276, 1985.

Liu, A., Stochastic modelling of rime icing, M.Sc. Thesis, University of Alberta, 150 pp., 1986.

Loset, S. and S. Vefsnmo, Computer modelling of ice accretion and control of icing on marine units, *Proceedings International Offshore and Navigation Conference and Exhibition* (POLARTECH 86), Helsinki, pp. 497-511, 1986.

Lozowski, E.P., J.R. Stallabrass, and P.F. Hearty, The icing of an unheated non-rotating cylinder in liquid water droplet-ice crystal clouds, National Research Council Canada Report LTR-LT-96, 1979.

Lozowski, E.P. and M.M. Oleskiw, Computer simulation of airfoil icing without runback, AIAA paper, AIAA 81-0402, 8 pp., 1981.

Lozowski, E.P. and M.M. Oleskiw, Computer modelling of time-dependent rime icing in the atmosphere, U.S. Army Cold Regions Research and Engineering Laboratory Report CRREL 83-2, 74 pp., 1983.

Lozowski, E.P., and E.M. Gates, Marine icing models: how do they work and how good are they? *Proceedings International Workshop on Offshore Winds and Icing*, T.A. Agnew and V.R. Swail, Eds., pp. 102-122, 1986.

Lozowski, E.P., E.M. Gates, and L. Makkonen, Recent progress in the incorporation of convective heat transfer into cylindrical ice accretion models, *Proceedings International Symposium of Cold Regions Heat Transfer*, K.C. Cheng, V.J. Lunardini, N. Seki, eds., pp. 17-24, 1987.

Ludlam, F.H., The heat economy of a rimed cylinder, *Quarterly Journal of the Royal Meteorological Society*, 77, pp. 663-666, 1951.

Ludlam, F.H., The hail problem, *Nubila*, 1, pp. 12-96, 1958.

Ludlam, F.H., *Clouds and Storms*, Pennsylvania State University Press, 405 pp., 1980.

MacDonald, K.A. and R.G. Jessup, Evaluation of a freezing spray forecast system, *Proceedings International Workshop on Offshore Winds and Icing*, T.A. Agnew and V.R. Swail, eds., pp. 267-277, 1985.

Macklin, W.C., The density and structure of ice formed by accretion, *Quarterly Journal Royal Meteorological Society*, 88, pp. 30-50, 1962.

Macklin, W.C., Heat transfer from hailstones, *Quarterly Journal of the Royal Meteorological Society*, 89, pp. 360-369, 1963.

Macklin, W.C., The characteristics of natural hailstones and their interpretation, In *Hail - a review of hail science and hail suppression*, G.B. Foote and C.A. Knight, eds., Meteorological Monographs, Vol. 16, No. 38, American Meteorological Society, pp. 65-88, 1977.

Makkonen, L., The heat balance of wet snow, *Meteorological Magazine*, 110, pp. 82, 1981.

Makkonen, L., Modelling ice accretion on wires, *Journal of Climate and Applied Meteorology*, 23, 929-939, 1984.

Makkonen, L., Atmospheric icing on sea structures, U.S. Army Cold Regions Research and Engineering Laboratory Monograph CRREL 84-2, 92 pp., 1984.

Makkonen, L., Heat transfer and icing of a rough cylinder, *Cold Regions Science and Technology*, 10, pp. 105-116, 1985.

Makkonen, L., The effect of conductor diameter on ice load as determined by a numerical icing model, *Proceedings Third International Workshop on Atmospheric Icing of Structures*, Vancouver, May 1986 (in press).

Makkonen, L., Salinity and growth rate of ice formed by sea spray, *Cold Regions Science and Technology* (in press). (1987)

Masters, C.O., The Federal Aviation Administration's Engineering and Development Aircraft Icing Program, AIAA paper, AIAA 85-0015, 8 pp., 1985.

McComber, P., Numerical simulation of cable twisting due to icing, *Cold Regions Science and Technology*, 8, pp. 253-259, 1984.

McComber, P. and G. Touzot, Calculation of the impingement of cloud droplets in a cylinder by the finite-element method, *Journal of the Atmospheric Sciences*, 38, pp. 1027-1036, 1981.

McComber, P. and G. Bouchard, The numerical calculation of the wind force coefficients on two-dimensional iced structures, *Proceedings Third International Workshop on Atmospheric Icing of Structures*, Vancouver, May 1986 (in press).

Mertins, H.O., Icing on fishing vessels due to spray, *Marine Observer*, 38, pp. 128-130, 1968.

Minsk, L.D., ed., *Proceedings First International Workshop on Atmospheric Icing of Structures*, U.S. Army Cold Regions Research and Engineering Laboratory Special Report CRREL 83-17, 1983.

Messinger, B.L., Equilibrium temperature of an unheated icing surface as a function of airspeed, *Journal of Aeronautical Sciences*, 20, pp. 29-41, 1953.

Narten, R., E.M. Gates and E.P. Lozowski, The influence of several factors on heat transfer from an isothermal cylinder, *Proceedings Third International Workshop on Atmospheric Icing of Structures*, Vancouver, Canada, May 1986 (in press).

Nelson, S.P., The influence of storm flow structure on hail growth, *Journal of the ATmospheric Sciences*, 40, pp. 1965-1983, 1983.

Norment, H.G., Calculation of water drop trajectories to and about arbitrary three-dimensional bodies in potential flow, NASA Contractor Report 3291, 82 pp., 1980.

Norment, H.G., Three-dimensional airflow and hydrometeor trajectory calculation with applications, AIAA paper AIAA 85-0412, 1985.

Obreiter, E., Physics of the marine ice accretion process. i. Spreading and freezing of salt water drops. ii. Mass loss due to splashing on a wet accretion, M.Sc. Thesis, University of Alberta, 145 pp., 1987.

Oleskiw, M.M., A computer simulation of time-dependent rime icing on airfoils, Ph.D. Thesis, University of Alberta, 302 pp., 1982.

Olsen, W. and E. Walker, Some experimental evidence for modifying the current physical model for ice accretion on aircraft surfaces, *Proceedings Third International Workshop on Atmospheric Icing of Structures*, (in press) (1986).

Overland, J.E., C.H. Pease, R.W. Preisendorfer, and A.L. Comiskey, A robust algorithm for prediction of vessel icing, *Proceedings International Workshop on Offshore Winds and Icing*, T.A. Agnew and V.R. Swail, eds., pp. 248-256, 1985.

Overland, J.E., C.H. Pease, and R.W. Preisendorfer, Prediction of vessel icing, *Journal of Climate and Applied Meteorology*, 25, pp. 1793-1806, 1986.

Parrish, J.L. and A.J. Heymsfield, A user guide to a particle-growth and trajectory model (using one-dimensional and three-dimensional wind fields), National Centre for Atmospheric Research, *Technical Note* NCAR/TN-259+1A, 69 pp., 1985.

Pearcey, T. and G.W. Hill, The accelerated motion of droplets and bubbles, *Australian Journal of Physics*, 9, pp. 19-30, 1956.

Personne, P. and J.-F. Gayet, Ice accretion on wires and anti-icing induced by the Joule effect, *Journal of Climate and Applied Meteorology*, (in press) (1987).

Pflaum, J.C., New clues for decoding hailstone structure, *Bulletin of the American Meteorological Society*, 65, pp. 583-593, 1984.

Pohlman, J.C. and D. Havard, Field research on the galloping of iced conductors - a status report, *Proceedings First International Workshop on Atmospheric Icing of Structures*, U.S. Army Cold Regions Research and Engineering Laboratory Special Report 83-17, pp. 319-325, 1983.

Potapczuk, M.G. and P.M. Gerhart, Progress in development of a Navier-Stokes solver for evaluation of iced airfoil performance, AIAA paper, AIAA 85-0410, 9 pp., 1985.

Proctor, F.H., Three-dimensional simulation of the 2 August CCOPE hailstorm with the terminal area simulation system, In *Notes for the International Cloud Modelling Workshop/Conference*, World Meteorological Organization Report WMO/TD-No. 57, 14 pp., 1985.

Prodi, F., G. Santachiara, and A. Franzini, Properties of ice accreted into two-stage growth, *Quarterly Journal Royal Meteorological Society*, 112, pp. 1057-1080, 1986.

Ranz, W.A. and W.R. Marshall, Evaporation from drops, *Chemical Engineering Prog.*, 48, pp. 141-180, 1952.

Rasmussen, R.M. and A.J. Heymsfield, Melting and shedding of graupel and hail: Part 1. Model physics, *Journal of the Atmospheric Sciences* (in press) (1987).

Roe, P.L., ed., *Numerical Methods in Aeronautical Fluid Dynamics*, Academic Press, 548 pp., 1982.

Roebber, P. and P. Mitten, Modelling and measurement of icing in Canadian waters. Final Report MT 86170 prepared for the Atmospheric Environment Service of Canada by the MEP Company, 7050 Woodbine Ave., Suite 100, Markham, Ontario, Canada L3R 4G8 (in press) (1987).

Rosen, K.M. and M.L. Potash, 40 years of helicopter ice protection experience at Sikorsky Aircraft, AIAA paper, AIAA 81-0407, 1981.

Sand, W.R., W.A. Cooper, M.K. Politovich, and D.L. Veal, Icing conditions encountered by a research aircraft, *Journal of Climate and Applied Meteorology*, 23, pp. 1427-1440, 1984.

Saunders, C.P.R. and C.C. Zhang, Measurements of the density of rime ice formed on a collecting rod attached to a rotor arm, *Atmospheric Research* (in press) (1987).

Schuepp, P.H. and R. List, Mass transfer of rough hailstone models in flows of various turbulence levels, *Journal of Applied Meteorology*, 8, pp. 254-263, 1969.

Schumann, T.E.W., The theory of hailstone formation, *Quarterly Journal of the Royal Meteorological Society*, 64, pp. 3-21, 1938.

Schwerdtfeger, P., The thermal properties of sea ice, *Journal of Glaciology*, 4, pp. 789-807, 1963.

Smith, S., Wind stress and heat flux over the ocean in gale force winds, *Journal of Physical Oceanography*, 10, pp. 709-726, 1980.

Soo, S.L., *Fluid Dynamics of Multiphase Systems*, Blaisdell, Waltham, Mass., 524 pp., 1967.

Stallabrass, J.R., Trawler icing - a compilation of work done at NRC National Research Council of Canada, Mechanical Engineering Report MD-56, NRC No. 19372, 103 pp., 1980.

Stallabrass, J.R., Aspects of freezing rain simulation and testing, *Proceedings First International Workshop on Atmospheric Icing of Structures*, U.S. Army Cold Regions Research and Engineering Laboratory, Special Report 83-17, pp. 67-74, 1983.

Szilder, K., E.P. Lozowski and E.M. Gates, Modelling ice accretion on non-rotating cylinders - the incorporation of time dependence and internal heat conduction, *Cold Regions Science and Technology*, 13, pp. 177-191, 1987.

Szilder, K., E.P. Lozowski and E.M. Gates, Some applications of a new time-dependent cylinder ice accretion model, *Atmospheric Research* (in press) (1988).

Tribus, M., Modern icing technology, Lecture notes, Engineering Research Institute, University of Michigan, 1952.

Van Fossen, G.J., R.J. Simoneau, W.A. Olsen, and R.J. Shaw, Heat transfer distributions around nominal ice accretion shapes formed on a cylinder in the NASA Lewis icing research tunnel, NASA Report TM-83557, 1984.

Wakahama, G., D. Kuroiwa, and K. Goto, Snow accretion on electric wires and its prevention, *Journal of Glaciology*, 19, pp. 479-487, 1977.

Welsh, L., ed., *Proceedings Third International Workshop on Atmospheric Icing of Structures* (in press) (1987).

Werner, J.B., Ice protection investigation for advanced rotary-wing aircraft, U.S. Army Air Mobility Research and Development Laboratory Technical Report 73-38, 1973.

Wirz, H.J. and J.J. Smolderen, eds., *Numerical Methods in Fluid Dynamics*, McGraw-Hill, 399 pp, 1978.

Xu, Jia-liu, Hail growth in a three-dimensional cloud model, *Journal of the Atmospheric Sciences*, 40, pp. 185-203, 1983.

Zakrzewski, W.P., Icing of ships. Part 1: Splashing a ship with spray, *NOAA Technical Memorandum*, EPL PMEL-66, 74 pp., 1986.

Zakrzewski, W.P. and E.P. Lozowski, The application of a vessel spraying model for predicting the ice growth rates and ice loads on a ship. *Proceedings Ninth International Conference on Port and Ocean Engineering under Arctic Conditions*, (POAC 87), Fairbanks, 3, pp. 591-603, 1987.

Zukauskas, A., Heat transfer from tubes in crossflow, *Advances in Heat Transfer*, 8, pp. 93-160, 1972.

Zukauskas, A. and J. Ziugzda, *Heat transfer of a cylinder in crossflow*, Hemisphere Publishing Co., 208 pp., 1985.

Chapter 20

Modeling and Forecasting Vessel Icing

W. PAUL ZAKRZEWSKI and E. P. LOZOWSKI
Division of Meteorology, Department of Geography, University of Alberta,
Edmonton, Alberta, Canada T6G 2H4

ABSTRACT

A comprehensive review of research on ship icing modelling and
forecasting is given. Effects of superstructure icing on ship safety
are discussed. Major research programs carried out world-wide for the
improvement of ship operations in icing-prone waters are briefly discussed. Numerical ship icing models are extensively reviewed. The model
calibration and the limitations of the model are discussed. Approaches
to the simulation of ship icing conditions in icing wind tunnels are
reviewed. Ship icing forecast techniques are assessed and the limitations of their use are given. The development of research on ice
removal from ships is discussed and the available techniques for anti-
icing and de-icing ships are reviewed.

CONTENTS

661

NOMENCLATURE

a	linearization coefficient = 4.4 W m^{-2} K^{-1}
C_d	droplet drag coefficient
c_a	specific heat capacity of dry air at constant pressure, J kg^{-1} K^{-1}
c_i	specific heat capacity of pure ice, J kg^{-1} K^{-1}
c_w	specific heat capacity at constant pressure of brine, J kg^{-1} K^{-1}
D	foremast diameter, m
d	droplet diameter, μ or m
E	collision efficiency
e	water vapor pressure over water at T_s, kPa
e_a	actual water vapor pressure in air at T_a, kPa
e_{as}	saturation vapor pressure in air at T_a, kPa
e_{is}	saturation vapor pressure over ice, kPa
e_s	water vapor pressure over water at T_a, kPa
e_T	water vapor pressure at surface of drop, kPa
F_o	source spray flux density, kg m^{-2} s^{-1}
F_{sn}	spray flux density normal to the impacting surface, kg m^{-2} s^{-1}
g	gravitational acceleration, 9.82 m s^{-2}
H	thickness, m
$\hat{H}$	incremental thickness of the saline spongy accretion, m h^{-1}
H_w	wave height, m
h	convective heat transfer coefficient, W m^{-2} K^{-1}
h_c	cylinder overall heat transfer coefficient, W m^{-2} K^{-1}
I	pure ice growth rate, kg m^{-2} s^{-1}
$\overline{I}$	time-averaged pure ice growth rate, kg m^{-2} h^{-1}
k	convective parameter
k_a	thermal conductivity of air, W m^{-1} K^{-1}
L	characteristic length scale, m
ℓ_f	latent heat of fusion of pure water, J kg^{-1}
ℓ_i	latent heat of sublimation, J kg^{-1}
ℓ_v	latent heat of vaporization, J kg^{-1}
Nu	Nusselt number
Nu_c	cylinder Nusselt number
Nu_d	droplet Nusselt number
n	freezing fraction
N_s	frequency of spraying the ship, min^{-1}
PR	predictor
Pr	Prandtl number for air
P_r	period between two sequential ship collisions with an oncoming wave, s
p	static pressure in the airstream, kPa
q	heat flux, J m^{-2} s^{-1}
R	direct spray flux, kg m^{-2} s^{-1}
$\overline{R}$	time-averaged spray flux, kg m^{-2} min^{-2}
$R*$	flux of shed brine, kg m^{-2} s^{-1}
$\overset{\circ}{R}$	incremental mass of brine entrapped in the accretion
$\hat{R}$	increment in the total mass of the spongy accretion
Re	Reynolds number
Re_c	cylinder Reynolds number
Re_d	droplet Reynolds number
$\overset{\circ}{S}$	salinity, parts per thousand (PPT)
$\hat{S}$	salinity of the entrapped brine, PPT
$\hat{S}$	overall salinity of the accretion, PPT
S_w	seawater salinity, PPT

S^* salinity of the shed water, PPT
Sc   Schmidt number for air
s   pure ice fraction
T   temperature, °C
T_a air temperature in the free airstream, °C
T_d spray droplet temperature at the moment of impact, °C
T_s equilibrium surface temperature, °C
T_w sea surface temperature, °C
t   time, s
U   wind speed, m s^{-1}
U_r relative wind speed, m s^{-1}
U_{10} wind speed at 10 m reference level
V_s ship speed, m s^{-1}
V_r ship speed relative to the waves, m s^{-1}
$\vec{V}_d$ droplet velocity
V_x, V_y, V_z droplet speed components in Cartesian coordinates, m s^{-1}
V_{zt} droplet terminal velocity, m s^{-1}
w   liquid water content, kg m^{-3}
X_t evaporative term
x,y,z   coordinates in the Cartesian system of reference fixed to the ship bow, m
z_{max} maximum height of the spray with respect to the ship deck m
z_0 height of the spray source above the ship deck, m
α ship heading with respect to the wind direction, deg
α_c angle between the ship heading and the relative wind velocity vector, deg
γ angle between the horizontal plane and the tangent to the droplet trajectory at the point of impact on the ship's surface, deg
Δt duration of spray, s
Δz range of droplet flight in the vertical direction, m
μ_a absolute viscosity of air, kg m^{-1} s^{-1}
ν_a kinematic viscosity of air, m^2 s^{-1}
ρ_a air density, kg m^{-3}
ρ_i pure ice density, kg m^{-3}
ρ_p pure water density, kg m^{-3}
ρ_s density of the salt in the seawater, kg m^{-3}
ρ_w seawater density, kg m^{-3}
$\hat{\rho}$ overall density of the accretion, kg m^{-3}
ε molecular weight ratio
τ duration of spray flight, s
θ equilibrium freezing temperature of brine, °C

Subscripts

a   air
c   cylinder
d   droplet
i   pure ice
p   pure water
s   salt in seawater or icing surface
w   seawater

1. INTRODUCTION

1.1 Background

Ship superstructure icing is the accumulation of ice on vessels as a result of the impingement and freezing of water (usually sea water from collision-generated spray). Icing is a menace to shipping and fisheries. A large number of small cargo vessels and small- and medium-sized fishing vessels and a few ocean-going tug boats and patrol vessels have been lost due to the loss of ship stability caused by super-structure icing (Figure 1). Almost all losses of ships also involved the loss of the ship crew. Shellard (1974) listed several known ship losses due to icing. Aksyutin (1979) discussed a few reported cases of losses of Soviet ships. DeAngelis (1974) listed a number of little-known losses of small ships operating in the Bering Sea and the Gulf of Alaska.

Larger ships (medium-cargo vessels, passenger or cargo ferries, large fishing vessels, and military vessels) are also prone to icing. Several cases of heavy icing on such vessels are listed by Shellard (1974), DeAngelis (1974), Panov (1976), and Aksyutin (1979). Also, the Overland et al., (1986) and Pease and Comiskey (1986) studies on the incidence of icing in Alaskan waters, show that large vessels are also susceptible to icing which may severely limit ship operations, but are unlikely to result in a ship capsizing. However, under conditions

Fig. 1   A crabber found in the Alaskan waters without the crew who perhaps abandoned the vessel when capsizing seemed imminent. They were never heard from.   (Official Coast Guard photo)

favourable to very heavy icing, several Soviet tankers and cargo vessels have been on the brink of icing disaster (Aksyutin, 1979). These vessels were only saved from capsizing through the bravery of their desperate crews, removing ice with axes and snow shovels or by sheltering the vessel in the lee of the land or in the ice field.

Over the last decade, ship disasters due to icing have not been as common as those in the 60s and 70s. The improvement of ship safety in cold oceans has been achieved by: (1) introducing larger vessels with better seaworthiness and performance under icing conditions; (2) altering the ship design - the great number of potential ice collectors (e.g., rigging) on modern vessels, designed for operation in icing-prone waters, were eliminated as a result of ship design studies initiated by the British Shipbuilding Research Association in 1957; (3) providing shipping and fisheries with more accurate icing warnings and forecasts as, for example, the NOAA ice accretion charts for the Gulf of Alaska and the Bering Sea.

However, despite the availability of ship icing forecasts, the modern large fishing vessel "Fjord West" was lost due to icing in the Bering Sea on January 19, 1989. In the cold waters off Newfoundland, a medium-sized trawler under the Phillipino flag capsized due to icing in February 1988 and only one survivor was rescued from the ocean. In the same area, two crewmen of the "Brendan" were swept overboard while removing ice from the deck of the heavily iced-up trawler. Very recently, on December 8, 1989, the 300-foot bulk carrier "Johanna B." and the 420-foot container ship "Captain Torres" capsized due to icing and high seas in the St. Lawrence Bay with 39 casualties. These recent icing disasters prove that icing must still be considered to be an extreme environmental hazard to shipping and fisheries. To date there are no effective de-icers and anti-icers readily applicable for ships.

In order to estimate the rate of ice growth on a ship surface, several models have been proposed. Some of them are based on empirical relationships between air-sea parameters and the icing rate (Sawada, 1966; Vasil'yeva, 1966; Mertins, 1968; Wise and Comiskey, 1980; Overland et al., 1986). Other models are aimed at specifying the icing rate in terms of physical and meteorological parameters. In order to predict the icing rate, all such models usually require the solution of the heat balance equation for the icing surface (Borisenkov and Panov, 1972; Kachurin et al., 1974; Panov, 1976; Stallabrass, 1980, Zakrzewski and Lozowski, 1987). For icing to occur two basic criteria have to be fulfilled: an object must be exposed to a water flux, and the heat flux from the wetted surface must be sufficiently large to result in the growth of ice. In the case of the most common type of icing of ships, spray icing, some of the heat balance equation components are related to the spray flux. Therefore, a knowledge of spray parameters is essential for determining the icing rate (Lozowski and Gates, 1985; Zakrzewski, 1987).

The goal of this paper is to outline the state-of-the-art modelling of ice accretion on ships. The paper is intended to provide a background and a synoptic perspective for those scientists, engineers and weather forecasters who may wish to focus their research objectives on any aspect of icing modelling, forecasting and icing prevention. Since icing has not yet been eliminated as a serious environmental hazard menacing ship operations in cold oceans, the intention of this paper is to assure anyone new to the field that there is still a lot to do to improve the safety of ships so that icing will not take a greater toll.

1.2 Synopsis of Ship Icing Research

Ship icing modelling is an international undertaking. The involvement in research on ship icing usually reflects either specific national weather service needs (Canada, USA, USSR) or ship safety requirements (Japan, USSR, UK) or, in the case of some research institutes, available funding (Norway, Canada). Table 1 lists the selected research centres and individuals involved in research on ship icing modelling and forecasting over the last few decades. Because of the scope of this paper, we cannot present all groups and individuals involved in ship icing modelling and forecasting. The Marine Icing Newsletter (1989), edited by Ross Brown of the Atmospheric Environment Service in Ottawa, Canada, can provide the reader with the most updated comprehensive review of research projects carried out world-wide in ship icing. In addition to the data listed in Table 1, a brief description of research on ship icing carried out internationally is given below.

Table 1   Selected research groups and individuals around the world involved in research on ship icing modelling and forecasting.

CANADA

Mr. R. Brown, Atmospheric Environment Service, 373 Sussex Drive, La Salle Academy, Block "E", Ottawa, Ontario K1A 0H3 (ship icing forecasting, validation of ship icing models).

Mr. R. Jessup, Atmospheric Environment Service, 4905 Dufferin Street, Downsview, Ontario M3H 5T4 (ship icing forecasting).

Dr. G.S.H. Lock, Prof. E.P. Lozowski, Dr. W.P. Zakrzewski, Dr. K. Szilder, Mr. R. Blackmore, University of Alberta, Edmonton, Alberta T6G 2H4 (ship spraying/icing modelling, ship icing forecasting, marine icing wind tunnel tests, field data collection on ships).

Dr. R. Gagnon, Institute for Marine Dynamics, National Research Council, P.O. Box. 12093, Station "A", St. John's, Newfoundland A1B 3T5 (ship icing modelling).

Mr. E. Guy, C-CORE, Memorial University of Newfoundland, Arctic Avenue, St. John's, Nfld. A1B 3X5 (new to the field).

Mr. P. Mitten, Compusult Limited, Suite 309, 198 Water Street, P.O. Box 607, Station C, St. John's, Nfld. (oil rig icing, data collection, development of icing measurement techniques).

Dr. M. Oleskiw, Low Temperature Laboratory, National Research Council of Canada, Montreal Road, Bldg. 22, Ottawa, Ontario K1A 0R6 (ship icing modelling and forecasting, ship icing data collection program, anti-icing, de-icing). The program was initiated by Mr. J. Stallabrass who retired in 1988.

FEDERAL REPUBLIC OF GERMANY

Dr. H.O. Mertins, German Weather Service, Hamburg, Germany (ship icing forecasting, field data collection).

(cont'd...)

FINLAND
Dr. L. Makkonen, Technical Research Scientist, Laboratory of
Structural Engineering, SF-12150 Espoo (modelling, ice accretion
properties).

JAPAN
Prof. T. Tabata, Dr. N. Ono, Mr. S. Iwata, Dr. S. Fukusako,
University of Hokkaido, Sapporo (field data collection, anti-icing,
de-icing).

Mr. T. Sawada, Meteorological Agency of Japan, Tokyo (ship icing
forecasting).

NORWAY
Prof. T. Carstens, Mr. I. Horjen, Norwegian Hydrotechnical
Laboratories, Trondheim, Klaebuveien 153, N-7034 (marine icing
modelling).

SWEDEN
Mr. J.E. Lundqvist, Mr. I. Udin, Swedish Meteorological and
Hydrological Institute, Norrkoping, S-601.

UNITED KINGDOM
British Shipbuilding Research Association, London, UK (trawler
model tests in wave tanks and cold rooms). Also, Mr. H.C.
Shellard, UK, Meteorological Service (review of ship icing research and forecasting techniques).

USA
Dr. L.D. Minsk, Dr. C.C. Ryerson, U.S. Army Cold Regions Research
and Engineering Laboratory, Hanover, New Hampshire, 03755-1290
(ship icing research, modelling, data collection, equipment
development).

Dr. J. Overland, Mrs. C. Pease, Pacific Marine Environmental
Laboratory, National Oceanic Atmospheric Administration, 7600
Sand Point N.E., Seattle, Washington 98115 (ship icing forecasting,
data collection).

Prof. W. Sackinger, University of Alaska - Fairbanks, Geophysical
Institute, Fairbanks, Alaska 99775-0800 (marine icing modelling,
de-icing).

USSR
Dr. Ye. Borisenkov, Dr. V.V. Panov (and several others), Arctic
and Antarctic Research Investigative Institute, Bering St. 58,
Leningrad (ship icing modelling, forecasting, field data collection,
anti-icing, de-icing).

Prof. L.G. Kachurin, Dr. L.G. Gashin, Leningrad State Meteorological Institute, Maldokhtinskii Pr. 98, Leningrad (ship icing
modelling and forecasting).

Based on an overview of available literature, one can state that
prior to 1957, icing was known to affect shipping and fisheries in the
North Atlantic and Arctic oceans (Aksyutin, 1979; Panov, 1976), but
little, if anything, was done to investigate it or to predict its
severity. Perhaps the first systematic research on ship icing was

carried out in the United Kingdom in 1957 (Trawler Ic. Res., 1957) in the wake of the icing disaster of the British steam trawlers "Lorella" and "Roderigo". These vessels were lost off Iceland on January 26, 1955 during an icing storm, and their crews were never heard from again.

In order to investigate the effects of ship heading and design on ship susceptibility to icing, the British Shipbuilding Research Association initiated a series of wave tank tests in cold rooms with 1:12 scale models of the trawlers (for details see Section 5). These tests resulted in the improvement of ship safety by issuing several recommendations on ship design to be implemented by the Inter-Governmental Maritime Consultative Organization. In addition, in the wake of these studies, more attention was paid to ship stability criteria under ice loads.

Similar to the British experience, Japan lost several tens of its fishing vessels in the cold waters of the Japan and Bering seas and the North Pacific Ocean. These losses resulted, in the early 60s, in the launching of a comprehensive Japanese research program (Tabata et al., 1963) focussed on (1) collection of ship icing field data, (2) investigation of the mechanisms of ice growth on ships, (3) examination of the effects of ice loads due to icing on ship stability, and (4) development of anti-icing and de-icing techniques for ships. The Japanese used several ships (fishing vessels, patrol boats) in their program and, based on the data collected (Sawada, 1968), produced nomograms for the icing rate on sea-going ships.

The Germans, fishing extensively in the North Atlantic Ocean, experienced several near-disaster and life-threatening situations, but their low speed and well-built trawlers had enough luck to avoid capsizing. The German Weather Service launched a data collection program (Mertins, 1968) targetted at the determination of an empirical relationship between the weather and sea conditions and the icing rate on ships. The well-known nomogram of Mertins (1968), based on approximately four-hundred cases of icing, has been used world-wide for forecasting the icing rate on ships since its publication.

However, the most comprehensive research program for the investigation of ship icing was launched by the Soviets in 1966 and was carried out to 1972-1973. The Soviet program was initiated in the wake of a major loss to the Soviet fishing fleet caused by icing. On one occasion, ten fishing trawlers (six Japanese and four Soviet) were lost to icing in the eastern part of the Bering Sea. The four Soviet medium-sized fishing trawlers lost with their crews on January 25, 1965 were the "Boksitogorsk", "Natchitchevan", "Sievsk", and "Siebeizh" (Figure 2). Only one crewman of the "Boksitogorsk" was miraculously rescued from the cold seas. The aim of the Soviet program (Panov, 1976) was to improve the safety of fishing vessels operating in cold oceans and to develop satisfactory techniques for forecasting ship icing and for removing ice from ships. More than two hundred research papers were produced under this program (for a detailed English review see Zakrzewski and Lozowski, 1989b).

Superstructure icing was considered by the US Navy as an environmental factor negatively affecting the operations of American ships in cold oceans (Fein and Freiberger, 1965). Several icing tests were carried out in cold rooms to test the de-icing capabilities of water-

Fig. 2   The "Piarnu" on her return to Tallin from the cold Baltic Sea
in 1969 (Aksyutin, 1979).

repelling coatings (Winegrad, 1987).   Several Soviet papers on ship
icing have been translated by the US Army Cold Regions Research and
Engineering Laboratory in Hanover and this was followed by a comprehensive review of research on ship icing (Minsk, 1977).   In 1985, 1987,
and 1989 the United States Navy organized a series of symposia on Arctic/
Cold Weather Operations of Surface Ships.   In order to improve the
operational capability of USN warships in cold oceans, a comprehensive
program (Barr and Kordenbrock, 1987) was proposed.

The National Oceanic and Atmospheric Administration (NOAA),
alerted by several ship icing disasters in Alaskan waters in the 70s,
launched a field data collection program for ship icing-related parameters (Wise and Comiskey, 1980; Pease and Comiskey, 1986).   Based on
the data collected, Overland et al. (1986) proposed a semi-empirical
algorithm that has been implemented by NOAA for ship icing forecasts
issued daily for the North Pacific Ocean, the Bering Sea, and the Gulf
of Alaska (Figure 3).

The icing of ships in cold waters off eastern Canada, where several
ships were lost to icing (Shellard, 1974; DeAngelis, 1974), was given
a lot of attention by the Low Temperature Laboratory of the National
Research Council (NRC) of Canada in Ottawa in the 70s.   A field data
collection program for icing-related parameters on ships was launched
by NRC.   Stallabrass (1980) analyzed the data from several tens of
ship icing reports distributed to sailors in this region, and produced
the first Canadian data set for icing rates on ships.   Moreover, a

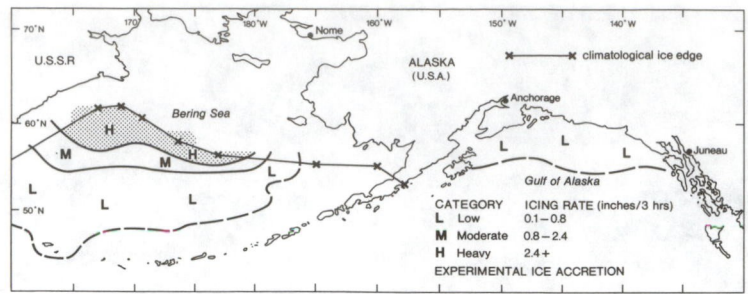

Fig 3 A fragment of the NOAA ice accretion chart (redrawn at the University of Alberta from a copy provided, courtesy of J. Overland.

numerical ship-icing model was developed by Stallabrass (1980) and verified with the data available. In addition, several techniques for ice removal from ships were tested in the NRC facility in Ottawa (Stallabrass, 1970).

A research group at the University of Alberta in Edmonton, led by Professors Edward P. Lozowski and E.M. Gates in the early 80s, has been in the forefront of icing research and has developed several models for freshwater icing (e.g., Lozowski et al., 1983; Lozowski and Gates, this volume). Another research objective of the University of Alberta icing group was the simulation of icing in wind tunnels and aircraft icing modelling. In 1985, a refrigerated marine icing wind tunnel was constructed in Edmonton and the group extended its research interests to marine icing (Lozowski and Gates, 1985; Zakrzewski and Lozowski, 1987a, b, 1989a,b; Szilder and Lozowski, 1989). For ship icing, the research objectives of the group are focussed on the development of a 3-D ship spraying/icing model, the collection of icing-related parameters on ships, the simulation of marine icing conditions in the MARINE icing tunnel (Marine Icing in Northern Environments), and most recently, on the development of anti-icers and de-icers for ships.

2. THE OCCURRENCE OF ICING EVENTS ON SHIPS

2.1 The Causes of Icing on Ships

The data collection by the Soviets under their research program in the 60s and 70s (Shekhtman, 1967, 1968; Borisenkov and Panov, 1972; Panov, 1976) show that icing due to freezing of collision-generated spray is the most common cause of ship icing. The Canadian data, reviewed by Brown and Roebber (1985), confirm the Soviet findings (Table 2). One can see, however, that the contribution of spray icing to all cases of icing on ships varies from region to region and it is the smallest for the Arctic seas. This is because of the more frequent incidence of conditions favourable to atmospheric icing and the presence of sea ice during most of the navigation season. Sea ice, by attenuating the waves, reduces and often completely stops the generation of spray,

Table 2   Causes of icing of Soviet ships compared to Canadian data

Region/ Sea	Period	Number of icing events	Spray	Spray and...				Atmospheric icing			Reference
				Fog	Rain	Frost	Snow	Fog	Rain	Snow	
All seas	1955-1967	~2000	89.8		6.4		1.1		2.7		Kravtsov and Stadnik in Borisenkov and Panov (1972)
All seas	1965-1966	~300	89.0	5.7	2.3		*		3.0		Shekhtman (1968)
All seas	1965-1967	~400	89.0		7.0				4.0		Shekhtman (1968)
Barents, Kara Laptiev, East Siberian, Chukchi seas	1967-1970	2269	50.0		41.0				9.0**		Kovrova et al. (1969)
Canadian Eastern Arctic	1973-1981	69	56.4	0.0	2.9	-	-	33.3	7.3	-	Brown & Roebber (1985, Table 4.5)
Canadian Western Arctic	1973-1982	40	60.0	7.5	0.0	-	-	22.5	10.0		Brown & Roebber (1985, Table 4.6)
Waters east of Canada	1970-1984	1041	93.1	1.3	1.5			1.2	2.8		Brown & Roebber (1985, Table 4.1)

*   We think that Shekhtman (1967) analyzed only those ship icing reports that did not include spray-snow mixed icing;
** Mostly due to supercooled fog.

and the ships can experience icing due to freezing of water only from
atmospheric sources (Zakrzewski et al., 1988a).

2.2 The Geographical Distribution of Icing Events on Ships

Borisenkov and Panov (1972) and later Panov (1976) presented
several maps showing the locations of icing events on Soviet ships
world-wide (Figure 4). A similar map was produced by Stallabrass
(1980) who showed the locations of the icing events on Canadian ships
participating in the NRC data collection program. One can easily notice
that the locations of the ship positions when icing was in progress are
concentrated in the main fishing zones. As the traffic along the Great
Northern Sea Route and in Antarctica has increased over the last decade,
one may suppose that more icing events have now been experienced in
these areas, although there is as yet no up-to-date analysis. However,
since Panov (1976) reports that approximately 3500 icing events on
Soviet ships alone were reported from 1967 to 1973, one might imagine a
very high incidence of icing on ships in the 80s. Based on the analyzed
data, the duration of the potential icing season, and the frequencies
of experiencing icing by ships during the icing season can be estimated
for several seas (Table 3).

Table 3   Period and frequency of Soviet ship icing (Borisenkov and
           Pchelko, 1975)

Region	No. of cases	Period of icing	Frequency (%)
NW part of Atlantic	85	15 Dec – 15 Mar	92
Norwegian and Greenland seas	109	15 Dec – 31 Mar	77
Northern part of Atlantic	63	15 Dec – 15 Mar	92
Barents Sea	390	1 Jan – 15 Mar	78
Baltic Sea	21	15 Dec – 29 Feb	85
Newfoundland region	15	1 Jan – 15 Mar	79
Bering Sea	185	1 Dec – 31 Mar	70
Sea of Okhotsk	337	1 Dec – 31 Mar	70
Sea of Japan	226	1 Dec – 29 Feb	85
NW part of Pacific	183	15 Dec – 31 Mar	79
Arctic seas (Kara, Laptev, East Siberian, and Chukchi)	71	15 Jun – 15 Nov	100

2.3 Weather and Oceanographical Conditions Favourable to Icing on Ships

Upon the completion of their research program, the Soviets concluded that icing of ships occurred at air temperatures from 0°C to
-26°C, sea surface temperatures from -1.8°C to +6°C, and for wind speeds
from 0 to 30 m/s (Borisenkov and Panov, 1972). Panov (1976) extended
the range of wind speeds recorded during the progress of icing of a
ship to 55 m/s. Shekhtman (1967) reported a few cases of ship icing
for which the sea surface temperatures were as high as 6-8°C. This
last finding is in agreement with Mertins (1968). The ranges of the
air-sea parameters involved in the growth of ice on the ships included
in the Atlantic data set of Stallabrass (1980) and the Alaskan data set
of Pease and Comiskey (1986), also fit those given above.

Fig. 4
  The positions of
  Soviet ships
  experiencing icing
  (Panov, 1979).

    For ship icing due to freezing of sea spray, one should consider
the reasons for the thresholds for air temperature and wind speed. The
threshold on air temperature occurs because it must be lower than the
freezing point of the impinging sea water. Of course, for ships exposed to freshwater spray in lakes or large rivers (e.g., the Great
Lakes or St. Lawrence River), the air temperature can be close to 0°C
provided that the heat flux from the wetted ship surface is large
enough to cause the growth of ice crystals. The threshold on wind
speed depends on ship design, and also on ship speed, heading and fetch.
For example, a wind speed of 8-9 m/s is sufficient to develop ocean
waves which can splash a medium-sized fishing trawler of Soviet type
with collision-generated spray (Panov, 1976; Zakrzewski and Lozowski,
1987a). Larger and newer ships of higher bow and freeboard and of a
modified design are known to be sprayed only in greater wind speeds.
For most of the conditions experienced in February 1988 by the large
stern trawler "Zandberg" of Catalina, Newfoundland, Canada (overall
length 65.1 m, gross tonnage 1,370 tonnes), the wind threshold for spraying was about 12-13 m/s. Large cargo vessels can be sprayed only in

even higher winds. Although they can eventually get iced-up, icing does not appear to threaten their stability.

3. PRINCIPLES OF SHIP ICING MODELLING

Ship icing models have been reviewed by Jessup (1985), and very recently by Lozowski (1988a,b) and Zakrzewski et al. (1988b,d). The first physical ship icing models were developed in the USSR (Borisenkov, 1969; Panov, 1971a; Borisenkov and Panov, 1972; Kachurin et al., 1974) and these were followed by the model of Stallabrass (1980). A new generation of ship icing models, which include a detailed spray flux calculation, has been developed at the University of Alberta (Zakrzewski, 1986; Zakrzewski and Lozowski, 1987a; Zakrzewski et al., 1988d; Blackmore et al., 1989). In this section we would like to address the following problems, which are applicable to ship icing in general, and to spray icing in particular:

(a) determination of the water flux to the ship surface,
(b) investigation of the water droplet trajectories,
(c) determination of the spray droplet temperature,
(d) water transport on the ship surface that determines the surface flow and water entrapment,
(e) determination of the heat balance equation components necessary to calculate how much of the delivered water freezes on the ship surface, and
(f) the physics of ice growth and the related accretion/brine parameters.

Because of the scope of this paper, all the above aspects of ship icing are discussed below in relation only to sea spray icing. For information about the modelling of atmospheric icing on ships, we recommend the Horjen (1981), Lozowski et al. (1983), and Makkonen (1984) models discussed by Zakrzewski et al. (1988a).

3.1 Water Flux Calculations

A knowledge of the spray flux to the icing surface of a ship is essential to determine the magnitude of the heat loss necessary to warm up the impinging spray to the surface equilibrium temperature T_s (or the heat gain in cooling the spray to T_s). In order to do this, one needs to calculate the droplet temperatures at impact. Perhaps the first attempt to incorporate this heat flux into the heat balance equation was made by Kachurin et al. (1974), who investigated the growth of ice on a cylinder oriented perpendicular to the impinging spray and located on a ship. The spray was assumed to be delivered to the target surface continuously. The liquid water content in the spray cloud was assumed to depend only on wave height H_w:

$$w = 10^{-3} \, H_w, \qquad kg \; m^{-3} \tag{1}$$

Equation (1) was recommended by Kachurin et al. (1974) for use on Soviet medium-sized fishing trawlers moving with a speed of $3.1 - 4.1$ m s^{-1} in head seas. The particular constant for the liquid water content in the spray cloud and the neglect of ship speed and heading limit the generality of this parameterization. Stallabrass (1980), who used Equation (1) in his ship icing model and tested it for the conditions

674

reported by Canadian ships, pointed out that in order to obtain a satisfactory correlation between the model predictions and the observed icing rates, a reduction of the constant in Equation (1) by a factor of six was necessary.

Borisenkov et al. (1975) analyzed the unique data set on the vertical profile of liquid water content in spray clouds generated by the trawler "Narva" in the Japan Sea in 1973, and proposed a relationship which, after making unit conversions, is:

$$w = 2.30 \times 10^{-2} \exp(- z/1.82), \qquad kg\ m^{-3} \qquad (2)$$

where z is the height above the ship deck (in meters). To date, this is the only source of data on collision-generated spray available in the open literature. The well-known equation of Preobrazhenskii (1973) considers wind-generated spray and should not be used for calculating the flux from the primary source of water delivery to ship collision generated spray. However, one can add the flux of wind-generated spray to that of collision-generated spray as introduced by Horjen and Vefsnmo (1984) who studied the icing of oil rigs. Because the magnitude of the water flux with wind-generated spray is typically several times smaller than that of collision-generated spray, neglecting the first factor simplifies the calculations and changes the results very little.

Zakrzewski (1987) argued that Equation (2) can be valid only for a medium-sized fishing trawler of Soviet type and that it should be used only for the weather, oceanographic, and ship motion parameters experienced during the data collection. In order to avoid the particularity of Equation (2), Zakrzewski (1987) attempted to generalize the coefficient in the formula of Borisenkov et al. (1975) by correlating the conditions during the data collection with the ship speed relative to the waves. In an attempt to generalize this result for arbitrary ship speeds, headings, and wave heights, the following equation was derived (Zakrzewski, 1987):

$$w = 6.15 \times 10^{-5} H_w V_r^2 \exp(- z/1.82), \qquad kg\ m^{-3} \qquad (3)$$

This formula is also applicable only to medium-sized fishing trawlers of Soviet type and only if the assumption of the proportionality factor $H_w V_r^2$ holds. There is an apparent need to verify Equation (3) and also to derive a suitable equation for other ships. However, because there is no other data available at present, the formula of Zakrzewski (1987) along with that of Borisenkov et al. (1975) should be used with care for the time being.

Spray splashes the ship in short pulses of a few seconds duration, separated by intervals of several tens of seconds during which no spray is delivered to the target. Panov (1971b) obtained the following empirical equation for the frequency of spraying a ship:

$$f = 15.78 - 18.04 \exp(- 4.26/P_r), \qquad min^{-1} \qquad (4)$$

where $1/P_r$ is the wave encounter frequency. Equation (4) is applicable to Soviet trawlers if $3.5\ s \leq P_r \leq 15\ s$. This suggests that vessels of this type are, on average, sprayed by every other wave. The frequency of spraying for ships of other types would presumably be different from that of the trawlers. For example, the measured frequency of spraying the ex-whaling ship "Andre Dyrøy" was only two splashes per minute (Horjen, 1988), while for the same conditions, Equation (4) predicts nearly seven splashes per minute for the Soviet trawler.

The duration of ship exposure to the impinging spray was measured on several occasions by Zakrzewski on the "Zandberg" in the Labrador Sea in February 1988. For wind speeds of 18 to 22 m/s, ship speeds of 2 to 4 m/s, and ship headings of 140-180°, each spray splashing the ship superstructure lasted, on average, 2.1 s, whereas the duration of spray cloud residence over the entire vessel was as long as 4.5 s. (The ship heading is an angle between the ship velocity vector and wave velocity vector. It is equal to 180° for the ship moving precisely into the waves and 0° for the ship sailing down the waves.) The limited number of measurements precludes any attempt to derive a relationship between the spray cloud parameters and the ship speed and wave height. The collected data, though scarce, suggest that the equation of Zakrzewski (1987) derived for the Soviet trawlers and based on only one significant measurement, overpredicts the duration of spraying by a factor of two or higher. Until some data from ships permit the derivation of more accurate formulae, we suggest that a value in the range of 1 to 3 s be used.

In order to calculate the time-averaged spray flux to the ship surface, Zakrzewski and Lozowski (1989a) used the equation:

$$F = F_o \sin(\alpha_c - 90) \cos \gamma, \qquad \text{kg m}^{-2} \text{ s}^{-1} \tag{5}$$

where α_c is the angle between the ship heading and the projection of the spray trajectory onto the deck of the moving ship (Figure 5), γ is the angle between the horizontal plane and the tangent to the droplet trajectory at the point of impact on the ship surface, and F_o is the spray flux density in the spray source. Because of the large drop

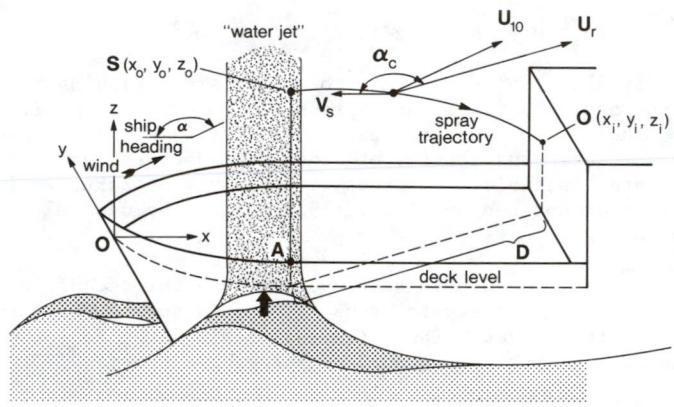

Fig. 5   Schematic view of the peculiar water jet generated by the wave impact onto the ship. The extent of the wave impact zone and the height of the water jet are ship specific. They are assumed to depend on the ship speed and heading and the wave height (for details see Zakrzewski and Lozowski, 1987a; Zakrzewski et al., 1988c).

diameters (>1 mm), the model of Zakrzewski and Lozowski (1987a) assumes
that the drops are not deflected in the airstream prior to hitting the
target. Consequently, the collision efficiency in Equation (5) is
assumed to be unity. For smaller drops, one might calculate the collision
efficiency by the well-known formulae of Langmuir and Blodgett (1946) or
Stallabrass (1980), or through the detailed calculation of individual
droplet trajectories if the airflow around the ship were known.

The source spray flux density is given by the formula:

$$F = w\,U_r, \qquad kg\ m^{-2}\ s^{-1} \tag{6}$$

where w is the liquid water content in the spray cloud formed above the
windward bulwark) and U_r is the wind speed relative to the ship. In the
model of Zakrzewski and Lozowski (1987a) the wind field around the ship
is represented by the wind velocity vector at a 10 m reference level.
The effects of turbulence are neglected and no variations of the wind
vector, either in the horizontal or in the vertical, due to flow around
the ship are allowed for. This model in question was intended for small
ships such as Soviet trawlers (overall length 39.5 m, freeboard 3.5 m,
tonnage 442 tonnes), which is the model's reference vessel. However,
these assumptions for the wind field should be checked for larger ships
whose influence on the airflow may be much greater. The University of
Alberta spraying model is capable of predicting the extent of the
spraying zone on a trawler (Figure 6) and the magnitudes of the spray
flux to several ship components (Figure 7).

3.2 Spray Drop Trajectory Calculations

In the ship icing models of Stallabrass (1980) and Panov (1976),
spray drop trajectory calculations are necessary to determine the duration of the spray drop flight and the drop temperature at impact. In
the model of Stallabrass (1980) the drop is assumed to be wind-rafted
with a speed equal to the relative wind speed, and to move in the air
over a fixed range of 20 meters above the vessel from the wave crest to
the target. The model of Panov (1976) is more sophisticated in this
matter since it takes into account the air drag and the droplet motion
relative to the airstream. In the ship icing models of Zakrzewski and
Lozowski (1987a, 1989a), the drop trajectory calculations are used not
only to determine the duration of the drop flight, but also to determine
if a given ship component is exposed to airborne spray. In the University of Alberta models, the spray flux calculations (see Section 3.1)
are being executed only for the cells located within the spray-splash
area. For cells located in the spray-free area, the ice growth calculations are performed only if these cells are affected by the brine shed
from the elevated ship components. These features distinguish the
University of Alberta ship icing models from the other available models.
A detailed discussion of the equations of motion used by Zakrzewski and
Lozowski (1987a, 1989a) is given below, along with information about a
numerical procedure that employs the drop trajectories to determine the
spray-splash area and the duration of the drop flight to the target.

The University of Alberta ship spraying/icing model determines the
horizontal extent of the water jet along the area of wave impact on the
ship hull. The water jet was defined by using several series of measurements under various weather and ship motion conditions (Zakrzewski et al.,

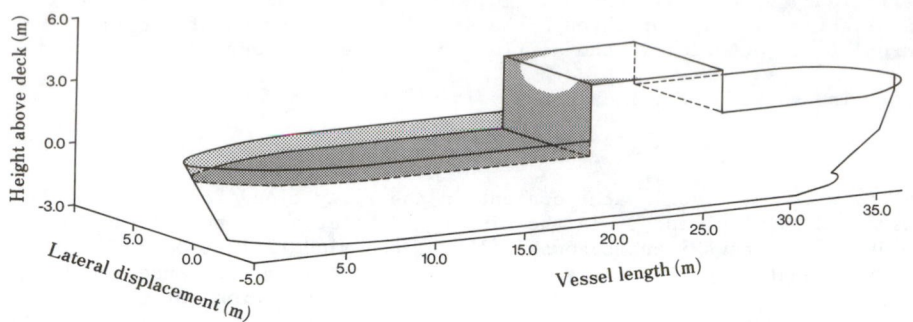

Fig. 6 The extent of the spray-receiving zone on a medium-sized fishing trawler for a wind speed of 12 m s^{-1}, a ship heading of 180°, and ship speed of 4 m s^{-1}.

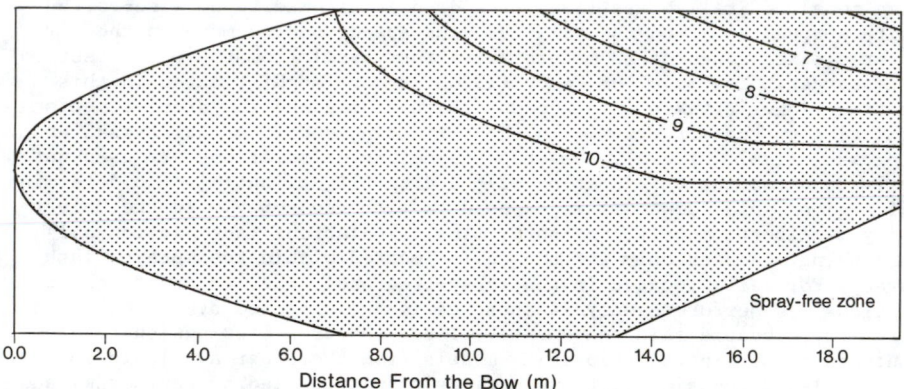

Fig. 7 The spray flux (in kg m^{-2} min^{-1}) to the ship deck for a wind speed of 18 m s^{-1}, ship speed of 2 m s^{-1}, heading of 150°, and fetch of 200 miles. The large zone of homogeneous spray in the fore part of the deck occurs because of the bulwark shadow effect on the spray trajectories (Zakrzewski and Lozowski, 1989a).

1988c). The maximum vertical extent of the spray was assumed to be equal to the maximum height of the water jet, and was assumed to be uniform along the area of wave impact (Figure 8). Zakrzewski et al. (1988c) assumed that the maximum height of the water jet was proportional to the ship speed relative to the wave. For medium sized fishing vessels of Soviet type, this is calculated by the formula:

$$z_{max} = 0.535\ V_r, \qquad m\ s^{-1} \tag{8}$$

The constant in Equation (8) was determined by using the data of Kuznietsov et al. (1971) on the upper limit of accretion on the foremast of such vessels as a function of the wind speed and ship heading. For each of the thirteen cases of icing on ships documented by Kuznietsov et al., Zakrzewski et al. (1988c) computed backwards the trajectories of the droplets that hit the upper limit of the ice build-up on the foremast.

In the University of Alberta model, the range of the droplet flight from the spray source to the target (SO in Figure 8) is equal to the range from the bulwark contour to the centre of a given grid cell of a numerical network superimposed on the ship. The contour of the bulwark is ship-specific, and in the case of our medium-sized Soviet fishing trawler, it is approximated by:

$$y = (x/0.5457)^{\frac{1}{2}} \qquad \text{for } 0 \leq x < 7.27, \quad m$$

$$y = \pm 3.65 \qquad \text{for } 7.27 \leq x < 27.0, \quad m \tag{9}$$

$$y = ((39.5 - x) / 0.356)^{\frac{1}{2}} \qquad \text{for } 27.0 \leq x < 39.5, \quad m$$

where x and y are the coordinates in a frame of reference fixed to the ship bow. In the same frame of reference, the equation of the line SO in Fig. 8 is given by:

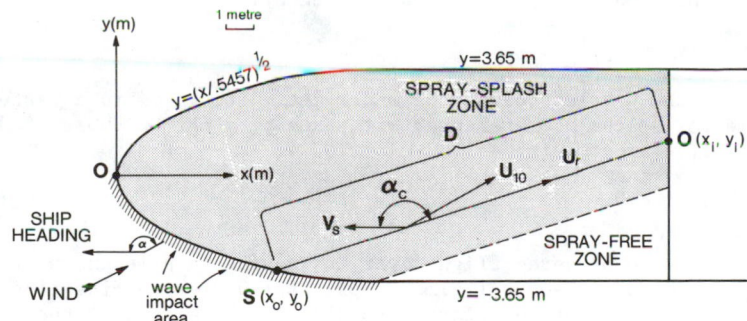

Fig. 8   Details of the spray droplet trajectory calculations for a medium-sized fishing trawler of Soviet type (Zakrzewski and Lozowski, 1987a, 1989a).

$$y - y_0 = (x - x_0) \tan(180 - \alpha_c) \tag{10}$$

where α_c is the angle between the ship heading and the relative wind velocity vector, u_r, (Figure 8). The angle α_c is given by:

$$\alpha_c = 180° - \sin^{-1} \frac{U_{10} \sin \alpha}{U_r} \tag{11}$$

The coordinates x_0 and y_0 are found analytically by solving equations (9) and (10) for the intersection point under the given air-sea and ship motion parameters. In the University of Alberta model, Equation (10) is also used to determine whether the cell is splashed with spray. The cell is exposed to spray only when the intersection point of equations (9) and (10) is located within the wave impact area for the given air-sea and ship motion parameters. Otherwise, the cell is considered to be in the spray-free zone (Figure 8).

In order to simplify the calculations of droplet flight duration, it is assumed that the wind velocity is constant with height within the spray zone, which extends up to 5-10 m above the ship deck. In the model, the wind speed at a 10 m reference level is used. No changes of the wind velocity vector are allowed for. Because of the large drop sizes, the effects of turbulence are neglected in the model. The water droplet is assumed to move according to the equation of motion (Zarling, 1980):

$$\frac{d\vec{V}_d}{dt} + \frac{3}{4} \frac{C_d}{d} \left(\frac{\rho_a}{\rho_w}\right) |\vec{V}_d - \vec{U}_{10}| (V_d - U_{10}) + \vec{g} \left(\frac{\rho_a}{\rho_w} - 1\right) = 0 \tag{12}$$

where $\vec{V}_d$ is the droplet velocity in a non-moving frame of reference, and C_d is the droplet drag coefficient given by (Langmuir and Blodgett, 1946):

$$C_d = 24/Re + 4.73/Re_d^{0.37} + 6.24 \times 10^{-3} Re_d^{0.38} \tag{13}$$

where Re_d is the droplet Reynolds number given by:

$$Re_d = d|\vec{V}_d - \vec{U}_{10}| / \nu_a \tag{14}$$

At time $t = 0$ the droplet leaves the water jet. Its speed in a Cartesian, non-moving system of reference, with axes parallel to those of the ship frame of reference, is:

$$V_x = V_x, \quad V_y = 0, \quad V_z = 0 \tag{15}$$

Once the range of spray flight over the moving ship is known, the droplet is followed for several time steps ($\Delta t = 0.01$ s) on its flight, using the equation of motion relative to the moving ship until the droplet completes the distance SO (Fig. 8) and hits the target. This makes it possible to numerically determine the duration of the spray flight, τ.

The height of the spray source in the water jet is computed by the formula:

$$z_o = z_i + \Delta z, \qquad m \tag{16}$$

where z_o is the height of the spray source with respect to the ship deck, z_i is the elevation of the centre of the cell above the ship deck, and Δz is the distance in the vertical direction travelled by the droplet on its way from the spray source to the target. Since the duration of the spray flight, τ, has already been determined, Δz can be found numerically by solving the equation for droplet motion in the vertical direction:

$$\frac{d\vec{V}_z}{dt} = -\vec{g}\left(\frac{\rho_w - \rho_a}{\rho_w}\right) - \frac{3\rho_a\, C_d V_z |\vec{V}_d - \vec{U}_{10}|}{4\rho_w d} \tag{17}$$

where V_z is the component of droplet velocity in the vertical direction, and the other terms are the same as those in Equation (12). No net vertical motion in the air stream is allowed for. When the droplet reaches its terminal velocity, V_{zt}, it is given by (Zarling, 1980):

$$V_{zt} = 2\sqrt{dg(\rho_w - \rho_a)/3\,\rho_a\, c_d}, \qquad m\ s^{-1} \tag{18}$$

Its falling speed in the vertical direction remains constant.

One should note that the solution of Equation (16) makes it possible to calculate for each grid cell the liquid water content in its spray source in the water jet according to Equation (3). When this parameter is known, the spray flux to the ship surface can readily be computed by using equations (5) and (6).

3.3 Spray Temperature

The spray impact temperature is required to calculate the heat loss necessary to warm up the impinging spray to the surface equilibrium temperature. When icing progresses in the wet growth regime and the spray flux to the target is large, a knowledge of the spray temperature at impact is essential to accurately calculate this heat flux and, consequently, the icing rate.

The spray droplets ejected from the wave, cool in the cold air. When an icing model is based on the assumption that the spray droplets fly over a fixed range, as it is in the model of Stallabrass (1980), the droplet temperature at impact is uniform over the target. If a ship icing model is capable of calculating the range of spray flight to the target as a function of the cell location with respect to the water jet (Zakrzewski and Lozowski, 1987a), the spray impact temperature varies over the ship surface, even though the air-sea and ship motion conditions remain unchanged.

When heat losses due to convection and evaporation are considered, the spray temperature, T_d, may be calculated from (Stallabrass, 1980):

$$T = T_a + (T_w - T_a)\exp(-kXt) \tag{19}$$

where t is the time after formation of the spray drop. X is the factor representing the effect of evaporation. It is given by:

$$X = 1 + \varepsilon \frac{\ell_v}{pc_a} \frac{e_a - e_T}{T_a - T} \tag{20}$$

where T is the instantaneous spray temperature, e_T is the saturation vapor pressure at temperature T, and e_a is the actual vapor pressure in the airstream.

The convective parameter in Equation (19), k, is given by:

$$k = 6 \, Nu_d \, k_a / \rho_w \, c_w \, d^2 \tag{21}$$

where the droplet Nusselt number, Nu_d, is given by (Michiev and Michieva, 1973):

$$Nu_d = \begin{cases} 0.49 \, Re_d^{0.50} & \text{for } Re_d < 10^3 \\ 0.24 \, Re_d^{0.60} & \text{for } Re_d > 10^3 \end{cases} \tag{22}$$

and Re_d is the droplet Reynolds number.

To date, none of the developed ship icing models takes into account the effect of evaporation on the reduction of the spray drop diameter and the increase of the drop salinity as the drop moves in the air. There are, however, several models developed for cloud droplets (Pruppacher and Klett, 1978) or for drops of wind-generated spray (Andreas, 1989) that consider the size and salinity evolutions of the drops.

3.4 Water and Salt Transport on the Ship Surface

The ship icing model of Zakrzewski and Lozowski (1987a) has been developed for medium-sized fishing vessels of Soviet type. These ships are known to be very susceptible to spray that is generated during every other wave-ship collision. Moreover, the duration of spray is usually long (4-6 s) and the high generation frequency of spray clouds results in short intervals between successive sprayings. With these facts in mind, we assume that the spray is delivered continuously to the target with a time-averaged water flux rate. Next, we will discuss the water and salt transport on the vertical face of the ship's superstructure.

As the icing progresses, brine is entrapped between the crystals of pure ice. Besides the brine pockets there are also entrapped air bubbles. The brine and air entrapments make the ice spongy and porous. We assume that over a time interval, Δt, the ice growth rate, I, is constant, and the physical properties of the ice (density, salinity, and sponginess/porosity) do not change. Since the spray flux varies over the ship's superstructure, both in the vertical and horizontal directions, we superimpose a numerical grid (0.5 x 0.5 m) on the superstructure face (Figure 9). The equation for the increment of the mass of the ice deposit over a given grid cell is:

$$\hat{R}_{ij} = \overset{o}{R}_{ij} + I_{ij} \tag{23}$$

Fig. 9

Schematic of the water and salt transport down the wall of the ship's superstructure in the University of Alberta ship icing model. (Notation the same as in the text.)

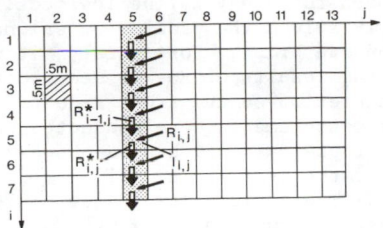

where $\overset{\circ}{R}_{ij}$ is the mass of brine entrapped within the cell ij over the time interval Δt, I_{ij} is the incremental mass of "pure ice" accreted during Δt, and $\hat{R}_{ij}$ is the increment in the total mass of the spongy ice deposit. $\hat{R}_{ij}$ can be specified in terms of the pure ice fraction, s_{ij} (Zakrzewski and Lozowski, 1987a):

$$s_{ij} = I_{ij} / \hat{R}_{ij} \tag{24}$$

In terms of s_{ij}, the incremental brine mass may be expressed as:

$$\overset{\circ}{R}_{ij} = I_{ij} (1 - s_{ij}) / s_{ij} \tag{25}$$

Since there is no salt trapped inside the pure ice mass, the salt balance equation for the ice deposit on the grid cell ij is:

$$\overset{\circ}{R}_{ij} \overset{\circ}{S}_{ij} = \hat{S}_{ij} \hat{R}_{ij} \tag{26}$$

where $\overset{\circ}{S}_{ij}$ is the salinity of the entrapped brine and $\hat{S}_{ij}$ is the overall accretion salinity. Consequently, $\overset{\circ}{S}_{ij}$ can be specified in terms of the ice fraction, s_{ij} and the overall ice salinity, $\hat{S}_{ij}$:

$$\overset{\circ}{S}_{ij} = \hat{S}_{ij} / (1 - s_{ij}) \tag{27}$$

For any grid cell the water mass balance equation is:

$$R_{ij} + R^*_{i-1,j} = \overset{\circ}{R}_{ij} + R^*_{ij} + I_{ij} \tag{28}$$

where R_{ij} is the direct spray flux, $R^*_{i-1,j}$ is the flux of shed brine from the cell immediately above, and R^*_{ij} is the flux of shed brine out of the grid cell. I_{ij} is the pure ice growth rate. The salt balance equation for the cell is:

$$R_{ij} S_o + R^*_{i-1,j} S^*_{i-1,j} = \overset{\circ}{R}_{ij} \overset{\circ}{S}_{ij} + R^*_{ij} S^*_{ij} \tag{29}$$

where S_o is the spray salinity (assumed to be equal to that of seawater at the sea surface), $S^*_{i-1,j}$ is the salinity of the shed brine flowing from the cell immediately above, $\overset{\circ}{S}_{ij}$ is the salinity of the entrapped brine, and S^*_{ij} is the salinity of the shed brine leaving the cell.

When salinity effects are considered, several assumptions are typically made in order to simplify the icing model due to the scarcity of field data. For example, Horjen and Vefsnmo's (1984) model uses Japanese data on salinity (Tabata et al., 1963). Even the recent approach to spongy saline ice modelling (Makkonen, 1986) has not finally solved the problem of saline ice deposits and the salinity of rundown brine. Fortunately, it was possible to apply to Equation (28) the results of Soviet measurements of the physical properties of ice accretion on the front side of the superstructure of their trawlers. These measurements were conducted while the ice growth was progressing. Using these data, it has been possible to determine the vertical profile of the overall ice salinity for any grid cell section in our model. The Soviet data (Panov, 1976) are well fitted by the following regression equation:

$$\hat{S}_{ij} = -3.85\, z_{ij} + 25.8, \qquad \text{PPT} \tag{30}$$

where z is the height (in meters) above the ship's deck. Equation (30) shows a similar tendency towards decreasing ice salinity with height above the ship's deck as does our own data from the research cruise on the RFT "Wilfred Templeman", in the waters east and south of Newfoundland, in February 1987 (Figure 10). The ice fraction in our ice deposit was measured by a calorimetric technique. Our results are fitted by the regression line:

$$s_{ij} = 0.047\, z_{ij} + 0.79, \qquad \text{PPT} \tag{31}$$

where z is the same as in Equation (30), and s is the ice fraction.

The solution of equations (28) and (29) with the aid of equations (26), (28), and (31) leads to the determination for each cell of the flux of rundown brine shed to the cell immediately below:

$$R^*_{ij} = (1 - n_{ij}/s_{ij})\,(R_{ij} + R^*_{i-1,j}), \qquad \text{kg m}^{-2}\,\text{s}^{-1} \tag{32}$$

The overall salinity of the moving water film is given as:

$$S^*_{ij} = \frac{R_{ij} S_o + R^*_{i-1,j} S^*_{i-1,j} - n_{ij}\hat{S}_{ij}(R_{ij} + R^*_{i-1,j})/s_{ij}}{(1 - n_{ij}/s_{ij})\,(R_{ij} + R^*_{i-1,j})}, \qquad \text{PPT} \tag{33}$$

where all terms are the same as those in equations (26) to (31), and n_{ij} is the freezing fraction given by:

$$n_{ij} = I_{ij}/(R_{ij} + R^*_{i-1,j}) \tag{34}$$

Three regimes exist - no icing, dry-growth, or wet-growth, according to whether the freezing fraction is 0, 1, or $0 < n_{ij} < 1.0$, respectively.

In order to estimate the ice growth rate and freezing fraction over the grid in the University of Alberta ship spraying/icing model, equations (32) and (33) are incorporated into the heat balance for the icing surface.

3.5 Heat Balance of the Icing Surface

Let us assume that spray droplets at temperature T_d hit an icing surface (Figure 11). The temperature of the moving water film is

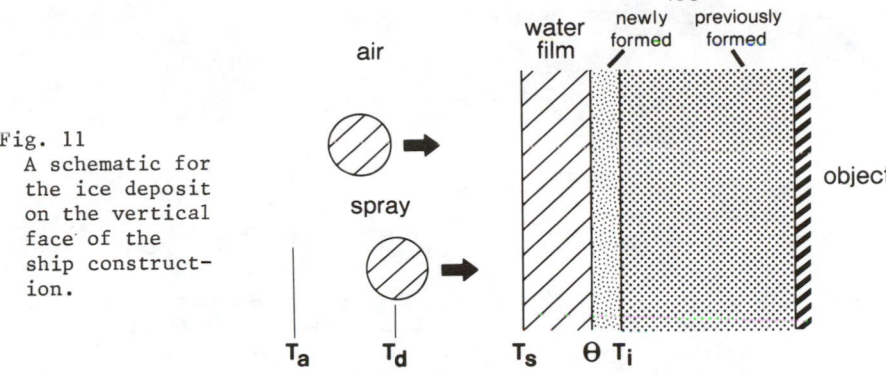

Fig. 11
A schematic for
the ice deposit
on the vertical
face of the
ship construct-
ion.

assumed to be the equilibrium surface temperature, T_s. This is assumed
to be homogeneous over the grid cell and throughout the thickness of
the water film. If icing progresses, it is equal to the freezing
temperature θ of the shed water of salinity S*. Icing is not in progress when $T_s > \theta$.

The freezing point of brine may be approximated by (Makkonen,
1987):

$$\theta_{ij} = 0.054 \ S^*, \quad °C \tag{35}$$

If one neglects the heat flux through the ice deposit due to the
presumably low thermal conductivity of spongy ice, the heat balance for
the icing surface is:

$$q_f + q_c + q_e + q_w + q_\ell + q_r = 0 \tag{36}$$

where q_f is the flux associated with the latent heat of freezing of the
pure ice crystals, q_c is the sensible heat flux between the icing surface and the air due to convection, q_e is the latent heat flux due to
forced evaporation, q_w is the net heat flux due to rundown of shed
water, q_ℓ is the heat loss necessary to warm up the impinging spray to
T_s (or the heat gain in cooling the spray to T_s), and q_r is the net
radiative heat flux. The detailed formulation of Equation (36) can be
found elsewhere for fresh water icing (e.g., Lozowski et al., 1983) and
for marine icing (Lozowski and Gates, this volume; Makkonen, 1988;
Zakrzewski et al., 1988a). The net heat flux due to rundown water for
the cell ij is:

$$q_w = R_{i-1,j}^* \ c_w \ \theta_{i-1,j} - R_{ij}^* \ c_w \ \theta_{ij}, \quad W \ m^{-2} \tag{37}$$

where $\theta_{i-1,j}$ and θ_{ij} are the equilibrium freezing temperatures of shed
water leaving the cells (i-1,j) and ij, respectively, and c_w is the
specific heat of brine within the temperature range from $\theta_{i-1,j}$ to
θ_{ij}.

When all the terms of Equation (36) are determined, one can find the icing rate I_{ij} for each cell as:

$$I_{ij} = h\ell_f^{-1}\left[\left(\frac{Pr}{Sc}\right)^{0.63}\left(\frac{\varepsilon\ell_v}{c_a p}\right)\left(e_{ij} - e_a\right) - \left(T_a - \theta_{ij}\right)\right]$$

$$+ R_{ij}\,\ell_f^{-1}\left[c_w\left(\theta_{ij} - (T_d)_{ij}\right)\right] - \frac{a}{\ell_f}(T_a - \theta_{ij})$$

$$- c_w\,\ell_f^{-1}(R_{i-1,j}^*\,\theta_{i-1,j} - R_{ij}^*\,\theta_{ij},\qquad kg\ m^{-2}\ s^{-1} \qquad (38)$$

where $Pr = 0.711$ and $Sc = 0.595$ are the Prandtl and Schmidt numbers for air, ℓ_v is the latent heat of vaporization, c_a and c_w are the specific heat capacities at constant pressure of air and seawater, ℓ_f is the latent heat of fusion of pure water, p is the air pressure, θ_{ij} is the freezing temperature of the brine in the cell ij, $R_{i-1,j}^*$ is the shed water flux from the cell immediately above, R_{ij} is the direct spray flux to the cell, T_a is the air temperature, T_d is the spray impact temperature, e_{ij} is the saturation vapor pressure over the brine at the temperature θ_{ij} and salinity S_{ij}, and e_a is the actual vapor pressure in the airstream.

The parameter ε is the ratio of the molecular weights of water vapor and dry air ($\varepsilon = 0.622$), and a is the linearization coefficient for the heat flux due to long-wave radiation ($a = 4.4\ W\ m^{-2}\ K^{-1}$).

The convective heat transfer coefficient, h, is related to the Nusselt number, Nu as:

$$h = k_a Nu/L, \qquad W\ m^{-2}\ K^{-1} \qquad (39)$$

where L is a characteristic length scale for the surface in question, and k_a is the thermal conductivity of the air. At present there is a complete lack of experimental data giving the Nusselt number for such objects as a ship's superstructure. Consequently, Nu for a flow parallel to a flat plate has been used (Kreith, 1969):

$$Nu = 0.036\ Pr^{0.33}\ Re^{0.8} \qquad (40)$$

where Pr is the Prandtl number for the air, and Re is the Reynolds number given by:

$$Re = LU_r/\nu_a \qquad (41)$$

where L is a characteristic length (here, we take L to be 4.47 m - the geometric mean dimension of the superstructure of height 3.5 m and breadth 5.72 m), U_r is the relative wind speed, and ν_a is the kinematic viscosity of air.

The above approach can also be used to determine the icing rate on other ship components (e.g., rigging, bulwarks, forecastle, etc.), provided that the specifics of heat transfer from these targets are taken into account and that empirical data on the accretion salinity are available. For example, for a ship's mast Lozowski and Zakrzewski (1988a) used the results of the measurements by Zakrzewski and Blackmore on the MT "Zogi" (February 13, 1988) and obtained the following regression equation:

$$\hat{S} = -7.5 \, z + 30, \qquad PPT \tag{42}$$

where z is the height above the ship deck. Equation (42) can be used when z does not exceed 3.5 m.

When the growth of ice on the ship mast is considered, the assumption for the growth of ice only on the windward face of the mast seems to be very realistic (Figure 12). The heat transfer from the ice-covered mast can be studied using the same heat balance equation as for the front face of the ship's superstructure (Equation 36). However, there are some important differences between the heat transfer coefficients for the ship mast and the ship's superstructure. The heat transfer coefficient, h, for the mast is given by the equation:

$$h = k_a \, Nu_c / D \tag{43}$$

where D is the mast diameter, and Nu_c is the Nusselt number for a cylinder. The latter is calculated for all wetted grid cells by the formula of Lozowski et al. (1983):

$$(Nu_c)_{ij} = Re_c^{0.5} \, (2.4 + 1.2 \sin (3.6 \, (\beta_{ij} - 25))) \tag{44}$$

where β_{ij} is the angle (in degrees) between the relative wind vector and the vector normal to the icing within the cell. The cylinder Reynolds number Re_c is:

$$Re_c = U_r D \rho_a / \mu_a \tag{45}$$

where ρ_a is the density of the air, and μ_a is absolute viscosity. It should be noted that equations (43) and (45) imply a time-dependence since the mast diameter increases as the icing progresses. The mast diameter, for a particular ring of the numerical grid network, at the beginning of the time step, k, can be estimated by the formula:

$$D_{jk} = D_o + 2 \, \overline{H}_{j,k-1} \tag{46}$$

where D_o is the diameter of the mast before icing ($D_o = 0.4$ m), and $\overline{H}_{j,k-1}$ is the angular mean thickness of the accretion within the ring j at the end of the previous time step.

3.6 Growth of Spongy Marine Accretions on the Ship Surface

The heat balance equation (see section 3.5) is used in many ship icing models (e.g., Stallabrass, 1980; Kachurin et al., 1974) to calculate the pure ice growth rate on ships. However, the accretion accumulated on sea-going ships is built of pure ice crystals and brine entrapments. This makes the marine ice accretion spongy and saline. The brine entrapments change not only the density of the accretion, but they apparently increase the thickness of the accretion and the ice loads due to icing on ships. Therefore, when the rate of growth of such an accretion is being calculated, especially for the purpose of ice load estimates, the model should take into account both the pure ice

a front view   b rear view

Fig. 12   The accretion on the foremast of the MT "Zogi" (Catalina,
           February 13, 1988). (Photo by W.P. Zakrzewski)

growth rate and the incremental mass of the entrapped brine between the
crystals of pure ice.

In order to determine the effect of the brine entrapments on the
icing rate and other accretion parameters associated with the icing
rate, one must have some knowledge of the contribution of the brine
entrapments to the total mass of the accretion. The term 'ice sponginess' is commonly used to express this characteristic feature of the
accretion. Ice sponginess can be readily determined by introducing
a physical parameter called the solid fraction, s, given below:

$$s = M_i / M \tag{47}$$

where M_i is the total mass of the pure ice crystals in the accretion,
and M is the total mass of the accretion sample.

Based on the measurements performed on the patrol boat "Chitose",
Tabata et al. (1963) estimated the solid fraction to be approximately
0.50. This implies that half of the accretion mass was entrapped brine.
The above value of the solid fraction is not questioned because of the
accurate measurement technique used in this Japanese study. However,
our own field experience suggests that the accretion of such a large

content of brine can be formed only on the ship deck. It is unlikely that such a spongy accretion would adhere to the vertical walls of the ship components. Even if this is not the case, the brine would creep down due to gravity, making the accretion porous. Our own data from the RFT "Wilfred Templeman" (February 1987) and the MV "Zandberg" (February 1988) show that the solid fraction of the newly formed accretion on vertical ship surfaces varies from 0.70 to 0.95 (see Figure 10).

When the solid fraction of the marine accretion s is known, the incremental thickness of the accretion can be calculated by the equation (Zakrzewski and Lozowski, 1987a):

$$\hat{H} = \bar{I} \, [1/\rho + (1 - s) / s\overset{\circ}{\rho}], \qquad m \ h^{-1} \tag{48}$$

where $\bar{I}$ is the time-averaged pure ice growth rate (in kg m^{-2} h^{-1}), ρ is the density of pure ice, and $\overset{\circ}{\rho}$ is the density of the brine entrapped. $\overset{\circ}{\rho}$ may be expressed in terms of the salinity of the brine $\overset{\circ}{S}$, the density of salt in seawater ρ_s, and the density of pure water ρ_p (Zakrzewski and Lozowski, 1987a):

$$\overset{\circ}{\rho} = 1000 \, \rho_s / [\overset{\circ}{S} + \rho_s \, (1000 - \overset{\circ}{S}) / \rho_p], \qquad kg \ m^{-3} \tag{49}$$

The density of the salt in seawater has been assumed to be $\rho_s = 3290$ kg m^{-3}. The density of pure water is taken to be $\rho_p = 1000$ kg m^{-3}.

A new approach to the calculation of the rate of saline spongy accretion has very recently been developed at the University of Alberta (Blackmore et al., 1989). This approach is based on the determination of the accretion equilibrium surface temperature and the pure ice growth rate. When implemented into the University of Alberta ship icing model, it has been demonstrated to be accurate and computer-time efficient.

4.   CALIBRATION OF THE SHIP ICING MODELS

In the opinion of Zakrzewski et al. (1988d), physical ship icing models demonstrate superiority over empirical relationships between the weather and sea parameters and the icing rate. The advantage of physical ship icing models is that they can accurately predict the mass of ice formed under given environmental heat transfer conditions, provided that, for a given ship and given component, (1) the primary heat fluxes from the icing surface are known, and (2) the model takes into account the target geometry, size, and orientation. These conditions are necessary since the icing rate is ship specific and also varies all around the vessel and over any ship component. Therefore, there is no 'generic' icing rate that would adequately determine the growth of ice on the ship. In our opinion, the icing rate is so strongly ship and target specific, that in order to accurately calculate the total ice load on the entire vessel, one must know the icing rate on all major ship components in order to add the ice load over the vessel.

If this is to be accomplished, one must have confidence in the magnitudes of several parameters involved in the growth of ice and the processes of water delivery to the ship surface and the salt and water transport on this surface. Unfortunately, the scarcity of field data from ships makes this task difficult. The use of data collected on one type or class of vessel is not recommended for use on another vessel type, as spray data are ship specific. Generally, one needs a list of icing-related parameters for the icing events experienced by several ships. These data sets should include the data on the icing rates and the meteorological and oceanographical parameters known to affect the progress of icing. In addition, data on the spray cloud parameters measured on the ship are required for model calibration, as the magnitude of the spray flux controls the heat transfer from the wetted ship surface.

A critical review of all available data sets on these parameters is given below.

4.1 Data Sets on the Spray Cloud Parameters

In order to calculate the spray flux to a target on ship board, the magnitudes of the following spray cloud parameters must be known:

(a) the extent of the spray envelope, both in the vertical and the horizontal,

(b) the liquid water content in the spray source in the spray cloud,

(c) the duration of the target exposure to spray and the frequency of spray cloud generation,

(d) the spray droplet size distribution.

For a given ship, all these parameters are functions of wind speed and wind direction with respect to the ship course, the ship speed, and the wave parameters (height, length, period, and steepness). The wave parameters depend on fetch, ocean depth and size, wind speed, wind direction with respect to the coastline, and the presence of sea ice.

The size of spray drops appears to be ship-dependent and to increase with ship size (Panov, 1976; Lozowski and Zakrzewski, 1988a; Horjen, 1988). In some cases, drops several millimeters in diameter are produced.

Based on a review of the literature, there is no data set on spray cloud parameters that lists the weather-sea conditions, along with data on ship motion and the above spray cloud parameters. However, some data on these parameters are available from miscellaneous sources, mostly from Soviet reports. In light of the recently reviewed Soviet publications on ship icing, Zakrzewski and Lozowski (1989a) concluded that the single equations for (1) the frequency of spray generation and (2) the liquid water content in the spray clouds, are available only for medium-sized fishing trawlers of Soviet type. The spraying module of the University of Alberta ship icing model is based on these data.

In the middle 80s, the spray cloud parameters were measured on Norwegian vessels during a few research cruises organized by the Norwegian Hydrotechnical Laboratories in Trondheim. The data collected on these ships are of proprietary nature and are not available in the open literature.

4.2 Data Sets on Icing Rates

In order to allow for the analysis of the effect of weather and sea parameters on the icing rate, data on the measured icing rates and on the weather and sea conditions involved in the growth of ice are required. Unfortunately, such data sets are scarce and often incomplete.

Two well-known data sets on ice growth on ships are available (Stallabrass, 1980; Pease and Comiskey, 1986). Stallabrass gives 39 documented cases of ship icing in the North Atlantic for which the icing rate "on the ship" has been determined as well as the air and sea surface temperatures, seawater salinity, and the relative wind speed. The data set of Pease and Comiskey presents 85 documented cases of ship icing in Alaskan waters for which the ice growth rates and the basic air-sea parameters are listed, along with the ship speed and course. Both data sets are based on ship icing reports. The icing rate was visually estimated, rather than measured. Moreover, the average thickness of the ice deposit was determined as the mean of the maximum ice thicknesses on the various ship components (Stallabrass, 1980). The data set of Pease and Comiskey (1986) gives no information as to which part of the ship the listed values of ice growth rates are based. Zakrzewski and Lozowski (1989b) do not recommend these data sets for model calibration.

Three very recent data sets on ship icing parameters are recommended for the calibration of ship icing models. The requirements are that the spray flux to the target be known, either from direct measurements carried out on the ships for which the calibration is being made, or, if necessary, by using the spray data from the Soviet trawlers. If this practice is acceptable, the Soviet trawlers can be used as the reference vessels until new data on spray cloud parameters are available for other types of ships.

The first data set, produced by Zakrzewski and Lozowski (1989b), has been compiled from miscellaneous Soviet reports on ship icing. Data on the weather-sea conditions and ship speed and heading are listed, along with the overall ship icing rate on the entire vessel (in tonnes per hour) or with a local icing rate (in cm per hour) on specified ship components for 111 documented icing events on Soviet ships. In most cases, these data were gathered on Soviet medium-sized fishing vessels. The Soviet data are considered to be better than those of Stallabrass (1980) and Pease and Comiskey (1986) as the Soviet data were based on measuring the icing rate, whereas the other data used data estimated visually.

The second data set of Lozowski and Zakrzewski (1988a) lists the measured data on the growth of ice on a stainless steel triangle (sides 0.60 m). This triangle was welded to the front face of the MV "Zandberg" in such a way that its centre was located 1.0 m above the deck and 2 m from the contour of the port side of the ship. These data are listed in Table 4.

The third data set of Horjen and Carstens (1989) lists the data gathered on the Norwegian vessel "Endre Dyrøy". The measurements of the ice growth rate were conducted on the railings, a panel, and pipes mounted to the ship mast at heights 5.1 to 10.1 m above the ship deck.

Table 4 The air-sea and ship motion parameters measured by Zakrzewski and Blackmore during icing on the "Zandberg" in the Labrador Sea and east of Newfoundland (February 8-26, 1988)

Day	Hour	Seawater salinity (PPS)	Air pressure (mb)	Air temperature (°C)	Sea sfc temperature (°C)	Wind speed (kts)	Wind direction (deg)	Ship speed (kts)	Ship course (deg)	Mean wave height (m)	Mean icing rate front of superstructure (cm h^{-1})
8	15	32.5	1028.0	-8.0	-0.3	33.0	290.0	0.9	329.0	1.3	1.0
8	16	32.5	1027.5	-5.1	-0.7	29.0	280.0	2.1	329.0	1.2	0.9
8	17	32.5	1026.9	-4.3	-0.8	18.0	285.0	2.1	329.0	1.1	0.5
8	18	32.5	1026.2	-3.0	-0.9	18.0	285.0	2.4	329.0	1.2	0.5
8	19	32.4	1025.4	-2.1	-1.0	18.0	285.0	2.4	329.0	1.3	0.4
8	20	32.0	1025.0	-2.0	-1.1	17.0	285.0	2.5	329.0	1.0	0.5
8	21	32.0	1025.0	-2.1	-1.1	18.0	285.0	2.0	329.0	0.9	0.4
8	22	32.0	1024.7	-2.1	-1.0	18.0	285.0	2.1	328.0	0.9	0.5
8	23	32.0	1024.5	-2.3	-1.0	18.0	285.0	2.0	328.0	0.9	0.5
11	14	32.8	1015.9	-2.1	-1.0	6.0	200.0	2.5	180.0	0.6	0.1
11	15	32.8	1015.5	-2.1	-1.0	7.0	210.0	8.5	260.0	0.6	0.5
11	16	32.9	1015.0	-2.2	-0.9	8.0	210.0	8.0	260.0	0.8	0.6
11	17	33.0	1014.5	-2.0	-1.0	7.0	210.0	8.0	260.0	0.8	0.7
11	18	32.9	1014.0	-2.0	-1.1	8.0	210.0	8.0	260.0	1.0	0.7
11	19	32.8	1013.3	-2.1	-1.1	8.0	210.0	8.0	260.0	1.0	0.7
11	20	32.7	1013.0	-2.8	-1.6	10.0	220.0	8.0	260.0	1.0	0.5
17	12	33.0	1009.1	-6.0	-1.6	15.0	280.0	10.5	245.0	0.3	1.0
17	13	33.0	1008.5	-6.0	-1.0	15.0	280.0	10.5	245.0	1.0	1.3
17	14	33.0	1007.5	-6.0	-1.0	15.0	280.0	10.5	245.0	1.0	1.5
17	15	33.0	1006.5	-6.0	-1.0	14.0	280.0	10.5	245.0	1.1	1.5
17	16	32.9	1006.0	-5.8	-1.0	17.0	280.0	9.8	325.0	0.9	1.7
17	17	32.3	1005.8	-5.5	-1.2	19.0	285.0	9.8	325.0	1.0	1.7
17	18	32.0	1005.8	-5.5	-1.2	20.0	285.0	9.8	325.0	1.0	1.8
17	19	32.0	1005.7	-5.6	-1.3	28.0	285.0	9.8	325.0	0.9	1.8
17	20	32.0	1005.0	-5.5	-1.7	30.0	285.0	9.8	325.0	0.9	1.9
17	21	32.0	1004.1	-5.5	-1.8	30.0	285.0	9.8	325.0	0.9	2.2
17	22	32.0	1003.8	-5.5	-1.8	30.0	285.0	9.8	325.0	1.0	2.3
17	23	32.0	1003.6	-5.5	-1.8	31.0	285.0	9.8	325.0	1.1	2.2
17	24	32.0	1003.1	-5.5	-1.8	30.0	285.0	9.8	325.0	1.2	2.3
18	01	32.0	1003.0	-5.5	-1.8	30.0	285.0	9.8	325.0	1.0	2.3
18	02	32.0	1002.8	-5.5	-1.8	29.0	285.0	9.5	325.0	1.1	2.2
18	03	32.0	1002.7	-5.5	-1.8	31.0	290.0	9.5	325.0	1.2	2.3
18	04	32.0	1002.3	-5.5	-1.8	35.0	290.0	9.5	325.0	1.0	2.1
18	05	32.0	1002.0	-5.5	-1.8	30.0	290.0	9.5	325.0	1.3	2.1
18	13	32.0	1001.5	-4.4	-1.4	32.0	290.0	12.8	255.0	1.0	0.8
18	14	32.0	1001.3	-4.1	-1.4	28.0	290.0	12.5	255.0	1.0	1.0
18	15	32.0	1001.1	-4.1	-1.5	30.0	290.0	12.5	255.0	1.0	1.0
18	16	32.0	1001.0	-4.0	-1.5	29.0	290.0	12.5	255.0	1.0	0.8
18	18	32.0	1001.0	-4.1	-1.4	22.0	295.0	12.8	230.0	1.0	0.1
18	19	32.0	1002.1	-4.3	-1.3	21.0	295.0	12.0	230.0	0.7	0.1
18	20	32.0	1003.1	-4.4	-1.3	20.0	295.0	12.0	230.0	0.7	0.1
18	21	32.0	1003.8	-4.3	-1.2	21.0	300.0	12.0	185.0	0.7	0.0
18	22	32.0	1003.8	-4.4	-1.1	20.0	300.0	12.0	185.0	0.7	0.0
18	23	32.0	1004.1	-4.4	-1.0	21.0	300.0	12.0	235.0	0.7	0.1

4.3 A Simple Procedure for Model Calibration

The goal of model calibration is to tune up a given model in such a way that after this procedure the model is capable of predicting the desired variables in reasonably good agreement with the available laboratory or field data representing the process approximated by this model. When ship icing models are being calibrated, field data from ships are required.

Let us assume that the ship icing model consists of two modules: the spray module and the ice growth module. The first module predicts the extent of the spray-receiving zone, and the time-averaged spray flux to the target. The ice growth module predicts the icing rate.

For a given ship component and the given air-sea and ship motion parameters, the model-predicted ice growth rates are affected by all discrepancies between the model-predicted and true magnitudes of (1) the extent of the spray zone, (2) the spray flux, and (3) the components of the heat balance equation. This implies an obvious sequence for model calibration:

-the extent of the spray receiving zone,
-the spray flux rate,
-the ice growth rate.

The model is run for conditions experienced during field measurements. A model-predicted magnitude of the variable V, known to affect the extent of the spray-receiving zone (e.g., the maximum height of the spray source), V_p, is compared with the available measurement V_m of this variable. A simple calibration factor S_1 for the first such variable is given by the equation:

$$S_1 = V_m / V_p \tag{50}$$

At this step, the model is corrected by introducing the following equation for 'calibrating' the variable in question:

$$V = S_1 V_p \tag{51}$$

where V_p is the model-predicted uncorrected magnitude of the variable and S_1 is the calibration factor from Equation (50). Next, another variable considered to affect the extent of the spraying zone (or the subsequent model-predicted parameter) is corrected in a similar way by again running the model for the conditions under which the field data were collected. This procedure is repeated until the last significant variable is calibrated by applying the S_n calibration factor as shown by Equation (51). As a product of the above simple operations, one may obtain a calibrated or tuned-up model. This would, for any air-sea and ship motion conditions, predict the icing rates correctly, provided that the applied empirical calibration factors $S_1 \ldots S_n$ cover all model deficiencies.

The University of Alberta ship spraying/icing model has been calibrated both for its spraying module (Zakrzewski et al., 1988c) and the ice growth module (Zakrzewski et al., 1988c). The data set of Zakrzewski and Lozowski (1989a) was used for this purpose (Figure 13).

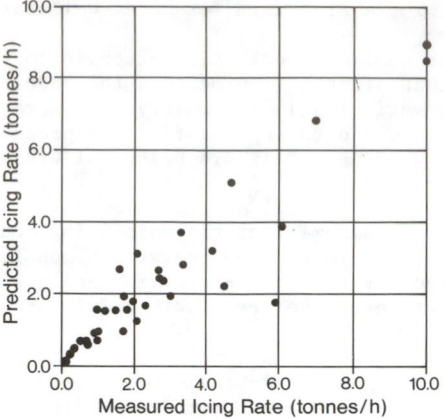

Fig. 13
Ship spraying/icing
model results for the
Soviet data set of
Zakrzewski and Lozowski
(1989a).

The calibration results in applying a calibration factor of 1.2 in
calculating the icing rates.

5.   LABORATORY INVESTIGATIONS OF SHIP ICING

Laboratory investigations of the icing of ships can be dated back
to the 1950s. These investigations were carried out world-wide (USA,
USSR, Canada, Japan), targetted either on the development of anti-icing
coatings for ships (e.g., Winegrad, 1987; Tabata, 1968; Semenova, 1972),
the improvement of ship design (Trawler Icing Research, 1957), or on the
simulation of the marine icing conditions on ship board in a wind icing
tunnel.

Many tests in icing wind tunnels in North America were carried out
in the 70s and 80s (e.g., Stallabrass, 1980; Lozowski et al., 1983;
Gates, 1985). Only a few of them were targetted at investigating the
growth of ice under simulated marine conditions. To 1985, there was no
facility that allowed for injecting saline spray into the cold environment of an icing wind tunnel. The MARINE (Marine Accretion Research in
Northern Environments) facility in the Department of Mechanical
Engineering at the University of Alberta was completed in 1986. To
date, it is one of two indoor refrigerated marine icing wind tunnels
in the world. The other is a smaller tunnel at the University of
Hokkaido. Its design, instrumentation and performance have been discussed by Foy et al. (1987). The tunnel has the capability to produce
freshwater and saline spray at temperatures to -30°C, and in wind speeds
up to 40 m s^{-1}. A general view of the tunnel is given in Figure 14.
In the spring of 1987, a series of tests was carried out in the MARINE
tunnel to examine the effect of the basic forcing parameters on the
icing rate of a 1" rotating cylinder (Figure 15). In order to simulate
the icing conditions on board ship in the tunnel, a special algorithm
was developed (Zakrzewski and Lozowski, 1987b) to transform ship icing
conditions into tunnel-controlled parameters. The tests in question
are relatively easy to run, though time consuming, and can be executed
in the lab throughout the year. Their cost is several times smaller
than the cost of a winter ship field survey targetted at collecting

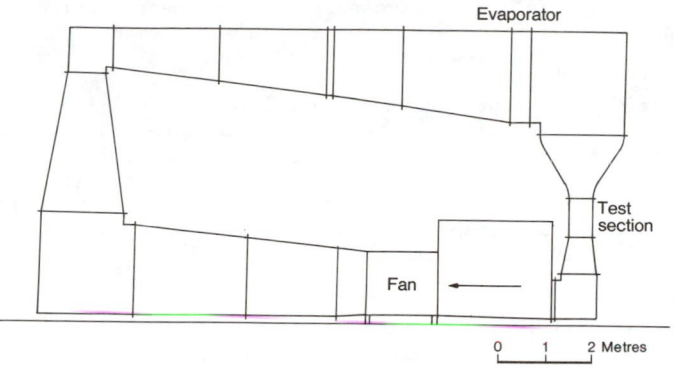

Fig. 14   The University of Alberta MARINE icing wind tunnel.

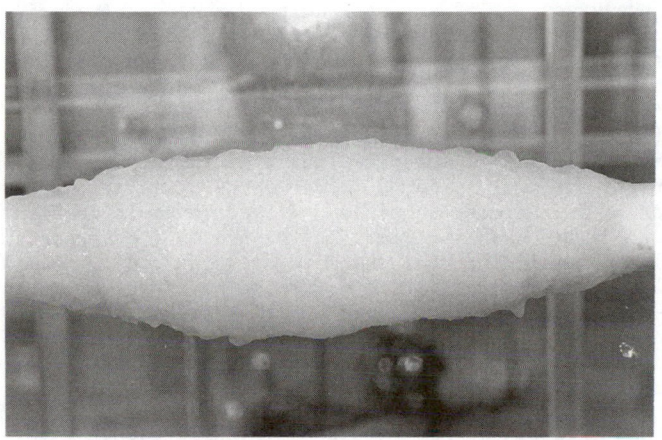

Fig. 15   Icing test in progress in the MARINE icing wind tunnel at
           the University of Alberta.

equivalent ship icing data at sea. Since the personnel and equipment
requirements of running tests in the already established facility are
smaller than those related to the launching of a ship expedition, the
possibility of simulating ship icing conditions in the cold environment
of the marine icing wind tunnel is very tempting and must be explored
in full to see if this hypothesis can be proven. Our data collected in
1987 and 1988 do not allow us to prove or disprove this hypothesis.

6.   SHIP ICING FORECAST TECHNIQUES

     National weather services base their ship icing forecasts mostly
on the empirical relationships (Japan – Sawada, 1966, 1967, 1968;
USA and UK – Overland et al., 1986; West Germany, Norway and Denmark –
Mertins, 1968). Weather services of Canada and the USSR use the model

of Kachurin et al. (1974). The Atmospheric Environment Service of Canada is considering implementation of a concise version of the University of Alberta ship icing model in the early 90s.

Ice accretion charts provide much more information about the ship icing conditions than verbal icing forecasts. To date, the NOAA ice accretion chart (see Figure 3) is the only map routinely issued by a weather service in the West. Two examples of various graphical representations of the icing rates on ice accretion charts for ships are given in Figure 16. These maps were computer-plotted by using the output from the University of Alberta ship icing model.

7. DE-ICING AND ANTI-ICING OF SHIPS

Among the techniques for anti-icing ships, one can distinguish the techniques based on (1) the application of an 'anti-icing' ship design (Figure 17), (2) the use of anti-icing coatings, mats (Figure 18), and paints, (3) the execution of an 'anti-icing' manoeuvre (reducing ship speed and/or heading down wind or sheltering the ship in the lee of land or in sea ice) by a ship experiencing icing or just alerted about the forecast of icing.

Panov (1976) and Winegrad (1987) reviewed the available techniques for ice removal from ships. Despite the large variety of trials and research carried out world-wide, a wooden or nylon mallet and an iron hammer appear to be the most common and effective de-icers used on ships. Snow shovels, used for removing the broken ice fragments off the deck, nicely compliment these antique de-icing tools. Similarly, urea pellets and salt are of little use on ships. However, several portable de-icers have been developed in the UK, USA, and the USSR. Pneumatic de-icers, pneumatic power hammers, vibrators, heaters, portable steam lances, and portable jet engines were tried on many occasions with various effects.

In the 80s, new approaches to the de-icing of ships, mostly USN warships, were developed. For example, Sackinger (1987) suggests the use of projectiles filled with two-component de-icing mixtures, and Winegrad (1987) describes the use of a very sophisticated portable de-icing system. This system operates on the principle that heated seawater is ejected by a water canon as the water is compressed to 3000 psi and given an ultrasonic resonance oscillation that is intended to destroy ice.

8. CONCLUSIONS AND RECOMMENDATIONS

The limited scope of this paper has allowed us merely to touch on several aspects of ship icing which we consider to be important. As a result of our discussion on ship icing modelling and forecasting, the following conclusions may be drawn.

1. Ice growth on ships is a complex environmental process shaped not only by the 'simple' growth of ice crystals, but also by (a) generation of spray due to wave-ship interaction, (b) spray droplet motion and cooling, and (c) transport of water and salt on the ship surface. The ship specifics must be considered in model development.

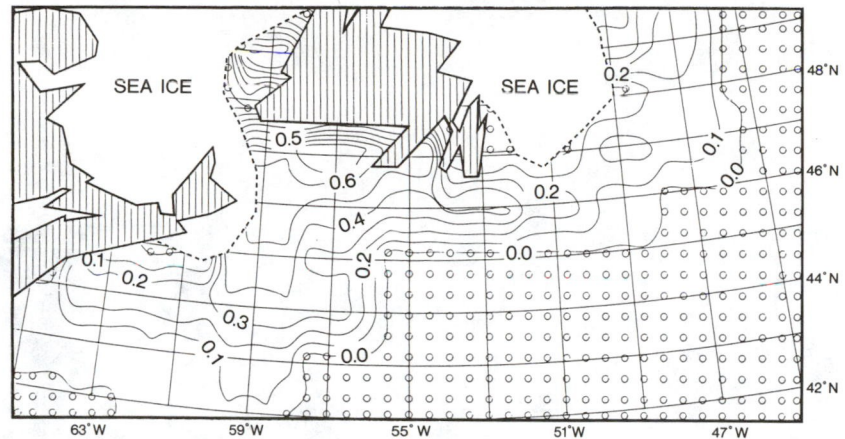

(a)

(b)

Fig. 16   Maximum potential icing rate on the foreface of a medium-
           sized fishing trawler (in cm h^{-1}) predicted by the University
           of Alberta ship icing model for the conditions of February 5,
           1985 (0000Z).

           a - isolines

           b - new graphical representation developed by Zakrzewski
               at the University of Alberta

Fig. 17 The MVs "Zandberg" and "Zandvort" of Fisheries Products
International in their home port of Catalina, Newfoundland,
Canada. Although these modern ships are susceptible to
spray and icing in moderate to high seas, they have proven
to be superior in anti-icing and performance over the older
trawlers.

(a)

Fig. 18
The installation of mats
covered with cryophobic
coatings on a Soviet trawler in the winter of
1968/69: (a) as given
by Panyushkin and Rozen-
berg (1972). What a
difference an effective
polyethylene coating
can make!; (b) applied on
the mast and rigging of the
ship (Panov, 1976) when
compared to the unprotected
ship in Fig. 2.

(b)

All the above processes take place under low- and high-frequency variations of ship motion parameters (six degrees of freedom) and low-frequency changes in the environmental parameters involved in spray generation and ice growth. A model attempting to consider all these changes of the model forcing parameters would be extremely complex.

2. Since brine movement depends on the shape, size and orientation of the target surface, we recommend development of several icing models for various components.

3. There is no 'generic' icing rate on a vessel. The only 'generic' icing rate for a vessel is the overall icing rate for the entire ship, which is the total of the incremental ice loads on all wetted ship components.

4. Reliable ship icing forecasts are sorely needed. These forecasts should be based on the output from a physics-based integrated ship icing/spraying model. Forecasts should be produced for several ship types/classes, rather than for a 'generic' ship. Ice accretion charts provide much more information about the icing intensity over the ocean than verbal messages.

5. Since ships can often avoid icing or reduce the icing rate by altering the ship speed and course, there is a strong potential for the development of a generation of tactical decision-making aids for ships.

6. The ship icing problem cannot be solved by icing forecasts alone, even if this accuracy is excellent. The ultimate goal is to develop and utilize an anti-icing and de-icing system for ships which is capable of ice removal under any condition. A change in the conventional ship design may be required.

7. In order to make model calibration possible, there is a need for generation of a reliable and comprehensive data base. This data base should be created by carrying out several laboratory and field ship icing and spraying experiments. Launching an international ship icing experiment is highly recommended here.

8. Boundary conditions and the environmental heat transfer from the ship surface need a new perspective and better approximations. Also, a new model for spray generation due to slamming and direct wave impacts is badly needed.

9. A series of international ship icing workshops would be helpful for the promotion of research on spraying and icing of ships.

REFERENCES

Aksyutin, L.R., 1979, Icing of ships, Sudostroenye Publishing House, Leningrad, 126 pp (in Russian).

Andreas, E.L., 1989, Thermal and size evolution of the spray droplets, CRREL Report 89-11, June 1989, 37 pp.

Barr, R.K. and J.U. Kordenbrock, 1987, U.S. Navy's Arctic/Cold Weather Program for Surface Ships, Proceedings U.S. Navy Symposium on Arctic/Cold Weather Operations of Surface Ships, Nov. 19-20, 1987, 31-39.

Blackmore, R.Z., Zakrzewski, W.P., and E.P. Lozowski, 1989, Growth of ice on a ship's mast, Proceedings of the 10th International Conference on Ports under Ocean and Arctic Conditions, Lulea, Sweden, June 12-16, 1989 (in press).

Borisenkov, Ye.P. and V.V. Panov, 1972, Primary results and prospects for investigating the hydrometeorological conditions of ship icing, Arkticheskii i Antarkticheskii Nauchno-Issledovatiel'skii Institut, Trudy, No. 298, Leningrad, 5-33 (in Russian).

Borisenkov, Ye.P. and I.G. Ptchelko (Editors), 1975, Indicators for forecasting ship icing, USA CREEL Draft Translation, No. 481.

Borisenkov, Ye.P., G.A. Zablockiy, A.P. Makshtas, A.I. Migulin, and V.V. Panov, 1975, On the approximation of the spray cloud dimension, Arkticheskii i Antarkticheskii Nauchno-Issledovatiel'skii Institut, Trudy No. 317, Gidrometeoizdat, Leningrad, 121-126 (in Russian).

Brown, R.D. and P. Roebber, 1985, The ice accretion problem in Canadian waters related to offshore energy and transportation, Canadian Climate Centre, Report No. 85013, Downsview, Ont., 295 pp.

DeAngelis, R.M., 1974, Superstructure icing, Mariners Weather Log, No. 18, 1-7.

Fein, N. and A. Freiberger, 1965, Survey of the literature on shipboard ice formation, Tech. Memorandum No. 2, Naval Applied Science Laboratory, Brookly, N.Y., 15 pp.

Foy, C.E., E.M. Gates, and E.P. Lozowski, 1987, Design, instrumentation and performance of a refrigerated marine icing wind tunnel, Proceedings International Symposium on Cold Regions Heat Transfer, 137-142, ASME.

Gates, E.M., 1985, Simulated marine icing in an icing wind tunnel, Department of Mechanical Engineering, University of Alberta, Report No. 49, 54 pp.

Horjen, I., 1988, Personal communication.

Horjen, I. and T. Carstens, 1989, Numerical modelling of sea spray icing on vessels, Proceedings of the 10th POAC Conference, Lulea, Sweden, June 12-16, 1989 (in press).

Horjen, I. and S. Vefsnmo, 1984, Mobile platform stability (MOPS) Subproject 02 - Icing, MOPS Report No. 15, Norwegian Hydrodynamic Labs, STF60 A 284002, 56 pp.

Jessup, R.G., 1985, Forecast techniques for ice accretion on different types of marine structures, including ships, platforms, and coastal facilities, Atmospheric Environment Service of Canada, Report presented to the 9th Session of the WMO Commission for Marine Meteorology, October 1984, 90 pp.

Kachurin, L.G., L.I. Gashin, and I.A. Smirnov, 1974, Icing rate of small displacement fishing vessels under various hydrometeorological conditions, Meteorologiya i Gidrologiya, No. 3, 50-60 (in Russian).

Kachurin, L.G., L.I. Gashin, and I.A. Smirnov, 1980, Icing of ships, Leningrad State Hydrometeorological Institute, 56 pp (in Russian).

Kennedy, W.P., 1987, Ice-phobic/low adhesion coatings for ship super-structure surfaces, Proceedings U.S. Navy's Symposium on Arctic/Cold Weather Operations of Surface Ships, November 19-20, 1987, 117-125.

Krasil'nikova, L.N. et al., 1969, Cryophobic coatings on organic-silicon foundation, In: Heat-Resistance Coatings, Leningrad, Izdatiel'stvo Nauka, 379-381 (in Russian).

Kreith, F., 1969, Principles of heat transfer, International Textbook Co., Scranton, Penn., 620 pp.

Kuznietsov, V.P., Ye.N. Kultashev, V.V. Panov, A.P. Tyurin, and A.V. Sharapov, 1971, Studying ship icing in the Sea of Japan in 1969, Theoretical and experimental investigations of conditions of icing on ships, Gidrometeorologicheskoye Izdatiel'stvo, Leningrad, 57-68 (in Russian.

Langmuir, I. and K.M. Blodgett, 1946, A mathematical investigation of water droplet trajectories, Collected Works of I. Langmuir, Pergamon Press, 10, 348-393.

Lozowski, E.P., 1988, Marine icing, Proceedings of the 9th IAHR Symposium on Ice, Sapporo, Japan, Aug. 23-28, 1988, Vol. 1, 43-66.

Lozowski, E.P. and E.M. Gates, 1990, On the modelling of ice accretion (this volume).

Lozowski, E.P. and E.M. Gates, 1985, An overview of marine icing modelling, Proceedings of the International Workshop on Offshore Winds and Icing, Halifax, Nova Scotia, Oct. 7-11, 1985, 102-122.

Lozowski, E.P., J.R. Stallabrass, and P.F. Hearty, 1983, The icing of an unheated, non-rotating cylinder, Journal of Climate and Applied Meteorology, 22, 2053-2074.

Lozowski, E.P. and W.P. Zakrzewski, 1988a, Forecasting of ice growth rates and ice loads due to icing on sea-going ships, Final Report to the Institute for Marine Dynamics, National Research Council, St. John's, Newfoundland, 227 pp.

Lozowski, E.P. and W.P. Zakrzewski, 1988b, Overview of Soviet research on ship icing, Milestone contract report to the U.S. Army CRREL, Hanover, N.H., 67 pp.

Lundquist, J.E. and I. Udin, 1977, Ice accretion on ships with special emphasis on Baltic conditions, Winter Navigation Research Board, Research Report No. 23.

Makkonen, L., 1984, Atmospheric icing, CRREL Monograph 84-2, 92 pp.

Makkonen, L., 1986, Salt entrapment in spray ice, Proceedings of the 8th IAHR Symposium on Ice, Iowa City, August 18-22, Vol. 2, 165-173.

Makkonen, L., 1987, Personal communication.

Makkonen, L., 1988, Formation of spray ice on offshore structures, Proceedings of the 9th IAHR Symposium on Ice, Sapporo, Japan, Vol. 2, 708-739.

Marine Icing Newsletter, 1989, Atmospheric Environment Service, Downsview, Ontario, Canada, Ross Brown (editor).

Mertins, H.O., 1968, Icing on fishing vessels due to spray, Marine Observer, Vol. 38, No. 221, 128-130.

Michieev, M.A. and I.M. Michieeva, 1973, Principles of heat transfer, Energiya Publishing House, Moscow, 319 pp (in Russian).

Minsk, L.D., 1977, Ice accumulation on ocean structures, CRREL Report No. 77-11, 42 pp.

Overland, J.E., C.H. Pease, R.W. Preisendorfer, and A.L. Comiskey, 1986, Prediction of vessel icing, Journal of Climate and Applied Meteorology, Vol. 25, No. 12, 1793-1806.

Panov, V.V., 1971a, Estimation of ice load due to spray icing of ships, Theoretical and experimental investigations of conditions of icing on ships, Gidrometeorologicheskoye Izdatiel'stvo, Leningrad, 26-48 (in Russian).

Panov, V.V., 1971b, On the frequency of splashing the medium fishing vessel with sea spray, Theoretical and experimental investigations of the conditions of the icing on ships, Gidrometeoizdat, Leningrad, 87-90 (in Russian).

Panov, V.V., 1976, Icing of ships, Arkticheskii i Antarkticheskii Nauchno-Issledovatiel'skii Institut, Trudy No. 334, Gidrometeoizdat, Leningrad, 263 pp (in Russian).

Pease, C.H. and A.L. Comiskey, 1986, Vessel icing in Alaskan waters 1979-1984 data set, NOAA Data Report, ERL PMEL-14, 16 pp.

Preobrazhenskii, L.Yu., 1973, Estimate of the content of spray-drops in the near water layer of the atmosphere, Fluid Mechanics - Soviet Research, Vol. 2, No. 2, 95-100 (in Russian).

Sackinger, W.M., 1987, New concepts for spray ice removal from ship superstructures, Proceedings U.S. Navy's Symposium on Arctic/Cold Weather Operations of Surface ships, Vol. 1, November 19-20, 1987, 297-303.

Sawada, T., 1966, A method of forecasting ice accretion in the waters off the Kurile Islands, Tokyo, Japan, Meteorological Agency, Journal of Meteorological Research, Vol. 18, 15-23.

Sawada, T., 1968, Ice accretion on ships in northern seas of Japan. Journal of the Meteorological Society of Japan, Vol. 46, No. 3, 250-254.

Semenova, Ye.P., 1972, Experimental investigations of chemical anti-freezers in the cold room at AANII. of ships, Arkticheskii i Antarkticheskii Nauchno-Issledovatiel'skii Institut, Trudy 298, Gidrometeoizdat, Leningrad, 97-104 (in Russian).

Sharapov, A.V., 1971, On the intensity of ice build-up on ships (of MFV type), Theoretical and experimental investigations of condit-ions of icing on ships, Gidrometeoizdat, Leningrad, 95-97 (in Russian).

Shekhtman, A.N., 1967, Hydrometeorological conditions of icing on ships, Nauchno-Issledovatiel'skii Institut, Aeroklimatologii, Trudy No. 45, Gidrometeoizdat, Moscow, 51-63 (in Russian).

Shekhtman, A.N., 1968, Probability and intensity of icing of sea-going ships, Nauchno-Issledovatiel'skii Institut, Aeroklimatologii, Trudy No. 50, Gidrometeoizdat, Moscow, 55-65 (in Russian).

Shellard, H.C., 1974, The meteorological aspects of ice accretion on ships, World Meteorological Organization, Marine Science Affairs Report No. 10 (WMO No. 397), 34 pp.

Stallabrass, J.R., 1970, Methods for the alleviation of ship icing, National Research Council, Ottawa, Canada, Report MD-51.

Stallabrass, J.R., 1980, Trawler icing, A compilation of work done at NRC, National Research Council, Mechanical Engineering Report, MD-56, NRC No. 19372, Ottawa, 103 pp.

Szilder, K. and E.P. Lozowski, 1989, Optimizing the growth thermo-dynamics of artificial floating ice platforms, Ocean Engineering, Vol. 16, No. 1, 99-115.

Tabata, T., 1968, Research on prevention of ship icing, Report to Hokkaido Prefectural Government, Institute of Low Temperature Science, Hokkaido University, Sapporo, Translation by the Defence Scientific Information Service of Canada, No. T95J, July 1968, 12 pp.

Tabata, T., 1969, Studies of ice accumulation on ships, III, Low Temperature Science, Series A, Part 27, 337-349.

Tabata, T., S. Iwata, and N. Ono, 1963, Studies of ice accumulation on ships I, Low Temperature Science, Series A, Part 21, 173-221.

Trawler Icing Research, 1957, British Shipbuilding Research Association, Report No. 221, London, England.

Tsvietukhin, A.S., 1977, Numerical algorithm with a small number of parameters for short-range ship icing forecasting, Problemy Arktiki i Antarktiki, No. 51, 6-12 (in Russian).

Vasil'yeva, G.V., 1966, Hydrometeorological conditions promoting the icing of sea-going ships, Rybnoye Khozyastvo, No. 12 (in Russian).

Winegrad, D.L., 1987, Shipboard ice removal techniques, Proceedings of of the 1987 U.S. Navy Symposium on Arctic/Cold Weather Operations of Surface Ships, November 19-20, 1987, 241-270.

Wise, J.A. and A.L. Comiskey, 1980, Superstructure icing in Alaskan waters, Pacific Marine Environmental Laboratory, Seattle, Washington, NOAA Special Report, 30 pp.

Wu, J., 1979, Spray in the atmospheric surface layer: review and analysis of laboratory and oceanic results, Journal of Geophysical Research, Vol. 84, No. C4, 1693-1704.

Zakrzewski, W.P., 1986, Icing of fishing vessels, Part II: Ice growth rates and simulation of icing, Proceedings 8th IAHR Symposium on Ice, Iowa City, August 18-22, 1986, 194-207.

Zakrzewski, W.P., 1987, Splashing a ship with collision generated spray, Cold Regions Science and Technology, Vol. 14, No. 1, 65-83.

Zakrzewski, W.P. and E.P. Lozowski, 1987a, The application of a vessel spraying model for predicting ice growth rates and ice loads on a ship, Proceedings of the 9th International Conference on Port and Ocean Engineering under Arctic Conditions, Fairbanks, Alaska, August 16-22, 1987, Vol. 3, 591-603.

Zakrzewski, W.P. and E.P. Lozowski, 1987b, Ice accretion of wires on ship board under simulated marine conditions, Proceedings of the 9th International Conference on Ports under Ocean and Arctic Conditions, Fairbanks, Alaska, Aug. 16-22, 1987, Vol. 3, 605-615.

Zakrzewski, W.P., E.P. Lozowski, and R.Z. Blackmore, 1988a, Atmospheric icing of ships and an overview of the research on atmospheric icing modelling applicable for ship icing, Proceedings of the 4th International Symposium on Atmospheric Icing on Structures, Paris, Sept. 5-7, 1988, 202-207.

Zakrzewski, W.P., E.P. Lozowski, R.Z. Blackmore, and R. Gagnon, 1988b, Recent approaches in the modelling of ship icing, Proceedings of the 9th IAHR Symposium on Ice, Sapporo, Japan, Aug. 23-27, 1988, Vol. 2, 458-476.

Zakrzewski, W.P., E.P. Lozowski, and D. Muggeridge, 1988c, The determination of the extent of the spraying zone on a ship, Ocean Engineering, Vol. 15, No. 5, 413-430.

Zakrzewski, W.P., R. Blackmore, and E.P. Lozowski, 1988d, Mapping the ice growth rates on sea-going ships, Journal of Meteorological Society of Japan, Vol. 66, No. 11, 661-675.

Zakrzewski, W.P., E.P. Lozowski, and R.Z. Blackmore, 1988e, Computer simulation of ship icing, Proceedings of the 9th IAHR Symposium on Ice, Sapporo, Japan, Aug. 23-27, 1988, Vol. 2, 436-457.

Zakrzewski, W.P. and E.P. Lozowski, 1989a, Modelling vessel icing, USA CRREL Report (in press).

Zakrzewski, W.P. and E.P. Lozowski, 1989b, Soviet marine icing data, Canadian Climate Centre Report, 89-2, 125 pp.

Zakrzewski, W.P., E.P. Lozowski, and I. Horjen, 1989: On the use of ship icing models for forecasting of the icing rates on ships, Proceedings of the 10th International Conference on Ports under Ocean and Arctic Conditions, Lulea, Sweden, June 12-16, 1989 (in press).

Zarling, J.P., 1980, Heat and mass transfer from freely falling drops at low temperatures, U.S. Army CRREL Report 80-18, 20 pp.

SPECIAL TOPICS (CASTING AND WELDING PROCESSES)

Chapter 21

Solidification Analysis of Castings

ITSUO OHNAKA
Department of Materials Science and Processing,
Faculty of Engineering, Osaka University,
Suitashi, Osaka 565, Japan

ABSTRACT

This paper presents the modeling of solidification with heat and mass transfer, and phase change, numerical techniques to solve the solidification models and software directing industrial application to the production of complicated shape castings. The discussed numerical techniques are the finite difference method, the direct finite difference method, the finite element method and the boundary element method. Numerical examples include the prediction of casting defects. Problems to be solved in the future are also discussed.

CONTENTS

NOMENCLATURE

a^*	coefficient in Eq.(51). $a^*=0$ and 1 corresponding to non-diffusion in solid and equilibrium condition, respectively.
a_s, a_{Lc}, a_{LT}	coefficient in Eq.(57)
A_e	area of an element (m^2)
B	magnetic flux density (T)
C_L, C_S	solute concentration in liquid and solid, respectively (mass%)
$\bar{C}$	mean solute concentration (mass%)
C_o	initial solute concentration (mass%)
$\bar{C}_S$	mean solute concentration in solid (mass%)
C_p	specific heat (kJ/(kg K))
d_{ij}	distance between nodal point i and j (m)
d_{ij}^i, d_{ij}^j	distance between nodal region boundary and nodal pint i and j, respectively (m)
D_d	dendrite arm spacing (m)
D_L, D_S	diffusion coefficient in liquid and solid, respectively (m^2/s)
Ei	exponential integral function
f_L, f_S	volume or mass fraction liquid and solid, respectively (-)
Δf_S	increase of fraction solid during a time step (-)
f_{sc}	a critical fraction solid above which the liquid can hardly move (-)
F	view factor in radiation heat transfer (-)
F	force or momentum (N)
g	acceleration of gravity vector (m/s^2)
h	heat transfer coefficient $(W/(m^2 K))$
H	specific enthalpy (kJ/m^3)
ΔH	latent heat of fusion (kJ/kg)
i	electric current density $(A/(m^2 s))$
k	equilibrium partition ratio (-)
K	permeability (m^2)
m_L	liquidus temperature slope on phase diagram (K/mass%)
n	outward normal vector or coordinate on element surface
N	interpolation function
P	pressure (Pa)
q	heat flux (W/m^2)
R_{ij}	thermal resistance between nodal points i and j $(m^2 K/W)$
r	distance (m)
S_{ij}	boundary surface area between nodal points i and j (m)
t	time (s)
t_f	local solidification time (s)
Δt	time step (s)
T	temperature (K)
T_f	melting point of pure metal on the liquidus line approximated to be linear (K)
T_L, T_S	liquidus and solidus temperature (K)
$\dot{T}$	temperature change, $\partial T/\partial t$ (K/s)
$\mathbf{u}$	velocity vector (m/s)

U	outward normal velocity on boundary surface of nodal region (m/s)
V	volume (m^3)
ΔV_p	pore volume increase during a time step (m^3)
ΔV_S	solid volume increase during a time step (m^3)
w	weighting function
x,y	x-,y-coordinates (m)
z	vertical coordinate (m)
α	thermal diffusivity (m^2/s)
ε	emissivity (-)
ε_p	porosity (-)
λ	thermal conductivity (W/(m K))
μ	viscosity (Pa s)
ρ	density (kg/m^3)
$\bar{\rho}$	element mean density (kg/m^3)
ρ_S	solid mean density (kg/m^3)
Γ	Stefan-Boltzmann constant (W/(m^2 K^4) or boundary

Subscripts

o	initial value
a	ambient
b,i,j,k,m	nodal point number
n,n1,n2	integration point number
ij	boundary between nodal points i and j
L	liquid
S	solid

Superscripts

B	known value at a time step before

1. INTRODUCTION

Since metal castings are used in many important industrial products, such as cylinder blocks of automobile engines and gas turbine blades, many researchers and engineers have been making an effort to improve the quality and productivity of metal castings and to reduce production costs. For these purposes it is necessary to predict casting defects, such as shrinkage porosity, and mechanical properties and to determine the optimal casting design and processing parameters. Computer simulation of solidification is the most powerful tool for solving these problems.

Analysis of solidification is very important not only in castings, but also in various solidification processes, such as the continuous casting of steel, welding, and the pulling-up of single crystals. However, the solidification phenomena are very complicated because they include heat and mass transfer, phase change and chemical reaction. Therefore it is necessary to solve different models depending on the process and purpose.

This paper will review and discuss the modeling of solidification, computer techniques to solve the models and software aiming industrial applications to complicated shape castings. The reader can also refer to previous reviews[1-13].

2. SOLIDIFICATION PHENOMENA IN CASTING PROCESS

2.1 Solidification Phenomena[14-17]

In the shape casting process a molten metal is poured into a mold and is partially cooled while filling the mold cavity(Fig.1)[*]. After filling the mold cavity, the heat of the melt is further removed through the mold.

In the melt, heat is transferred by thermal conduction and convection, though thermal radiation and phase change are also important in some molds. When the melt heats the mold surface gas is evolved and increases the thermal resistance between the melt and the mold. Oxide film formed on the melt also increases the thermal resistance in some alloys. In addition, the evolved gas may be absorbed in the melt, resulting in a gas porosity defect.

When the temperature of the melt falls below a liquidus temperature, crystals form by heterogeneous nucleation. Usually the undercooling at the nucleation stage is neglected. The morphology of the crystal varies

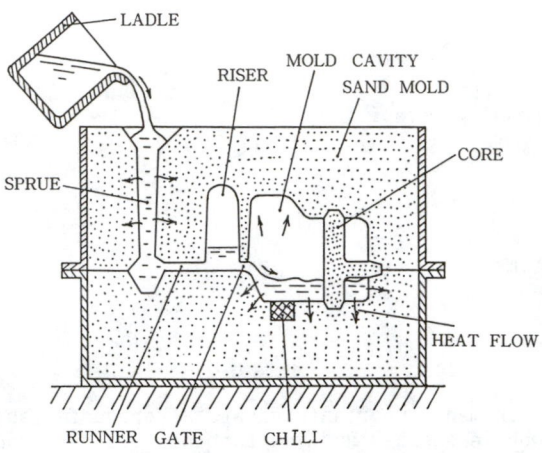

FIGURE 1. Metal casting in a sand mold

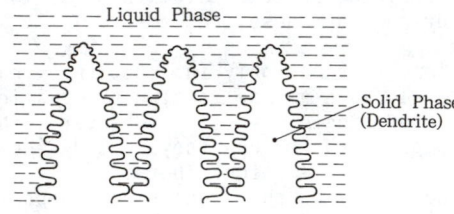

FIGURE 2. Dendrites in a solidifying alloy.

---

[*]In pressure die-casting the melt is injected into the mold by a piston, resulting in a high turbulent flow which is not easy to simulate.

with alloy composition and cooling conditions. Except for highly pure metals and unidirectional solidification under large temperature gradients, the shape of the solid-liquid interface is not planar but dendritic in many alloys(Fig.2). This causes a mushy region to be formed, where solid and liquid phases coexist. Therefore, it is difficult to employ the techniques proposed for solving so-called Stefan problem, where the planar solid-liquid interface is usually assumed. Further, information about the mushy region is very important, because many casting defects originate there. The required data includes temperature, volume fraction solid, solute concentration, pressure, shape and size of the solid phase and so on.

Along the growth of the crystal, the liberation of latent heat and, in some cases, solidification contraction, solute redistribution, and chemical reaction should be considered.

An increase in solid volume,ΔV_s, corresponds to liberation and removal of the latent heat $\rho_s \Delta V_s \Delta H$.

Since the solid density, ρ_s, is greater than the liquid density, ρ_L, in many alloys, the balance,$(\rho_s - \rho_L)\Delta V_s$, should be compensated. This density change forces the liquid to move(solidification contraction) resulting in a pressure decrease in the mushy region, which may cause the shrinkage porosity. The flow in the mushy region can be assumed to be the D'Arcy flow (see Section 4).

If the partition ratio,k, is less than unity, the solute concentration in the solid phase is less than that in the liquid phase and the balance is rejected into the liquid (microsegregation). The partition ratio and liquid concentration at the solid-liquid interface are known from equilibrium phase diagrams.

The melt contains gases such as hydrogen, nitrogen and oxygen, and their solubility decreases with decreasing temperature and with the phase change. This is one of main causes of porosity formation. Further, oxygen may react with solute elements, resulting in the formation of non-metallic inclusions or gas bubbles.

During solidification the liquid, and even the solid in many cases, move by thermal and solutal convections, and solidification contraction. This flow causes the macrosegregation and affects the macrostructure.

The temperature distribution in the solid brings about strain and stress, and deformation of the casting and mold. The deformation of the casting can cause liquid flow, resulting in porosity or macrosegregation. In addition, the thermal resistance between the casting and mold varies with the deformation and time.

Micro- and macro-structures, non-metallic inclusions, porosity and residual stresses, which determine the mechanical properties, are greatly affected by the cooling rate during solidification and phase transformation in the solid state.

For solidification analysis, it is necessary to predict temperature change and distribution, cooling rate, solid fraction change and distribution, and the place where solidification is delayed(hot spot). Further, information about solute concentration, pressure, and the degree of casting defects are also desired.

2.2 Solidification Model

The solidification phenomena sketched above should be modeled adequately, depending on the purpose, using one of the following:

(1) A thermal conduction model which considers only the thermal conduction heat transfer and the liberation of latent heat. It is the most simple, but practical, basic model. Most conventional solidification analyses employ this model.

(2) A D'Arcy flow model, which considers the D'Arcy flow in the mushy region and hence can calculate the pressure. If the solute transfer is also considered, the macrosegregation can be predicted. The convection heat transfer is often neglected.

(3) A laminar or turbulent flow model, where the laminar or turbulent flow is assumed in the liquid region and the flow in the mushy region is not considered. It is mainly used for continuous casting.

(4) D'Arcy-laminar flow model which considers the D'Arcy and laminar flows in the mushy and liquid regions, respectively. The flow effect can be analyzed more realistically.

(5) A D'Arcy-laminar-turbulent model. Since the flow in the liquid region is usually turbulent, the turbulent flow should be considered. However, it is more important to consider the heterogeneous distribution and movement of the solid phase and remelting. Further, the flow in the low fraction solid region is not well understood.

2.3 Relation between Fraction Solid and Temperature

In conventional casting processes it is thought that the local equilibrium condition holds at the solid-liquid interface and the temperature in the mushy region changes along the liquidus line on the equilibrium phase diagram. The liquidus temperature is given by a function of liquid concentration:

$$T_L = T_f - F(C_L) \tag{1}$$

If the function is linear and the alloy is binary, it becomes:

$$T_L = T_f - m_L C_L \tag{2}$$

The liquid concentration can be calculated by considering the solute redistribution. The following equation can be used: In the case of complete mixing in the liquid and complete diffusion in the solid, i.e. the equilibrium condition;

$$C_L = C_0\{ 1 + (k-1)f_S \}^{-1} \tag{3}$$

In the case of no diffusion in the solid(Scheil's equation[18]);

$$C_L = C_0(1-f_S)^{k-1} \tag{4}$$

In the case that a finite diffusion in the solid is considered[19];

$$C_L = C_0\{1-(1-\psi)f_S\}^{(k-1)/(1-\psi)}) \tag{5}$$

where $\Psi = 2\alpha k/(1+2\alpha)$, $\alpha \equiv 4D_S t_f/D_d^2$

From Eq.(2)-(5) the relation between the temperature and the fraction solid in the mushy region can be obtained:

$$T_L = F_s(f_S) \tag{6}$$

However, since engineering alloys are usually multicomponent, most of their phase diagrams have not been determined. Therefore, a linear or quadratic relationship is often assumed:

$$T_L = T_L^o - (T_L^o - T_S^o)f_S \tag{7-a}$$

$$T_L = T_L^o - (T_L^o - T_S^o)f_S^2 \tag{7-b}$$

where T_L^o and T_S^o are the measured liquidus and solidus temperatures, respectively. Further, it should be noted that the liquid concentration is affected by the liquid flow as mentioned in 2.1 and this effect is not considered in these equations.

2.4 Energy Balance Equation and Liberation of Latent Heat

The differential equation for the energy balance equation in the mushy region is :

$$\rho C_p \frac{\partial T}{\partial t} = Q_d + Q_s \tag{8}$$

where

$$Q_d = \nabla(\lambda\nabla T) + \rho C_p\nabla(f_L \mathbf{u}T) \tag{8-a}$$

$$Q_s = \rho\Delta H\partial f_S/\partial t \tag{8-b}$$

Q_d denotes the conduction and convection heat transfer and Q_s denotes the liberation of the latent heat due to fraction solid increase. The latent heat term is treated as follows:

Equivalent specific heat method. Rearranging Eq.(8) we obtain:

$$\rho C_{pE}\frac{\partial T}{\partial t} = Q_d \tag{9}$$

where

$$C_{pE} = C_p - \Delta H\partial f_S/\partial T \tag{10}$$

and $\partial f_S/\partial T$ can be calculated from Eqs.(6) or (7). Therefore, Eq.(9) can be solved as a non-linear transient heat transfer problem and conventional software packages can be used. However, if the alloy has a narrow solidification interval and a large time step is selected, the temperature may jump past the solidification interval, resulting in the latent heat being neglected. Special techniques are required to avoid this. In addition, the high non-linearity of C_{pE} requires special care in the numerical calculation(See 3.3). Further, since the fraction solid should be calculated, this method is not as simple as it appears.

Enthalpy method. The enthalpy of a material having the fraction solid,f_S, is defined by:

$$H = \int_{T_S}^{T} \rho C_p dT + \int_{0}^{1-f_S} (\rho \Delta H) df_S + H_o \qquad (11)$$

where H_o is the enthalpy at $T=T_S$. Assuming C_p and ρ to be constant and differentiating this equation with respect to time, we obtain:

$$\frac{\partial H}{\partial t} = \rho C_p \frac{\partial T}{\partial t} - \rho \Delta H \frac{\partial f_S}{\partial t} \qquad (12)$$

Therefore Eq.(8) becomes:

$$\frac{\partial H}{\partial t} = Q_d \qquad (13)$$

This equation is usually solved by the explicit method, and the temperature after a time step, $T^{t+\Delta t}$, is determined from a known enthalpy-temperature relation(Fig.3). A full enthalpy method is also proposed[20] where the temperature on the right hand side of Eq.(13) is converted to enthalpy.

<u>Temperature recovery method</u>. In this method, first, the temperature, $\tilde{T}$, after a time step is calculated from Eq.(8) by assuming $Q_s=0$. If T is in the solidification interval, the enthalpy change should be the sum of the latent heat and sensible heat changes:

$$\rho C_p(T^t - \tilde{T}) = C_p(T^t - T^{t+\Delta t}) + \rho \Delta H\{f_S(T^{t+\Delta t}) - f_S(T^t)\} \qquad (14)$$

The temperature after a time step, $T^{t+\Delta t}$, is calculated from this equation.

If the time step is small, we can use a simpler method [36]. Namely, if the temperature $\tilde{T}$ falls below the liquidus temperature $T_L(f_S^t)$, the fraction solid increase, Δf_S, is calculated by Eq.(15) and the temperature is recovered from $\tilde{T}$ to $T_L(f_S^t)$.

$$\Delta f_S = C_p(T_L(f_S^t) - \tilde{T})/\Delta H \qquad (15)$$

In the next time step the liquidus temperature $T_L(f_S^t)$ is changed to $T_L(f_S^t + \Delta f_S)$ and this procedure is repeated until $f_S = \sum \Delta f_S = 1$ is satisfied.

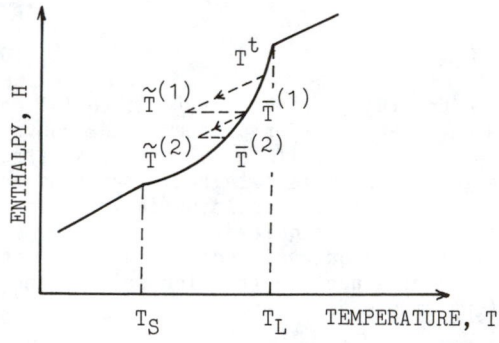

FIGURE 3. Relation between enthalpy and temperature(solid line). Broken line indicates the procedure of the fictitious heat flow method.

<u>Heat generation method</u>. The methods mentioned above are used mainly for the explicit method in the thermal conduction model. Since the implicit method usually employs a longer time step and results in a larger temperature change, it may lead to a larger error. Further, when temperature and solute fields are coupled, these methods are difficult to use.

"Fictitious Heat Flow Method" is one of the heat generation methods and can be applied to longer time steps. In this method the liberation of the latent heat is first estimated in a similar manner as in the enthalpy method, and the heat balance equation(Eq.(8)) is solved directly by the following procedure for each time step (Fig.3)[72,77,81]:

(a) Solve Eq.(8) by assuming $Q_s = 0$ and calculate $\tilde{T}^{(1)}$.

(b) From Eq.(14) calculate $\bar{T}^{(1)}(\equiv T^{t+\Delta t}$ in Eq.(14)) and $\Delta f_S^{(1)}$:

$$\rho C_p (T^t - \tilde{T}^{(1)}) = \rho C_p (T^t - \bar{T}^{(1)}) + \rho \Delta H \Delta f_S^{(1)} \qquad (16\text{-}a)$$

$$\Delta f_S^{(1)} = f_S(\bar{T}^{(1)}) - f_S(T^t) \qquad (16\text{-}b)$$

(c) Calculate the latent heat generation rate by :

$$Q_S^{(i)} = \rho \Delta H \Delta f_S^{(i)} / \Delta t, \quad i=1 \qquad (17)$$

(d) Calculate $\tilde{T}^{(i+1)}$ from Eq.(8) by setting $Q_S = Q_S^{(i)}$.

(e) Calculate $\bar{T}^{(i+1)}$, $\Delta Q_S^{(i+1)}$ and $\Delta f_S^{(i+1)}$ from :

$$\rho C_p (\bar{T}^{(i)} - \tilde{T}^{(i+1)}) = \rho C_p (\bar{T}^{(i)} - \bar{T}^{(i+1)}) + \rho \Delta H \Delta f_S^{(i+1)} \qquad (18\text{-}a)$$

$$\Delta f_S^{(i+1)} = f_S(\bar{T}^{(i+1)}) - f_S(\bar{T}^{(i)}) \qquad (18\text{-}b)$$

$$\Delta Q_S^{(i+1)} = \rho \Delta H \Delta f_S^{(i+1)} / \Delta t \qquad (18\text{-}c)$$

(f) If $\varepsilon_q = |Q_S^{(i+1)}/(\rho \Delta H)|$ is larger than a given small value, ε_{qo}, then

set $Q_S^{(i+1)} = Q_S^{(i)} + \Delta Q_S^{(i+1)}$, $i=i+1$ and repeat from (d) to (f) until

$\varepsilon_q < \varepsilon_{qo}$.

Although Dalhuijsen et al.[87] reported that the fictitious heat flow method is less accurate than the enthalpy method, it is still considered practical[81].

The other method of the heat generation method is " Coupling Method ". Essentially Eqs.(6),(8) and the liquid concentration C_L are coupled and they should be solved simultaneously(see 5). For example, from Eq.(6) and (8) we obtain:

$$\partial f_S/\partial t = - Q_d \{\rho \Delta H - \rho C_p \cdot dF_S/df_S \}^{-1} \qquad (19)$$

Equations (8) and (19) can be solved by the Runge-Kutta method. Alternatively, if Q_d is evaluated by using known values a time step before, then we can calculate $\Delta f_S = \Delta t \partial f_S/\partial t$ and $T^{t+\Delta t}$ from Eqs.(19) and (8), respectively.

3. SOLIDIFICATION ANALYSIS BY THE THERMAL CONDUCTION MODEL

Various methods can be grouped into the fixed domain or fixed grid method and the front tracking method. The latter calculates the phase change boundary position directly and includes the variable space grid method [21], the isotherm migration method[22], the solid phase-fitted curvilinear coordinate transformation method[23,24] and the deforming or moving element FEM[74,76,86]. Although both methods are useful for pure metal or single crystals, the fixed domain method is more practical for shape castings because of the existence of the mushy region described in 2.1. Therefore, in this paper only the fixed domain method is presented.

3.1 Finite Difference Method(FDM)

The FDM discretizes the differential equations by using Taylor expansion. Although many works using the FDM have been reported on solidification analysis of casting[25-34], they are usually limited to simple-shaped castings. Therefore, the FDM is not used in practice. Further, it should be noted that the FDM originally did not have the concept of the nodal region and it is not convenient for calculating the fraction solid as discussed in 3.3.

Nevertheless, the FDM with the boundary-fitted coordinate transformation method[35] has a potential applicability to shape castings. In this method, computational domains of arbitrary shapes are mapped onto simply shaped domains in the transformed coordinate system. The resultant differential equation is solved by the FDM. The difference between this method and the conventional curvilinear coordinate transformation method is that it uses the automatic numerical generation of a general curvilinear coordinate system with coordinate lines coincident with all boundaries of the arbitrarily shaped domains. This enables automatic meshing and is convenient for CAD. However, it is not clear whether suitable coordinate control functions applicable to all practical complicated castings exist or not. Further, comparisons of accuracy and applicability with other methods which use a similar automatic meshing algorithm are necessary.

3.2 Direct Finite Difference Method(DFDM)

In this method the discretized equations ,i.e. finite difference equations, are derived directly from physical phenomena[36-43,103,104].

Derivation of finite difference equation. The geometry concerned is divided into finite elements, and nodal regions and nodal points are designated. The nodal point is a representative point of the nodal region. There are two techniques in the DFDM, depending on the definition of the nodal point and region. In one technique, the inner nodal point technique, the nodal region is defined as the finite element itself and the nodal point is designated at the circumcenter of the element. In the other technique, the outer nodal point technique,* the nodal points are designated at the vertexes of the finite elements and the nodal region is

---

*An outer nodal point technique where the nodal region is formed by sides of quadrilateral elements and lines connecting midpoints of the sides has also been developed[111].

defined as the region formed around the nodal point by the bisectional surfaces of each element's edge. However, the inner and outer nodal point techniques are essentially the same.

The finite difference equation for the thermal conduction model is derived as follows(Fig.4). Assuming the linear distribution of temperature between the nodal points, the net rate of heat across the element surface S_{ij} by thermal conduction is:

$$\Sigma \lambda_{ij} S_{ij} (T_j - T_i)/d_{ij} \qquad (20\text{-}a)$$

where the thermal conductivity λ_{ij} is evaluated for the mean temperature $T_{ij} = (T_i + T_j)/2$.

If S_{ib} are boundary surfaces, the heat entering the nodal region is:

$$\Sigma S_{ib} (T_b - T_i)/R_{ib} \qquad (20\text{-}b)$$

where

$$R_{ib} = 1/h_{ib} + d_{ib}^i/\lambda_i + d_{ib}^b/\lambda_b \qquad (20\text{-}c)$$

If S_{im} are radiative boundary surfaces, the radiation heat transfer coefficient can be used for h_{ib} in Eq.(20-c). Alternatively, by assuming a uniform temperature in the nodal region, the following heat transfer can be considered:

$$\Sigma (\varepsilon \Gamma FS)_{im} \{ (T_m)^4 - (T_i)^4 \} \qquad (20\text{-}d)$$

The heat entering the nodal region(element),i, and the latent heat are balanced to the rate of accumulation of internal energy within the element:

$$(\rho C_p V)_i \frac{\partial T}{\partial t} = Q_{Ti}^+ + (\rho \Delta HV)_i \frac{\partial f_S}{\partial t} \qquad (21)$$

where

$$Q_{Ti}^+ = \Sigma (\lambda S)_{ij} (T_j - T_i)/d_{ij} + \Sigma (S/R)_{ib} (T_b - T_i) + \Sigma (\varepsilon \Gamma FS)_{im} \{ (T_m)^4 - (T_i)^4 \}$$

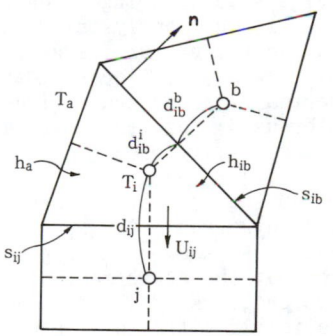

FIGURE 4. Energy conservation in nodal region(element),i.

This equation can be solved by the forward or backward difference schemes, Crank-Nicolson's scheme or other schemes. If the forward difference scheme is used, the time step is calculated by:

$$\Delta t < (\rho C_p V)_i / a_i \tag{22}$$

$$a_i = \Sigma (\lambda S/d)_{ij} + \Sigma (S/R)_{ib} + \Sigma (\varepsilon \Gamma FS)_{im} (T_m)^4 / (T_i)^4$$

This condition leads to all positive coefficients in the finite difference equation and satisfies the stability condition of the explicit method.

Characteristics of the DFDM. Similar methods to the DFDM have been proposed[44-60] where conservation laws are applied in finite elements or control volumes. They include the thermal resistance network method [44], the PIC method[46], the FLIC method[47], the subdomain method[50], the control volume method[53-60] and the finite element method. These methods, excluding the thermal resistance network method, use integration of governing differential equations, while the DFDM derives the discrete equations directly from physical phenomena. Therefore, those methods excluding the DFDM might be referred to as " Finite Volume Integration Methods (FVIM)". The DFDM has the following characteristics against the FVIM:

(a) The DFDM does not depend on coordinates.

(b) The DFDM is clearer in physical meaning and more reliable. Although the FVIM integrates the differential equations, it is very approximate and not always correct. For example, often the integration of a gradient, $\int (\partial T/\partial n) ds$, is assumed to be $S\Delta T/\Delta r$, where S is the element side length and Δr is the distance between nodal points, which are usually defined as being at the center of gravity of each element. This is not correct because the orthogonality between S and the heat flux, $\lambda \Delta T/\Delta r$, does not hold if equilateral triangular or rectangular elements are not used. This may result in an error.

(c) The DFDM has the potential to handle complex, non-linear, heterogeneous problems with ease.

However, the FVIM will become more reliable if its resultant finite difference equations are compared with those derived by the DFDM.

Numerical examples. As one of numerical examples, temperature contours and the solidified region(hatched region) at 101 s after pouring a cast iron melt in a composite mold are shown in Fig.5. In this analysis, the DFDM with a forward finite difference scheme and the temperature recovery method were employed. The mushy region(unhatched region) is located at the junction and some shrinkage defects are expected if an adequate riser is not used. The temperature distribution can be used for the stress analysis in order to estimate the lifetime of the metallic mold.

Another example is shown in Fig.6, which is a 3-D analysis of a steel wheel drum. Because Eq.(21) can easily be applied to 2- and 3-D problems, it is convenient for developing practical programs. As shown in Fig.6, the wheel drum is roughly rotationally symmetric, but the attachment of a riser requires 3-D analysis. Figure 6-b shows the progress of solidification at the symmetric center for two cases of slightly different shapes of riser base. In (b)-1 the hot spot is located at point P and porosity has actually been observed there(inside of the product). By changing the riser base as shown in (b)-2 , the hot spot moved outside of the product.

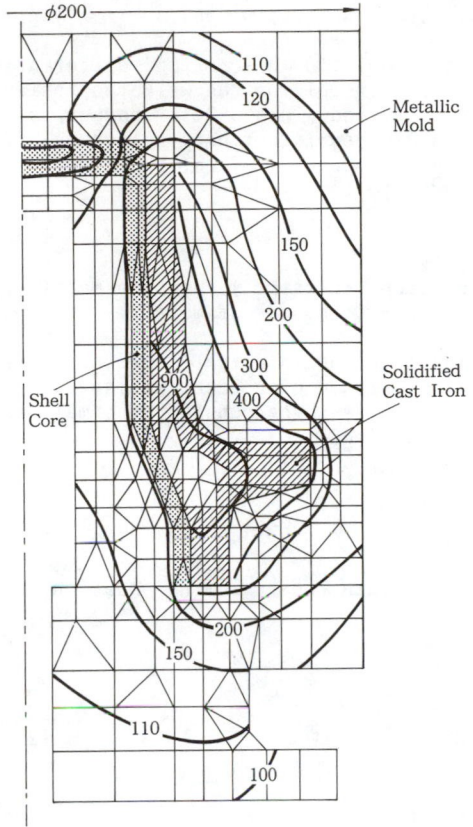

FIGURE 5. Temperature contours in a cast iron and in a composite mold constructed of metallic mold and shell core at 101 s after cooling.

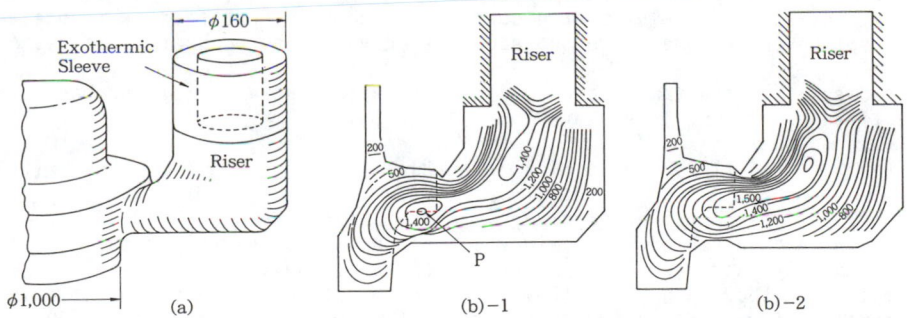

FIGURE 6. Progress of fraction solid 0.7 in steel drums cast in a sand mold. (b)-1 and (b)-2 differ in the riser bottom shape. A shrinkage defect was observed at P.

3.3 Finite Element Method (FEM)

The FEM is also practical for complicated castings[61-90].

Derivation of finite difference equation. To discretize differential equations the FEM uses the variational principle or the weighted residual method. The Galerkin's procedure is preferred. In a similar manner, as in the DFDM, the geometry concerned is divided into finite elements(2-D problems are considered to facilitate discussions). Then, the temperature distribution in the elements is assumed to be:

$$T(x,y,t) = [N(x,y)]^t [T(t)] = \sum N_i T_i \tag{23}$$

where $[N]$ is referred to as interpolation (or shape) function matrix, $[T]$ nodal temperature matrix and $[\ ]^t$ denotes transpose of matrix.

In Galerkin's procedure the governing equation is weighted by the interpolation function matrix, $[N]$, and integrated over each element. From Eq.(8), excluding the convection term and assuming constant thermal conductivity, the following equation is obtained:

$$\int_\Omega [N]\{\rho C_p \frac{\partial T}{\partial t} - \lambda \frac{\partial^2 T}{\partial x^2} - \lambda \frac{\partial^2 T}{\partial y^2} - \rho \Delta H \frac{\partial f_s}{\partial t}\}d\Omega = 0 \tag{24}$$

where Ω is the element's interior region.

Using Green's theorem and Eq.(23), we obtain the following equation:

$$[c][\frac{\partial T}{\partial t}] = [k][T] + [h] + [q] \tag{25}$$

where

$$[c] = \int_\Omega \rho C_p [N][N]^t d\Omega \tag{25-a}$$

$$[k] = -\int_\Omega \lambda (\frac{\partial [N]}{\partial x} \cdot \frac{\partial [N]^t}{\partial x} + \frac{\partial [N]}{\partial y} \cdot \frac{\partial [N]^t}{\partial y})d\Omega \tag{25-b}$$

$$[h] = \int_\Gamma [N]\lambda \frac{\partial T}{\partial n} ds \tag{25-c}$$

$$[q] = \int_\Omega [N]\rho \Delta H \frac{\partial f_s}{\partial t} d\Omega \tag{25-d}$$

where Γ denotes the boundary. The term $\lambda \partial T/\partial n$ in Eq.(25-c) is equal to the heat flux on the element's surface. Since this heat flux is canceled except on the boundary, we consider it only for the boundary surface.

To discretize the time space we can use various schemes. If the backward difference scheme is employed, Eq.(25) becomes:

$$\{[c]/\Delta t - [k]\}[T^{t+\Delta t}] = [c][T^t]/\Delta t + [h] + [q] \tag{26}$$

Numerical calculation procedure. Equation(25) is calculated for each element and assembled. The values corresponding to the same nodal points are simply added in the appropriate space of the global matrix. If an implicit method such as Eq.(26) is employed, it should be performed efficiently since this method requires a lot of memory. Prescribed boundary conditions, such as constant temperature and heat flux, are inserted into the final assembled matrix. Finally, the resulting equation is solved by a suitable method such as the Gauss-elimination method, the modified Cholesky method, etc.

Treatment of the latent heat. If the equivalent specific heat method is employed, more attention, in addition to the possibility of neglecting latent heat(see 2.4), should be taken in integrating Eq.(25-a), since a serious instability might be caused[64] if the specific heat is interpolated simply by:

$$C_p = \Sigma N_i C_p(T_i) \tag{27}$$

Therefore the following techniques are proposed:

In " Enthalpy-equivalent specific heat technique", instead of directly integrating the specific heat, the enthalpy defined by Eq.(11) is interpolated because it is a smooth function of temperature even in the phase change zone. Specifically

$$H = \Sigma N_i H_i(t) \tag{28}$$

is used, where $H_i(t)$ are the values at the nodal points.

Since $\rho C_p = dH/dT$, the following averagings are proposed:

$$\text{(a)} \quad \rho C_p = \frac{1}{2} \left(\frac{\partial H}{\partial x} / \frac{\partial T}{\partial x} + \frac{\partial H}{\partial y} / \frac{\partial T}{\partial y} \right) \qquad [65] \tag{29}$$

$$\text{(b)} \quad \rho C_p = \left[\frac{(\partial H/\partial x)^2 + (\partial H/\partial y)^2}{(\partial T/\partial x)^2 + (\partial T/\partial y)^2} \right]^{1/2} \qquad [66,71] \tag{30}$$

$$\text{(c)} \quad \rho C_p = \left(\frac{\partial H}{\partial x} \cdot \frac{\partial T}{\partial x} + \frac{\partial H}{\partial y} \cdot \frac{\partial T}{\partial y} \right) / \left[\left(\frac{\partial T}{\partial x}\right)^2 + \left(\frac{\partial T}{\partial y}\right)^2 \right] \qquad [67] \tag{31}$$

$$\text{(d)} \quad \rho C_p = \left(H^{t+\Delta t} - H^t \right) / \left(T^{t+\Delta t} - T^t \right) \tag{32}$$

Thomas et al.(80) reported that technique (b) is the best, though they differ slightly.

Besides these techniques, the enthalpy method, the temperature recovery method and the heat generation method described in 2.4 can be used without difficulties, though " nodal region (node-associated area) " should be reasonably defined in most of these techniques(see below).

Discussions. (1) Treatment of heat capacity matrix. Many difficulties in the non-linear transient heat conduction problems are caused by the treatment of the heat capacity matrix. For example, Eq.(25-a) results in the following matrices for constant physical properties:

In case of linear triangular elements with 3-nodes(TEM-3):

$$[c] = \frac{1}{12} C_p A_e \begin{vmatrix} 2 & 1 & 1 \\ 1 & 2 & 1 \\ 1 & 1 & 2 \end{vmatrix} \tag{33-a}$$

In case of bi-linear square isoparametric elements with 4-nodes(ISEM-4):

$$[c] = \frac{1}{36} C_p A_e \begin{vmatrix} 4 & 2 & 1 & 2 \\ 2 & 4 & 2 & 1 \\ 1 & 2 & 4 & 2 \\ 2 & 1 & 2 & 4 \end{vmatrix} \tag{33-b}$$

Where A_e is the element area. Therefore, the heat accumulation term for node 1 in Fig.7 (the first equation of $[c][\partial T/\partial t]$ is:

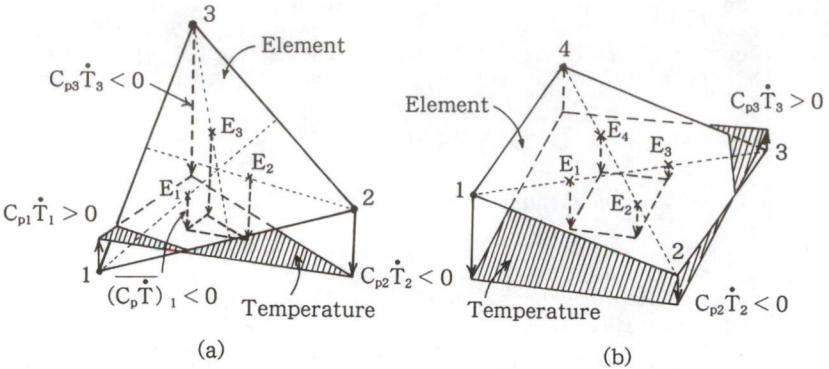

(a)　　　　　　　　　　　　(b)

FIGURE 7. Representative nodal value and distribution of C_pT for the linear distribution assumption in FEM. (a) Linear triangular element, (b) Bi-linear isoparametric element.

$$\overline{(C_p\dot{T})}_1 = \frac{1}{12} C_p A_e (2\partial T_1/\partial t + \partial T_2/\partial t + \partial T_3/\partial t) \qquad \text{for TEM-3} \quad (34)$$

$$\overline{(C_p\dot{T})}_1 = \frac{1}{36} C_p A_e (4\partial T_1/\partial t + 2\partial T_2/\partial t + \partial T_3/\partial t + 2\partial T_4/\partial t) \quad \text{for ISEM-4} \quad (35)$$

These results show that the heat capacity change is evaluated at point E_i (mid- and tri-sectional points on median and diagonal lines) and the nodal temperature changes are extrapolated from values at E_i. Therefore, even if the mean value $(C_p\dot{T})_i$ is negative, the values at nodal points $(C_p\dot{T})_i$ could be positive as illustrated in Fig.7. This is why an unreasonable temperature rise is often obtained in cooling problems. Further, this may cause a serious error if the equivalent specific heat method is employed, where C_p shanges suddenly[37]. Therefore, such smoothing techniques as described in 3.3 are required.

In addition, if $\partial T_i/\partial t$ are equal, then Eq.(34) becomes:

$$(C_p\dot{T})_1 = \frac{1}{3} C_p A_e \, \partial T/\partial t \tag{36}$$

This means that the nodal area is always $A_e/3$ and explains why the temperatures at P_1 and P_2 in Fig.8(a) differ for the same cooling condi-

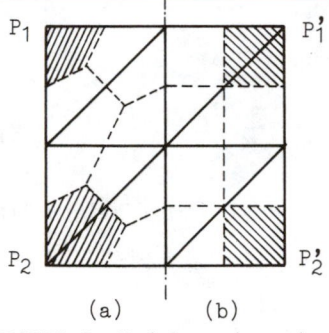

(a)　　　　(b)

FIGURE 8. Nodal regions in FEM with triangular element with 3-nodes. Hatched regions in (a) and (b) are the nodal regions of P_1, P_2 and P'_1, P'_2 in cases of linear and concentrated distribution of C_p, respectively.

tion. In this case, the resultant nodal area of P_1 and P_2 is $A_e/3$ and $2A_e/3$, respectively, while the net outflow of heat is the same[37].

These difficulties can be eliminated by employing the " lumped mass technique" [37,77,81,89,91] where the specific heat is concentrated at the nodes.   For example, the nodal area might be computed by:

$$A_i = A_e \iint N_i dxdy \tag{37}$$

where N_i is the interpolation function. However, it does not always work. For example, for quadratic 8-node elements negative value appears.   For TEM-3 it becomes $A_i = A_e/3$ and the same problem mentioned above occurs. For ISEM-4, Eq.(37) does not have this problem, hence it is recommended.

The reasonable nodal region area might be determined by considering the physical meaning in a similar manner as in the DFDM. Namely, if we choose the nodal region described in 3.2, the problems mentioned above do   not occur. For example, the following heat capacity matrix for TEM-3 is obtained by using the outer nodal technique in the DFDM[36]:

$$[c] = \begin{vmatrix} C_p A_1 & 0 & 0 \\ 0 & C_p A_2 & 0 \\ 0 & 0 & C_p A_3 \end{vmatrix} \tag{38}$$

where A_i is the nodal region area, which is determined by constructing a perpendicular bisector on each side of the triangle(Fig.8(b)).

(2) Linear and higher order elements. For non-linear problems, higher order elements such as 8-node quadratic isoparametric elements, does not always result in an improvement in accuracy and computing time. Further, from an engineering point of view, it is not so important to look for rigorous solutions for the simple model. It is enough if the accuracy of the solutions is almost the same as that of the physical model.   Since the computing cost is more important, the simple method, particularly the ISEM-4 or TEM-3 FEM, might be more practical.

(3) Implicit and explicit schemes. Most of conventional FEM employs implicit schemes, such as the backward finite difference scheme, Crank-Nicolson's scheme, Dupont's three-level scheme, Lee's three-level scheme and so on. Thomas et al.[80] reported that the Dupont's three-level scheme is superior in both accuracy and stability. However, if the memory size is taken into account, the backward difference scheme might be more economical. Further, since many industrial problems require a lot of elements and shorter computing time, the explicit scheme, typically the forward difference scheme, is often more practical. As for the computing time, if the inverse matrix $\{[c]/\triangle t+[k]\}^{-1}$ is stored in the backward difference scheme with ISEM-4 and the fictitious heat flow technique, the computing time can be greatly reduced[81].

(4) Comparison with DFDM. For the thermal conduction model there is not much difference between FEM and DFDM in accuracy, computing time and flexibility for shape. Rather it may depend on whether the user has a good preprocessor or not. However, for more complicated solidification models the DFDM might be more practical.

3.4 Boundary Element Method (BEM)

The BEM has a potential applicability to the analysis of complicated shape castings[92-95].

Derivation of finite difference equation. The BEM is also based on the governing differential equations and Eq.(8) is integrated over the whole region concerned, while integration over finite elements is used in the FEM. Further, a time integration from t_1 to t_2 is also performed. Namely from Eq.(8), excluding the convection and fraction solid terms:

$$\int_{t_1}^{t_2} \int_\Omega w \, \{ \rho C_p \, \frac{\partial T}{\partial t} - \lambda \nabla^2 T \} d\Omega \, dt = 0 \tag{39}$$

where w is a weight function. Applying Green's theorem and partial integration, and introducing a weighting function, $w=T^*$(fundamental solution), we obtain:

$$T_i + \frac{\alpha}{\lambda} \iint qT^* dsdt = \frac{\alpha}{\lambda} \iint Tq^* dsdt + \int TT^* d\Omega \big|_{t1} \tag{40}$$

where $q = -\partial T/\partial n$ and $q^* = -\partial T^*/\partial n$.

Assuming that the variation of T and q is much smaller than that of T^* and q^*, and hence T and q are constant for a small time step $\Delta t (= t_2 - t_1)$, we finally obtain the following equation for 2-D problems:

$$c_i T_i + \frac{1}{4\pi\lambda} \int_\Gamma q Ei(-a)d\Gamma = \frac{1}{2\pi} \int_\Gamma T \, \frac{|r|}{r^2} \, \exp(-a)d\Gamma + \frac{1}{4\pi\alpha\Delta t} \int T \big|_{t1} \exp(-a)d\Omega \tag{41}$$

where the integration on the boundary is evaluated in Cauchy's principal value sense, $a=r^2/\{4\pi(t_2-t_1)\}$ and $c_i=1$ for inner points i and $c_i=1/2$ for points i on smooth boundary. Further, the following boundary conditions are used(Fig.9):

$\lambda \partial T/\partial t = h(T_a - T)$ then $q = h(T - T_a)$

$\lambda \partial T/\partial t = - \bar{q}$ (constant) then $q = \bar{q}$

$T = \bar{T}$ (constant) then $q = q$ (unknown)

If we divide the boundary into boundary elements having N nodal points and the inner domain into rectangular elements having M nodal points, the following discretized equations are obtained:

$$\sum_j^N H_{ij} T_j^{t+\Delta t} = \sum_j^N G_{ij} B_j^{t+\Delta t} + \sum_k^M P_{ik} T_k^t \tag{42}$$

where

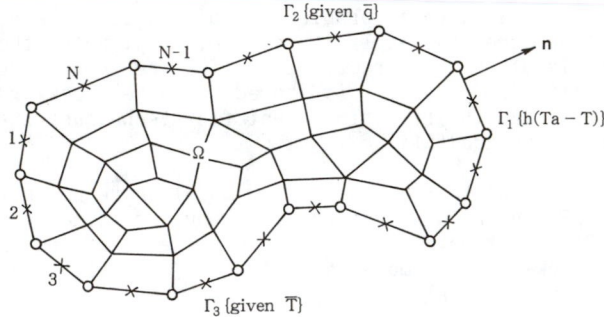

FIGURE 9. Boundary elements(1,2,3,---n) and internal elements in BEM. Γ_1, Γ_2 and Γ_3 denote boundaries with variable temperature, constant heat flux and constant temperature, respectively.

$B_j^{t+\Delta t}$ is $q_j^{t+\Delta t}$ or $h_j(T_j-T_a)$ (42-a)

$$G_{ij} = \frac{1}{4\pi\lambda}\int_{\Gamma}Ei(-a)d\Gamma = \frac{1}{4\pi\lambda}\Sigma\ Ei(-a)J_Bw_n \qquad (42\text{-}b)$$

$$H_{ij} = \frac{1}{2\pi}\int_{\Gamma}\frac{|r|}{r^2}\exp(-a)d\Gamma = \frac{1}{2\pi}\Sigma\ \frac{|r|}{r^2}\exp(-a)J_Bw_n \qquad (42\text{-}c)$$

$$P_{ij} = -\frac{1}{4\pi\alpha\Delta t}\int_{\Omega}\exp(-a)d\Omega = -\frac{1}{4\pi\alpha\Delta t}\Sigma\Sigma\exp(-a)J_Dw_{n1}w_{n2} \qquad (42\text{-}d)$$

J_B and J_D are Jacobian for the transformation of coordinates and w_n, w_{n1}, w_{n2} are the weighting factors for the numerical integration.

The formulation described above is for transient thermal conduction problems and the liberation of latent heat is considered by the temperature recovery method[93]. The enthalpy method may also be applicable.

Discussions. For steady state thermal conduction problems the BEM has a clear benefit. Namely, enmeshing the interior domain is not necessary and computing time is shorter. However , for the solidification analysis the interior domain should also be divided into elements. Further, the BEM has the following demerits. First, the resultant equations are implicit and require a lot of memory to be solved. Second, it is not easy to apply it to advanced models where fluid flow is considered.

The greatest advantage of the BEM may be that the shape of the interior domain element is not so limited as in the DFDM or the FEM where remarkably deformed elements decrease the accuracy. If this has a considerable benefit over the DFDM and FEM, and if memory size reduction is possible, the BEM may be used practically.

4. D'ARCY MODEL AND ESTIMATION OF SHRINKAGE DEFECT

The shrinkage defect in castings results from solidification contraction and gas evolution. The deformation of the casting and mold due to non-uniform temperature distribution and melt static pressure also contribute to the defect in some cases. However, since in many cases the pressure decrease from solidification contraction is the main cause for the shrinkage defect, i.e. pore formation, the estimation of the defect requires the knowledge of the amount of pressure in the mushy region. This can be found by using the D'Arcy model[42,43,96-98].

4.1 D'Arcy Model

Governing finite difference equation. The energy balance equation is given in Eq.(21). The convection heat transfer by the D'Arcy flow can usually be neglected because it is small. Further, if a pore is formed in an element, the mass of the element should be changed.

The mass balance equation is easily derived by applying the DFDM to the nodal region(element),i, shown in Fig.4:

$$\{(\rho_S-\rho_L)\Delta f_S + (\rho_L-\rho_L^B)f_L\}_iV_i/\Delta t - (\rho_L\Delta V_p)_i/\Delta t = -\Sigma\ (\rho_Lf_LSU)_{ij} \qquad (43)$$

The left hand side consists of the solidification contraction, liquid density change and pore formation terms. The liquid density change can be usually neglected for the shrinkage defect estimation.

Applying D'Arcy's law to the surface element shown in Fig.10, we obtain

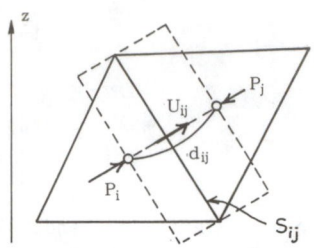

FIGURE 10. Equilibrium of forces in the surface element drawn by broken line.

the following equation:

$$P_i - P_j + \Delta\rho g(z_i - z_j) = (\mu d f_L/K)_{ij} U_{ij} \qquad (44)$$

where P_i is the pressure variation due to fluid flow and $\Delta\rho = \rho_0 - \rho_L$ is the liquid density difference from the initial condition. Substituting Eq.(44) into Eq.(43), we obtain:

$$\Sigma(\frac{\rho_L KS}{\mu d})_{ij}(P_i-P_j)=(\rho_L-\rho_S)V_i\Delta f_s/\Delta t + \rho_L\Delta V_p/\Delta t - \Sigma(\frac{\rho_L\Delta\rho g KS}{\mu d})_{ij}(z_i-z_j) \qquad (45)$$

This equation can be easily solved if the right hand side value is known.

Calculation procedure. First, $T^{t+\Delta t}$ and f_S are calculated by solving Eq.(21) and employing the temperature recovery method. The right hand side of Eq.(45) is determined by those values. The method to calculate ΔV_p will be mentioned below in 4.2. Then Eq.(45) is solved by an appropriate method, for example, the Gaussian elimination procedure, considering the boundary condition. This procedure is repeated until a given time. In these calculations the following element status index, Id, is designated and used for determining boundary conditions and equations to be used: Id=0 for vacant elements which does not have liquid and solid, 12 for free surface element where the liquid can flow, i.e. $f_S < f_{sc}$, Id=14 for free surface element where the liquid cannot move, i.e. $f_S \geq f_{sc}$, Id= 2 for flow element where $f_S < f_{sc}$, Id=3 for pore elements where pores are forming, and Id=4 for solid elements where $f_S \geq f_{sc}$.

4.2 Estimation of Shrinkage Defect

Application of the thermal conduction model. The formation of the shrinkage defect is often estimated by the parameters obtained from the thermal conduction model. In particular, the temperature gradient method and the fraction solid gradient methods are widely used[12,55]. It is thought that the pore forms easily if these gradients are small, because the smaller the gradient, the wider the mushy region and the more difficult it becomes to make up the solidification contraction. However, since the the critical value for the pore formation changes with alloys, shapes and dimensions of castings, the critical value should be determined empirically.

Application of the D'Arcy model. The next simple method is to assume that the defect forms in elements below a critical absolute pressure. However, the accuracy of this method depends largely on the accuracy of the permeability data, which is difficult to measure. Therefore, it is not practical at present.

Alternatively, the defect may be evaluated by the pressure gradient in the mushy region. It is thought that the shrinkage defect may form in an element where the pressure gradient is larger than a critical value, because the pressure may be low in the element.

However, it is desirable to simulate the pore formation more realistically and to evaluate the pore more quantitatively. If we can calculate the shrinkage pore volume in each element, the shrinkage defect can be evaluated by the porosity:

$$\varepsilon_{pi} = V_{pi}/V_i, \quad V_{pi} = \Sigma \Delta V_{pi} \tag{46}$$

A simple method for calculating the pore volume is to assume a critical pressure for the pore formation. After the formation of pores at the critical pressure, the melt in the element is thought to be used to compensate the solidification contraction. Therefore, the pore volume increase is calculated by:

$$\Delta V_{pi} = \Delta t \Sigma (f_L SU)_{ij} \tag{47}$$

The elements having a free surface facing the atmosphere are also treated as pore elements, because the free surface is lowered due to the solidification contraction. If a flow element(Id=2) is surrounded by a solidified shell, it is assumed that the pore is formed in the following order: (i) elements with the lowest head and $f_S < f_{sc}$, and (ii) elements having the lowest fraction solid.

In a more realistic model, to calculate the pore volume we assume: (i) the equilibrium condition holds for the gas element, (ii) the gas in the pore is ideal one, (iii) the pores grow with a constant radius of curvature which is assumed to be proportional to residual liquid thickness between dendrites or grains which are determined mainly by the cooling rate in the mushy region, and (iv) the solid phase does not move. From these assumptions we can calculate the pore volume[106].

4.3 Examples and Discussions

Figure 11 shows the flow field in tapered steel ingots, demonstrating that the downward flow is mainly due to solidification contraction and the upward flow is due to solutal convection from microsegregation. In steels the liquid density usually decreases with the progress of solidification[42]. It can be seen that the negative pressure at the late stage of solidification decreases with the taper. This is why the taper is important in producing steel ingots without porosity. Further, the upward flow may result in the A-segregation[99,100].

Figures 12 and 13 show examples[43] where the critical pressure of the pore formation was assumed to be a constant. These results show that the shrinkage porosity can be simulated if proper permeability and critical pressure are used. However, it is not easy to apply this method to practical casting processes, because this simulation requires large memory size, high computing costs, and accurate data of permeability and critical pressure. However, with the development of computer hardware such simulation could be used practically in the near future.

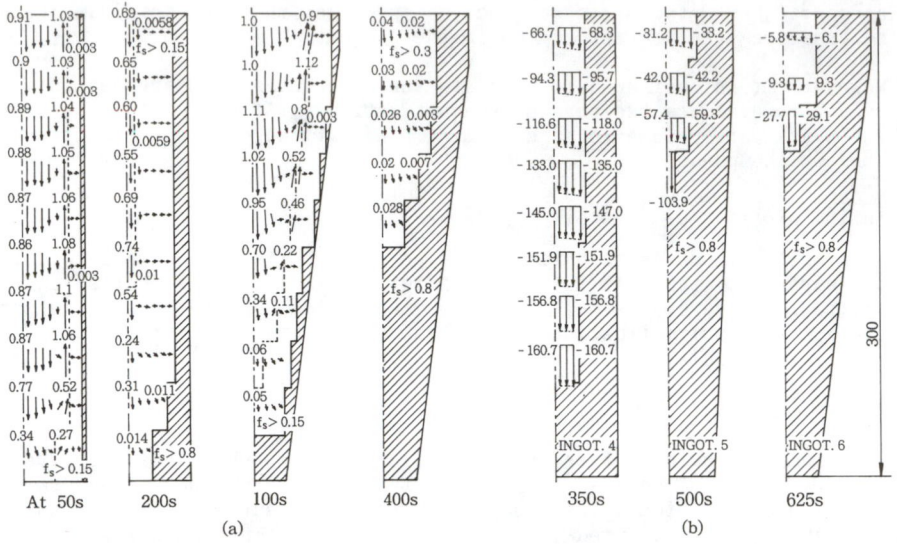

FIGURE 11. Example of analysis of tapered steel ingots by the D'Arcy
model. (a) Flow field in mm/s, (b) Pressure field in kPa.

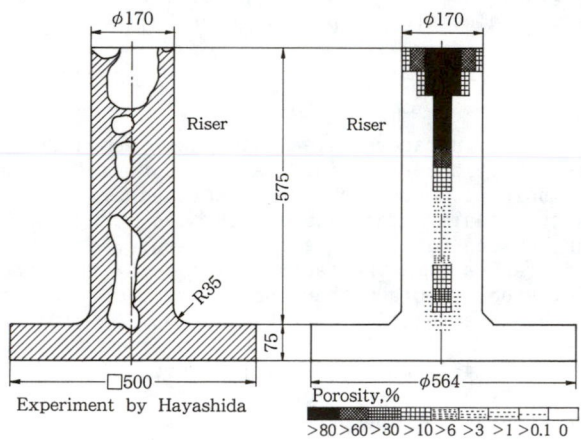

FIGURE 12. Comparison of measured and simulated shrinkage porosity in a
steel casting (D'Arcy model)

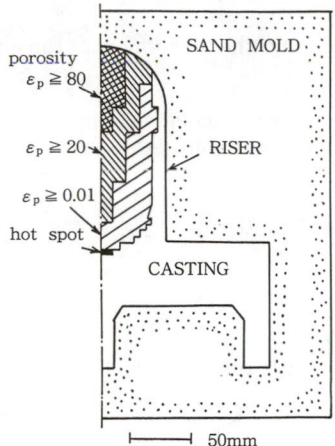

FIGURE 13. Simulated shrinkage porosity in a spheroidal graphite cast iron (D'Arcy model)

5. SIMULATION OF MACROSEGREGATION

Although some macrosegregation such as V- or A-segregation might be estimated by parameters calculated by the thermal conduction model, the parameters do not contain the essential information about concentration and fluid flow. Therefore, it is necessary to simulate the fluid flow and solute concentration directly.

There are two approaches to estimating the flow and concentration. One approach considers only the flow in the liquid region and the other considers the flows in both the liquid and the mushy regions. In the former approach, the liquid concentration is calculated by using an effective diffusivity[102] or mass transfer coefficient between the solid and liquid region. This method can easily employ turbulent models and is suitable for single crystals or for alloys with a narrow solidification interval. However, the fluid flow in the mushy region is often more important for macrosegregation and it is not easy to determine the effective diffusivity and the mass transfer coefficient for various alloys. In this section the latter approach, i.e. a D'Arcy-laminar flow model is presented, where D'Arcy and laminar flows are considered in the mushy and liquid regions, respectively, and the concentration field is also solved by the DFDM.

5.1 Governing Finite Difference Equations

Energy balance equation. The energy balance equation, Eq.(21), is used by adding the convection term:

$$T_i = T_i^B + \Delta t Q_T / (\rho C_p V)_i + \Delta H (\Delta f_S / C_p)_i \qquad (48)$$

where $Q_T = Q_T^+ - \Sigma (\rho_L f_L C_p SU)_{ij} (\overline{T} - T_i^B)$ and $\overline{T} = T_i^B$ for $U > 0$ and $\overline{T} = T_j^B$ for $U \leq 0$.

This convection term is derived from

$$(C_pV)_i \, \partial \bar{\rho} T/\partial t = (\bar{\rho}C_pV)_i \partial T_i/\partial t + (C_pTV)_i \partial \bar{\rho}/\partial t,$$

and mass balance equation, Eq.(43), by neglecting the pore formation and assuming $C_{pij} \simeq C_{pi}$.

<u>Solute mass balance equation</u>. The solute mass balance equation is derived in a similar manner as in the derivation of the energy balance equation:

$$V_i \partial (\bar{\rho}\bar{C})_i/\partial t = - \Sigma (\rho_L f_L SUC_L)_{ij} + \Sigma (\rho_L D_L S)_{ij}(C_{Lj}-C_{Li})/d_{ij} \qquad (49)$$

where

$$(\bar{\rho}\bar{C})_i = (\bar{\rho}_S f_S \bar{C}_S + \rho_L f_L C_L)_i \qquad (50)$$

If we assume non-diffusion in a solid or equilibrium condition, the following equation is obtained:

$$\partial (\bar{f}_S \bar{C}_S)_i/\partial t = k(C_L \partial f_S/\partial t + a^* f_S \partial C_L/\partial t)_i \qquad (51)$$

Combining these equations and employing the forward finite difference scheme we obtain the following solute mass balance equation:

$$C_{Li} = \{ 1 + \rho_S(1-k)\Delta f_S/G_{fi}\}C_{Li}^B + \Delta t Q_{ci}/(G_f V)_i \qquad (52)$$

where

$$Q_{ci} = - \Sigma (\rho_L f_L SU)_{ij}(\bar{C}_L - C_{Li}^B) + \Sigma (\rho_L D_L S)_{ij}(C_{Lj}^B - C_{Li}^B)/d_{ij} \qquad (53)$$

$$\bar{C}_L = C_{Li}^B \quad \text{for } U \geq 0 \text{ and } \bar{C}_L = C_{Lj}^B \text{ for } U < 0,$$

$$G_{fi} = (\rho_L f_L + a^* \rho_S k f_S^B)_i$$

The solute concentration in the solid and the mean concentration are calculated by:

$$\bar{C}_S = (\bar{\rho}_S^B f_S^B \bar{C}_S^B + \rho_S \Delta f_S C_S)/(\bar{\rho}_S f_S) \qquad (54)$$

$$\bar{C} = (\bar{\rho}_S f_S \bar{C}_S + \rho_L f_L C_L)/(\bar{\rho}_S f_S + \rho_L f_L) \qquad (55)$$

where the mean solid density is:

$$\bar{\rho}_S = (\bar{\rho}_S^B f_S^B + \rho_S \Delta f_S)/f_S \qquad (56)$$

The solid and liquid densities are assumed to be:

$$\rho_S = \rho_{So} + a_s k C_L \qquad (57\text{-}a)$$

$$\rho_L = \rho_{Lo} + a_{Lc} C_L + a_{LT} T \qquad (57\text{-}b)$$

<u>Momentum balance equation</u>. The net momentum flowing in the element, i, (Fig.14) through the element surface area, S_{ij}, per unit time is:

$$F_m = - \Sigma (\rho_L f_L uUS)_{ij} \qquad (58\text{-}a)$$

The shear force due to viscosity acting on the element surface is:

$$F_v = \Sigma (\mu f_L S)_{ij}(u_j - u_i)/d_{ij} \qquad (58\text{-}b)$$

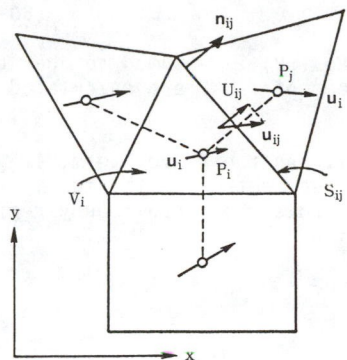

FIGURE 14. Momentum balance in element, i.

The melt pressure force acting on the element surface is:

$$F_p = - \Sigma (f_L \ nPS)_{ij} - (\rho_0 f_L V)_i \cdot \mathbf{g} \qquad (58\text{-}c)$$

where P is the pressure variation from the initial static pressure.
As the body force we consider the gravity and the electromagnetic forces:

$$F_b = (\rho_L f_L V)_i \ \mathbf{g} + \mathbf{i} \times \mathbf{B} \ V_i \qquad (58\text{-}d)$$

The force acting in the element, i, due to the D'Arcy flow may be proportional to the nodal velocity, viscosity, a mean liquid surface area,
$\overline{f_L S}$, and flow distance, $\overline{L}$, normal and parallel to the nodal velocity,
respectively:

$$F_d \propto - (\mu \mathbf{u})_i \overline{f_L S} \cdot \overline{L}$$

Here if we assume $\overline{f_L S} \cdot \overline{L} = (f_L V)_i$, then we obtain the following equation by
introducing the permeability, K, which is a function of fraction solid
[105]:

$$F_d = - (\mu f_L^2 \ uV/K)_i \qquad (58\text{-}e)$$

These forces should be balanced with the momentum accumulation per unit
time, i.e.:

$$V_i \frac{\partial \rho_L f_L \mathbf{u}}{\partial t}\Big|_i = - \Sigma (\rho_L f_L \mathbf{u} US)_{ij} + \Sigma (\mu f_L S)_{ij} (\mathbf{u}_j - \mathbf{u}_i)/d_{ij} + (\rho_L - \rho_0)_i (f_L V)_i \cdot \mathbf{g}$$

$$- \Sigma (f_L \ nPS)_{ij} - (\mu f_L^2 \ uV/K) + \mathbf{i} \times \mathbf{B} V_i \qquad (59)$$

If we employ the explicit method and assume $\partial \rho_L f_L / \partial t = 0$, we finally obtain the following momentum balance equation:

$$\mathbf{u}_i = \tilde{\mathbf{u}}_i - \frac{\Delta t}{(\rho_L f_L V)_i E_k} \Sigma (f_L \ nPS)_{ij} \qquad (60\text{-}a)$$

$$\tilde{\mathbf{u}}_i = \mathbf{u}_i^B / E_k + \frac{\Delta t}{(\rho_L f_L V)_i E_k} \{ - \Sigma (\rho_L f_L \mathbf{u} US)_{ij}^B + \Sigma (\mu f_L S)_{ij} (\mathbf{u}_j^B - \mathbf{u}_i^B)/d_{ij} \}$$

$$+ (\rho_L - \rho_0) \Delta t \ \mathbf{g}/(\rho_L E_k) + \mathbf{i} \times \mathbf{B} \Delta t/(\rho_L E_k) \qquad (60\text{-}b)$$

where $E_k = 1 + \{\mu f_L \Delta t/(\rho_L K)\}_i$ (60-c)

and the velocity **u** in the D'Arcy term of Eq.(59) is assumed to be the value at time t+Δt. The element surface values of f_L, P are calculated by a linear interpolation.

The greatest advantage of Eq.(59) is that it has the D'Arcy term. If we apply the D'Arcy and laminar flow equations separately in the mushy and liquid regions[100,101], it is not easy to handle the narrow mushy region appearing in the early stage of solidification.

5.2 Calculation Procedure

The calculation procedure is as follows:

(a) Data such as element volume, node distance, and initial and boundary conditions are read from a file which is generated by a preprocessor.
(b) Q_T in Eq.(48) and Q_c in Eq.(53) are calculated by using known values a time step before. The fraction solid increase can be calculated by solving simultaneously Eqs.(2),(48) and (52):

$$\Delta f_S = \{ F_1 - (F_1^2 - 4F_2 F_3)^{1/2} \}/(2F_2)$$ (61)

where

$$F_1 = (\bar{\rho}V)_i[b^* H + \rho_S C_p m_L (1-k) C_L^B]_i/\rho_L + F_4 , \qquad F_2 = (\bar{\rho}V)_i \Delta H$$

$$F_3 = \Delta t \cdot m_L (\bar{\rho} C_p Q_c)_i/\rho_L + b^* F_4/\rho_L$$

$$F_4 = (\bar{\rho} C_p V)_i (T_f - m_L C_{Li}^B - T_i^B) - \Delta t \cdot Q_{Ti}$$

$$b^* = \rho_L f_L^B + a^* \rho_S k f_S^B$$

(c) T_i, C_{Li} and C are calculated by Eq.(48),(52) and (55), respectively.
(d) The velocity field is calculated by using values obtained in the step (c). Here the transient velocity, $\tilde{u}$, is first calculated by Eq.(60-b). Then the pressure field is calculated so that the mass conservation equation, Eq.(43), is satisfied. Namely, applying Newton's law on the element surface(Fig.10):

$$(\rho_L f_L dS)_{ij} \Delta U_{ij}/\Delta t = f_L S_{ij}(P_i - P_j) - \mu \Delta U_{ij}(dS)_{ij} f_L^2/K$$ (62)

then substituting U=U + ΔU, calculated from Eq.(62) and from a linear interpolation of **u**, into Eq.(43), we obtain the pressure equation:

$$\sum \frac{\Delta t (f_L S)_{ij}}{d_{ij} E_k} (P_i - P_j) = - \sum (\tilde{U} \rho_L f_L S)_{ij} - \{(\rho_S - \rho_L)\Delta f_S$$
$$+ (\rho_L - \rho_L^B) f_L\} V_i/\Delta t + (\rho \Delta V_p)_i/\Delta t$$ (63)

(e) Results are printed out at a prescribed time interval.
(f) The above steps are iterated until a prescribed time.

5.3 Examples and Discussions

<u>Flow field in a heavy steel ingot</u>. Figure 15 demonstrates that the fraction solid distribution obtained by this model is very different from that obtained by the thermal conduction model. If the convection is not

considered, the fraction solid at the ingot center is always lowest. On the other hand, in this example, the thermal convection dominates at the early stage of solidification(Fig.15(a)) and the solutal convection becomes stronger with time because the temperature difference in the melt becomes smaller and microsegregation occurs. The solutal convection, shown in Fig.15(b) and (c)(i.e. the upward flow near the mold), transports low temperature melt into the ingot center, resulting in the higher fraction solid(Fig.15(c)). This result suggests that the fluid flow greatly changes the solidified structure: the transition from columnar to equiaxed structures may occur at the low fraction solid region between the mold and ingot center.

Macrosegregation in a steel ingot having stepwise thickness change.
Figure 16 shows a simulated flow at 50 s after cooling of the bottom of a columnar steel specimen having a stepwise change. The change of flow direction at the lower step resulted in the macrosegregation shown in Fig.17. In this analysis it was assumed that the bottom of the specimen was cooled by convection heat transfer. The simulated results agreed with the measured ones.

Macrosegregation in Al-Cu ingots solidified horizontally. Figure 18 shows the velocity field at 400 s after cooling in a horizontally unidirectionally solidifying Al-4.4 mass % Cu ingot. Thermal and solutal convections were observed in the liquid and mushy regions, respectively. The velocity in the mushy region was smaller than that in the liquid region by about one order. This flow field resulted in the macrosegregation shown in Fig.19, which agreed with the measured ones. This macrosegregation varied mainly with the solutal convection in the mushy region and was affected by the permeability[104,107].

Although this model can simulate various types of macrosegregation, it is not easy to simulate the V- and A-segregations*, because the former is related to the solid phase movement and the latter to the inhomogeneous solid phase distribution.

6. SOFTWARE PACKAGES FOR SOLIDIFICATION ANALYSIS

Several software packages are currently used in foundries(Table 1). Most of them are based on the thermal conduction model and employ the FEM and DFDM. Although these software packages are useful, the following should be further improved:

(1) Preprocessor. Current preprocessors are not sufficient for the input data generation of complicated castings in terms of time for shape input and of 3-D enmeshing. More user friendly preprocessors should be developed.
(2) Heat transfer analysis during filling. The heat transfer and gas absorption or inclusion pick-up during the filling of mold cavity are very important for estimating the casting defects. Although approximate analysis is possible [112], accurate flow analyses including gas and inclusion pick-up are not easy.

---

* Quite recently Bennon et al.[110] simulated the A-segregation by using a super computer.

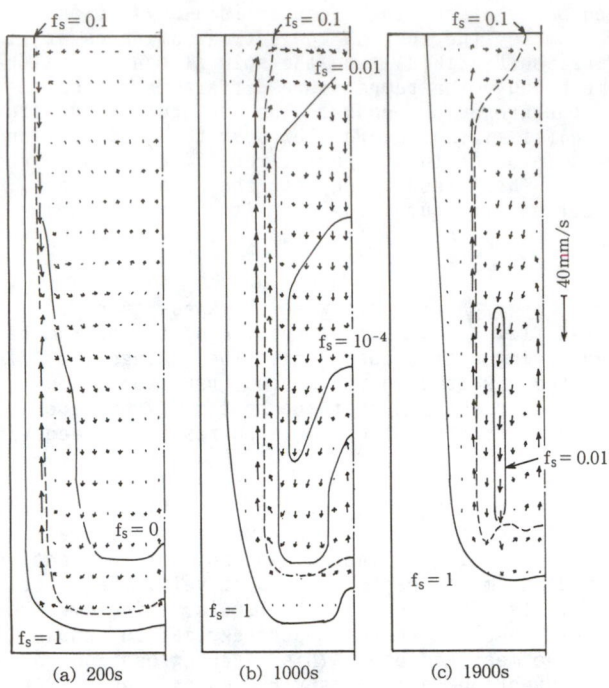

FIGURE 15. Fluid flow and solid fraction distribution in a two-dimensional steel ingot of 500 mm in width and 1000 mm in height. The top surface was heat insulated.

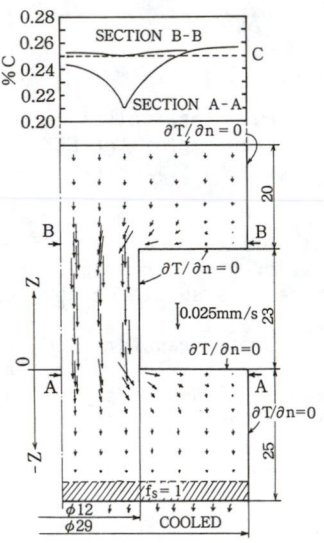

FIGURE 16. Simulated fluid flow at 50 s after cooling and carbon segregation in a unidirectionally solidified Fe-0.25 mass% C specimen.

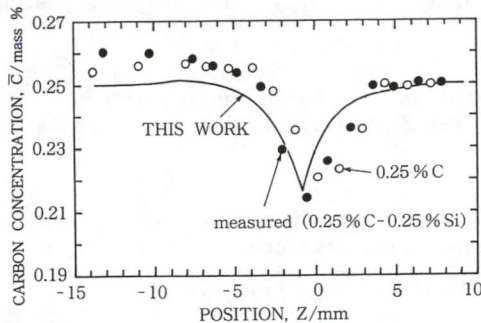

FIGURE 17 Comparison of simulated solute distribution(mean value over cross-section) with measured ones by Nomura et al.[108] in the specimen shown in Fig.16.

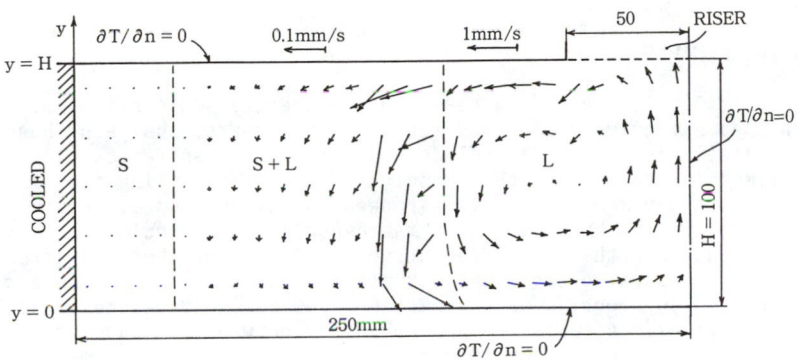

FIGURE 18. Simulated fluid flow at 400 s after cooling in a horizontally unidirectionally solidifying Al-4.4 mass % Cu ingot.

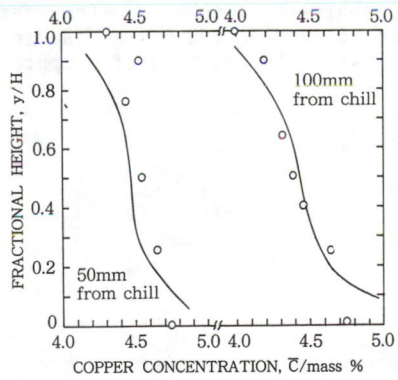

FIGURE 19. Comparison of simulated solute distribution with measured one by Mehrabian et al.[109] in Al-4.4 mass % Cu ingot shown in Fig.18.

(3) Quantitative evaluation of casting defects. It is necessary to evaluate quantitatively defects such as the porosity. This requires such models as described in 4.

Table 1. Typical software packages for solidification analysis
(FEM : the finite element method, DFDM : the direct finite difference method, FDM* : not the FDM by Taylor expansion but rather the thermal resistance network method).

Software	Method	Organization developed the software
BACCAS	DFDM	Kawasaki Steel Work(Japan)
CAST	FDM*	Abex Co.(USA)
CASTEM	FEM	COBELCO(Japan)
CASTS	FEM	Technische Uni. Aachen(BRD)
HICASS	FDM*	Hitachi Electric Co.(Japan)
SOLAN	DFDM	Sokeizai Center, Osaka Univ.(Japan)
SOLDIA	DFDM	Komatsu Co.(Japan)

7. CONCLUDING REMARKS

Although solidification analysis based on the thermal conduction model have been used extensively in foundries in recent years, analyses based on more realistic models which include heat and mass transfer should be developed. Models considering the turbulent flow while filling the mold, and solid phase movement in particular, need to be developed. Further, microscopic models combined with the macroscopic model should also be developed to estimate the solidified structure[113]. The thermal stress analysis[114,115] is also important. It affects not only crack defects and the mechanical properties of the casting, but also dimensional accuracy. Further, it also affects the heat transfer between the casting and mold, hence casting defects such as porosity.

At the same time, physical properties such as thermal conductivity, specific heat, latent heat, density change with temperature, and permeability of various practical alloys and molds should be measured more accurately, because more accurate data are required to achieve accurate models. Further, it is necessary to combine such analyses with CAE (Computer Aided Engineering) of casting processes and to improve the understanding of the complicated solidification phenomena with the progress of computer hardware and software.

8. ACKNOWLEDGMENT

The author acknowledges coworkers and students for their useful discussion and help in the development of software and computation.

9. REFERENCES

1. Ockendon, J.R. and Hodgkins, W.R., Moving Boundary Problems in Heat Flow and Diffusion, Clarendon Press, Oxford, 1975.

2. Erickson, W.C., Computer Simulation of Solidification, AFS Int. Cast

Metals J., vol.5, pp.30-40, 1980.

3. Brody, H.D. and Stoehr, R.A., Computer Simulation of Heat Flow in Castings, J.Metals, vol.32, Sept., pp.20-27, 1980.

4. Brody, H.D. and Apelian, D., Modeling of Casting and Welding Processes, The Metallurgical Society of AIME, Warrendale, 1981.

5. Crank,J., in:Numerical Methods in Heat Transfer, ed. R.W.Lewis, K.Morgan and O.C.Zienkiewicz, pp.177-199, John Wiley & Sons, 1981.

6. Shih, T.M., A Literature Survey on Numerical Heat Transfer, Numerical Heat Transfer, vol.5, pp. 369-420, 1982.

7. Ohnaka, I., Computer Simulation of Solidification, Bulletin of Japan Inst. Metals, vol.22, pp. 802-808, 1983

8. Dantzig, J.A. and Berry, J.T., Modeling of Casting and Welding Processes II, The Metallurgical Society of AIME, Warrendale, 1984

9. Sahm, P.R. and Hansen, P.N., Numerical Simulation and Modelling of Casting and Solidification Processes for Foundry and Cast-House, Int. Committee of Foundry Technical Associations, 1984

10. Ohnaka, I., Introduction to Computer Analysis of Heat Transfer and Solidification, Maruzen, Tokyo, 1985

11. Niyama, E. and Aizawa, T., R & D Trend on Computer Aided Analysis in Solidification Castings, Sokeizai, vol. 26, no. 5, pp. 7-12, 1985.

12. Fredriksson, H., State of the Art of Computer Simulation of Casting and Solidification Process, J. de physique, Les Ulis Cedex, 1986.

13. Kou,S. and Mehrabian, R., Modeling and Control of Casting and Welding, The Metallurgical Society of AIME, Warrendale, 1986

14. Chalmers,B., Principles of Solidification, John Wiley & Sons, 1964

15. Flemings,M.C., Solidification Processing, McGraw-Hill, New York, 1974

16. Kurz, W. and Fisher, D.J., Fundamentals of Solidification, Trans Tech Publications, Switzerland, 1984

17. Minkoff, I., Solidification and Cast Structure, John Wiley & Sons, Chichester, 1986

18. Scheil,E., Bemerkungen zur Schichtkristallbildung, Z.Metallkd, vol.34, pp. 70-72, 1942.

19. Ohnaka,I., Mathematical Analysis of Solute Redistribution during Solidification with Diffusion in Solid Phase, Trans. Iron and Steel Inst. Japan, vol.26, pp.1045-1051, 1986.

20. Rogers,J.C.W., Berger,A.E. and Ciment,M., The alternating phase truncation method for numerical solution of a Stefan problem, SIAM J. Numerical Analysis, vol. 16, pp. 563-587, 1979.

21. Murray, W.D. and Landis, F., Numerical and Machine Solutions of Transient Heat-Conduction Problems Involving Melting and Freezing,

Trans. ASME, J. Heat Transfer, vol. 81, pp. 106-112, 1959.

22. Crank, J. and Crowley, A.B., On an implicit scheme for the isotherm migration method along orthogonal flow lines in two dimensions, Int. J. Heat Mass Transfer, vol. 22, pp. 1331-1337, 1979.

23. Sparrow, E.M. Ramadhyani,S. and Patankar,S.V., Effect of Subcooling on Cylindrical Melting, Trans. ASME, J. Heat transfer, vol.100, pp.395-402, 1978.

24. Saitoh,T., Numerical Method for Multidimensional Freezing Problems in Arbitrary Domains, ibid, vol.100, pp. 294-299, 1978.

25. Sarjant, R.J. and Slack,M.R., Internal temperature distribution in the cooling and reheating of steel ingots, J.Iron Steel Inst.,vol.177, pp. 428-444, 1954.

26. Muzikar,E.A., Mathematical Heat Transfer Model for Solidification of Continuously Cast Steel Slabs, Trans. Metall. Soc. AIME, vol.239, pp.1747-1753, 1967

27. Narita,K. and Mori,T., Study on Solidification of Heavy Steel Ingots, Tetsu-to-Hagane, vol.56, pp.1323-1341, 1970.

28. Marrone,R.E., Wilkes,J.O. and Pehlke,R.D., Numerical Simulation of Solidification, Part 1: Low Carbon Steel Casting-"T" Shape, Part 2: "L" Shape, AFS Cast Metals Res.J., vol.6, pp.184-192, 1970.

29. Davies,V. de L, Stokke,S. and Westby,O., Numerical Computation of Heat and Temperature Distribution in Castings, The British Foundry-man, vol.66, pp.305-313, 1973.

30. Pehlke,R.D., Kirt,M.J., Marrone,R.E. and Cook,D.J., Numerical Simula-tion of Casting Solidification, AFS Cast Metals Res. J., vol.9, pp.49-55, 1973.

31. Ebisu,Y., Computer Simulations on the Macrostructure in Centrifugal Castings, AFS Trans., vol.85, pp.643-654, 1977.

32 Jeyarajan,A. and Pehlke,R.D., Application of Computer-Aided Design to a Steel wheel Casting, AFS Trans., vol.86, pp.457-464, 1978

33. Hsiao,J.S., An Efficient Algorithm for Finite-Difference Analyses of Heat Tranfer with Melting and Solidification, Numerical Heat Transfer, vol.8, pp.653-666, 1985.

34. Pham,Q.T., A Fast, Unconditionally Stable Finite-Difference Scheme for Heat Conduction with Phase Change, Int. J. Heat Mass Transfer, vol.28, pp.2079-2084, 1985.

35. Uchikawa,S. and Takeda,R., Use of a Boundary-Fitted Coordinate Trans-formation for Unsteady Heat Conduction Problems in Multiconnected Regions with Arbitrarily Shaped Boundaries, Trans. ASME, J. Heat Transfer, vol.107, pp.494-498, 1985.

36. Ohnaka,I. and Fukusako,T., Calculation of Solidification of Casting by a Matrix Method, Trans.Iron and Steel Inst. Japan, vol.17, pp.410-418, 1977.

37. Ohnaka,I. and Fukusako,T., Finite-Element Method and A Matrix Method in Transient Heat-Conduction Problems, Sixth Int. Heat Transfer Conf., vol.3, pp.251-256, 1978.

38. Ohnaka,I., Classification of Numerical Methods for Transient Heat Transfer Problems and Improved Inner Nodal Point Method, Tetsu-to-Hagane, vol.65, pp.1737-1746, 1979.

39. Ohnaka,I., Yashima,Y. and Fukusako,T., Numerical Analysis of Solidification of Steel Casting with L-Junction in Sand Mold, Trans. Japan Foundrymen's Soc., vol.1, pp.7-11, 1982.

40. Ohnaka,I., in Innovative Numerical Analysis in Applied Engineering Science, ed.R. Shaw et al., Univ. Press of Virginia, pp.555-565,1981

41. Ohnaka,I., Nagasaka,Y., Fukusako,T. and Yoshioka,J., Three-Dimensional Solidification Analysis of Castings by Inner Nodal Point Method, Imono(J. Japanese Foundrymen's Society), vol.53, pp.376-382, 1981.

42. Ohnaka,I. and Fukusako,T., Solidification Analysis of Steel Ingots with Consideration on Fluid Flow, Trans.Iron and Steel Inst. Japan, vol.21, pp.485-494, 1981.

43. Ohnaka,I., Mori,Y., Nagasaka,Y. and Fukusako,T., Numerical Analysis of Shrinkage Cavity Formation without Solid Phase Movement, Imono(J. Japanese Foundrymen's Society), vol.53, pp.673-679, 1981.

44. MacNeal,R.H., An Asymmetrical Finite Difference Network, Quarterly Applied Mathematics, vol.11, pp.295-310, 1953.

45. Campbell,D.J. and Vollenweider,D.B., Unusual Techniques Employed in Heat Transfer Problems, Proc. Eastern Joint Computer Conf., vol.16, pp.143-147, 1959.

46. Evans,M.E. and Harlow,F.H., The Particle-in-Cell Method for Hydrodynamic Calculations, Los Alamos Scientific Lab., Rept.No.LA-2139,Los Alamos, New Mexico, 1957.

47. Gentry,R.A., Martin,R.E. and Daly,B.J., An Eulerian Differencing Method for Unsteady Compressible Flow Problems, J. Computational Physics, vol.1, pp.87-118, 1966.

48. Saeki,S. and Iwaki,T., Analytical Method for Thermal Load of Piston Crown, J. Marine Engine Society Japan, vol.4, pp.51-58, 1969.

49. Henzel Jr,J.G. and Keverian,J., Comparison of Calculated and Measured Solidification Patterns in a Variety of Steel Castings, Trans. AFS, vol.74, pp.661-679, 1966.

50. Finlayson,B.A., The Method of Weighted Residuals and Variational Principles, Academic Press, 1972.

51. Hodgkins,W.R. and Waddington,J.F., in Moving Boundary Problems in Heat Flow and Diffusion, ed. J.R.Ockendon and W.R.Hodgkins, Clarendon Press, Oxford, pp.26-37, 1975.

52. Weatherwax,R.B. and Riegger,O.K., Computer-Aided Solidification Study of a Die-Cast Aluminum Piston, AFS Trans., vol.85, pp.317-322, 1977.

53. Patankar,S.V., Numerical Heat Transfer and Fluid Flow, Hemisphere Publishing Co. New York, 1980.

54. Ueda,S. and Ono,S., Application of Computer in Determination of Casting Plans, J. Japanese Foundrymen's Society(Imono), vol.47, pp.658-666, 1981.

55. Niyama,E., Uchida,T., Morikawa,M. and Saito,S., Predicting Shrinkage in Large Steel Castings from Temperature Gradient Calculations, AFS Int. J. Cast Metals, vol. 6, pp.16-22, 1981.

56. Ohtsuka,Y., Mizuno,M. and Yamada,J., Application of a Computer Simulation System to Aluminum Permanent Mold Castings, AFS Trans., vol.90, pp.635-646,1982.

57. Raw,M.J. and Schneider,G.E., A New Implicit Solution Procedure for Multidimensional Finite-Difference Modeling of the Stefan Problems, Numerical Heat transfer, vol.8, pp.559-571, 1985.

58. Hamer,R., Numerical Simulation in Precision Castings, Int. J. Numer. Methods Engng., vol.24, pp.219-229, 1987.

59. Voller,V.R. Cross,M. and Markatos,N.C., An Enthalpy Method for Convection/Diffusion Phase Change, Int. J. Numer. Method Engng., vol.24, pp.271-284, 1987

60. Voller,R.V. and Prakash,C., A Fixed Grid Numerical Modeling with Methodology for Convection-Diffusion Mushy Region Phase-Change Problems, Int. J. Heat Mass Transfer, vol.30, pp.1709-1719, 1987.

61. Soliman,J.I. and Fakhroo,E.A., Finite Element Solution of Heat Transmission in Steel Ingots, J. Mech. Eng. Sci., vol.14, pp.19-24, 1972.

62. Hwang,C.T., Murray,D.W. and Brooker,E.W., A Thermal Analysis for Structures on Permafrost, Canadian Geotechnical J., vol.9, pp.33-46, 1972.

63. Zienkiewicz,O.C., Parekh,C.J. and Wills,A.J., The Application of Finite Element to Heat Conduction Problems Involving Latent Heat, Rock Mechanics, vol.5, pp.65-76, 1973.

64. Ohnaka,I. and Fukusako,T., Calculation of Solidification of Molten Metal by Finite Element Method, Tech. Rept. Osaka Univ., vol.24, pp.461-475, 1974.

65. Comini,G., Guidice,S.Del, Lewis,R.W. and Zienkiewicz,O.C., Finite Element Solution of Non-Linear Heat Conduction Problems with Special Reference to Phase Change, Int. J. Numer. Methods Engng., vol.8, pp.613-624, 1974.

66. Morgan,K., Lewis,R.W. and Zienkiewicz,O.C., An Improved Algorithm for Heat Conduction Problems with Phase Change, Int. J. Numer. Method Engng, vol.12, pp.1191-1195, 1977.

67. Giudice,S.Del and Comini,G., Finite Element Simulation of Freezing Processes in Soils, Int. J. Num. Anal. Methods in Geomechanics, vol.2, pp.223-235, 1978.

68. Tashiro,K., Watanabe,S., Kitagawa,I. and Tamura,I., Influence of Mold

Design on the Solidification and Soundness of Heavy Forging Ingots, Tetsu-to-Hagane, vol.67, pp.103-112, 1981.

69. Hsu,T.R. and Pizey,G., On the Prediction of Fusion Rate of Ice by Finite Element Analysis, J. Heat Transfer, vol.103, pp.727-732, 1981.

70. Imafuku,K., A Study on the Estimation of Generated Shrinkage Cavity in Steel Casting, J. Japan Soc. Mech. Engi.(C), vol.47, pp.918-926, 1981, ibid. vol.48, pp.1959-1967, pp.1968-1976, 1982.

71. Lemmon,E.C., in: Numerical Methods in Heat Transfer ed. R.W.Lewis, K.Morgan and O.C.Zienkiewicz, John Wiley & Sons, pp.201-213, 1981.

72. Rolph III,W.D. and Bathe,K-J., An Efficient Algorithm for Analysis of Nonlinear Heat Transfer with Phase Changes, Int. J. Numerical Methods Engng., vol.18, pp.119-134, 1982.

73. Makimura,M., Sakai,K., Nishimura,Y. and Tanaka,M., Axisymmetrical and Three-Dimensional Thermal Analysis of Solidification of Cast Iron Specimens with Various Thickness, J. Japanese Foundrymen's Soc. (Imono), vol.55, pp.736-741, 1983.

74. Yoo,J. and Rubinsky,B., Numerical Computation Using Finite Elements for the Moving Interface in Heat Transfer Problems with Phase Transformation, Numerical Heat Transfer, vol.6, pp.209-222, 1983.

75. Lewis,R.W., Morgan,K. and Roberts,P.M., Application of An Alternating-Direction Finite-Elements Method to Heat Problems Involving A Phase of Change, Numerical Heat Transfer, vol.7, pp.471-482, 1984.

76. Tsai,H.L. and Rubinsky,B., A Numerical Study Using "Front Tracking" Finite Elements on the Morphological Stability of a Planar Interface during transient Solidification Processes, J. Crystal Growth, vol.69, pp.26-46, 1984.

77. Roose,J. and Storrer,O., Modelization of Phase Changes by Fictitious Heat Flow, Int. J. Numer. Methods Engng., vol.20, pp.217-225, 1984.

78. Blanchard,D. and Fremond,M., The Stefan Problem: Computing Without the Free Boundary, ibid., vol.20, pp.757-771, 1984.

79. Borshukova,S. and Konovski,P., Transformation of Dependent Variable in the Finite Element Solution of Some Phase Change Problems, ibid., vol.20, pp.1815-1821, 1984.

80. Thomas,B.G., Samarasekera,I.V. and Brimacombe,J.K., Comparison of Nummerical Modeling Techniques for Complex, Two-Dimensional Heat Conduction Problems, Metall. Trans. B., vol.15B, pp.307-318, 1984.

81. Nakagawa,T., Hirose,K. and Takebayashi,Y., Solidification Simulation of Castings by the Finite Element Method, Proc. Beijin Int. Foundry Conf., vol.1, pp.674-698, 1986.

82. Samonds,M., Morgan,K. and Lewis,R.W., Finite element modelling of solidification in sand castings employing an implicit-explicit algorithm, Appl. Math. Modelling, vol.9, pp.170-174, 1985.

83. Abis,S., Numerical Simulation of Solidification in an Aluminum Casting, Metall. Trans. B., vol.17B, pp.209-216, 1986.

84. Weiss,K., Wendt,J. and Sahm,P.R., Berechnung der Erstarrung und Abkühlung realer Gussstücke-ein neues Werkzeug für die Giesserei, Giesserei, vol.73, pp.345-348, 1986.

85. Kumar,T.S.P., Pathak,S.D and Prabhakar,O., Finite Element Formulations for Estimating Feeding Efficiency Factors, AFS Trans., vol.93, pp.789-800, 1985

86. Albert,M.R. and O'Neill,K., Moving Boundary-Moving Mesh Analysis of Phase Change Using Finite Elements with Transfinite Mappings, Int. J. Numer. Methods Engng., vol.23, pp.591-607, 1986.

87. Dalhuijsen,A.J. and Segal,A., Comparison of Finite Element Techniques for Solidification Problems, Int. J. Numer. Methods Engng., vol.23, pp.1807-1829, 1986.

88. Crivelli,L.A. and Idelsohn,S.R. A Temperature-Based Finite-Element Solution for Phase-Change Problems, Int. J. Numer. Methods Engng., vol.23, pp.99-119, 1986.

89. Pham,Q.T., The use of lumped capacitance in the finite-element solution of heat conduction problems with phase change, Int. J. Heat Mass Transfer, vol.29, pp.285-291, 1986.

90. Yoo,J., and Rubinsky,B., A Finite Element Method for the Study of Solidification Processes in the Presence of Natural Convection, Int. J. Numer. Methods Engng., vol.23, pp.1785-1805, 1986.

91. Hromadka II,T.V. and Guyman,G.L., Mass-Lumping Numerical Models of Three-Dimensional Heat Conduction, Numerical Heat Transfer, vol.6, pp.367-375, 1983.

92. Wrobel,L.C. and Brebbia,C.A., in Numerical Methods in Heat Transfer, ed. R.W.Lewis et al., John Wiley & Sons, pp.91-113, 1981.

93. Hong,C.P., Umeda,T. and Kimura,Y., Numerical Models for Casting Solidification, Met.Trans.B., vol.15B, pp.91-99, pp.101-107, 1984.

94. Coleman,C.J., A Boundary Integral Approach to the Solidification of Dilute Alloys, Int. J. Heat Mass Transfer, vol.30, pp.1727-1732, 1987.

95. Zabaras,N. and Mukherjee,S., An Analysis of Solidification Problems by the Boundary Element Method, Int. J. Numer. Method Engng., vol.24, pp.1879-1900, 1987.

96. Flemings,M.C. and Nereo,G.E., Macrosegregation: Part I, Trans. Metall. Soc.AIME, vol.239, pp.1449-1461, 1967.

97. Flemings,M.C., Mehrabian,R. and Nereo,G.E., Macrosegregation: Part II, Trans. Metall. Soc. AIME, vol.242, pp.41-49, 1968.

98. Mehrabian,R., Keane,M. and Flemings,M.C., Interdendritic Fluid Flow and Macrosegregation; Influence of Gravity, Metall. Trans., vol.1, pp.1209-1220, 1970.

99. Asai,S. and Muchi,I., Theoretical Analysis and Model Experiments on the Formation Mechanism of Channel-type Segregation, Trans.Iron and Steel Inst. Japan, vol.18, pp.90-98, 1978.

100. Szekely,J. and Jassal,A.S., An Experimental and Analytical Study of the Solidification of a Binary Dendrite System, Metall. Trans.B., vol.9B, pp.389-399, 1978.

101. Ridder,S.D., Kou,S. and Mehrabian,R., Effect of Fluid Flow on Macro-segregation in Axi-Symmetric Ingots, Metall. Trans.B., vol.12B, pp.435-447, 1981.

102. Tacke,K.H., Grill,A., Miyazawa,K. and Schwerdtfeger,K., Macrosegre-gation in Strand Cast Steel: Computation of Concentration Profiles with a Diffusion Model, Arch. Eisenhüttenwes., vol.52, pp.15-20,1981.

103. Ohnaka,I. and Kobayashi,K., Flow Analysis during Solidification by the Direct Finite Difference Method, Trans.Iron and Steel Inst. Japan, vol.26, pp.781-789, 1986.

104. Ohnaka,I. and Matsumoto,M., Computer Simulation of Macrosegregation in Ingots, Tetsu-to-Hagane, vol.73, pp.1698-1705, 1987.

105. Poirier,D.R., Permeability for Flow of Interdendritic Liquid in Columnar-Dendritic Alloys, Metall. Trans.B, vol.18B, pp.245-255, 1987.

106. Zhu,J-D. and Ohnaka,I., Interdendritic Microporosity in Al-4.5%Cu Ingots and Its Computer Simulation, J. Japanese Foundrymen's Soc. (Imono), vol.59, pp.542-547, 1987.

107. Ohnaka,I., Microsegregation and Macrosegregation, Metals Handbook, Ninth Ed.,Vol.15 Casting, ASM Int. pp.136-141, 1988

108. Nomura,H., Tarutani,Y. and Mori,K., Formation of Macrosegregation during Unidirectional Solidification of Iron Alloys, Tetsu-to-Hagane, vol.67, pp.88-92, 1981

109. Mehrabian,R.,Kean,M. and Flemings,M.C., Experiments on Macrosegrega-tion and Freckle Formation, Metall. Trans., vol.1, pp.3238-3241,1970

110. Bennon,W.D. and Incropera,F.P., The Evolution of Macrosegregation in Statically Cast Binary Ingots, Metall.Trans.B, 18B, pp.611-616, 1987

111. Ohnaka,I., Solidification Analysis by An Outer Nodal Point Method, Proc.116th Grand Meeting of Japanese Foundrymen's Society,pp.90,1989

112. Ohnaka,I. and Kaiso,M., in Modeling and Control of CASTING AND WELD-ING PROCESSES IV, ed. A.F.Giamei adn G.J.Abbaschian,pp.141-150, TMS, Warrendale, USA, 1988

113. Rappaz,M., Modelling of microstructure formation in solidification processes, Int. Materials Reviews, vol.34, no.3, pp.93-123, 1989.

114. Kelly,J.E.,Michalek,K.P.,O'Connor,T.G.,Thomas,B.G. and Dantzig,J.A., Initial Development of Thermal and Stress Fields in Continuously Cast Steel Billets, Metall. Trans. A., 19A, pp.2589-2602, 1988

115. Ohnaka,I. and Y.Yashima, in:Modeling and Control of CASTING AND WELDING PROCESSES IV, ed. A.F.Giamei adn G.J.Abbaschian,pp.385-394, TMS, Warrendale, USA, 1988

Chapter 22

Heat Transfer Phenomena in Welding Processes

CHON L. TSAI
Department of Welding Engineering, The Ohio State University,
Columbus, Ohio 43210, USA

ABSTRACT

This paper presents an overview of the welding heat transfer phenomena, the modeling of heat transfer, and its applications. Mathematical formulations of the heat source and the surface heat loss are used in conjunction with the heat conduction equation to solve for the temperature field in the weldment. Various methods for obtaining analytical solutions are discussed. Numerical methods are also briefly discussed, though the specific details are left to the referred references. The use of these solutions for practical welding problems is also addressed.

CONTENTS

Applications of the Solutions
       Modifications of the Analytical Solutions
       Temperature History and Temperature Distribution
       Weld Pool Growth and Solidification Behaviour
       Calculation of Peak Temperature and Cooling Rate
       Effect of Welding Conditions on Material Thermal Response

Concluding Remarks

NOMENCLATURE

B	characteristic surface dimension (mm)
C	shape constant for finite source distribution (mm^{-2})
C_1, C_2	constants
C_p	specific heat (J/kg-°C)
d	bead width (mm)
D	diameter of electrode (mm)
E	welding voltage (V)
F	concentration factor
h	surface heat loss coefficient $(W/cm^2-°K)$
h_{nc}	natural convection heat loss coefficient $(W/cm^2-°K)$
h_{rad}	radiation heat loss coefficient $(W/cm^2-°K)$
h_{uw}	convection heat loss coefficient under water $(W/cm^2-°K)$
H	plate thickness (mm)
ΔH	change of enthalpy due to latent heat
I	welding current (A)
I_o	modified Bessel function of the first kind of zero order
K	thermal diffusivity (mm^2/s)
K_o	modified Bessel function of the second kind of zero order
K_1	modified Bessel function of the second kind of first order
l_w, l_y, l_z	direction cosines of the boundary surfaces

L	latent heat (J/mm^3)
$\underline{n}$	surface normal vector
$\dot{q}$	heat input flux (W/mm^2)
$\dot{q}_e$	internal heat generation rate (W/mm^3)
$\dot{q}_o$	heat flux at the source center (W/mm^2)
Q	total heat input rate (W)
r	radial distance from the source center (mm)
r_a	radius of heat input area on surface (mm)
r_B	distance between the source center and the maximum pool width
S	location of the melting front (mm)
t	time (s)
t_o	welding time (s)
t_1	time after welding source termination (s)
V	welding speed (mm/s)
x,y,z	global Cartesian coordinates (mm)
w,y,z	moving Cartesian coordinates (mm)
W	pool width (mm)
ε	emissivity
θ	temperature $(^\circ C)$
θ_0	initial temperature $(^\circ C)$
θ_∞	environmental temperature $(^\circ C)$
θ_ℓ	liquidus temperature $(^\circ C)$
θ_s	solidus temperature $(^\circ C)$
$\bar{\theta}$	temperature in the previous time step
β	weight constant
σ	Boltzeman constant $(5.76 \times 10^{-12}\ W/cm^2 - {}^\circ K^4)$
λ	thermal conductivity $(W/mm-{}^\circ C)$
ρ	density (kg/mm^3)
Ψ	angular coordinate (rad)
∇	differential operator

INTRODUCTION

During welding, the thermal cycles produced by the moving heat source cause a metallurgical change in the heat affected zone, transient thermal stress, and metal movement, resulting in the creation of residual stress and distortion in the finished product. In addition, during the cooling cycle defects may form due to excessively rapid weld solidification. In order to analyze these problems, it is essential to start with an analysis of thermal behaviour.

Welding Thermal Process

Figure 1 represents a physical model of the welding system. The welding heat source moves at a constant speed along a straight path. The end result from either the initiation or the termination of the heat source is the formation of a transient thermal state in the weldment. Some time after the initiation and before the termination of the heat source, the temperature distribution is stationary, or in equilibrium, with respect to the moving coordinates, the origin of which coincides with the center of the heat source. The intense welding heat melts the metal and forms a molten pool. Some of the heat conducts into the base metal and some is lost from the arc column or the metal surface to the environment surrounding the plate. Three metallurgical zones,

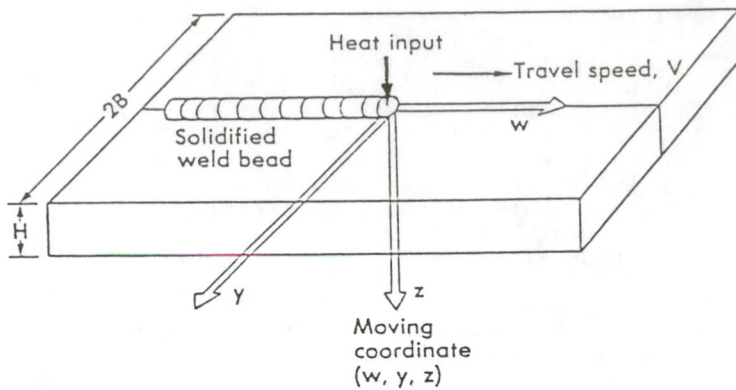

Figure 1   A schematic diagram of the welding thermal model.

weld metal (WM), heat-affected zone (HAZ) and base metal
(BM), are formed in the plate upon the completion of the
thermal cycle. The peak temperature and the subsequent
cooling rates determine the HAZ structures. The thermal
gradients, the solidification rates and the cooling rates at
the liquid and solid pool boundary determine the
solidification structure of the WM zone. The size of the pool
determines the amount of dilution and weld penetration. The
material response in the temperature range near melting is
primarily responsible for the metallurgical changes.

Two thermal states, quasi-stationary and transient, are
associated with the welding process. The transient thermal
response occurs during the source initiation and the
termination stages of welding, the latter of which is of
greater metallurgical interest. Hot cracking usually begins
in the transient zone due to the non-equilibrium
solidification of the base material. A crack formed in the
source initiation stage may propagate along the weld if the
solidification strains sufficiently multiply in the wake of
the welding heat source. During source termination, the weld
pool solidifies several times faster than the weld metal in
the quasi-stationary state. Cracks usually appear in the weld
crater and may propagate along the weld. Another dominant
transient phenomenon occurs when a short repair weld is made
to a weldment. Rapid cooling results in a brittle HAZ
structure and causes either cracking problems or creates a
site for fatigue crack initiation.

The quasi-stationary thermal state represents a steady
thermal response of the weldment with respect to the moving
coordinates, the origin of which coincides with the welding
heat source. The majority of the thermal expansion and
shrinkage in the base material occurs during the

quasi-stationary thermal cycles. Residual stress and weld
distortion cause the remaining thermal stress and strain in
the weldment after completion of the thermal cycle.

Relation to Welding Engineering Problems

To model and to analyze the thermal process, an
understanding of thermally induced welding problems is
important. A simplified modeling scheme, with adequate
assumptions for specific problems, is possible for practical
applications without using complex mathematical
manipulations. The relationship between the thermal behaviour
of weldments and welding metallurgy, welding control and
welding distortion is summarized as follows.

Welding Metallurgy

Defective metallurgical structures in HAZ and cracking
in WM usually occurs under the transient thermal condition.
The weld pool solidifies and cools several times faster than
at the quasi-stationary state. Cracks usually appear in the
weld crater as a result of the termination of the heat
source, and can propagate along the weld. The rapid cooling
may also initiate HAZ embrittlement. The same problems may
also occur in short repair welds. A transient thermal model
is needed to analyze cracking and embrittlement problems.

To evaluate the various welding conditions for process
qualification, the quasi-stationary thermal responses of the
weld material need to be analyzed. The minimum required
amount of welding heat input within the allowable welding
speed range must be determined in order to avoid rapid
solidification and cooling of the weldment. Preheating may be
necessary if the proper thermal conditions cannot be obtained
under the specified welding procedure. A quasi-stationary
thermal model is adequate for this type of analysis.

Hot cracking is formed as a result of a combination of
"force" and "metallurgy" effects. The "force" effect results
from weld metal displacement at near-melting temperatures
due to solidification shrinkage and weldment restraint. The
"metallurgy" effect relates to the segregation of alloying
elements and the formation of the eutectic during the high
non-equilibrium solidification process. Using metallurgical
theories, it is possible to determine the chemical
segregation, the amounts and distributions of the eutectic,
the magnitudes and directions of grain growth, and the weld
metal displacement at high temperatures. Using the heating
and cooling rates, and the retention period predicted by
modeling and analysis, hot cracking tendencies can be
determined. To analyze these tendencies, it is important to
employ a more accurate numerical model which considers finite
welding heat distribution, latent heat and surface heat loss.

Welding Control

In-process welding control has been studied recently.
Many of the investigations are aimed at the development of
sensing and control hardware. A link between weld pool

geometry and weld quality, however, has not been fully established. A transient heat flow analysis needs to be used to correlate the melted surface, which is considered the primary control variable, to the weld thermal response in a time domain.

Pulsing the welding current during welding can change the energy input mode on a real time scale and formulates additional parameters for more flexible and accurate heat input control. Because of the time dependency of the welding current, a transient thermal model and analysis must be used to study the effect of pulsing parameters on weld pool formation and solidification.

Welding Distortion

The temperature history and distribution, caused by the welding thermal process, creates non-linear thermal strains in the weldment. Thermal stresses are induced if any incompatable strains exist in the weld. Plastic strains are formed when the thermal stresses are higher than the material yield stress. Incompatable plastic strains accumulate over the thermal process and result in residual stress and final weldment distortion. The material response in the lower temperature range during the cooling cycle is responsible for the residual stresses and weldment distortion. For this type of analysis, the temperature field away from the welding heat source is needed for modeling the heating and cooling cycle during and after welding. A quasi-stationary thermal model with a concentrated moving heat source can predict the temperature information for the subsequent distortion analysis with reasonable accuracy.

Literature Review

Many investigators have analytically, numerically and experimentally studied welding heat flow modeling and analysis.[1-18]. The majority of the studies were concerned with the quasi-stationary thermal state. Lance and Martin[1], Rosenthal and Schmerber[2] and Rykalin[3] independently obtained an analytical temperature solution for the quasi-stationary state using a point or line heat source moving along a straight line on a semi-infinite body. A solution for plates of finite thickness was later obtained by many investigators using the imaged heat source method[3,4]. Tsai[5] developed an analytical solution for a model which incorporated a welding heat source with a skewed Gaussian distribution and finite plate thickness. It was later called the "Finite Source Theory".[6]

With the advancement of computer technology and the development of numerical techniques like the finite difference and finite element methods, more exact welding thermal models were studied and additional phenomena were considered, including non-linear thermal properties, finite heat source distributions, latent heat and various joint geometries. Tsai[5], Pavelic[7], Kou[8], Kogan[9] and Brody[10] studied the simulation of the welding process using the finite difference scheme. Hibbitt and Marcal[11],

Friedman[12] and Paley[13] made some progress in welding simulation using the finite element method.

Analytical solutions for transient welding heat flow in a plate were first studied by Naka[14], Rykalin[3], Masubuchi and Kusuda[15] in the 40's and 50's. A point or line heat source, constant thermal properties and adiabatic boundary conditions were assumed. Later, Tsai[16] extended the analytical solution to incorporate Gaussian heat distribution using the principle of superposition. The solution was used to investigate the effect of pulsed conditions on weld pool formation and solidification without the consideration of latent heat and non-linear thermal properties.

The analysis of the transient thermal behaviour of weldments using numerical methods has been the focus of several investigations since 1980. Friedman[17] discussed the finite element approach to the general transient thermal analysis of welding processes. Brody[10] developed a two-dimensional transient heat flow model using a finite difference scheme and a simulated pulsed current gas tungsten arc welding process. Tsai and Fan[18] modeled the two-dimensional transient welding heat flow using a finite element scheme to study the transient thermal behaviour of the weldment.

The General Approach

The various modeling and analysis schemes, summarized in the previous section, are capable of investigating the thermal process of different welding applications. Analytical solutions for the simplified model, with adequate assumptions, can be used to analyze welding problems which show a linear response to the heat source if the solutions are properly calibrated by experimental tests. Numerical solutions which incorporate non-linear thermal characteristics of weldment are usually required for investigating the weld pool growth or solidification behaviour. Numerical solutions may also be necessary for metallurgical studies in the weld HAZ if the rapid cooling phenomenon is significant under an adverse welding environment, such as welding under water.

Thermally related welding problems can be summarized into three categories: solidification rates in the weld pool, cooling rates in the HAZ and its vicinity and thermal strains in the general domain of the weldment. The domain of concern in weld pool solidification is within the molten pool area, in which the arc (or other heat source) phenomena and the liquid stirring effect are significant. A convective heat transfer model with a moving boundary at the melting temperature is needed for studying the first category. Numerical schemes are usually required.

The HAZ is always bounded, on one side, by the liquid and solid interface during welding. This inner boundary condition is the solidus temperature of the material. The liquid weld pool might be eliminated from thermal modeling if the interface can be identified. A conduction heat transfer model would be sufficient for the analysis of HAZ. Numerical methods are often employed and very accurate results may be obtained.

The thermal strains caused by welding thermal cycles are caused by the non-linear temperature distribution in the general domain of the weldment. Since the temperature in the material near the welding heat source is high, very little stress could be accumulated from the thermal strains due to low rigidity (i.e. small elastic modulus). The domain for thermal strain study is less sensitive to the arc and fluid flow phenomena and needs only a relatively simple thermal model. Analytical solutions with minor manipulations often provide satisfactory results.

In this paper, only the heat flow models are addressed. The convective models for fluid flow in a molten weld pool is not presented.

MATHEMATICAL FORMULATIONS

The Conduction Equation

Figure 1 shows a schematic diagram of the welding thermal model. The origin of the moving coordinates (w,y,z) is fixed at the center of the welding heat source. The coordinates move with the source at the same speed. The conduction equation for heat flow in the weldment is

$$\nabla \cdot (\lambda \nabla \theta) + \rho C_p V \frac{\partial \theta}{\partial t} + \dot{Q} = \rho C_p \frac{\partial \theta}{\partial t} \qquad [1]$$

The initial condition is

$$\theta = \theta_0 \qquad \text{at } t=0 \qquad [2]$$

The general boundary condition is

$$\lambda \{ \frac{\partial \theta}{\partial w} l_w + \frac{\partial \theta}{\partial y} l_y + \frac{\partial \theta}{\partial z} l_z \} - \dot{q} + h (\Theta - \Theta_\infty) = 0 \qquad [3]$$

The volumetric heat source, $\dot{Q}$, represents the Joule heating in the weldment due to the electric current flow within that conducting medium. The total energy of such heating in welding is usually minimal compared to the arc heat input. The majority of the energy is concentrated in a very small volume beneath the arc [5]. In other words, a very high energy density generation exists in the weld pool and it may have a significant effect on transient pool growth and solidification.

The Heat Source Formulation

The direction cosines on the surface which receives the heat flux from the welding source $(z=0)$ are $l_w = l_y = 0$ and $l_z = -1$. Within the significant heat input area (to be defined later in this section), the heat loss coefficient h is zero.

The distribution of the welding heat flux on the weldment surface can be characterized, in a general form, by a skewed Gaussian function [19]

$$\dot{q}\ (r,w) = \dot{q}_0 \exp\{\ -Cr^2 - \frac{\beta V}{2K}\ w\} \qquad [4]$$

The weight constant, β, indicates the significance of the welding travel speed. A normal distribution of the welding heat flux is obtained if the weight constant is zero.

In general, the total energy input to the weldment, which is a fraction of the total welding power generated by the welding machine, is the sum of the concentrated heat and the diffused heat [20]. The concentrated heat is carried by the core of the energy transmission medium, for example, the arc plasma column. The diffused heat reaches the weld surface by radiation and convection energy transport from the core surface. The heat flux distribution is a function of the proportional values between these two types of energy. The fraction of the total welding power reaching the weldment indicates the heating efficiency of the welding process and the fraction percentage is defined as "welding heat efficiency, η".

The shape constant, C, can be obtained in terms of the core diameter, D, and the concentration factor, F. The concentration factor is defined as the ratio of the concentrated heat to the net energy reaching the weldment. The core diameter may be assumed as the diameter of the plasma column in the arc welding process. The concentration factor and welding heat efficiency are not fully understood and have been subjected to manipulation during the mathematical analyses in order to obtain a better correlation with the experimental data.

Assuming a normal heat flux model, two conditions are required to determine the shape constant and the heat flux at the source center, $\dot{q}_0$. By integrating Eq. [4] over the core heating area and the entire heat input domain ($r = 0 \rightarrow \infty$), the shape factor can be determined by dividing the two integrals. The heat flux at the source center can then be determined from the second integral. The two constants are expressed as follows

$$C = 4/D^2 \cdot \ln\ \{1/(1-F)\} \qquad [5]$$

$$\dot{q}_0\ /\dot{Q} = C/\pi \qquad [6a]$$

In the case of arc welding,

$$\dot{q}_0\ /\eta EI = C/\pi \qquad [6b]$$

where E is the welding arc voltage and I is the welding current.

For practical purposes, the welding heat source can be considered to be restricted within a circle of radius r_a, where the heat flux drops to 1/100 of the center flux $\dot{q}_0$. The radius of the significant heat input area can be written as

$$r_a = (\ln 100/C)^{0.5} \qquad [7]$$

Surface Heat Loss

The heat loss coefficient, h, represents both radiation and convection heat losses from the boundary surfaces outside the significant heat input area. The formulation for both heat loss mechanisms can be written as

Radiation heat loss coefficient (in air)

$$h_{rad} = \varepsilon\sigma (\theta_w + \theta_\infty) (\theta_w^2 + \theta_\infty^2) \qquad [8]$$

Natural convection heat loss coefficient (in air)

$$h_{nc} = 0.00042 \{ (\theta_w - \theta_\infty)/B \}^{0.25} \qquad [9]$$

Convection heat loss coefficient (in water)

$$h_{uw} = 0.442 (\theta_w - \theta_\infty)^{0.25} \qquad [10]$$

Natural convection is dominant at a temperature below 550 °C, while radiation becomes more important beyond it. The total heat loss coefficient is the sum of Eqs. [8] and [9]. The characteristic surface dimension is the effective distance from the source beyond which the temperature rises insignificantly during welding. The characteristic dimension for steel is about 150 mm[5]. In underwater welding, heat losses are mainly due to heat transfer from the surface to the moving water environment. This motion is created by the rising gas column in the arc area.[21]

For an insulated surface, no heat transfer into or out of the surface is assumed. The temperature gradient normal to the surface is zero and can be represented by

$$\underset{\sim}{n} \cdot \nabla \theta = 0 \qquad [11]$$

where n is an unit vector normal to the surface and equals $(l_w^2 + l_y^2 + l_z^2)^{0.5}$.

Other Conditions

There are several other possible boundary conditions in welding heat flow modeling which depend on the assumptions used for model simplification.

(a) Condition at Infinity

$$\theta = \theta_\infty \text{ or } \lim_{r \to \infty} \frac{\partial \theta}{\partial r} = 0 \qquad [12]$$

(b) Condition Near Heat Source

Line source for a thin plate

$$-2\pi\lambda H \lim_{r \to 0} r \frac{\partial \theta}{\partial r} = \eta EI \qquad [13]$$

Point source for a thick plate

$$-2\pi\lambda \lim_{r\to 0} r^2 \frac{\partial\theta}{\partial r} = \eta\ EI \qquad [14]$$

Finite source for a thick plate

$$\underset{\sim}{n} \cdot (-\lambda \nabla\theta) = \dot{q}; \quad r \leq r_a \qquad [15]$$

(c) Conditions at Solid-Liquid Interface

$$\theta_\ell = \theta_s = \theta_m \qquad [16a]$$

$$\underset{\sim}{n} \cdot (\lambda \nabla\theta)_s - \underset{\sim}{n} \cdot (\lambda \nabla\theta)_\ell = \pm \rho_s L \frac{ds}{dt} \qquad [16b]$$

where + indicates the melting process and - indicates the
solidification process. The subscripts s and ℓ indicate the
temperature and the properties in a solid and liquid,
respectively. $\underset{\sim}{n}$ is a normal vector on the boundary surface or
interface, r_a is the radius of the heat input area, L is the
latent heat of the base material, and the subscript m
represents the melting temperature of the base material.

THE ANALYTICAL SOLUTIONS

Two general analytical methods, which have been used to
solve welding heat conduction problems are Green's Function
Integration, and the Separation of Variables. The basic
assumptions associated with these solution methods are
homogeneity and linearity of the thermal model. In most
cases, the boundary surfaces, except for the heat input area,
are assumed adiabatic, and the thermal properties are
independent of temperature.

Green's Function Integration Method

The use of Green's functions in the theory of potential
is well known[22]. For a closed surface, the function is most
conveniently defined as the potential which vanishes over the
surface, and is infinite at the point of source or sink
inside the surface. In welding heat conduction, Green's
function is taken as the temperature at (x,y,z) at the time t
due to an instantaneous point source of unit strength
generated at the point (x',y',z') at the time t', the surface
being kept at zero temperature.

The general solutions for the instantaneous point source
which satisfies the linear parabolic equation of heat
conduction in two- and three-dimensional solids, are

The equation of heat conduction

$$\nabla^2\theta = \frac{1}{K} \frac{\partial\theta}{\partial t} \qquad [17]$$

The 3-D solution

$$\theta - \theta_0 = \frac{Q}{4\rho C_p \{\pi\ K(t-t')\}^{1.5}} \exp \frac{-\{(x-x')^2+(y-y')^2+(z-z')^2\}}{4\ K\ (t-t')} \qquad [18a]$$

The 2-D solution

$$\theta - \theta_o = \frac{Q}{4\pi H\lambda(t-t')} \exp\frac{-\{(x-x')^2+(y-y')^2\}}{4 K (t-t')} \qquad [18b]$$

where Q is the instantaneous energy discharge of unit strength at the point of a stationary source.

By integrating the equation with respect to time, the solution for the continuous point source corresponding to the release of heat at a given point at a prescribed rate is obtained. By integrating the solution further, with respect to the appropriate space variables, the moving source solutions which correspond to the successive release of heat at a given point and time at a prescribed rate, is obtained. The space integration can also be used to treat the distributed heat source or sink in finite areas. The distributed source incorporates the finite welding heat source distribution or simulates latent heat at the solidifying weld pool interface. The heat sink distributed over the boundary surfaces simulates boundary heat loss or latent heat at the melting front of the weld pool. Therefore, various welding heat conduction problems can be solved by employing the principle of superposition.

Point Source Theory

For welding an infinitely thick plate, the temperature solution at (w,y,z) in the moving coordinates at the time t due to a point heat source of strength Q, moving at a constant speed V along a straight path (i.e. y=o) on the plate surface (i.e. z=o), is

$$\theta - \theta_o = \frac{Q}{\pi^{1.5}\lambda r} \exp\{\frac{-Vw}{2K}\} \int_{\frac{r}{2(Kt)^{\frac{1}{2}}}}^{\infty} \exp\{-\zeta^2-(\frac{Vr}{4K\zeta})^2\} \, d\zeta \qquad [19a]$$

where $r^2 = w^2 + y^2 + z^2$ and $\zeta = r/[2\sqrt{(t-t')}]$. $\dot{Q}$ is the welding heat input rate and equals ηEI.

If $t\to\infty$, a quasi-stationary state is established and the temperature solution at (w,y,z) becomes

$$\theta - \theta_o = \frac{\dot{Q}}{2\pi\lambda r} \exp\{\frac{-Vw}{2K}\} \exp\{\frac{-Vr}{2K}\} \qquad [20]$$

Line Source Theory

For welding a thin (i.e. 2 dimensional) plate, the temperature solution at (w,y) in the moving coordinates at the time t due to a distributed line heat source of strength $\dot{Q}/H$ through the plate thickness H moving along a straight path (i.e. y=0) at a constant speed V is

$$\theta - \theta_o = \frac{\dot{Q}}{4\pi H \lambda} \exp\{\frac{-Vw}{2K}\} \int_o^{\frac{V^2 t}{4K}} \frac{\exp\{-\zeta - (\frac{Vr}{4K})^2/\zeta\}}{\zeta} \, d\zeta \qquad [21a]$$

where $r^2 = w^2 + y^2$ and $\zeta = V^2 (t-t')/(4 K)$.

If $t \to \infty$, a quasi-stationary state is established, and the temperature at (w,y) become

$$\theta - \theta_o = \frac{\dot{Q}}{2\pi H \lambda} \exp\{\frac{-Vw}{2K}\} K_o(\frac{Vr}{2K}) \qquad [22]$$

where K_o is the modified Bessel function of the second kind of the zero order. The integral form of K_o. $\dot{Q}$ is the welding heat input rate.

Temperature Changes at the Start and End of Weld

To approximate the transient temperature changes at the start and end of weld, Figure 2 shows a global coordinate system (x,y,z), the origin of which is fixed at the point of source initiation, with t_o being the welding time and t_1 being the time after the welding heat source termination. The temperature solutions at t_o and t_1 can be derived, respectively, from Eqs. [19a] and [21a].

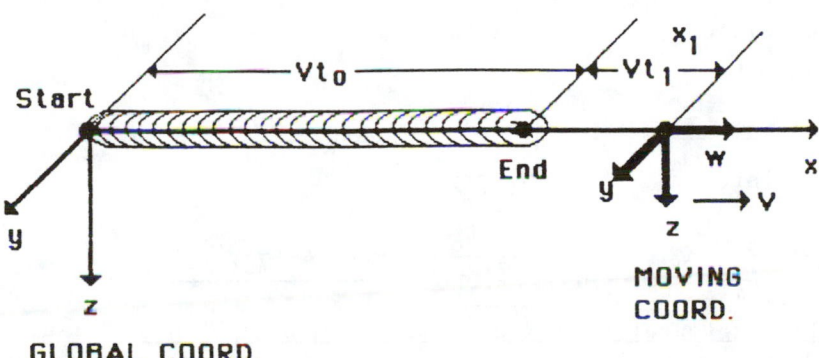

Figure 2   Global and moving coordinate systems for welding heat conduction.

At the starting point $x=y=z=0$, $t_o \to 0$ and $t_1=0$. The location of the starting point with respect to the moving coordinates becomes $(-Vt_o, 0, 0)$. The temperature change for infinitely thick and thin plates can be derived, respectively, as

Thick Plate:

$$\theta - \theta_o = \frac{\dot{Q}}{2\pi\lambda\ Vt_o} \qquad [23]$$

Thin Plate:

$$\theta - \theta_o = \frac{\dot{Q}}{4\pi\lambda\ H}\ \exp\{\frac{V^2\ t_o}{2K}\}\ K_o(\frac{V^2\ t_o}{2K}) \qquad [24]$$

At the welding termination point $x=Vt_o$ and $y=z=0$, $t_o \to \infty$ and $t_1 \to 0$. In the moving coordinates, the end point is at $(-Vt_1,0,0)$. The temperature solutions are

Thick Plate:

$$\theta - \theta_o = \frac{\dot{Q}}{2\pi\lambda\ V\ t_1} \qquad [25]$$

Thin Plate:

$$\theta - \theta_o = \frac{\dot{Q}}{4\pi\lambda\ H}\ \exp\{\frac{V^2 t_1}{2K}\}\ K_o(\frac{V^2 t_1}{2K}) \qquad [26]$$

The general solutions for the time period between the start and end of welding (i.e. $0 \le t' \le t_o$) can be modified from Eqs. [19a] and [21a] by substituting t by t_o+t_1.

Thick Plate:

$$\theta - \theta_o = \frac{\dot{Q}}{\pi^{1.5}\lambda\ r}\ \exp\{\frac{-Vw}{2K}\}\int_{\frac{r/2}{\sqrt{K(t_o+t_1)}}}^{\frac{r/2}{\sqrt{Kt_1}}}\ \exp\{-\zeta^2-(\frac{Vr}{4K})^2\}\ d\zeta \qquad [19b]$$

Thin Plate:

$$\theta - \theta_o = \frac{\dot{Q}}{2\pi\lambda H}\ \exp\{\frac{-Vw}{2K}\}\ \int_{\frac{V^2 t_1}{4K}}^{\frac{V^2(t_o+t_1)}{4K}}\frac{\exp\{-\zeta-(\frac{Vr}{4K})^2/\zeta\}}{\zeta}\ d\zeta \qquad [21b]$$

<u>Spatial Integration for a Distributed Heat Source</u>

The general temperature solution for a welding heat source with a Gaussian flux distribution (i.e. Eq. [4]) can be obtained by integrating Green's function over the heating area. The solution can be written as

$$\theta - \theta_o = \frac{1}{4\rho C_p(\pi K)^{1.5}}\int_o^t\frac{1}{(t-t')^{1.5}}\int_o^\infty\int_o^{2\pi}\dot{q}\ r'\ \cdot$$

$$\exp\{\frac{-\{\{w+V(t-t')-r'SIN\Psi\}^2+\{y-r'COS\Psi\}^2+z^2\}}{4\ K\ (t-t')}\}d\Psi\ dr'\ dt' \qquad [27]$$

Figure 3 defines the radial and angular coordinates used in Eq. [27].

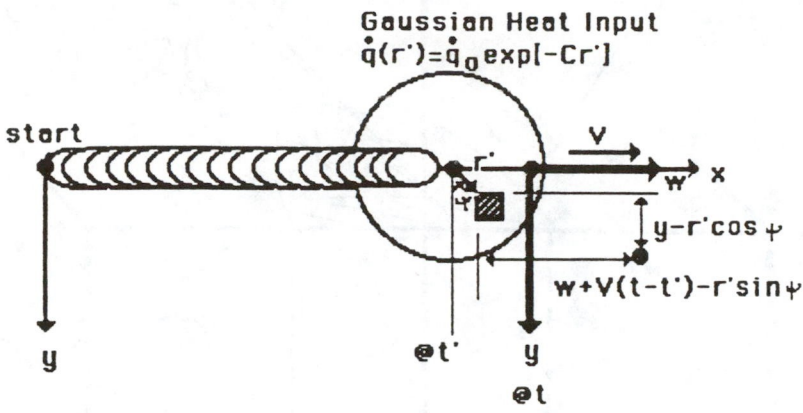

Figure 3   The radial and angular coordinates for welding heat input model.

Separation of Variables Method

Finite Source Theory

For a quasi-stationary state, the time derivative of Eq. [1] vanishes and the spatial variables can be separated by the product solutions of each variable. Since linearity and homogeneity are assumed, it follows that any linear combination of such solutions corresponding to different eigenvalues will also satisfy the heat conduction equation. Assuming a skewed Gaussian heat flux distribution (i.e. Eq. [4]) with the weight constant being unity, the product equation can be written as

$$\theta - \theta_o = \exp\{\frac{-Vw}{2K}\} \cdot \phi(r,z) \qquad [28]$$

where the three variables are w, z and r. The ϕ function is an axisymmetrical temperature function which can be obtined for the cylindrical column beneath the source area ($r \leq R$) and the exterior region ($r > R$) separately. Figure 4 defines the mathematical domains of the solution. The inner and exterior solutions can be derived and presented in reference (6).

Thin Plates with a Constant Heat Loss Coefficient

The general equation [1] can be simplified by ignoring the variation of thermal conductivity against the temperature, the heat conduction in the direction of thickness, and the time derivative. The boundary heat loss

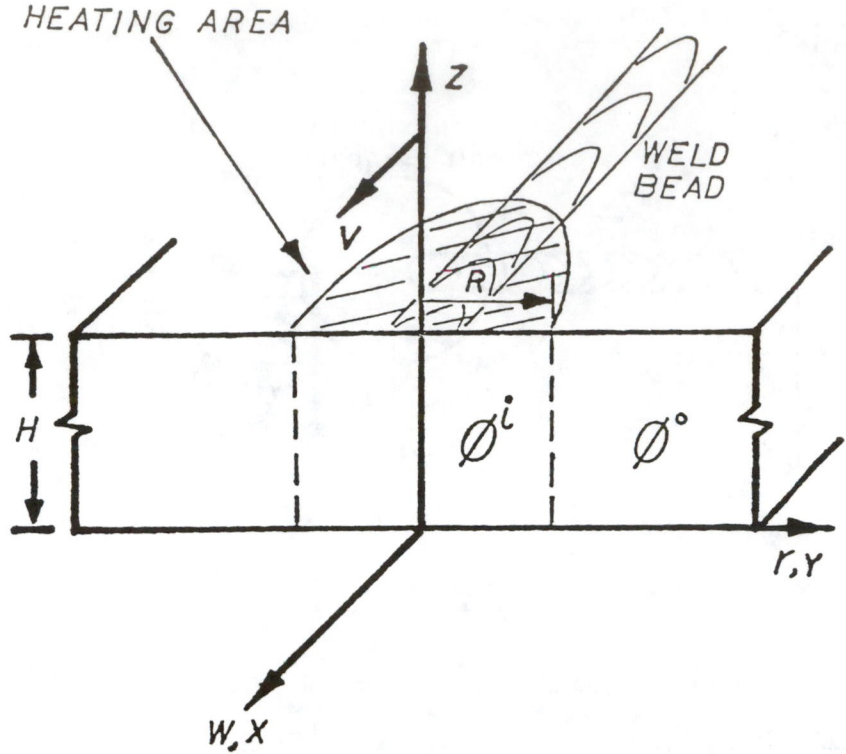

Figure 4 The mathematical domains of the inner and exterior
solutions in the finite source theory.

can be treated as a volumetric heat sink and the
two-dimensional heat flow equation becomes

$$\nabla^2\theta + 2\delta\ V\ \frac{\partial\theta}{\partial w} - \frac{h}{\lambda\ H}\ (\theta - \theta_o) = 0 \qquad\qquad [29a]$$

where $\delta = 1/2K$ and H is the characteristic thickness of
the thin plate. Given $r^2 = w^2 + y^2$, the product temperature
solution can be written as

$$\theta - \theta_o = \exp\ (-\ \delta V w)\cdot\ \phi(r) \qquad\qquad [30]$$

where $e^{-\delta v w}$ is due to the moving heat source and $\phi(r)$ is a
symmetric function with respect to the moving coordinates.

By substituting Eq. [30] into Eq. [29a], a linear differential equation involving only the symmetric function $\phi(r)$ can be derived:

$$\frac{\partial^2 \phi}{\partial w^2} + \frac{\partial^2 \phi}{\partial y^2} + \zeta^2 \phi = 0 \qquad [29b]$$

where $\quad \zeta = [(\delta V)^2 + h/(\lambda H)]^{0.5}$.

By transferring Eq. [29b] into a cylindrical coordinate system, it becomes

$$r \frac{d\phi^2}{dr^2} + \frac{1}{r} \frac{d\phi}{dr} - \zeta^2 \phi = 0 \qquad [29c]$$

Given $r' = \zeta r$, Eq. [29c] can be rewritten in the form of Bessel's modified differential equation of the zero order:

$$r'^2 \frac{d\phi^2}{dr'^2} + r' \frac{d\phi}{dr'} - r'^2 \phi = 0 \qquad [29d]$$

The general solution of this equation is

$$\phi(r) = C_1 I_0(\zeta r) + C_2 K_0(\zeta r) \qquad [31]$$

where C_1 and C_2 are arbitrary constants and I_0 and K_0 are modified Bessel's functions of the first and second kind of the zero order, respectively.

If a point source moves at a constant speed through an infinitely large medium (i.e. thin plate), the two constants can be determined by applying the boundary conditions of Eqs. [12] and [13]. The final solution is similar to Eq. [22] from the line source theory, except that the argument of K_0 is replaced by ζr, which is now a dependent of surface heat loss. The solution can be written as

$$\theta - \theta_0 = \frac{\dot{Q}}{2\pi \lambda H} \exp\{\frac{-Vw}{2K}\} K_0(\zeta r) \qquad [32]$$

Image Method for Plates of Finite Thickness

The image method enables the investigator to superimpose the solutions for an infinitely thick plate, the source of which is placed on imaginary surfaces until the proper boundary conditions on the plate surfaces are obtained. This method is based on the premise that if a solution satisfies the governing equation and the boundary conditions, the solution must be not only a correct solution, but the only solution. (i.e. Uniqueness of Solution).

Using the image method, the solution for plates of finite thickness with adiabatic surfaces can be modified from the respective temperature solutions described in the previous sections.

763

Let $\phi_0(w,y,z,t)$ be the initial solution for an infinitely thick plate. The temperature solution for a finite thick plate can be obtained by superimposing the imaginary solutions $\phi_{mn}(w,y_m,z_n,t)$ and $\phi_{mn}'(w,y_m',z_n',t)$ on the initial solution and this can be written in a general form as

$$\theta - \theta_o = -\phi_o(w,y,z,t) + \sum_{m=o}^{\infty} \sum_{n=o}^{\infty} \{\phi_{mn}(w,y_m,z_n,t) + \phi'_{mn}(w,y'_m,z'_n,t)\}$$

[33]

where

$$y_m = 2mB - y; \quad z_n = 2nH - z$$
$$y_m' = 2mB + y; \quad z_n' = 2nH + z$$

[34]

where B is the half width and H is the thickness of the plate. The subscripts m and n are integers which vary from zero to infinity.

For a plate with sufficient width, the subscript m is zero. The solution will converge and reach the correct adiabatic surface condition in six to ten superposition steps, depending upon the thickness of the plate. The two-dimensional solution (i.e. thin plate) is generally used for any solution which requires more than ten superposition steps.

Treatment of Temperature Dependent Thermal Properties

Practically all analytical studies on heat flow in weldments are performed using the linear theory, in which thermal properties are assumed to be constants. In a situation where temperature variation is the central focus of the investigation, such an approach is obviously not satisfactory. When the actual temperature dependencies are taken into account, however, the problem becomes non-linear and the mathematical analysis is then extremely complex.

With the advent of sophisticated computer programs for the simulation and analysis of the mechanism of the welding process, the solution of non-linear problems is gaining increasing significance. Apart from solving the non-linear problem using a computer in an iterative scheme, very limited efforts have been expended to advance the solutions for non-linear problems.

In the 1950s, a conscientious effort was made by Grosh et al [23] to account for the variable properties of non-linear problems. The method they employed hinges on the assumption that thermal properties change linearly with temperature:

$$\lambda = \lambda_o \, f'(\theta)$$

[35a]

$$\rho C_p = (\rho \, C_p)_o \, f'(\theta)$$

[35b]

where λ_0 and $(\rho C_p)_0$ are constants and $f'(\theta)$ is the temperature derivative of some linear function of temperature $f(\theta)$. $f(\theta)$ and $f'(\theta)$ are assumed to be different and $f'(\theta)$ is assumed to be greater than zero. Equations [35a] and [35b] imply that both λ and ρC_p can be expressed by the same temperature dependent function of $f'(\theta)$.

While this method was applicable for stainless steel, it appears to be one of the very few construction materials that possesses the desired linear properties. However, with a slight relaxation of the required condition, properties with non-linear temperature dependencies can eventually be treated using this method. The details of this method are presented in reference (24).

A substitution relation originating from the chain rule in differential calculus with general validity is given by:

$$\frac{\partial f(\theta)}{\partial \xi} = f'(\theta) \frac{\partial \theta}{\partial \xi} \tag{36}$$

With the substitution relation, Eq. [36], the quasi-stationary welding heat conduction equation, Eq. [1], ignoring volumetric heat generation, can be transformed into

$$\nabla^2 f(\theta) + \frac{V}{K} \frac{\partial f(\theta)}{\partial w} = 0 \tag{37}$$

where K_0 is a constant thermal diffusivity and equals to $\lambda/\rho C_p$ or $\lambda_0/(\rho C_p)_0$. The associated boundary conditions are

$$\lim_{r \to \infty} \frac{\partial f(\theta)}{\partial r} = 0 \tag{38a}$$

and

$$\lim_{r \to 0} \left\{ -2\pi r^2 K_0 \frac{\partial f(\theta)}{\partial r} \right\} = \dot{Q} \tag{38b}$$

The solution of the above boundary value problem, as elaborated in the previous sections, is given by

$$f(\theta) - f(\theta_0) = \frac{\dot{Q}}{2\pi K_0 r} \exp\left\{\frac{-Vw}{2K_0}\right\} \exp\left\{\frac{-Vr}{2K_0}\right\} \tag{39}$$

If $f'(\theta)$ is expressed as

$$f'(\theta) = (a + b\theta)^2 \tag{40}$$

that is, both the thermal conductivity and the product of density and specific heat are quadratic functions of temperature.

$$f(\theta) = \frac{(a+b\theta)^3}{3b} + C' \qquad [41]$$

From Eqs. [39] and [41], the final temperature solution can be derived as

$$\theta = \frac{1}{b} \left\{ \left\{ 3b \left(\frac{\dot{Q}}{2\pi K_o H} \exp\left\{ \frac{-Vw}{2K_o} \right\} \exp\left\{ \frac{-Vr}{2K_o} \right\} + N \right) \right\}^{\frac{1}{3}} - a \right\}$$

$$[42]$$

where $N = (a + b\,\theta_0)^3 / (3\,b)$.

The three-dimensional case for a plate of finite thickness can be evaluated in the usual manner by adding an infinite series of mirror images of the point source to satisfy the adiabatic boundary condition on the top and bottom surface of the plate.

Analogous to the development of the three-dimensional case, the two-dimensional solution for a thin plate can be derived as

$$\theta = \frac{1}{b} \left\{ \left\{ 3b \left(\frac{\dot{Q}}{2\pi K_o H} \exp\left\{ \frac{-Vw}{2K_o} \right\} K_o \left(\frac{Vr}{2K_o} \right) + N \right) \right\}^{\frac{1}{3}} - a \right\}$$

$$[43]$$

The solution for a plate with finite thickness and width can also be evaluated by the image superposition method previously described.

THE NUMERICAL SOLUTIONS

Most recent efforts in the numerical treatment of welding heat flow problems incorporated not only a variation of mechanical and physical properties with temperature, but also metallurgical transformations, into computer programs and computer-aided studies. Clearly, these efforts will be greatly enhanced if efficient methods are available to solve the basic non-linear heat conduction problem. In this section, two common numerical schemes, the finite difference method and the finite element method, as well as a general approach to latent heat, are described.

The Finite Difference Method

The finite difference method approximates the derivatives of a temperature function by Taylor's series expansion about a space point, also called the grid-point, and a time instance. The heat conduction equation and its boundary conditions can then be translated into a discrete difference form.

In the quasi-stationary heat flow case, the time derivative is ignored. The Gauss-Seidel point iterative method can be used to solve the difference equations effectively. This method uses the latest iterative values as soon as they are available and scans the mesh points systematically, that is to say, the order in which one solves for components must be established before hand. The iterative formula can be found in reference (25).

The non-linear points can be treated as known quantities by using their already calculated values of the n-th iteration. As the number of iterations increases, the difference between the values of the function of the n-th and n+1 iterations approches zero and a unique solution of the problem is obtained.

For transient welding heat flow analysis in an alternating direction, the implicit method may be used[25]. A tridiagonal coefficient matrix is usually formulated. The linear equations with a maximum of three variables per equation is readily solved using the Gaussian elimination method and the solution can be expressed very concisely. The implicit method converges with the solution of the heat conduction equation as the time and space increments approach zero, regardless of the value of the ratio $\Delta t/(\Delta X)^2$. The value of the ratio $\kappa/[V \Delta X]$ greater than or equal to one is the necessary condition for the stability of the method[26].

The Finite Element Method

Using Galerkin's method and integrating the heat conduction equation using Green's theorem, the finite element method idealization yields a set of simultaneous equations for each element at each time step. These are designated as the thermal stiffness equations. They relate the "thermal forces" at the nodal points of the element to the nodal temperatures at the end of the time step, and to the distributed "thermal loads" acting on the element. In matrix notation,

$$[H]\{T\} + [P]\{\dot{T}\} + \{F\} = 0 \qquad\qquad [44]$$

where $[H]$ is the thermal stiffness matrix, $[P]$ is the specific heat matrix, $\{F\}$ is the thermal force vector, $\{T\}$ is the unknown temperature vector, and $\{\dot{T}\}$ is the derivative of the temperature vector with respect to time.

For a linear heat conduction model, the elements of the coefficient matrices would all be known. With the effects of non-linearity, including latent heat, they now depend on temperature so that an iterative process is employed merely to evaluate the matrix coefficients for i-th iteration as a function of the temperature solutions computed for iteration i-1. Acceleration parameters may also be introduced to speed convergence. A hypothesis is made for the temperature distribution to evaluate the matrix coefficients.

The global coefficient matrices can be assembled from the element matrices. The time variable is treated in the same way as a spatial variable using the finite element scheme. A quadratic interpolation function could be used to treat the time variable for the case which is associated with large time derivatives (i.e. high cooling rate). This method would result in a more accurate numerical solution compared to the linear interpolation used in the finite difference method.

A detailed description of the finite element heat conduction analysis is presented in reference (18).

Numerical Treatment of Latent Heat

In welding, latent heat is liberated or absorbed as a consequence of phase transformation over a finite temperature range. The phase change front is specified by the temperature calculated from the analysis. The analysis is virtually a moving boundary problem in a two- or three-dimensional space. This complexity requires the employment of a numerical method of heat conduction analysis.

A common means of incorporating phase change into numerical solution methods is to treat the latent heat as an increase in the specific heat between the solidus and liquidus temperatures.

A very large increase in heat capacity over a relatively small temperature range requires that, for a given time increment, the proper magnitude of latent heat involved in the transformation is dependent upon temperature both at the beginning and at the end of the time step. An iterative method is thus employed to directly account for the latent heat. Although the method is iterative, convergence is usually quite rapid. Time increments larger than those required for the non-iterative increased specific heat procedure may generally be used, even though the inaccuracies inherent in a procedure which does not account for the temperature at the end of the increment are avoided.

To summarize the numerical method, with the derivation details described in reference (17), the change of enthalpy due to latent heat is expressed in the form:

$$\Delta H = C_p(\bar{\theta}) \; (\theta - \bar{\theta}) + \{A\theta - (B_1 \; \theta_s + B_2 \theta_\ell + B_3 \bar{\theta}\;)\} \cdot$$

$$\{\frac{L}{\theta_\ell - \theta_s} - C_p(\bar{\theta})\}$$

[45]

where $\bar{\theta}$ is the previous temperature, θ is the present temperature under calculation, θ_s is the solidus temperature θ_ℓ is the liquidus temperature and the four phase change coefficients are given in reference (17). The latent heat is treated in the same way as other non-linear properties by the iterative procedure described previously.

APPLICATIONS OF THE SOLUTIONS

Modifications of the Analytical Solutions

The analytical solutions tend to predict higher temperatures than those actually recorded near the heat source as a result of the assumed point (or line) source and linear temperature responses. In addition, the predicted temperatures near the source suffer from problems inherent in the complex behaviours of metals at the molten state, the phase change between the solid and liquid boundary, the heat recovery from metal transfer and, of the greatest consequence, the unknown energy transfer from the arc to the base material. There is no simple way to consider these factors in the analytical treatment.

Fortunately, there exists a physical relationship between heat input and the weld puddle geometry. The thermal model can be simplified to a pure heat conduction problem by eliminating the puddle. The temperature at the point of the maximum pool width on the plate surface, when plotted, is a straight line passing through this point and parallel to the welding direction.

Figure 5 shows the surface geometry of a weld pool at a quasi-stationary state. Two geometric conditions are well defined at the pool boundary of maximum width (point B):

$$(\frac{d\theta}{dw}) \text{ at point B} = 0 \qquad\qquad [46a]$$

$$(\frac{dy_m}{dw_m}) \text{ at point B} = 0 \qquad\qquad [46b]$$

where y_m and w_m represent the melting contour of the weld pool.

The temperature solution defines the pool boundary if the temperature is at the melting point. If the temperature is a function proportional to the spatial variables and the heat input coefficient is an unknown factor, the heat input coefficient can can be solved, as well as the relative distance of point B from the source center, wB, using Eqs. [46a] and [46b]. The coefficient can be expressed in an implicit form.

For a Thin Plate

$$\theta - \theta_o = B_2 \exp\{\frac{-Vw}{2\kappa}\} K_o(\frac{Vr}{2\kappa})$$

$$[47a]$$

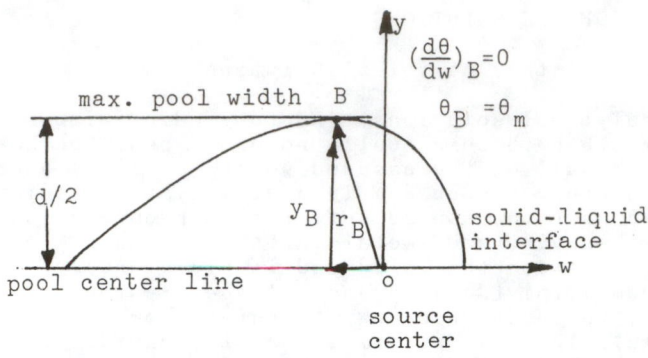

Figure 5 The source geometry of a weld pool at a quasi-stationary state.

$$B_2 = \frac{(\theta_m - \theta_o)\ \exp\{\frac{-Vr_B}{2\kappa}\}\ \dfrac{K_o(\frac{Vr_B}{2\kappa})}{K_1(\frac{Vr_B}{2\kappa})}}{K_o(\frac{Vr_B}{2\kappa})} \qquad [47b]$$

Bead Width, $\qquad\qquad\qquad\qquad\qquad\qquad\qquad$ [47c]

$$d = 2\ r_B\ \{1 - \{\frac{K_o(\frac{Vr_B}{2\kappa})}{K_1(\frac{Vr_B}{2\kappa})}\}^2\ \}^{\frac{1}{2}}$$

For an Infinitely Thick Plate

$$\theta - \theta_o = B_2\ \exp\{\frac{-Vw}{2\kappa}\}\ \exp\{\frac{Vr}{2\kappa}\}\ /r \qquad [48a]$$

$$B_2 = (\theta_m - \theta_o)\ r_B\ \exp\{\frac{-Vr_B}{2\kappa}\}\{\frac{1}{1 + \frac{2\kappa}{Vr_B}} - 1\} \qquad [48b]$$

Bead Width, $d = 2\, r_B \{ 1 - \dfrac{1}{1 + \dfrac{2\kappa}{V\, r_B}} \}^{\frac{1}{2}}$ [48c]

The modification described in the above equations forces the temperature solution to be compatable with the melting condition at the weld fusion boundary. This constitutes a calibration procedure for better results. This method has been used to inversely calculate the welding heat efficiency from the measurement of the weld bead width.

Temperature History and Temperature Distribution

To demonstrate the applicability of the aforementioned temperature solutions, calculated results from different solutions are presented in this section.

Figure 6 shows the predicted temperature history, using the Green's function integral solution, at a point 3 mm from the weld centerline of a pulsed-current welding process on a 10 mm-thick 304L stainless steel plate. With a 2 Hz frequency, this point goes through a remelting process. Using the same solution, Fig. 7 shows the temperature history at points along a line 13 mm away from and parallel to the weld centerline. The thermal, initial transient, quasi-stationary and final transient states, are clearly shown.

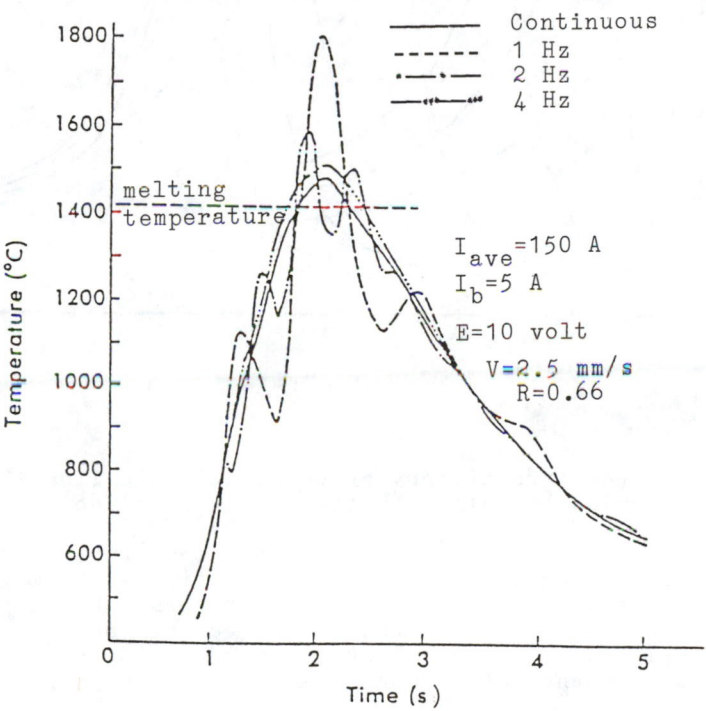

Figure 6   The predicted temperature history of a pulsed-current GTAW process.

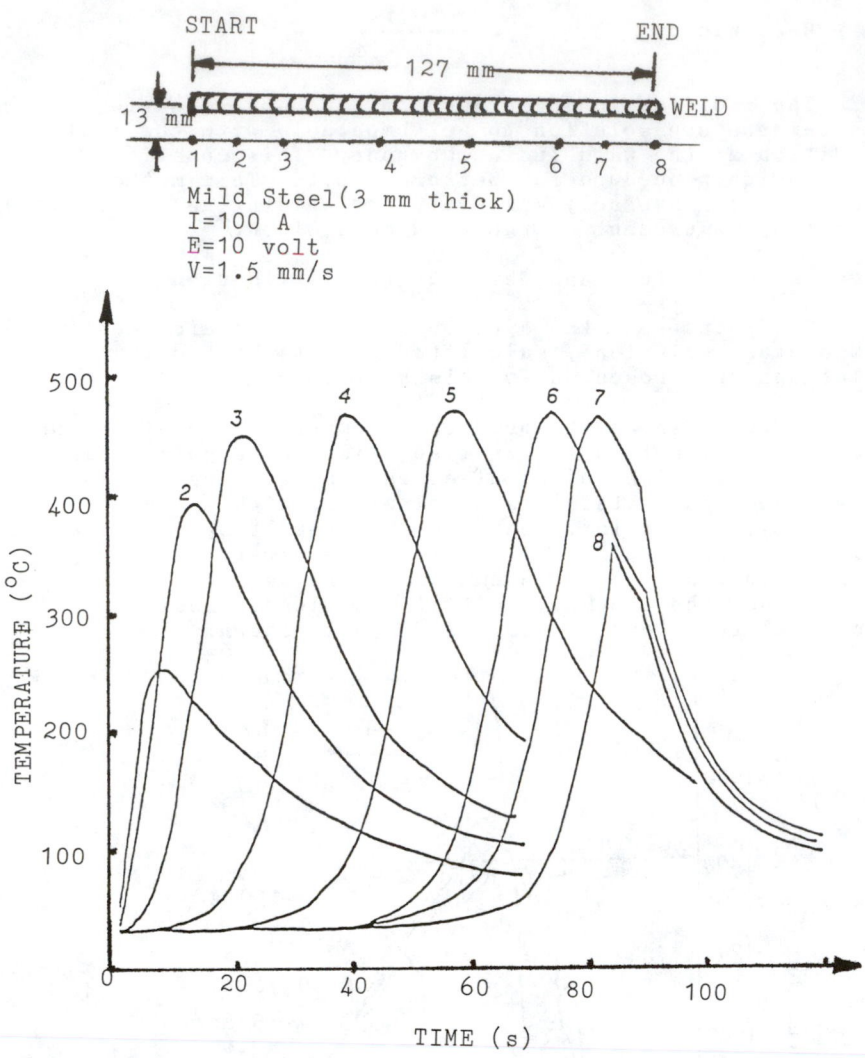

Figure 7   Temperature history at points along a line 13 mm
            from and parallel to the weld centerline.

Figure 8 shows the quasi-stationary temperature
distributions using the point source solution with image
superpositions on the top and bottom surfaces, the
cross-sectional plane under the source center, and the
longitudinal central plane along the weld. Weld pool surface
dimensions and penetration can be seen from the graphic plot.

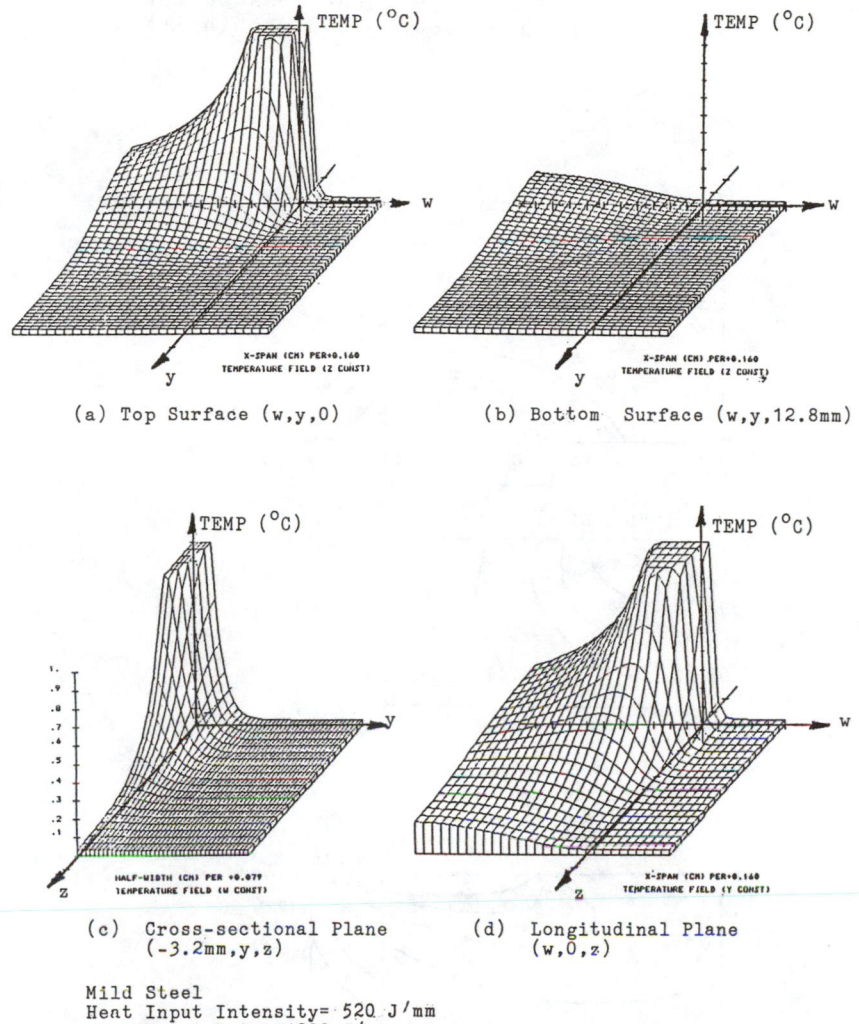

(a) Top Surface (w,y,0)

(b) Bottom Surface (w,y,12.8mm)

(c) Cross-sectional Plane
     (-3.2mm,y,z)

(d) Longitudinal Plane
     (w,0,z)

Mild Steel
Heat Input Intensity= 520 J/mm
Heat Input Rate= 4200 J/s

Figure 8   Quasi-stationary temperature distributions (a) top
           surface; (b) bottom surface; (c) cross-sectional
           plane; (d) longitudinal plane. plate thickness=16mm
           welding speed=3.8 mm/s

Figure 9 shows the isothermal contours in three planes in an underwater weldment. The solution was obtained from the finite difference calculations, with a melting inner boundary condition and the heat loss coefficient prescribed by Eq. [10]. As indicated by the isothermal contours, the critical surface area responsible for larger heat loss to the water environment is behind the heat source and within its vicinity.

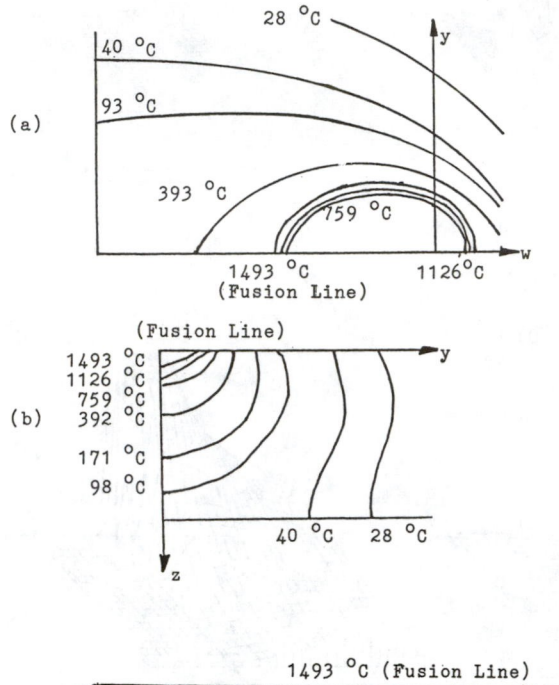

Mild Steel (Underwater Welding); I=200 A; E=12 volt
Plate Thickness=16 mm; Welding Speed=3.8 mm/s

Figure 9   Isothermal contours in three planes in an underwater weldment. (a) top surface; (b) cross-sectional plane; (c) longitudinal plane.

Weld Pool Growth and Solidification Behaviour

Figure 10 shows the weld pool growth in a thin aluminum plate, predicted by the transient, finite element analysis. For the analysis of pool solidification, Fig. 11 shows the solidifying solid-liquid interface of a thin 30Cu-70Ni alloy plate, plotted at different time increments, growing towards the center of the molten puddle. The solidification direction and the solidification rate at every point on the solidifying front can be determined. Figure 12 shows the predicted solidification rates, the temperature gradients across the interface in the solid phase and the cooling rates at the melting temperature.

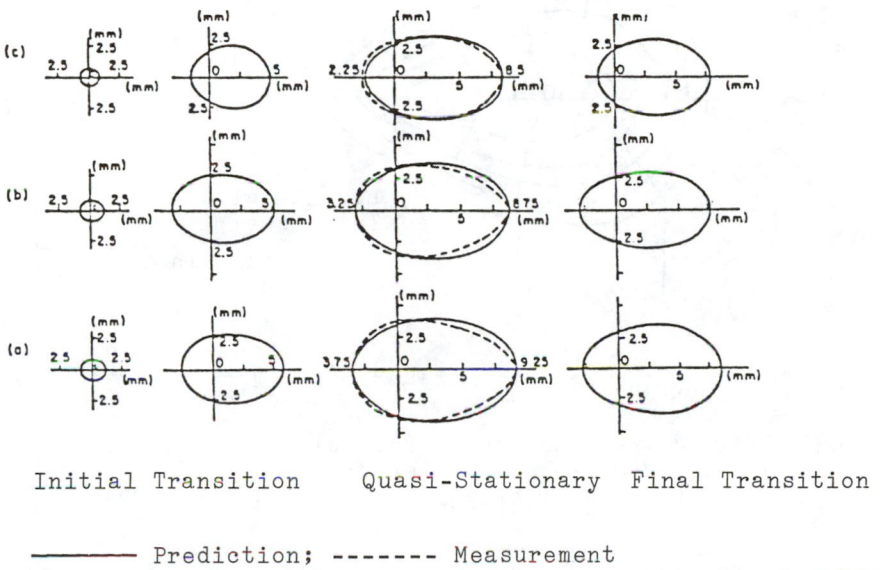

Initial Transition   Quasi-Stationary   Final Transition

————— Prediction; ------- Measurement

Figure 10   Weld pool growth in a thin aluminum plate. Transient changes of the weld pool shape and comparison between the predictions and the measurements. I=165 A; E=11 volts
(a) V=5.8 mm/s; (b) V=6.5 mm/s; (c) V=7.3 mm/s.

Calculation of Peak Temperature and Cooling Rate

Frequently it is desirable to know the peak temperature and cooling rate experienced at any given location in a weldment, to make the prediction of the metallurgical structure in that area possible. They can be calculated from numerical methods which incorporate the non-linearity in the solution. For industrial applications, simple equations can

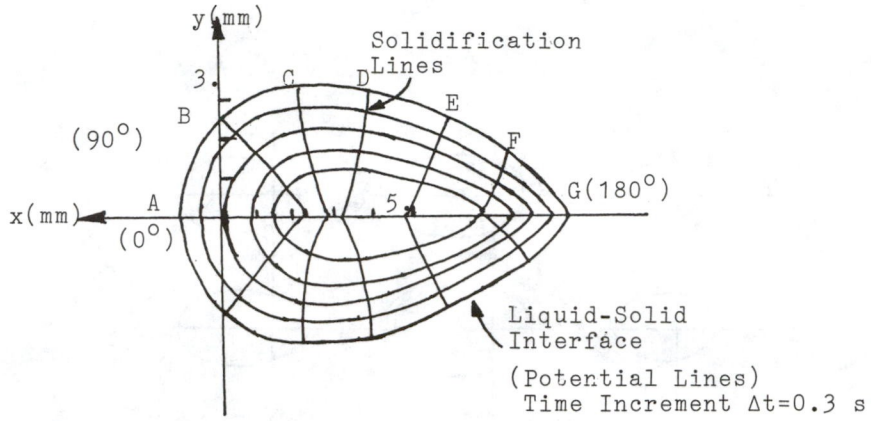

70Ni-30Cu (1.6 mm thickness)
I= 84 A
E= 10 volt
V= 2 mm/s

Solidification Lines

y(mm)

(90°)

x(mm)

(0°)

G(180°)

Liquid-Solid Interface

(Potential Lines)
Time Increment Δt=0.3 s

Figure 11   Solidifying solid-liquid interface of a thin
            70Ni-30Cu plate.

also be derived from the analytical solutions. The
derivations are described in reference (27). Reference (27)
also presents the practical applications of the simple
equations.

Effect of Welding Conditions on Material Thermal Responses

Figure 13 shows the effect of material properties and
welding conditions on the surface pool geometries in the
quasi-stationary state. The analysis was conducted and based
on the point source solution for an infinitely thick plate.
Aluminum, carbon steel and stainless steel were simulated to
explain the effect of thermal conductivity on the thermal
response of the base materials. More energy input is required
to obtain the same amount of volumetric melting in high
conductive materials than for low conductive materials. The
effect of welding travel speed is also explained. For a given
heat input per unit weld length, an increase of the
welding speed elongates the pool towards the back of the
welding source. The expansion of the isothermal contours
transverse to the welding direction is insignificant.

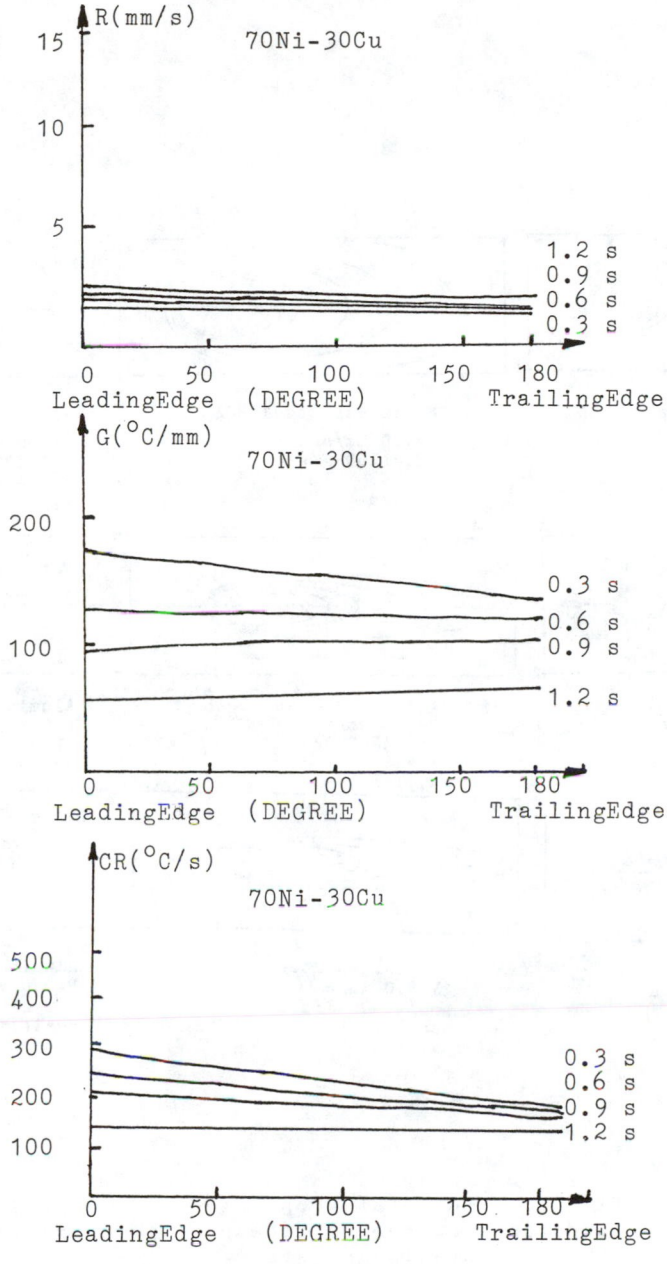

I=84 A; E=10 volt; V=2 mm/s
Plate Thickness=1.6 mm)

Figure 12   The predicted solidification rates (R, mm/s),
            temperature gradients (G,°C/mm) and cooling
            rates (CR,°C/s).

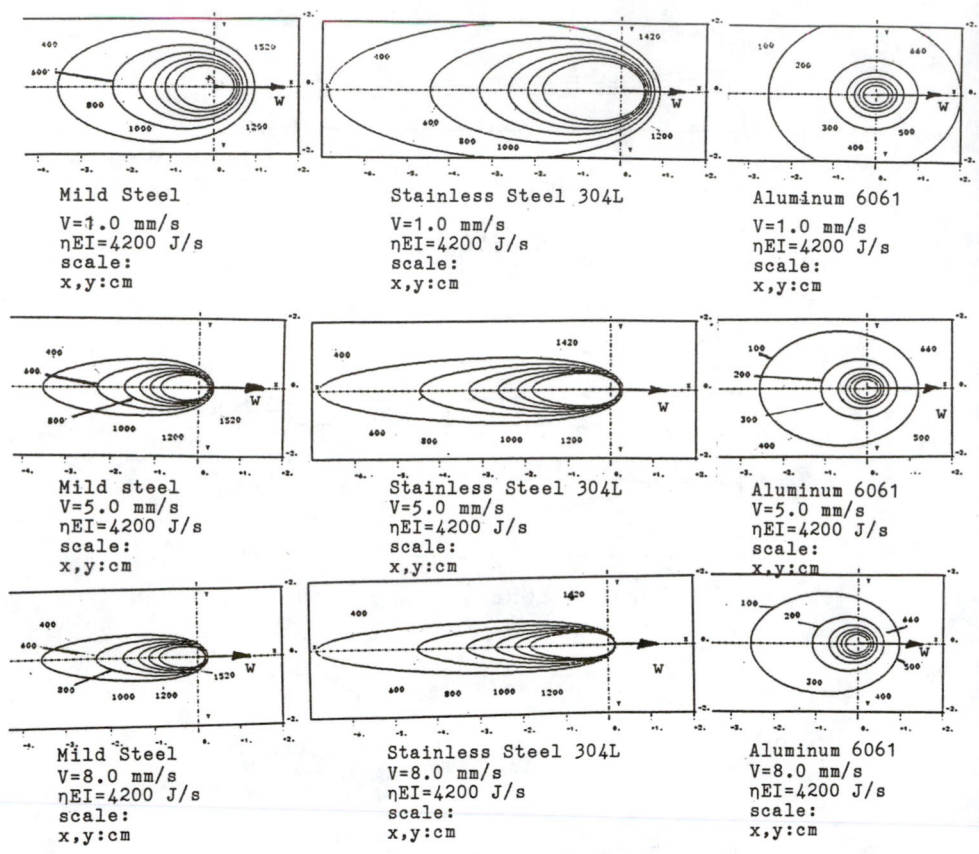

Figure 13   Effect of Material properties and welding
            conditions on the surface pool geometry in
            the quasi-stationary state.

CONCLUDING REMARKS

Mathematical formulations and analyses are useful in understanding the basic heat flow mechanisms associated with the welding processes. Because of unknown factors or the complexity of the non-linear nature of the physical phenomena, the primary objective of welding heat flow modeling is to provide a mathematical tool for thermal data analysis, design iterations, or the systematic investigation of the thermal characteristics of any welding parameters. Exact comparison with an experimental measurement may not be feasible, unless some calibration through experimental verification procedure is conducted.

REFERENCES

1. Boulton, N. S. and Lance-Martin, H. E., "Residual Stresses in Arc Welded Plates," _Proc. Inst. Mech. Engr._, 33, 1986, p. 295
2. Rosenthal, D. and Schmerber, R., "Thermal Study of Arc Welding," _Weld. J._, 17(4), 1983, p. 2s
3. Rykalin, N. N., "Calculations of Thermal Processes in Welding," Moscow, Mashgiz, 1951
4. Masubuchi, K., "Analysis of Welded Structures," Pergamon Press, 1980
5. Tsai, C. L., "Parametric Study on Cooling Phenomena in Underwater Welding," Ph.D Thesis, MIT (1977)
6. Tsai, C. L., "Finite Source Theory," _Modeling of Casting and Welding Processes II_, AIME, p. 329, Eds. Dantzig and Berry
7. Pavelic, R., Tanakuchi, R., Czehara, O. and Myers, P., "Experimental and Computed Temperature Histories in Gas Tungsten Arc Welding in Thin Plates," _Weld. J._, 48(7), 1969, p. 295s
8. Kou, S., "3-Dimensional Heat Flow During Fusion Welding," _Modeling of Casting and Welding Processes_, 1981, p. 129
9. Kogan, P. G., "The Temperature Field in the Weld Zone," Ave. Svarka, (9), 1979, p. 8
10. Ecer, G. M., Downs, H. D., Brody, H. D. and Gokhale, A., "Heat Flow Simulation of Pulsed Current Gas Tungsten Arc Welding," _Modeling of Casting and Welding Processes_, Engr. Foundation, 1981, p. 139
11. Hibbitt, H. and Marcal, P., "A Numerical Thermomechanical Model for Welding and Subsequent Loading of a Fabricated Structure," _Computer and Structures_, 3, 1973, p. 1145
12. Friedman, E., "Thermomechanical Analysis of the Welding Process Using Finite Element Methods," _J. Pressure Vessel_, Trans. of ASME, _97_, Seires J, (3), August, 1975, p. 206
13. Paley, Z. and Hibbert, P., "Computation of Temperature in Actual Weld Design," _Weld. J._, 54(11), 1975, p. 385.s
14. Naka, T., "Temperature Distribution During Welding," _J. Japan Weld. Soc._, 11(1), 1941, p. 4
15. Masubuchi, K. and Kusuda, T., "Temperature Distribution of Welded Plates," _J. Japan Weld. Soc._, 22(5), 1953, p. 14

16. Tsai, C. L. and Hou, C. A., "Theoretical Analysis of Weld Pool Behaviour in the Pulsed Current GTAW Process," _Transport Phenomena in Materials Processing_, ASME Winter Annual Meeting, 1983
17. Friedman, E., "Finite Element Analysis of Arc Welding," DOE R & D Report, WAPD-TM-1438, 1980
18. Fan, J. S. and Tsai, C. L., "Finite Element Analysis of Welding Thermal Behavior in Transient Conditions," 84-HT-80, ASME
19. Tsai, C. L., "Welding Heat Source Model," to be published
20. Apps, R. L. and Milner, D. R., "Heat Flow in Argon-arc Welding," _British Weld. J._, 2(10), 1955, p. 475
21. Tsai, C. L., "An Investigation of Heat Transport Phenomena in Underwater Welding," Winter Annual Meeting, 85-WA/HT-37, ASME, 1985
22. Carslaw, H. S. and Jaeger, J. C., "Conduction of Heat in Solids," Oxford Press
23. Grosh, R. J., Trabant, E. A. and Hawkins, G. A., "Temperature Distribution in Solids of Variable Thermal Properties Heated by Moving Heat Source," _Quarterly of Applied Mathematics_, 13(2), 1955, p.161
24. Tsai, C. L., "Treatment of Non-Lineat Thermal Properties," WE 723 Class Notes, Dept. of Welding Engr., The Ohio State University
25. Carnahan, B., Luther, H. A. and Wilkes, J. O., Applied Numerical Methods, Chapter 7, John Wiley & Sons, Inc.
26. Levy, E., Tsai, C. L. and Groover, M. P., "Analytical Investigation of the Effect of Tool Wear on the Temperature Variations in a Metal Cutting Tool," _J. Engr. for Industry_, Tran. ASME, 98(1) Series B, 1976
27. Tsai, C. L., WE 620 Class Notes, 1986, Department of Welding Engineering, The Ohio State University

Index